T0135265

Advances in Intelligent Systems and Computing

Volume 871

The series "Advances in Intelligent Systems and Computing" contains publications on theory, applications, and design methods of Intelligent Systems and Intelligent Computing. Virtually all disciplines such as engineering, natural sciences, computer and information science, ICT, economics, business, e-commerce, environment, healthcare, life science are covered. The list of topics spans all the areas of modern intelligent systems and computing such as: computational intelligence, soft computing including neural networks, fuzzy systems, evolutionary computing and the fusion of these paradigms, social intelligence, ambient intelligence, computational neuroscience, artificial life, virtual worlds and society, cognitive science and systems, Perception and Vision, DNA and immune based systems, self-organizing and adaptive systems, e-Learning and teaching, human-centered and human-centric computing, recommender systems, intelligent control, robotics and mechatronics including human-machine teaming, knowledge-based paradigms, learning paradigms, machine ethics, intelligent data analysis, knowledge management, intelligent agents, intelligent decision making and support, intelligent network security, trust management, interactive entertainment, Web intelligence and multimedia.

The publications within "Advances in Intelligent Systems and Computing" are primarily proceedings of important conferences, symposia and congresses. They cover significant recent developments in the field, both of a foundational and applicable character. An important characteristic feature of the series is the short publication time and world-wide distribution. This permits a rapid and broad dissemination of research results.

More information about this series at http://www.springer.com/series/11156

Natalia Shakhovska · Mykola O. Medykovskyy
Editors

Advances in Intelligent Systems and Computing III

Selected Papers from the International Conference on Computer Science and Information Technologies, CSIT 2018, September 11–14, Lviv, Ukraine

Editors
Natalia Shakhovska
Lviv Polytechnic National University
Lviv, Ukraine

Mykola O. Medykovskyy
Institute of Computer Science and
Information Technologies
Lviv Polytechnic National University
Lviv, Ukraine

ISSN 2194-5357 ISSN 2194-5365 (electronic)
Advances in Intelligent Systems and Computing
ISBN 978-3-030-01068-3 ISBN 978-3-030-01069-0 (eBook)
https://doi.org/10.1007/978-3-030-01069-0

Library of Congress Control Number: 2014951000

This Springer imprint is published by the registered company Springer Nature Switzerland AG
The registered company address is: Gewerbestrasse 11, 6330 Cham, Switzerland

Contents

Applied Linguistics

Decision Support Systems

Software Engineering

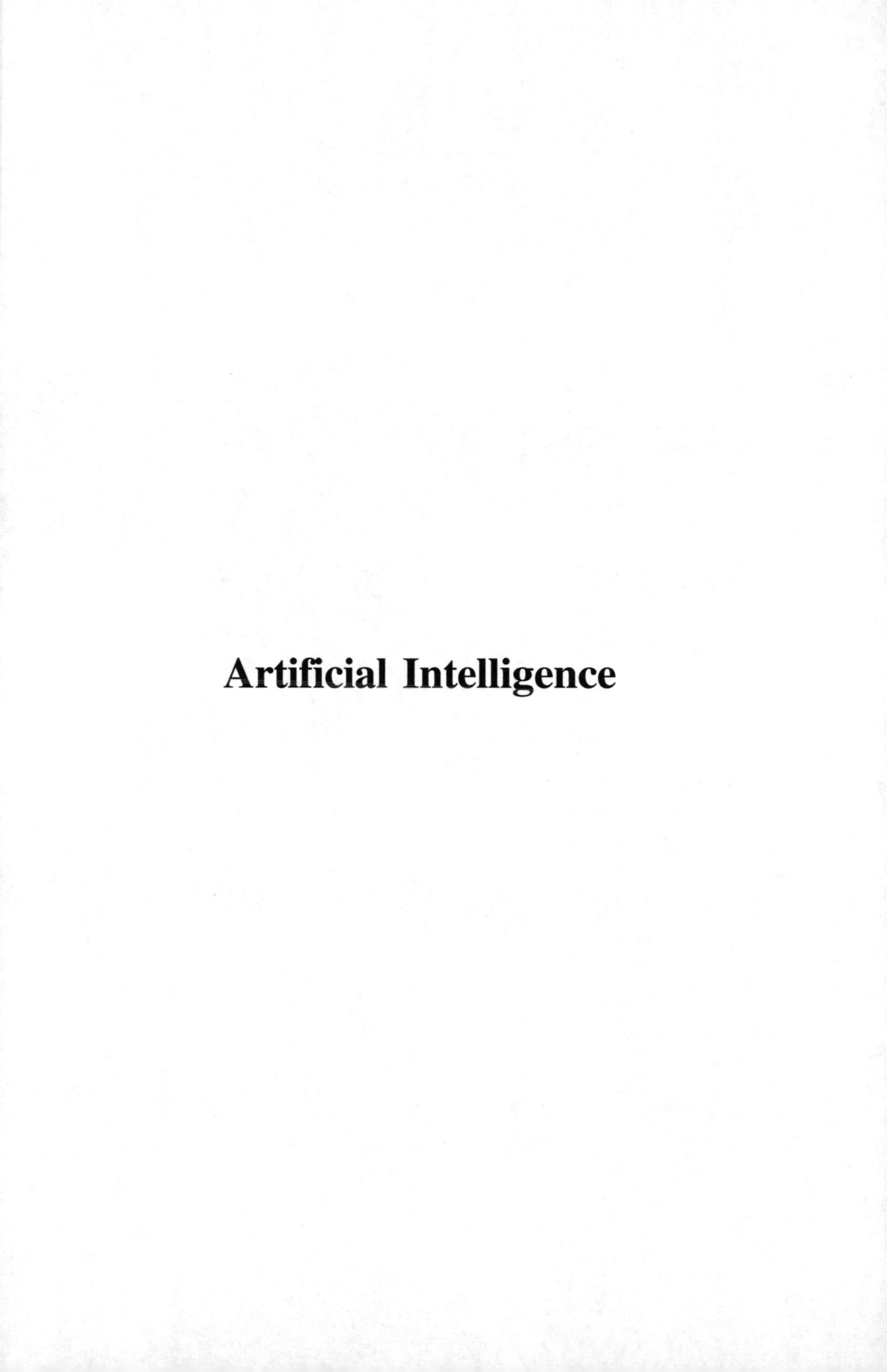

Artificial Intelligence

Fast Coordinate Cross-Match Tool for Large Astronomical Catalogue

Volodymyr Akhmetov(✉), Sergii Khlamov, and Artem Dmytrenko

V. N. Karazin Kharkiv National University, Svobody Sq. 4, Kharkiv 61022, Ukraine
akhmetovvs@gmail.com

Abstract. In this paper we presented the algorithm designed to efficient coordinate cross-match of objects in the modern massive astronomical catalogues. Preliminary data sort in the existed catalogues provides the opportunity for coordinate identification of the objects without any constraints with the storage and technical environment (PC). Using the multi-threading of the modern computing processors allows speeding up the program up to read-write data to the storage. Also the paper contains the main difficulties of implementing of the algorithm, as well as their possible solutions.

Keywords: Database · Data mining · Parallel processing Astronomical catalogue · Cross-match

1 Introduction

In recent years in astronomy the development of telescope- and instrument-making has led to an exponential growth of the observational data. The modern astronomical catalogues are the 2D-spreadsheets that contain the various information about the celestial bodies. An each row of this table corresponds to the data of one object.

There is a lot of information about the object in this row, such as:

- position in spherical coordinate system;
- errors in determining of the coordinates;
- stellar magnitude in the different photometric bands (brightness of the object);
- standard errors of stellar magnitude;
- proper motions and other useful information.

The number of objects in the modern astronomical catalogues reaches to the several billion objects, and the size of tables that contain information about these objects varies from hundreds of GB to the several TB.

So, the knowledge extraction from such data will be the most complicated challenge for researchers and scientists.

In the near future the following telescopes will be launched: Large Synoptic Survey Telescope (LSST) [1] (Fig. 1) and Thirty Meter Telescope (TMT) [2] (Fig. 2). Both telescopes will give about 30 TB of data for one observational night.

N. Shakhovska and M. O. Medykovskyy (Eds.): CSIT 2018, AISC 871, pp. 3–16, 2019.
https://doi.org/10.1007/978-3-030-01069-0_1

Fig. 1. Large Synoptic Survey Telescope (LSST).

Fig. 2. Thirty Meter Telescope (TMT).

Even today, the scientists of the world are faced with the problem of the large data during the following missions:

- ESA GAIA space mission [3, 4]: 3D-map of Milky Way with collecting of about 1 PB of data in 5 years for 1.3 billion objects;
- Pan-STARRS [5]: collecting of more than 100 TB of data for more than 2 billion objects;
- ESA Euclid space mission: collecting of more than 200 TB of data (less than 800 GB/day over at least 6 years).

The data mining techniques and intelligent management technologies of data analysis are rapidly evolving, but the cross-matching is still one of the main step of any standard modern pipeline for data analysis or reduction.

For example, a pipeline of the following software includes estimation of the objects position (data analysis), astrometry and photometry reduction: CoLiTec (Collection Light Technology) software (http://www.neoastrosoft.com) [6, 7], Astrometrica [8] and others.

One of the main step of comparing and analyzing the data is the coordinate identification of common objects in the modern massive astronomical catalogues, Big Data or any large data sets or streams that contain useful information about celestial objects. For this purpose the different databases are used, but all of them are based on MSSQL for Windows and PostgreSQL for UNIX systems.

This approach is very convenient for the storing and obtaining the quick access to data from various tables (catalogues). Such approach allows developing the software for analysis of the data from the different tables (catalogues) of database. Also, there is an opportunity in database for coordinate identification of the objects in the different catalogues.

The example of such database of astronomical catalogues is VizieR (http://vizier.u-strasbg.fr). It is a joint effort of CDS (Centre de Données astronomiques de Strasbourg) and ESA-ESRIN (Information Systems Division).

VizieR has been available since 1996, and was described in a paper published in 2000 [9]. VizieR includes more than 18 thousand catalogues that are available from CDS. 17 629 catalogues from all of them are available online as full ASCII or FITS files. 17 342 catalogues are also available through the VizieR browser [9].

Using the online access to the different astronomical catalogues provided by VizieR, the different software, such as CoLiTec [6, 10] and Astrometrica [8], can perform data analysis using different data mining techniques and intelligent management technologies.

The estimation accuracy of the object's position or of the object's brightness in both software is in the direct ratio with accuracy of the used astronomical astrometric and photometric catalogues and their fullness.

The comparison of statistical characteristics of positional measurements with CoLiTec and Astrometrica software has demonstrated that the accuracy of the objects position in the specified catalogue is a key factor for the catalogue selection for the astrometric and photometric reduction [11].

In addition to the obvious advantages of using such databases (different astronomical catalogues) there are a number of disadvantages that need to be corrected. In this paper we presented the one of available algorithms for the quick coordinate identification of common objects (intersection) in the modern large astronomical catalogues without using of the algorithms that implemented only to database using.

2 Cross-Identification

The modern catalogues include values of stellar magnitude in different photometric bands that are also can be obtained at the various epochs. In this case, we could not use photometry for cross-identification of these catalogues. Therefore, we had to perform cross-identification using only coordinates of objects. Such cross-identification is not necessarily an exact identification.

Let's represent the data of various astronomical catalogues in the form of sets *A* and *B*. So, the result of cross-identification of these astronomical catalogues will be represented as one of the combinations of join types in the Fig. 3.

In the first example (Fig. 3a) set *A* completely belongs to set *B*. In this case the result of cross-identification of sets *A* and *B* will be all objects from set *A* and the number of objects can be predicted.

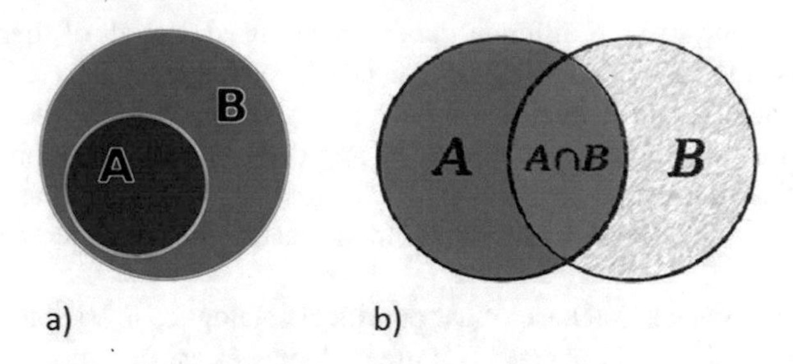

Fig. 3. Combinations of join types.

We perform the intersection of objects with a specified circular radius of search. For this we create a full set of vectors between the objects of the first and second catalogues. Also we select the distance, which is less than the specified circular radius of search. In theory, the number of such vectors depends on the number of objects in the catalogues under investigation. So, the computational complexity of the algorithm can be represented as $O(N * M)$, where N, M are the numbers of objects in the first A and the second B catalogues.

The size of modern astronomical catalogues is from several hundred millions to a billions of objects. Thus, even for the modern computers, the intersection of the large astronomical catalogues with billions of objects will take several days. Therefore, it is necessary to optimize the crossing process so that the matching of catalogues with billions records can be performed in a few hours instead of days.

To perform the intersection of common objects in our paper we sorted the data of all catalogues according to the declination of stars. The algorithm has preprocessing cost of $O(N * logN + M * logM)$ using the fast sorting with inserts. Typically, this sorting is performed once when loading and unpacking the given catalogue. Then we store and use only the file with sorted data from the catalogue.

According to our estimates the proposed intersection algorithm costs $O((N + M) * log(k * M))$ for processing, where k is a coefficient depending on the size of intersection window. Different tasks use the different intersection windows: from 0.1 to 10 and more arc seconds. For the intersection of modern high-precision catalogues in close observation epochs, the search radius does not exceed 1 arc second, so the coefficient k is usually equal to 1.

Usually, the situation with the coordinate identification (intersection) of objects in large astronomical catalogues looks much more complicated. In general, this complication is due to the large random and systematic errors in determining of the objects coordinates. Also the complication can be caused by the significant difference between the epochs of observations and the stellar proper motions.

Because of the described reasons the coordinates of object in various catalogues can be also different. So, the result of intersection of two catalogues will be only objects whose coordinates do not exceed the optimal search radius. This intersection result can be also represented as subset of common objects and as combinations of sets A and B (Fig. 3b).

In PMA catalogue the windows with various sizes ranging from 0.1 to 15 arc seconds with a step 0.1 arc seconds are used because of a very large difference of stellar density at the different galactic latitudes.

We counted the increment of a number of stars *dN* (Fig. 4, blue points), which fell into the circular zones with radii *R* and *R* + *dR*. This increment is a function of the circular radius and can be represented by a sum of two independent functions (Fig. 4, green points).

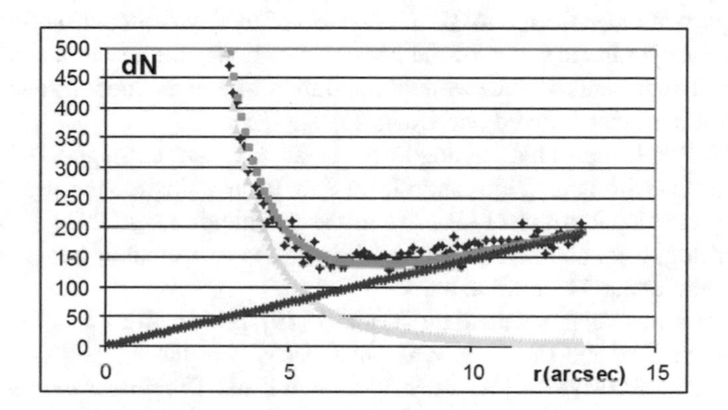

Fig. 4. The increment of a number of stars as a function of the circular radius.

The first function describes uniform density distribution (Fig. 4, red points) of stars over the sky pixel and is directly proportional to the radius of window.

The second function is the density distribution function of angular distances for the nearest neighbors.

The distribution function was calculated for the random (Poisson) distribution (Fig. 4, yellow points) of star positions.

The intersection point of these two functions allowed us to establish the optimal window size for cross-identification of catalogues. This point corresponds to such radius where the probability of misidentification reaches the probability of omitting a star with a considerable proper motion.

The described algorithm does not guarantee a correct identification for all objects from the catalogues, but according to our research and analysis the almost all objects have been identified correctly [12].

3 Data Preparation

In this paper we described some important steps for developing of the cross-identification method. It was used for creating of the catalogue with about 420 million positions and absolute proper motions of stars (PMA) [12] and for cross-identification of the following catalogues: UCAC4 [13], UCAC5 [14], Tycho-2 [15], TGAS [16], PPMXL [17], HSOY [18], 2MASS [19], Pan-STARRS (PS1) [20], ALLWISE [21] and Gaia DR2 catalogues [4, 22].

PMA Catalogue. This catalogue contains about 420 million absolute proper motions and stellar magnitudes of stars that were combined from Gaia DR1 [3] and 2MASS [19] catalogues. Most of the systematic zonal errors inherent in the 2MASS catalogue were eliminated before deriving the absolute proper motions. The absolute calibration procedure (zero-pointing of the proper motions) was carried out using about 1.6 million positions of extragalactic sources.

The mean formal error of the absolute calibration is less than 0.35 mas/yr. The derived proper motions cover the whole celestial sphere without gaps for a range of stellar magnitude range from 8 to 21. The system of the PMA proper motions does not depend on the systematic errors of the 2MASS positions, and in the magnitude range from 14 to 21 represents an independent realization of a quasi-inertial reference frame in the optical and near-infrared wavelength range [23].

UCAC4 Catalogue. This catalogue is an all-sky star catalogue, which covers mainly the stars with magnitude range from 8 to 16 in a single band pass between V and R [13]. Positional errors of all stars in the magnitude range from 10 to 14 from UCAC4 catalogue are about 15 to 20 mas for stars. Proper motions have been derived for most of the about 113 million stars.

These data are supplemented by 2MASS [19] photometric data for about 110 million stars and 5-band (B, V, g, r, i) photometry from the APASS (AAVSO Photometric All-Sky Survey) [24] for over 50 million stars. The proper motions of bright stars are based on about 140 catalogs, including Hipparcos and Tycho [25], as well as all catalogues used for the Tycho-2 [15] proper motion construction.

UCAC5 Catalogue. This catalogue contains the new astrometric reductions of the all-sky observations from US Naval Observatory CCD Astrograph Catalog (UCAC) [14]. These observations were performed using the TGAS stars in the magnitude range from 8 to 11 as a reference star catalogue.

UCAC5 catalogue has the significant improvements in the astrometric solutions as compared with UCAC4 [13]. The UCAC5 positions on the Gaia [4] coordinate system provide additional data of similar quality to the Hipparcos mission and Tycho catalogue. Using UCAC5 catalogue the TGAS proper motions will be improved.

Tycho-2 Catalogue. This is an astrometric reference catalogue that contains positions and proper motions for the 2.5 million brightest stars in the sky [15]. The Tycho-2 positions and magnitudes are based on precisely the same observations as the original Tycho catalogue, which was collected by the star mapper of the ESA Hipparcos satellite [25]. But Tycho-2 catalogue is much bigger and precise because of using the more advanced reduction technique.

Novelty of this technique is including the components of double stars with separations down to 0.8 arc seconds. Proper motions are determined in comparison with the Astrographic Catalogue and 143 other ground-based astrometric catalogues. Tycho-2 proper motions were reduced to the Hipparcos celestial coordinate system [25].

TGAS Catalogue. The first data release of GAIA astrometric satellite mission (Gaia-DR1) [4] contains three parameters: celestial positions (α, δ) and G-band magnitudes for about 1.1 billion objects that based on the observations during only the first 14 months of its operational phase. The TGAS catalogue is the first large five-parameter astrometric catalogue of celestial positions, parallaxes and proper motions that were obtained in combination of GAIA data and positions from Hipparcos [25] and Tycho-2 [15] catalogues for 2 million brightest stars.

PPMXL Catalogue. The USNO-B1.0 [26] and 2MASS [19] catalogues are the most widely used full-sky surveys. However, the 2MASS catalogue does not have the proper motions of objects at all, and the USNO-B1.0 catalogue published only relative, not absolute proper motions of objects. PPMXL catalogue [17] determines the mean positions and proper motions of objects in the ICRS system by combining USNO-B1.0 and 2MASS astrometry.

PPMXL catalogue contains about 900 million objects including 410 million with 2MASS photometry, and is the largest collection of ICRS proper motions. The resulting typical individual mean errors of the proper motions are within range from 4 mas/yr to more than 10 mas/yr. The mean errors of the objects positions at epoch 2000 are from 80 to 120 mas when 2MASS astrometry is used. Otherwise, the mean errors are from 150 to 300 mas.

HSOY Catalogue. The "Hot Stuff for One Year" (HSOY) catalogue [18] was created on basis of measurements from PPMXL [17] and Gaia DR1 [3] catalogues using the weighted least squares method. The last one was applied to derive the PPMXL catalogue itself. The HSOY catalogue contains 583 million stars with positions of the Gaia DR1. The accuracy of the objects proper motions is from 1 to 5 mas/yr and depends on the object's magnitude and coordinates in the sky.

2MASS Catalogue. The "Two Micron All Sky Survey" (2MASS) catalogue [19] contains the data with uniformly scanning of the entire sky in three near-infrared bands. This information was very important for the detection and characterizing of the point objects that are brighter than 18 stellar magnitudes in each band, with signal-to-noise ratio (SNR) greater than 10.

2MASS project was designed to close the gap between the current technical capability and knowledge of the near-infrared sky. By using the obtained context for the interpretation of results obtained at infrared and other wavelengths, the 2MASS project helped with the clarification of the large-scale structure of the Milky Way and the Local Universe.

Pan-STARRS (PS1) Catalogue. The Panoramic Survey Telescope and Rapid Response System (Pan-STARRS) is a system for wide-field astronomical observations [20]. The Pan-STARRS (PS1) was developed and operated by the Institute for Astronomy at the University of Hawaii. The PS1 survey is the first completed part of Pan-STARRS and is the basis for Data Release 1 (DR1). All sky observations were made by 1.8 m telescope and its 1.4 Gigapixel camera (GPC1) in 5-band filters (g, r, i, z, y). The PS1 took approximately 370 thousands exposures from 2010 to 2015 for more than 1.9 billion stars.

ALLWISE Catalogue. The Wide-field Infrared Survey Explorer (WISE) [21] was a NASA Medium Class Explorer mission. The main goal of WISE was a digital imaging survey of the entire sky in the 3.4, 4.6, 12 and 22 um mid-infrared band filters. The ALLWISE program extends the work of the WISE mission by combining data from the cryogenic and post-cryogenic survey phases.

The results are presented in the form of the most comprehensive view of the mid-infrared sky. The ALLWISE program provides a new catalogue and an image atlas with improved accuracy compared with earlier WISE data releases. ALLWISE catalogue includes the two complete sky coverage. Advanced data processing using for ALLWISE can be performed for the measuring of proper motions, and to compile a massive database of light curves.

Gaia DR2 Catalogue. This catalogue contains the five-parameter astrometric solution including the positions on the sky (right ascension, declination), parallaxes and proper motions for more than 1.3 billion objects with a magnitude up to 21 and a bright limit approximately equals to 3.

Parallaxes deviations in Gaia DR2 catalogue are the following: up to 0.04 mas for the objects with magnitude up to 15; approximately 0.1 mas for the objects with magnitude equals to 17; about 0.7 mas for the objects with magnitude equals to 20. The corresponding deviations in the proper motions are the following: up to 0.06 mas/yr (for <15 mag); 0.2 mas/yr (for 17 mag) and 1.2 mas/yr (for 20 mag) [22].

4 Astronomical Database

During research the special database has been developed. It provides a quick and simple access to the modern astronomical catalogues that contain data of millions or even billions celestial objects including stars, galaxies, quasars and others data.

All of these modern astrometric catalogues were collected in the database using MSSQL (Windows) server. This database contains about 50 catalogues with data more than 2 TB.

The list of some modern astronomical catalogues that were obtained for the last several years is provided in the Table 1. It also contains appropriate size of catalogues and amount of objects that are included in the catalogue.

Table 1. General data about catalogues from database.

Catalogue name	Size, Mb	Amount of objects
Gaia DR1	73 173.688	1 142 679 769
Gaia DR2	155 599.203	1 692 919 135
GPS1	38 110.445	341 469 435
GSC23	42 717.320	940 464 379
HSOY	51 757.977	583 001 652
PMA	33 944.477	421 454 398
PPMXL	184 754.203	910 468 710
PS1	263 357.275	1 919 106 885
2MASS(PSC)	41 814.016	470 992 970
TGAS	131.742	2 057 050
UCAC4	15 325.109	113 773 555
UCAC5	10 142.953	107 758 513
USNOB1	98 337.547	1 044 738 050
WISE	100 128.141	563 921 584
XPM	29 167.617	313 610 083
XPM2	117 749.281	1 077 651 504

Also the special web-interface has been created to facilitate access to the all available astronomical data of these catalogues. The web-interface was written using PHP programming language and available by the following address: http://astrodata.univer.kharkov.ua/astrometry/db.

The database allows the user to carry out data selection that containing in a small region of the celestial sphere from the large astronomical catalogues by using: a local network, an internet browser, special scripts and programs.

5 Cross-Match Astronomical Catalogue by Mean of Database

The CDS cross-match service is a new tool, which allows astronomers to efficiently cross-identify sources between very large catalogues (up to 1 billion rows) or between a user-uploaded list of positions and a large catalogue (http://cdsxmatch.u-strasbg.fr/xmatch).

In this service you can perform a cross-match based only on the positions lying at an angular distance less than radius R (1), or a cross-match based on positions taking into account error uncertainties.

$$R = 3 * sqrt(eRA_1^2 + eDEC_1^2 + eRA_2^2 + eDEC_2^2), \quad (1)$$

where eRA_1, $eDEC_1$ – the uncertainties in right ascension and declination of objects in the first catalogue; eRA_2, $eDEC_2$ – the uncertainties in right ascension and declination in the second catalogue.

Cross-match can be performed on all sources of both tables, or can be restricted to a cone around a given position or object name, or to a given HEALPix cell [27].

It turns out that when using CDS cross-match service, in addition to advantages, there are several serious disadvantages that can greatly affect to the results of the intersection.

The main disadvantage of this service is the incorrect identification of objects in the areas with high density, where more than one object falls into the search window. The processing result of this service is the selecting of all possible pairs of objects, but the total number of crossed objects may exceed 100% of objects in set A (Fig. 3a).

For example, Tycho-2 catalogue contains 2 539 913 objects in total [15]. Cross-match between itself (Tycho-2 × Tycho-2) should give the result equals to 2 539 913 – the same number of objects in total of Tycho-2 catalogue.

But when we performed cross-match between itself only for the small search window with size at 1.5 arc sec, the CDS cross-match service has returned the number of common objects equals to 2 543 009.

So, for the catalogues that contain more than billion objects the number of false-identified objects is from 5 to 10% depending on the objects density in the area of the sky (the number of objects per one square degree).

Otherwise, using of the CDS cross-match service for the very large astronomical catalogues that contain up to the billion objects the number is not directly possible due to the timeout for limitation of execution time. Therefore, the whole catalogue should be broken into the several hundred sections using pixilation HEALPix cell [27] and with intersection of each pixel separately. Then they need to be combined into one file. This approach does not allow the correct cross-identification at the edges using HEALPix especially for the large search radii.

6 Fast Cross-Match of Catalogues Without Using Database

The cross-match between catalogues that include hundreds millions of data units is a big technical problem. Mainly because these data cannot be kept in RAM. Of course, the disk storage also should be very large for this purpose.

Usually the input data of the astronomical catalogues are organized in format of tables using the following databases: MSSQL, MySQL or PostgreSQL. But the storage engine is more suitable to the fast reading after carry out of indexing.

The output data of the catalogues cross-match are written to the new table. But as shown the results of testing, actually the data writing to the database rather than the data reading was a bottleneck. In the current research we did not use any database for the cross-match of astronomic catalogues.

All data are stored in a text files that are sorted by declination. The source code of the program was written using C++ language with supporting of the CPU multi-threading. The optimization was a compromise between CPU usage and RAM limitations. In our case source code was optimized for the performance on the server with 32 GB RAM, Intel(R) Core(TM) i7 CPU with 4 cores at 3.2 GHz with hyper threading for a total of 8 CPUs.

In the developed method the calculations are performed in RAM and the input data for both the first and the second catalogues are read from previously sorted files by small areas of declination (several millions objects). In addition, the input data of catalogue are divided into several (depend from amount of hyper threading) declination strips and the calculations of objects pairs for the different strips are run in parallel.

The developed tool for cross-match uses C data structures as some "string data" with RA and DEC double type and boolean flag (true or false). This "flag" equals to "true" if the object does not have any pair from the second catalogue. The "flag" equals to "false" when the object is have a pair and cannot used for other pairs calculation. In case, when several objects are in the optimal radius of search, the nearest object will be selected.

The developed method for cross-match of catalogues includes only one object from the area with high density. After cross-match of objects for the current area the result will be presented as two "string data" from the first and second catalogues. This data will be written to the text file as one object and then the tool moves on to the next area and read it and so on.

It should be noted that the time for reading and writing the data result of cross-match more than 70% from total time in case of using 8 CPUs (threading) for catalogues with several millions objects.

A dependence of the intersection time (finding common objects) for different catalogues is provided in the Fig. 5. For the intersection in all cases the PMA catalogue (421454398 objects) was taken as a first catalogue, as the second catalogue the following catalogues were taken: TGAS (2 057 050 objects), ALLWISEAGN (1 354 775 objects), WSGAL (20 416 142 objects), UCAC5 (107 758 513 objects).

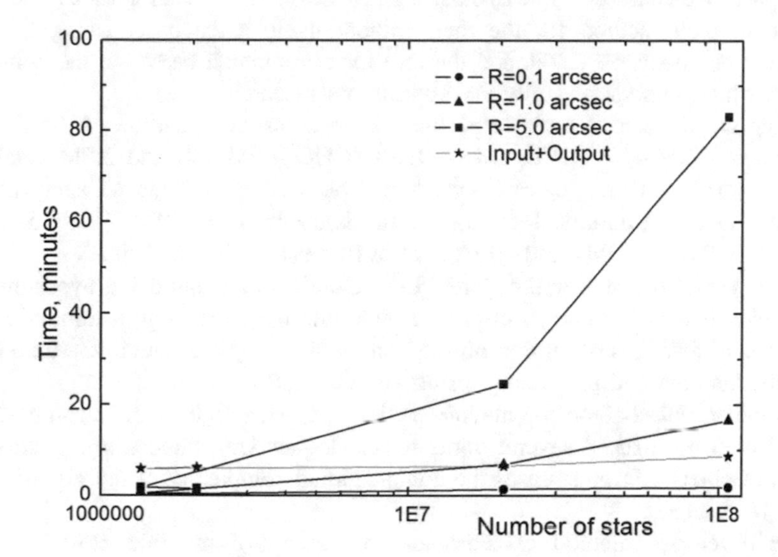

Fig. 5. Dependence of intersection time on the number of objects in the catalogues.

The plots in Fig. 5 are presented for the following different search radii: 0.1, 1.0 and 5.0 arc seconds.

As seen from the Fig. 5, the cross-match of catalogues (identification of common objects) with a small search window of 0.1 arc second takes the time comparable with time for the reading of data, even for catalogues with hundreds millions of objects.

The intersection of catalogues with a large search window of 5.0 mas, are usually used to derive objects proper motions on large difference between epochs. In this case the intersection takes a lot of time and is significantly increases with increasing of the search radius. But only for high-density catalogues with hundreds millions of objects, the crossing time exceeds the time of reading + writing.

The Table 2 contains the following information: amount of objects in the catalogues for intersection with PMA catalogue, time in minutes of reading + writing the data and time in minutes of matching for the different search radii and catalogues.

Table 2. Preprocessing time (minutes).

Catalogue name	Amount of objects	Input + output	Matching 0.1 arcsec	Matching 1.0 arcsec	Matching 5.0 arcsec
TGAS	2 057 050	5.8 + 0.4	0.07	0.41	1.52
ALLWISEAGN	1 354 775	5.6 + 0.3	0.18	0.48	1.72
WSGAL	20 416 142	6.1 + 0.4	1.38	6.86	24.47
UCAC5	107 758 613	6.9 + 1.8	1.85	16.57	83.06

7 Conclusions

In this paper we analyzed some of existed large astronomical catalogues and developed fast cross-match method for the then without using a database. Using C ++ programming language we developed the tool for cross-match based on the appropriate method, which was successfully used during our research.

Using the developed tool we performed the cross-identification of the following catalogues: UCAC4, PPMXL, USNOB1, TYCHO2, 2MASS and XPM catalogues. Also the developed tool for cross-match will be used for the further analysis of the following large astronomical catalogues in Gaia era: ALLWISE, UCAC5, HSOY, GPS1, PS1, PMA, XPM2 and all Gaia Data Release (1, 2, 3 and final).

With help of the preliminary sorted by declination data and using hyper threading the developed tool can carry out cross-match data from the large astronomical catalogues (up to few tens of million objects) on the time approximate to the time of read and write the data and processing results from/to HDD.

For the proposed algorithm the intersection time even on the simple computers (not servers) does not exceed several hours for catalogues with hundreds of millions and billions of objects. In most cases the computational complexity of our algorithm is $O((N + M) * logM)$.

The developed method excludes an opportunity of multiple cross-matches of objects in areas with high density. In this case only one (nearest) object can be presented as a result of cross-match. This approach is very important particularly for the modern massive astronomical catalogues that contain data with more than several billion objects. Also the developed tool is very effective for preparation of the reference data with the highest accuracy for astrometric and photometric reduction during pipeline processing using the different software, such as CoLiTec, Astrometrica, etc.

Acknowledgment. The authors thank CDS (Strasbourg, France) who provided online access to the different astronomical catalogues by VizieR (http://vizier.u-strasbg.fr) [9] and cross-match service (http://cdsxmatch.u-strasbg.fr/xmatch). We especially thank all creators of astronomical catalogues that described in the paper. We are grateful to the reviewer for their helpful remarks that improved our paper and, in particular, for the suggestion “to calculate the complexity of the algorithm”.

References

1. Tuell, M., Martin, H., Burge, J., Gressler, W., Zhao, C.: Optical testing of the LSST combined primary/tertiary mirror. In: Modern Technologies in Space- and Ground-based Telescopes and Instrumentation, p. 77392V (2010)
2. Skidmore, W.: Thirty meter telescope detailed science case. Res. Astron. Astrophys. **15**(12), 1945–2140 (2015)
3. Gaia Collaboration: Gaia Data Release 1 - Summary of the astrometric, photometric, and survey properties. Astron. Astrophys. **595**(A2), 23 (2016)
4. Gaia Collaboration: The Gaia mission: Astron. Astrophys. **595**(A1), 36 (2016)

5. Denneau, L., Jedicke, R., Grav, T., Granvik, M., Kubica, J.: The Pan-STARRS moving object processing system. Publ. Astron. Soc. Pac. **125**(926), 57 (2013)
6. Khlamov, S., Savanevych, V., Briukhovetskyi, O., Pohorelov, A.: CoLiTec software – detection of the near-zero apparent motion. In: Proceedings of the International Astronomical Union, vol. 12(S325), pp. 349–352. Cambridge University Press (2017)
7. Savanevych, V., Khlamov, S., Vavilova, I., Briukhovetskyi, A., Pohorelov, A., Mkrtichian, D., Kudak, V., Pakuliak, L., Dikov, E., Melnik, R., Vlasenko, V., Reichart, D.: A method of immediate detection of objects with a near-zero apparent motion in series of CCD-frames. Astron. Astrophys. **609**(A54), 11 (2018)
8. Raab, H.: Astrometrica: astrometric data reduction of CCD images. In: Astrophysics Source Code Library, record ascl:1203 (2012)
9. Ochsenbein, F., Bauer, P., Marcout, J.: The VizieR database of astronomical catalogues. Astron. Astrophys. **143**(1), 23–32 (2000)
10. Khlamov, S., Savanevych, V., Briukhovetskyi, O., Oryshych, S.: Development of computational method for detection of the object's near-zero apparent motion on the series of CCD–frames. East. Eur. J. Enterp. Technol. **2**(9), 41–48 (2016)
11. Savanevych, V., Briukhovetskyi, A., Ivashchenko, Yu., Vavilova, I., Bezkrovniy, M., Dikov, E., Vlasenko, V., Sokovikova, N., Movsesian, Ia., Dikhtyar, N., Elenin, L., Pohorelov, A., Khlamov, S.: Comparative analysis of the positional accuracy of CCD measurements of small bodies in the solar system software CoLiTec and Astrometrica. Kinemat. Phys. Celest. Bodies **31**(6), 302–313 (2015)
12. Akhmetov, V., Fedorov, P., Velichko, A., Shulga, V.: The PMA Catalogue: 420 million positions and absolute proper motions. MNRAS **469**(1), 763–773 (2017)
13. Zacharias, N., Finch, C., Girard, T., Henden, A., Bartlett, J., Monet, D., Zacharias, M.: The fourth U.S. Naval Observatory CCD Astrograph Catalog (UCAC4). Astron. J. **145**, 44 (2013)
14. Zacharias, N., Finch, C., Frouard, J.: UCAC5: new proper motions using Gaia DR1. Astron. J. **153**, 166 (2017)
15. Hog, E., Fabricius, C., Makarov, V., Urban, S., Corbin, T., Wycoff, G., Bastian, U., Schwekendiek, P., Wicenec, A.: The Tycho-2 catalogue of the 2.5 million brightest stars. Astron. Astrophys. **500**, 583–586 (2009)
16. Fedorov, P., Akhmetov, V., Velichko, A.: Testing stellar proper motions of TGAS stars using data from the HSOY, UCAC5 and PMA catalogues. MNRAS **476**(2), 2743–2750 (2018)
17. Roeser, S., Demleitner, M., Schilbach, E.: The PPMXL catalog of positions and proper motions on the ICRS. Combining USNO-B1.0 and the two Micron All Sky Survey (2MASS). Astron. J. **139**(6), 2440–2447 (2010)
18. Altmann, M., Roeser, S., Demleitner, M., Bastian, U., Schilbach, E.: Hot Stuff for One Year (HSOY). A 583 million star proper motion catalogue derived from Gaia DR1 and PPMXL. Astron. Astrophys. **600**, 4 (2017)
19. Cutri, R., Skrutskie, M., Van, D., Beichman, C., Carpenter, J., Chester, T., Cambresy, L., Evans, T., Fowler, J., Gizis, J., Howard, E., Huchra, J., Jarrett, T., Kopan, E., Kirkpatrick, J., Light, R., Marsh, K., McCallon, H., Schneider, S., Stiening, R., Sykes, M., Weinberg, M., Wheaton, W., Wheelock, S., Zacarias, N.: The 2MASS all-sky catalog of point sources. CDS/ADC Collection of Electronic Catalogues 2246 (2003)
20. Chambers, K., Magnier, E., Metcalfe, N., Flewelling, H., Huber, M., Waters, C., Denneau, L., Draper, P., Farrow, D., Finkbeiner, D.: The Pan-STARRS1 surveys. arXiv: Astrophysics – Instrumentation and Methods for Astrophysics, 1612.05560 (2016)

21. Wright, E., Eisenhardt, P., Mainzer, A., Ressler, M., Cutri, R., Jarrett, T., Kirkpatrick, J., Padgett, D., McMillan, R., Skrutskie, M., Stanford, S., Cohen, M., Walker, R., Mather, J.: The Wide-field Infrared Survey Explorer (WISE): mission description and initial on-orbit performance. Astron. J. **140**, 1868–1881 (2010)
22. Gaia Collaboration: Gaia Data Release 2: observations of solar system objects. Astron. Astrophys. **614**, 13 (2017)
23. Akhmetov, V., Fedorov, P., Velichko, A.: The PMA Catalogue as a realization of the extragalactic reference system in optical and near infrared wavelengths. In: Proceedings of the IAU, vol. 12(S330), pp. 81–82. Cambridge University Press (2018)
24. Henden, A., Templeton, M., Terrell, D., Smith, T., Levine, S., Welch, D.: AAVSO Photometric All Sky Survey (APASS) DR9. American Astronomical Society, AAS Meeting, vol. 225 (2015)
25. European Space Agency: The hipparcos and tycho catalogues. European Space Agency SP-1200, pp. 1–17 (1997)
26. Monet, D., Levine, S., Casian, B., Ables, H., Bird, A., Dahn, C., Guetter, H., Harris, H., Henden, A., Leggett, S., Levison, H., Luginbuhl, C.: The USNO-B Catalog. Astron. J. **125**, 984–993 (2003)
27. Gorski, K., Hivon, E., Banday, A., Wandelt, B., Hansen, F., Reinecke, M., Bartelman, M.: HEALPix: a framework for high-resolution discretization and fast analysis of data distributed on the sphere. Astron. J. **622**(2), 759–771 (2005)

Implementation of the Face Recognition Module for the "Smart" Home Using Remote Server

Kazarian Artem(✉), Vasyl Teslyuk, Ivan Tsmots, and Tykhan Myroslav

Lviv Polytechnic National University, Lviv 79013, Ukraine
artem.kazarian@gmail.com, vasyl.m.teslyuk@lpnu.ua

Abstract. This article presents the use of the remote computing resources for the face recognition process as a part of a "smart" home security system. Such approach allows us to optimize the load of the computation resources and to reduce the price of security system by using non-powerful hardware and to run high load face recognition calculations on the remote server of the service provider. Article describes different cases of face recognition usage, combined with the manual user interactions for better reliability of the security system.

Keywords: Face recognition · Remote computing · Security system

1 Introduction

The development of the modern technical systems is impossible without using intelligent information technologies [1, 2]. Such technologies are used in medicine [3, 4], household devices [5], engineering, military equipment [6], and others. They allow us to improve the technical and operational system parameters. Intelligent devices in the field of medicine follow the indicators of patient health with high precision, in military equipment - improves the accuracy of causing the opponent's lesions. In addition, intelligent information technology applications include "smart" home systems that allow the owner to provide a high level comfort and save energy significantly. One of the most important subsystems of a "smart" home is the subsystem of protection [7]. Process of face detection is one of the main functions in such security systems [8, 9].

Nowadays the most frequently used technology for house protection is based on the motion detection in the room by motion sensors [10], on the windows integrity and windows and doors open events. It is proposed to get the motion sensor signal as a command for face recognition execution. Such approach will prevent the mistaken alarm from triggering in cases that are not caused by the penetration of outsiders inside. The list of similar situations includes objects falling down from household surfaces caused by public transport vibrations, or repair work carried out in the sibling apartment or house. In addition, the use of security system based on reaction to movement within the apartment, causes the restrictions on keeping domestic animals within the protected area. The use of motion sensors and the human faces recognition functionality simultaneously makes it possible to ascertain that movement within the protected

N. Shakhovska and M. O. Medykovskyy (Eds.): CSIT 2018, AISC 871, pp. 17–27, 2019.
https://doi.org/10.1007/978-3-030-01069-0_2

area is caused by human actions of the house inhabitants or the intruders, which increases the reliability of the security system.

2 Architecture Implementation with Using Remote Server

Processes of data receiving from the sensors and processes of commands sending to the house devices can be based on popular microcomputers and microcontrollers. One of the most popular microcomputers Raspberry Pi 3 Model B has such technical specifications that can perform the communication between the elements of the system [11] and not complicated calculations.

The process of face recognition requires powerful computation resources, so the systems of "smart" homes need to include hardware that can provide such resources. This causes the increasing of system price. In order to prevent this, it is proposed to use architectural solution based on remote server using for calculation of tasks that require powerful computation resources, including face recognition processes. The communication with remote server is provided by global Internet network [12]. The client side hardware performs the actions of gathering information from the sensors [13], the processes of commands being sent to the house devices and communication with remote server related to processes that need large computation resources (Fig. 1).

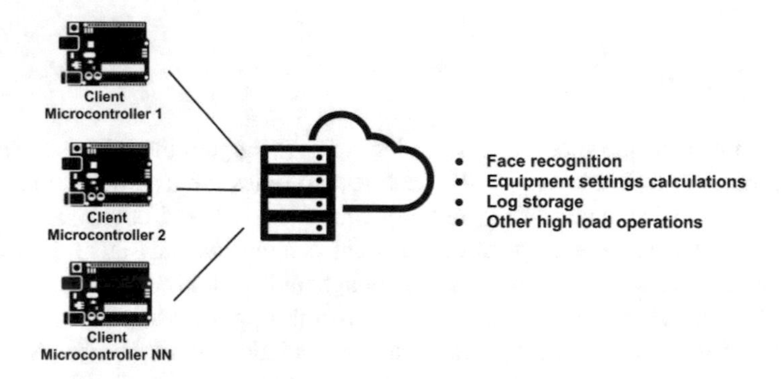

Fig. 1. Function distribution scheme

The remote server after receiving the command from client side performs computations using automated balancing of computation load between the client's requests and sends the results of calculations back to the respective client. The results of remote server calculations can be such data as the results of face recognition, new parameters of the house devices, etc. In the case of lost connection with the server – "smart" home system switches to the manual control mode and shows the notification for users about that.

The result is the developed structure of information system based on client-server architecture and an optimization of the load on the system client parts [14] by performing complex computations on a powerful server of the service provider [15].

3 Technical Stack

The server part for the implemented system is based on Node.js technology [16]. Node.js is an open source platform for executing high-performance network applications written on JavaScript. There are a lot of ready to use software solutions for face recognition implementation. One of the most popular software is the OpenCV image computation library [17]. OpenCV is a library that presents functions and algorithms for image processing, computer vision and numerical open source general-purpose algorithms. The library provides the tools for analysis and processing image content, including the object tracking, image transformation, recognition of objects in photographs (for example, figures of people, text, etc.), application of machine learning methods and the identification of common elements on different images.

Face recognition methods initially process the input image for identification and distinguish the characteristic parts of the face such as the eyes, mouth, nose, etc., and then it calculates the geometric interconnections between them, creating a vector of geometric peculiarities from the image of the person's face. Standard statistical methods are used to find faces based on measured data.

MongoDB is used to store data about the inhabitants inside the "smart" home, the results of user faces recognition and the historical data on user's movement inside. It is a document based database management system (DBMS) with open source code that does not require a table schema description. MongoDB takes a niche between fast and scalable systems that uses key/value data and relational DBMS, functional and user-friendly in query formation.

As a result, usage of the modern technology stack in development process make the system easy to maintain and scalable for connection of the new clients [18].

4 Face Recognition Process

The system must have preloaded user face image samples for comparison with received image to recognize him. Before starting a system training, there is a need to collect graphic files of users whose faces are depicted. For each image, we need to select an area centred near the user's face. The CascadeClassifier class of the OpenCV library is used to identify the face [19]. CascadeClassifier can be used to detect objects as it is created from an XML file that contains the representation of the trained model. OpenCV provides some premade models for various usage cases, such as face detection, eye detection, full body detection, etc. The HAAR_FRONTALFACE_ALT2 is used to identify the face. Receiving a black and white image, the detectMultiScale function returns the bounding rectangles of the potential boundaries of the face. For processing, the first best result is taken and the separated part of the image that is in this rectangle is rotated. Image files are processed alternately by creating a training sample. The next step is to resize the images. We need this process to bring all samples from the test sample to one size requiring recognition functions for further processing. Then the system marks the image according to the given user and begins the training process on the received sample of images. For system training, it is necessary to prepare an array of images and an array of appropriate user names for training function. As soon as the data is prepared, the process of recognizers initializations is performed. Logically, the method of training expects to receive arrays of images and corresponding users of the same length at the input.

Person is identificated by the webcam, will be saved in a separate frame for recognition. For this process, a similar sequence of actions is performed, as in the process of training the system. In other words, there are separation of some image area with the area of the person's face, the transformation of the resulting zone into black and white, the size of the isolated zone is converted to the size of the images from the training sample. Then the resulting converted image area is transmitted to the recognition function to compare it with the reference images. The result of this procedure is the coincidence coefficients for each user of the system that was added to the training sample.

The result of the OpenCV library using for face recognition tasks is the high rate of correct user identifications with the small size of the training sample.

5 Functions of the Developed System

5.1 Allow the Access for Recognized User

The granting and restricting access functions work by means of the face recognition process and compare the recognized image with the images of users who have been granted access. This functionality can be used, both for providing access to a private house or adjoining territory, as well as for commercial enterprises to provide access to individual rooms for group workers with the appropriate access level.

System will be waiting for the user face in the area of webcam vision. After the face detection process, a face recognition process will be performed in scope of the webcam vision, with the usage of the existing users information. When a high degree of accuracy matches the resulting image with one of the existing user images - the system decides whether to grant or restrict access to the current user according to the level of access users from the database.

If we get the average index of the accuracy of the current image with one of the existing images of users for which access is granted - the system will offer the user to enter a PIN code. If the correct pin-code is entered corresponding by to the user with the highest coincidence coefficient - the system checks the level of access of the user entered in the database and according to this data gives or refuses access to the user and

Fig. 2. Example of the interface with pin code panel

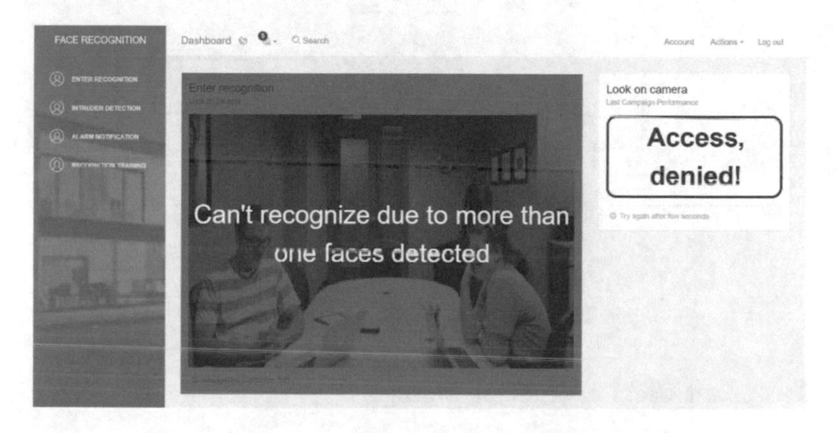

Fig. 3. Example of interface with multiple faces recognition

displays the corresponding message. The images below show the user interface of implemented face recognition module of “smart” home (Fig. 2).

If an incorrect pin code is entered corresponding by to the user with the highest coincidence coefficient - the system forbids the user to access and displays a message to the user. If it detected more than two faces in the webcam area or got a low index of the matching image accuracy with one of the existing images of users, the system issues a message to the user that the user identification and access procedure can not be performed (Fig. 3).

The algorithm of the presented function is given below (Fig. 4).

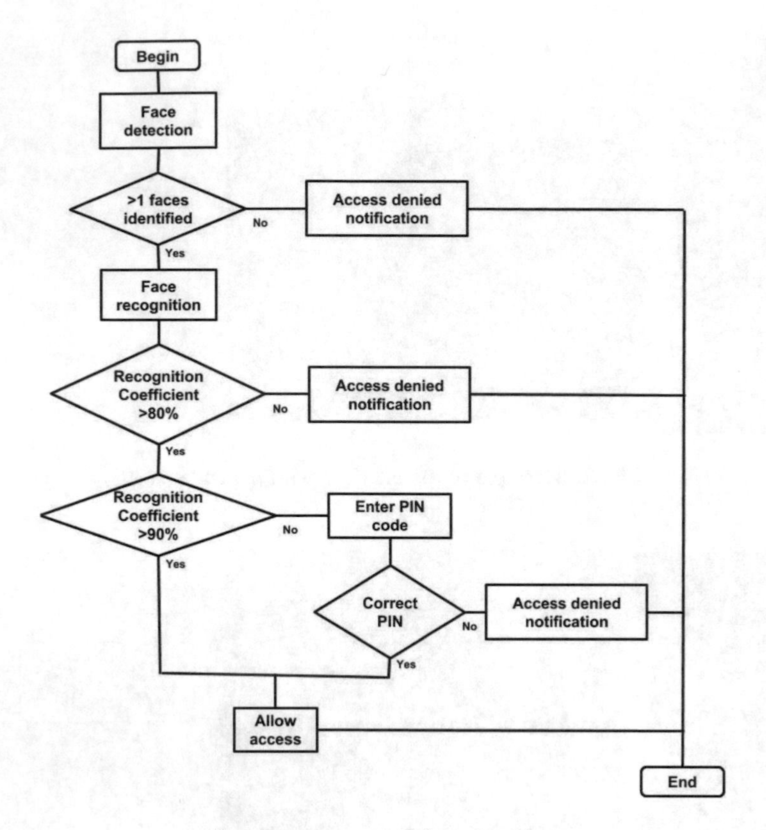

Fig. 4. Access providing algorithm

5.2 Recognized Users Information Storage

Functionality for collecting information about the presence of users in the house works by means of the face recognition process and the comparison of the recognized image with the images of users, whose data is entered into the system. This functionality can be applied for the creation of a test database that will be used to train the neural network, which is part of the "smart" home appliances control system, which will be used to automate the settings of devices in accordance with the wishes of the intellectual house inhabitants. Also, this function will be useful in monitoring elderly people, children and to provide this information to law enforcement bodies in case of situations of unlawful penetration into the house.

When human faces will be identified in the area of the camera vision, the system launches the face recognition process and compares the received images with the existing images of system users from the database. If a high degree of accuracy matches the resulting image with one of the existing user images, the system records information about the user name, the date and time of the successful identification process into the database, and displays the recent records of the successful presence identification as a list using the user interface system (Fig. 5).

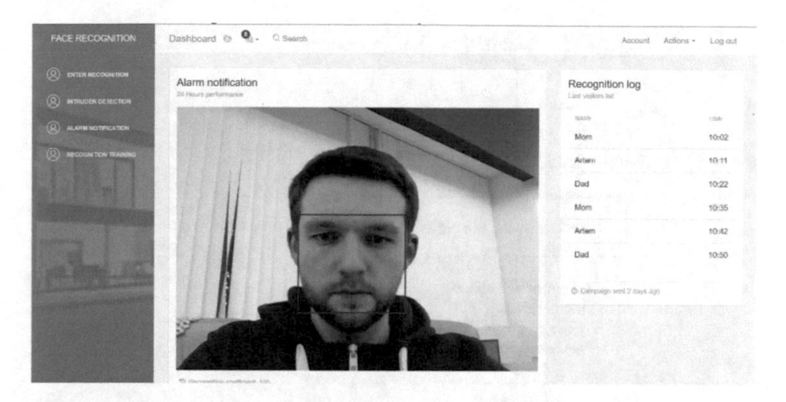

Fig. 5. Example of interface with users recognition stored information panel

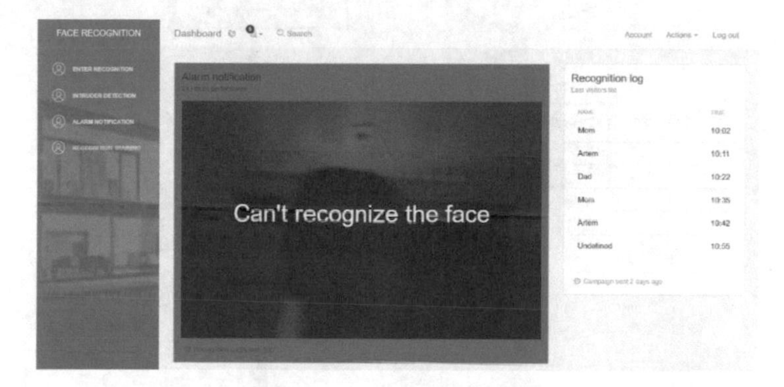

Fig. 6. Example of the interface with failed recognition

In case of several human faces identification in the camera vision area, the system launches the face recognition process and compares the received images with the existing images of system users from the database for each identified person and records information about the name of the recognized user, the date and time of the procedure of successful identification to the database.

If a low indicator of the received image coincidence with one of the existing user images for each identified face within the camera's actions is received, the system records the date and time of the identification procedure into the database separately for each image named "Undefined person", which is accordingly reflected in the list using the system user interface (Figs. 6 and 7).

5.3 Detection of Unrecognized Persons Inside the House, Notifications of the Users

Activating "Detection of unrecognized persons" function, the system triggers the process of detected face recognition and compares them with faces of system users. With low rate of comparison the system will send SMS message on defined numbers

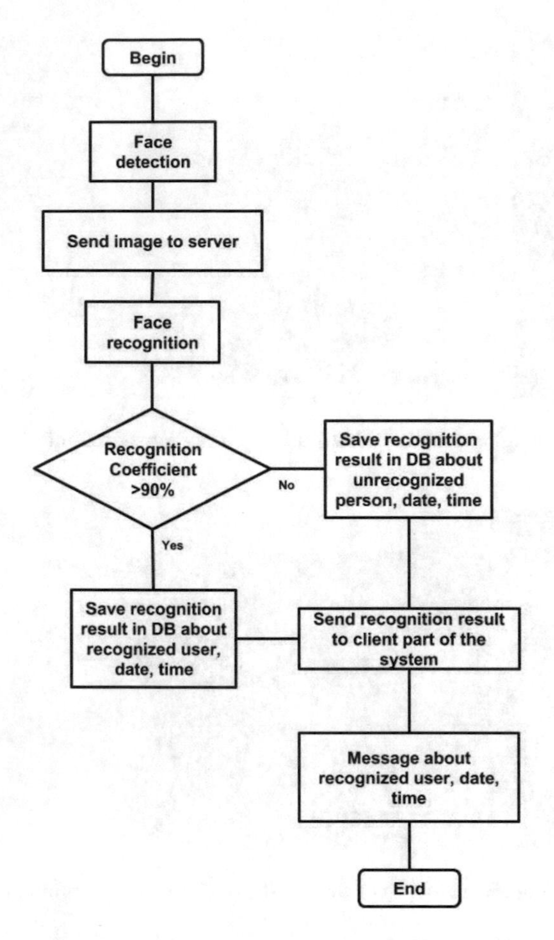

Fig. 7. Recognition results storage process algorithm

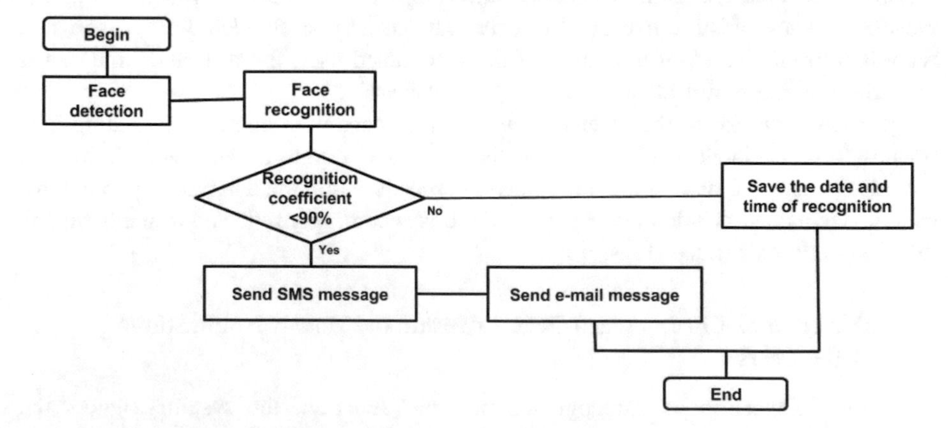

Fig. 8. Notification sending algorithm

with notification about intruders inside the house and send the image from the camera on the defined addresses of e-mails. Information about sending of such notifications will be displayed in system user interface (Fig. 8).

As a result of the developed algorithms software implementation, the information system performs functions of providing access to the house for the system users based on the operation of human face recognition; users are notificated about identification of an unknown person by sending a message by SMS and e-mail; storage of historical data about users presence at house.

6 System Training Based on User Faces Images

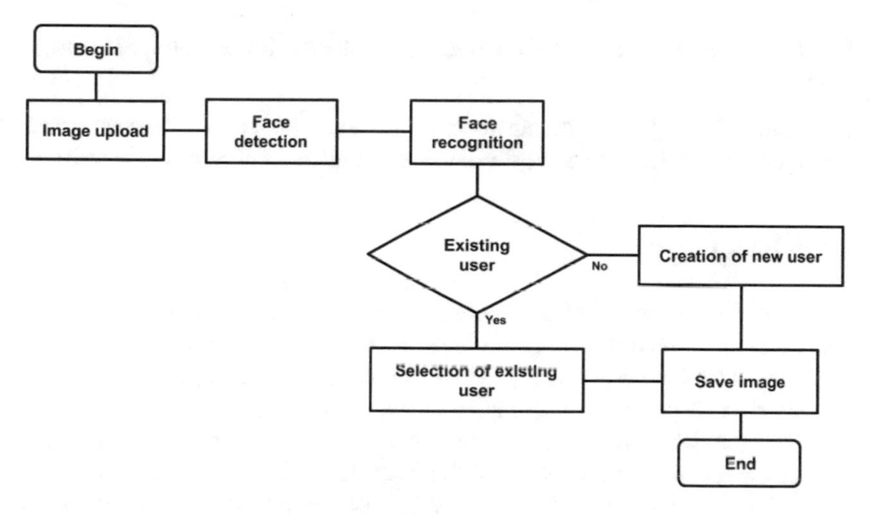

Fig. 9. The system training algorithm

For the process of user face recognition, the system has to work with the set of user faces images for self training [20]. Identification of the user depends on the count of uploaded test images, their quality and user emotions on this images. Adding new images for face recognition accuracy increasing provided by the user interface that allows to upload files with images from the memory of the computer. After uploading, the user can see the result of the identification, select the name of the user from the list of users created before or to select the new user by filling of data inside the form. After the confirmation of the data, the system adds the uploaded image of the user inside database, which will be used in future for the system face recognition training (Fig. 9).

The developed algorithm is based on the usage of machine learning techniques and allows to improve the results of the system users recognition by learning the system on additionally loaded user face images. Dependency graph of the successful face recognition, on the size of the training sample with uploaded system users face images is presented below (Fig. 10):

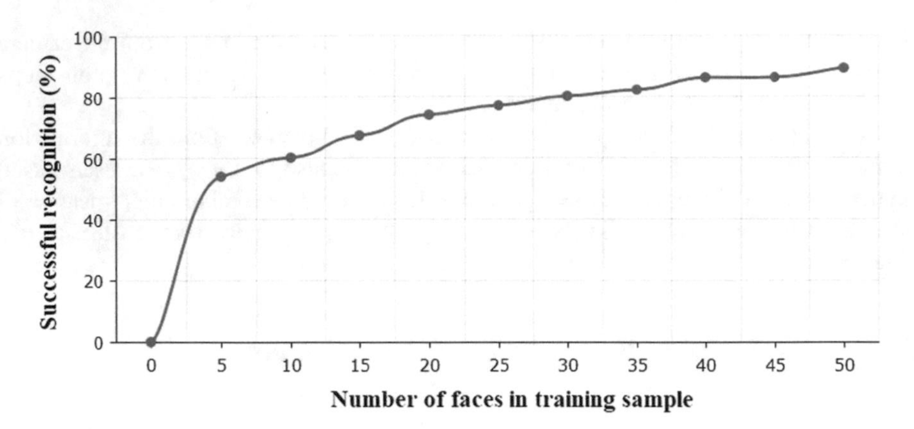

Fig. 10. Dependence of successful recognition on the number of downloaded images

The result of the system training function use is an increase of the successful user face recognitions, which affects positively the overall reliability of the system.

7 Conclusion

As a result of the system implementation, the real time face recognition method was improved thanks to the usage of a remote server based on the technology NodeJS. The developed information system can be used as a separate module of the “smart” home system, which will perform functions to enhance the security of a “smart” home or commercial establishment. Moreover, this module can perform data collection functions to use them in the learning stage of the machine learning algorithms for automated control of devices inside “smart” homes or commercial institutions, depending on the users preferences.

References

1. Lombardi, R., Dumay, J., Trequattrini, R., Lardo, A.: Modern trends for the strategic use of Intellectual Property rights: dynamic IP portfolio management, open innovation and collaborative organizations. In: Lombardi, R., Dumay, J., Trequattrini, R., Lardo, A. (eds.) Managing Globalisation: New Business Models, Strategies, and Innovation, pp. 114–137 (2016)
2. Bonomi, F., Milito, R., Zhu, J., Addepalli, S.: Fog computing and its role in the internet of things. In: Proceedings of the 1st Edition of the MCC Workshop on Mobile Cloud Computing, Helsinki, Finland, 13–17 August 2012, pp. 13–16
3. Collins, T., Crosson, J., Peikes, D., McNellis, R.: Using Health Information Technology to Support Quality Improvement in Primary Care, 19 p. AHRQ Publication, Princeton (2015). (15-0031-EF)

4. Berezsky, O., Melnyk, G., Datsko, T., Verbovy, S.: An intelligent system for cytological and histological image analysis. In: Proceedings of the 13th International Conference on Experience of Designing and Application of CAD Systems in Microelectronics, CADSM 2015, Polyana-Svalyava (Zakarpattya), Ukraine, 24–27 February 2015, pp. 28–31 (2015)
5. Lobaccaro, G., Carlucci, S., Löfström, E.: A review of systems and technologies for smart homes and smart grids. Energies **9**, 348–381 (2016)
6. Khizhnaya, A.V., Kutepov, M.M., Gladkova, M.N., Gladkov, A.V., Dvornikova, E.I.: Information technologies in the system of military engineer training of cadets. Int. J. Environ. Sci. Educ. **13**, 6238–6245 (2016)
7. Isa, E., Sklavos, N.: Smart home automation: GSM security system design & implementation. J. Eng. Sci. Technol. Rev. JESTR. **10**(3), 170–174 (2017). (1791–2377)
8. Sahani, M., Subudhi, S., Mohanty, M.: Design of face recognition based embedded home security system. KSII Trans. Internet Inf. Syst. TIIS **10**(4), 1751–1767 (2016). (1976–7277)
9. Teslyuk, V., Beregovskyi, V., Denysyuk, P., Teslyuk, T., Lozynskyi, A.: Development and implementation of the technical accident prevention subsystem for the smart home system. Int. J. Intell. Syst. Appl. (IJISA) **10**(1), 1–8 (2018). https://doi.org/10.5815/ijisa.2018.01.01
10. Peleshko, D., Ivanov, Y., Sharov, B., Izonin, I., Borzov, Y.: Design and implementation of visitors queue density analysis and registration method for retail videosurveillance purposes. In: IEEE First International Conference on Data Stream Mining & Processing (DSMP), Lviv, pp. 159–162 (2016). https://doi.org/10.1109/dsmp.2016.7583531
11. Kazarian, A., Teslyuk, V., Tsmots, I., Mashevska, M.: Units and structure of automated smart house system using machine learning algotithms. In: Proceeding of the 14th International Conference on the Experience of Designing and Application of Cad Systems in Microelectronics, CADSM 2017, Polyana, Lviv, Ukraine, 21–25 February 2017, pp. 364–366 (2017)
12. Choy, S., Wong, B., Simon, G., Rosenberg, C.: A hybrid edge-cloud architecture for reducing on-demand gaming latency. Multimed. Syst. **20**, 503–519 (2014)
13. Vujović, V., Maksimović, M.: Raspberry pi as a wireless sensor node: performances and constraints. In: Proceedings of the 2014 37th International Convention on Information and Communication Technology, Electronics and Microelectronics (MIPRO), Opatija, Croatia, 26–30 May 2014, pp. 1013–1018
14. Hajji, W., Tso, F.P.: Understanding the performance of low power Raspberry Pi Cloud for big data. Electronics **5**(2), 29 (2016)
15. Tso, F., White, D., Jouet, S., Singer, J., Pezaros, D.: The glasgow raspberry pi cloud: a scale model for cloud computing infrastructures. In: Proceedings of the 2013 IEEE 33rd International Conference on Distributed Computing Systems Workshops (ICDCSW), Philadelphia, PA, USA, 8–11 July 2013, pp. 108–112
16. Johanan, J.: Building Scalable Apps with Redis and Node.js, vol. 1, 297 p. Packt Publishing Ltd., Birmingham (2014). ISBN 978-1-78398-448-0
17. Viraktamath, S.V., Katti, M., Khatawkar, A., Kulkarni, P.: Face detection and tracking using OpenCV. SIJ Trans. Comput. Netw. Commun. Eng. (CNCE) **1**(3), 45–50 (2013)
18. Shah, H., Soomro, T.: Node.js challenges in implementation. Glob. J. Comput. Sci. Technol. E Netw. Web Secur., 0975–4350 (2017)
19. Attaullah, M., Dhere, S., Hipparagi, S.: Real time face detection and tracking using OpenCV. Int. J. Res. Emerg. Sci. Technol., 39–43 (2017)
20. Pinto, N., DiCarlo, J., Cox, D.: How far can you get with a modern face recognition test set using only simple features? In: IEEE Conference on Computer Vision and Pattern Recognition, Miami, FL, pp. 2591–2568

Improved Multi-spiral Local Binary Pattern in Texture Recognition

Nihan Kazak[1] and Mehmet Koc[2(✉)]

[1] Computer Engineering, Bilecik Seyh Edebali University, Bilecik, Turkey
nihan.kazak@bilecik.edu.tr
[2] Electrical and Electronics Engineering, Bilecik Seyh Edebali University, Bilecik, Turkey
mehmet.koc@bilecik.edu.tr

Abstract. Local Binary Pattern (LBP) is a well-known appearance-based local feature descriptor. Since it is successfully applied to many pattern recognition applications such as texture recognition, face recognition, and so on, many variants of LBP are proposed by researchers. It is known that edges carry important discriminative information about the geometric structure and content of the image. In this paper, the discrimination ability of the feature descriptors derived from the edges of an image using Spiral Local Binary Patterns (S1BLP) and its two variants, namely two Spiral LBP (S2LBP) and four Spiral LBP (S4LBP) are investigated. We also combine this descriptor with S1LBP, S2LBP, and S4LBP features which are derived from the whole image. Linear Regression Classification (LRC) and χ^2 test are used to investigate the performance of the proposed descriptors in terms of classification accuracy. The classification tests conducted on two different texture datasets, namely CURet and UIUC show that the proposed feature descriptor has important discriminative information which improves the classification accuracy.

Keywords: Texture recognition · Spiral Local Binary Pattern
Edge detection · Linear Regression Classification

1 Introduction

The recognition problems consist of two main steps. Firstly, features are extracted from images and secondly obtained features are classified using a suitable classification method.

Local Binary Pattern (LBP) is one of the well-known appearance based local feature extraction methods which is firstly used for texture recognition in [1]. Conventional LBP calculates the intensity difference with the reference pixel and its neighbors on square topology. To overcome the limitations of square topology, sampling points are chosen from the neighbors of a center pixel over circular topology in [2]. In [3], elliptical topology is used for the assignment of neighbors' location and in this way anisotropic structural information is extracted for a face recognition system. Different geometric shapes such as ellipse, circle, parabola, hyperbola, and Archimedean spiral are tested for neighborhood topology in [4]. Also, different encodings are used for the

N. Shakhovska and M. O. Medykovskyy (Eds.): CSIT 2018, AISC 871, pp. 28–37, 2019.
https://doi.org/10.1007/978-3-030-01069-0_3

calculation of LBP code. According to this work, proposed quinary encoding and using elliptic neighborhood performs the best with the medical image databases. In [5], Archimedean spiral topology (SLBP) is analyzed for texture recognition and they propose the topologies with two (S2LBP) and four spirals (S4LBP) in [6].

In the classification phase, linear discriminant analysis (LDA), linear regression classifier (LRC), support vector machines (SVM), Chi-square test, and G-test are used with LBP and some variants of LBP features in [5, 6]. LRC [7] is one of the popular subspace based method which is mostly used in face recognition [8–11].

Edges are generally the result of sharp lightning, color, illumination, or grey level changes. Edge is one of the important visual information to define the image. Edges represent the geometric structure of an image and the edges in the images are of great importance for people to recognize objects. For this reason, it is used with LBP and its variants in many studies [12–15]. We used Canny edge detector which is one of the most widely used edge detection algorithms to obtain edges in an image.

In this work, we have investigated the effect of combining the LBP codes obtained from the edges of an image with the conventional LBP codes for recognition performance. We use several topologies, namely LBP, S1LBP, S2LBP, and S4LBP. LRC, and χ^2 test are used for classification purposes. Performances of the features evaluated using different topologies are tested on UIUC [16] and CURet [17] texture databases.

The rest of the paper is organized as follows. Section 2 introduces the LBP operator. LBP codes obtained using spiral topology and its variants are explained in Sect. 3. The proposed feature descriptor extension is given in Sect. 4. The experimental work is given in Sect. 5 and finally conclusions are presented in Sect. 6.

2 Local Binary Pattern

LBP is a well-known feature descriptor which is used to extract local properties of texture patterns in an image. LBP assigns a label to each pixel in a 3×3 neighborhood by comparing the grey scale intensity values with the center-pixel in order to generate a binary code. If the neighbor pixel's grey scale value is lower than center pixel it is assigned a zero otherwise a one. This operation will yield an 8-bit binary code. After the application of LBP to each pixel in the image, the obtained 8-bits numbers are converted to decimal values and accumulated in a 256-bin histogram. Histogram is the LBP descriptor of the image and used as feature vector.

In Fig. 1, a sample calculation of original LBP is given. After the thresholding the neighbor pixels with the center pixel, the binary code is evaluated as 01101010. Finally, when converted to the decimal, LBP code for the center pixel is 106.

The use of 3×3 square neighborhood restricts the LBP when acquiring the structural information of the texture image. To overcome this limitation, the LBP operator is extended to circular topology with different radius and number of sampling points. Figure 2 represents circular topology for LBP for radius 1 with 8 sampling points and radius 2 with 16 sampling points. As it is seen from the figure that most of the sampling points are not at a neighbor pixel's center. In this case, the bilinear interpolation is used the intensity value of the sampling point.

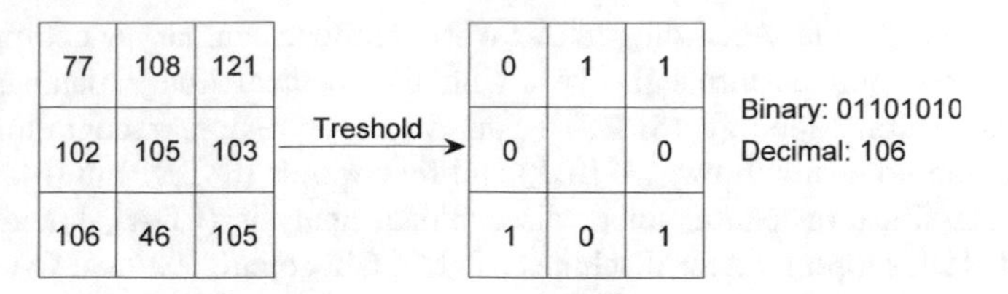

Fig. 1. A sample application of the original LBP.

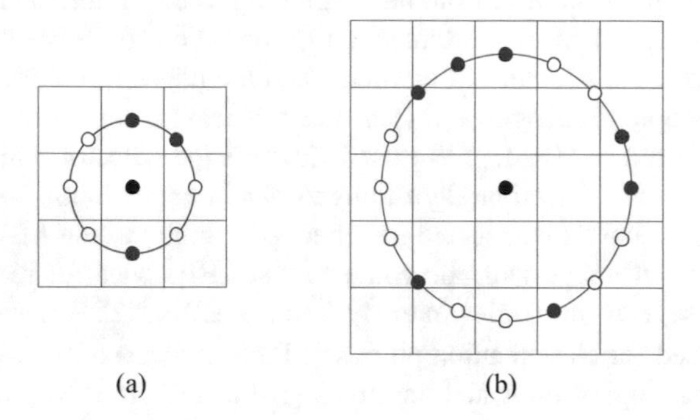

Fig. 2. (a) 8 samples representation with 1-pixel radius (b) 16 samples representation with 2-pixel radius.

Let (x_c, y_c) be the center pixel, then the LBP code of center pixel for P sampling points with radius R is calculated as follows.

$$LBP_{P,R} = \sum\nolimits_{p=0}^{P-1} s(g_p - g_c)2^p, s(x) = \begin{cases} 1, & \text{if } x \geq 0 \\ 0, & \text{otherwise} \end{cases} \tag{1}$$

Here g_p and g_c are the grey scale values of the pth sampling point and the center pixel respectively.

3 Spiral Local Binary Pattern

In this section, we describe local binary pattern with spiral topology (SLBP) and its two variants. In the original LBP, the neighbor sampling points are chosen over a circular neighborhood and the LBP code is calculated by thresholding the sampling points with the grey level value of the center pixel. In spiral LBP, sampling points are chosen over an Archimedean spiral neighborhood topology (see Fig. 3).

The Archimedean spiral can be expressed, in polar coordinates, by the following equation:

$$r = a + b\theta \tag{2}$$

Here a and b are real numbers, r is the radial distance, and θ is the polar angle. In SLBP, Archimedean spiral starts from the center pixel as illustrated in Fig. 3 and stops after 2π radians. The sample points are chosen over the Archimedean curve with $\theta = \pi/4$ intervals for 8-point neighborhood. Since the most of the sampling points do not overlap with a neighbor pixel center, the grey level intensity value of the sampling point is calculated using bipolar interpolation.

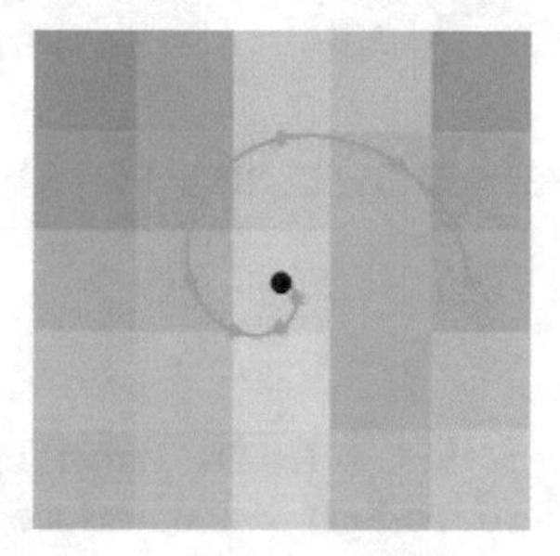

Fig. 3. An illustration of Spiral Local Binary Pattern (SLBP) topology

After calculating the grey level values of the sampling points, the computation of the LBP code of the center pixel for SBLP is the same as the basic LBP. The sample pixel values are thresholded with the grey level value of the center pixel. If the value of the sampling point is smaller than the center pixel, it is coded as a zero, otherwise it is coded as a one. Then the binary code is converted to decimal. This procedure is repeated for each pixel and the resultant values are accumulated in a histogram with 256 bin which is to be used as feature descriptor of the image.

In addition to S1LBP, there are other variants proposed in [6]. Local Binary Pattern with two spirals topology (S2LBP) is one of the methods. As understood from its name, in this topology we use two spirals. Firstly, S1LBP is applied to whole image and a histogram is obtained. Then the second spiral is obtained by adding π to θ in Eq. 2. This topology is shown in Fig. 4a. The locations of the neighbors of the center pixel can be easily found by adding π to θ in Eq. 2. S1LBP is applied to the whole image using the second spiral and feature descriptor of the image is obtained. Finally, the histograms obtained from the two spirals are concatenated to get the feature descriptor for S2LBP.

Another variant of S1LBP is Local Binary Patterns with four spirals (S4LBP). In this topology we use four spirals which are placed with equally angular distances on an interval of 2π length. S1LBP is applied to the image by using each of the spirals separately and the final feature vector is obtained by concatenating the histograms evaluated from the spirals. The spirals can be found by adding $\pi/2$, π, and $3\pi/2$ to θ in Eq. 2. The topology for S4LBP is illustrated in Fig. 4b.

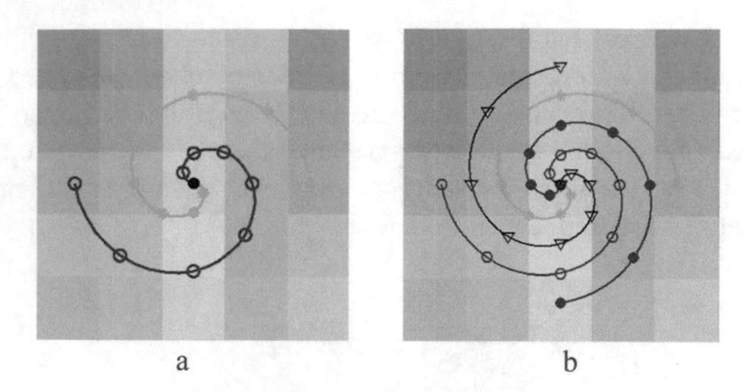

Fig. 4. Illustration of (a) S2LBP and (b) S4LBP topologies

4 Proposed Method

In this section, we give a brief review of feature extraction method, namely S1LBP_EI (one spiral LBP with edge information) and introduce its variants S2LBP_EI (two spirals LBP with edge information) and S4LBP_EI (four spirals LBP with edge information).

S1LBP_EI is firstly introduced in [18]. In this work, authors successfully applied the method to texture recognition problem. It is known that edges of an image may contain discriminative information that can be used for classification purposes. In S1LBP_EI, edge maps are generated using Canny edge detector. The main aim of S1LBP_EI is to combine the conventional S1LBP and S1LBP codes derived from the edges of the image. To form S1LBP_EI descriptor of an image, firstly S1LBP is applied to the whole image and histogram for the LBP codes is obtained. Secondly, S1LBP is applied to the pixels on the edges of the image then histogram of the evaluated codes is obtained. Finally, two histograms are concatenated in a single histogram to be used as feature descriptor.

In this study, we extend this idea to S2LBP and S4LBP to get more discriminative feature descriptors. Basically, we apply S1LBP_EI to each of the spirals in multi-spiral LBP topologies then we concatenate the obtained feature descriptors into single feature vector. Figure 5 illustrates S2LBP with edge information (S2LBP_EI). Since each of the histograms is 256 dimensional, the length of the final feature vector for S2LBP_EI is 1024 and for S4LBP is 2048.

In Fig. 5, two histograms for S2LBP are calculated using the green and the red spirals separately. Then, the edge maps are generated using Canny edge detector. S2LBP is applied to the pixels in the edge map using the blue and the light blue spirals. Finally, four histograms obtained from the spirals concatenated to get 1024-dimensional feature vector.

The procedure to obtain feature vector for S4LBP_EI is the same as S2LBP_EI. In this case we need to calculate S1LBP codes for eight histograms. Then we have 2048-dimensional feature vector to describe the image.

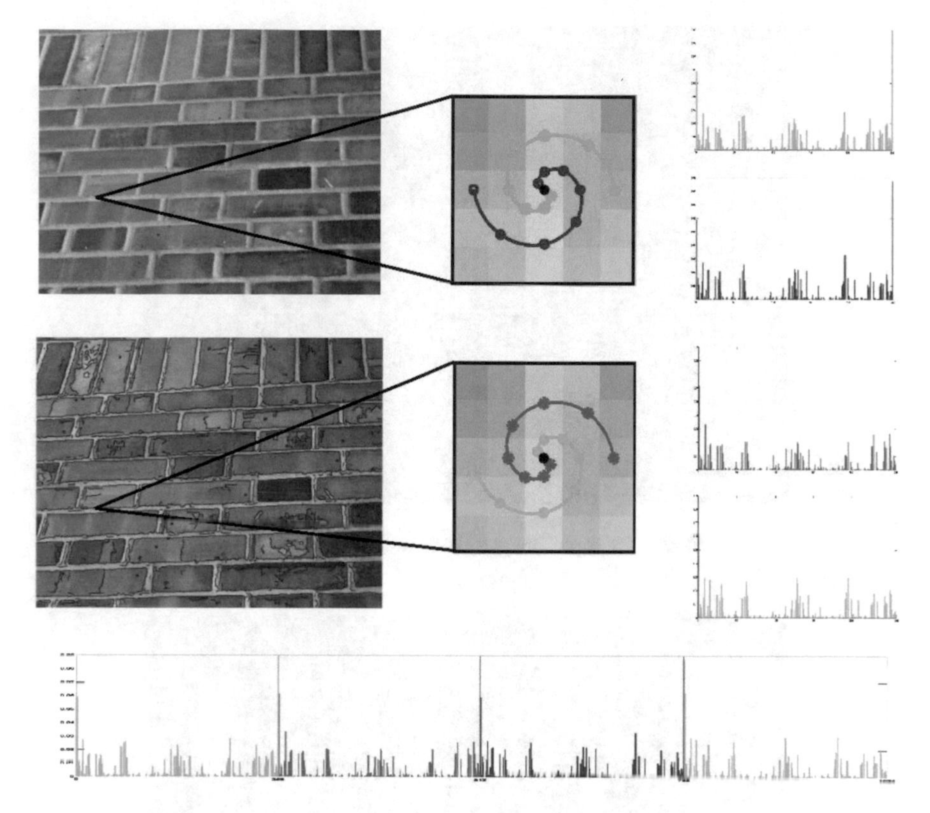

Fig. 5. An illustration of S2LBP_EI feature vector generation

5 Experimental Results

The performances of LBP, S1LBP, S2LBP and S4LBP are compared with LBP_EI, S1LBP_EI, S2LBP_EI and S4LBP_EI on UIUC [16] and CURet [17] texture databases.

UIUC texture database consists of 40 texture classes and each class involves 25 texture images which have strong viewpoint and scale differences. Furthermore, the dataset includes non-rigid deformations, illumination changes and viewpoint-dependent appearance variations. The resolution of each sample is 640×480 pixels. Figure 6 shows example images from each class.

CURet texture database contains 61 samples. Each class has 205 images which acquired at different viewpoints and illumination conditions. 92 images from which a sufficiently large region could be cropped (200×200) across all texture classes are selected [19]. We converted all the cropped regions to gray-scale. Figure 7 shows example images from each class.

In all variants of topologies, neighbor number is set to 8. The grey level values of the sampling points that do not overlap with a pixel center is calculated using bilinear interpolation. For S1LBP, Archimedean spiral starts from the center pixel and stops

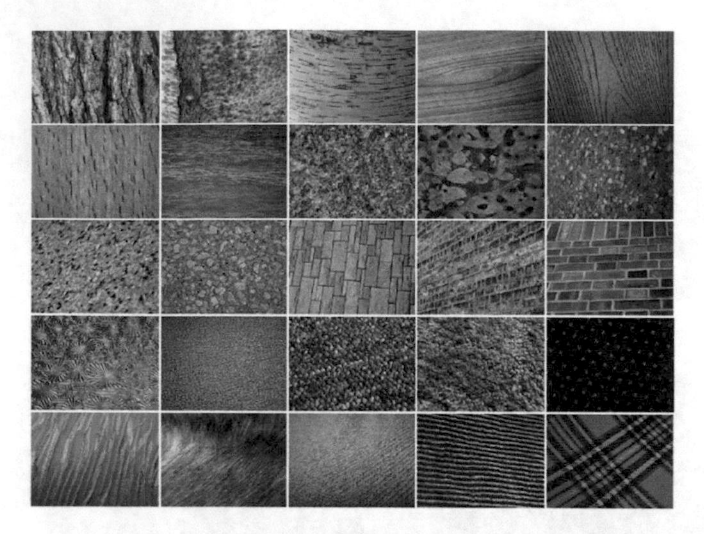

Fig. 6. Sample images from each class of UIUC

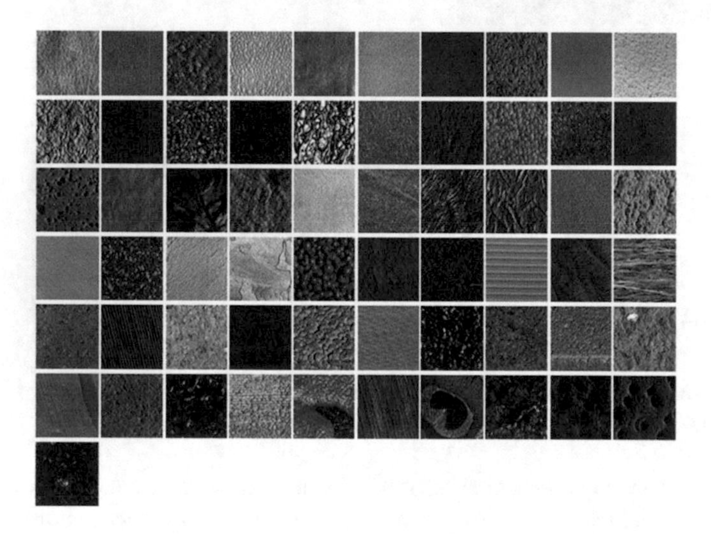

Fig. 7. Sample images from each class of CURet

after 2π radians angular movement along the curve. Sampling points are located with $\theta = \pi/4$ angular intervals on the spiral. The locations of the sampling points around the center pixel are determined by adding π to θ in Eq. 2 for the second spiral in S2LBP. Similarly, we add $\pi/2$, π, and $3\pi/2$ to θ in Eq. 2 to get the sampling points on the other three spirals in S4LBP. Then we apply S1LBP and S1LBP_EI to each of the spirals. Finally, we concatenate the obtained histograms in order to get feature vector.

In the classification stage we use Chi-square test and Linear Regression Classification (LRC) as classifier. Chi-square is a well-known statistical test mostly used to measure similarity between two histograms. The latter is a subspace-based method generally used in face recognition problems. To get statistically significant experimental results, N training images are randomly selected from each class while the remaining $(40 - N)$ images per class for UIUC texture database and M training images are randomly selected from each class while the remaining $(92 - M)$ images per class for CURet texture database are used as the test set. The average accuracies over 10 randomly splits and the standard deviations are listed in Tables 1 and 2 for UIUC and CURet databases respectively. Tables 3 and 4 represent the recognition rates of LRC method in UIUC and CURet databases respectively. The best recognition rates are highlighted in bold.

Table 1. Recognition rates for UIUC texture database with Chi-square

Method	30	20	15	10
LBP	64.60 ± 2.61	58.72 ± 1.77	53.66 ± 1.69	47.86 ± 1.88
LBP_EI	65.72 ± 2.66	62.02 ± 1.75	58.53 ± 1.71	52.82 ± 1.20
S1LBP	63.84 ± 2.87	57.04 ± 1.94	52.19 ± 1.67	46.46 ± 1.18
S1LBP_EI	66.06 ± 2.58	**63.68 ± 1.92**	58.70 ± 1.64	52.24 ± 1.73
S2LBP	64.44 ± 2.76	57.46 ± 2.04	52.49 ± 1.84	46.73 ± 1.21
S2LBP_EI	66.12 ± 2.84	63.05 ± 2.11	58.71 ± 1.72	53.65 ± 1.97
S4LBP	63.84 ± 2.77	57.02 ± 1.96	52.32 ± 1.85	46.69 ± 1.38
S4LBP_EI	**67.01 ± 2.45**	63.56 ± 1.82	**58.72 ± 1.79**	53.67 ± 2.03

Table 2. Recognition rates for CURet texture database with Chi-square

Method	46	23	12
LBP	91.09 ± 0.43	85.28 ± 0.51	77.83 ± 0.54
LBP_EI	92.86 ± 0.20	85.97 ± 0.45	78.91 ± 0.60
SLBP	91.73 ± 0.36	86.20 ± 0.42	78.61 ± 0.64
SLBP_EI	93.61 ± 0.63	87.02 ± 0.51	79.08 ± 0.41
S2LBP	92.11 ± 0.32	86.66 ± 0.43	78.98 ± 0.64
S2LBP_EI	94.80 ± 0.89	87.98 ± 0.48	81.54 ± 0.56
S4LBP	92.33 ± 0.38	87.03 ± 0.42	79.45 ± 0.59
S4LBP_EI	**96.34 ± 0.58**	**89.59 ± 0.46**	**82.17 ± 0.99**

Table 3. Recognition rates for UIUC texture database with LRC

Method	30	20	15	10
LBP	85.60 ± 2.01	78.14 ± 2.36	72.59 ± 2.42	61.94 ± 2.06
LBP_EI	88.80 ± 2.30	81.42 ± 1.84	76.22 ± 2.88	65.84 ± 2.46
SLBP	85.92 ± 1.76	80.08 ± 2.20	74.48 ± 2.24	64.01 ± 1.77
SLBP_EI	89.08 ± 1.33	83.44 ± 1.90	78.62 ± 2.02	68.25 ± 1.85
S2LBP	87.28 ± 2.56	81.48 ± 1.96	75.47 ± 1.89	64.44 ± 1.89
S2LBP_EI	88.85 ± 1.92	82.90 ± 2.07	78.69 ± 2.16	68.96 ± 1.98
S4LBP	88.04 ± 1.88	81.92 ± 2.39	75.72 ± 2.19	64.61 ± 2.08
S4LBP_EI	**89.84 ± 1.65**	**83.50 ± 1.67**	**78.80 ± 1.79**	**69.08 ± 2.20**

Table 4. Recognition rates for CURet texture database with LRC

Method	46	23	12
LBP	97.14 ± 0.36	95.26 ± 0.52	89.52 ± 0.92
LBP_EI	97.76 ± 0.24	95.31 ± 0.38	89.11 ± 0.69
SLBP	97.80 ± 0.17	95.98 ± 0.36	90.54 ± 0.93
SLBP_EI	97.92 ± 0.24	95.58 ± 0.31	89.71 ± 0.65
S2LBP	98.61 ± 0.13	96.68 ± 0.49	91.14 ± 0.88
S2LBP_EI	98.72 ± 0.56	96.84 ± 0.47	91.40 ± 0.72
S4LBP	98.82 ± 0.17	96.83 ± 0.40	91.32 ± 0.86
S4LBP_EI	**99.24 ± 0.70**	**97.59 ± 0.42**	**91.97 ± 0.88**

Experimental results in the both databases show that LRC has better classification performance than chi-square test. Tables show that, S4LBP is the best among the all topologies. We can conclude from the tables that adding edge information improves the classification accuracy. Another finding from the tables that increasing the number of training samples increases the recognition accuracy as expected.

6 Conclusions

LBP is one of the most well-known feature extraction method which reveal the local properties of an image. Also, it is known that edges in an image carry discriminative information that can be used for classification purposes. In this study, the edge information is used to increase the discrimination ability of the histograms obtained as a result of LBP and its variants with various topologies, in texture recognition. We extend the previously proposed S1LBP_EI feature extraction algorithm to two-spirals and four-spirals case. We use LRC and χ^2 test to investigate the performance of the proposed descriptors in terms of classification accuracy. The classification tests show that the proposed extensions (S2LBP_EI and S4LBP_EI) have important discriminative information which improves the classification accuracy.

Experimental results show that LRC has better classification performance than chi-square test. The other note according to results is that LBP with spiral topology defines texture images better than LBP with circular topology. The last and most significant improvement is that edge information is efficiently used with LBP variants in both databases.

Increasing the feature vector size increases the computational cost which is the major drawback of our proposed method. As a future work, we aim to propose a feature selection approach which is not only decreases the feature vector size but also improves the recognition performance.

References

1. Ojala, T., Pietikainen, M., Harwood, D.: A comparative study of texture measures with classification based on feature distributions. Pattern Recogn. **29**(1), 51–59 (1996)
2. Ojala, T., Pietikainen, M., Maenpaa, T.: Multiresolution gray-scale and rotation invariant texture classification with local binary patterns. IEEE Trans. PAMI **24**(7), 971–987 (2002)
3. Liao, S., Chung, A.: Face recognition by using elongated local binary patterns with average maximum distance gradient magnitude. In: Computer Vision, ACCV 2007. Lecture Notes in Computer Science, vol. 4844, pp. 672–679. Springer, Berlin (2007)
4. Nanni, L., Lumini, A., Brahnam, S.: Local binary patterns variants as texture descriptors for medical image analysis. Artif. Intell. Med. **49**, 117–125 (2010)
5. Kazak, N., Koc, M.: Performance analysis of spiral neighbourhood topology based local binary patterns in texture recognition. Int. J. Appl. Math. Electron. Comput. **4**, 338–341 (2016)
6. Kazak, N., Koc, M.: Some variants of spiral LBP in texture recognition. IET Image Proc. **12** (8), 1388–1393 (2018)
7. Naseem, I., Togneri, R., Bennamoun, M.: Linear regression for face recognition. IEEE Trans. Pattern Anal. Mach. Intell. **32**(11), 2106–2112 (2010)
8. Naseem, I., Togneri, R., Bennamoun, M.: Robust regression for face recognition. Pattern Recogn. **45**, 104–118 (2012)
9. Koc, M., Barkana, A.: Application of linear regression classification to low dimensional datasets. Neurocomputing **131**, 331–335 (2014)
10. Suruliandi, A.: Local binary pattern and its derivatives for face recognition. IET Comput. Vis. **6**, 480–488 (2012)
11. Ergul, S., Koc, M.: Analysis of LRC performance with LBP features in face recognition. In: 6th International Conference on Advanced Technologies (ICAT 2017), Istanbul, Turkey (2017)
12. Sun, J., Fan, G., Wu, X.: New local edge binary patterns for image retrieval. In: Proceedings of the 20th IEEE International Conference on Image Processing, pp. 4014 4018 (2013)
13. Lin, J., Chiu, C.T.: LBP edge-mapped descriptor using MGM interest points for face recognition. In: IEEE International Conference on Acoustics, Speech and Signal Processing (ICASSP), New Orleans, LA, pp. 1183–1187 (2017)
14. Vipparthi, S.K., Murala, S., Gonde, A.B., Wu, Q.M.: Jonathan: local directional mask maximum edge patterns for image retrieval and face recognition. IET Comput. Vision **10**(3), 182–192 (2016)
15. Kaur, H., Dhir, V.: Local maximum edge cooccurance patterns for image indexing and retrieval. In: International Conference on Signal and Information Processing (IConSIP), Vishnupuri, pp. 1–5 (2016)
16. Lazebnik, S., Schmid, C., Ponce, J.: A sparse texture representation using local affine regions. IEEE Trans. Pattern Anal. Mach. Intell. **27**(1), 1265–1278 (2005)
17. Dana, K., Van-Ginneken, B., Nayar, S., Koenderink, J.: Reflectance and texture of real world surfaces. ACM Trans. Graph. (TOG) **18**(1), 1–34 (1999)
18. Kazak, N., Koc, M.: Improving the Performance of spiral local binary pattern using edge information. In: XIII International Scientific and Technical Conference Computer Science and Information Technologies (CSIT), pp. 364–367. Lviv, Ukraine (2018)
19. Varma, M., Zisserman, A.: A statistical approach to texture classification from single images. Int. J. Comput. Vis. **62**(1–2), 61–81 (2005)

An Overview of Denoising Methods for Different Types of Noises Present on Graphic Images

Oleh Kosar(✉) and Nataliya Shakhovska(✉)

Lviv Polytechnic National University, Lviv 79013, Ukraine
oleh.kosar94@gmail.com, nataliya.b.shakhovska@lpnu.ua

Abstract. It has been established that one of the reasons for the complication of the decision-making process is the deterioration of the quality of the input information obtained on the basis of various images due to overlaying noise on them, which may have different origin and characteristics. Studying a certain class of noise in the context of considering it as a function allows you to focus on determining its parameters, the degree of influence of these parameters and the artificial noise generation. An overview of the noise of different types and their effects was performed for further evaluation of the quality of recognition systems. Noises that arise in this case, are subject to classification in order to study, formalize and further eliminate or minimize their harmful effects. Studying a certain class of noise in the context of considering it as a function allows you to focus on determining its parameters, the degree of influence of these parameters and the artificial noise generation. Research shows that there are many types of noise that negatively affect the processing and analysis of images. An overview of various types of noise - Gaussian noise, shot noise (Poisson noise), noise type "salt and pepper" (impulse noise), noise of grains of a film, speckle noise, noise, giving a blur effect (they can be imposed with different degree of transparency); the features of overlaying such noise are determined. Also, the listed types of noise can be superimposed on each other. The method of logical generalization, overlaying of image noise using the Matlab environment is used. Comparison of several noise that creates the effect of blurriness when applied to images with varying degrees of transparency. Generating different noise leads to further overlay on real images of special noise masks with given parameters values - such as the intensity and size of the noise, the law of distribution of their centers, and so on.

Keywords: Donoising · ICA filters · Mean filter · Median filter
Wiener filter

1 Introduction

In process of images processing and recognition in particular, there is a problem of identifying the features that represent the images and can subsequently be treated as input signals for recognition devices. "Low-level" processing of each pixel of digital images has limited usage and is rather computationally expensive. Instead, there are

N. Shakhovska and M. O. Medykovskyy (Eds.): CSIT 2018, AISC 871, pp. 38–47, 2019.
https://doi.org/10.1007/978-3-030-01069-0_4

several types of "higher-level" features. Among such features is the structure of the histograms of the images brightness with different amount of quanta - for analysis of the color spectrum of images.

An analysis of the edges shape is also used with using so-called chain codes. An object outline can be detected using a special convolution filter. Such detection is based on the brightness difference that takes place on the verge of two areas collision. The Harris detector makes it possible to determine the angles of objects as features.

Gabor filters and Independent Component Analysis filters (ICA filters) are used to detect the features basing on the image texture.

Extremely wide range of digital images areas of usage has led to further work and studies related to noises removal (minimization of their harmful effects). Such noises, as one of the main factors of image quality impairment, distort the features and accordingly reduce the correctness of subsequent processing operations - classification, segmentation and other.

2 State of Arts

Based on the analysis of literary sources, it has been established that noise is an overlay image of a pixel mask of random color and brightness. Regarding the origin of noise, it is worth noting the possibility of their occurrence due to the unessential nature of the equipment used, the adverse external conditions (e.g., weather, level of illumination, sensor temperature), obstacles in the transmission channel, etc.

Noises that arise in this case, are subject to classification in order to study, formalize and further eliminate or minimize their harmful effects. Studying a certain class of noise in the context of considering it as a function allows you to focus on determining its parameters, the degree of influence of these parameters and the artificial noise generation. Such generation results in a further overlay on real images of special noise masks with given parameter values - such as the intensity and size of the noise, the law of distribution of their centers, etc.

In practice, various methods of denoising and, consequently, image quality improvements are used, each of which has its advantages/disadvantages and application areas. These groups of methods include linear, nonlinear, and morphological filters; noise removal using wavelets; regularization; multichannel, multiframe, and iterative image recovery [1].

One of the most widespread types of noise is the Gaussian noise that has zero mathematical expectation for the distributed values. Accordingly, this noise can be eliminated by averaging the pixels values of a certain unit range (mean filtering). Such unit range is usually called window. In this case the size of window is usually picked as square of 33, 55, and so on. On the other hand, this averaging leads to blurring of the image edges and loss of details. At the same time, one pixel, which is very different from the others, will noticeably affect the resulting values for the entire range. This noise cancellation method is also often used for other noises such as "salt-pepper" noise, but it is not always effective. There is also a modification of this method using k-means filtering.

To remove "salt-pepper" noise a median filtering is often used that involves choosing a single pixel value range that is in the middle of its brightness level row [2]. Such a filter requires pre-sorting of the range values, that increases its computational complexity. This filter does not leave problems of edges blurring.

A filter of alpha-trimmed means can be also applied to eliminate noise on the images. It is a combination of a mean filter and median filter. It uses the average of the ascending ordered values of the range, but with the truncation of a certain part (α) of the ordered range at both ends. This allows you to get rid of those points of the range that are the most different from the others and often allow you to increase the adequacy of the calculated mean. In [3], we also refer to the interrelation of the α-truncated means filter and median filter.

To remove the Gaussian noise, a Wiener filter is also used, for which the point-scattering function is selected in terms of minimizing the mean square error between the original (non-distorted) and the restored image [1, p. 131]. The Wiener filter can have three types of implementations: non-causal systems (requiring an unlimited number of past and future data), causal systems (requiring an unlimited amount of past data), and implementation of a finite-pulse response [4]. In [5] it is also said that the disadvantage of the Wiener filter is that its application in practice requires certain a priori knowledge of the original image spectrum and noise to select parameters and thresholds. The usage of this filter is also often worsened by having it blurred the image edges.

Another technique for noise removing is the nonlinear technique of bilateral filtering. This technique was also designed for tasks such as texture removal, outline highlighting, and video correction [6]. The essence of this technique is to calculate the pixel brightness value as a weighted average value of the bright nesses of the neighboring pixels. The weights are based not only on the Euclidean distance from the target pixel (as for the Gaussian convolution), but also on the radiometric characteristics - the intensity difference, the depth of the color, and so on. The bi-lateral filter can be defined as:

$$I^{filtered}(x)\frac{1}{W_p}\sum_{x_i\in\Omega} I(x_i)f_r(\|I(x_i)-I(x)\|)g_s(\|x_i-x\|). \tag{1}$$

Where the normalization coefficient W_p:

$$W_p=\sum_{x_i\in\Omega} f_r(\|I(x_i)-I(x)\|)g_s(\|x_i-x\|). \tag{2}$$

In this case: I and $I^{filtered}$ are the original image and image after filter applying respectively; x is the position of the corresponding pixel; Ω - is the window centered at x; f_r - window kernel to smooth the difference in intensity; g_s - a spatial kernel for smoothing the difference of coordinates (may be a Gaussian function).

The bilateral filter works well with preserving the edges with a stepwise change in intensity. However, this filter is not very effective to maintain a sharp change in intensity for ridge- and valley-like edges on the image [7]. Proceeding from this, the extensions of the bilateral filter are a trilateral filter, a symmetric bilaterial filter, a regression filter.

It is also known about the usage of morphological operations to remove noises and obstacles. Mathematical morphology is often used to extract image properties, that are useful for its representation and description, such as outlines extracting. Morphological operations are also used for pre-processing or final processing of images - for example, thickening or thinning. It uses additional input data - a structural element that has some form and is much smaller than the main image.

Speaking about the noise removal, morphological operation of erosion will replace the central pixel by the darkest pixel of the range, when the morphological operation of extension will replace it by the brightest pixel. The complex operation of opening (erosion with the following extension) is well suited for removing "salt" and the operation of closing (extension and subsequent erosion) is suitable for removing "pepper". [8] deals with operations of mathematical morphology and proposes an algorithm for detecting the edges of the noised image (with some specific noise type) in context of medical images and task of human organs recognition.

Image averaging method is also used to eliminate noise - it can bring good results if the level of illumination or other external conditions affecting the image construction are changing. However, in this case the angle must be unchanged, as well as the scene in the image. At the same time, assuming that the average noise value at a given point is 0, we can say that with increasing number of images used for averaging, the resulting image will go to noiseless [1, p. 31]. The additional advantage of this is that averaging can increase the color depth of the image to such an extent that it could not be achieved with a single shot. A similar task is to detect changes in video by intercepting video images at short intervals and finding their "differences."

Next, let's look at a few specific methods for removing noises of other types

[9] contains information about the specifics of the Poisson noise removing using a nonlocal mean value filter, a bilateral filter and the BM3D algorithm.

In [10] you can find information about the specifics of the film grain noise eliminating. The authors emphasize on the importance of computing high order statistics for such images based on the nonlinear relation of such noises to the original image. Also, calculation of such statistics will contribute to a more reliable evaluation of the parameters of grain film noise.

A number of articles consider the removal of speckle noise for images of ultrasound diagnostics. Thus, the method description in [11] involves a wavelet-based definition of the threshold, and the method described in [12, 13] requires logarithmic transformations and a nonlinear diffusion tensor.

3 Evaluating the Effectiveness of Noise Elimination Techniques

When eliminating noise we can evaluate the retrieval efficiency, which corresponds to the degree of proximity of the restored image to the original (ideal, noiseless). This characteristic can be estimated as the ratio of the number of pixels whose values coincide with the original (with a certain allowed margin of error) to the number of pixels of the original image. The noise level, respectively, can be estimated as the ratio of the number of noise pixels to the number of pixels of the original image.

Also, in order to evaluate the quality of noise elimination algorithms, higher-level metrics of image quality estimation are used. Such metrics can be divided into 3 groups according to the originality of the original (noiseless) image:

- with the original image present;
- with the original image partly present (there are some parts of the original image);
- with the missing original image.

Quite often peak signal-to-noise ratio (PSNR) is used. This characteristic assumes the presence of an original image, is determined using logarithms and measured in decibels (dB):

$$PSNR = 10\log_{10}\left(\frac{MAX_i^2}{MSE}\right). \tag{3}$$

With MAX_i^2 to be the maximum value taken by the pixel of the image, and MSE is mean square error for the original image i and noisy image K:

$$MSE = \frac{1}{mn}\sum_{i-0}^{m-1}\sum_{j=0}^{n-1}|I(i,j) - K(i,j)|^2. \tag{4}$$

Figure 1 presents the original image. After that we add different types of noise and try to use denoising methods. The developed Matlab script imposes filters on a black and white image to clear the noise of different types. After that we calculate PSNR value for each case.

For a noiseless image, the value of the mean square error is 0, and the PSNR goes to infinity. The default value for a 256-tone image is from 30 to 50 dB. Generally, increasing the PSNR means improving image quality. Next on Figs. 2, 3, 4 and 5 the results of noise different types overlaying are given along with their removing based on some of listed methods.

Fig. 1. Original image

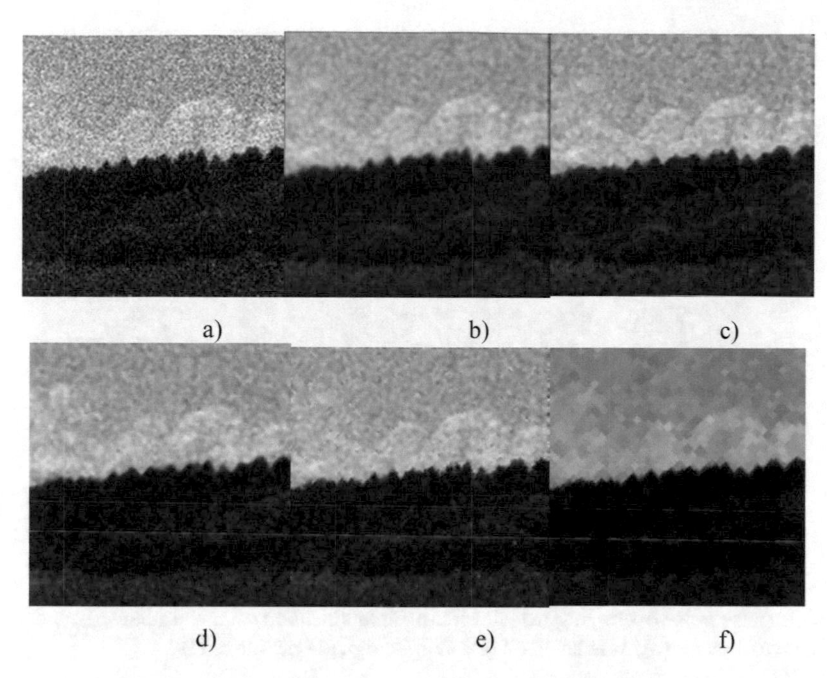

Fig. 2. Gaussian noise overlaying (a) with following application of: mean filter (b), median filter (c), α-trimmed mean (d), Wiener filter (e), morphological operations (f)

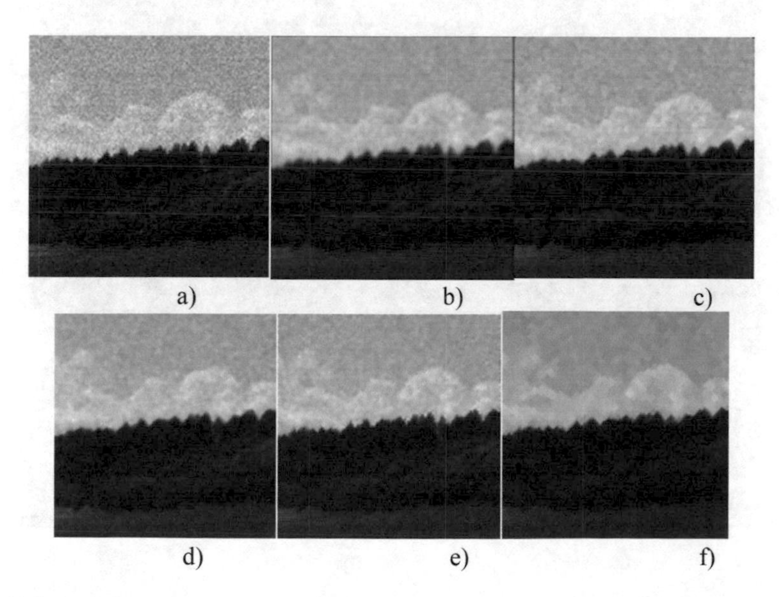

Fig. 3. Poisson noise overlaying (a) with following application of: mean filter (b), median filter (c), α-trimmed mean (d), Wiener filter (e), morphological operations (f)

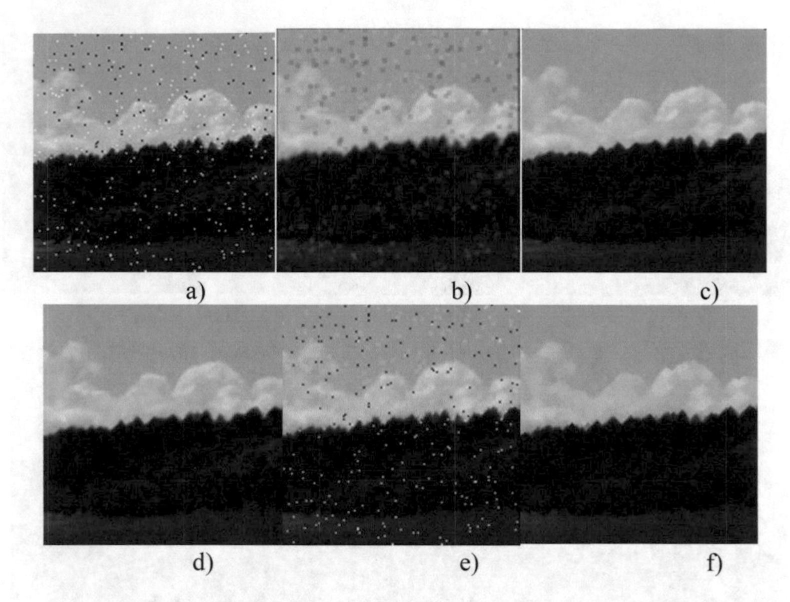

Fig. 4. Impulse noise overlaying (a) with following application of: mean filter (b), median filter (c), α-trimmed mean (d), Wiener filter (e), morphological operations (f)

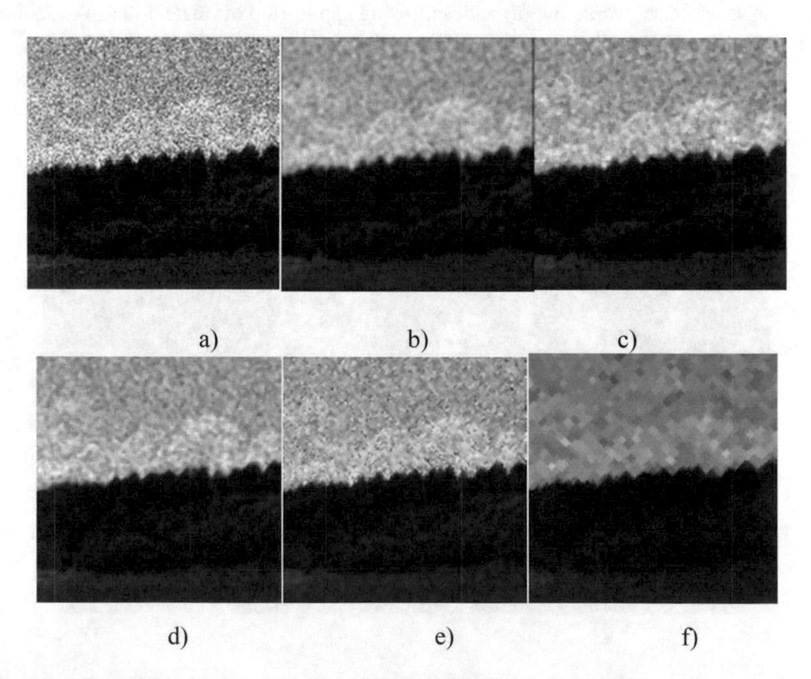

Fig. 5. Speckle-noise overlaying (a) with following application of: mean filter (b), median filter (c), α-trimmed mean (d), Wiener filter (e), morphological operations (f)

The table after that contains information about PSNR value for each case (Table 1).

Table 1. Noise removing methods comparison for different types of noises based on PSNR

	Gaussian noise	Shot noise (Poisson noise)	Impulse noise ("salt-pepper")	Speckle noise
Without removing	20.365	27.760	21.871	19.991
Mean filter	26.802	28.996	27.453	26.650
Median filter	26.757	32.055	35.171	26.321
Wiener filter	27.804	34.110	22.969	24.808
Morphological operations (with "diamond" as structural element)	20.437	27.102	32.792	19.890
α-trimmed mean filter	27.711	33.138	36.424	27.393

As can be seen from the conducted research: the Wiener method and the α-trimmed mean filter proved to be most effective to eliminate Gaussian noise; the result for the medium filter and the median filter was slightly worse. Almost all noise has remained on the image with the use of morphological operations.

There are clear trends in eliminating impulse noise ("salt-pepper"), where the α-trimmed mean filter, the median filter and morphological operations were most effective. The worst removal results are for the Wiener filter and the mean filter.

With regard to Poisson's noise, one can see that it is better to remove it with the Wiener filter, the α-trimmed mean and the median filter; and the speckle noise - with the α-trimmed mean filter, the mean filter, and the median filter (Wiener filter application has significantly worse result).

So, it should be noted that, in most cases, the α-trimmed mean filter was very effective, while Wiener filter effectively eliminated Gaussian noise and the median filter was very effective for "salt-pepper" noise.

However, the results of this study are limited to this type of image and the simultaneous application of only one noise and filter. Also, as a limitation, we can consider the noise measuring metric used. It is possible to move towards eliminating this restriction in further researching.

4 Discussion

As mentioned above, currently the technology of denoising, computer vision and tools for image recognition are actively developing. For qualitative recognition, it is important to identify elements without noise. It is huge problem for images processing taken from a drone or from space. Small parts of the image due to the presence of noise may not be recognized. Global technology leaders invest in research and development a lot of money because this problem has wide opportunities for use. The direction for the development of denoising technologies is primarily the protection and increased protection systems. Currently the most used in entertainment, I believe this trend will grow also. Proposed approach can be used for big data processing [13, 14].

5 Conclusions

An overview of the noise of different types and their effects was performed for further evaluation of the quality of recognition systems. Research shows that there are many types of noise that negatively affect the processing and analysis of images. The article describes and describes their characteristics and origin. Among these noises are Gaussian noise, Poisson noise, pulse noise, film noise, speckle noise, impulse noise and others.

Various approaches to the removal of the three main types of noise: Gaussian noise, impulse noise, Poisson noise, combined noise. An adaptive method for removing these noise. On its basis, an algorithm was developed and implemented in Matlab environment. A system was created that removes the listed noises more effective than existing systems. In addition, this system is universal and itself is adjusted according to the available noise. A comparative analysis was carried out.

In the future, it is planned to adapt these approaches for processing not only black and white, but also color images. It is planned to create hybrid algorithms, combining the effective components of new and standard algorithms.

In the context of evaluating the quality of image recognition systems, it's worth noting that noise is not the only type of interference. Performing such actions on the image as, for example, purposeful modification, rotation or zooming of the image will also have a negative effect on the image resolution. However, in this paper we consider only the overlay of noise.

References

1. Bovik, A.C.: The Image and Video Processing Handbook, 1st edn. Academic Press, Boston (2000)
2. Bovik, A.C., Huang, T.S., Munson D.C.: The effect of median filtering on edge estimation and detection. IEEE Trans. Pattern Anal. Mach. Intell. PAMI **9**(2), 181–194 (1987)
3. Bednar, J., Watt, T.: Alpha-trimmed means and their relationship to median filters. IEEE Trans. Acoust. Speech Signal Process. **32**(1), 145–153 (1984)
4. Xiao-dan, Z., Xuan-chi-cheng, L., Mei, L.: The implementation of wiener filtering deconvolution algorithm based on the pseudo-random sequence. Am. J. Circuits, Syst. Signal Process. **2**(1), 1–5 (2016)
5. Vijaykumar, V.R., Vanathi, P.T., Kanagasabapathy P.: Fast and efficient algorithm to remove gaussian noise in digital images. IAENG Int. J. Comput. Sci. https://pdfs.semanticscholar.org/31aa/963c66d55e78f1489d3a09d52eaa97bc79fe.pdf
6. Rashkevych, Y., Peleshko, D., Vynokurova, O., Izonin, I., Lotoshynska, N.: Single-frame image super-resolution based on singular square matrix operator. In: 2017 IEEE First Ukraine Conference on Electrical and Computer Engineering
7. Paris, S., Kornprobst, P., Tumblin, J., Durand, F.: Bilateral filtering: theory and applications. Found. Trends Comput. Graph. Vis. **4**(1), 1–73 (2009)
8. Yu-qian, Z., Wei-hua, G., Zhen-cheng, C.: Medical images edge detection based on mathematical morphology. In: IEEE Engineering in Medicine and Biology 27th Annual Conference, Shanghai, pp. 6492–6495 (2005)

9. Thakur, K.V., Damodare, O.H.: Poisson noise reducing bilateral filter. Procedia Comput. Sci. **79**, 861–865 (2016)
10. Yan, J.C.K.: Statistical Methods for Film Grain Noise Removal and Generation. http://www.nlc-bnc.ca/obj/s4/f2/dsk2/ftp01/MQ28858.pdf
11. Sudha, S., Suresh, G., Sukanesh R.: Speckle Noise Reduction in Ultrasound Images by Wavelet Thresholding Based on Weighted Variance. http://www.ijcte.org/papers/002.pdf
12. Benzarti, F., Amiri, H.: Speckle Noise Reduction in Medical Ultrasound Images. https://arxiv.org/ftp/arxiv/papers/1305/1305.1344.pdf
13. Veres, O., Shakhovska, N.: Elements of the formal model big date. In: XI International Conference on Perspective Technologies and Methods in MEMS Design (MEMSTECH), Lviv, Ukraine, pp. 81–83 (2015)
14. Melnykova, N.: Semantic search personalized data as special method of processing medical information. In: Advances in Intelligent Systems and Computing, pp. 315–325. Springer, Cham (2017)

Using Multitemporal and Multisensoral Images for Land Cover Interpretation with Random Forest Algorithm in the Prykarpattya Region of Ukraine

Olha Tokar[1(✉)], Serhii Havryliuk[2], Mykola Korol[2], Olena Vovk[1], and Lubov Kolyasa[1]

[1] Lviv Polytechnic National University, Lviv, Ukraine
tokarolya@gmail.com

[2] Ukrainian National Forestry University, Lviv, Ukraine

Abstract. Random Forest (RF) classification algorithm was used for investigation of land cover dynamics in the Prykarpattya region of Ukraine. This approach was applied for two types of images – Landsat and Sentinel-2 with different spatial resolution and obtained in different time (multitemporal). For correct comparison of classifications resulting from the Landsat and Sentinel-2 images the same ground truth data were used for forming the signatures for image interpretations. All classifications were done using a script developed for R software involving a special library realizing the Random Forest algorithm for image interpretation. For land cover investigation main land cover types (coniferous forest, deciduous forest, water, urban territory, grasslands, and additionally areas under clouds) were interpreted using different multitemporal images with different (similar) spatial resolutions. For comparison of the results, all images were formed with similar spatial resolution. Accuracy of classification was estimated using a few indexes, including OOB (Out-of-Bag) and Kappa. Using obtained land cover classifications, thematic maps of land cover were formed, and land cover change was analyzed.

Keywords: Land cover · Landsat · Sentinel · Random Forest algorithm Image interpretation · Accuracy assessment · Multitemporal images

1 Introduction

Land cover is very dynamic system, and for its investigation it is important to have objective data. It especially is important for estimation of changes in land cover. It is a very expensive business to use only ground researches (for example geodesy surveying, ground laser scanning, etc.), this takes much time and resources. One of the possible ways of land cover investigation is using remote sensing data. For these tasks, we can use remote sensing data, where we have a lot of possibilities to use. There are different types of remote sensing data – greyscale images, multispectral, hyperspectral, LIDAR, RADAR data, etc. For land cover interpretation multispectral data from different types of sensors with different spatial and spectral resolution are used most often. This is because the multispectral remote sensing market now is most useful for these tasks.

N. Shakhovska and M. O. Medykovskyy (Eds.): CSIT 2018, AISC 871, pp. 48–64, 2019.
https://doi.org/10.1007/978-3-030-01069-0_5

Moreover, for these images be implemented a lot of algorithms and special software for interpretation.

There are a lot of algorithms and approaches for image interpretation. It depends from kinds of images, spectral ranges of image receiving etc. The main approach to automated image interpretation (also known as computer-based classification or digital classification) is spectral characteristic of pixel (coefficient of spectral brightness), where different algorithms are analyzing all pixels to refer them to their corresponding classes. These classes are chosen depending on the task and on the formed signatures.

In practice, it is important to obtain better result with minimum outgoings, in other words with minimum ground data for classification. Its important condition is the possibility to use data base (data set) for image interpretation many times for different kinds of image and for multitemporal images. For some practical tasks it is important to choose algorithm which will satisfy most of conditions.

In general, there are distinguished two main approaches to image interpretation [1, 2]: unsupervised and supervised classifications, additional hybrid classification, which combines some fundamentals from both. These approaches now are dominated in automated image interpretation and have different detections known as algorithms. They differ only in mathematical set, that dividing pixels in spectral range to some classes. For land cover interpretation all these approaches and algorithms could be used.

Different researches prefer different algorithms for land cover interpretation. Baumann et al. [3] used Support Vector Machine (SVM) algorithm for land cover interpretations on Landsat images in Post-Soviet European zone of Russia with next post-classification comparison to monitor forest area, disturbance, and reforestation. They used winter images for increasing accuracy of classification and detecting changes in land cover and land use. The similar approach was used for forest cover changes and detecting of illegal loggings in Ukrainian Carpathians [4].

Kuemmerle et al. [5] used a hybrid classification technique, combining advantages from supervised and unsupervised methods, to derive a land cover map for Polish, Slovak and Ukrainian Carpathians Mountains. For different classes of land cover interpretation they used different algorithms. For water, dense settlement, open settlement, shrubland, grassland, and agricultural land were used hybrid classification, for forest cover land (broadleaf, mixed and coniferous) were used algorithm ISODATA and for high mountains land – knowledge-based image analysis. There were used different approaches depending of distributions of pixels depending to different classes, which often have some overlapping and different algorithms will be in different ways to refer pixels to different classes. Usage of one algorithm can give a lot of misclassification pixels that will give low accuracy statistics and wrong imagine about land cover.

The similar approach was used by Carreiras et al. [6]. They used supervised classification with different algorithms, maximum likelihood (parametric supervised classification method), classification tree (non-parametric method) and *k*-nearest neighbors (*k-NN*) (non-parametric classification). Using these techniques, they investigated agriculture/pasture and secondary succession forest in Brazil. Kimes et al. [7] were used neural networks and linear analysis techniques for mapping primary forest, forest regeneration age classes and deforested areas on SPOT images. Neural networks and linear analysis were employed to single out secondary forest age. The unsupervised classification approach was used for investigation of changes in forest cover with Landsat images for Carpathians [8]. Havryliuk et al. [9] used different band combinations of Landsat images for land cover interpretations with different algorithms of supervised classification: maximum likelihood, minimum distance and Mahalanobis distance.

Now, there are widely used comparatively new approaches in remote sensing data interpretation, which adapted to image classification. Powell et al. [10] classified multitemporal Landsat images for estimation of how forest disturbance and regrowth processes are influencing in carbon dynamics. For these reasons they used three statistical techniques: Reduced Major Axis regression, Gradient Nearest Neighbor imputation and Random Forests regression trees. In their study, the best results were obtained from using Random forest algorithm for biomass estimation. The similar conclusions received Chan and Paelinckx [11]. They were comparing Adaboost and Random Forest algorithms for land use/land cover classification of airborne hyperspectral images. They emphasize that random forest algorithm is "faster in training and more stable". The best results of land cover classification with random forest algorithm has been obtained by Rodriguez-Galiano et al. [12]. This is why we choose Random Forest algorithm for our investigations of land cover.

2 Methods

2.1 Study Area

Our investigations were focused on a small part of Ukraine named Prykarpattya [13]. Prykarpattya is a historical and ethnographic district of Ukraine, which spreads close to Ukrainian Carpathians like a thin strip of territory (Fig. 1).

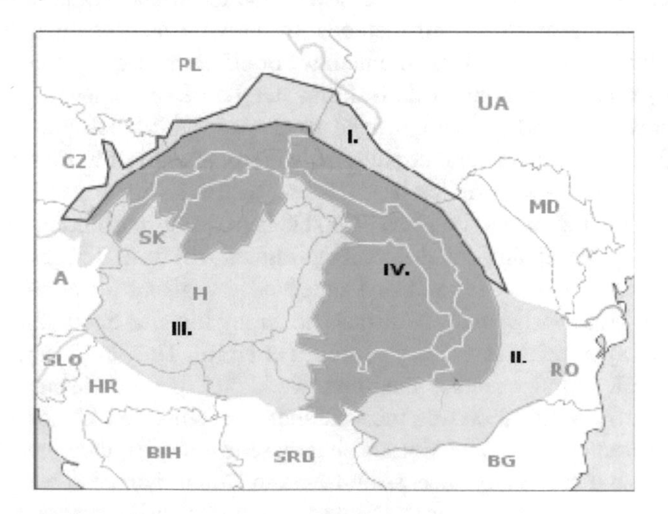

Fig. 1. Map of the Carpathian region, Subcarpathian depressions (I.) outlined in red (downloaded from https://en.wikipedia.org/wiki/Subcarpathia).

Prykarpattya includes parts of two administrative regions of Ukraine: Lviv region and Ivano-Frankivsk region [13]. Prykarpattya belongs to Basin of Dniester River and has different types of land cover including those of flat area and mountain specificity.

For the area of interest, there were chosen multitemporal and multisensoral remote sensing data.

2.2 Remotely Sensed Data

The investigation of land cover calls for using of remote sensing data. Nowadays, there are a lot of different kinds of images which we can receive free of charge or for some payments. There are few global sites for image downloading like USGS Explorer (https://earthexplorer.usgs.gov/). There are a lot of remote sensing data like archive and actual.

For land cover investigation of Prykarpattya region, we used multitemporal and multisensoral remote sensing data, downloaded from USGS Explorer. For these aims, we used the following images:

- Landsat-8 (date of acquisition 2015/05/26);
- Landsat-8 (date of acquisition 2018/05/28);
- Sentinel-2 (date of acquisition 2018/05/29).

We choose these remote sensing data because they have almost similar season of receiving and differ by three years. Landsat images and images from Sentinel have different types of sensors, differ in their spectral ranges of surveying and have different spatial resolutions.

Satellite Landsat-8 has 2 different types of sensors – Operational Land Imager (OLI) and the Thermal Infrared Sensor (TIRS) [14]. Landsat bands have different spatial and spectral resolutions. Those which are from 1 to 7 and 9 bands have 30 m; 8 – 15 m; 10 and 11 bands – 100 m. Spatial resolutions are change with the increase in bands and differ from ultra-blue (band 1) to thermal infrared (band 10 and 11) [15]. Mostly, for land cover investigations are using image bands in optical range of wavelength with the same spatial resolution.

For stacking images, we used these bands – 1 to 7 and 9 with 30 m spatial resolution. Additionally, for we did resolution merging with panchromatic band (8 band of Landsat-8 with 15 m spatial resolution). For it we used the Principal component's method with the Nearest neighbor technique [1]. This method are using for merging data with multiple to 2 input data set, that is spatial resolution of multispectral stacked image is equal to 30 m, panchromatic band resolution is equal to 15 m, it means 30/15 = 2. In results we obtained multispectral image with 15 m spatial resolution.

For Prykarpattya, region we used multitemporal Landsat images with 185 path and 26 row (Fig. 2), which were clipped in overlapping with Sentinel image.

Sentinel-2 is a comparatively new European satellites, which have on their boards multispectral sensor (MSI) with high spatial and spectral resolution. It produces 13 spectral bands with different spatial resolutions: four bands at 10 m, six bands at 20 m and three bands at 60 m [16]. At once it surveys the path 290 km width. The spectral range of Sentinel images covers range from visible to Short wave Infrared spectrum similar to Landsat-8.

The parts of these images were clipped on overlapping for next investigations of land cover in Prykarpattya region (see Fig. 2).

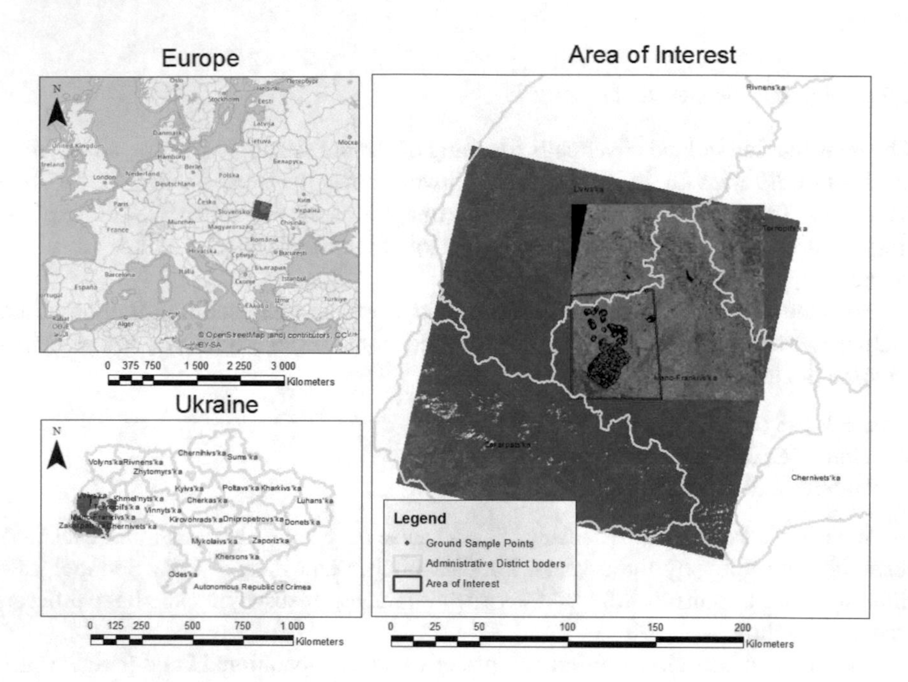

Fig. 2. Area of interest for land cover investigation in Prykarpattja region.

2.3 The Ground Truth Data

Image interpretations with the supervised techniques require having data set with class labels. Mostly these data could be ground checked points/areas or points chosen as a result of photointerpretation. The field sample plots are more precise, but often it is too expensive and resource consuming. Therefore, for increasing data set it is useful to apply photointerpretation approach for gathering ground truth data.

For our investigations, different approaches for gathering ground truth data were used. The first stage was to randomly generate points on area of interest with its next checking on the ground. For non-forest classes these points have been checked on the ground using their coordinates with GPS system. For forest cover lands were conducted sample plots using base of point cruising. The compositions of tree species were estimated on the circle sample plots with relascope of Bitterlich. Additionally, the squared and rectangular sample plots with the wholesale measuring of all trees on sample plots were used. Using these data all points and areas were separated on 2 classes – coniferous and deciduous lands. Because, the forest cover lands are usually separating from other land cover with difficulty, extra points from two sources were additionally chosen. Firstly, the digital forest inventory maps (in shape format with georeferencing) for this territory and separated points with coniferous and deciduous labels were used. Secondly, high resolution images were used on Google Earth platform and these classes were interpreted on QuickBird images. Totally 2637 points were used, where 1020 are coniferous and 856 – deciduous labels. For non-forest classes ground truth data were interpreted on QuickBird images on Google Earth software. An approximately equal number of sample points were chosen for all classes. Less representative points were chosen for well separated classes, and for worse separated ones – more representative points.

All sample points for all classes were merged in one file and separated on two data sets – 70% for training data and 30% for the test data. This was done randomly using algorithm described on Technical Support site from ESRI [17]. In results we received two data set: one is for classification and another one – for accuracy assessment of classification.

Peculiarities of forming data set were class of clouds, which were defined according to every image. Multispectral sensors are very sensitive to clouds cover and often it is not possible to receive images without clouds. Our input images were received in different times when clouds were different and we were forced to form representative points for this class separately for every image.

Generally, signatures for 6 classes of land cover were formed: deciduous forest, coniferous forest, water land, urban land, grassland land and clouds for next classification with Random Forest algorithm.

2.4 The Random Forest Algorithm

The algorithm of Random Forest classification now is widely used in different types of investigation [10–12, 18]. This algorithm can be successfully used in image classification including land cover.

The basic of RF algorithm as statistical method is described by Breiman [19] and some overview for image classification – by Richards [20]. The idea of RF algorithm by Breiman [19] describes the following: "Random forests are a combination of tree predictors such that each tree depends on the values of a random vector sampled independently and with the same distribution for all trees in the forest". Algorithm is working the following way: we have *N* pixels in training set and algorithm selecting first pixel randomly, conducting calculations and returning this pixel to the training set. The next one selects second pixel and so on. These calculations are conducted for training tree in random forest and splitting nodes in tree using Gini index. The peculiarity of RF algorithm is that it does not need to have some additional data for testing performance of tree forming and splitting nodes, but uses the original data set with one third of set for estimation. The accuracy estimation of algorithm training is conducted on OOB (Out-of-Bag) value.

Richards [20] emphasizes two requirements for working algorithm successfully: the randomly generated trees have to be uncorrelated and individual trees should be strong classifiers. The RF algorithm has a few advantages, such as internal estimates of error, strength, correlation and variable importance, is very fast, simply and easy parallelized [20] and gives high precise of classification (using RMSE estimator) for biomass interpretation [10] and does not need a lot of training data set.

Taking into account these advantages we choose this algorithm for land cover investigation. In *R* software a script for classification was formed. We used it in the previous investigation [21] where not only classification process was realized but accuracy assessment and visualization of classification. Below this script is presented.

```
library(sp)
library(rgdal)
library(raster)
library(randomForest)
sdata <- readOGR("G:/...", "Merge_70_Clip_Sent")
tdata <- readOGR("G:/...", "Merge_30_Clip_Sent")
r <- stack(paste("G:/...","T34UGV_20180529T092029_B0stack_10m_Clip_Z35.tif",
sep="/"))
sdata@data <- data.frame(sdata@data, extract(r, sdata))
tdata@data <- data.frame(tdata@data, extract(r, tdata))
plot(sdata@data,col=rainbow(1000))
print(sdata@data[1:10,])
print(tdata@data[1:10,])
rf.mdl <- randomForest(x=sdata@data[,4:ncol(sdata@data)],
y=sdata@data[,"CLASS"], ntree=200, importance=TRUE,confusion=TRUE, prox-
imity=TRUE,do.trace=10)
plot(rf.mdl)
varImpPlot(rf.mdl, type=1)
print(rf.mdl,confusion,proximity)
plot(outlier(rf.mdl), type="h")
save(rf.mdl,file="G:/.../2018-05-29_Sent-10.RData")
predict(r, rf.mdl, filename="G...", type="response", overwrite=FALSE, pro-
gress="window")
```

2.5 The Accuracy Assessment

As a result of image classification we can receive thematic maps with determined classes. For assessment result of classification different approaches for accuracy assessment are used: different indexes, distances etc. The best results could be received from comparing thematic maps with more precise maps which have been created before. But very often it is not possible to have these materials or their precise is not sufficient. That is why for accuracy assessment different indexes or distances are usually used.

The accuracy assessment can be divided on two parts:

(1) assessment of training data set for possibility of separating and usage for next classification (pre-classification accuracy assessment);
(2) post classification accuracy assessment using ground truth samples, that did not use for classification.

In both cases the users' and producers' accuracy, overall accuracy, index Kappa (also known as Cohens' kappa index) and for RF algorithm OOB (out-of-bag) are the most useful for accuracy assessment of image classification.

Users' accuracy shows variability of some class classification and calculating as:

$$P'_{users} = \frac{n_{ii}}{n_i} \times 100 \tag{1}$$

where n_{ii} – pixel counts of correctly classified to class i; n_i – pixel counts classified to class i.

Producers' accuracy is probability that pixel that belong to some class was correctly classified and calculated as:

$$P'_{produsers} = \frac{n_{ii}}{m_i} \times 100 \tag{2}$$

where m_i – pixel counts of class i from the test data.

Overall accuracy, in contrast to the users' and producers' accuracies which are estimating accuracy for individual class, shows quality of classification for all classes in total and are calculated as:

$$P_{overall} = \frac{\sum_{i-1}^{k} n_{ii}}{N} \times 100 \tag{3}$$

where k – count of classes included into classification (including unclassified class); N – total count of pixels.

Index Kappa shows quality of classification for all classes in total as overall accuracy. But it is more informative about quality of classification, because in contrast to overall accuracy, this index the total counts of pixels in different classes from test data and from classified data and taken into account and are calculated as:

$$k = \frac{N\sum_{i=1}^{r} x_{ii} - \sum_{i=1}^{r}\left(x_{i+} \cdot x_{+i}\right)}{N^2 - \sum_{i=1}^{r}\left(x_{i+} \cdot x_{+i}\right)} \tag{4}$$

where N is a the total number of sites in the confusion matrix, r is a the number of rows in the confusion matrix, x_{ii} is a number in row i and column i, x_{+i} is a total for row i, and x_{i+} is a total for column i.

OOB (Out-of-Bag) estimation was used for the pre-classification accuracy assessment of the training and test data and shows aggregated pixels (in percentage) from the data set, which was not included to the training algorithm.

2.6 The Change Detection

The usage of multitemporal and multisensoral images for classification allows to make changes using detection analysis. It is a very informative instrument for land cover investigation because we can study changes occurred in different time periods.

For this analysis we used the simple instrument in ArcGIS named "Image analysis" that formed thematic maps with all differences between two thematic maps.

In this case Coppin et al. [22] focus on the image acquisition dates and the change interval length (temporal resolution). They propose to use multitemporal images multiple to annual cycles (1, 2, 3… years) for the change detection because than the peculiarities in reflectance caused by seasonal vegetation fluxes and Sun angle differences will be minimized. For our investigation we obtained accomplishment of this recommendation, because we have different images with the same data of acquisition (26/05, 28/05 and 29/05) with differentiations in three years.

Additionally, we conducted analysis of land cover changes by area for better understanding of regularities of changes.

3 Results and Discussion

The land cover investigation was conducted using the RF algorithm in *R* software. For classification the ground truth data was prepared. It was separated into two parts: 70% of points for classification (the raining data) and 30% for checking accuracy (the test data). All classification was spent for all four images (two Landsat-8 and Sentinel-2 with different spatial resolution).

In the performed script not only classification and received thematic maps of interpretation were provided, but also some additional actions for estimation accuracy. In particular, for classification certain libraries are needed. In our script four libraries were expected:

(1) sp – package for importing, manipulations and exporting spatial data in *R* (this package is a collaborative effort of Edzer Pebesma, Roger Bivand, Barry Rowlingson and Virgilo Gomez-Rubio);
(2) rgdal – multifunctional package for reading out coordinate systems and working with spatial data;
(3) raster – package for creation of raster object and commonly works with raster data (the functions in this package were written by Robert J. Hijmans);
(4) randomForest – package for preparing and spending of the RF classification and includes a lot of operators.

These libraries are needed for the correct working for whole script.

On the first stage of this script after announcing libraries we show the training and test data. Using their coordinates the algorithm from raster image selects spectral characteristics of all points. It is needed for training algorithm and building trees with splitting nodes. We can present and analyze all spectral characteristics with 'print' command.

For better analysis of spectral characteristics in training data it is possible to visualize these data on different kinds of plots. It is possible after training of the algorithm with 'randomForest' command. On this stage the building of random forest with different trees is fulfilled, in other words all statistical operations described by Breiman are conducted [20].

After training algorithms we can analyze the training data in the RF model on graphs. The prepared graphs for different images are presented in Fig. 3.

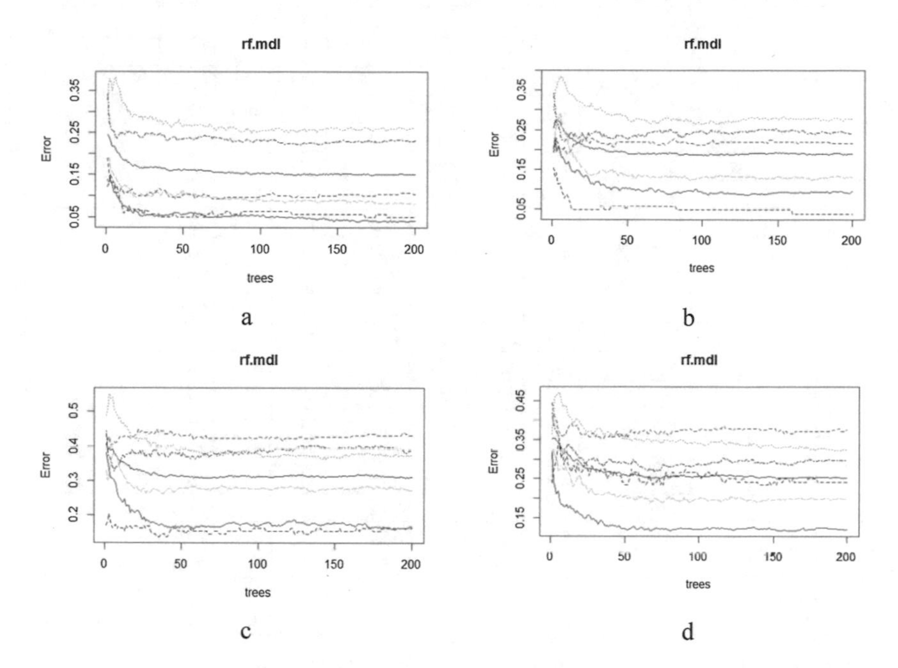

Fig. 3. The OOB estimation of different land cover classes in the RF models depending on a number of trees: a – for Landsat-8 (date of acquisition 2015/05/26); b – for Landsat-8 (date of acquisition 2018/05/28); c – for Sentinel-2 (10 m resolution); d – for Sentinel-2 (20 m resolution).

As you can see in Fig. 3, for Landsat images OOB is smaller than for Sentinel images (see Y axes). For Landsat images building of trees starting from OOB equals 0,35 and with increasing of numbers of trees decreases. For Sentinel images the situation is similar but algorithm starts building trees from 0,45–0,50 OOB.

For some classes OOB is decreases with the increasing numbers of trees faster (pink, red and cyan lines on graphs), but for others – decreasing is less. It means that these classes will be classified with a lot of mistakes and for our opinion it happens for classes with similar spectral characteristics to others classes.

Totally from these graphs we can get information about optimal numbers of trees for correct classification – if a line with the increasing numbers of trees does not decrease, it means that it is enough for classification of this class. Additionally, if a line looks direct but does not have many peaks, the algorithm finds an optimal distribution for this class.

Numeric information about OOB for different classes and totally for all can be received in time of the algorithm training in the table format (see Table 1).

Table 1. OOB error depending from number of trees.

No. of trees	Landsat-8 (2015/05/26)							Sentinel-2 (20 m resolution)						
	OOB, %	Classes[a]						OOB, %	Classes					
		1	2	3	4	5	6		1	2	3	4	5	6
10	19	8,4	32,8	24,8	12	8,4	11,7	31,8	29,8	43,8	32,8	28,1	17,9	36,1
20	17	6,3	29	25	9,6	6,4	10,4	29,2	26,6	39,4	30,4	25,9	15,8	36,6
30	16,6	5,6	28,2	24,9	9,8	5,5	10,4	27,2	27,4	37,3	29,6	20,9	14,2	35,8
40	16,1	5	27	23	10,9	5,3	10,4	27	26,6	36,6	30,1	21,1	13,2	36,2
50	16,1	5	26,3	23,4	11,1	5,7	10,4	25,9	23,4	35,2	29,2	20,6	12,3	35,8
60	16,1	5	27	24	10	5,7	10	25,8	24,2	35,2	28,1	20,2	12,3	37,1
70	15,7	5,6	26,1	24	9,6	5	9,2	25,3	24,2	34,6	26,9	20	12,1	37,1
80	15,7	6,3	26,5	23,5	9,1	5,2	9,6	25,6	25	34,6	28,1	20,4	11,9	36,6
90	15,4	6,3	25,6	23	9,3	5,5	9,6	26	26,6	34,7	28,7	20,2	12,3	37,9
100	15,3	6,3	25,4	23	8,9	5,2	10	25,6	25,8	34	28,9	19,8	12,1	37,5
110	15,4	6,3	26,1	23	8,7	5,2	10,4	25,9	25,8	34,5	29,2	19,6	12,5	37,9
120	15	5,6	25,6	22,4	8,7	5	9,6	25,9	25	34,5	29,7	19,6	12,1	37,9
130	15	5,6	25,5	22,5	8,9	4,6	9,6	25,6	25	33,3	29,6	19,8	12,3	37,9
140	15	5,6	25,8	22,4	8,7	4,6	10	25,3	23,4	33,2	29,2	19,1	12,1	37,9
150	15,2	5,6	25,9	22,5	8,9	4,5	10,4	25,5	24,2	33,2	29,6	19,6	12,1	37,9
160	15,1	5,6	25,9	22,7	8,7	4,3	10,4	25,2	25	33,2	28,6	19,8	11,7	36,6
170	15,1	5	25,9	23	8,7	4,1	10	25,4	24,2	32,9	29,4	20	12,3	37,1
180	15	5,6	25,9	23	8,1	4,1	10,4	25,4	25	32,6	29,6	19,8	12,5	37,1
190	15,2	5	26,2	23,2	8,5	4,1	10,4	25,3	24,2	32,5	29,7	19,6	12,3	37,1
200	15	5	26,1	23	8,1	4,1	10,4	25,3	24,2	32,2	29,7	20	12,1	37,5

[a]Classes are numbered by alphabet: 1 – Clouds; 2 – Coniferous forest; 3 – Deciduous forest; 4 – Grassland; 5 – Urban lands; 6 – Water lands.

Table 1 shows similar information as Fig. 3. Estimation was prepared with a step of number of trees equal 10. For Landsat-8 (15%) OOB is totally less than for Sentinel image (25,3) for number of trees equal 200. It means that the training data for Landsat image are better separated than for Sentinel. As you remember, we used the same training data for all images excluding data for the "Clouds" class. The best results with Landsat image we receive for the interpretation classes 1, 5, 4 and 6 (i.e. clouds, urban, grassland and water), worse – coniferous and deciduous forests. With interpretation the Sentinel image results for each classes will be different. The best result will be for the urban territory (OOB = 12,1%), for other classes OOB is very high. It is interesting, that the "Clouds" and "Water" classes are poorly classified. In our opinion it happened because these classes are similar in spectral distribution (for the "Clouds" class similarities are present in dark shadows from clouds). Additionally, they have similarities with old coniferous forests.

Our suppositions demonstrating outliers for different images in Fig. 4. These graphs show the outlier of the training data set that means how the spectral characteristics of pixels differ from the spectral characteristics of the training set. If we have some peaks on a graph it means that these pixels are very different from the others. In statistics, it means that they are lying too far from the others and could be wrong for this class or noise.

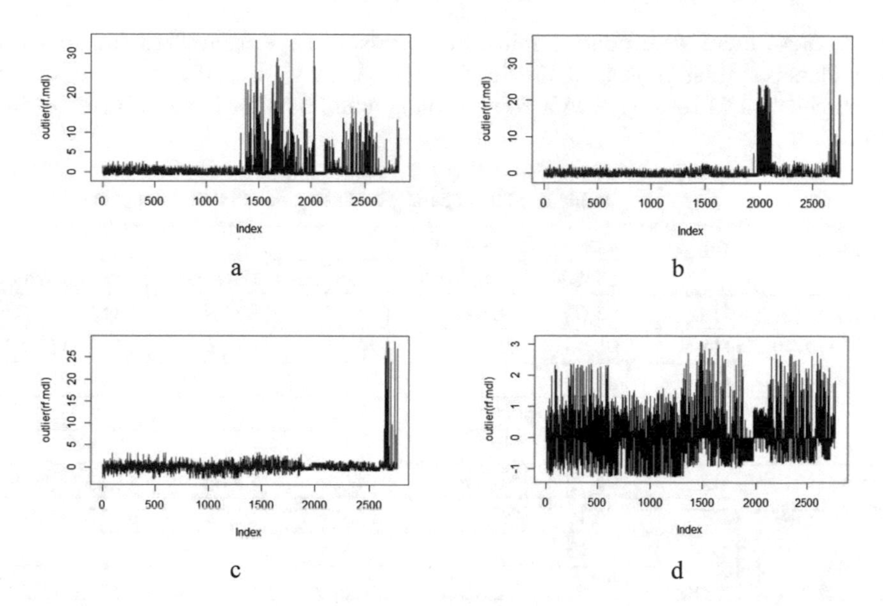

Fig. 4. Outlier graphs for training data sets in the RF models: a – for Landsat-8 (date of acquisition 2015/05/26); b – for Landsat-8 (date of acquisition 2018/05/28); c – for Sentinel-2 (10 m resolution); d – for Sentinel-2 (20 m resolution).

For Landsat image (date of acquisition 2015/05/26) these peaks are presented in the second part of a graph from approximately 1400 and more. It means that these pixels do not resent some class in a good way and the result of its classification could be wrong. For Landsat image from 2018 these outliers are present in pixels near 2000 and near 2700. The similar situation is for Sentinel with 10 m resolution. The outlier graph for Sentinel with 20 m resolution is more homogeneous and does not have these peaks. It could be a result of the spectral ranges of surveying images where different land cover in different spectrum has differences.

Taking into account all these pre-classification analysis, we conducted accuracy assessment of the RF model with the input data set (Table 2).

For Landsat image from 2015 the Users' and Producers' accuracy is more than 90% for all classes excluding coniferous and deciduous forest. The last classes have accuracy more than 73%. It is naturally because these classes are similar in spectral distribution. The assessment for other Landsat image shows less precision for the whole indexes. For these images Index Kappa is equal to 0,81 and 0,76 accordingly that indicates satisfactory result of classification.

The accuracy assessment of Sentinel images shows worse results. The worst results (according to the Users' and Producers' accuracy) are for coniferous and deciduous forests class and, that is strange, for water class. The detail reviewing of confusion matrices for these classes shows the following: those coniferous and deciduous classes are mixing among themselves. With a class of water mixing is deeper, because pixels are mixing with a few classes, such as urban and grassland ones. To our minds, it is linked with a water regime in this territory because before the date of acquisition

(2018/05/24) there were pouring rains. For Landsat image from 2018 situation with water class is similar. In general, the Sentinel images showed satisfactory results (index Kappa equals 0,61 for image with 10 m resolution and 0,68 for image with 20 m resolution).

Table 2. The accuracy assessment.

Image	Index[a]	Classes names					
		Clouds	Coniferous	Deciduous	Grassland	Urban	Water
Landsat (2015/05/26)	UA	95,60	76,97	75,33	92,59	90,59	94,33
	PA	95,00	73,95	76,96	91,89	95,91	89,62
	OA	84,96					
	Kappa	0,81					
	OOB	15,04					
Landsat (2018/05/28)	UA	97,12	78,57	74,35	83,06	81,83	95,29
	PA	96,19	72,41	75,96	86,98	90,57	78,45
	OA	81,15					
	Kappa	0,76					
	OOB	18,85					
Sentinel (10 m)	UA	92,04	64,23	61,33	74,95	69,23	83,54
	PA	83,87	62,89	61,44	72,83	83,27	56,90
	OA	69,07					
	Kappa	0,61					
	OOB	30,93					
Sentinel (20 m)	UA	88,68	70,25	70,64	78,37	76,83	77,96
	PA	75,81	67,79	70,28	80,00	87,90	62,50
	OA	74,68					
	Kappa	0,68					
	OOB	25,32					

[a]UA – Users' accuracy; PA – Producers' accuracy, OA – Overall accuracy; Kappa – Index Kappa; OOB – Out-of-Bag

The decreasing accuracy of classification from the Landsat to Sentinel images has possibly a few explications: (1) the accuracy of image georeferencing that has an influence on the stage of reading spectral characteristics from the training data set; (2) different spectral ranges of image bands for stacking.

The classification was done and respective thematic maps were obtained for all images (some of the thematic maps are presented in Fig. 5). For better visualization of thematic maps and for reducing noises the kernel majority filter with size 3×3 pixels was used.

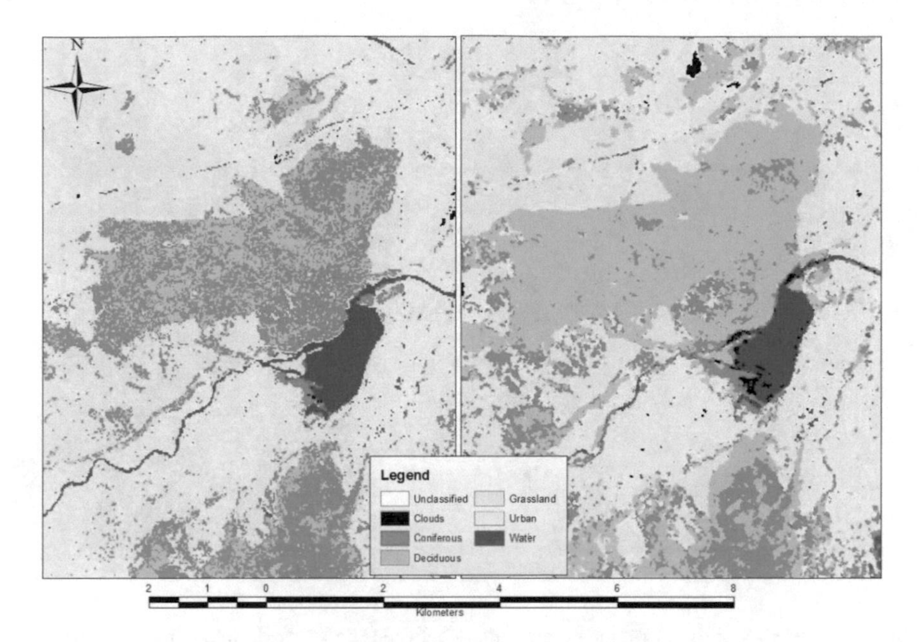

Fig. 5. Fragments of thematic maps of classification of Landsat-8 (date of acquisition 2018/05/28) (left) and Sentinel-2 (with spatial resolution 20 m) (right).

The visual analysis of thematic maps shows considerable difference between classes, especially between coniferous and deciduous forests (forest massive above lake on thematic maps). Some changes are visible with class of clouds close to lake which is possible wrong classified.

The changes of land cover in 2015 and 2018 were detected using the thematic maps obtained from the images of respective years (ArcGIS Image analysis tools were applied). The results are presented in Fig. 6.

Differentiation was performed as the subtractions between sequence number of class on initial and final images. Since we have sequence numbers of classes equal to the sorted ones by alphabet names of classes, the sequence numbers of classes are the following: clouds – 1, coniferous – 2 and so on. So, on the difference maps subtracting between these numbers is presented. Zero on these maps means no changes between initial and final images. Other values show changes between classes. For example, value '–1' shows that these pixels were lower classes and became higher classes (coniferous to deciduous, deciduous to grasslands and so on). The main changes are presented near lake (compare Figs. 5 and 6 where the same territory is presented) and on the edges of forest missives.

Visual analysis of the difference maps is very subjective and not full. Therefore, we compared changes using areas in thousands hectares of all classes on different thematic maps. The results are presented in Table 3.

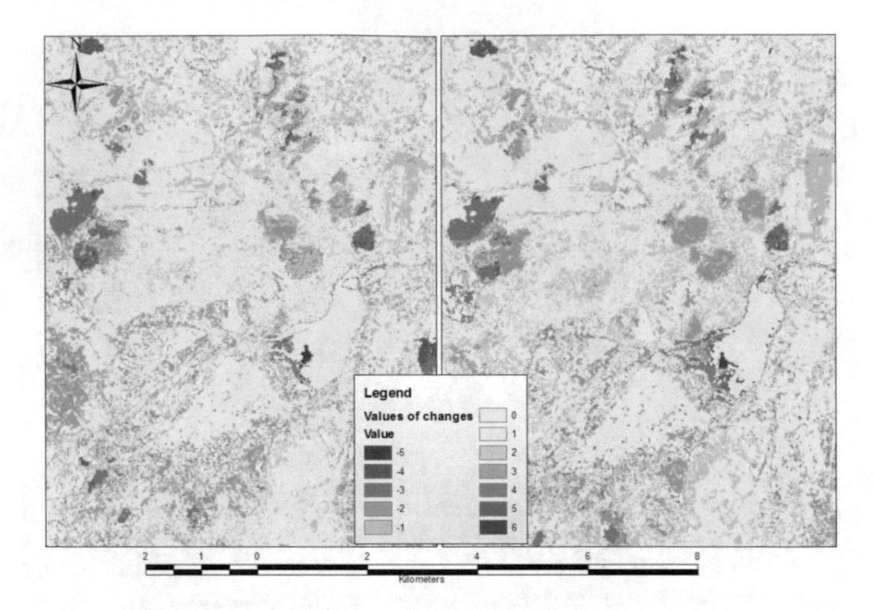

Fig. 6. Fragments of the difference maps: left – between the thematic maps from Landsat 2015 and 2018; right – between the thematic maps from Landsat 2015 and Sentinel (20 m spatial resolution).

Table 3. The change detection between the thematic maps.

Class name	Area, thousand ha[a]				Difference in pairs[b]					
	1	2	3	4	ha			%		
					1–2	1–3	1–4	1–2	1–3	1–4
Clouds	5,2	1,2	1,9	2,3	3990	3286	2917	77	63,4	56,3
Coniferous	118,2	112,0	89,2	110,9	6228	29009	7275	5,3	24,5	6,2
Deciduous	77,0	67,0	102,6	72,0	9924	−25640	5006	12,9	−33,3	6,5
Grassland	51,9	58,9	40,4	39,8	−6978	11585	12151	−13,4	22,3	23,4
Urban	34,6	49,6	53,5	63,5	−14970	−18870	−28826	−43,2	−54,5	−83,2
Water	6,1	4,3	5,5	4,6	1800	634	1476	29,4	10,3	24,1

[a]Area, thousand ha: 1 – Landsat (date of acquisition 2015/05/26); 2 – Landsat (date of acquisition 2018/05/28); 3 – Sentinel (spatial resolution 20 m); 4 – Sentinel (spatial resolution 10 m).

[b]Pairs are assembled according to image number (see above).

The estimated area has a rich forest territory, so the biggest classes on the thematic maps (by area) are coniferous and deciduous forests. Considerable areas are for urban territory and grasslands. For the thematic map from Sentinel with 20 m resolution, a considerable part of coniferous turned for deciduous forests. The large divergences are in area of water class, where areas are avoiding nearly twice on different thematic maps. We do not analyze the class of clouds in detail because it is variable on different images.

The differences on different pairs of images have different directions and different absolute values. Water and coniferous areas comparing with Landsat from 2015 are growing in all cases. In all cases urban class is decreasing.

Comparing differences between multitemporal Landsat images shows increasing of areas of coniferous and deciduous forests and water, however decreasing in grassland and urban areas. For Sentinel images with different spatial resolutions differences vary where some classes are decreasing and some – increasing. The values in percentages of differences are very high and reach 83,2%. In general, differences between Landsat images are smaller than those with Sentinel images.

4 Conclusions

The prepared script for Random Forest classification in *R* software showed a few advantages, in particular reusable, speed of classification and possibilities of receiving additional information about accuracy of classification and separabilities of data sets. For classification the same data sets for training and testing excluding representative pixels for the 'Clouds' class were prepared because in multitemporal images this class is not the same. For the next classification a few common classes of land cover were chosen: coniferous, deciduous, grassland, urban and water areas on multitemporal and multisensoral images from Landsat and Sentinel satellites. For objective comparing of the results these images were chosen in different years but in the same season.

The accuracy assessment of the testing data showed higher quality for Landsat images and lower for Sentinel. Considering it for the next classification and analyzing Sentinel images in all cases showed lower results. The differences for Sentinel images are very high and have not some fixed direction. These results we link with differentiations in spectral ranges of image receiving for Landsat and Sentinel and with an accuracy of images geocoding.

References

1. Richards, J.A., Jia, X.: Remote Sensing Digital Image Analysis: An Introduction. Springer, Berlin (1999)
2. Aronoff, S.: Remote Sensing for GIS Managers, 1st edn. ESRI Press, Independent Publishers Group (IPG), Redlands (2005)
3. Baumann, M., Ozdogan, M., Kuemmerle, T., Wendland, K.J., Esipova, E., Radeloff, V.C.: Using the Landsat record to detect forest-cover changes during and after the collapse of the Soviet Union in the temperate zone of European Russia. Remote Sens. Environ. **124**, 174–184 (2012)
4. Kuemmerle, T., Chaskovskyy, O., Knorn, J., Radeloff, V.C., Kruhlov, I., Keeton, W.S., Hostert, P.: Forest cover change and illegal logging in the Ukrainian Carpathians in the transition period from 1988 to 2007. Remote Sens. Environ. **113**, 1194–1207 (2009)
5. Kuemmerle, T., Radeloff, V.C., Perzanowski, K., Hostert, P.: Cross-border comparison of land cover and landscape pattern in Eastern Europe using a hybrid classification technique. Remote Sens. Environ. **103**, 449–464 (2006)
6. Carreiras, J.M.B., Pereira, J.M.C., Campagnolo, M.L., Shimabukuro, Y.E.: Assessing the extent of agriculture/pasture and secondary succession forest in the Brazilian Legal Amazon using SPOT VEGETATION data. Remote Sens. Environ. **101**, 283–298 (2006)
7. Kimes, D.S., Nelson, R.F., Salas, W.A., Skole, D.L.: Mapping secondary tropical forest and forest age from SPOT HRV data. Int. J. Remote Sens. **20**(18), 3625–3640 (1999)

8. Kozak, J., Estreguil, C., Vogt, P.: Forest cover and pattern changes in the Carpathians over the last decades. Eur. J. For. Res. **126**, 77–90 (2007)
9. Havrylyuk, S.A., Myklush, S.I.: Classification of forest cover types of Western Steppe of Ukraine using remote sensing data. Sci. Bull. Ukr. Natl. For. Univ. **17**(3), 26–35 (2007)
10. Powell, S.L., Cohen, W.B., Healey, S.P., Kennedy, R.E., Moisen, G.G., Pierce, K.B., Ohmann, J.L.: Quantification of live aboveground forest biomass dynamics with Landsat time-series and field inventory data: a comparison of empirical modeling approaches. Remote Sens. Environ. **114**, 1053–1068 (2010)
11. Chan, J.C.-W., Paelinckx, D.: Evaluation of Random Forest and Adaboost tree-based ensemble classification and spectral band selection for ecotope mapping using airborne hyperspectral imagery. Remote Sens. Environ. **112**, 2999–3011 (2008)
12. Rodriguez-Galiano, V.F., Chica-Olmo, M., Abarca-Hernandez, F., Atkinson, P.M., Jeganathan, C.: Random Forest classification of Mediterranean land cover using multi-seasonal imagery and multi-seasonal texture. Remote Sens. Environ. **121**, 93–107 (2012)
13. Smolij, V.A. (ed.): Encyclopedia of History of Ukraine, vol. 8. Naukova dumka, Kyiv (2011)
14. Landsat 8. https://landsat.usgs.gov/landsat-8. Accessed 21 June 2018
15. What are the band designations for the Landsat satellites? https://landsat.usgs.gov/what-are-band-designations-landsat-satellites. Accessed 21 June 2018
16. Overview. https://sentinel.esa.int/web/sentinel/missions/sentinel-2/overview. Accessed 21 June 2018
17. How To: Select random points from an existing point feature layer. https://support.esri.com/en/technical-article/000013141. Accessed 22 June 2018
18. Stumpf, A., Kerle, N.: Object-oriented mapping of landslides using Random Forests. Remote Sens. Environ. **115**, 2564–2577 (2011)
19. Breiman, L.: Random forests. Mach. Learn. **45**, 5–32 (2001)
20. Richards, J.A.: Remote Sensing Digital Image Analysis: An Introduction, 5th edn. Springer, Berlin (2013)
21. Tokar, O., Vovk, O., Kolyasa, L., Havryliuk, S., Korol, M.: Using the Random Forest classification for land cover interpretation of Landsat images in the Prykarpattya region of Ukraine. In: 13th International Scientific and Technical Conference "Computer Science and Information Technologies", Lviv, Ukraine (2018, in press)
22. Coppin, P., Jonckheere, I., Nackaerts, K., Muys, B., Lambin, E.: Digital change detection methods in ecosystem monitoring: a review. Int. J. Remote Sens. **25**(9), 1565–1596 (2004)

A New Approach to Image Enhancement Based on the Use of Raw Moments for Subranges of Brightness

Sergei Yelmanov[1(✉)] and Yuriy Romanyshyn[2,3]

[1] Special Design Office of Television Systems, Lviv, Ukraine
sergei.yelmanov@gmail.com
[2] Lviv Polytechnic National University, Lviv, Ukraine
yuriy.romanyshynl@gmail.com
[3] University of Warmia and Mazury, Olsztyn, Poland
yuriy.romanyshyn@uwm.edu.pl

Abstract. The problem of improving the quality of complex low-contrast images in automatic mode with an acceptable level of computational costs is considered in this paper. The task of increasing the contrast for complex low-contrast images with a wide dynamic range and multi-modal distribution of brightness is considered. The purpose of this work is to improve the efficiency of increasing the overall contrast for complex images with a wide dynamic range and multi-modal distribution of brightness. A new approach to increase the contrast of complex low-contrast image by its adaptive non-linear contrast stretching is proposed based on the measuring of raw moments for subranges of image brightness. The proposed approach is based on measuring the ratios between the values of the raw moments for different subranges of brightness. A new technique of contrast enhancement for complex monochrome images based on measuring the mean values for subranges of image brightness is also proposed. The research of the various known and proposed techniques of image contrast enhancement in the automatic mode was carried out using the no-reference metrics of overall image contrast. The proposed technique provides an efficient redistribution of the contrast of objects in the image regardless of their size and enables to effectively increase the overall contrast of the image in automatic mode with an acceptable level of computational costs.

Keywords: Image enhancement · Contrast · Non-linear stretching
Raw moment · Brightness subrange

1 Introduction

The widespread use of gadgets with image sensors (photo and video cameras) and applying of new technologies in imaging and image processing requires the solution of the problem of operative (in real-time) improving the quality of the images in automatic mode with minimal level of computational costs.

There are various approaches to increasing the objective quality of the image [1].

Increasing the contrast is the most effective, rapid, high-performance and cost-effective approach to image enhancement [2, 3]. Contrast enhancement enables

N. Shakhovska and M. O. Medykovskyy (Eds.): CSIT 2018, AISC 871, pp. 65–84, 2019.
https://doi.org/10.1007/978-3-030-01069-0_6

significantly enhance the visual perception of the image and improve its objective quality and is the basic technology in image pre-processing [3–5].

Various techniques for image contrast enhancement are known [1–3, 6]. However, the known methods of increasing the contrast of the image have a number of disadvantages that substantially limit their practical use for image processing in an automatic mode. Their main disadvantages are low efficiency of contrast enhancement for complex images with multi-modal distribution of brightness and a wide dynamic range, an overly increase in the contrast of large-sized objects, a decrease in contrast and the possible disappearance of small-sized objects in the image, a relatively high level of computational costs. To address these disadvantages, we propose a new approach to contrast enhancement by adaptive non-linear stretching of dynamic range of image brightness.

The problem of improving the quality of complex low-contrast images in automatic mode with an acceptable level of computational costs is considered in this paper (Sect. 2). The object of study is the process of improving the quality for complex images. The task of increasing the contrast for complex low-contrast images with a wide dynamic range and multi-modal distribution of brightness is considered. The purpose of this work is to improve the efficiency of increasing the overall contrast for complex low-contrast images with multi-modal distribution of brightness and a wide dynamic range. The subject of the study is techniques of increasing the contrast of the image in the automatic mode by adaptive nonlinear stretching of the dynamic range of its brightness.

A new approach to increase the contrast of complex low-contrast image by its adaptive non-linear contrast stretching is proposed based on the measuring of raw moments for subranges of image brightness. The proposed approach is based on measuring the ratios between the values of the raw moments for different subranges of brightness (Sect. 3).

A new technique of contrast enhancement for complex monochrome images based on measuring the mean values for subranges of image brightness is also proposed.

The proposed technique of adaptive non-linear stretching provides an efficient redistribution of the contrast of objects in the image regardless of their size and enables to effectively increase the overall contrast of the image in automatic mode, while maintaining the contrast of small-sized objects in the image.

The research of the effectiveness of the known methods and proposed technique of contrast enhancement in the automatic mode was carried out using no-reference metrics of overall contrast for sixth groups of complex low-contrast images with multi-modal distribution of brightness and a wide dynamic range (Sects. 4 and 5).

2 Contrast Enhancement Techniques

Currently, various approaches to increase the contrast of the image in the automatic mode are known [7–9].

Known techniques for enhancing the contrast in images can be conditionally subdivided into three main groups, namely, methods of piecewise linear contrast stretching, nonlinear transformations of brightness levels, and methods based on the histogram specification and its modifications.

Methods of piecewise-linear contrast stretching (e.g. min-max linear stretching, partial linear stretching, percentage stretching, piecewise linear stretching, etc.) are based on a piecewise linear transformations of the dynamic range of the brightness.

Min-max linear stretching, at which the dynamic range of the original image is stretched to the range of possible values of brightness, is the standard procedure of image processing.

The min-max linear stretching is most often defined as [3]:

$$y_i = y_{\min} + (y_{\max} - y_{\min}) \cdot \frac{x_i - x_{\min}}{x_{\max} - x_{\min}}, \tag{1}$$

where x_i, y_i – brightness values for pixels of original and processed image; x_{min}, x_{max}, y_{min}, y_{max} – minimum and maximum brightness values; $x, y \in [0, 1]$.

It is most often assumed that $y_{min} = 0$ and $y_{max} = 1$.

Another widely used method of increasing the contrast is partial linear stretching, which is usually defined as [3, 10]:

$$y_i = \begin{cases} 0, & if\ x_i \le a \\ y_{\min} + (y_{\max} - y_{\min}) \cdot \frac{x_i - a}{b - a}, & if\ a < x_i < b \\ 1, & if\ x_i \ge b \end{cases} \tag{2}$$

where a, b – lower and upper threshold values, parameters.

The threshold values of a and b are most often determined from the equations (the method of percentage linear stretching) [10]:

$$\int_0^a p_x(z)\,dz = \delta \ \wedge \ \int_0^b p_x(z)\,dz = 1 - \delta, \tag{3}$$

where $p_x(\bullet)$ – probability density of brightness x of original image; δ – parameter, threshold value.

Parameter δ most often takes a value from the range 0,005–0,03.

Piecewise linear contrast stretching is usually defined as [3, 10]:

$$y_i = \begin{cases} s \cdot \frac{x_i}{t}, & if\ x_i \le t \\ s + (1 - s) \cdot \frac{x_i - t}{1 - t}, & if\ x_i > t \end{cases}, \tag{4}$$

where t – threshold value, $0 < t < 1$; s – stretching value, $0 < s < 1$.

In [3, 10], a method of piecewise linear contrast stretching is proposed based on the brightness transformation in the form:

$$y_i = \begin{cases} s_1 \cdot \frac{x_i}{t_1}, & if\ x_i \le t_1 \\ s_1 + (s_2 - s_1) \cdot \frac{x_i - t_1}{t_2 - t_1}, & if\ t_1 < x_i \le t_2 \\ s_2 + (1 - s_2) \cdot \frac{x - t_2}{1 - t_2}, & if\ t_2 < x_i \end{cases}, \tag{5}$$

where t_1, t_2 – lower and upper threshold values; s_1, s_2 – lower and upper stretching values, and:

$$0 < t_1 < t_2 < 1 \wedge 0 \leq s_1 \leq s_2 \leq 1. \quad (6)$$

The effectiveness of the techniques of piecewise linear contrast stretching depends significantly on the values of the parameters of transformation and is relatively low for complex images with a multimodal distribution of brightness and a wide dynamic range.

Another widely used approach to image enhancement is nonlinear contrast stretching. Nonlinear stretching techniques are widely used in image pre-processing.

Nonlinear contrast stretching is based on nonlinear gray-level transformations (e.g. logarithmic, exponential and power-law transformations, sigmoid stretching, etc.) and on techniques of histogram specification (e.g. histogram equalization, histogram hyperbolization, etc.) and their modifications (BBHE [11], DSIHE [12], RSMHE [13], etc.).

A known method of nonlinear stretching is logarithmic transformation, where is often defined as [1, 3]:

$$y_i = \alpha \cdot \log(\beta \cdot x_i + 1), \quad (7)$$

where α, β – parameters of transformation.

The exponential transformation has the form [1, 3]:

$$y_i = \alpha \cdot \log\left(e^{\beta \cdot x_i} - 1\right). \quad (8)$$

Another widely used technique for contrast enhancement is power law transformation, also known as gamma correction [1, 3]:

$$y_i = y_{\min} + (y_{\max} - y_{\min}) \cdot \left(\frac{x_i - x_{\min}}{x_{\max} - x_{\min}}\right)^{\gamma}, \quad (9)$$

where γ – parameter, exponent.

At $\gamma = 2$ is square law, similar to exponential transformation, at $\gamma = 1/3$ is cubic root, similar to logarithmic transformation.

The parameter γ is often set so that the mean value y_{mean} of brightness of the processed image Y was equal to 1/2.

A method of nonlinear contrast stretching on the basis of use of sigmoid function is known [14].

$$y_i = y_{\min} + (y_{\max} - y_{\min}) \cdot \frac{1}{1 + e^{-\tau(x_i - \omega)}}, \quad (10)$$

where τ, ω – parameters of the sigmoid function.

The results of nonlinear contrast stretching using (7)–(10) essentially depend on the values of the parameters of the transformation function, which are usually set in interactive mode.

The histogram equalization is the most well known and widely used standard procedure in image pre-processing and is widely used to enhance the image contrast in automatic mode.

The histogram equalization is defined as [15]:

$$y_i = y_{\min} + (y_{\max} - y_{\min}) \cdot \int_0^{x_i} p_x(z)\, dz, \tag{11}$$

where $p_x(\bullet)$ – probability density of brightness x.

Another widely used technique for non-linear contrast stretching is the histogram hyperbolization, which is defined as [16]:

$$y_i = y_{\min} + (y_{\max} - y_{\min}) \cdot x_{\min} \cdot (x_{\max}/x_{\min})^{\int_0^{x_i} p_x(z)dz}. \tag{12}$$

The main disadvantages of the techniques, based on the histogram specification and its modifications, are an overly increase in the contrast of large-sized objects, a decrease in contrast and the possible disappearance of small-sized objects in the image. To address these disadvantages, in this paper we propose a new approach to image enhancement by nonlinear contrast stretching based on the use of raw moments for subranges of brightness.

3 Proposed Approach to Image Enhancement

The article deals with the problem of improving the quality of complex images in automatic mode with an acceptable level of computational costs.

The purpose of this work is to improve the efficiency of increasing the overall contrast for complex low-contrast images with multi-modal distribution of brightness and a wide dynamic range.

This paper suggests a new approach to increase the contrast of complex low-contrast image by its adaptive non-linear contrast stretching based on the measuring of raw (crude) moments for subranges of brightness.

The proposed approach is based on measuring the ratios between the values of raw moments for different subranges of brightness of the original image.

To increase the contrast in complex low-contrast image, we propose adaptive nonlinear statistical no-inertial transformation for image brightness in the form:

$$y_i = \begin{cases} y_{\min}, & \text{if } x_i \le x_{\min} \\ y_{\min} + \xi \cdot (y_{\max} - y_{\min}) \cdot \frac{\mu_x^k(0,x_i)}{\mu_x^n(0,x_i)}, & \text{if } x_{\min} < x_i < x_{\max}, \\ y_{\max}, & \text{if } x_i \ge x_{\max} \end{cases} \tag{13}$$

where $\mu_x^k(a, b)$ – k^{th} raw moment of brightness x in the range [a, b] of original image X; ξ – normalizing coefficient; k, n – exponents, parameters; where:

$$\mu_x^k(a,b) = \int_a^b z^k \cdot p_x(z)\, dz, \tag{14}$$

$$\xi = \left[\frac{\mu_x^k(0,\ 1)}{\mu_x^n(0,\ 1)}\right]^{-1} = \frac{\mu_x^n(0,\ 1)}{\mu_x^k(0,\ 1)}. \tag{15}$$

Expressions (13)–(15) define the proposed approach in a general form.

To demonstrate the possibilities of the proposed approach to adaptive improvement of complex images, assume that $k = 1$ and $n = 0$.

In this case the following equation holds:

$$\frac{\mu_x^1(a,b)}{\mu_x^0(a,b)} = \int_a^b z \cdot p_x(z)dz \,/\, \int_a^b p_x(z)dz = \mathrm{mean}_x(\mathrm{a}, b), \tag{16}$$

where mean_x (a, b) – the mean value of brightness of the original image in the range [a, b].

In this case, the definition (13) taking into account (16) takes the form:

$$y_i = \begin{cases} y_{\min}, & \mathit{if}\ x_i \le x_{\min} \\ y_{\min} + (y_{\max} - y_{\min}) \cdot \dfrac{\mathrm{mean}_x(0, x_i)}{\mathrm{mean}_x(0, 1)}, & \mathit{if}\ x_{\min} < x_i < x_{\max} \\ y_{\max}, & \mathit{if}\ x_i \ge x_{\max} \end{cases} . \tag{17}$$

The expression (17) defines the proposed technique of image enhancement by adaptive nonlinear stretching using the raw moments of the 0th and 1st degree.

The proposed technique provides an effective increase in the overall contrast for complex low-contrast images with multi-modal distribution of brightness and wide dynamic range.

4 Research

The research is carried out by measuring the overall contrast for the six groups of test images and by expert estimates of their quality.

Each group of test images consists of original images and the results of their processing using five known methods of image enhancement and the proposed technique, namely:

- min-max linear stretching (1);
- percentage linear stretching (2) and (3) for δ = 0,01;
- nonlinear stretching using sigmoid function (10);
- gamma correction (9) for y_{mean} = 1/2,
- histogram equalization (11);
- proposed technique of adaptive non-linear stretching (17).

Appearance of the six original images and their histograms are shown in Fig. 1 [17], Fig. 2 [17], Fig. 3 [17], Fig. 4 [18], Fig. 5 [17] and Fig. 6 [17]. The results of processing for the six original images are shown in Figs. 7, 8, 9, 10, 11 and 12.

Fig. 1. The first image [17] and its histogram.

Fig. 2. The second image [17] and its histogram.

Measurement of the overall contrast for the six groups of test images was carried out using known no-reference metrics of contrast, namely [19–21]:

(1) generalized (complete integral) weighted contrast [19]:

$$C_{gen}^{wei} = \int_0^1 \int_0^1 \frac{|x_k - x_n|}{x_k + x_n} \cdot p_x(x_k) \cdot p_x(x_n)\, dx_k dx_n, \tag{18}$$

(2) generalized linear contrast [20]:

$$C_{gen}^{lin} = \frac{1}{x_{\max} - x_{\min}} \cdot \int_0^1 \int_0^1 |x_k - x_n| \cdot p_x(x_k) \cdot p_x(x_n)\, dx_k dx_n, \tag{19}$$

(3) incomplete integral weighted contrast [21]:

$$C_{inc}^{wei} = \int_0^1 \frac{|x_k - \mathrm{mean}_x(0,\ 1)|}{x_k + \mathrm{mean}_x(0,\ 1)} \cdot p_x(x_k) dx_k, \tag{20}$$

Fig. 3. The third image [17] and its histogram.

Fig. 4. The fourth image [18] and its histogram.

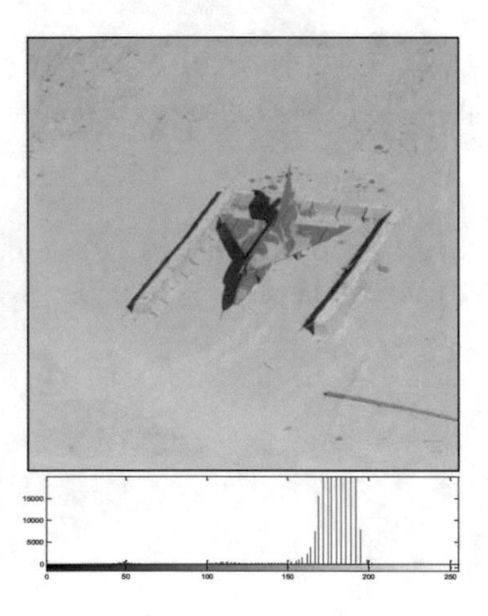

Fig. 5. The fifth image [17] and its histogram.

Fig. 6. The sixth image [17] and its histogram.

(4) incomplete integral linear contrast [21]:

$$C_{inc}^{lin} = \frac{1}{x_{\max} - x_{\min}} \cdot \int\limits_{0}^{1} |x_k - \text{mean}_x(0,\ 1)\,| \cdot p_x(x_k) dx_k. \tag{21}$$

The results of contrast measurements for six groups of test images using known no-reference metrics of overall contrast (18)–(21) are shown in Table 1.

(a) min-max linear stretching.

(b) percentage linear stretching, δ=0,01.

(c) sigmoid non-linear stretching.

(d) gamma correction, $\gamma_{mean} = 0,5$.

(e) histogram equalization.

(f) proposed technique.

Fig. 7. The results of processing for the first original image (Fig. 1, [17]).

(a) min-max linear stretching.

(b) percentage linear stretching, δ=0,01.

(c) sigmoid non-linear stretching.

(d) gamma correction, $y_{mean} = 0,5$.

(e) histogram equalization.

(f) proposed technique.

Fig. 8. The results of processing for the second original image (Fig. 2, [17]).

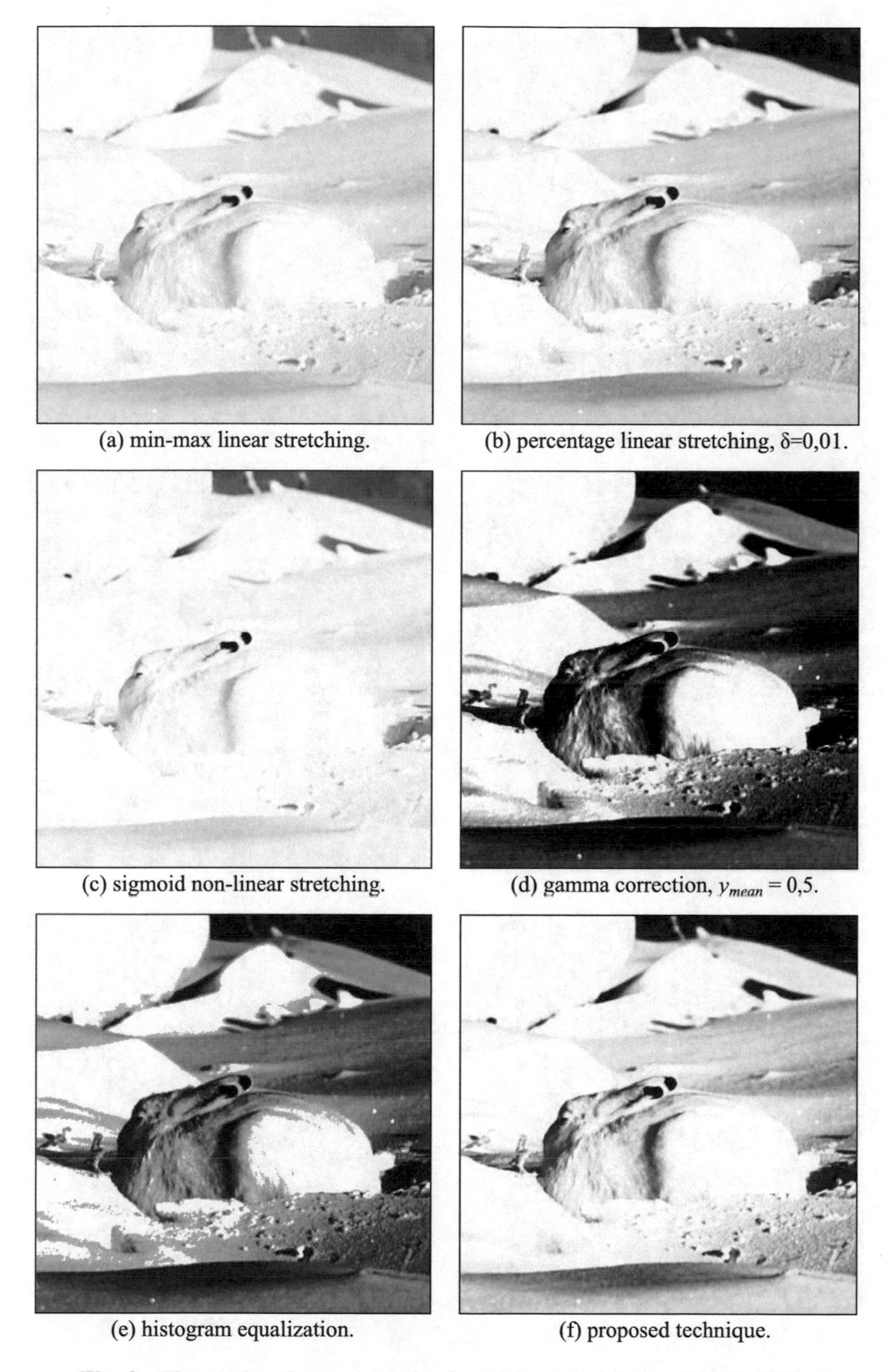

(a) min-max linear stretching.

(b) percentage linear stretching, δ=0,01.

(c) sigmoid non-linear stretching.

(d) gamma correction, $y_{mean} = 0{,}5$.

(e) histogram equalization.

(f) proposed technique.

Fig. 9. The results of processing for the third original image (Fig. 3, [17]).

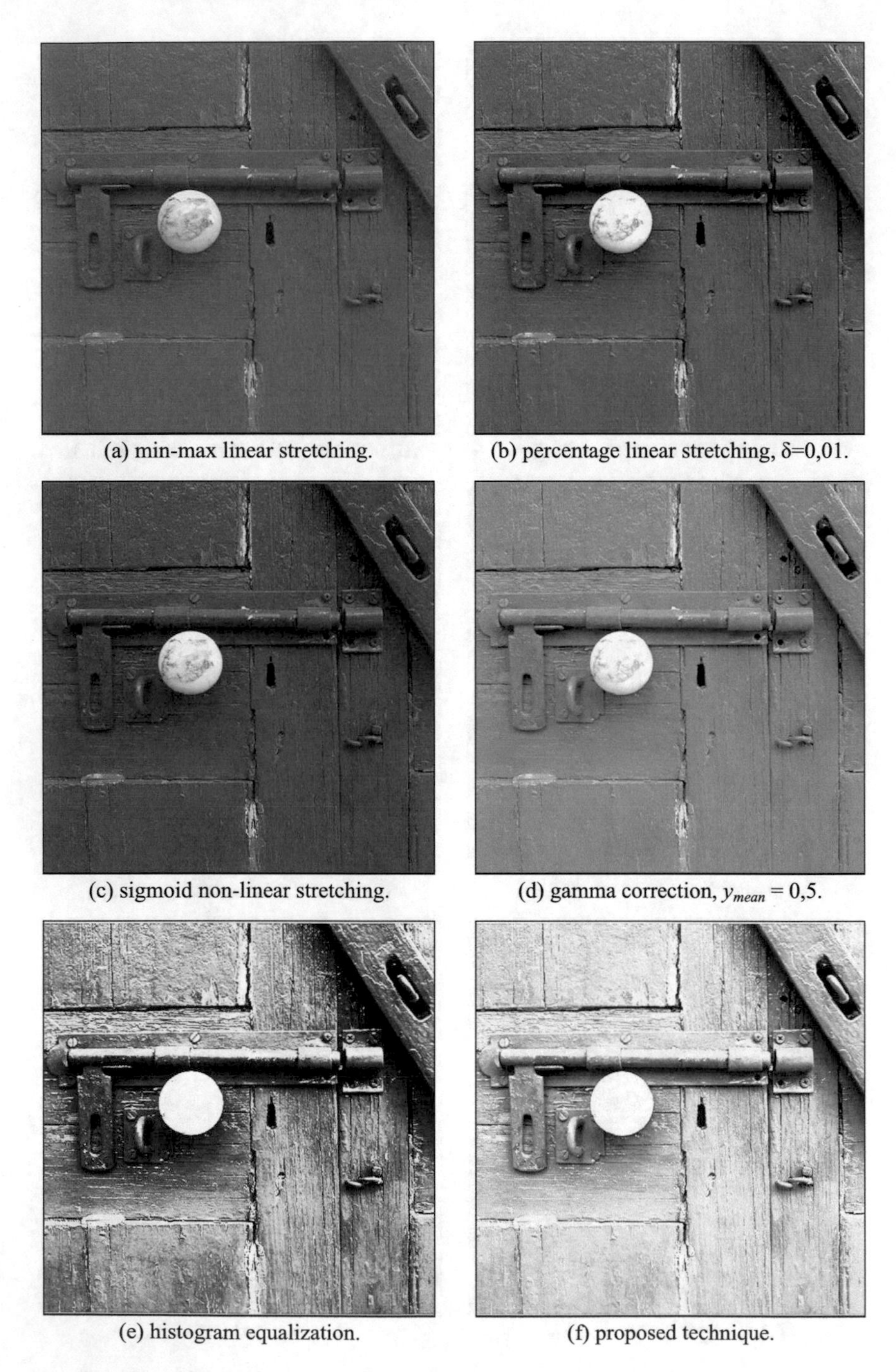

(a) min-max linear stretching.

(b) percentage linear stretching, δ=0,01.

(c) sigmoid non-linear stretching.

(d) gamma correction, y_{mean} = 0,5.

(e) histogram equalization.

(f) proposed technique.

Fig. 10. The results of processing for the fourth original image (Fig. 4, [18]).

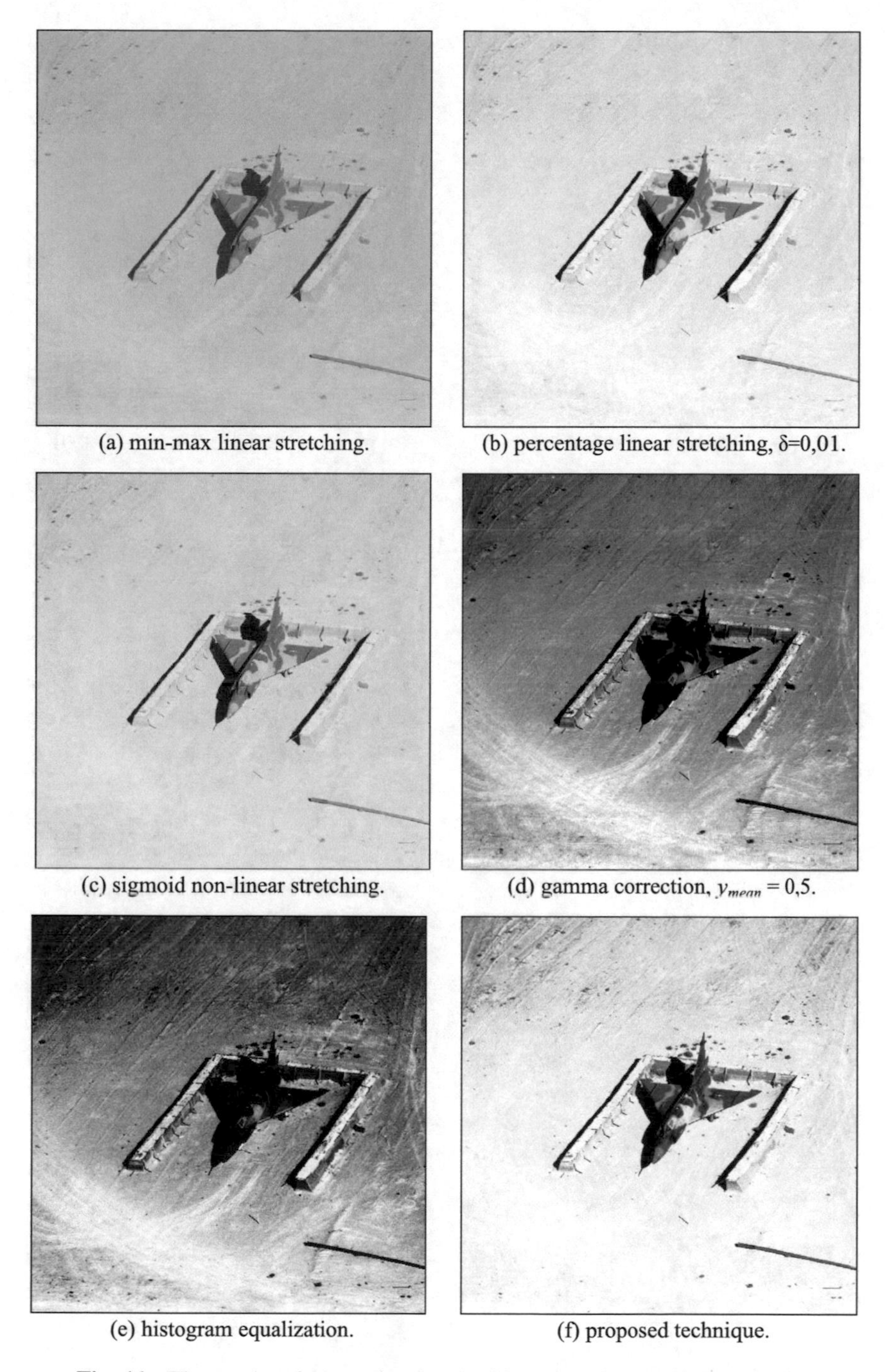

(a) min-max linear stretching.

(b) percentage linear stretching, δ=0,01.

(c) sigmoid non-linear stretching.

(d) gamma correction, $y_{mean} = 0{,}5$.

(e) histogram equalization.

(f) proposed technique.

Fig. 11. The results of processing for the fifth original image (Fig. 5, [17]).

(a) min-max linear stretching.

(b) percentage linear stretching, δ=0,01.

(c) sigmoid non-linear stretching.

(d) gamma correction, $y_{mean} = 0{,}5$.

(e) histogram equalization.

(f) proposed technique.

Fig. 12. The results of processing for the sixth original image (Fig. 6, [17]).

Table 1. The results of measurements of contrast for test images

Id of images	C^{wei}_{gen}	C^{lin}_{gen}	C^{wei}_{inc}	C^{lin}_{inc}
1	0.154	0.227	0.117	0.181
7(a)	0.186	0.227	0.141	0.181
7(b)	0.218	0.260	0.163	0.208
7(c)	0.217	0.259	0.168	0.219
7(d)	0.483	0.361	0.355	0.274
7(e)	0.389	0.335	0.289	0.251
7(f)	0.274	0.307	0.201	0.237
2	0.190	0.273	0.146	0.222
8(a)	0.246	0.273	0.185	0.222
8(b)	0.271	0.301	0.202	0.244
8(c)	0.283	0.307	0.218	0.262
8(d)	0.530	0.395	0.410	0.319
8(e)	0.388	0.334	0.289	0.251
8(f)	0.353	0.352	0.259	0.271
3	0.092	0.142	0.065	0.104
9(a)	0.092	0.142	0.065	0.104
9(b)	0.166	0.211	0.113	0.155
9(c)	0.078	0.113	0.055	0.086
9(d)	0.552	0.452	0.458	0.383
9(e)	0.410	0.394	0.315	0.306
9(f)	0.221	0.262	0.156	0.203
4	0.113	0.076	0.070	0.047
10(a)	0.113	0.076	0.070	0.047
10(b)	0.194	0.117	0.120	0.073
10(c)	0.178	0.096	0.112	0.059
10(d)	0.131	0.116	0.082	0.074
10(e)	0.389	0.334	0.290	0.251
10(f)	0.223	0.253	0.158	0.192
5	0.057	0.069	0.037	0.046
11(a)	0.059	0.069	0.038	0.046
11(b)	0.087	0.116	0.055	0.078
11(c)	0.071	0.088	0.046	0.060
11(d)	0.294	0.260	0.205	0.185
11(e)	0.385	0.332	0.289	0.251
11(f)	0.132	0.169	0.091	0.127
6	0.119	0.133	0.085	0.098
12(a)	0.133	0.133	0.095	0.098
12(b)	0.200	0.231	0.140	0.171
12(c)	0.193	0.196	0.138	0.147
12(d)	0.374	0.300	0.264	0.220

(*continued*)

Table 1. (*continued*)

Id of images	C_{gen}^{wei}	C_{gen}^{lin}	C_{inc}^{wei}	C_{inc}^{lin}
12(e)	0.388	0.334	0.289	0.251
12(f)	0.254	0.287	0.181	0.217

The graphs for the contrast values (Table 1) for each of the six groups of test images are shown in Figs. 13, 14, 15, 16, 17 and 18.

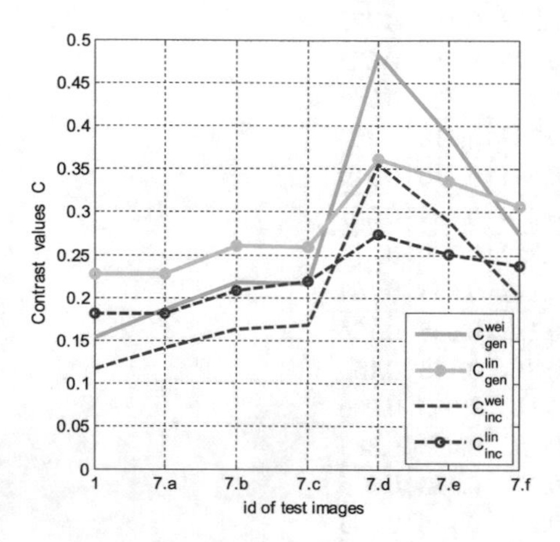

Fig. 13. Contrast for the first group of images.

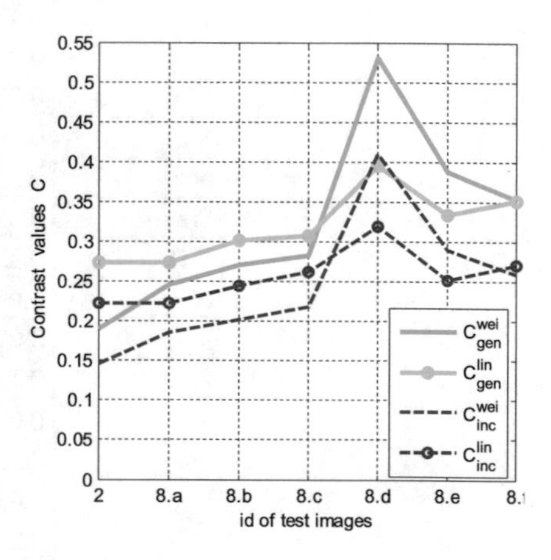

Fig. 14. Contrast for the second group of images.

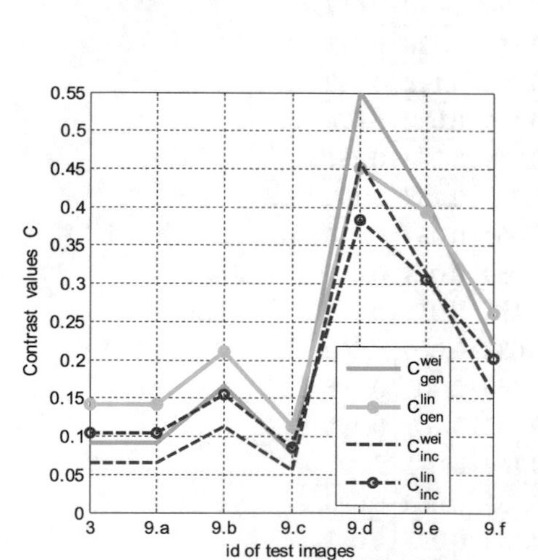

Fig. 15. Contrast for the third group of images.

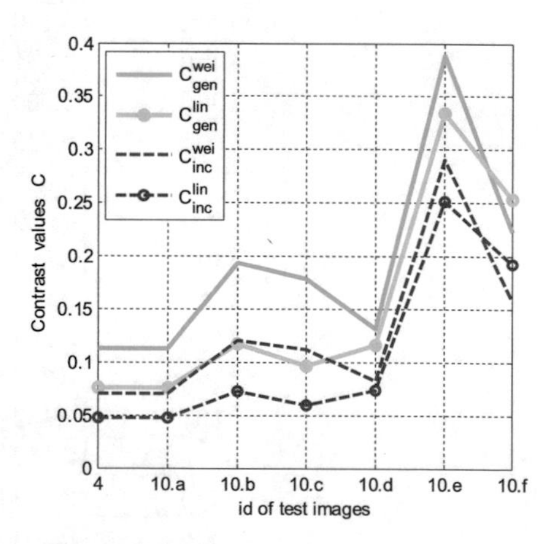

Fig. 16. Contrast for the fourth group of images.

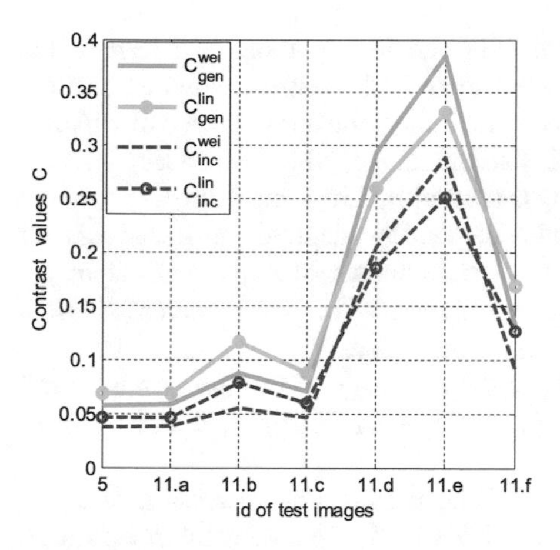

Fig. 17. Contrast for the fifth group of images.

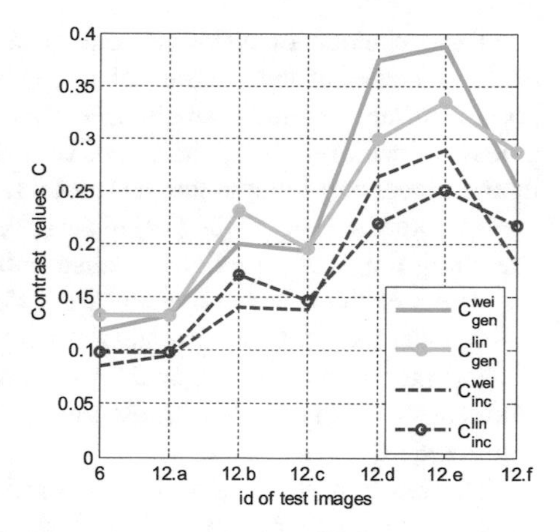

Fig. 18. Contrast for the sixth group of images.

5 Discussion

The study shows that the methods of the piecewise linear and nonlinear contrast stretching make it possible to increase the overall contrast in image and is widely used in image pre-processing for image enhancement.

However, the known techniques for contrast enhancement have a number of significant disadvantages, which substantially limit their practical use for processing of complex images in an automatic mode.

The study shows that the minimum and percentage linear stretching (1), (2) enable to increase the contrast for images with a narrow dynamic range of brightness (Figs. 11(b) and 12(b)). However, the effectiveness of min-max and percentage linear stretching (1), (2) depends significantly on the distribution of brightness and is very low for images with a wide dynamic range (Figs. 7(a), (b), 8(a), (b), 9(a), 10(a), 11(a) and 12(a)).

The results of image processing using piecewise linear contrast stretching (4) and (5), logarithmic and exponential transformation (7) and (8), gamma correction (9) and sigmoid stretching (10) depends significantly on the values of parameters of the transformation, which are usually set in an interactive mode (Figs. 7(c), (d), 8(d), 9(c), (d), 10(b), (c), (d), 11(d), 12(b) and (d)).

The efficiency of image processing with using these methods (4), (5) and (7)–(10) is relatively low for complex low-contrast images with a multi-modal distribution of brightness and a wide dynamic range.

The most effective increase in image contrast is provided by methods based on the technology of histogram specification and its modifications ((11), (12), [11–13]).

The techniques based on the histogram equalization (11) provide the most effective contrast enhancement, however, expert estimates show that high values of the assessment of overall contrast for histogram equalization are the result of an overly increase in the contrast of large-sized objects.

The technique of histogram equalization (11) and its variations can lead to the overly increase in the contrast of large-sized objects, a decrease in contrast and the possible disappearance of small-sized objects in the image, the consequence of which is a possible deterioration in the objective quality of the image, which is unacceptable for image processing in an automatic mode (Figs. 8(e), 9(e), 11(e) and 12(e)).

The proposed technique (17) of image enhancement by adaptive non-linear contrast stretching provides an efficient redistribution of the contrast of objects in the image regardless of their size and enables to effectively increase the overall contrast, while maintaining the contrast of small-sized objects in the image.

The proposed technique enables to improve the efficiency of increasing the overall contrast for complex low-contrast images with multi-modal distribution of brightness and a wide dynamic range.

The research shows that the proposed technique provides an increase in the efficiency of contrast enhancement (on average by 84%–129%) in automatic mode for all original images and an effective increase in the contrast of objects in image regardless of their size.

6 Conclusion

In this paper, the problem of improving the quality of complex low-contrast images in automatic mode with an acceptable level of computational costs was considered. The task of effectively increasing the contrast for complex low-contrast images with small-sized objects, a wide dynamic range and multi-modal distribution of brightness was considered. The purpose of the work is to improve the efficiency of increasing the overall contrast for complex images with multi-modal distribution of brightness and a wide dynamic range.

A new approach to increase the contrast of a complex low-contrast image by adaptive non-linear stretching of its dynamic range was proposed based on measuring of the raw moments for the subranges of image brightness. The proposed approach is based on measuring the ratios between the values of the raw moments for different subranges of brightness. The proposed approach provides an efficient redistribution of the contrast of objects in the image regardless of their size and enables to effectively increase the overall contrast, while maintaining the contrast of small-sized objects in the image. The proposed approach enables to improve the efficiency of increasing the overall contrast for complex low-contrast images with multi-modal distribution of brightness and a wide dynamic range.

A new technique of contrast enhancement for complex monochrome images based on measuring the mean values for subranges of image brightness was proposed.

The proposed technique enables to improve the efficiency of increasing the overall contrast for complex low-contrast images in an automatic mode with an acceptable level of computational costs.

The research of the various known and proposed techniques of image contrast enhancement in the automatic mode was carried out using the no-reference metrics of overall contrast for six groups of test images. The results of the research confirmed the high efficiency of the proposed technique for increasing the contrast of complex images. The proposed technique has a low computational cost and can be recommended for operative (in real-time) image pre-processing in automatic mode in various modern applications.

References

1. Pratt, W.K.: Digital Image Processing: PIKS Inside, 3rd edn. Wiley, New York (2001)
2. Weeks, A.R.: Fundamental of Electronic Image Processing. The International Society for Optical Engineering, Bellingham, Washington, SPIE, USA (1996)
3. Gonzalez, R.C., Woods, R.E.: Digital Image Processing, 3rd edn. Prentice Hall, Upper Saddle River (2010)
4. Wang, Z., Bovik, A.C.: Modern image quality assessment. In: Synthesis Lectures on Image, Video, and Multimedia Processing, vol. 2, no. 1, pp. 1–156. Morgan and Claypool Publishers, New York (2006)
5. Peli, E.: Contrast in complex images. J. Opt. Soc. Am. A **7**(10), 2032–2040 (1990)
6. Ketcham D.J.: Real time image enhancement technique. In: Proceedings SPIE/OSA Conference on Image Processing, Pacific Grove, USA, vol. 74, pp. 120–125 (1976)
7. Kaur, M., Kaur, J., Kaur, J.: Survey of contrast enhancement techniques based on histogram equalization. Int. J. Adv. Comput. Sci. Appl. **2**(7), 138–141 (2011)
8. Raju, A., Dwarakish, G.S., Reddy, D.V.: A comparative analysis of histogram equalization based techniques for contrast enhancement and brightness preserving. Int. J. Signal Process. Image Process. Pattern Recognit. **6**(5), 353–366 (2013)
9. Kotkar, V.A., Gharde, S.S.: Review of various image contrast enhancement techniques. Int. J. Innov. Res. Sci., Eng. Technol. **2**(7), 2786–2793 (2013)
10. Mokhtar, N.R., et al.: Image enhancement techniques using local, global, bright, dark and partial contrast stretching for acute leukemia images. In: Proceedings of the World Congress on Engineering (WCE), London, U.K., vol. 1, pp. 807–812 (2009)
11. Kim, Y.T.: Contrast enhancement using brightness preserving bi-histogram equalization. IEEE Trans. Consum. Electron. **43**(1), 1–8 (1997)
12. Wang, Y., Chen, Q., Zhang, B.: Image enhancement based on equal area dualistic sub-image histogram equalization method. IEEE Trans. Consum. Electron. **45**(1), 68–75 (1999)
13. Chen, S., Ramli, A.R.: Contrast enhancement using recursive mean-separate histogram equalization for scalable brightness preservation. IEEE Trans. Consum. Electron. **49**(4), 1301–1309 (2003)
14. Hassan, N., Akamatsu, N.: A new approach for contrast enhancement using sigmoid function. Int. Arab. J. Inf. Technol. **1**(2), 221–225 (2004)
15. Hummel, R.: Image enhancement by histogram transformation. Comp. Graph. Image Process. **6**, 184–195 (1977)
16. Frei, W.: Image enhancement by histogram hyperbolization. Comput. Graph. Image Process. **6**(3), 286–294 (1977)
17. http://sipi.usc.edu/database/database.php?volume=misc
18. http://optipng.sourceforge.net/pngtech/corpus/
19. Nesteruk, V.F., Sokolova, V.A.: Questions of the theory of perception of subject images and a quantitative assessment of their contrast. Opt.-Electron. Ind. **5**, 11–13 (1980)

20. Yelmanova, E., Romanyshyn, Y.: No-reference metric of generalized contrast for complex images, In: Proceedings of the 2017 IEEE First Ukraine Conference on Electrical and Computer Engineering, Kyiv, Ukraine, pp. 1088–1093 (2017)
21. Yelmanov, S., Romanyshyn, Y.: A method for rapid quantitative assessment of incomplete integral contrast for complex images. In: Proceedings of IEEE 14th International Conference on TCSET 2018, Lviv-Slavske, Ukraine, pp. 915–920 (2018)

A New Approach to Measuring Perceived Contrast for Complex Images

Sergei Yelmanov[1(✉)] and Yuriy Romanyshyn[2,3]

[1] Special Design Office of Television Systems, Lviv, Ukraine
sergei.yelmanov@gmail.com
[2] Lviv Polytechnic National University, Lviv, Ukraine
yuriy.romanyshynl@gmail.com
[3] University of Warmia and Mazury, Olsztyn, Poland
yuriy.romanyshyn@uwm.edu.pl

Abstract. The problem of assessing the overall contrast of complex images is considered. The problem of increasing the accuracy of the quantitative evaluation of the perceived contrast for the complex multi-element images is being solved. This paper proposes a new approach to measuring the overall contrast of a complex image based on an estimate of the perceived contrast of the all pairs of elements of image (objects and background). A new approach to estimating the contrast of two elements of a complex image is proposed based on measuring the perceived contrast of a simple two-element image of these elements relative to the adaptation level for this simple image. The proposed approach is based on measuring the ratios between the values of mean absolute deviation of the brightness of simple image relative to the level of adaptation and the level of adaptation. A possible implementation of the proposed approach to measuring the perceived contrast for a weighted contrast kernel is considered. A new technique of measuring the perceived contrast of multi-element images for weighted contrast is proposed. The paper presents the results of research of the known no-reference metrics of contrast and proposed technique of measuring the perceived contrast of image for two groups of test images. The proposed technique of measuring the contrast enables to increase the accuracy of the estimated contrast for complex images.

Keywords: Image quality · Measuring contrast · No-reference assessment
Overall contrast · Complex image

1 Introduction

Wide use of new technologies for digital image processing requires addressing the problem of the no-reference quantitative assessment of image quality [1].

There are various approaches to assessing the objective quality of the image [2, 3].

Objective quality of image is usually determined based on an analysis of its basic quantitative characteristics [1, 2].

The contrast of the image is its basic quantitative characteristic [1, 2, 4].

Contrast to a large extent determines the objective quality of the image and the effectiveness of its visual perception, the effectiveness and reliability of its subsequent

N. Shakhovska and M. O. Medykovskyy (Eds.): CSIT 2018, AISC 871, pp. 85–101, 2019.
https://doi.org/10.1007/978-3-030-01069-0_7

processing and analysis [2]. Currently, the use of new techniques for measuring the perceived contrast for complex images is relevant as never before [1].

The overall contrast of a complex multi-element image is usually estimated based on the values of contrast of its elements (objects and background). The contrast of two image elements (two objects or an object and a background) determines the difference in their objective characteristics. The contrast of the image elements is usually determined on the basis of the difference between their values of brightness [4–6].

There are various approaches to assessing the contrast of an image [1, 4–6].

However, the known methods for measuring the contrast of an image have a number of disadvantages that significantly reduce the effectiveness of their practical use. The main disadvantage of the known methods is their low accuracy of measuring the contrast for complex multi-element images with a wide dynamic range and uneven illumination.

To address these disadvantages, we propose a new approach to measuring the overall contrast of a complex image based on an estimate of the perceived contrast of the pairs of elements of image.

The problem of increasing the accuracy of the quantitative evaluation of perceived contrast for complex multi-element images is considered in this paper (Sect. 2). A study of various approaches to measuring the overall contrast for complex multi-element images is carried out. The object of the study is the process of measuring the objective quality of the image. The problem of increasing the accuracy measuring the contrast of image elements is considered. The purpose of this work is to improve the accuracy of the no-reference quantitative assessment of the contrast of a complex image based on the measurement of contrast values for pairs of elements of this image (objects and background). The subject of the study is techniques of measuring the contrast of elements of complex multi-element images.

A new approach to estimating the contrast of two elements of a complex image is proposed based on measuring the perceived contrast of a simple two-element image of these elements relative to the adaptation level for this simple image (Sect. 3). The proposed approach is based on measuring the ratios between the values of mean absolute deviation of the brightness of simple image relative to the level of adaptation and the level of adaptation. A new technique of measuring the perceived contrast of multi-element images for weighted contrast is proposed.

The proposed technique for measuring the perceived contrast provides an effective increase in the accuracy of the estimate of integral contrast for complex multi-element images with a wide dynamic range and uneven illumination.

The research of the known methods and proposed technique for no-reference measuring the perceived contrast of complex low-contrast images was carried out for two groups of test images with a wide dynamic range and uneven illumination (Sects. 4 and 5).

2 Overall Contrast in Complex Images

The overall contrast of a complex multi-element image is usually based on the measurement of contrast values for all pairs of its elements (objects and background). The contrast of two image elements (two objects or an object and a background) determines the distinction in their quantitative characteristics.

There are various approaches to measuring the overall contrast of complex multi-element images.

The overall (generalized) contrast of a complex multi-element image is defined as the mean value of contrast for all pairs of image elements in original image [4]:

$$C_{gen} = \int_0^1 \int_0^1 | C_{ij} | \cdot h(C_{ij})\, dC_{ij}, \tag{1}$$

where C_{ij} – contrast of a pair (i, j) of image elements; $h(C_{ij})$ – probability density function for contrast C_{ij}.

However, estimating the distribution $h(C_{ij})$ of contrast of image elements is in itself a rather complex task and requires a significant level of computational costs.

Therefore, to simplify the calculation, expression (1) is usually represented in the form [4]:

$$C_{gen} = \int_0^1 \int_0^1 | C_{ij} | \cdot p(B_i, B_j)\, dB_i dB_j, \tag{2}$$

where B_i, B_j – brightness of two image elements i and j; $p(B_i, B_j)$ – two-dimensional distribution of brightness for image elements.

For the practical implementation of these approaches (1) and (2), it is necessary to solve the problem of choosing the definition of contrast for two image elements and estimate the two-dimensional distribution of brightness. For this case, it is assumed that the contrast of the two image elements is equal to the contrast of a simple two-element image. The contrast of two elements of a simple image is usually defined on the basis of the difference between the values of their brightness.

Various definitions for the contrast of simple two-element image are known.

In [4] the definition of weighted contrast of the image elements on the basis of the contrast law of light perception has been proposed:

$$C_{ij}^{wei_1} = \frac{B_i^2 - B_j^2}{B_i^2 + B_j^2}. \tag{3}$$

Currently, the most widely used definition of the contrast of a simple two-element image is the weighted contrast in the form [2]:

$$C_{ij}^{wei_2} = \frac{B_i - B_j}{B_i + B_j}. \tag{4}$$

Another known definition of the contrast of image elements is the relative contrast, which is most often defined as [5]:

$$C_{ij}^{rel_1} = \frac{B_i - B_j}{\max(B_i,\ B_j)}, \tag{5}$$

$$C_{ij}^{rel_2} = \frac{B_i - B_j}{1 - \min\left(B_i,\ B_j\right)}. \tag{6}$$

The definition of linear contrast of the two image elements is known, which has the form [6]:

$$C_{ij}^{lin} = \frac{B_i - B_j}{B_{\max} - B_{\min}}, \tag{7}$$

where B_{min}, B_{max} – minimum and maximum values of brightness in image.

The definitions (3)–(5) of the contrast for simple two-element image are called the contrast kernels and are the basis for the metrics of contrast for complex multi-element images.

Estimation of the two-dimensional distribution $p(B_i,\ B_j)$ of brightness is also quite a challenge.

To simplify the calculations, it is usually assumed that the elements in image are independent and do not affect each other. For this case, the estimate of the two-dimensional distribution $p(B_i,\ B_j)$ of brightness takes the form [4]:

$$p\left(B_i, B_j\right) = p(B_i) \cdot p\left(B_j\right), \tag{8}$$

where $p(B_i)$ – probability density for brightness of image.

In this case, the generalized contrast (2) for (4), (5) and (7) using (8) takes the form [6]:

$$C_{gen}^{wei} = \int_0^1 \int_0^1 \frac{\left|B_i - B_j\right|}{B_i + B_j} \cdot p(B_i) \cdot p\left(B_j\right) dB_i dB_j, \tag{9}$$

$$C_{gen}^{rel} = \int_0^1 \int_0^1 \frac{\left|B_i - B_j\right|}{\max\left(B_i,\ B_j\right)} \cdot p(B_i) \cdot p\left(B_j\right) dB_i dB_j, \tag{10}$$

$$C_{gen}^{lin} = \int_0^1 \int_0^1 \frac{\left|B_i - B_j\right|}{B_{\max} - B_{\min}} \cdot p(B_i) \cdot p(B_j) dB_i dB_j. \tag{11}$$

Expressions (9)–(11) are no-reference histogram-based metrics of generalized contrast for complex multi-element images.

Another known approach to assessing the overall contrast in a complex image is based on measuring the contrast of image elements relative to a known level of adaptation.

To simplify the calculations, in [4] the estimate of complete integral contrast as the mean value of contrast for all pairs of image elements relative to a preset adaptation level has been proposed:

$$C_{com} = \int_0^1 \int_0^1 | C_{ij0} | \cdot p(B_i, B_j)\, dB_i dB_j, \tag{12}$$

where C_{ijO} – contrast of an elementary two-element image with brightness B_i and B_j relative to the preset value of adaptation level B_O.

The contrast of the two image elements relative to the adaptation level is defined on the basis of the law of perception of an elementary two-element image relative to the preset level of adaptation and has the form [4]:

$$C_{ij0} = \frac{C_{i0} + C_{j0}}{1 + C_{i0} \cdot C_{i0}}, \tag{13}$$

where C_{iO}, C_{jO} – contrast values of image element i and j relative to the value B_O of adaptation level.

Expression (13) for weighted contrast kernel (4) has the form [4]:

$$C_{ij}^{wei_3} = \frac{B_i \cdot B_j - B_0^2}{B_i \cdot B_j + B_0^2}. \tag{14}$$

In this case, the complete integral contrast (12) for weighted contrast (4) using (14) and (8) takes the form [4]:

$$C_{com}^{wei} = \int_0^1 \int_0^1 \frac{B_i \cdot B_j - B_0^2}{B_i \cdot B_j + B_0^2} \cdot p(B_i) \cdot p(B_j)\, dB_i dB_j. \tag{15}$$

The value B_O of the adaptation level is usually assumed to be equal to the mean brightness in the image [4]:

$$B_0 = \int_0^1 B \cdot p(B)\, dB. \tag{16}$$

Expression (15) is the definition of the complete integral contrast using the kernel (4) of weighted contrast [4]:

In [4] another approach to the estimation of the two-dimensional distribution of brightness of the image elements was proposed.

To reduce computational costs, the estimation of the distribution of two-dimensional brightness was defined as [4]:

$$p(B_i, B_j) = p(B_i) \cdot \delta(B_i - B_j), \tag{17}$$

where $\delta(\cdot)$ – delta function.

In [4] on the basis of (12), (13) and (17) the definition of incomplete integral image contrast was proposed, which has the form:

$$C_{inc} = 2 \cdot \int_0^1 \left| \frac{C_{i0}}{1 + C_{i0}^2} \right| \cdot p(B_i)\, dB_i. \tag{18}$$

For definition (4) of weighted contrast, the expression (18) takes the form [4]:

$$C_{inc}^{wei_1} = \int_0^1 \left| \frac{B^2 - B_0^2}{B^2 + B_0^2} \right| \cdot p(B)\, dB. \tag{19}$$

Another known definition of incomplete integral contrast has the form [6]:

$$C_{inc} = \int_0^1 |\, C(B, B_0) \,| \cdot p(B)\, dB. \tag{20}$$

The definition of the incomplete integral contrast for the weighted and linear contrast kernels (4) and (7) takes the form [6]:

$$C_{inc}^{wei_2} = \int_0^1 \frac{|B - B_0|}{B + B_0} \cdot p(B)\, dB, \tag{21}$$

$$C_{inc}^{lin} = \frac{1}{B_{\max} - B_{\min}} \int_0^1 |B - B_0| \cdot p(B)\, dB. \tag{22}$$

The definitions (19), (21) and (22) are no-reference metrics of incomplete integral contrast for rapid estimation of contrast for multi-element images.

It should be noted that the choice of the value B_0 of adaptation level is a very difficult problem and largely determines the accuracy of the measuring of complete and incomplete integral contrast using (15), (19), (21) and (22).

The main disadvantage of the known methods (15), (19), (21), (22) is the low accuracy of measuring the contrast for different pairs of the image elements relative to the constant value of adaptation level. The value of the adaptation level does not change when calculating the contrast for different pairs of image elements. In addition, when choice the value of the adaptation level, the brightness values and the sizes of the image elements for their different pairs are not taken into account.

Increasing the accuracy of the estimate of integral contrast for a complex image is possible by increasing the accuracy of the contrast measurement for the pairs of its elements. The choice of the adaptation level for each pair of image elements makes it possible to significantly improve the accuracy of measuring their contrast. When calculating the contrast of a pair of elements of simple image, for choosing the value of adaptation level it is also necessary to take into account the brightness values and the dimensions of image elements of this pair. The analysis of the ratio of the sizes of elements of a simple image enables increasing the accuracy and reliability of measuring its level of adaptation and perceived contrast.

To improve the accuracy of estimating the integral contrast, we propose a new approach to measuring the perceived contrast for pairs of image elements.

In this work, we propose an estimate of the perceived contrast of two elements of a complex image based on the contrast measurement of a simple two-element image relative to the adaptation level, which is equal to the mean value of the brightness of a simple image.

3 Proposed Method

In this paper, the problem of increasing the accuracy of the quantitative evaluation of the integral contrast for the complex multi-element images is being solved.

An effective increase in the accuracy of the estimation of contrast for complex images is possible by measuring the contrast of a pair of image elements relative to the adaptation level for a simple image of the elements of this pair.

In this paper, we propose a new approach to measuring the overall contrast of complex image based on assessing the perceived contrast of all pairs of elements of this image.

A new approach to estimating the contrast of image elements is proposed based on measuring the contrast of a simple two-element image of these elements relative to the adaptation level for this simple image.

Consider a simple image which consists of two elements i and j of a complex multi-element image (Fig. 1).

In this paper, we propose a new definition for the perceived contrast of a simple image of two elements i and j of a multi-element image (Fig. 1) in the form:

$$C^{per} = \Delta / L, \tag{23}$$

$$\Delta = \int_0^1 |B^* - L| \cdot p(B^*)\, dB^*, \tag{24}$$

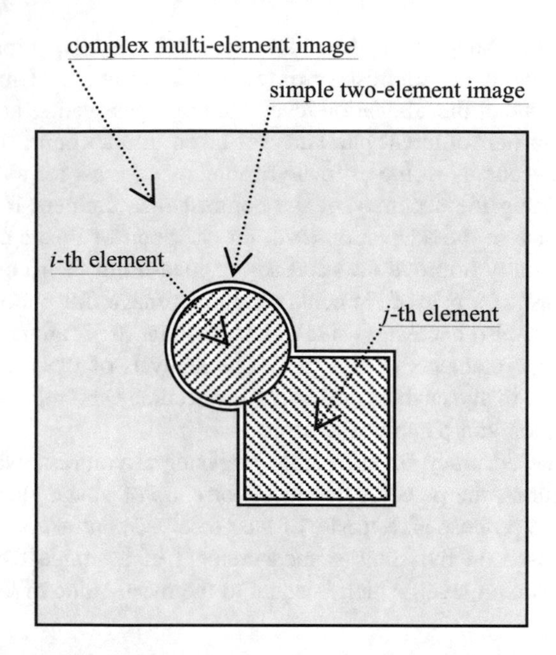

Fig. 1. Simple two-element image for two elements of a complex multi-element image.

where B^* – brightness of simple image; L – value of adaptation level for simple image; Δ – mean absolute deviation of the brightness of simple image relative to the value L of adaptation level.

Expressions (23) and (24) describe the proposed definition for perceived contrast of a simple two-element image.

Assume that the value of the adaptation level L for weighted contrast (4) is defined as:

$$L^{wei} = \int_0^1 B^* \cdot p(B^*)\, dB^*, \tag{25}$$

where $p(B^*)$ – probability density for brightness of simple image.

The proposed definition of perceived contrast of a simple image using (23)–(25) for weighted contrast takes the form:

$$C^{per} = \int_0^1 |B^* - L| \cdot p(B^*)\, dB^* / \int_0^1 B^* \cdot p(B^*)\, dB^*. \tag{26}$$

The expression (26) defines the proposed approach to measuring perceived contrast for pairs of image elements for a weighted contrast kernel.

Assume that the brightness of the elements i and j of a simple image (Fig. 1) is equal to B_i and B_j (Fig. 2).

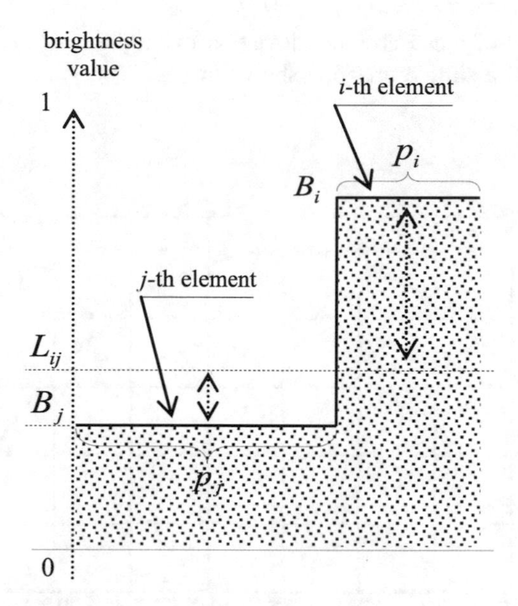

Fig. 2. The values of brightness for two elements of a simple image.

In this case, for a simple two-element image, mean absolute deviation Δ of brightness relative to the value L of adaptation level is equal to (Fig. 2) ($B_i > L > B_j$, $p(B_i) > 0$, $p(B_j) > 0$):

$$\Delta_{ij} = \frac{p_i}{p_i + p_j} \cdot \left(B_i - L_{ij}\right) + \frac{p_j}{p_i + p_j} \cdot \left(L_{ij} - B_j\right), \tag{27}$$

where L_{ij} – adaptation level for simple image of pair (i, j) of image elements; Δ_{ij} – mean absolute deviation of brightness for simple image of elements i and j.

Expression (26) for simple two-element image of the elements i and j has the form:

$$C_{ij}^{per} = \frac{\frac{p_i}{p_i + p_j} \cdot \left(B_i - L_{ij}\right) + \frac{p_j}{p_i + p_j} \cdot \left(L_{ij} - B_j\right)}{L_{ij}}. \tag{28}$$

Assume that the adaptation level L_{ij} for a simple two-element image for weighted contrast (4) is equal to the average brightness of simple image (Fig. 2):

$$L_{ij} = \frac{p_i}{p_i + p_j} \cdot B_i + \frac{p_j}{p_i + p_j} \cdot B_j. \tag{29}$$

In this case, we represent expression (27) in the form:

$$\Delta_{ij} = 2 \cdot \frac{p_i \cdot p_j}{\left(p_i + p_j\right)^2} \cdot \left(B_i - B_j\right). \tag{30}$$

The dependence of mean absolute deviation of brightness (30) on the ratio of the sizes of elements of a simple image is shown in Fig. 3.

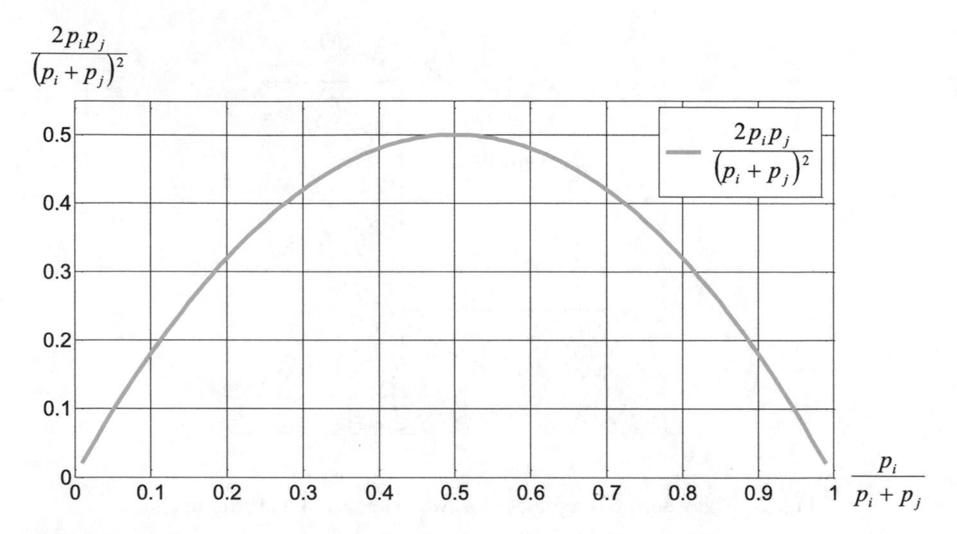

Fig. 3. Dependence of value of mean absolute deviation of brightness on the ratio of sizes of elements of simple image.

The perceived contrast (23) of a simple image for weighted contrast using (29) and (30) is equal to:

$$C_{ij}^{per} = \frac{2p_i \cdot p_j \cdot (B_i - B_j)}{(p_i + p_j) \cdot (p_i \cdot B_i + p_j \cdot B_j)}. \tag{31}$$

Expression (31) describes the proposed definition of the perceived contrast of a simple two-element image for weighted contrast kernel (4).

The complete integral contrast of a multi-element image in accordance with (2) is defined as the average value of the estimates of contrast for all pairs of its elements:

$$C_{int} = \int_0^1 \int_0^1 \left| C_{ij}^{per} \right| \cdot \omega(B_i, B_j)\, dB_i dB_j, \tag{32}$$

where $\omega(B_i, B_j)$ – two-dimensional distribution of brightness for pairs of image elements.

For the practical implementation of the proposed approach (32), it is necessary to solve the problems of estimating the two-dimensional distribution of brightness for pairs of image elements.

To demonstrate the possibilities of the proposed approach, we assume that by analogy with (8) the distribution $\omega(B_i, B_j)$ has the form:

$$\omega\left(B_i, B_j\right) = \omega(B_i) \cdot \omega\left(B_j\right), \tag{33}$$

where

$$\omega(B_i) = \beta \cdot p(B_i)^{\gamma}, \quad \beta = \left(\int_0^1 p(B_k)^{\gamma} dB_k \right)^{-1}, \tag{34}$$

where β – normalizing coefficient; γ – exponent, parameter.

The proposed definition (32) for complete integral contrast using (31), (33) and (34) takes the form:

$$C_{int}^{wei} = 2\beta^2 \cdot \int_0^1 \int_0^1 \frac{p_i \cdot p_j \cdot \left| B_i - B_j \right|}{\left(p_i + p_j\right) \cdot \left(p_i \cdot B_i + p_j \cdot B_j\right)} \cdot p_i^{\gamma} \cdot p_j^{\gamma} \, dB_i dB_j. \tag{35}$$

Expression (35) describes the implementation for the proposed approach (23), (24), (32) to the estimation of the integral contrast of a complex image on the basis of measuring the perceived contrast of simple images for pairs of its elements.

The known (9)–(11), (15), (19), (21), (22) and proposed (35) definitions are no-reference metrics for measuring of integral contrast for complex images.

The research of known and proposed techniques for measuring the integral contrast of complex images was carried out in Sects. 4 and 5.

4 Research

The research was carried out by measuring the integral contrast using contrast metrics (9), (11), (15), (19), (21), (22) and (35) for the two groups of test images.

The first group of test images [7] consists of twelve images with a complex structure and different levels of contrast and illumination.

Test images of the first group are shown in Fig. 4 [7].

Test images of the second group are shown in Fig. 5.

The second group of test images consists of the original image (Fig. 5(a), [7]) and the results of its processing using known methods of image enhancement, namely:

(1) original image (Fig. 5(a), [7]);
(2) percentage stretching [2, 3] (Fig. 5(b));
(3) piecewise linear stretching [8] (Fig. 5(c));
(4) sigmoid stretching [9] (Fig. 5(d));
(5) gamma-correction [3], γ = 0,75 (Fig. 5(e));
(6) gamma-correction [3], γ = 0,50 (Fig. 5(f));
(7) gamma-correction [3], for a value 0,5 of the average brightness of the processed image (Fig. 5(g));
(8) histogram equalization [10] (Fig. 5(h));
(9) histogram hyperbolization [11] (Fig. 5(i));

(a) [7] (b) [7] (c) [7]

(d) [7] (e) [7] (f) [7]

(g) [7] (h) [7] (i) [7]

(j) [7] (k) [7] (l) [7]

Fig. 4. The first group of test images.

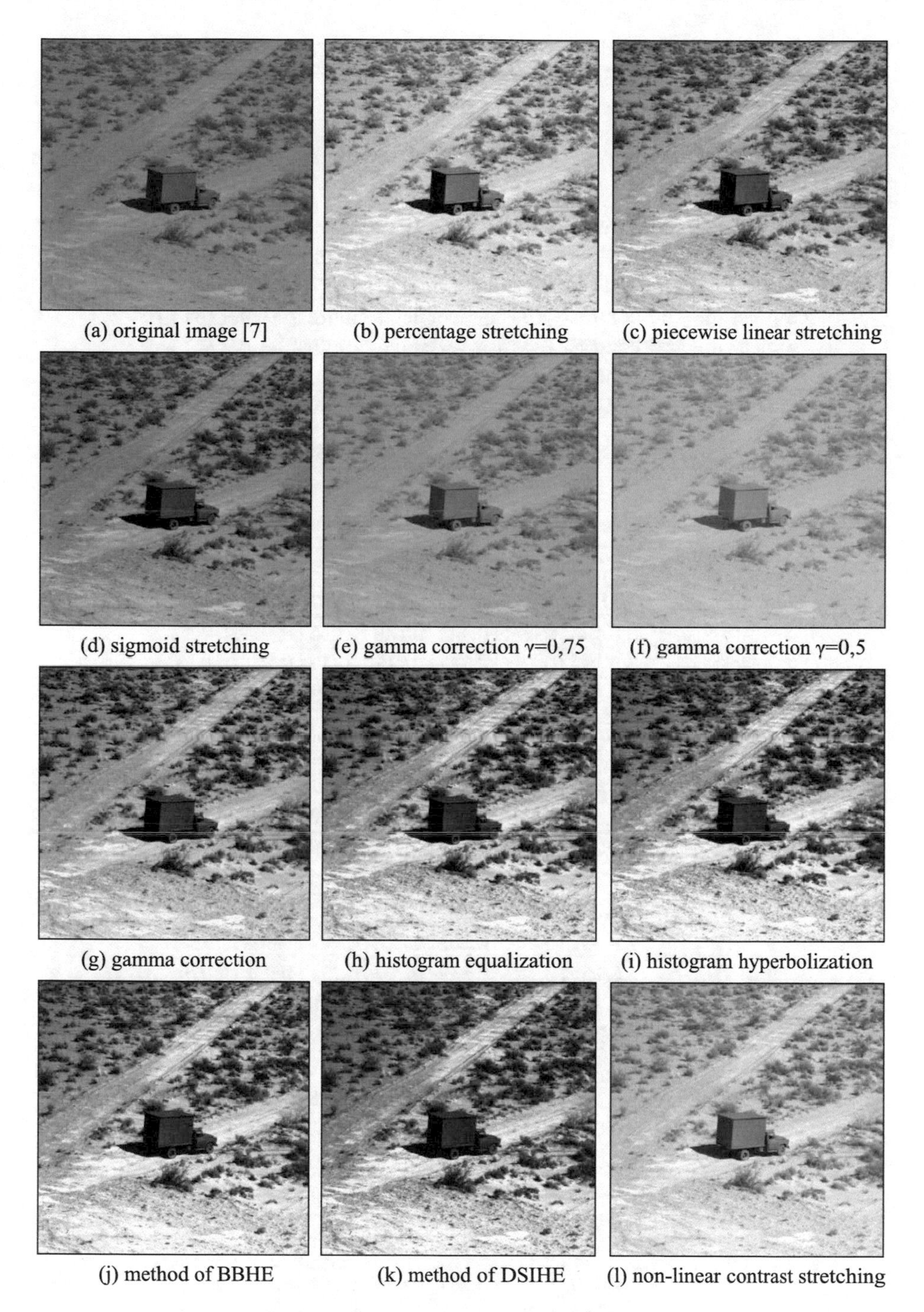

Fig. 5. The second group of test images.

(10) method of BBHE [12] (Fig. 5(j));
(11) method of DSIHE [13] (Fig. 5(k));
(12) non-linear contrast stretching [14] (Fig. 5(l)).

Research was carried out by measuring of complete and incomplete integral contrast using known and proposed definitions of contrast of image elements, namely:

(1) complete integral contrast (15) for the weighted contrast kernel (3) [4];
(2) generalized contrast (9) for weighted contrast kernel (4) [6];
(3) generalized contrast (11) for linear contrast kernel (7) [6];
(4) incomplete integral contrast (19) for weighted contrast kernel (3) [4];
(5) incomplete integral contrast (21) for weighted contrast kernel (4) [6];
(6) incomplete integral contrast (22) for linear contrast kernel (7) [6];
(7) proposed metric of integral contrast for weighted contrast (35).

The values of integral contrast for two groups of test images are shown in Table 1.

Table 1. The values of integral contrast for two groups of test images

Id of images	C_{com}^{wei}	C_{gen}^{wei}	C_{gen}^{lin}	$C_{inc}^{wei_1}$	$C_{inc}^{wei_2}$	C_{inc}^{lin}	C_{int}^{wei}
4(a)	0.062	0.057	0.069	0.068	0.037	0.046	0.021
4(b)	0.102	0.092	0.142	0.122	0.065	0.104	0.027
4(c)	0.182	0.154	0.227	0.218	0.117	0.181	0.066
4(d)	0.186	0.176	0.176	0.250	0.132	0.134	0.100
4(e)	0.186	0.175	0.226	0.226	0.124	0.165	0.101
4(f)	0.257	0.232	0.196	0.278	0.162	0.140	0.138
4(g)	0.229	0.222	0.188	0.272	0.153	0.131	0.152
4(h)	0.235	0.229	0.233	0.314	0.170	0.174	0.191
4(i)	0.253	0.246	0.244	0.325	0.178	0.172	0.218
4(j)	0.249	0.238	0.249	0.316	0.173	0.181	0.216
4(k)	0.278	0.262	0.267	0.363	0.200	0.201	0.241
4(l)	0.441	0.368	0.263	0.422	0.281	0.205	0.333
5(a)	0.165	0.155	0.119	0.207	0.111	0.088	0.079
5(b)	0.220	0.206	0.234	0.259	0.146	0.173	0.116
5(c)	0.281	0.264	0.248	0.326	0.189	0.183	0.158
5(d)	0.268	0.252	0.171	0.321	0.182	0.127	0.137
5(e)	0.131	0.123	0.115	0.168	0.088	0.085	0.063
5(f)	0.090	0.084	0.099	0.117	0.060	0.073	0.041
5(g)	0.362	0.341	0.286	0.403	0.246	0.211	0.200
5(h)	0.407	0.388	0.334	0.465	0.289	0.251	0.223
5(i)	0.405	0.385	0.345	0.464	0.287	0.260	0.215
5(j)	0.389	0.369	0.336	0.444	0.273	0.253	0.208
5(k)	0.420	0.401	0.341	0.489	0.304	0.258	0.235
5(l)	0.216	0.205	0.235	0.261	0.145	0.171	0.104

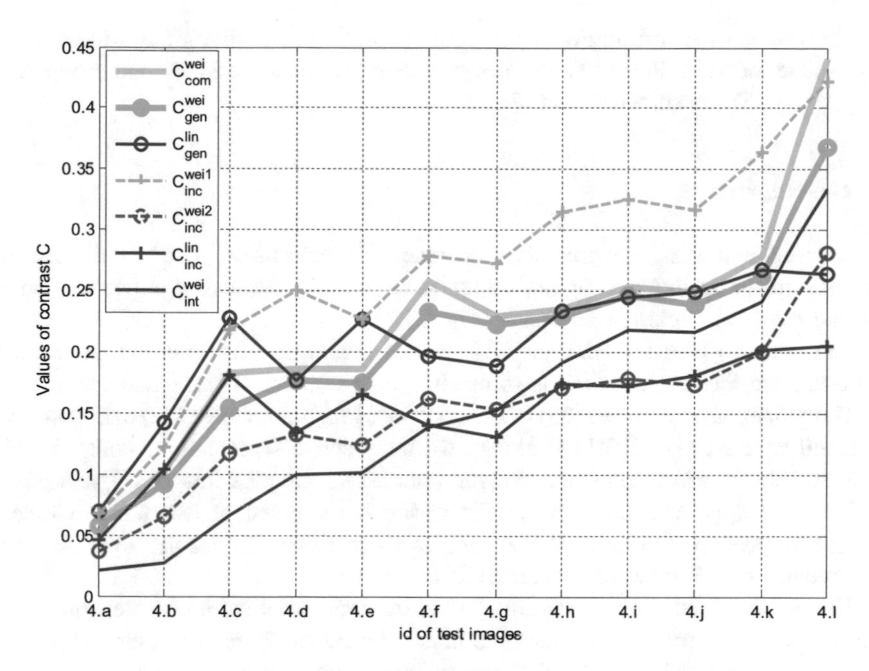

Fig. 6. The values of contrast for the first group of test images.

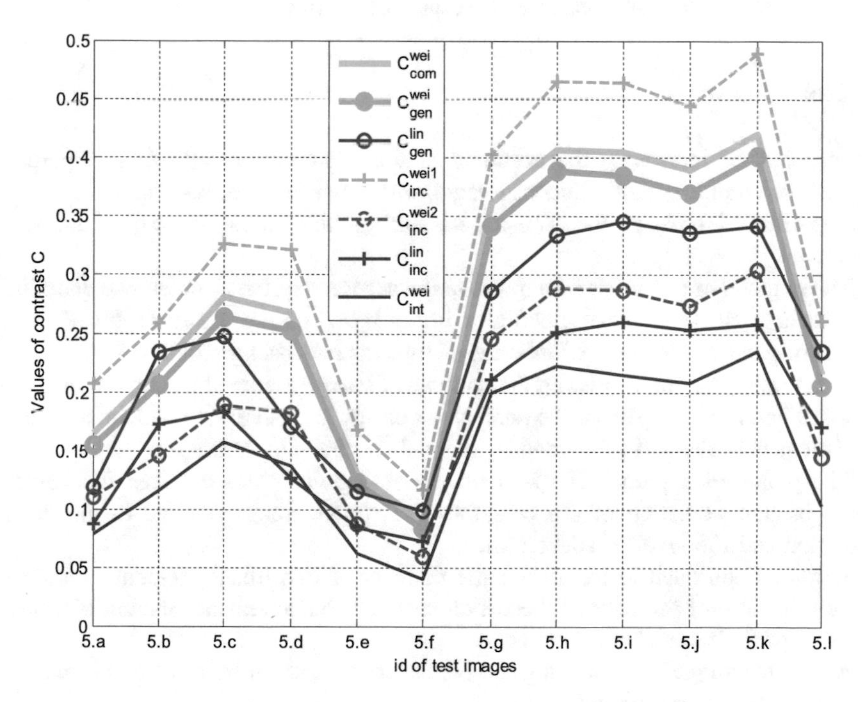

Fig. 7. The values of contrast for the second group of test images.

The results of measuring of the integral contrast for the first group of test images (Fig. 4) are shown in Fig. 6. The values of integral contrast for the second group of test images (Fig. 5) are shown in Fig. 7.

5 Discussion

Research show that assessments of the complete (15) and incomplete (19) contrast of a complex image using a weighted contrast kernel (3) depend significantly on the average value of brightness in image (Fig. 7).

The known estimates (19) and (21) of the incomplete integral contrast are proportional, and the estimate (19) is somewhat overestimated (Figs. 6 and 7).

The proposed assessment (35) of the complete integral contrast (32) using proposed weighted contrast kernel (31) is closest to the expert assessments of contrast and is most suitable for assessment of perceived contrast of complex images (Figs. 4 and 6).

Proposed approach (23)–(24) to measuring the contrast of two image elements, taking into account the ratio of the sizes of these elements, enables to enhance the effectiveness of estimates of its perceived contrast.

The proposed definitions (26) and (31) of the contrast of the image elements enable to improve the accuracy of the estimation of contrast of simple two-element images.

The proposed technique (35) of non-reference measurement of the complete integral contrast enables to increase the accuracy and reliability of the estimation of overall contrast for complex images and is most suitable for measuring the complete integral contrast of multi-element images with complex structure.

6 Conclusion

The problem of increasing the accuracy of the quantitative evaluation of perceived contrast for complex multi-element images was considered in this paper.

The task of increasing the accuracy measuring the contrast of image elements was considered.

The purpose of this work is to improve the accuracy of the no-reference quantitative assessment of the contrast of a complex image based on the measurement of contrast values for pairs of elements of this image (objects and background).

A new approach to estimating the contrast of two elements of a complex image was proposed based on measuring the perceived contrast of a simple two-element image of these elements relative to the adaptation level for this simple image.

The proposed approach is based on measuring the ratios between the values of mean absolute deviation of the brightness of simple image relative to the level of adaptation and the level of adaptation.

Proposed approach to measuring the contrast of two image elements, taking into account the ratio of the sizes of these elements, enables to enhance the effectiveness of estimates of their perceived contrast.

A new technique of measuring the perceived contrast of multi-element images for weighted contrast was proposed.

A new definition of kernel for weighted contrast was proposed based on measuring the perceived contrast of a simple two-element image for weighted contrast.

The research of the known methods and proposed technique for no-reference measuring the perceived contrast of complex low-contrast images was carried out for two groups of test images with a wide dynamic range and uneven illumination.

The proposed technique for measuring the perceived contrast provides an effective increase in the accuracy of the estimate of integral contrast for complex multi-element images with a wide dynamic range and uneven illumination.

The proposed assessment of complete integral contrast on the basis of proposed definition of perceived contrast of the image elements is closest to the results of expert assessments of contrast for test images and is most suitable for measuring the complete integral contrast of multi-element images with complex structure.

References

1. Wang, Z., Bovik, A.C.: Modern image quality assessment. In: Synthesis Lectures on Image, Video, and Multimedia Processing, vol. 2, no. 1, pp. 1–156. Morgan and Claypool Publishers, New York (2006)
2. Pratt, W.K.: Digital Image Processing: PIKS Inside, 3rd edn. Wiley, New York (2001)
3. Gonzalez, R.C., Woods, R.E.: Digital Image Processing, 3rd edn. Prentice Hall, Upper Saddle River (2010)
4. Nesteruk, V.F., Sokolova, V.A.: Questions of the theory of perception of subject images and a quantitative assessment of their contrast. Opt.-Electron. Ind. **5**, 11–13 (1980)
5. Peli, E.: Contrast in complex images. J. Opt. Soc. Am. A **7**(10), 2032–2040 (1990)
6. Yelmanova, E., Romanyshyn, Y.: No-reference contrast metric for medical images. In: Proceedings of the 2017 IEEE 37th International Conference on Electronics and Nanotechnology (ELNANO), Kyiv, Ukraine, pp. 338–343 (2017)
7. http://sipi.usc.edu/database/database.php?volume=misc
8. Yelmanov, S., Romanyshyn, Y.: Image contrast enhancement in automatic mode by nonlinear stretching. In: Proceedings of the 2018 XIV-th International Conference on Perspective Technologies and Methods in MEMS Design, Lviv, Ukraine, pp. 104–108 (2018)
9. Hassan, N., Akamatsu, N.: A new approach for contrast enhancement using sigmoid function. Int. Arab. J. Inf. Technol. **1**(2), 221–225 (2004)
10. Hummel, R.: Image enhancement by histogram transformation. Comp. Graph. Image Process. **6**, 184–195 (1977)
11. Frei, W.: Image enhancement by histogram hyperbolization. Comput. Graph. Image Process. **6**(3), 286–294 (1977)
12. Kim, Y.T.: Contrast enhancement using brightness preserving bi-histogram equalization. IEEE Trans. Consum. Electron. **43**(1), 1–8 (1997)
13. Wang, Y., Chen, Q., Zhang, B.: Image enhancement based on equal area dualistic sub image histogram equalization method. IEEE Trans. Consum. Electron. **45**(1), 68–75 (1999)
14. Yelmanov, S., Romanyshyn, Y.: Automatic contrast enhancement of complex low-contrast images. In: Proceedings of IEEE 14th International Conference on TCSET 2018, Lviv-Slavske, Ukraine, pp. 952–957 (2018)

Applied Linguistics

Authorship and Style Attribution by Statistical Methods of Style Differentiation on the Phonological Level

Iryna Khomytska(✉) and Vasyl Teslyuk

Lviv Polytechnic National University, Lviv 79013, Ukraine
Iryna.khomytska@ukr.net, vasyl.m.teslyuk@lpnu.ua

Abstract. A novel approach to authorship and style attribution and differentiation on the phonological level has been suggested. Each style is considered a statistical system the elements of which are mean frequencies of groups of consonants chosen as a style attribution and differentiation criterion. Statistical analogues of the phonological subsystems of style systems have been obtained by mathematical statistical methods (the hypothesis, ranking and style distance determination methods). Interrelations of style, language and individual manner of writing factors as well as the style-differentiating capability of eight groups of consonants (labial, forelingual, mediolingual, backlingual, nasal, constrictive, occlusive and sonorant) have been established. The results of the research show that only the three methods combined above allow to fully characterize each style (belles-lettres, colloquial and scientific) under study and establish authorship of a text. The closeness and distance established between the compared styles have been shown in the three models proposed.

Keywords: Consonant phonemes · Mean frequency
Style-differentiating capability · Authorship attribution

1 Introduction

The phonostatistical characteristics of a text help identify with mathematical accuracy its stylistic peculiarities and relate it to a particular type of style. It is typical to apply methods of mathematical statistics to style attribution and differentiation as these methods allow to avoid any ambiguity which may be caused by the use of different criteria for singling out a style's distinctive features. However, the results obtained by these methods have a probabilistic, stochastic nature. A number of linguistic and non-linguistic factors cause frequency fluctuations in groups of consonants. Thus the way groups of consonants function abides by different phonological laws depending on factors such as a phoneme's position, neighboring phonemes and intonation patterns. In other words, the effect of the language factor is stronger on the phonological level than on other language levels.

The novel approach of the present paper is to determine the effect of three factors: the author's manner of writing, style and language in statistical systems of the belles-lettres, colloquial and scientific styles. The elements of these statistical systems are

N. Shakhovska and M. O. Medykovskyy (Eds.): CSIT 2018, AISC 871, pp. 105–118, 2019.
https://doi.org/10.1007/978-3-030-01069-0_8

mean frequencies of the occurrence of groups of consonants and relations among them (statistical structures) that have been established by the hypothesis (HM), ranking (RM) and style distance determination (DR) methods. The aim of our study is to improve efficiency and define phonostatistical peculiarities of the researched styles caused by the interrelation of three factors: the author's manner of writing, style and language. The suggested approach to style attribution and differentiation is new when compared with other authors' studies [1–4, 8, 9, 12–15, 18–21] in which the hypothesis method and style distance determination method were used separately to analyse phoneme functioning in texts of different functional styles.

In contrast with the above-mentioned studies, we suggest the application of a ranking method to solve style differentiation problems. The method should be combined with the hypothesis and style distance determination methods because only all three of them, for each phonemes position, give the most exact and reliable data. We consider the problem of closeness and distance of the compared texts from the angle of the effect of the three factors: the author's manner of writing, style and language. The effect of a strong style factor has been indicated by the many groups of phonemes where the hypothesis method has identified essential differences, by the ranking method that has established a great difference of rank indices, and by the Interrelation of style and language factors has been studied in four main aspects. In the first aspect we determine the specifics of an author's individual manner of writing. The texts under study (poems by two poets, George Gordon, Lord Byron, and Thomas Moore) belong to the same historical period and literary trend—Romanticism. The second aspect involves comparison of texts of different styles of the belles-lettres style (poetry by Byron and Moore on the one hand and emotive prose by Byron on the other hand) from the same historical period. In the third aspect we study the interrelation of different styles from the same historical period (colloquial and scientific styles). Styles are represented by their statistical structures which are compared in pairs to analyse their interrelation. The fourth aspect deals with the texts of different historical periods representing different styles. These are the scientific and colloquial styles on the one hand and the belles-lettres style as represented by poetry by Byron and Moore, emotive prose by Byron, and drama by George Bernard Shaw on the other hand. In the second, third and fourth aspects, the style dominance prevails over the language dominance. It is seen more vividly in the fourth aspect of the studies in which the texts of different styles belonging to different historical periods do not differ as much as those of the same historical period. Here similarities are caused by common stylistic elements of the texts of the colloquial and belles-lettres styles.

2 Main Part

Dwelling briefly upon each of the methods used, we should note that the hypothesis method involves fulfilment of the following tasks:

1. an authorship and a style attribution and differentiation criterion is consonant group mean frequency;

2. consonant phoneme mean frequencies $\bar{X}$ must be distributed according to normal distribution;
3. the investigated texts of different authors and styles must be differentiated by the obtained difference of the values of the type $\bar{x}_1^{\alpha} - \bar{x}_2^{\alpha}$ for three phoneme's position;
4. three models must be built which determine the degree of style and an author's manner of writing factors.

The present study's results establish that a sample size of 31000 phonemes is sufficient. To obtain information about the criterion of differentiation $\bar{x}_r^{\alpha}$ on the basis of the data of the sample, we use the chi-square criterion x^2 [5–7, 10, 11]. In this investigation we use 5% level of significance. The calculated value x^2 is such that phoneme group frequencies are distributed according to the equation:

$$f(x) = \frac{1}{\sigma\sqrt{2\pi}} \cdot e^{-\frac{(X-\mu)^2}{2\sigma^2}} \tag{1}$$

where $f(x)$ is density of normal distribution, μ is mean frequency of a general population and σ^2 is variance of frequency in a general population.

Information about mean occurrence frequencies of consonant groups has been obtained from Liapunov's theorem according to which they follow normal distribution [2, 10, 11, 17]. With the use of Student's t-test, it has been proved that the values of differences $\bar{x}_1^{\alpha} - \bar{x}_2^{\alpha}$; $\bar{x}_2^{\alpha} - \bar{x}_3^{\alpha}$ for the given level of significance 0.05 are essential. To estimate the difference of the type $\bar{x}_1^{\alpha} - \bar{x}_2^{\alpha}$, a double-sided significance criterion 2Q has been used. The chosen level of significance is 0.05. The results of the calculations have shown that the difference $\bar{x}_1^{\alpha} - \bar{x}_2^{\alpha}$ is essential if 2Q < 0.05 and the value $\bar{x}_r^{\alpha}$ can be considered a criterion of differentiation of the researched styles on the level of phonology [1, 12, 16]. The data obtained is represented in 63 tables, one of which is given in this article (Table 1).

Table 1. Comparison of the colloquial style and poetry by Thomas Moore: unidentified position of a phoneme

Groups of phonemes	Scientific style		Colloquial style		S	t	2Q	Type of the value
	$\bar{X}$	$\sum (X_i - \bar{X})^2$	$\bar{X}$	$\sum (X_i - \bar{X})^2$				$\bar{X}_1 - \bar{X}_2$
Labial	137.9	4156.56	131.9	7611.48	14.00	1.69	>5%	Slight
Forelingual	425.0	8178.00	362.9	32500.3	26.04	9.39	<0.1%	Essential
Mediolingual	5.9	143.58	18.6	5175.36	9.42	5.31	<0.1%	Essential
Backlingual	59.3	3242.26	72.6	5157.36	11.83	4.43	<0.1%	Essential
Nasal	82.9	1902.71	76.8	4202.84	10.09	2.38	<5%	Essential
Sonorant	233.9	4890.01	226.9	14575.5	18.01	1.56	>10%	Slight
Constrictive	210.3	8529.18	158.9	7948.71	16.57	12.2	<0.1%	Essential
Occlusive	182.7	10670	226.5	5725.74	1653	10.4	<0.1%	Essential

The capability of style differentiation of each group of consonant phonemes determines its place in the phonological subsystem of a style system. It differs for each functional style and that is why it is expedient, along with the hypothesis method referred to above, to give the data obtained with the help of the ranking method [2, 10, 11]. This method consists of putting in order empirical data (consonant phoneme mean frequencies groups for the investigated styles) and building a decreasing row of values $\bar{x}_r^\alpha$. The rank index of the value $\bar{x}_r^\alpha$ is a number of the place it occupies in the row.

We have considered ranking of two types:

(1) building for each style a decreasing row of consonant phoneme group mean frequencies;
(2) building for each group of consonant phonemes a decreasing row of consonant phoneme group mean frequencies.

SDDM gives important information about style differentiation [10]:

$$l = \frac{t - t_0}{t}. \tag{2}$$

Here the parameter t of the Student's t-test corresponds to the value of the type $\bar{X}_1 - \bar{X}_0$ and t_0 stands for significance criterion 0.05. The values $\bar{X}_1$ and $\bar{X}_0$ are consonant phoneme group mean frequencies that our study compares in paired styles. In our investigation the number of degrees of freedom is $\nu = 60$. Consequently, $t_0 = t_{0.05}$; 60 = 2.00 [2, 10].

The developed methods have been coded on the java programming language. The developed system program of differentiation of phonostatistical structures of styles has the following software classes: Main, Window, PanelFile, ExtFileFilter, PanelTranscription, PortionDistribution, GroupDistribution, Pearson'sTest, Student's t-Test. For Main jFrame from javax.swing is used. To develop tabs for Window, jtabbedPane is used. Every tab is named. To add a new tab, the add method is used. Panel File consists of two text areas: jTextAreaRight jTextAreaLeft. To choose a file jFileChooser is used. ExtFile is used to select files of the type. txt. To read a text, BufferedReader from java. io. is used. To divide the TextArea into jTextAreaRight and jTextAreaLeft, jSplitPane is used. To place the text area on jSplitPane, two panels are made with the help of jPanel. Every text area has it scroll from jScrollPane. the components jTextArea, jSplitPane, jPanel, jScrollPane are from javax.swing.

The algorithm of functioning of the developed program system of differentiation of phonostatistical structures of styles is as follows (Fig. 1).

The Transcription tab is realised in PanelTranscription. The program system works independantly from the internet. To fill the Transcription .txt file, ArrayList <String> is used. Javax.swing and java.io are used for transformation the text into its transcriptional variant. In Fig. 2 the transcriptional variant of the text files is given.

The sample with the size of 31000 phonemes is divided into 31 portions with 1000 phonemes in each. In Fig. 3 phoneme distribution in groups is given.

To develop Pearson's test tab and Student's t-test tab, the mathematical library Colt of java.util has been used. The results obtained by Pearson's test are positive (normal distribution). The two texts chosen from Byron and Moore's poems differ in sonorant,

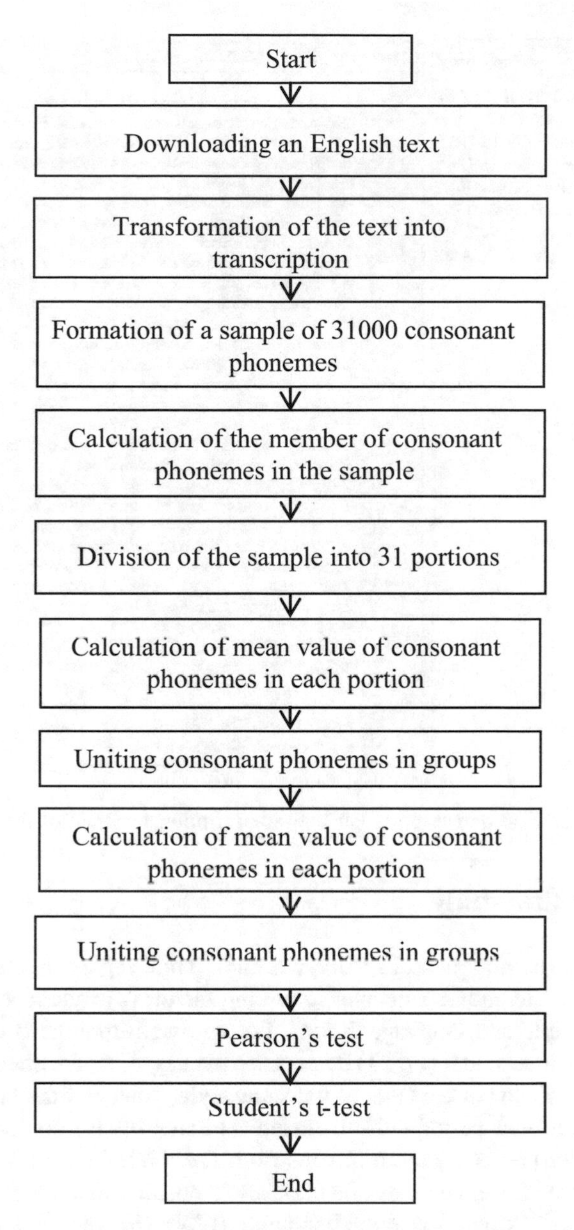

Fig. 1. Block-scheme of algorithm of functioning of the program system, of differentiation of phonostatistical structures of styles

constrictive and occlusive phoneme groups. Consequently, authorship attribution has been done by phonostatistical parameters – mean frequencies of consonant phoneme groups.

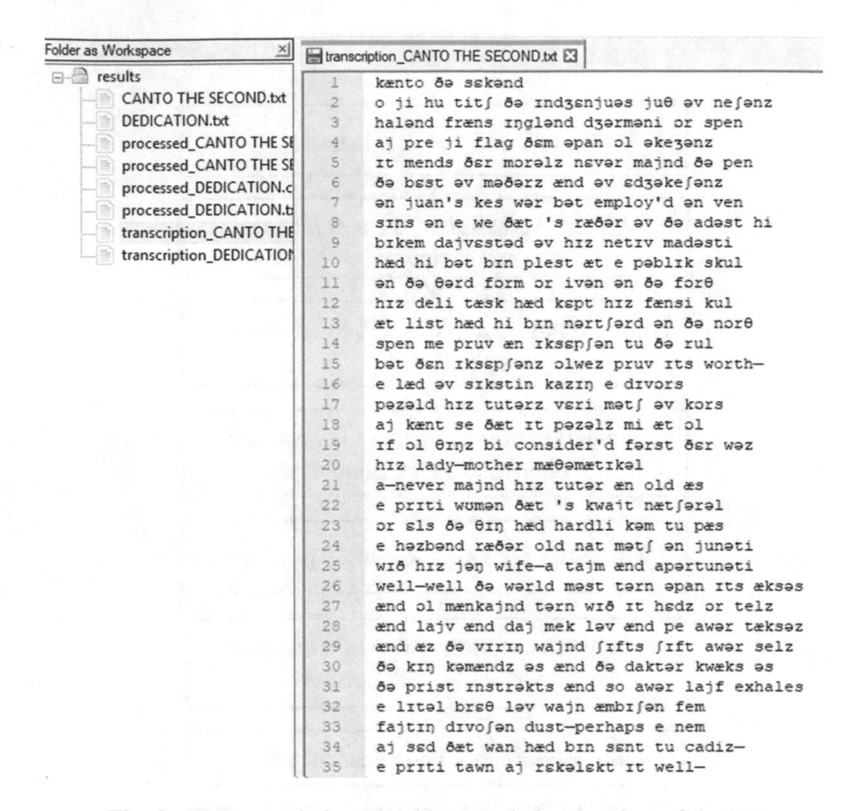

Fig. 2. The transcription tab with transcriptional variant of the text

3 Results of the Study

Three models are the new aspect of this research. They represent the interrelation of (1) the style factor and author's manner of writing factor; (2) style factors of the belles-lettres style; (3) style and language factors. Let us dwell upon each of the suggested three models. The first model (Fig. 1) determines the degree of the effect of the author's manner of writing factor in the texts of the same style, poetry of the belles-lettres style and the same historical period—the Romantic. Poems by Byron (*The Corsair, The Prisoner of Chillon*) and by Moore (poems from *Lalla Rookh: An Oriental Romance*) have been compared for three cases of a phoneme's position in a word: (1) when it is an unidentified (UP); (2) when it is at the beginning (BW); (3) when it is at the end (EW).

The position of a phoneme in a word must be taken into account because it makes it possible to fully analyse specific phoneme functioning. It is not a surprise that for different cases of phoneme position in a word we can obtain different results. Thus relative similarity has been established comparing the poetry by Byron with poetry by Moore. The samples differ considerably only for two groups of consonants (backlingual and occlusive for BW; labial and constrictive for EW). For the case of UP there is less closeness as the samples differ considerably in three groups of consonants (constrictive, occlusive, sonorant). Here the factor of an author's individual manner of

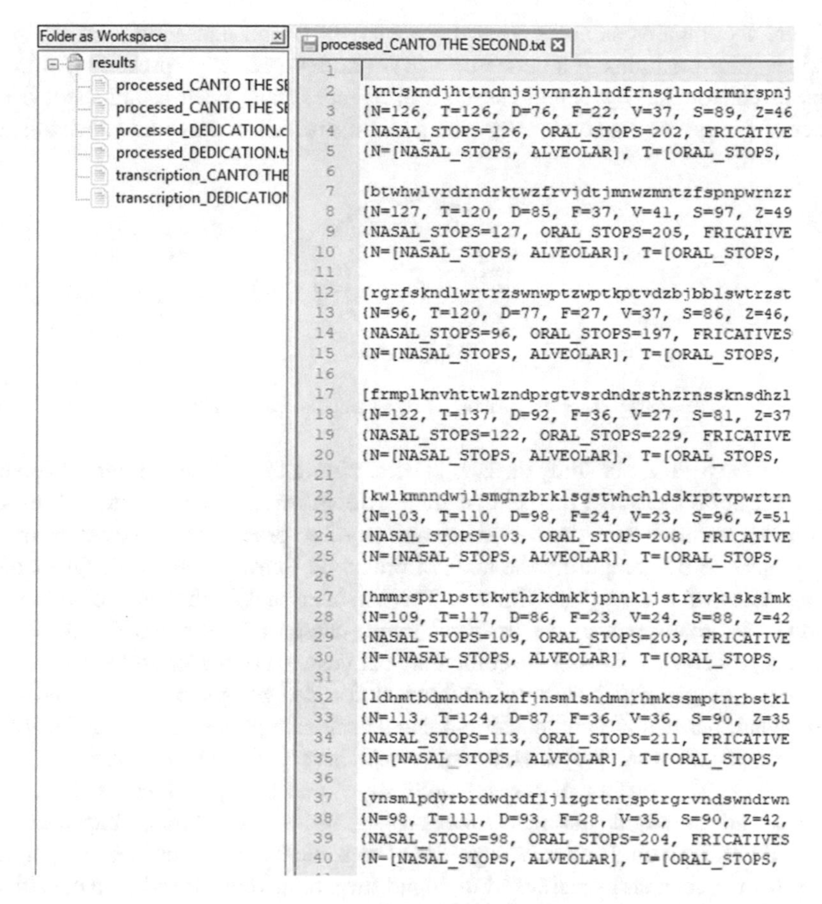

Fig. 3. Phoneme distribution in groups

writing is determined by these three groups of consonants. The effect of the style factor is significant. For the cases of UP, BW and EW groups of constrictive and occlusive phonemes have high style-differentiating capability. Similarity has been revealed for six (BW, EW) and five (UP) groups of consonants. Stylistic peculiarities determined by the specific features of Romantic poetry prevail over features of the author's individual style. Figure 1 represents interrelation of the style and the author's individual manner of writing factors as established by the hypothesis method.

The effect of an author's individual style has been established by the MR. Ranking indices differ in three units for groups of backlingual and occlusive phonemes. For groups of labial, forelingual and nasal phonemes the effect is slight—a difference in one unit.

According to the first type of ranking, the highest ranking index has the group of forelingual phonemes. Quantitative analysis of the texts from the poems The Corsair and The Prisoner of Chillon by Byron gives the same result—the group of forelingual phonemes has the highest frequency of occurrence.

Figure 4 compares samples from the poetry by Byron and Moore on the phonological level for positions of a phoneme at BW and EW. Here 2 represents a number of groups of consonants by which essential differences have been established; 6 is the number of groups of consonants by which slight differences have been established.

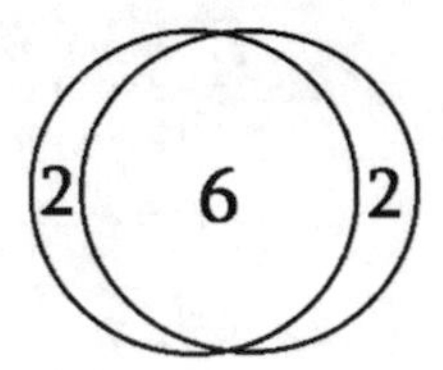

Fig. 4. Comparison of two samples from poetry.

The next aspect of our study deals with determining the degree of interrelation of the effect of the style factors of the belles-lettres style. From the comparison of three styles of the belles-lettres style—drama, poetry and emotive prose—we have established the style dominance by comparing the texts of drama by Shaw (John Bull's Other Island) with the texts of poetry (The Corsair, The Prisoner of Chillon) and emotive prose (introductory prose texts to the first and fourth cantos of the poem Childe Harold's Pilgrimage) by Byron. Thus in the comparison of the texts of drama and emotive prose a strong style dominance has been established for six groups of consonants (labial, mediolingual, nasal, sonorant, constrictive and occlusive phonemes in the case of BW; forelingual, backlingual, nasal, sonorant, constrictive and occlusive phonemes in the case of EW). The results of the research show that groups of constrictive, nasal, occlusive and sonorant phonemes have high style-differentiating capability. The essential differences obtained for the groups of consonants have been caused by a lexical layer of Irish speech and specifics of dialogue speech (as represented by a considerable amount of auxiliary verbs in interrogative and negative sentences).

The following data have been obtained by the ranking method in the comparison above: strong style dominance in the case of UP—difference in six units for groups of labial and sonorant phonemes. A strong style dominance has been established for the case of BW—difference in six units for group of mediolingual phonemes. For the case of EW, strong style dominance has been revealed—difference in six units for group of sonorant phonemes. Summing up the results obtained by the hypothesis and ranking methods, we can maintain that the group of sonorant phonemes has high style-differentiating capability for the cases of UP and EW.

In a comparison of the texts of drama by Shaw with texts of poetry by Byron for the case EW a style dominance has been established for five of eight groups of consonants (labial, forelingual, backlingual, sonorant and constrictive phonemes). The style dominance is represented by three groups of consonants (mediolingual, backlingual and occlusive).

The results obtained by the ranking method in this comparison for the case of UP have shown medium style dominance—difference in four units for groups of labial, mediolingual and sonorant phoneme. The following results have been obtained for the case BW: medium style dominance—difference in four units for group of forelingual phonemes. For the case EW, medium style dominance—difference in five units has been established for groups of nasal and constrictive phonemes. Summing up the results, we can draw the conclusion that there is an equal effect of style and style factors —difference in four of eight units has been revealed by the ranking method. Hypothesis and ranking methods have shown that groups of labial, backlingual, nasal and constrictive phonemes reflect the effect of the style factor. Phonological features of the compared samples have been caused by both style (penetration of colloquial elements into the texts of the researched drama by Shaw and poetry by Byron) and style (phonological characteristics of the samples from the belles-lettres style) factors.

The researched styles of the belles-lettres style have been compared with the colloquial style represented in the sample from a phrase-book (texts on everyday topics). This is a literary variant of the colloquial style which is to the least extent marked by stylistic expressive means and can be considered to some extent neutral.

Having compared the results obtained by the hypothesis method for the texts of the colloquial style with the texts of Moore's poetry for the case of UP, we have established a strong style dominance—an essential difference for six groups of consonants (forelingual, mediolingual, backlingual, nasal, constrictive, occlusive). For the case of BW, medium style dominance has been revealed—difference for five groups of consonants (forelingual, mediolingual, nasal, constrictive, occlusive). For the case of EW, medium style dominance has been established—difference for four groups of consonants (forelingual, backlingual, constrictive, occlusive). Analysis of the results obtained has shown that for all three cases—UP, BW and EW—groups of forelingual and occlusive phonemes have high style-differentiating capability. Results of the comparison for the case of UP are represented in Table 1.

The results obtained by the ranking method in the comparison of the texts of the colloquial style and the texts of Moore's poetry have shown for the case of UP strong style dominance: difference in six units for groups of mediolingual, constrictive and occlusive phonemes. For the case of BW, there is strong style dominance: difference in six units for group of constrictive phonemes. For the case of EW: medium style dominance—difference in five and four units respectively for groups of constrictive and forelingual, sonorant phonemes. The results obtained show that for the cases of UP and BW the group of constrictive phonemes has high style-differentiating capability. Table 2 presents the results of the comparison for the case of UP. The upper left index represents the first type of ranking, lower right represents the second one.

Table 2. Two types of ranking for poetry by Moore (PM)

Groups of phonemes	PM
Labial	$^{5}\,138_{\,2}$
Forelingual	$^{1}\,425_{\,2}$
Mediolingual	$^{8}\,5.9_{\,7}$
Backlingual	$^{6}\,59.3_{\,7}$
Nasal	$^{6}\,82.9_{\,4}$
Sonorant	$^{2}\,234_{\,2}$
Constrictive	$^{3}\,210_{\,1}$
Occlusive	$^{4}\,183_{\,7}$

Table 3. Distances between the colloquial style and poetry by Moore

Phoneme groups	Position of a phoneme in a word is not taken into account	A phoneme at the BW	A phoneme at the EW
Labial	–	–	–
Forelingual	0.79	0.55	0.81
Mediolingual	0.62	0.58	–
Backlingual	0.55	–	0.20
Nasal	0.16	0.14	–
Sonorant	–	–	–
Constrictive	0.84	0.80	0.75
Occlusive	0.81	0.73	0.29

Having applied the method of the determination of style distance to the compared texts of the colloquial style and Moore's poetry, we have established for the case of UP strong style dominance (great distance) 0.84, 0.81 and 0.79 respectively for groups of constrictive, occlusive and forelingual phonemes. For the case of BW, there is strong style dominance (great distance) and 0.80 and 0.73 have been revealed respectively for groups of constrictive and occlusive phonemes. For the case of EW, there is strong style dominance (great distance) and 0.81 and 0.75 have been established respectively for groups of forelingual and constrictive phonemes. There is a weak style dominance (little distance), 0.20, for group of backlingual phonemes. Analysis of the results is presented in Table 3 which shows that for all three cases—UP, BW and EW—the group of constrictive phonemes has high style-differentiating capability.

Before any linguistic interpretation of the numerical data represented above, it should be noted that the strong style dominance prevails in the results of the comparison of the colloquial style and Moore's poetry as obtained by the three methods of research (the hypothesis, ranking and style distance determination methods). Strong style dominance has been caused by the features of Romantic literary trends on the one hand (archaic, folklore and colloquial elements, expressive means of poetry, peculiar features of verse arrangement) and, on the other hand, the relative neutrality of the literary variant of the colloquial style under study. The strong style dominance has been established for groups of forelingual, constrictive and occlusive phonemes for the cases of UP, BW and EW in comparison of the texts of the colloquial style and poetry by Byron and Moore.

For the case of BW, the distance to poetry by Byron, 0.79, and by Moore, 0.80, is great with strong style dominance for group of constrictive phonemes. For the case EW, the distance to poetry by Byron and Moore is great—0.81—strong style dominance for group of forelingual phonemes. For two cases of the three (UP, EW) great distance—a strong style dominance—has been established between the compared samples for group of forelingual phonemes. Having analysed the results obtained by three methods (hypothesis, ranking and style distance determination methods), we can draw the conclusion that the group of forelingual phonemes has the highest style-differentiating capability of the styles under study.

We have established by the ranking method a strong style dominance in the comparison of the samples from the colloquial and scientific styles for the case of UP for group of forelingual phonemes. For the case of BW, we have revealed strong style dominance, difference in six units, for group of occlusive phonemes. For the case of EW, we have obtained medium style dominance: difference in five units for group of backlingual phonemes. The results of the comparison have shown that the group of forelingual phonemes in the case of UP, and the group of occlusive phonemes in the case of BW, have high style-differentiating capability.

Analysis of the results obtained with the style distance determination method in comparing the scientific and colloquial styles has shown for the case of UP strong style dominance (great distance): 0.78 for groups of forelingual and constrictive phonemes; medium style dominance (medium distance): 0.57, 0.52, 0.52 and 0.49 respectively for groups of occlusive, mediolingual, backlingual and nasal phonemes; weak style dominance (little distance): 0.11 for group of labial phonemes. For the case of BW, strong style dominance has been revealed (great distance): 0.87, 0.74 and 0.71 respectively for groups of sonorant, occlusive and labial phonemes; and a medium style dominance (medium distance): 0.47 for group of nasal phonemes. For the case of EW, strong style dominance has been established (great distance): 0.84 and 0.70 respectively for groups of occlusive and backlingual phonemes. Analysis of the style-differentiating capability of the given groups of consonants has shown that it is high for groups of forelingual and constrictive phonemes in the case of UP, and for groups of labial, occlusive and sonorant phonemes in the case of BW.

The results above of the comparison of the texts representing the colloquial style and drama by Shaw have shown an equal effect of style and language factors. Therefore, we can suppose that the texts of the scientific style differ from the texts of the colloquial style and Shaw's drama in a similar way. The supposition has proved to be true. Thus a strong style dominance has been established in the comparison of the samples of the scientific style and Shaw's drama: for the case of UP, the texts differ in six groups of consonants (forelingual, mediolingual, backlingual, constrictive, occlusive, sonorant); for the case of BW, in seven groups (labial, forelingual, mediolingual, backlingual, nasal, occlusive, sonorant); for the case of EW, in six groups of consonants (labial, forelingual, backlingual, constrictive, occlusive, sonorant). In the given comparison we should note that groups of forelingual, backlingual, occlusive and sonorant phonemes have high style-differentiating capability.

Having compared the samples from the scientific style and drama by Shaw, we have obtained the following results: for the case of UP, medium style dominance, difference in five units for groups of forelingual and sonorant phonemes; and weak style dominance, difference in one unit for groups of labial and nasal phonemes. For the case of BW: strong style dominance, difference in six units for groups of nasal and sonorant phonemes; medium style dominance, difference in five and four units respectively for groups of mediolingual and backlingual phonemes and group of occlusive phonemes; weak style dominance, difference in one unit, for groups of labial and constrictive phonemes. For the case of EW: medium style dominance, difference in four units for groups of labial, forelingual, backlingual, occlusive and sonorant phonemes; weak style dominance, difference in one unit for group of nasal phonemes. The above results show that in this comparison groups of nasal and sonorant phonemes have high style-differentiating capability. For group of nasal phonemes the same results have been obtained by the hypothesis method.

We have established a strong style dominance in the comparison of the texts representing emotive prose by Byron and the scientific style. The samples compared differ in an essential way for six of eight groups of consonants. These are: for the case of UP, groups of labial, forelingual, constrictive, nasal, occlusive and sonorant phonemes; for the case of BW, groups of labial, forelingual, backlingual, constrictive, nasal and sonorant phonemes; for the case of EW, groups of labial, forelingual, constrictive, nasal, occlusive and sonorant phonemes. Here we should maintain that strong capability of style differentiation has been revealed for three cases of the position of a phoneme in a word (UP, BW, EW) for five of eight groups of consonants—labial, forelingual, constrictive, nasal and sonorant phonemes. Figure 5 represents interrelation of style and language factors established by the hypothesis method.

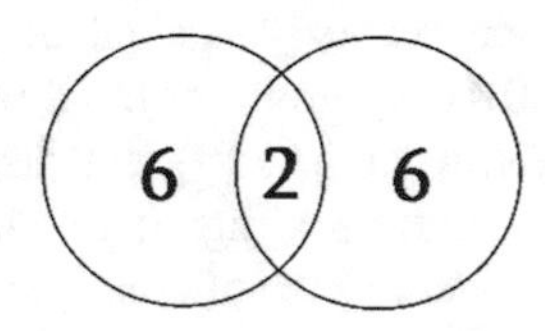

Fig. 5. Comparison of samples from the scientific style and emotive prose.

Figure 5 shows the comparison of samples from the scientific style and emotive prose by Byron for all three positions of a phoneme in a word (UP, BW, EW). Pr—emotive prose by Byron; SS—scientific style; 2—the number of groups of consonants by which slight differences have been established; 6—the number of groups of consonants by which essential differences have been established.

Having analysed the data obtained by the ranking method for the compared samples from the texts of the scientific style and emotive prose by Byron for the case of UP, we have established medium style dominance—difference in five units for group of labial phonemes, in four units for groups of nasal and occlusive phonemes; weak style dominance—difference in one unit for group of sonorant phonemes; zero style dominance—no difference of rank indices—they are equal for group of backlingual phonemes. For the case of BW—weak style dominance—difference in one unit for groups of mediolingual, occlusive and sonorant phonemes. For the case of EW—strong style dominance—difference in six units for groups of labial and nasal phonemes; weak style dominance—difference in one unit for group of backlingual phonemes. Consequently, high style-differentiating capability has been established for groups of labial and nasal phonemes in the case of EW.

From a lexical point of view, the difference obtained between the scientific style on one hand, and the colloquial style, emotive prose by Byron, and drama by Shaw on the other hand, has been found by the use of specific terms from technical physics. For group of occlusive phonemes in the case of BW high style-differentiating capability has been obtained by the hypothesis, ranking and style distance determination methods.

4 Conclusions

A statistical analogue (statistical system) of a phonological subsystem of a style system has been suggested to differentiate styles on the level of phonology for authorship and style attribution. The elements of the statistical system are consonant phoneme group mean frequencies, which is a style differentiation criterion. The relations between the elements have been established by the three methods of mathematical statistics referred to above. The suggested combination of three methods made it possible to reveal similarities and differences between the texts representing the belles-lettres (poetry, emotive prose and drama), colloquial and scientific styles. On the basis of the obtained results three models proposed. The first model represents the effect of author's manner of writing factor in comparison of poetry by Byron and Moore (PB–PM). It shows high style-differentiating capability of backlingual and occlusive phonemes. The second model represents an interrelation of style factors in comparisons of poetry by Byron and drama by Shaw (PB–Dr); drama by Shaw and emotive prose by Byron (Dr–Pr); and poetry and emotive prose by Byron (PB–Pr). Style-differentiating capability is high for groups of labial, forelingual, constrictive and sonorant phonemes in comparison PB–Dr; for group of sonorant phonemes in comparison Dr–Pr; for group of labial phonemes in comparison PB–Pr. The third model represents an interrelation of style and language factors in the comparison of the scientific style and emotive prose by Byron (SS–Pr). The model shows high style-differentiating capability of groups of labial and nasal phonemes. Therefore, the results of the investigation have shown regularities and specifics of functioning of consonant groups in the belles-lettres, colloquial and scientific styles. The results obtained allow to improve efficiency of author and style attribution and can be used in forensic stylometry and plagiarism testing. Future research will deal with authorship-attributing capability of each phoneme group.

References

1. Alamothoda, S.M.: Phonostatistics and phonotactics of the syllable in modern Persian. Series: Studia Orientalia. Finnish Oriental Society, Helsinki (2000)
2. Altmann, G., Levickij, V., Perebyinis, V.: Problems of Quantitative Linguistics. Ruta, Chernivtsy (2005)
3. Argamon, S., Koppel, M., Pennebaker, J., Schler, J.: Automatically profiling the author of an anonymous text. Commun. ACM **52**(2), 119–123 (2009)
4. Bisikalo, O.V., Vysotska, V.A.: Sentence syntactic analysis application to keywords identification Ukrainian texts. Radio Electron. Comput. Sci. Control **3**(38), 54–65 (2016)
5. Everitt, B.S.: A Handbook of Statistical Analyses Using R. Chapman and Hall/CRC, London/Boca Raton (2009)
6. Gomez, P.C.: Statistical methods in language and linguistic research. University of Murcia, Spain (2013)
7. Gries, Th.S.: Statistics for linguistics with R. Mouton Textbook (2009)
8. Juala, P.: Authorship attribution, foundations and trends(R) in information retrieval. **1**(3), 233–334 (2008)

9. Kapociute-Dzikiene, J., Utka, F., Sarkute, L.: Authorship attribution and author profiling of Lithuanian literary texts. In: Proceedings of the 5th Workshop on Balto-Slavic Natural Language Processing, Hissac, Bulgaria, pp. 96–105 (2015)
10. Khomytska, I., Teslyuk, V.: The method of statistical analysis of the scientific, colloquial, belles-lettres and newspaper styles on the phonological level. In: Advances in Intelligent Systems and Computing, vol. 512, pp. 149–163 (2016)
11. Khomytska, I., Teslyuk, V.: Modelling of phonostatistical structures of the colloquial and newspaper styles in English sonorant phoneme group. In: Proceedings of the XIIth Scientific and Technical Conference, CSIT, Lviv, pp. 67–70 (2017)
12. Kingston, J., Baayen, H., Clopper, C.G.: Statistical analyses: statistics in laboratory phonology. Mixed-effects models clustering and classification methods. Oxford (2011)
13. Koppel, M., Schler, J., Argamon, S.: Computational methods in authorship attribution. J. Assoc. Inf. Sci. Technol. **60**(1), 9–26 (2009)
14. Kornai, A.: A Mathematical Linguistics. Springer, Heidelberg (2008)
15. Lytvyn, V.: Development of a method for the recognition of author's style in the Ukrainian language texts based on linguometry, stylemetry and glottochronology. East. Eur. J. Enterp. Technol. **4/2**(88), 10–18 (2017)
16. Mines, M.A., Hanson, B.F., Shoup, J.E.: Frequency of occurrence of phonemes in conversational English. Lang. Speech **21**(3), 221–241 (1978)
17. Roberts, A.H.: A Statistical Linguistic Analysis of American English. Mouton, The Hague (1965)
18. Shakhovska, N.B., Noha, R.Y.: Methods and tools for text analysis of publications to study the functioning of scientific schools. J. Autom. Inf. Sci. **47**(12)
19. Sovkowiak, W.: On the phonostatistics of English onomatopoeia. Adam Mickiewicz University, Poznan (1990)
20. Stamatatos, E.: A survey of modern attribution methods. J. Assoc. Inf. Sci. Technol. **60**(3), 538–556 (2009)
21. Zhezhnych, P., Markiv, O.: A linguistic method of web-site content comparison with tourism documentation objects. In: Proceedings of the XIIth Scientific and Technical Conference, CSIT, Lviv, pp. 340–343 (2017)

Attitudes Toward Feminism in Ukraine: A Sentiment Analysis of Tweets

Olena Levchenko(✉) and Marianna Dilai

Applied Linguistics Department, Lviv Polytechnic National University, Lviv, Ukraine
levchenko.olena@gmail.com, mariannadilai@gmail.com

Abstract. This paper presents a sentiment analysis of Ukrainian tweets on feminism. In order to carry out a computational study of opinions, we have adjusted the SentiStrength algorithm to the Ukrainian language by replacing the English term lists in the program files with the Ukrainian ones. The main contribution is an attempt to compile a social domain sentiment lexicon for Ukrainian (3,736 words). The SentiStength output has shown a prevailing negative sentiment of the analyzed tweets. The program performance was evaluated in terms of accuracy, precision, recall, error, fallout and F1 Score. In addition, we found a number of common attributes of a feminist, which also predominantly express negative attitude. Overall, the findings show that a direct support of a key feminist goal, i.e. equality of women and men in society, by the Ukrainian Tweeter users couples with misconception about the concept of feminism and unwillingness to be called a feminist.

Keywords: Sentiment analysis · SentiStrength · Ukrainian sentiment lexicon
Feminism

1 Introduction

We can speak about the perception of feminism by various social strata of the population, different political groups and public institutions, but this paper addresses the "naive" perception of feminism, i.e. without analyzing and penetrating into the whole spectrum of ideologies, political and social movements. Hence, the aim of this research is to carry out sentiment analysis of Ukrainian tweets on feminism in order to reveal the affective attitude of Ukrainians to the feminist ideas.

In recent decades, social media platforms have become a valuable resource of ample research data for the study of public opinions on different controversial topics from various domains. The development of the algorithms and tools that can automatically extract opinions over the Web in real-time and determine the sentiments they convey has become crucial not only for businesses, which want to know customers' feedbacks and preferences, but also for social science as well as government to collect data on public attitude toward different notions and events, and to monitor and analyze social behavior.

Since early 2000 multiple methods for measuring sentiments have been designed, including lexicon-based approaches and supervised machine learning methods.

N. Shakhovska and M. O. Medykovskyy (Eds.): CSIT 2018, AISC 871, pp. 119–131, 2019.
https://doi.org/10.1007/978-3-030-01069-0_9

However, most of them are limited to the analysis of English texts only. Some contributions to the development of sentiment analysis of Ukrainian texts were made by the works of Romanyshyn [1, 2].

This paper attempts at the Ukrainian localization of a SentiStrength program [3]. Unlike most opinion mining algorithms (General Inquirer, WordNet Affect, SentiWordNet, QWordNet etc.), the SentiStrength program attempts to detect both the polarity (positive, negative and neutral) and the strength of sentiment in text. First and foremost, SentiStrength is adjusted to the Ukrainian language by compiling a term list EmotionLookupTable.txt consisting of sentiment-bearing words.

Although sentiment-mining offers numerous research challenges (e.g. figurative expressions, irony and lie detection, etc.), it promises insight into the study of opinions and attitudes expressed in discussions surrounding feminist ideas in social media.

A corpus-driven approach to sentiment analysis allows processing a large number of texts and automatically retrieving information on their positive/negative polarity, emotions, and evaluations. The analyzed corpus is made up of 923 tweets.

2 Previous Research on Sentiment Analysis

Sentiment analysis is the study of subjectivity (neutral vs. emotionally loaded) and polarity (positive vs. negative) of a text [4]. It deals with the computational study of opinions, sentiments and emotions expressed in text [5]. It is the process of algorithmically identifying and categorizing opinions expressed in text to determine the user's attitude toward the subject of the document (or post). This process relies on sentiment vocabulary, i.e. large collections of words, each marked with a positive or negative orientation. The overall sentiment of a text is identified based on the polarity of its words.

Sentiment analysis has become one of the most dynamic research areas in natural language processing, data mining, Web mining, and text mining. Due to a wide set of applications, particularly in business and social sphere, a number of sentiment analysis tools are available. Some programs based on a lexicon of positive and negative words, such as General Inquirer [6], WordNet Affect [7], QWordNet [8] or SentiWordNet [9], mainly count their frequency. More advanced modifications presuppose the detection of words that intensify sentiment in other words and overall sentence structures [10]. Another approach is based on identifying text features that could potentially be subjective in certain contexts and then use contextual information to determine whether they are subjective in every new context [11]. A primarily linguistic approach to opinion mining makes use of simple rules based upon compositional semantics (information about probable meanings of a word based upon the surrounding text) to detect the polarity of an expression [12].

3 Adjusting SentiStrength to Ukrainian

SentiStrength (Thelwall et al. 2012) is a state-of-the-art, lexicon-based classifier that exploits a sentiment lexicon built by combining entries from different linguistic resources. The SentiStrength algorithm is used to identify the polarity of sentiment in text and detect the strength of the sentiment expressed [13]. SentiStrength outputs both positive and negative sentiment scores, namely −1 (not negative) to −5 (extremely negative), 1 (not positive) to 5 (extremely positive) for any input text written in English, based on the research from psychology, which has revealed that we process positive and negative sentiment in parallel (mixed emotions). Based on their algebraic sum, SentiStrength can also report the overall trinary score, i.e. the overall positive (score = 1), negative (score = −1) and neutral (score = 0).

The SentiStrength algorithm relies on the information given in various files, including:

- EmotionLookUpTable which is a list of emotion-bearing words, each one with the word, then a tab, then an integer 1 to 5 or −1 to −5. Strengths of +1 and −1 have no effect on the program;
- NegatingWordList.txt which reverses the polarity of subsequent words, e.g. not happy is negative;
- BoosterWordList.txt which increases sentiment intensity, e.g. very happy is more positive than happy;
- IdiomLookupTable.txt which overrides the sentiment strength of the individual words in the phrase

In order to adjust SentiStrength for the Ukrainian language, first of all, we should replace the English term list EmotionLookupTable.txt with the Ukrainian one. The word list is compiled by translating the English word list and adding the most common sentiment-bearing words from the analyzed tweets. Each word in EmotionLookupTable.txt should be marked with a sentiment score that indicates the typical polarity and the strength of the sentiment expressed using the following scheme:

[−5] Very strong negative sentiment (e.g., *жахливий*)
[−4] Strong negative sentiment (e.g., *страшний*)
[−3] Moderate negative sentiment (e.g., *прикрий*)
[−2] Mild negative sentiment (e.g., *незручний*)
[2] Mild positive sentiment (e.g., *задоволений*)
[3] Moderate positive sentiment (e.g., *добрий*)
[4] Strong positive sentiment (e.g., *чудовий*)
[5] Very strong positive sentiment (e.g., *фантастичний*)

In order to get unbiased results, 50 labelers of different age and gender were asked to rate the words. As sentiment of a word is hugely influenced by context and depends on the domain in which it is used, the rating is performed taking into account the typical use of a word referring to gender issues in informal situations. For instance, the word *домагатися* has a positive sentiment in general context, however, it has opposite orientation when used in reference to a woman (1):

(1) Дилема: якщо вона феміністка - то може образитись. А може й подцмати, що я ***домагаюся*** *її..))*

Creation of domain-specific sentiment lexicons is crucial to computational social science research, which can be misled without domain-specific lexicons by sentiment assignments biased towards domain-general contexts, neglecting factors like genre, community-specific vernacular, or demographic variation [14, 15].

Once the labeling is complete, each word was given the average score of the 50 coders. We enter one word per line, followed by a tab, followed by the sentiment score: −5, −4, −3, −2, 2, 3, 4, 5. (−1, 0, and 1 are not used). The words with different word endings that give it the same sentiment meaning are truncated and the endings are replaced with a star *. For instance, *добр** would match all grammatical forms of the nouns *добро, доброта* as well as the adjective forms of *добрий*, and the adverb *добре*, so that separate entries are not needed for all of these words. After that the word list was incorporated in the program. In total, the list consists of 3,736 words.

The following files are also replaced with a Ukrainian equivalent: NegatingWordList.txt, BoosterWordList.txt, and QuestionWords.txt. BoosterWordList.txt contains a list of words that tend to increase or decrease the sentiment of the word that follows, such as *дуже, аж, суттєво*, etc. Each word has a booster score that indicates typical increase or decrease in sentiment strength:

[−2] Large decrease in sentiment (e.g., *мало*)
[−1] Moderate decrease in sentiment (e.g., *не дуже*)
[1] Moderate increase in sentiment (e.g., *дуже*)
[2] Large increase in sentiment (e.g., *надзвичайно*)

NegatingWordList.txt contains a list of words that almost always indicate that the sense of a sentence, word of phrase is negated, such as *не, ніколи, ні*, etc. QuestionWords.txt contains a list of words that almost always indicate that the sentence is a question, such as *який, чому, коли* etc.

We use the Java version that needs the utf8 option to read the input files.

The SentiStrength analysis output is a copy of the text file with overall comment assigned with the most positive of its sentence emotions and the most negative of its sentence emotions. Table 1 presents the results in Excel.

In order to test the results, the research proceeds with building a manually annotated dataset which also allows taking into account the context determining emotional loading of words and utterances.

4 SentiStrength Results

The SentiStrength findings prove that tweets on feminism are emotionally loaded (40.2% of segments are either positive or negative) (see Fig. 1, Table 2). The obtained corpus strength scale shows that negative tweets (−4 to −1) prevail the positive ones (1 to 4), they make up about 23%, and include very strong negative texts (−4–2.9%).

The average sentiment of the Ukrainian tweets on feminism is −0.2.

Table 1. SentiStrength output

Positive	Negative	Sum	EmotionRationale
3	−5	−2	*висновок[0] [[Sentence = −1, 1 = word max, 1–5]] Я[0] не [0] расистка[−2][NegatedDueToPreviousWord] я[0] просто[0] не[0] люблю[4][NegatedDueToPreviousWord] негрів[0] [[Sentence = −5, 3 = word max, 1–5]] і[0] не[0] феміністка[0] [[Sentence = −1, 1 = word max, 1–5]] просто[0] думаю[0] що[0] жінки[0] рівні[0] у[0] правах [0] з[0] чоловіками[0] [[Sentence = −1, 1 = word max, 1–5]][[[3, −5 max of sentences]]]*
1	−4	−3	*Якась[0] маразматична[−3] феміністка[0] написала[0] не[0] читайте[0] витратите[0] даремно[0] час[0] [[Sentence = 4, 1 = word max, 1–5]][[[1, −4 max of sentences]]]*
4	−2	2	*Джоан[0] слово[0] ШИКАРНА[1] це[0] синонім[0] її[0] імені[0] сильна[3] жінка[2][+1 MultiplePositiveWords] феміністка[0] йде[0] проти[−1] суспільства[0] приклад[0] для[0] мене[0] [[Sentence = −2 ,4 = word max, 1–5]][[[4, −2 max of sentences]]]*
4	−1	3	*Бейонсе[0] Я[0] –[0] сучасна[3] феміністка[0] Бейонсе [0] запевнила[0] що[0] їй[0] подобається[1] бути[0] жінкою[0] й[0] пояснила[0] чому[0] назвала[0] га[0] [[Sentence = −1, 4 = word max, 1–5]][[[4, −1 max of sentences]]]*
1	*−4*	*−3*	*Відкрив[0] двері[0] та[0] пропустив[0] дівчину[0] вперед[0] а[0] вона[0] замість[0] дякую[0] подивилася[0] як[0] на[0] ідіота[−3] [[Sentence = − 4,1 = word max, 1–5]] Феміністка[0] мабуть[0] [[Sentence = −1, 1 = word max, 1–5]][[[1,−4 max of sentences]]]*
3	*−5*	*−2*	*навіть[0] при[0] тому[0] що[0] можу[0] назвати[0] себе[0] феміністкою[0] я[0] іноді[0] вживаю[0] в[0] розмові[0] щодо[0] жінок[0] те[0] чого[0] сама[0] не [0] люблю[4][NegatedDueToPreviousWord] [[Sentence = −5, 1 = word max, 1–5]] Але[0] все[0] з[0] часом[0] я[0] себе[0] пробачаю[2] [[Sentence = −1, 3 = word max, 1–5]][[[3, −5 max of sentences]]]*
2	*−3*	*−1*	*То[0] феміністка[0][+0.6 EmphasisInPunctuation] [[Sentence = −1, 2 = word max, 1–5]] Вона[0] образиться[−2] [[Sentence = −3, 1 = word max, 1–5]] [[[2, −3 max of sentences]]]*

Most words in the analyzed texts are neutral (marked [0]), they make up 66%. Negatively marked words prevail over the positive ones, 20% and 14% correspondingly (see Fig. 2). Words marked [−4] mostly include swear words.

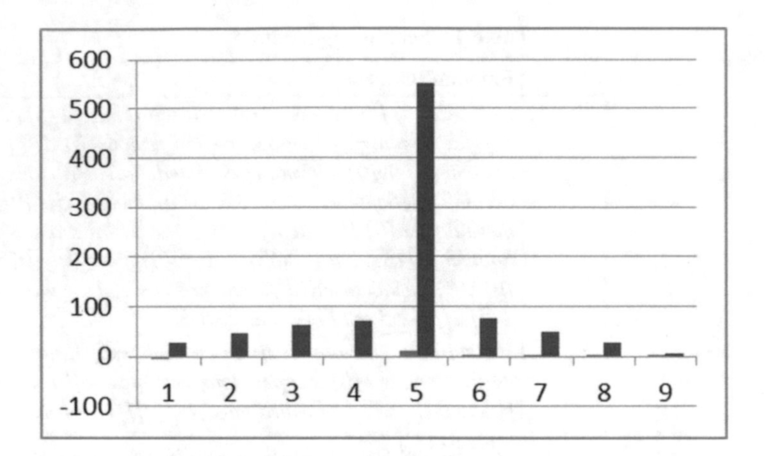

Fig. 1. SentiStrength results

Table 2. SentiStrength results

Sentiment strength scale	Frequency	Percentage, %
−4	27	2.9
−3	48	5.2
−2	63	6.8
−1	71	7.7
0	552	59.8
1	76	8.2
2	50	5.4
3	27	2.9
4	7	0.8

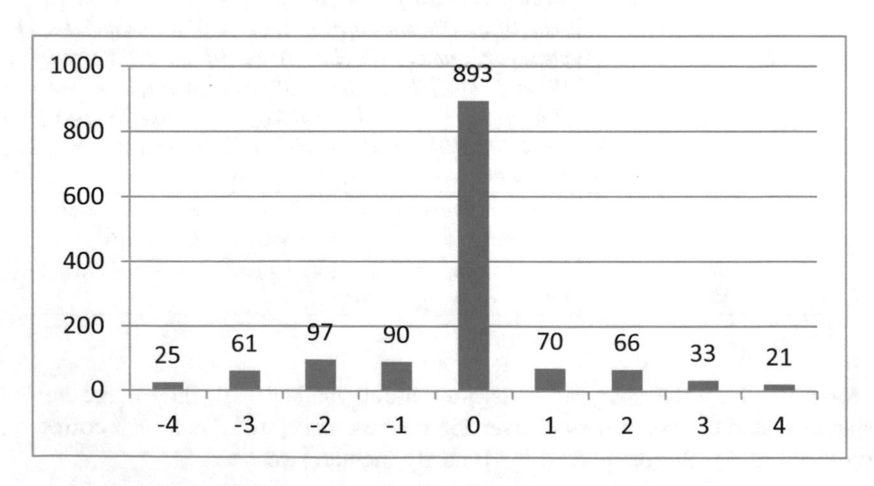

Fig. 2. Polarity of words in Ukrainian tweets on feminism

5 Evaluation of Program Performance

When validating a sentiment analysis program, the testing methodology is crucial. The data source and how it is scored, subject matter and volume of data tested are all significant variables that can dramatically affect results. The evaluation of sentiment analysis system performance shows how well it agrees with human judgments. Table 3 shows agreement between manual annotation (how a human performs, which is seen as a gold standard) and SentiStrength results in terms of true positive (57), false positive (34), true negative (72) and false negative (21) tweets. The volume of data tested is 184 tweets.

Table 3. Agreement between manual labeling and SentiStrength

		Manual labeling	
		Positive	Negative
SentiStrength	Positive	57	34
	Negative	21	72

It is common to analyze the following measures that go into determining how well a sentiment detection program works: accuracy, precision, error, fallout and recall [16]. Accuracy (1) tracks how many of the tweets that were rated to have negative and positive sentiment were rated correctly:

$$Accuracy = \frac{tp + tn}{tp + tn + fn + fn} \tag{1}$$

Precision (2) is also referred to as positive predictive value; it is a number of tweets correctly identified as belonging to certain (positive or negative) category over the total number of tweets belonging to that category:

$$Precision = \frac{tp}{tp + fp} \tag{2}$$

Recall (3) is a measure of how many tweets with sentiment were rated as sentimental. This could be seen as how accurately the system determines neutrality. Recall is also referred to as the true positive rate or sensitivity.

$$Recall = \frac{tp}{tp + fn} \tag{3}$$

Error is calculated using expression (4):

$$Error = \frac{fp + fn}{tp + tn + fp + fn} \tag{4}$$

Fallout is a harmonic mean of precision and recall (5):

$$Fallout = \frac{fp}{fp + tn} \tag{5}$$

F1 Score, also called F-Score or F-Measure, this is a combination of precision and recall. It is commonly used amongst experts and researchers in the linguistics and natural language processing fields and is considered one of the most important measures describing how the system is performing. F1 score reaches its best value at 1 (perfect precision and recall) and worst at 0. The formula for calculating F1 Score is (6):

$$F1 = \frac{2}{\frac{1}{recall} + \frac{1}{precision}} \tag{6}$$

The outcomes of the calculations shown in Table 4 prove that SentiStrength for Ukrainian texts provide quite valuable results or results that we can trust (accuracy 0.70, recall 0.73).

Table 4. Evaluation of the program performance

	Accuracy	Precision	Recall	Error	Fallout	F1
SentiStrength	0.70	0.63	0.73	0.3	0.32	0.68

The analysis of false positive and false negative tweets shows that the sentiment detection system fails to correctly interpret complicated language structures. Metaphorical expressions, irony and sarcasm are other challenges that make accurate automated analysis of natural language quite difficult.

Most of the incorrectly interpreted tweets are sarcastic or ironical. Although tweets (2–5) are obviously negative mockery, they were marked as positive.

(2) Напівгола[0] феміністка[0] зірвала[0] святкову [2] месу[0] на[0] честь [3] Різдва[0] Христового[0] у[0] Кельні[0] [[Sentence = −1, 4 = word max, 1–5]] [[[4, −1 max of sentences]]]

(3) ok[0] [[Sentence = −1, 1 = word max, 1–5]] тепер[0] я[0] як[0] справжня [3] феміністка[0] ушло[0] читати[0] теоретичні[0] тексти[0] [[Sentence = −1, 4 = word max, 1–5]][[[4, −1 max of sentences]]]

(4) Отак[0] спілкуєшся[0] з[0] здавалося[0] б[0] розумною [2] людиною[0] а[0] потім[0] дізнаєшся[0] що[0] вона[0] феміністка[0] [[Sentence = −1, 3 = word max, 1–5]][[[3, −1 max of sentences]]]

(5) @alinakarban[0] Щось[0] тема[0] емансипації[0] і[0] т[0] [[Sentence = −1, 1 = word max, 1–5]] д[0] часто[0] проскакує [1] в[0] тебе[0] [[Sentence = −1, 2 = word max, 1–5]] ще[0] не[0] дай[0] боже[0] феміністкою[0] виявишся[0] [[Sentence = −1, 1 = word max, 1–5]][[[2, −1 max of sentences]]]

Example (6) is falsely labelled negative:

(6) якщо[0] баба[−1] жирна[−3] то[0] вона[0] феміністка[0] с[0] цитати[0] убогих[0] людей[0] [[Sentence = −4,1 = word max, 1–5]][[[1, −4 max of sentences]]]

6 Attributes of a Feminist in Ukrainian Tweets

The content analysis of the Internet posts shows that Ukrainian women do not often admit belonging to feminism: I'm a feminist – 172, I'm not a feminist – 80. Twitter posts imply that Ukrainian women in this social network most often deny their belonging to feminists: not a feminist – 68, I'm a feminist – 59 (8), (9).

(8) Я не феміністка, ні дай Бог.

"I'm not a feminist, God forbid"

(9) …ну чоловіки та жінки рівні я не якась там феміністка але це справді так.

"Well, men and women are equal. I'm not any kind of feminist, but it's really so."

The analysis of tweets has revealed a number of attributes inherent in the linguistic image of a feminist in the minds of Ukrainian speakers (see Table 5). It has been found out that 53.4% of the attributes have a negative connotation, 25% are neutral and 21.6% are positive.

In 2006, Oksana Kis claimed that a stereotyped view on a feminist as a manlike, rough, sexually disgruntled, angry and aggressive, predisposed to sexual deviation man-hater still characterizes the public and women's discourse of modern Ukraine [17]. Remarkably, the scholar singles out, in fact, similar attributes referring to the stereotyped idea of a feminist.

A frequent use of the attribute *внутрішня (inner)* (10–14) as well as the construction *Я не феміністка, але … (I'm not a feminist, but …)* (19–21) implies that Ukrainian women are reluctant to state directly that they are feminists, however, they do support a key feminist idea that is advocacy of women's rights on the grounds of political, social and economic equality of the sexes.

(10) Чомусь і 8 березня вже не державний вихідний, моя внутрішня феміністка невдоволена.

"The 8th of March is not a public holiday anymore; still my inner feminist is not satisfied."

(11) Не скажу, що моя внутрішня феміністка повністю задоволена після перегляду диво-жінки, але це вже хоч щось.

"I can't say that my inner feminist is completely satisfied after watching a wonder-woman, but still it is something."

(12) Бо баби всі роззяви, вони всі однакові!! Окей, моя внутрішня феміністка (вона в мене, як виявилося, є) страшенно обурена.

"It is because all women are absent-minded, they are all the same! Ok, my inner feminist (it appears I have it) is furious."

(13) Після перегляду фільму про викрадення, маньяка і т.д моя внутрішня феміністка зненавиділа мужиків.

"After watching the film about kidnapping, maniac and etc. my inner feminist started hating men."

Table 5. Attributes of a feminist

Attribute	Frequency
Внутрішня "inner"	*7*
Радикальна "radical"	*6*
Головна "main", затята "busy", справжня "real", страшна "ugly", жирна "fat"	*4*
Горда "proud", кончена "fucked", сучасна "modern"	*3*
Гетеросексуальна "heterosexual", маленька "small", оголена "naked", чергова "alternate", заядла "confirmed", махрова "entrenched", п'яна "drunk", побутова "domestic", самодостатня "self-sufficient", чокнута "crackpot", шалена "crazy", якась там "a kind of", яра "rampant"	*2*
*Академічна "academic", гола [nude], гомосексуальна [homosexual], дорога [dear], лгбт [LGBT], лезбі [lesbi], лезбійська [lesbian], майбутня [future], мала [small], напівгола [half-naked], незворушна [insensitive], перша [first], православна [orthodox], рафінована [refined], рижа [ginger], ситуативна [situational], типова [typical], атгресивна [aggressive], безплідна [acyetic], бидлувата [swinish], бідна [poor], бісова [devilish], відома [famous], войовнича [militant], жахлива [terrible], жостка [rough], заклята [sworn], залізна [iron], заслужена [well-deserved], збс [f***ing], зла [evil], змучена [tired], клята [swearing], крута [cool], лютіша [furious], маразматична [mad], найбожевільніша [the most crazy], найвидатніша [the most prominent], найзнаменитіша [the most famous], найперша [first], настояща [true], неголена [unshaven], недотр***[underf***], незавісіма [independent], нереалізована [unrealized], несвідома [unaware], нещасна [unhappy], нормальна [normal], озлоблена [angry], паршива [lousy], переконана [convinced], повернута [twisted], поїхавша [nuts], прибацана [bananas], провідна [leading], самостійна [independent], свідома [conscious], сильна [strong], скромна [modest], срана [shitty], тлуста [fat], тупа [dull], х *** [f***ing],чорна [black], шизоїдна [schizoid]*	*1*

(14) Я це прочитала, і моя внутрішня феміністка аж встрепенулася від щастя (навіть не знала, що вона в мені живе).

"I have read it and my inner feminist is over the moon (I didn't know that I have it).

In addition, the analyzed tweets reveal misconception of feminism and, as a result, negative attitudes, mockery and rejection. The analysis of in-depth searching revealed 8 main reoccurring categories of attributes of a feminist in Ukrainian tweets:

- They look shabby and are ugly (15–17).

(15) Нашо тобі бути феміністкою? ти ж гарна.
"Why are you a feminist? You are beautiful"
(16) Будь_сильною_і_красивою_а_не_феміністкою.
"Be strong and beautiful, but not a feminist"
(17) Ти жирна, як феміністка.
"You are as fat as a feminist"

- A feminist is an offence (16–18).

(16) Не даремно Жанна Д'Арк була жіночої статі)) Тільки не подумайте, що я феміністка. Просто думки ПОходу…

"For a good reason Jeanne D'Arc was female)) Just do not think I'm a feminist. Just the thoughts…"

(17) Лол. Феміністка — головна образа наших часів.

"Loll. The word "a feminist" is the greatest offence today."

(18) Сьогодні моя подруга сказала, що я виглядаю як 30тирічна феміністка… Обідка.

"Today my friend said that I look like a 30 year old feminist. Offence"

- Supporting feminism, but denying having anything to do with the concept (19–21).

(19) Зокрема, манчестерському боксеру приписують сексистські слова про те, що "найкраще місце жінки - або на кухні, або лежачи на спині", а також критику гомосексуалізму й абортів". І от звідки пішов цей сексизм?? ЧОМУ так багато чоловіків стверджують, що жінки це "нижча" раса. Я, звісно, не феміністка, але такі слова обурюють!

"In particular, the Manchester boxer is believed to say sexist words that" the best place for women is either in the kitchen or lying on the back, "as well as criticism of homosexuality and abortion". And where did this sexism come from? WHY so many men claim that women are a "lower" race. Of course, I'm not a feminist, but such words are outrageous!"

(20) Я не феміністка, я люблю, коли є можливість сісти в транспорті, коли чоловічу роботу роблять чоловіки і т.д., але будь-які справи, які хоча б трішки стосуються мене, люблю вирішувати сама.

"I'm not a feminist, I love when it's possible to take a seat in public transport when male work is done by men, etc. But I do like to deal with any matters that concern me on my own."

(21) Я не феміністка, але сама з таким зіткнулася.

"I'm not a feminist, but I faced it myself."

- Feminists are stupid (22–23).

(22) Отак спілкуєшся з, здавалося б, розумною людиною, а потім дізнаєшся що вона феміністка…

"So, it seems you communicate with an intelligent person, but then you find out that she is a feminist…"

(23) Феміністка тупа, бісить.

"A stupid feminist, she drives me nuts"

- Feminists do not marry, hate or despise men (24–26).

(24) препод с психологии говорит одногруппнице: -хочете вийти заміж? - нє - шо, ще одна феміністка??

"a psychology teacher said to my groupmate: Do you want to marry? No – another feminist?"

(25) Сьогодні Міжнародний день боротьби за ліквідацію насильства над жінками. Я не феміністка, але зневажаю чоловіків, які дозволяють собі фізичне чи моральне насильство над жінкою.

"Today is the International Day of the Elimination of Violence Against Women. I'm not a feminist, but I despise men who dare practice physical or moral violence against women".

(26) Я як феміністка, тільки навпаки. тупо ненавиджу баб.

"I'm like a feminist, but simply hate women"

- A feminist is psychologically unstable (27–28).

(27) Одна дотепна жінка якось сказала: "назватися феміністкою – це навісити на себе бирку « зла, істерична, неадекватна...

"One witty woman once said: "To call oneself a feminist is to bear a label of evil, hysterical, inadequate…"

(28) Якась маразматична феміністка написала, не читайте, витратите даремно час.

"A marasmic feminist wrote it, don't read it, it's a waste of time."

- Feminists get no help (29–30).

(29) Ситуація з чоловіками на сьогодні плачевна. Потрібно написати собі на спині: "Ей, я не феміністка і не відмовляюся від допомоги понести валізу"

"The situation with men today is appalling. We need to write on our backs: "Hey, I'm not a feminist and will not turn down the help if you offer to carry a suitcase".

Не можна одночасно бути феміністкою і вимагати, щоб тобі поступались чоловіки місцем у транспорті:)

"You can't be a feminist and at the same time want men to let you take their seat on a bus"

(30) Відкрив двері та пропустив дівчину вперед, а вона замість "дякую" подивилася як на ідіота. Феміністка, мабуть

"He opened the door and let the girl go forward, and she instead of saying 'thank you' looked at him as at an idiot. The feminist, perhaps".

- Feminists are treated aggressively (31).

*(31) Господи, як я чекаю на ті часи, коли за слова "я феміністка" будуть п*** і ніх** за це не буде....*

"Lord, I am waiting so much for the time when for the words "I'm a feminist" they are f *** and not a f ** thing is done for that".

7 Conclusions

The research findings show that feminism has become an array of clashing opinions of Ukrainian Twitter users. With the help of the adjusted SentiStrength algorithm we automatically detected prevailing negative and often strong negative sentiment of the Ukraine tweets on feminism. The results of the manual annotation have also shown overall negative attitude. Nevertheless, we revealed a general public support for a key feminist goal, i.e. equality of women and men in society, coupled with unwillingness to consider oneself a feminist by women and some misconception about the term *feminism*, often used synonymously with man-hating.

Further studies on detecting sentiment and sentiment strength in Ukrainian tweets, as well as other types of text, is promising as they help understand the role of emotion in on-line communication and, particularly, identify utterances associated with discrimination, aggression and threatening behavior.

References

1. Romanyshyn, M.: Rule-based sentiment analysis of Ukrainian reviews. Int. J. Artif. Intell. Appl. (IJAIA) **4**(4), 103 (2013)
2. Lobur, M., Romaniuk, A., Romanyshyn, M.: Defining an approach for deep sentiment analysis of reviews in Ukrainian. Visnyk Natsionalnogo Universytetu Lvivska Politehnika. Komputerni systemy proektuvannia, Teoria i praktyka 747, pp. 124–130 (2012)
3. SentiStrength. http://sentistrength.wlv.ac.uk
4. Pang, B., Lee, L.: Opinion mining and sentiment analysis. Found. Trends Inf. Retr. **2**(1–2), 1–135 (2008)
5. Liu, B.: Sentiment Analysis and Opinion Mining. Morgan & Claypool Publishers, San Rafael (2012)
6. Stone, P.J., Dunphy, D.C., Smith, M.S., Ogilvie, D.M.: The General Inquirer: A Computer Approach to Content Analysis. The MIT Press, Cambridge (1966)
7. Strapparava, C., Valitutti, A.: Wordnet-affect: an affective extension of wordnet. In: Proceedings of the 4th International Conference on Language Resources and Evaluation, Lisbon, pp. 1083–1086 (2004)
8. Agerri, R., García-Serrano, A.: Q-WordNet: Extracting polarity from WordNet senses. http://www.lrec-conf.org/proceedings/lrec2010/pdf/2695_Paper.pdf
9. Baccianella, S., Esuli, A., Sebastiani, F., SentiWordNet 3.0: An enhanced lexical resource for sentiment analysis and opinion mining. http://www.lrec-conf.org/proceedings/lrec2010/pdf/2769_Paper.pdf
10. Turney, P.D.: Thumbs up or thumbs down? Semantic orientation applied to unsupervised classification of reviews. In: Proceedings of the 40th Annual Meeting of the Association for Computational Linguistics (ACL), Philadelphia, PA, pp. 417–424 (2002)
11. Wiebe, J., Wilson, T., Bruce, R., Bell, M., Martin, M.: Learning Subjective Language. Computat. Linguist. **30**(3), 277–308 (2004)
12. Choi, Y., Cardie, C.: Learning with compositional semantics as structural inference for subsentential sentiment analysis. In: Proceedings of the Conference on Empirical Methods in Natural Language Processing, pp. 793–801 (2008)
13. Thelwall, M., Buckley, K., Paltoglou, G., Cai, D., Kappas, A.: Sentiment strength detection in short informal text. J. Am. Soc. Inf. Sci. Technol. **61**(12), 544–2558 (2012)
14. Hamilton, W. L., Clark, K., Leskovec, J., Jurafsky, D.: Inducing domain-specific sentiment lexicons from unlabeled corpora. In: Empirical Methods in Natural Language Processing (EMNLP) (2016)
15. Yang, Y., Eisenstein, J.: Putting things in context: community-specific embedding projections for sentiment analysis. https://www.semanticscholar.org/paper/Putting-Things-in-Context%3A-Community-specific-for-Yang-Eisenstein/17e7efbb17dd1e13c6def85cb6eacf6d0d803e8b
16. Olson, D.L., Dursun, D.: Advanced Data Mining Techniques, 1st edn. Springer, Heidelberg (2008)
17. Kis, O.: Who is not protected by the Berehynya, or matriarhy as a male invention. **4**(16), 11–16 (2006)

Method for Determining Linguometric Coefficient Dynamics of Ukrainian Text Content Authorship

Victoria Vysotska[1(✉)], Vitor Basto Fernandes[2], Vasyl Lytvyn[1,3], Michael Emmerich[4], and Mariya Hrendus[1]

[1] Lviv Polytechnic National University, Lviv, Ukraine
{Victoria.A.Vysotska, Vasyl.V.Lytvyn, Mariya.H.Hirnyak}@lpnu.ua
[2] University Institute of Lisbon, Lisbon, Portugal
[3] Silesian University of Technology, Gliwice, Poland
[4] Leiden Institute of Advanced Computer Science, Leiden University, Leiden, The Netherlands

Abstract. The article describes the peculiarities of linguometry information technologies usage to determine the linguometric coefficients dynamics of the text content authorship. The linguistic and statistical analysis of the author texts within a certain time period takes advantage of the text content-monitoring based on the NLP methods to determine the set of stop words and to study n-grams. The latter is used in the methods of linguometry and stylometry to determine the linguometric coefficients dynamics of the ownership of the analyzed text to a specific author in percentage points. There is proposed a formal approach to the definition of the author's style of the Ukrainian text in the article. The experimental results of the proposed method for determining the ownership of the analyzed text to a particular author upon the availability of the reference text fragment are obtained. The study was conducted on the basis of the Ukrainian scientific texts of a technical area.

Keywords: NLP · Content · Content monitoring · Stop words
Content analysis · Statistical linguistic analysis · Quantitative linguistics
Statistical linguistics · Linguometry

1 Introduction

Today, the task of statistical linguistics are to determine the statistical structure of the text for solving problems of computational linguistics, in particular, of linguometry, stylometry and glottochronology [1–4]. For example, these tasks are to computerize the lexicographic processes, to compare dictionaries, to create the stenography systems, to automatically define the language or the authenticity of the work, and the like. In addition, the results of statistical linguistics are used in cryptolinguistics. The beginning of the quantitative and statistical methods application to language and speech reach the antiquity age, and a new impetus to its development in modern times was, in particular, the growing popularity of foreign language learning in the middle of XIX - early XX

N. Shakhovska and M. O. Medykovskyy (Eds.): CSIT 2018, AISC 871, pp. 132–151, 2019.
https://doi.org/10.1007/978-3-030-01069-0_10

century. The theorists and practitioners of language teaching have realized that it is impossible to fully master the language in a few years with several hours of classes a week, so they decided to limit the dictionary to the most frequently used words. The real "explosion" of the statistical research took place with the advent and development of computer technology [5–12]. Computers have helped to simplify the mechanical work, in particular, such as the calculation of word tokens in the text to determine its authorship [13–20]. In contrast to the only possible manual calculation formerly, now the text transformation in electronic form and the application of the appropriate software allows you to obtain these data automatically, promptly and with minimal deviation [21–26]. Today, linguistic and statistical research is carried out in each country with a well-developed linguistics: in Germany, Austria, USA, Australia, Czech Republic, Slovakia, Poland, Russia and the like [27–33]. There are international societies and magazines: IQLA (International Quantitative Linguistics Association)., Journal of Quantitative Linguistics, "Quantitative Linguistics" series, Computational Linguistics and Intelligent Systems (http://colins.in.ua). A significant contribution to the development of this branch of linguistics was made by Gabriel Altmann, Reinhard Köhler (Germany)., Peter Grzybek (Austria)., Geiza Wimmer (Slovakia)., Adam Pawłowski, Jadwiga Sambor (Poland)., Valentyna Perebyinis, Nataliya Darchuk, Dmytro Lande, Nataliya Sharonova (Ukraine)., Yukhan Tuldava (Estonia)., Raimund Piotrovskyy (Russia)., etc. [33–47]. Linguistic and statistical research in Ukraine began in the 1950s, and at first they were related to the selection of lexical minimum for foreign languages, but subsequently their scope of application was extended. So, in the early 1960s at the A. A. Potebnya Institute of Linguistics of the USSR Academy of Sciences a group of structural and mathematical linguistics was organized. This group began a targeted statistical study of the Ukrainian texts of fiction, scientific and technical, and socio-political functional styles, which allowed the identification of their statistical parameters. At the same time, a project on the formation of a series of frequency dictionaries began, in particular: dictionaries of fiction, drama, poetry, journalism, scientific prose, to which the laboratory of computer linguistics of the Taras Shevchenko National University of Kyiv was also involved. Apart from the above-mentioned research centers, one can also consider Kyiv National Linguistic University, Yuriy Fedkovych Chernivtsi National University, Ivan Franko National University of Lviv, Lviv Polytechnic National University and others.

2 Literature Data Analysis and Problem Statement

A significant concept in linguistic statistics is the unit distribution in the text, the presence of units in various (usually equal). subsamples (fragments). If the analyzed unit functions in only one of subsamples, albeit with high frequency, then such a sample is not representative towards this unit. It is important that the unit under study is evenly distributed in the main entity, so it is present in the text of the absolute majority of subsamples. To determine the equitability of units in the text, the prevalence ratio is analyzed, the ratio of the subsamples number with a certain linguistic unit to the total number of subsamples. However, the characteristics obtained by the sample material may differ from the actual characteristics of the main entity, as the relative study inaccuracy is possible. The frequency distribution of language units in the text has a certain regularity and forms its statistical (frequency, probabilistic). structure. It is different for different language elements – phonemes, morphemes, lexemes and the like. Statistical parameters of the styles that are set at different levels have different style distinction power for different pairs of styles; more related styles are clearly distinguished at the syntactic level, less related – at the lexical [21–26].

The statistical text structure is described in the form of models and theoretical formulas, for example, in the form of benefits law, Zipf's law and Mandelbrot's law [21–26]. The difference between the statistical structures of the different texts is a visual representation of the differences between these texts.

A lot of attention is devoted to the statistics of phonological units, in particular, it is revealed that it has a direct impact on the statistical structure of the lexical units [21–26]. The statistical text structure at the lexeme level is usually determined by the data of the frequency dictionary. Frequency dictionary (FD) gives the frequency of a certain linguistic unit (syllable, word, word form, word combination, idiom, phraseological unit). in the examined texts (sample) of a certain volume. According to the nature of the sample, the frequency dictionaries are divided into the FD of: the whole language, a certain functional style, a writer and/or a particular work. To determine the authorship of the Ukrainian text, it is usually used the frequency dictionary of a particular work or the whole language, but it is desirable to use the frequency dictionary of a certain functional style and writer. This will give more accurate results, but it takes a lot of time for the preparatory stages (collection and analysis of the original profile/ thematic author texts for a long period of time). In addition, such texts are not always available either because of their absence or because of the source uncertainty, or because of source absence (or the small number)., or because of the lack of access to them because of the property rights protection.

Usually, the frequency dictionary includes absolute and relative frequency of language units use. The form of the material representation is essential for the frequency dictionary, in particular, dictionary entries are placed in descending frequency order. It actually visualizes the statistical structure of the text that is the material for its formation, and also allows you to calculate the degree of text coverage. According to the calculation unit, there are distinguished such frequency dictionaries as: of sounds, syllables, morphemes, words (the most common). or word combinations, idioms and phraseological units. To determine the Ukrainian text authorship, it is usually used FD of morphemes, words (the most common). or word combinations of a particular work. It is desirable then to compare the results with the FD of morphemes, words (the most common). or word combinations of a certain functional style of a particular writer.

Frequency dictionaries are monolingual and translated according to the number of languages, by means of material recording – oral and written speech FD. Although depending on the material peculiarities and the author idea, each FD is formed differently, however, we can distinguish the general technique of the frequency dictionary formation:

1. the determination of an enumeration unit,
2. the identification of the representative sources of FD formation,
3. the determination of the FD formation principles, in particular,

(a) what statistical information will be provided by the FD (absolute/relative/medium frequency, the degree of text coverage, the units distribution in the text, etc.).,
(b) the word lemmatization schemes,
(c) the determination of the list amount (words in descending frequency order, word forms in descending frequency order, words in alphabetical order, a separate list of proper names, etc.).,

4. the determination of FD formation approach,
5. the homonymy (ambiguity). elimination,
6. the determination of the FD formation milestones.

An important problem of the frequency dictionary formation is the selection of sources. For the FD of a particular writer, the sources are all his works, and the most authoritative is the last lifetime edition. For the FD of the whole language, for example, modern Ukrainian literary language, ideally it is taken into account all its functional and stylistic nature: fiction, journalistic, colloquial, scientific, official, epistolary, and confessional. Within each of the styles, it is also taken into account different themes, areas, etc. The problem of quantitative correlation of styles in FD of the idioethnic language is open. In the FD of different languages, this issue was solved in different ways. Since the unit of calculation in modern FD is usually the word, so the studies use lemmatization (bringing the word form of the text in question to its initial dictionary form). For the correct implementation of this operation, in the initial stages there is the identification of the lexical (for example, *мука* (UA).). and grammatical homonyms (for example, *коси* (UA).). – the imperative mood of the 2 singular person, nominative case of the plural noun and genitive case of the singular noun. Sometimes we distinguish between the single meanings of the polysemantic words, however, it is impossible to make it fully automatically.

Words and word forms are arranged in descending frequency order. The number of unit in sequence in the list in descending frequency order is called the rank.

There are the following key features of the FD:

- The text volume, the number of word tokens in a text (N). is the total number of words in the text. For example, in the sentence *квітка квітці посміхається* there are 3 word tokens.
- The volume of the dictionary of word forms, the number of word forms in the text ($W_ф$). is the number of words in the text in a certain form. For example, in the text *людина людину повинна поважати, адже вона людина* there are 6 word forms since the first and last word – *людина* – is a noun in the same form (nominative case, singular).
- The volume of the dictionary of lexemes, the number of words in the text (W). is the number of lemmatized words in the text. For example, in the previous sentence, there are 5 words, because of the case forms (*людина, людину, людина*). of one word *людина*.

It is interesting to compare the statistical characteristics of the styles that are rendered by the FD. There are different approaches to the methods of the FD comparing. For example, V. Perebiynis proposed for this purpose the concept of *zero style* [21–26],

M. Arapov and his co-authors put forth the theory of determining *the quantitative distance* between dictionaries [34–43], etc. However, for a correct comparison, the frequency dictionaries must be formed:

- on the material of the same volume, because the lengthening of the text differently affects the increase in the number of words and word forms: the number of word forms grows slightly faster than the number of words;
- on the same principles (enumeration unit, word lemmatization scheme, homonymy elimination, etc.).

According to the FD data, such characteristics as the wealth of the dictionary, the *diversity index* (K_l). are calculated. K_l is the ratio of the lexeme dictionary volume (*W*). to the text volume (*N*)., that is, $K_l = W/N$. In Table 1 and hereinafter the asterisk (*). denotes the numbers obtained as a result of scaling. The table shows that the most diverse and rich vocabulary (lexis). is in the poetry speech, further in descending order – in fiction, colloquial and journalistic styles. There are the least different words in the scientific and official speech.

Table 1. Results of the vocabulary diversity, indexes of uniqueness and concentration according to the FD of the Ukrainian functional styles

Dictionaries	W/N	W_1/W	W_1/N	W_{10t}/N	W_{10}/W
FD of the poetry	0,103	0,052	0,495	0,789	0,098
FD of the fiction	0,083* (0.067)	0,038* (0,029)	0,455* (0,430)	— (0,821)	— (0,149)
FD of the colloquial style	0,073	0,034	0,465	0,789	0,161
FD of the journalism	0,070	0,031	0,450	0,804	0,121
FD of the scientific prose style	0,059	0,025	0,427	0,890	0,189
FD of the official style	0,030	0,0085	0,280	0,935	0,303

The average repeatability of a word in the text (*A*). is the ratio of the text volume (*N*). to the volume of the lexeme dictionary (*W*).; reciprocal of the diversity index, that is $A = N/W$. According to the FD data, every word in colloquial style is used on average 14 times, and in the scientific style – 17. *Hapax legomena* parameter is also analyzed, i.e. words that happened once in the sample to be studied, that is, have a frequency of 1.

The uniqueness index is calculated separately for the dictionary and for the text (Table 1). It characterizes the vocabulary variability, that is, the proportion of the text (dictionary)., which is covered by words that happened once. *The uniqueness index for the dictionary* (I_{wt}). is the ratio of the number of lexemes with frequency of 1 (W_1). to the total number of lexemes: $I_{wt} = W_1/W$. *The uniqueness index for the text* (I_t). is the ratio of the number of lexemes with frequency of 1 (W_1). to the text volume (*N*).: $I_t = W_1/N$.

The opposite to the uniqueness index is *the concentration index of the dictionary and the text*, indicating the proportion of the text (dictionary)., which is covered by words that have happened 10 times or more (Table 1). *The concentration index of the dictionary* is the ratio of the number of words in the dictionary with an absolute frequency of 10 (W_{10}). to the total number of words in the dictionary (W).: I_{kt} = W_{10}/W. *The concentration index of the text* is the ratio of the sum of absolute word frequencies with an absolute frequency of 10 or more (W_{10t}). to the text volume (N).: I_{tn} = W_{10t}/N.

As can be seen from the frequency dictionary, a speech prefers a small number of frequently used units. They form the core of any speech subsystem, while the overwhelming number of units is of low-frequency. This conformity was noticed by the scientist Dewey in the early XX century, calling it *the benefits law*. This conformity was further investigated by the German linguist J. Zipf, formulating the Zipf's law, which establishes the dependencies:

- of word frequency and dictionary rank: the more frequent the word, the higher its rank, as $F \times i = const$, where F is the word frequency in the frequency dictionary, i is the word rank;
- of word frequency and its length: the more frequent the word, the shorter it is, as $k = C \lg r$, where k is the word length in phonemes, C is constant, r is a rank;
- of word frequency and number of its meanings: the more frequent the word, the more meanings it has, as $m = C\sqrt{f}$, where m is the number of word meanings, C is constant, f is a word frequency;
- of word frequency and its origin: the older the word, the more frequent it is.

According to the law of the German linguist P. Mentserat, the length of the language pattern (word, word combination, superphrasal unity, sentence). is inversely related to the length of its components (syllables, words, word combinations, etc.)., that is, the longer the language pattern, the shorter its components. It was mathematically formulated by G. Altmann: $y = ax^b$, where y is the average length of the components, x is the length of the language pattern, b is an indicator characterizing the dynamics of the components length (the law is in force if $b < 0$). Krylov's law establishes a relationship between the number of polysemantic words and their frequency: $p_x = 1/2^x$, $px = (\omega - 1)^{x-1} / \omega^x$, where p_x is the probability of a word usage that has x meanings, ω is the average number of the word meanings in the dictionary.

Some basic quantitative characteristics of a language are very simple. For example, the difference between the number of words (10^4–10^5)., the number of morphemes (several thousand)., the number of syllables (from several hundred to several thousand). and the number of phonemes (10 to 80). It is suggested that such relations are associated with the property of human memory. It should be also noted that the more frequent the word, the faster the person will be able to remember it.

But there is no research in the field of change dependency in the coefficients of the author's lexical speech during the period of his work.

3 Purpose and Objectives of the Research

The purpose of this work is to develop a method of determining the dynamics of the text content authorship coefficients in Ukrainian texts on the basis of linguometry.

To achieve this goal, the following objectives are formulated:

(1) to develop a method for determining the text author based on the analysis of the coefficients of the author's lexical speech in the reference text;
(2) to develop a method for determining the dynamics of the linguometric coefficients of the text content authorship;
(3). to develop content monitoring software to determine the dynamics of the linguometric coefficients of the text content authorship in Ukrainian texts based on the linguometric analysis of certain stop words of the text content;
(4) to obtain and analyze the results of experimental trialability of the proposed content monitoring method to determine the dynamics of the linguometric coefficients of the text content authorship in Ukrainian scientific texts of the technical area.

4 Method for Determining the Text Content Author

Linguometry is a branch of applied linguistics, it detects, measures and analyzes quantitative characteristics of the units of different levels of language or speech [21–26]. Applying the apparatus of mathematical statistics, the important tasks of linguistics, which can be solved by linguometry, are the creation and comparison of dictionaries (including frequency and statistical dictionaries)., the creation of the automatic dictionaries, thesauri, development of stenography systems, automatic language detection, information retrieval, and the like. One way to characterize the literary text wealth is to assess the nature of the linguistic unit use at all language levels. This allows us to identify the concepts of *wealth* and *diversity* of speech [21–26]. Calculation of the linguistic diversity coefficients should assume the relationship of such coefficients as *lexical diversity, degree (measure). of syntactic complexity* [21–26], *cohesion, indexes of the text uniqueness and concentration* [34–43]. Since the coefficient is an absolute value, the length of the compared texts can be neglected within certain limits. The theoretical interest is also the study of the internal text "dynamics" in terms of comparing the coefficients of different parts of the text with each other and with a common coefficient for the whole text (Table 2).

It was found out [21–26] that the Ukrainian fairy tale text has K_z = 0,77, and the text of the Ukrainian scientific article – 3,0, that is, the coherence in the second text is 3,9 times stronger than in the first one. There are no official standards for the coefficients of speech diversity for K_l and K_s [21–26], but the reference point for the comparison and evaluation of a text in a homogeneous group of texts is the average norm of the coefficient for equal length fragments. The minimum size (length). of the fragment will be 100 words, we will assume that the coefficients here are already stable, reflecting the real features of the author's language. The proximity or distance of a single individual coefficient from average serves as the basis for assessing the speech diversity in the relevant text. Texts are considered sufficient if their diversity coefficients fall within the range of mean square deviations from a certain average. As a reminder, a mean square deviation is calculated as: $D = \sqrt{\frac{\sum_{i=1}^{n} x_i^2}{n} - (\bar{x})^2}$, where D is the

Table 2. Text diversity coefficients [21–26]

Coefficient	Definition
Lexical diversity	The ratio of the number of words to the total number of word forms in the text, that is: $K_l = W/N$, where K_l is the coefficient of lexical diversity, W is the number of words in a certain text, N is the total number of words in this text. The coefficient value lies in the range of [0; 1]. The higher the decimal fraction, the higher the lexical diversity of the text
Syntactic complexity	The ratio of the number of sentences to the number of words in a particular text: $K_s = 1 - P/W$, where K_s is the coefficient of syntactic complexity, P is the number of sentences, W is the number of words in the entire text. The greater the fraction (within [0; 1]), the wordier the whole sentence of such text is, and therefore, there is a higher probability of the diversity of syntactic relations between words in a separate sentence
Cohesion coefficient	The ratio of the number of prepositions and conjunctions to the number of separate sentences (the coefficient equals to one, when in one sentence there are three connective elements (prepositions and conjunctions).: $K_z = (Z + S)./(3P)$., where Z is the number of prepositions, S is the number of conjunctions, P is the number of separate sentences
Uniqueness index	The variability of the vocabulary, that is, the proportion of the text, which is words that happened 1 time, that is, $I_{wt} = W_1/W$, where I_{wt} is the text uniqueness index, W_1 is the number of words with frequency of 1, W is the number of words in the entire text
Concentration index	The text proportion covered with words that have happened 10 times or more, that is, $I_{kt} = W_{10}/W$, where I_{kt} is the text concentration index, W_{10} is the number of words with a frequency of 10 and more, W is the number of words in the entire text

mean square deviation from the average, x is each specific coefficient of such manifold, n is the total number of coefficients (of fragments). The standard deviation zone boundaries are calculated as a range within $\bar{x} \pm D$. It bears reminding that $\bar{x}$ is the arithmetic mean value, which is calculated by the formula: $\bar{x} = \frac{x_1 + x_2 + \ldots + x_N}{N} = \frac{\sum_{i=1}^{N} x_i}{N} = \frac{1}{N}\sum_{i=1}^{N} x_i$, where N is the number of experiments, x_i is the individual evaluation of the experiment.

Table 3 provides the most basic stages of the analysis and interpretation of the stylistic features and regularities of the writer's style of a certain author (or certain literature of a certain era). at the linguistic level [21–26, 34–43].

According to the outlined text study scheme, it is possible to solve the author attribution, which can be formulated, for example, in the following way. Let there be a statistically processed artistic legacy of the author (reference). It is necessary to assess the belonging of certain fragments to the reference using appropriate methods. Figure 1 shows the graphic representation of the relative frequency of function words occurrence in fragment 4 and in reference. The correlation index for function words, in this case, equals to R_{e-U4} = 0,7326. We also give the correlation indexes for each of the function words for fragments 1–4 (Table 4). Analyzing the correlation indexes for the function

Table 3. The analysis and interpretation of the stylistic features and regularities of the writer's style of a certain author at the linguistic level

№	Name	Explanation
1.	Text selection	The selection method and the amount of text sample are important. To determine the characteristics, it should be at least 18 thousand words
2.	Text unit lemmatization	Combining word forms under the lemma of the language
3.	Elimination of text unit heterogeneity	The solution to the problem of the text unit heterogeneity, for example, according to their relation to different types of speech (author's, not author's, etc.)
4.	System construction, organization of the statistical distributions in the necessary scales of frequency dictionaries	Frequency dictionary is a type dictionary, which shows the number of uses (frequency). of a particular language unit (syllable, word, word form, word combination, idiom, phraseological unit). in various texts of a certain volume. Usually, the absolute and relative frequency of the language unit use is given, dictionary entries are divided in descending order of frequencies
5.	Search for parameters that adequately reflect the structure of the frequency dictionary	The number of parameters can be different, for example, to describe the French texts of the XVII century, 51 parameters are proposed. The obtained common parameters allow you to formulate some of the basic linguistic and statistical methods of text research: • the method of prop words (counting the total use frequency and finding the percentage composition of function words: prepositions, conjunctions, particles); • the method of punctuation marks (counting only the number of internal and external punctuation marks); • word method (counting only words of a certain length); • the method of sentences (counting only sentences of a certain length); • syntactic method (counting the punctuation marks, words and sentences of a certain length); • mixed method (a combination of the prop word method and the syntax method)

(*continued*)

Table 3. (*continued*)

№	Name	Explanation
6.	Parameter checkout for efficiency	Application of common validation methods of the selected effectiveness parameters
7.	Mathematical modeling of lexical and statistical distributions	Application of common methods of mathematical apparatus of lexical and statistical distributions modeling
8.	Construction of statistical classifications	Construction of the author reference texts that reflect the stylistic patterns within the works of a particular author or a particular literary epoch (or the sequence of literary epochs)
9.	Interpretation of results	Interpretation of the results obtained from the standpoint of historical and literary ideas, general and historical stylistics

words, we conclude that the probability of belonging the fragments to the studied reference is the greatest for Fragment 4, followed by Fragment 2, Fragment 1, Fragment 3. Note that for all four fragments there are consistently high correlation indexes for particles. This implies that there is no impact of particles on the author's style. Additionally, for fragments, we analyze the occurrence frequency of solely prepositions and conjunctions, find the corresponding correlation indexes and compare the results (Table 4).

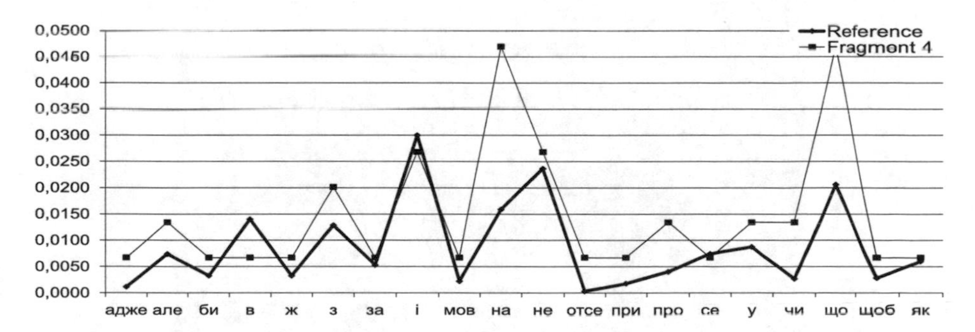

Fig. 1. The relative frequency of the function words occurrence in Fragment 4 and Reference

Fragment 4 remained the most likely candidate to be the reference, and the next by a slight margin was Fragment 1, then – Fragment 2. Fragment 3, as in the previous study, has the least probability of belonging to the reference. To confirm the results, let us turn to [21–26], from which fragments for the study are taken.

Table 4. Correlation indexes for the function part of speech and for each fragment

Fragment	Preposition	Conjunction	Particle	R_{e-U}	R'_{e-U}
1	0,72	0,79	1	0,6076	0,6900
2	0,4928	0,5714	0,9580	0,7066	0,4913
3	0,1517	0,1624	0,8800	0,2810	0,2254
4	0,5639	0,9544	0,9594	0,7326	0,6905

5 Results of Studies on Determining the Author's Style in the Ukrainian Texts Based on the Technology of Statistical Linguistics

To achieve the research goal, a system with the possibility of selecting the language/languages of the analyzed content was developed, which is implemented on the Victana Web-resource [20]. The information resource contains such fields (Fig. 2).

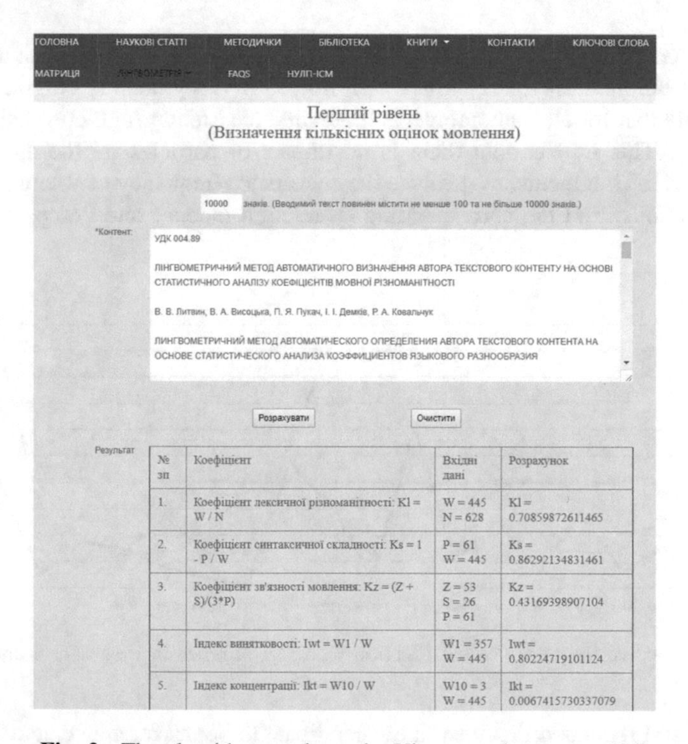

Fig. 2. The algorithm result on the Victana web-resource [34]

- Characters. The user enters the number of characters (the size of the text to be produced). and the text to analyze in the corresponding fields. The text to be entered must contain at least 100 and no more than 10000 characters. The maximum content size is set.

- Content field, where the studied text is copied from the buffer.
- Calculate field runs the calculation of the text diversity coefficients of the author's text under study.
- Delete field clears the entered data.

On the server, after going the calculation of the text diversity coefficients, the algorithm of text analysis is run:

1. Check of the text length, the excess is cut off.
2. Determination of the number of sentences.
3. Cleaning of the studied text (from numbers, special characters).
4. Determination of the total number of words in the text N.
5. Determination of the number of words W (word stems without repetition).
6. Determination of the number of prepositions Z.
7. Determination of the number of conjunctions S.

Analyzing the components of the formulas that allow evaluating the work wealth, it can be concluded that it is necessary to find such variables: the number of sentences, the number of words and word forms, the number of prepositions and conjunctions, the number of words with a frequency of 1 and a frequency of at least 10. For convenience, we will enter the obtained data in the table. On the website page the generated table is presented (Table 5). and the results of the study are displayed on the screen.

Table 5. Example of the generated table as the algorithm result on Victana [20]

N	Coefficient	Input data	Calculation
1.	Lexical diversity coefficient: $K_l = W/N$	$W = 184$ $N = 295$	$K_l = 0.62372881355932$
2.	Syntactic complexity coefficient: $K_s = 1 - P/W$	$P = 18$ $W = 184$	$K_s = 0.90217391304348$
3.	Cohesion coefficient: $K_z = (Z + S)./(3*P)$	$Z = 20$ $S = 28$ $P = 18$	$K_z = 0.88888888888889$
4.	Uniqueness index: $I_{wt} = W_1/W$	$W_1 = 141$ $W = 184$	$I_{wt} = 0.76630434782609$
5.	Concentration index: $I_{kt} = W_{10}/W$	$W_{10} = 2$ $W = 184$	$I_{kt} = 0.010869565217391$

Proceeding from the above, we assess the wealth of the work fragments of individual scientific articles of the technical area of the Lviv Polytechnic National University Bulletin, Information Systems and Networks series for the period 2001–2017 [20] using the coefficients of diversity and cohesion, indexes of the text uniqueness and concentration. For the analysis, we choose the first part (10000 characters). of each article. Statistical analysis of the system operation for the detection of stop words set from 215 scientific articles in the technical area was conducted in 3 stages.

Stage 1. The analysis of 100 scientific articles was carried out to determine the range of the optimal size of the studied text. At first, the texts were analyzed in full volume, and then these texts were analyzed for different numbers of characters. The results showed that the optimal range of texts is [100; 10000] characters. Less than 100 characters are uninformative extracted information, often the coefficient values of different authors are similar, and of the same author on different texts – differ significantly. If more than 10000 characters, the coefficients do not change significantly, but the analogs for the study have different lengths, and due to the lack of large length analogs diversity, the maximum number for the analysis was chosen 10000.

Stage 2. The analysis of more than 200 individual works in the technical area of over 50 different authors for the period 2001–2017 to define whether the text diversity coefficients by these authors change and how at different time periods.

Stage 3. The analysis of more than 200 individual works in the technical area by about 100 different authors for the period 2001–2017 to define whether the text diversity coefficients by these authors change and how at different time periods.

Analyzing the components of the formulas that allow evaluating the work wealth, it can be concluded that it is necessary to find such variables: the number of sentences, the number of words and word forms, the number of prepositions and conjunctions, the number of words with a frequency of 1 and a frequency of at least 10. For convenience, we will enter the obtained data in the table. To maintain the research integrity, it is necessary to analyze, whether the work publication time influences on the text diversity coefficients, that is, whether these coefficients do not change over time on the sample of the same authors and their texts. We will first analyze how the total amount of words in the equal-sized fragments varies in the range of 2001–2017. As you can see, over time, those same authors often use short words (Fig. 3).

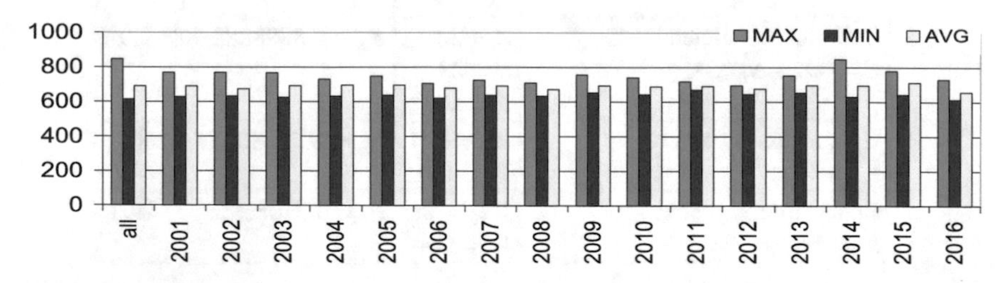

Fig. 3. The distribution of words in equal-sized fragments in the range of 2001–2017

Over time, the coefficient of lexical diversity K_l does not significantly change (Fig. 4). Similarly, the syntax complexity coefficient K_s does not significantly change over time. But the cohesion coefficient K_z decreases over time within 16 years, although not significantly. At the beginning (2001). it is variable in the range of [0,5; 1,2], and at the end of the period – in the range of [0,4; 0,9] (Fig. 5).

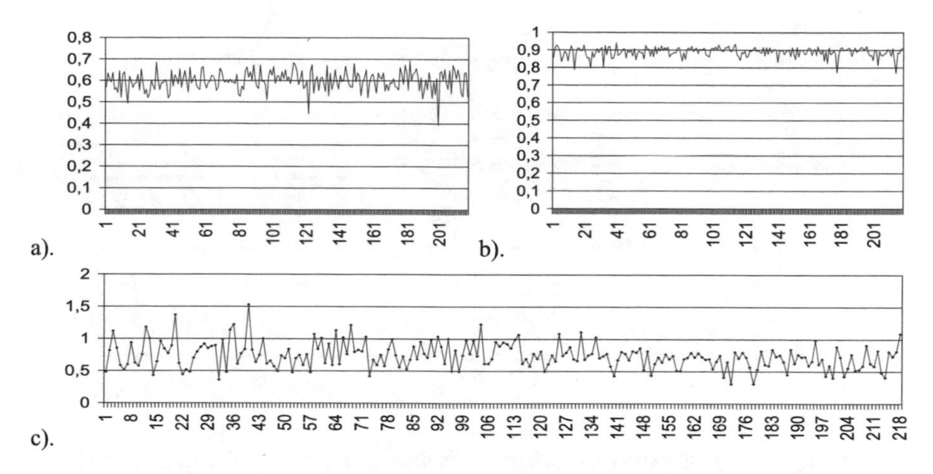

Fig. 4. The distribution of speech coefficients for the equal-sized fragments in the range of 2001–2017: (a). K_l; (b). K_s; (c). K_z

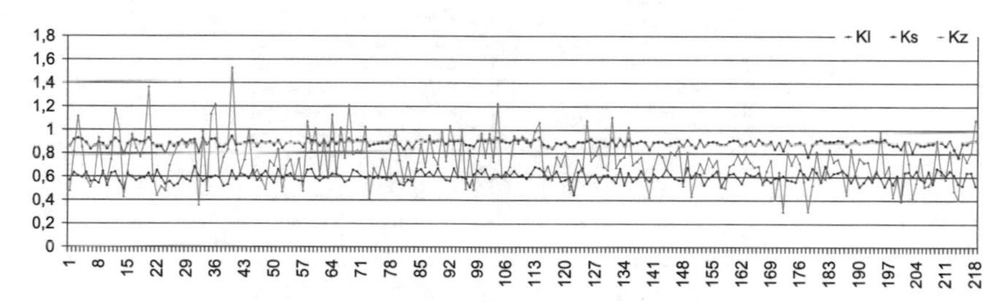

Fig. 5. Comparison of the speech coefficients distribution for the equal-sized fragments in the range of 2001–2017

Similarly, let us compare the distributions of the uniqueness and concentration indexes (Fig. 6). If the size of the distribution does not change significantly for I_{wt} over time, then there are fixed significant changes for I_{kt}. Eventually, the author of these works increasingly repeats some of the terms in his work more than 10 times, narrowing the scope of the research. Figure 7 shows the analysis result of speech coefficients for the equal-sized fragments in the range of 2001–2017 as the minimum, maximum and average values for this period (determination of the fluctuation in this time interval). A more significant fluctuation is observed in K_z.

Separately, we analyze the distribution of the use of all word forms, words used 1 time, words used more than 10 times, used in the studied texts for the equal-sized fragments in the range of 2001–2017 (Fig. 8).

Figure 9 provides the analysis of the use of prepositions, conjunctions and separate sentences in the studied texts for the equal-sized fragments in the range of 2001–2017, where Z is the number of prepositions, S is the number of conjunctions, P is the number of separate sentences.

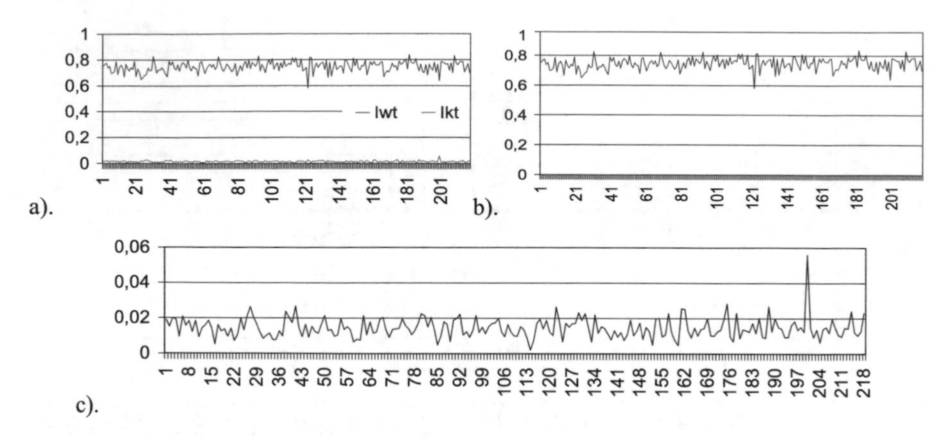

Fig. 6. Comparison of the speech coefficients distribution for the equal-sized fragments in the range of 2001–2017: (a). of both indexes; (b). of I_{wt}; (c). of I_{kt}.

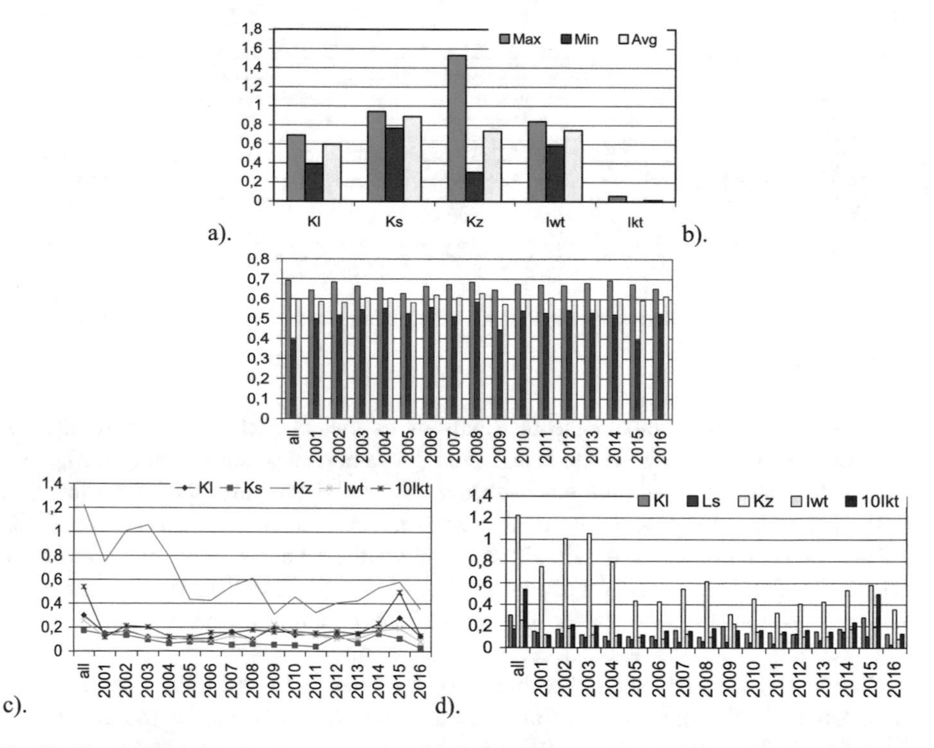

Fig. 7. The analysis result of speech coefficients for the equal-sized fragments in the range of 2001–2017: (a). minimum, maximum and average value for this period for all coefficients; (b). minimum, maximum and average value for this period for K_l; (c). diagram of all coefficients dynamics for a certain period; (d). histogram of dynamics of all coefficients for a certain period

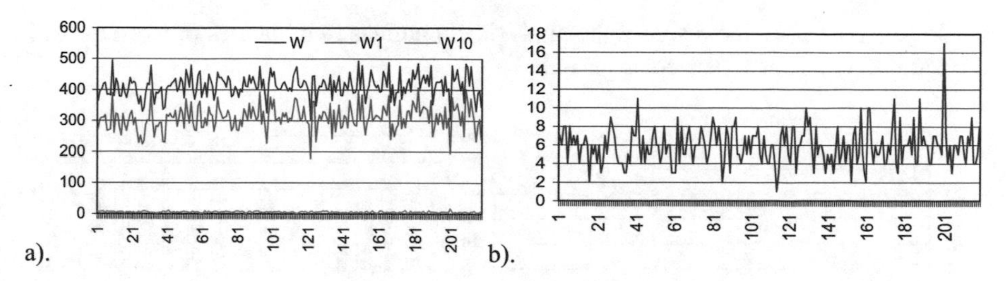

Fig. 8. The distribution of: (a). all word forms, words used 1 time, words used more than 10 times, (b). W_{10} used in the studied texts for the equal-sized fragments in the range of 2001–2017

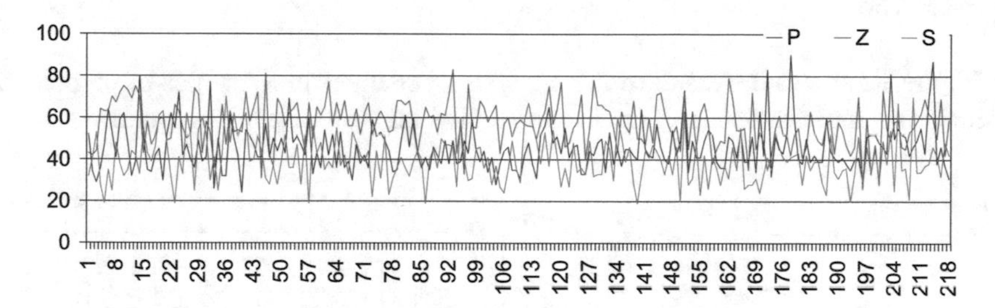

Fig. 9. The analysis of the use of prepositions, conjunctions and separate sentences in the studied texts

Figure 10a shows that over time, the authors use shorter sentences to describe the subject area than at the beginning of the study period. If the number of prepositions decreases, the distribution of conjunctions does not decrease significantly (Fig. 10b).

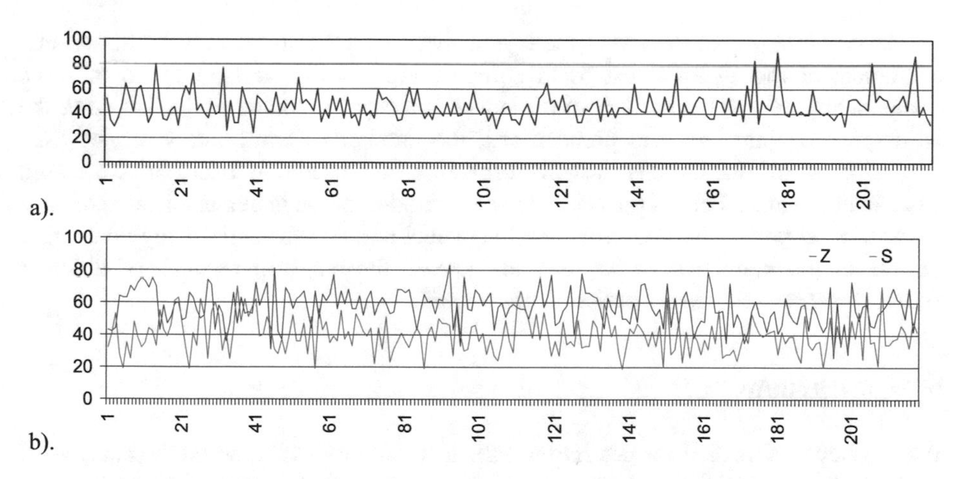

Fig. 10. In-depth analysis of the use of prepositions, conjunctions and separate sentences in the texts under study for a certain period

Figure 11 shows the result of the analysis of changes in dynamics of word usage in the studied texts for a particular period.

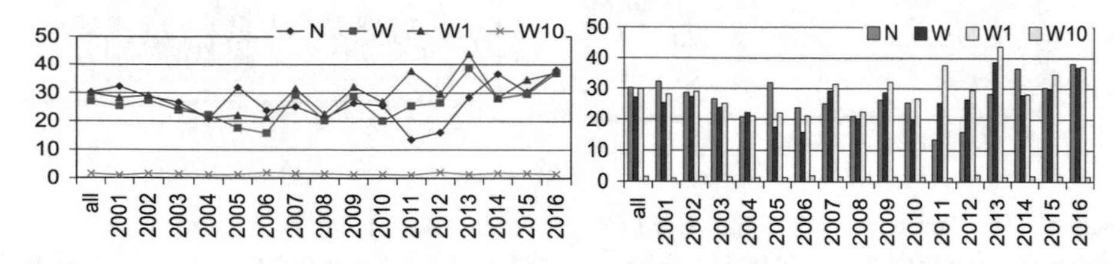

Fig. 11. In-depth analysis of changes in dynamics of word usage in the studied texts for a particular period

Figure 12 shows the result of the analysis of changes in dynamics of prepositions, conjunctions and sentences usage in the studied texts for a particular period.

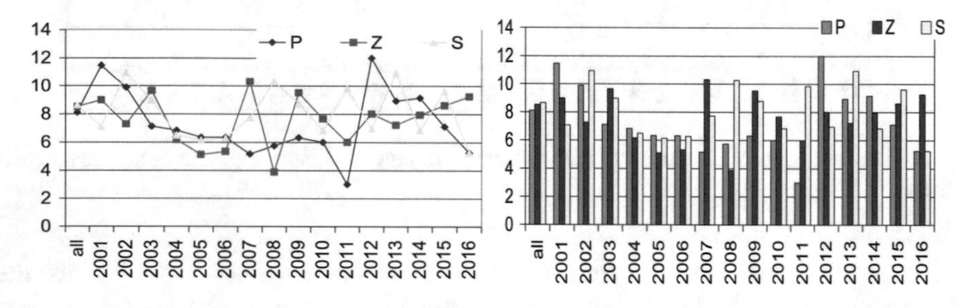

Fig. 12. In-depth analysis of changes in dynamics of prepositions, conjunctions, and sentences used in the studied texts for a particular period

As a result, we can say that there is a dynamic pattern not only of the speech coefficient of the author's text for a certain period of his work, but also of some components, such as the number of sentences in a certain volume of the fragment, the number of conjunctions and prepositions, the number of word forms for the total number of words, the number of word forms that were used only once, and were used more than 10 times. For a more accurate determination of the incremental value of each of the studied parameter, it is necessary to conduct a more substantial study on a larger sample of the individual works, and to increase the range of research of different authors' works for a longer period of their work.

6 Conclusions

We developed a method for determining the text author style based on the analysis of the lexical coefficients of the author speech in the reference fragment of the author text. It is developed a method for determining the dynamics of the linguometric coefficients of the text content authorship based on the analysis of the coefficients of the author lexical speech in the reference fragment of the author text. We developed the algorithm for determining the stop words of text content on the basis of text content linguistic analysis. We developed also the algorithm of the lexical analysis of Ukrainian texts and the algorithm of the text content syntactic analyzer. Its distinctive feature is the adaptation of morphological and syntactic analysis of lexical units to the peculiarities of the Ukrainian words/texts constructions. The article provided the theoretical and experimental justification of the content monitoring method and the determination of stop words of the Ukrainian text. This method is aimed at the automatic detection of the significant stop words in the Ukrainian text due to the proposed formal approach to the implementation of content parsing. We proposed an approach to the development of content monitoring software to determine the dynamics of the linguometric coefficients of the text content authorship in the Ukrainian texts based on Web Mining. We studied the results of the experimental testing of the proposed method of content monitoring to determine the dynamics of the linguometric coefficients of the text content authorship in the Ukrainian scientific texts of the technical area. We studied more than 200 individual publications of the technical area by about 100 different authors for the period 2001–2017 from more than 20 issues of the Lviv Polytechnic National University Bulletin of the Information Systems and Networks series. Further experimental research requires testing of the proposed method to determine the dynamics of the linguometric coefficients of the text content authorship from other categories of texts – scientific humanitarian, belles-lettres, journalistic, etc.

References

1. Khomytska, I., Teslyuk, V.: The method of statistical analysis of the scientific, colloquial, belles-lettres and newspaper styles on the phonological level. In: Advances in Intelligent Systems and Computing, vol. 512, pp. 149–163 (2017)
2. Khomytska, I., Teslyuk, V.: Specifics of phonostatistical structure of the scientific style in English style system. In: Proceedings of the XIth International Conference on Computer Science and Information Technologies, Lviv, pp. 129–131 (2016)
3. Kowalska, K., Cai, D., Wade, S.: Sentiment analysis of polish texts. Comput. Commun. Eng. **1**(1), 39–41 (2012)
4. Kotsyba, N.: The current state of work on the Polish-Ukrainian Parallel Corpus. In: Organization and Development of Digital Lexical Resources, pp. 55–60 (2009)
5. Mobasher, B.: Data mining for web personalization. In: The Adaptive Web, pp. 90–135. Springer, Heidelberg (2007)
6. Dinucă, C.E., Ciobanu, D.: Web content mining. University of Petroşani, Economics, 85 (2012)
7. Xu, G., Zhang, Y., Li, L.: Web content mining. In: Web Mining and Social Networking, pp. 71–87. Springer (2011)
8. Ganesh, J.A.: A comparative study of stemming algorithms. Int. J. Comp. Tech. Appl. **2**(6), 1930–1938 (2011)
9. McGovern, G., Norton, R.: Content Critical. FT Press, Upper Saddle River (2001)
10. McKeever, S.: Understanding web content management systems: evolution, life cycle and market. Ind. Manag. Data Syst. **103**(9), 686–692 (2003)
11. Rockley, A., Norton, R.: Managing Enterprise Content: A Unified Content Strategy. New Riders Press, Reading (2002)

12. Mishler, A., Crabb, E.S., Paletz, S., Hefright, B., Golonka, E.: Using structural topic modeling to detect events and cluster Twitter users in the Ukrainian crisis. In: Communications in Computer and Information Science, vol. 528, pp. 639–644 (2015)
13. Davydov, M., Lozynska, O.: Information system for translation into Ukrainian sign language on mobile devices. In: Proceedings of the International Conference on Computer Science and Information Technologies, Lviv, pp. 48–51 (2017)
14. Davydov, M., Lozynska, O.: Mathematical method of translation into Ukrainian sign language based on ontologies. In: Advances in Intelligent Systems and Computing, vol. 689, pp. 89–100 (2018)
15. Davydov, M., Lozynska, O.: Linguistic models of assistive computer technologies for cognition and communication. In: Proceedings of the International Conference on Computer Science and Information Technologies, Lviv, pp. 171–175 (2017)
16. Mykich, K., Burov, Y.: Uncertainty in situational awareness systems. In: Proceedings of the International Conference on Modern Problems of Radio Engineering, Telecommunications and Computer Science, pp. 729–732 (2016)
17. Mykich, K., Burov, Y.: Algebraic framework for knowledge processing in systems. In: Advances in Intelligent Systems and Computing, pp. 217–228 (2017)
18. Mykich, K., Burov, Y.: Research of uncertainties in situational awareness systems and methods of their processing. East. Eur. J. Enterpr. Technol. **1**(79), 19–26 (2016)
19. Mykich, K., Burov, Y.: Algebraic model for knowledge representation in situational awareness systems. In: Proceedings of the International Conference on Computer Sciences and Information Technologies, Lviv, pp. 165–167 (2016)
20. Victana. http://victana.lviv.ua
21. Lytvyn, V., Vysotska, V., Uhryn, D., Hrendus, M., Naum, O.: Analysis of statistical methods for stable combinations determination of keywords identification. East. Eur. J. Enterpr. Technol. **2/2**(92), 23–37 (2018)
22. Lytvyn, V., Pukach, P., Bobyk, I., Vysotska, V.: The method of formation of the status of personality understanding based on the content analysis. East. Eur. J. Enterpr. Technol. **5/2** (83), 4–12 (2016)
23. Lytvyn, V., Vysotska, V., Pukach, P., Brodyak, O., Ugryn, D.: Development of a method for determining the keywords in the slavic language texts based on the technology of web mining. East. Eur. J. Enterpr. Technol. **2/2**(86), 4–12 (2017)
24. Lytvyn, V., Vysotska, V., Pukach, P., Bobyk, I., Uhryn, D.: Development of a method for the recognition of author's style in the Ukrainian language texts based on linguometry, stylemetry and glottochronology. East. Eur. J. Enterpr. Technol. **4/2**(2), 10–18 (2017)
25. Lytvyn, V., Vysotska, V., Pukach, P., Vovk, M., Ugryn, D.: Method of functioning of intelligent agents, designed to solve action planning problems based on ontological approach. East. Eur. J. Enterpr. Technol. **3/2**(87), 11–17 (2017)
26. Lytvyn, V., Vysotska, V., Chyrun, L., Chyrun, L.: Distance learning method for modern youth promotion and involvement in independent scientific researches. In: Proceedings of the IEEE First International Conference on Data Stream Mining & Processing (DSMP), pp. 269–274 (2016)
27. Kravets, P.: The game method for orthonormal systems construction. In: The Experience of Designing and Application of CAD Systems in Microelectronics (2007)
28. Kravets, P., Kyrkalo, R.: Fuzzy logic controller for embedded systems. In: Proceedings of the 5th International Conference on Perspective Technologies and Methods in MEMS Design (2009)
29. Kravets, P.: Game model of dragonfly animat self-learning. In: Perspective Technologies and Methods in MEMS Design, pp. 195–201 (2016)
30. Kravets, P.: The control agent with fuzzy logic. In: Perspective Technologies and Methods in MEMS Design, pp. 40–41 (2010)

31. Basyuk, T.: The main reasons of attendance falling of internet resource. In: X-th International Conference on Computer Science and Information Technologies, Lviv, pp. 91–93 (2015)
32. Maksymiv, O., Rak, T., Peleshko, D.: Video-based flame detection using LBP-based descriptor: influences of classifiers variety on detection efficiency. Int. J. Intell. Syst. Appl. **9** (2), 42–48 (2017)
33. Peleshko, D., Rak, T., Izonin, I.: Image superresolution via divergence matrix and automatic detection of crossover. Int. J. Intell. Syst. Appl. **8**(12), 1–8 (2016)
34. Bazylyk, O., Taradaha, P., Nadobko, O., Chyrun, L., Shestakevych, T.: The results of software complex OPTAN use for modeling and optimization of standard engineering processes of printed circuit boards manufacturing. In: TCSET, pp. 107–108 (2012)
35. Bondariev, A., Kiselychnyk, M., Nadobko, O., Nedostup, L., Chyrun, L., Shestakevych, T.: The software complex development for modeling and optimizing of processes of radio-engineering equipment quality providing at the stage of manufacture. In: TCSET, p. 159 (2012)
36. Teslyuk, V., Beregovskyi, V., Denysyuk, P., Teslyuk, T., Lozynskyi, A.: Development and implementation of the technical accident prevention subsystem for the smart home system. Int. J. Intell. Syst. Appl. **10**(1), 1–8 (2018)
37. Pasichnyk, V., Shestakevych, T.: The model of data analysis of the psychophysiological survey results. In: Advances in Intelligent Systems and Computing, vol. 512, pp. 271–282 (2016)
38. Zhezhnych, P., Markiv, O.: Linguistic comparison quality evaluation of web-site content with tourism documentation objects. In: Advances in Intelligent Systems and Computing, vol. 689, pp. 656–667 (2018)
39. Vysotska, V., Rishnyak, I, Chyrun, L.: Analysis and evaluation of risks in electronic commerce. In: 9th International Conference on CAD Systems in Microelectronics, pp. 332–333 (2007)
40. Chernukha, O., Bilushchak, Y.: Mathematical modeling of random concentration field and its second moments in a semispace with erlangian distribution of layered inclusions. Task Q. **20**(3), 295–334 (2016)
41. Shakhovska, N., Vysotska, V., Chyrun, L.: Features of e-learning realization using virtual research laboratory. In: XIth International Conference on Computer Sciences and Information Technologies, Lviv, pp. 143–148 (2016)
42. Shakhovska, N., Medykovsky, M., Stakhiv, P.: Application of algorithms of classification for uncertainty reduction. Przeglad Elektrotechniczny **89**(4), 284–286 (2013)
43. Schahovs' ka, N., Syerov, Y.: Web-community ontological representation using intelligent dataspace analyzing agent. In: 10th International Conference on the Experience of Designing and Application, CADSM 2009, pp. 479–480 (2009)
44. Shakhovska, N., Vovk, O., Hasko, R., Kryvenchuk, Y.: The method of big data processing for distance educational system. In: Conference on Computer Science and Information Technologies, pp. 461–473. Springer, Cham (2017)
45. Shakhovska, N., Vovk, O., Kryvenchuk, Y.: Uncertainty reduction in Big data catalogue for information product quality evaluation. East. Eur. J. Enterp. Technol. **1**(2), 12–20 (2018)
46. Rashkevych, Y., Peleshko, D., Vynokurova, O., Izonin, I., Lotoshynska, N.: Single-frame image super-resolution based on singular square matrix operator. In: IEEE 1th Ukraine Conference on Electrical and Computer Engineering (UKRCON), pp. 944–948 (2017)
47. Tkachenko, R., Tkachenko, P., Izonin, I., Tsymbal, Y.: Learning-based image scaling using neural-like structure of geometric transformation paradigm. In: Studies in Computational Intelligence, vol. 730, pp. 537–565 (2018)

Decision Support Systems

Analysis of the Activity of Territorial Communities Using Information Technology of Big Data Based on the Entity-Characteristic Mode

Nataliya Shakhovska[1,2](✉), O. Duda[3], O. Matsiuk[3], Yuriy Bolyubash[4], and Roman Vovnyanka[4]

[1] Lviv Polytechnic National University, Lviv 79013, Ukraine
nataliya.b.shakhovska@lpnu.ua
[2] University of Economy, Bygdoszcz, Poland
[3] Ternopil Ivan Puluj National Technical University, Ruska str., 56, Ternopil, Ukraine
oleksij.duda@gmail.com, oleksandr.matsiuk@gmail.com
[4] Zolochiv College of Lviv Polytechnic National University, Zolochiv, Ukraine
bol_jura@ukr.net

Abstract. The definition of Big data is given and the main characteristics are described. There are analyzed mathematical means submission and processing of Big data and their limitations are defined. Software for working with Big data is given. The formal description of Big data is defined. There are posted patterns of associations between entities and characteristics for various categories NoSQL databases. The analysis of the objects of the model of Big data in objects of NoSQL databases is presented. For distributed storage of large amounts of data there is proposed model used in the system of Google BigTable which is characteristic for incomplete relational data model and support dynamic control over the layout data.

Keywords: Big data · Territorial community · Database
Entity-characteristics model

1 Introduction

Ukraine has deployed full-scale reform of local government. That is why, local communities and authorities have essential reorganization. New territorial entities are formed, endowed with a wide range of authorities. An innovative system of local self-government is created, which is designed to ensure the effective development of territorial communities and provide a wide range of high-quality public services to the population. In the process of functioning of newly formed territorial communities, powerful information flows are generated that require operational real-time processing in order to prepare and adopt well-considered effective management decisions.

N. Shakhovska and M. O. Medykovskyy (Eds.): CSIT 2018, AISC 871, pp. 155–170, 2019.
https://doi.org/10.1007/978-3-030-01069-0_11

2 State of Arts

The development of information technologies, methods, tools, models and meta-language, processing of various types of unconverted data types, mostly relational, have been engaged in research since the 70s of the XX century. Usually this contributed to solving a wide range of partial tasks for processing multi-type data, such as indexing, to speed up search procedures.

Modern researchers develop information technologies and methods of processing various types of data, which ensure the effective implementation of procedures for integrating data into the data warehouse. Existing methods for data integration by their functionality are divided into two main classes: integration with the use of web applications (Lahoze K., Van de Zompel H.) and data warehousing integration (Kosman D., Helevi A.). The analysis of the research results in the given field confirms the necessity in many cases of the combination of both types of integration and improvement of existing data models in connection with the formation of new requirements for existing data sources and their dynamic growth.

In spite of significant theoretical and experimental work aside researchers are still problems as the development of effective tools for processing data from multi-source and developing the principles and criteria for evaluating the quality of the integrated data in order to create effective support and management decision-making procedures.

2.1 The Primary Subject of Local Self-government

The territorial community is the primary subject of local self-government, the main bearer of its functions and authority, and is an integral territorial part of a country characterized by the continuity of the territory, the definition of boundaries on the external borders of the jurisdiction of the councils of territorial communities, which united by the availability of local governments, a certain structure of production, internal communications, the population, production and social infrastructure [1, 2].

The Law of Ukraine on the voluntary association of territorial communities of villages, towns, cities emphasizes that the formation of these structures provides for compliance with certain conditions, including important one as "in deciding on a voluntary association of local communities are taken into account historical, natural, ethnic, cultural and other factors influencing the socio-economic development of the united territorial community" [Law of Ukraine Voluntary association of territorial communities].

The formation of models of territorial communities, which predict effective procedures for processing their information resources, should provide a number of features as (Fig. 1):

- large sets of entities that characterize the population, institutions, natural resources, recreational funds, legislation, reports, etc.;
- databases that contain: information resources that play the role of the basis for data mining processes, data dictionaries that are intended for information binding of objects in difficult structured problem areas, etc.

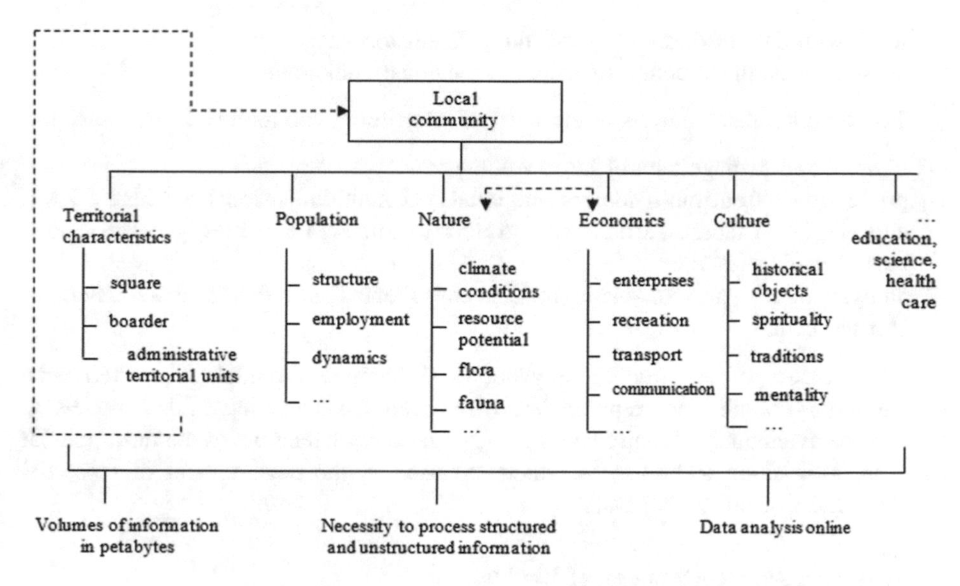

Fig. 1. Component structure of the territorial community

Analysis and modeling of territorial communities as socio-economic systems should be conducted taking into account a number of characteristic features [3]:

– the territorial community is considered as a complex, weakly structured system, the methodology of which is a systematic analysis taking into account the following features:
  - the presence of a large number of complex interconnected causal relationships among the factors under analysis;
  - need for analysis of processes of interaction of elements of a complex system;
  - analysis of the processes of events and consequences of actions, which are not always obvious at the time of the adoption of appropriate management decisions;
  - the study of stochastic processes in systems operating under uncertainty and ambiguity;
– the territorial community is a complex of dynamic system, the analysis of which involves in-depth study of the dynamics of development processes, the implementation of scientifically sound procedures for analyzing the processes of change in the complex of the relevant indicators, taking into account the features of the overall life cycle, and its components as the population, enterprises, residential fund, etc. and their adaptive evolution.

While modeling regional systems we should take into account the number of integral features that are necessary for:

– processing both detailed and aggregated data;
– obtaining data from different types of sources both at random request and according to certain regulations;

- work with data of different types and different formats;
- data analysis, the structure of which is previously unknown.

For complex data analysis at the level of a territorial community is necessary to:

- to store and manage data of large volumes;
- process data from multi-source: relational and multidimensional databases, XML and NoSQL databases, structured and poorly structured text files, geospatial databases, media files, etc.;
- analyze multi-type data using both a consolidated and federated approach for storing them.

The process of constructing a generalized (complex) model of the territorial community is complicated, in particular, by the need to use a variety of data models, as well as the availability of many levels of aggregation of data. One of the most popular modern information technologies which are used in the development of territorial management systems is Big Data.

2.2 Information Technology of Big Data

Big data information technology involves the use of sets of methods and tools for processing structured and unstructured multi-type dynamic large volumes of data for the purpose of their analysis and use in support decision-making processes. Big data is an alternative to traditional database management systems and the Business Intelligence class. The technology class includes tools for parallel processing of data, which include, in particular, tools such as NoSQL, the MapReduce, Hadoop and several other algorithms [4].

The significant characteristics for Big data are volume (in the sense of the value of physical volume), speed (in the sense of the speed of data growth, and the need for high-speed processing and obtaining results), diversity (in the sense of the possibility of one-time processing various types of structured, semi-structured and unstructured data). Due to the heterogeneity and processes of continuous growth of Big data, the latter require unconventional approaches to their storage and processing. To work effectively with such data arrays, complex solutions are needed for monitoring, filtering, structuring, and searching for hierarchical links. The use of information technology Big Data can provide monitoring of a large number of variables, and allows them to identify global trends and formulate system recommendations for the prospects of the development of complex weakly structured systems.

One of the effective technologies that is expedient to use for processing Big data of a territorial community is, in particular, information technology based on spatial data, block vectors of which contain a plurality of information products of the domain, divided into three blocks: structured data (databases and data warehouses), semi-structured data (XML, spreadsheets) and unstructured data (including text). To operate the data space and its individual elements, the corresponding systems of operations and predicates are formed, which provide: the mutual transformation of various elements of the vector into each other; combining elements of the same type; search for relevant items per keyword.

2.3 Analysis of the Problems of Processing Various Types of Information

Historically, the research of well-known and development of new means of rapid receipt of various types of data (tabular, text documents, charts, etc.) in order to formulate operative solutions based on them appeared during the 2nd World War. The results were used in the field of atomic technology, missile systems, naval and air navigation, and combat control systems.

Especially rapid development of means of operative collection of diverse types of data, loading them into data warehouses, analyzing and forecasting is observed in the branches of power engineering and administration, large-scale scientific and technological projects [5]. Scheme for information authorities involves the creation of statistical reports by industry objects predetermined shape with a turn of aggregation (Fig. 2). In this case, a system of equal sharing of data is usually formed.

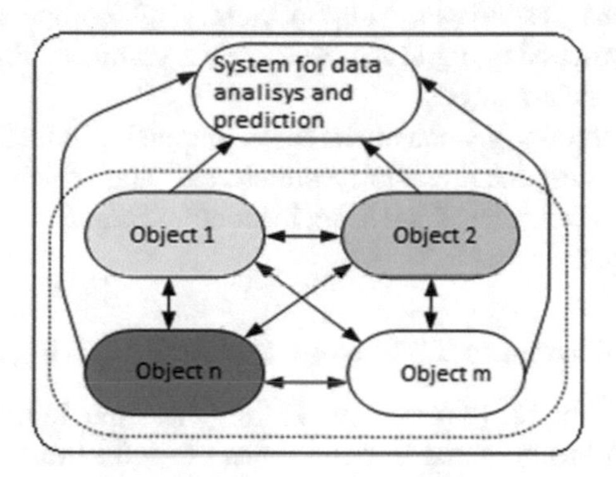

Fig. 2. Scheme of equal sharing of data

This leads to the fact that the person who decides to receive information in aggregated form, in accordance with certain criteria grouping, and get detailed information with a delay. Therefore, solutions usually do not take sufficient account of the peculiarities of the course of the analyzed processes.

Consolidation of data intended for analysis and forecasting of the process of development of territorial communities pursues the following objectives:

- increasing the efficiency of obtaining, analyzing and using the information necessary to support decision-making on management of the territorial community;
- improving the quality and efficiency of managerial decisions through the management of reliable information which is received directly from the relevant information source;
- identifying new aspects of the territorial community through the analysis of data that was not included in traditional reports and therefore usually not taken into account in decision-making;
- timely detection of negative trends with a view to their further elimination.

The rapid development of the information society results in a significant increase in the amount of information accumulated in various subject areas of social life. Such industries include, in particular, state and municipal government. The volumes of information that require operational processing grow exponentially. According to the IDC Digital Universe Study [6], conducted by EMC, the world's total volume of data in 2013 amounted to 4.4 zettabytes. The 2020 forecast for the same study suggests an increase in digital data up to 44 zettabytes, which will increase the volumes of individual databases to the petabytes of volumes. Most of the generated data are generally not analyzed, or only pass-through processing is carried out [7].

The main problems that arise when processing Big data is caused by the lack of high-tech and complex analysis methods. Existing techniques and methods are mostly narrow-minded, can be used in narrow-profile areas of data analysis that accumulate in a territorial community, in particular, it concerns numerical and geo-data, tools for analysis of poorly structured reports, etc. Variety of methods and tools which are used requires a lot of human resources to support them. High computational complexity of algorithms for data processing, rapid increase of data volumes, leads to an extension of the time required for data analysis.

These features naturally generated the need to develop effective methods for analyzing multi-type, heterogeneous, poorly structured data, which are rapidly accumulated in various subject areas of such socioeconomic entities as modern Ukrainian territorial communities.

2.4 Analysis of Information Technology Working with Big Data

Big data is at the stage of active formation, at the same time, information technology is actively used both in business and in other areas of social life. The analysis of actual researches became the basis for the allocation of the following basic aspects of this innovative information technology concept. [8]: sources of Big data, hardware and infrastructure, software for collection and storage, information technologies (methods and means of data processing), practical aspects of the use of Big data and technologies for their analysis.

Experts come in agreement with saving and processing Big data, it is expedient to use cloud computing technologies, which is an innovative, economically and technologically efficient paradigm for solving problems of Big data cluster placement and providing a wide range of services on the network. Cloud hosting of Big data clusters allows customers to store and perform the necessary computations based on huge volumes of data that are in a suitable cloud environment.

The volume of individual databases that characterize the activities of territorial communities (in particular, the databases present in social networks) are growing so rapidly that their processing using OLAP class technologies is practically impossible (Fig. 3) [9].

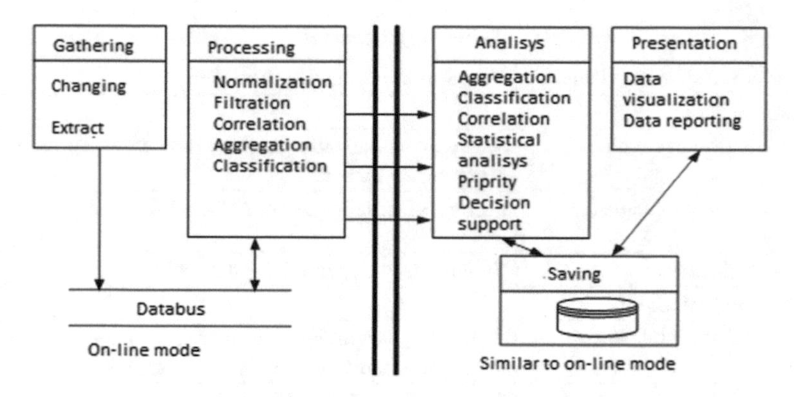

Fig. 3. Functional characteristics of OLAP information technology and Big data

In Table 1 [10] is given information about software tools for processing Big data with using the cloud computing infrastructure. Most of these tools are implemented with the orientation of the Apache license. Products are grouped according to the types of tasks that arise in the processing of Big data.

Table 1. Tools for processing Big data with open source [10]

Tool for processing Big data	Description of functionality
Data analysis tools	
Ambari http://ambari.apache.org	Apache Hadoop cluster service delivery, management, and monitoring tool
Avro http://avro.apache.org	Data serialization system
Chukwa http://incubator.apache.org/chukwa	Data collecting system for managing large distributed systems
Hive http://hive.apache.org/	Data warehouse infrastructure that provides aggregation of data
Pig http://pig.apache.org	High-level data stream language and executable framework for parallel computing
Spark http://spark.incubator.apache.org	A fast and general calculator for Hadoop data that provides a simple and distinct programming model that supports a wide range of applications, such as ETL, machine learning, video streaming
ZooKeeper http://zookeeper.apache.org/	High-performance coordination service for distributed applications
Actian http://www.actian.com/about-us/#overview	The system of storage of raw data and their preparation for further analysis
HPCC http://hpccsystems.com	System of rapid conversion and parallel processing of data

(*continued*)

Table 1. (*continued*)

Tool for processing Big data	Description of functionality
Data mining tools	
Orange http://orange.biolab.si	Means of visualization and analysis for the novice and professional experts
Mahout http://mahout.apache.org	Library of machine learning and data mining
KEEL http://keel.es	Evolutionary algorithm for data mining
Business analysis tools	
Talend http://www.talend.com	A tool for data integration, management, application integration, tools and services for large data
Jedox http://www.jedox.com/en	Analysis system, reporting, planning
Pentaho http://www.pentaho.com	System of data integration, business analysis, data visualization, forecasting
Rasdaman http://rasdaman.eecs.jacobs-university.de/	The system of storing multidimensional spatial data (array) without size limits with a specialized query language
Data search tools	
Apache Lucene http://lucene.apache.org	Full-text indexing and search system
Apache Solr http://lucene.apache.org/solr	System of full-text, spatial and facet search, dynamic clustering
Elasticsearch http://www.elasticsearch.org	Distributed full-text search engine with a web-based interface
MarkLogic http://developer.marklogic.com	NoSQL and XML database
mongoDB http://www.mongodb.org	Cross-platform document-oriented database management system and dynamic schemes
Cassandra http://cassandra.apache.org	Scalable multi-tool database
HBase http://hbase.apache.org	Scalable distributed database with storing structured data of a large volume
InfiniteGraph http://www.objectivity.com	Distributed graph database

3 Models of Big Data Presentation

Table 2 presents comparison of Big data representation.

Table 2. Comparison of models of representation of Big data

Name	Authors	Advantages	Disadvantages
Multidimensional model	Maté [11]	It is used for solving problems of data visualization and analysis	In connection with the heterogeneity of the hypercube with heterogeneous data, their volume increases, which is unacceptable for Big data
Object model	Chang [12], Papakonstantinou [13]	For a particular modification, it can be used to represent Big data	Unresolved is the task of displaying certain types of data in an object model
Graph model	Feng [14]	Convenient when analyzing links between objects	Computational complexity of search algorithms provided a large number of objects

3.1 Formal Description of the Structure of Big Data

One of the clearest examples of using the concept of Big data is the process of creating and analyzing an array of data, which provides information on the functioning of the territorial community, which reaches several petabytes.

This is an urgent need for processing structured (database departments of health, tax inspection, etc.) and unstructured information (statistical reporting (data on population and migration status and dynamics of the labor market, data from the fields of education, health, information regarding income and living conditions, level of social protection, etc.)).

The processing of data which characterize the territorial communities usually involves data analysis in real time.

Consequently, the informational model of the territorial community implies:

- the availability of a large set of entities that are: persons (physical, legal), organizations, events, natural resources (rivers, forests, lakes) and minerals, recreation fund (historical monuments, sanatoria), legislative acts and financial and economic reports etc.
- large in scope of the database of features: documents requiring intellectual processing, ontological systems, data dictionaries that allow the formation and maintenance of information links between object descriptions.

Based on this information, the need for technological tracking of existing and associative connections between entities is naturally formed.

We will present the informational model of the territorial community using the information technology of Big data in the following form: $BigD = \langle e, f, a \rangle$, where entities $e \in E$, characteristics $f \in F$, association $a \rightarrow n_{e,f}$ between entities e and characteristic f. Formally we can divide all objects into the following categories: entities e, characteristic f, and association between entities e and characteristic f.

Let also be defined: the set of entities E, set of characteristics F, for each e and f the relevant association number is given as $n_{e,f}$. The total number of entities is defined as $|E|$, the total number of characteristics is the capacity of the set F: $|F|$. Also describe:

- for each f set $e(f) = \{e \in E : n_{e,f} > 0\}$ all associated f entities;
- for each e set $f(e) = \{f \in F : n_{e,f} > 0\}$ all associated from e characteristics.

We will supply qualitative signs in quantitative form.

To do this, we will use an analogue of the description of the TF-IDF's measure of text-specific documents.

In similar situations, when we have several features related to the characteristic, we use the quantitative representation of information, that is, the number of binary questions (yes or no) to be answered in order to find the desired object. In general, if it is known that some object belongs to a set containing elements, then we can divide this set in half and, by asking binary questions, find out to which half the object belongs to. So, then the number of elements among which the search will be made $\frac{N}{2}$. We will continue the specified procedure: ask the second question, for which we divide the selected half in half. So, after two questions in the subset, we will have a search $\frac{N}{4}$ objects, among which there is the desired object. After three questions we will have $\frac{N}{8}$. In general, after responding to q binary questions we have a set consisting of $N \cdot 2^{-q}$ elements that necessarily contain the required object [15, 16].

When the corresponding result set consists of one element, we confidently and correctly identify the alternative we need. The number of binary questions that should be asked to find the characteristics for N alternatives: $N \cdot 2^{-q} = 1$, or $q = \log_2(N)$.

Similarly, you can describe processes over entities. We have $|E|$ entities with the amount of information $\log_2(|E|)$. When we know that a certain entity is associated with a characteristic (we have $|e(f)|$ entities associated with the characteristic f), then the number of questions you need is equal $\log_2(|e(f)|)$. Thus, the fact that the essence is related to the characteristic f, allows you to reduce the number of questions to

$$k = \log_2(|E|) - \log_2(|e(f)|) = \log_2\left(\frac{|E|}{|e(f)|}\right) \quad (1)$$

In addition, the effect of several associations can be described by counting the number of additional binary questions that need to be asked in order to form correct knowledge about the association of a certain characteristic with the corresponding entity - denote by $n_{e,f}$. The answer to each binary question reduces the number of objects in the target search set by half; accordingly issues reduce this amount to $n_{e,f} \cdot 2^{-q}$. We will continue to analyze the association until the appropriate number of objects is made ≥ 1. The largest quantity q, for which the association is determined to support as well $n_{e,f} \cdot 2^{-q} = 1$, consequently $q = \log_2(n_{e,f})$. Adding another additional request is defined as $1 + \log_2(n_{e,f})$ and means the absolute availability of the association.

The general importance of the characteristics f for entity e defines as $\log_2\left(\frac{|E|}{|e(f)|}\right)$ with a factor of importance $1 + \log_2(n_{e,f})$. The resulting number of questions is defined as

$$I(e,f) = \left(1 + \log_2(n_{e,f})\right) \cdot \log_2\left(\frac{|E|}{|e(f)|}\right) \tag{2}$$

This formula is one of the options for representing the frequency of terms - the so-called inverse frequency of a document tf-idf [17, 18]. For each entity we have a number of questions $I(e,f)$ for different characteristics f. The value of importance is necessary to normalize:

$$V(e,f) = \frac{\left(1 + \log_2(n_{e,f})\right)\right) \cdot \log_2\left(\frac{|E|}{|e(f)|}\right)}{\sqrt{\sum_{j \in f(e)} \left(\left(1 + \log_2(n_{e,j})\right) \cdot \log_2\left(\frac{|E|}{|e(j)|}\right)\right)^2}} \tag{3}$$

For each e available weight $V(e,f)$. Thus, as a measure of the proximity of two objects E_1 i E_2, we can take the distance between the corresponding vectors $(V(e_1,f), V(e_2,f), \ldots)$. So, for every weight $V(e,f)$, which represents the number of "yes" or "no" answers, we will have

$$d(e_1, e_2) = \sum_{f \in F} |V(e_1,f) - V(e_2,f)|. \tag{4}$$

This distance depends on the number of characteristics: for example, if, in addition to the documents, we store their copies, the distance increases accordingly twice. To avoid this dependence, distance $d(e_1, e_2)$, as a rule, normalizes in the interval [0,1] by dividing by as much as possible the value of this distance [3].

3.2 Models of Associations Between Entities and Characteristics in Different Categories of NoSQL Databases

The data carrier in the key-value model (another name - column database) is described by tuples of the form:

$$KV = \{<k, v>\} \tag{4}$$

where k – key, which takes unique values in each pair, v – value corresponding to this key.

Keys can be presented as set (major or minor), values maintain virtually unlimited semantics, $e \leftrightarrow k; f \leftrightarrow v$.

The model's signature is presented as:

$$O = \langle \pi, \sigma \rangle \quad (5)$$

where π – attribute projection operation (key or value), σ - operation of selection of attributes (choice of value by key, keys by value, keys by the value of ancestors, etc.). The listed operations belong to the category of operations "reading" [19, 20]. One of the brightest examples of a column-type DBMS is the Cassandra software product.

For distributed storage of large amounts of data proposed model used in the system of Google BigTable which is characteristic for incomplete relational data model and support dynamic control over the layout data. The main elements of the Bigtable data model are quite simple: rows, columns, and timestamps.

$$BigTable = \{<r, c, t>\} \quad (6)$$

In the search database, the names of the strings can be Internet document addresses, and the column names - the features of these documents (for example, the contents of the document can be stored in the column "content:", and links to child pages - in the "anchor:"). Another example is Google maps, which consist of billions of images, each of which details a particular geographic area of the planet. In Bigtable, Google maps are structured as follows: each line has one geographic segment, and the columns are the images from which this segment is composed, the different columns store images submitted with varying degrees of detail:

$$f \leftrightarrow r(t), e \leftrightarrow c \quad (7)$$

If several columns contain data of the same type, these columns, according to the Bigtable model, form a family: $colF = \{c_i, c_j | dom(c_i) \in T \wedge dom(c_j) \in T\}$. The convenience of using column families is, in particular, the compression of homogeneous data, which in turn allows you to significantly reduce the amount of data stored in the repository. In the specified model family of columns there are units of access to data.

The content of the Internet pages is constantly changing. In order to take account of such changes, each copy of the data stored in a particular column is assigned a time stamp. Bigtable's temporary label is a 64-bit number, which can, in particular, be encoded by the time and date that are required by the parameters of a wide range of client programs. For example, timestamp for copies of a web page in the content: column is the date and time when the specified copies were created. Using Bigtable temporary labels in applications can only search for the most recent copies of data.

One of the main advantages of this approach is that the database generated under this rule can be easily decomposed and distributed to a given set of servers. Alphabetized bands are divided into ranges called "Tablets" (tablets) - non-independent tables. Since the rows in each tablet are sorted by the key name, then the client applications easily find the desired tablet, and in it the desired tape.

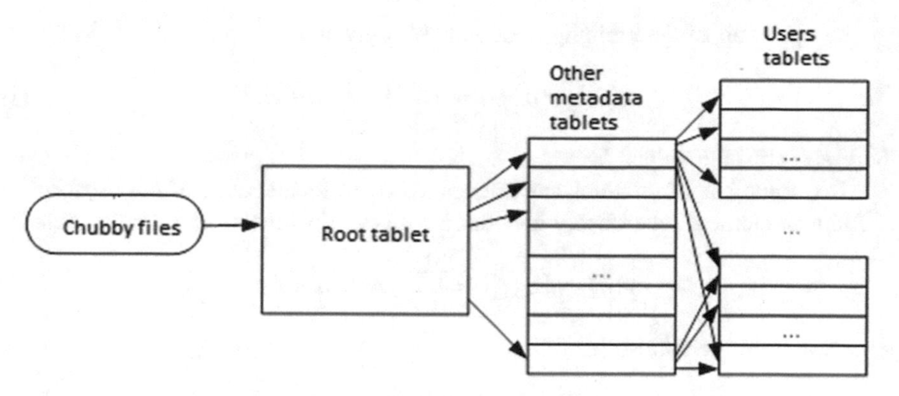

Fig. 4. The process of finding custom tablets

In this model, the key identifies a tape that contains data stored in one or more column families. Within these families, each tape can have many column values. The values in each column contain a timestamp, so some of the matching values between the row and the column may be within the same column family.

Bigtable is a large distributed system for submitting and processing synchronized objects and for which the distributed lock service, called Google, is called Chubby, and its role in Bigtable is to commensurate with the role of transactions in traditional DBMSs [21].

For each tablet-server, Chubby creates a special chubby file, which makes it easy for Bigtable Master to know which servers are working. Another chubby file contains a link to the Root-Table Placements that contains the location of all other tablets. This file contains information about which server is managing the tablets.

Of course, the use of Chubby's service in Bigtable solves in a certain way the problem of maintaining data incompatibility in a distributed environment with a large amount of replica. It is believed that Bigtable is perhaps the first attempt to achieve a realistic balance between system performance, its scalability and non-consistency of data. The result is the support of the so-called weak consistency, which, in principle, satisfies the requirements of most employees working with Bigtable services.

Figure 4 shows how to search for a custom tablet [22].

The carrier of the "object-document" model is represented by tuples of the form:

$$OD = \{ <f_0, <f_1 : e_1, f_2 : e_2, f_n : e_n, f_{n+1} : d_1, f_2 : d_2, f_{n+l} : d_l > \} \tag{8}$$

where f_0 – document ID, $f_1..f_m$ – characteristics (attributes) of the document, $e_1..e_m$ – atomic values of characteristics $f_1..f_m$, $d_1..d_l$ – links to other documents, $d_i = e(f_i)$.

Operation manipulations in this model are object-oriented, in particular, the operation of determining the nodes of an element is provided

$$v(f_i) = \{C\} \cup \{f_{0_i} | i = \overline{1,n}\} \cup \{e(f_i) | i = \overline{0, n+l}\} \tag{9}$$

where C is collection of documents f_{0_i}.

The operation of determining node values is given as

$$v(f_i) = \{n_{e_j,f_i} | i = \overline{1,n}, j = \overline{0, m+l}\} \tag{10}$$

where e_j – attribute values f_i.

Also, a number of relations are determined over the elements of the carrier.

Element-element relationship is defined between documents and a collection:

$$v(f_i) = \{n_{e_j,f_i} | i = \overline{1,n}, j = \overline{0, m+l}\}; \tag{11}$$

element-attribute relationship:

$$f_i \times OD \rightarrow EA; \tag{12}$$

link-element relationship:

$$f_i \times d_j \rightarrow ER; \tag{13}$$

the data-element relation is defined as:

$$f_i \times e_j \rightarrow ED. \tag{14}$$

Examples of this type of DBMS are MongoDB and CouchDB.s.

The graph data model is presented in the form:

$$O = \langle ID, A, z, r \rangle \tag{15}$$

where ID-set of id, graph nodes; A – a set of marked oriented arcs (p,l,c), $p, c \in ID$, l – "line-label", record $(p,1,c)$ means that there is a connection l between the nodes p and c; z – a function that represents each node $n \in ID$ in the concrete meaning of a compound or atomic type, $z : n \rightarrow v$; V– special root node of the graph.

The structure of the XML-document consists of nested elements tags is well known. Difference between above graph model is mainly in interpretation tags and labels to boxes labels which are used to designate relationships between elements of these schemes, and the labels are not needed to mark an item, and in XML document-oriented model it is necessary that each (non-text) data item has an identifying sign. An XML document is translated into a tree structure, which is a partial case of a graph model.

In the graph model for semi structured XML data is necessary to use specialized types of attributes, such as *ID*, *IDREF, IDREFS.* These types allow you to organize cross-reference storage in XML-type elements < *eid,vahie* > (<element identifier, value>) and attributes of the form <*label, eid* > (<label, value>).

There are several species of *RDF(*Resource Description Framework) which can be submitted in the form of a graph model: *RDF/XML, N3, Turtle, RDF/JSON.*

Description of information resources in the form *RDF* – are "Subject" - "predicate" - "object", that is, for the set U (Universal R*esource Identifier, URI*) – elements f of set *{Black nodes,* empty nodes) of set *L {Literal, RDF*), $B \in e, L \in e$, is determined by the set (*f*, *e*(*f*), *e*), where *f*– «subject»; *e*(*f*)– «predicate»; *e*– «object».

RDF graph the data model is then given in the following way: let $t = (f, e(f), e)$ is *RDF* data element, where $(f, e(f), e) \in (UB) \times U \times (UBL)$, moreover t call main, if it does not contain nodes that do not have identifiers. RDF graph G is a set of $T \supseteq t$ [23].

Consequently, the information technology of Big data combines various data models, which generates the urgent need to develop effective methods of their transformations with minimal loss in data.

4 Conclusion

The need to develop methods, models and tools for processing multi-valued data for information technologies of Big data has become relevant in the context of the creation of high-tech tools, information systems oriented to meet the needs of territorial communities and regions. It is systematically substantiated the necessity of creating effective high-tech tools for integrating various types of information resources using information technologies of Big data in territorial management systems both at the level of individual territorial communities and regions as a whole.

In order to conduct complex data analysis at the level of the territorial community, it is necessary to store and manage data of large volumes; to process data from multi-source: relational and multidimensional databases, XML and NoSQL databases, structured and poorly structured text files, geospatial databases, media files, etc.; to analyze various types of data, using both a consolidated and federated approach for their storage. To ensure these conditions, the authors form an original model of Big data, which contains constructive tools for their processing and determines the set of restrictions that are inherent in these processes.

The analysis of the objects of the model of Big data in objects of NoSQL databases is analyzed. For distributed storage of large amounts of data proposed model used in the system of Google BigTable which is characteristic for incomplete relational data model and support dynamic control over the layout data. One of the main advantages of this approach is that the database generated under this rule can be easily decomposed and distributed to a given set of servers.

It is anticipated that further research will be conducted to develop tools that will allow the use of concepts of relational databases in the parallel processing of Big data.

References

1. About Voluntary Association of Territorial Communities. Bulletin of the Verkhovna Rada 13 (2015)
2. Chistov, S.M., Nikiforov, A.Y., Kutsenko, T.F.: State regulation of the economy. KNEU, Kyiv (2000)
3. Kut, V., Kunanets, N., Pasichnik, V., Tomashevskyi, V.: The procedures for the selection of knowledge representation methods in the “Virtual University” distance learning system. In: Hu, Z., Petoukhov, S., Dychka, I., He, M. (eds.) ICCSEEA 2018. AISC, vol. 754, pp. 713–723. Springer, Cham (2018)
4. Bolyubash, Y.Y.: Methods and tools for processing Big data in territorial administration systems. Sci. Bull. NLTU Ukraine Collect. Sci. Tech. Work **26**(4), 341–354 (2016)

5. Zacharias, G.L., MacMillan, J., Van Heme, S.B.: National Research Council. Behavioral Modeling and Simulation: From Organizations to Societies, From Individuals to Societies. The National Academies Press, Washington (2008)
6. Digital universe of opportunities. http://www.emc.com/leadership/digital-universe/index.htm?pid=landing-digitaluniverse-131212
7. Beyer, M.A., Laney, D.: The importance of "Big Data": A definition. https://www.gartner.com/doc/2057415/importance-big-data-definition
8. Shakhovska, N.: The method of Big data processing. In: XIIth International Scientific and Technical Conference Computer Sciences and Information Technologies (CSIT), pp. 122–125, Lviv (2017)
9. Hilbert, M.: Big Data for development: From information- to knowledge societies. http://papers.ssrn.com/abstract=2205145
10. Fang, L., Sarma, A.D., Yu, C., Bohannon, P.: Rex: explaining relationships between entity pairs. Proc. VLDB Endowment **5**(3), 241–252 (2011)
11. Maté, A.: A novel multidimensional approach to integrate Big Data in business intelligence. J. Database Manag. (JDM) **26**(2), 14–31 (2015)
12. Chang, F., Dean, J., Ghemawat, S., Hsieh, W.C., Wallach, D.A., Burrows, M., Chandra, T., Fikes, A., Gruber, R.E.: Bigtable: a distributed storage system for structured data. http://static.googleusercontent.com/media/research.google.com/ir/archive/bigital-osdi06.pdf
13. Bomba, A., Kunanets, N., Nazaruk, M., Pasichnyk, V., Veretennikova, N.: Information technologies of modeling processes for preparation of professionals in smart cities. In: Advances in Intelligent Systems and Computing, pp. 702–712 (2018)
14. Zhou, F., Hsu, W., Lee, M.L.: Efficient pattern discovery for semistructured data. In: Proceedings of the 17th IEEE International Conference on Tools with Artificial Intelligence (ICTAI 2005), pp. 301–309, Hon Kong (2005)
15. Nekrasov, V.: Basic concepts and approaches in creation of contextual search engines based on relational databases. http://www.citforum.ru/database/articles/search_sys.shtml
16. Novitsky, A.V., Reznichenko, V.A., Proskudin, G.Y.: An example of constructing scientific archives using Eprints. In: RCDL 2006, pp. 154–161, Cudzal, Russia (2006)
17. Veres, O., Shakhovska, N.: Elements of the formal model big date. In: XI International Conference Perspective Technologies and Methods in MEMS Design (MEMSTECH), pp. 81–83 (2015)
18. Shakhovska, N., Syerov, Y.: Web-community ontological representation using intelligent dataspace analyzing agent. In: X-th International Conference the Experience of Designing and Application of CAD Systems in Microeletronics (CADSM-2009), pp. 479–480 (2009)
19. SDMX initiative: New approaches to the exchange of statistical data. http://citforum.univ.kiev.ua/internet/xml/sdmx/
20. Inmon, B.: Productivity of data warehouse systems. Performance in the data warehouse environment **4**, 41–48 (2000)
21. Pedrycz, W., Chen, S.-M.: Information granularity, Big Data, and computational intelligence. https://books.google.com/books?id
22. Di Ciaccio, A., Coli, M., Angulo Ibanez, J.M.: Advanced Statistical Methods for Analysis of Big Data. Springer, Berlin (2012)
23. Shakhovska, N., Veres, O., Bolubash, Y., Bychkovska-Lipinska, L.: Data space architecture for Big Data managering. In: Xth International Scientific and Technical Conference Computer Sciences and Information Technologies (CSIT), pp. 184–187 (2015)

Model of Innovative Development of Production Systems Based on the Methodology of Optimal Aggregation

Taisa Borovska[1(✉)], Inna Vernigora[1], Victor Severilov[1], Irina Kolesnik[1], and Tetiana Shestakevych[2]

[1] Vinnitsa National Technical University, Vinnytsia, Ukraine
taisaborovska@gmail.com, shulganinna@gmail.com, severilovvictor0@gmail.com, iraskolesnyk@gmail.com

[2] Lviv Polytechnic National University, Lviv, Ukraine
tetiana.v.shestakevych@lpnu.ua

Abstract. The mathematical model of the integrated system "innovation, development, production" has been developed, software implemented and studied. The development is based on methodologies of applied system analysis and optimal aggregation of production systems. Analogues are analyzed: a model of a three-level ecological system, a model of the production system "production, development, innovation" on the basis of a ternary operator of optimal aggregation. The new development uses the binary operator of optimal aggregation "production, development", the essence of which is the optimal distribution of the "resource quantum" between the subsystems "production" and "development", the "resource quantum" is allocated for a finite time interval. Optimal control is obtained as a function of the duration of the interval and the magnitude of the resource quantum.

Keywords: Optimal aggregation · Methodology · Production function
Task variation · Development strategy · A binary operator · Production system
The integration of the subsystems

1 Introduction

Today, sustainable development and positions of the enterprise on the market are complicated by the acceleration of changes in all segments of socio-technical-ecological systems. The manufacturer of a certain product must constantly and taking into account the dynamics of the market and its enterprise, optimally allocate resources between production, development of production and innovations. However, the objects themselves – "production", "development", "innovations" do not have clear definitions and boundaries for the financier, the technologist, the manager and the mathematician – the "model designer". However, the creation of satisfactory mathematical models does not obligatory solve the problem of optimal control of the system "production, development, innovation" (PDI). We have a dynamical system with parametric connections between elements, that does not satisfy the limitations of known methods.

N. Shakhovska and M. O. Medykovskyy (Eds.): CSIT 2018, AISC 871, pp. 171–181, 2019.
https://doi.org/10.1007/978-3-030-01069-0_12

To solve the problem of optimal allocation of resources between the subsystems of the PDI system, we use the methodology of optimal aggregation. The object of optimal control is specified. Production: material production is considered as more simple for formalization. We leave the production of intellectual products and services for a generalized model. The main characteristic is the production function (PF). Development – all changes in production, increasing efficiency and production capacity: mastering, modernization of the main production, replacement of means and models of production. The development function (DF) is more complex model of the "costs, output" class in comparison with the production function (PF). Innovation: between "development" and innovation there is no clear boundary. Classical examples of innovations: the transition from sailboats to steamships, from radio tubes to semiconductors, from glass containers to plastic bottles and back. Innovations affect much, up to the development of civilizations. In this paper, we consider the local aspect of the impact of innovation – production. Modeling is not closed by a pair: (object, model), modeling of production systems is a triad: object, model creation purpose, model. Such an approach guarantees the obtaining of new and useful theoretical and practical results.

To solve the problems of modeling and optimization, the optimal aggregation methodology was chosen [1, 2, 9]. The obvious advantages of this methodology in the field of solving multidimensional optimization problems: – the absence of constraints such as linearity, convexity, continuity for the functions "costs, output"; – linear dependence of computational costs on the dimension of the problem. Qualitative advantages of the methodology: - the result of optimal aggregation is not a point, but a function of the class "costs, output"; – when solving the global optimization problem, the optimization tasks of all subsystems are solved. New in this article: the methodology of modeling, the class of the problem, the method of solving the problem. Analogues and prototypes in these areas: methodology of optimal aggregation [1, 9], analogue of the mathematical model of aggregation as middleware, a similar methodology – "open management" [3, 10], models based on generating mechanisms in Forrester [4], functions "costs, output", the interbranch balance of Leontiev [5]. Analogues appeared of fundamental problems of modeling and management of modern production systems. In [6, 7] the model is considered as a tool for innovative concepts search. In [8] the modern mass high-tech production is considered in the aspects of modeling, optimization, decomposition of production processes and environmentally friendly waste recycling based on Toyota's car manufacturing experience. Disadvantages of the analogues studied: limitations on the type of production functions, inefficient multidimensional optimization, inadequate production and development models. Figure 1 shows two effective examples of modeling three-level systems with parametric resource links [2, 9], on the basis of which it is possible to develop a generalized model of development.

Figure 1a presents a model of the classical model of the ecosystem with the addition of the subsystem "grass". The model takes into account the resource feedbacks FB1, FB2, FB3 and the inter-level functions of the influence of the IF21, IF31, IF32. In the modeling example, the graphics of the dynamics of subsystems "grass" and "wolves" are combined. The possibilities of using the model for the study of socio-technical-ecological systems are obvious. Figure 1b presents a continuous model of the

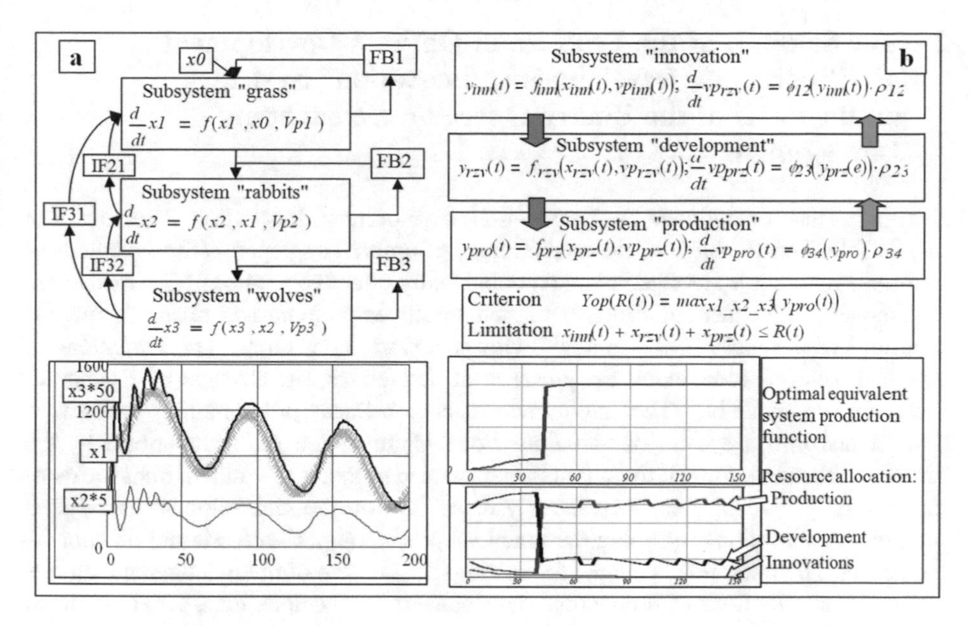

Fig. 1. Models-analogues of a two-level resource system with parametric constraints

system "innovation, development, production". In the upper part of Fig. 1b presents a continuous model of the dynamics of such a system. The model takes into account direct and reverse inter-level resource links. In the lower part of Fig. 1b – an example of modeling: at high development costs, the state is achieved "to close existing production and switch to new technologies and products of production". The drawback of the model is two-dimensional optimization (see "criterion" in Fig. 1b). Based on the analysis of analogs and prototypes, the goals and objectives of the research are formulated. The aim of the research is to establish and solve the problem of optimal aggregation of the system "production, development, innovation" on the basis of the binary operator of optimal aggregation "production, development".

Objectives of the study:

- development of mathematical model of optimization of the system "production, development, innovation";
- analysis of the binary operator of optimal aggregation "production, development";
- solving the problem of optimal aggregation of the system "production, development, innovation".

2 The Solution of the Problem of Optimal Development of "Production, Development, Innovation" System on the Basis of the Binary Operator "Production, Development"

Setting the task of optimization "production, development, innovation" structure. The task is solved on the basis of the methodology of optimal aggregation. The methods of optimal aggregation are algebraic, computationally effective methods. The mathematical model is a sequential structure of non strictly monotonic and positive bounded functions of the class "costs, output" with parametric constraints. The limitations of classical methods (convexity, the presence of derivatives, the absence of discontinuities) are not available. The multidimensional non-linear programming problem is transformed into a system of one-dimensional optimization problems solved by the direct search method, with the possibility of vectorization. The computational costs of the method grow not more than linearly depending on the dimension of the applied problem. Assumptions regarding the functions "costs, output" and criterial functions – are absent. The main points (steps) for solving the problem of optimal aggregation are: obtaining the functions of production, development, innovations for a certain segment of production (for example: agricultural sector and waste processing, automotive industry, microelectronics, power engineering) using applied system analysis; obtaining the functions of resource links; building the resource structure of the production system; isomorphic mapping of the resource structure of the production system into a binary tree of optimal aggregation, which is an algebraic formula for solving the optimization problem. The form of the solution of the optimization problem is the optimal equivalent function of the class "costs, output" of the aggregate production system. Control variables – the allocation of resources between the subsystems of the production system. All these points are given in the formulas and graphs of this article.

The effectiveness of optimal aggregation methods is due to the fact that the multidimensional optimization problem is isomorphically mapped to a system of one-dimensional optimization problems. It is the search and formation of such a mapping that is the goal of the study. Figure 2 presents the scheme for solving the optimization problem based on the optimal aggregation methodology: obtaining an object model as a resource structure from generalized technological converters (Fig. 2a); mapping this structure into a binary tree of optimal aggregation (Fig. 2b); the result of aggregation is an extended solution of the problem of non-linear programming: optimal equivalent production function of the system (OEPF): $Fs(Xs, Mp)$. As a result, we obtain an aggregated model of the optimal system (Fig. 2c) – a convenient tool for research on "virtual reality". Such research is impossible or unacceptable to perform on a real system.

In the scheme (Fig. 2a), the connections of the subsystems "innovation" and "development", "development" and "production" are highlighted. The essence of these links: the costs of research and development in the "innovation" subsystem lead to an increase in the efficiency of the "development" subsystem, the costs of modification and replacement of production capacities in the subsystem "development" lead to an increase in the efficiency of the subsystem "production". Formally: for all the

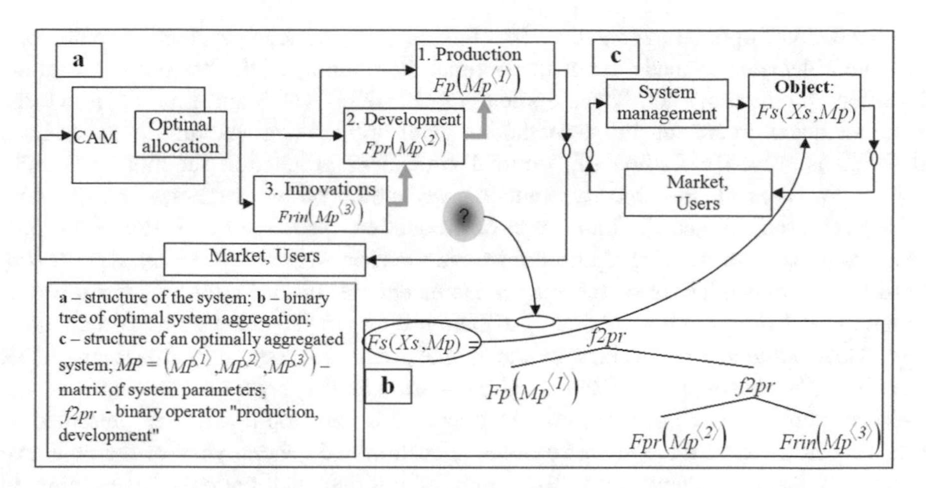

Fig. 2. Setting the task of optimizing "production, development, innovation" structure

subsystems, the functions "costs, output" are defined (Fig. 2a), and the "output" of the subsystem of the lower level changes the parameters of the upper-level subsystem. Thus, parametric links take place in the subsystem "production, development, innovation". A new solution to the problem (Fig. 2b) assumes a two-fold use of the operator *f2pr*. The problem of the optimal distribution of the resources of the production system between "production" and "development" was solved in [1] for the case of one-level optimal aggregation "production, development". Consider the features of a binary operator *f2pr* at the visual level (Fig. 3).

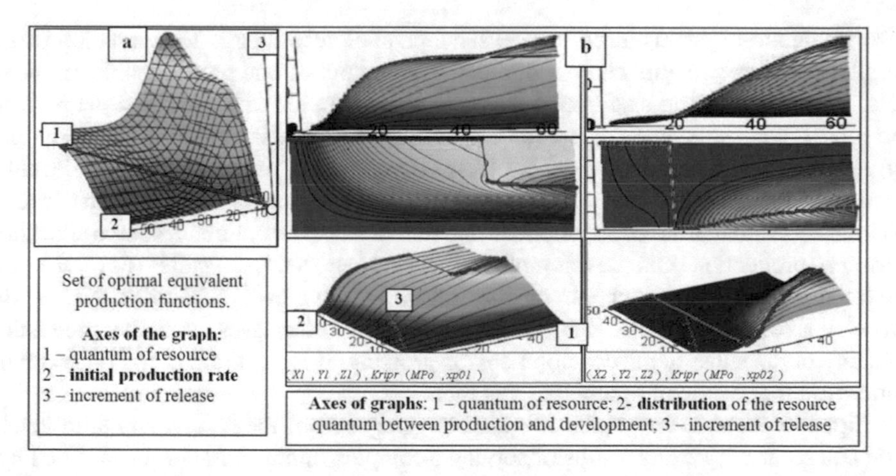

Fig. 3. Analysis of the operator of optimal aggregation of the structures "production, development"

Scenario of optimal aggregation: in a functioning production system, at some point of time a decision is made about the optimal distribution of the "resource quantum" between "production" and "development", $(\Delta X, \Delta T)$ – value and time of use of the resource quantum are set. In the methodology of optimal aggregation, the generalized problem is solved – finding of "optimal equivalent production function" – OEPF. Figure 3a represents the objective function, depending on two variables: the quantum of the resource ΔX and the initial rate of production *xp01*. Figure 3b shows the projections for the optimal aggregation objective function *Kripr(MPo, xp01)*, *Kripr(MPo, xp02)* – the dependences of the output increment on the magnitude of the resource quantum and the proportion of its distribution between the subsystems "production" and "development". The values of the initial production rate *xp01*, *xp02* are fixed. Graphs – the hodographs of the maxima – are superimposed on the graphs of the objective functions. Projections of 3D graphs (Fig. 3b, top-down) are presented: – OEPF, – function of the optimal resource allocation and general view of the objective functions. Data sets with such discontinuous optimal distributions: "everything in production, proportionally between production and development" and "everything in production, everything in development, is proportional to production and development" are selected. Recall: the difference between the methods of optimal aggregation from analogs is the equivalent replacement of the multidimensional non-linear programming problem by a system of one-dimensional optimization problems that can be executed in parallel. For optimization, a direct search method is used that allows vectorization and removes all search problems in a multidimensional phase space.

3 Analysis of the Problem of Optimal Aggregation of a Parallel Structure

Real production systems at each level have parallel subsystems. Interpretation of the parallel resource structure can be different – the output of one product, different products. Let us consider an example of the optimal aggregation of a parallel structure from the subsystems "production, development". Figure 4 presents: – the formula for optimal aggregation of the production system (PS); – production functions (PF) and development functions (DF) of the subsystems; – results of aggregation of the first level – optimal equivalent production functions and functions of the optimal resource allocation between production and development. Figure 4 shows four pairs (PF, DF), the parameters of which are chosen in the following way: from subsystem 1 to 4, the production capacity of the subsystem "production" decreases, and the production capacity of the subsystem "development" is growing. It is not difficult to notice certain patterns in the results of optimal aggregation.

Figure 5 presents the results of optimal aggregation of the second and third levels. You can see the logic and results of solving a complex optimization problem. The logic of the operations in Fig. 5 is simple. As a result of aggregation of subsystems (1, 2) and (3, 4), two objects with some optimal functions of the class "costs, output". These functions are complemented by functions of the optimal resource allocation between the parts of these subsystems. The result of the optimal aggregation of these functions into the optimal equivalent function is presented below *Fopps(PR1, PR2)*.

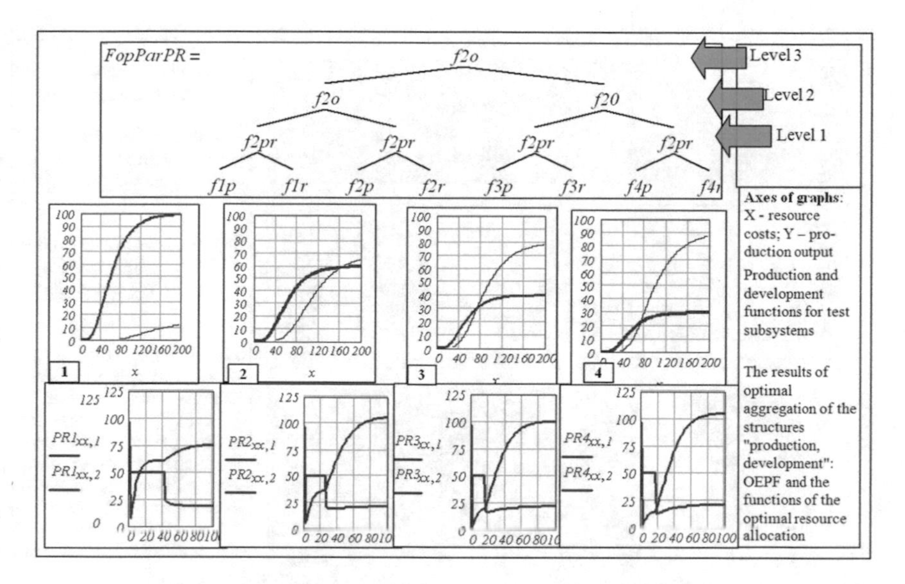

Fig. 4. Optimal aggregation of the system from the subsystems “production, development”. Level 1

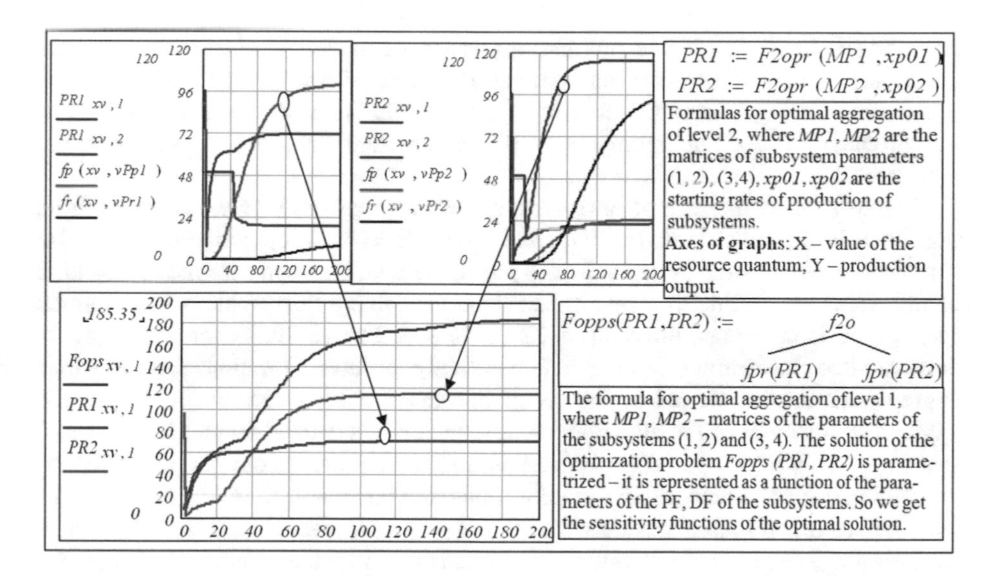

Fig. 5. Optimal aggregation of the second and third level for the test system

Figure 6 shows the graphs of optimal distribution of the resource for the test task.

Let us compare the graphs in Figs. 3, 4 and 5. In general, it is an algebra of optimal aggregation, similar to the usual algebra of numbers. In this regard, the methods of optimal aggregation are simple for the practical application of the obtained solutions of optimization problems, but are difficult to understand: – optimization of resource

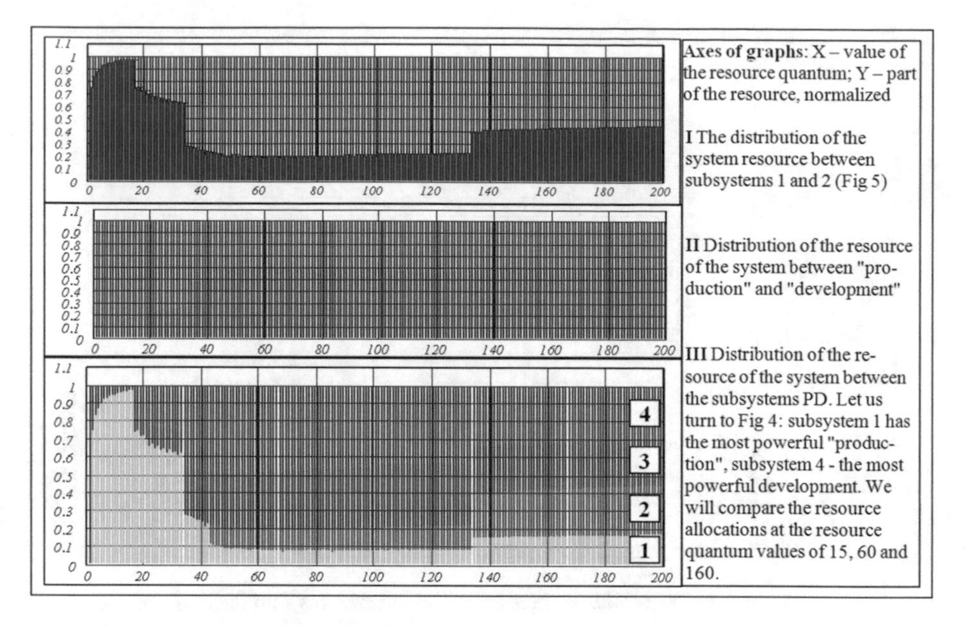

Fig. 6. Optimal resource allocations for the test system

allocation is "built-in" in the optimal aggregation operator, – operands of optimal aggregation – matrix "constructions" similar to records of databases. The nature of the distributions in Fig. 6 does not contradict the empirical rules of business. Solution of the problem of optimal aggregation of the structure "production, development, innovation".

Figure 7 shows the scheme of optimal system aggregation. Figure 7a is taken from Fig. 2a it is an initial structure with parametric links that represents an enterprise similar to environmental systems. In this paper, the goal is to optimize the aggregation of this structure. Figure 7b shows the scheme for solving this problem. The optimal aggregation is performed three times. The difference between this aggregation procedure and that considered in Sect. 3 is that not only resource allocations from previous aggregations are transferred to the result operand, but also new parameters of the aggregated subsystems. Initially, optimal aggregation was defined for parallel and serial structures [1], where the result operands did not contain data about the parameters of the "costs, output" functions. At present, optimal aggregation methods for structures with parametric resource links are developed and investigated [2], for which it is necessary to include in the structure of operands not only the "memory" of previous aggregations. The solution obtained is an important part of the adaptive CAM. Figure 7c is a diagram of the main functional modules of the CAM. The use of optimal aggregation methods substantially changes even the statements of standard tasks of the CAM. The continuation of this work is the development of a management module, functioning model and development of the production system, an environment model – consumers, competitors, suppliers. The scheme, shown in Fig. 7b is program-realized. As it was already noted, close analogues of the theory and practical application have not been found. Therefore, identifying errors and debugging of the interfaces is a small part of the research on the model of functioning and development (Fig. 7c). A comprehensive testing of the system "production, development, innovation" started. The system model has 12 parameters, characterizing the subsystems and the system as a whole. In addition, it is necessary to obtain statistics of "virtual reality" when considering the stochasticity of the functions of innovation, development and production.

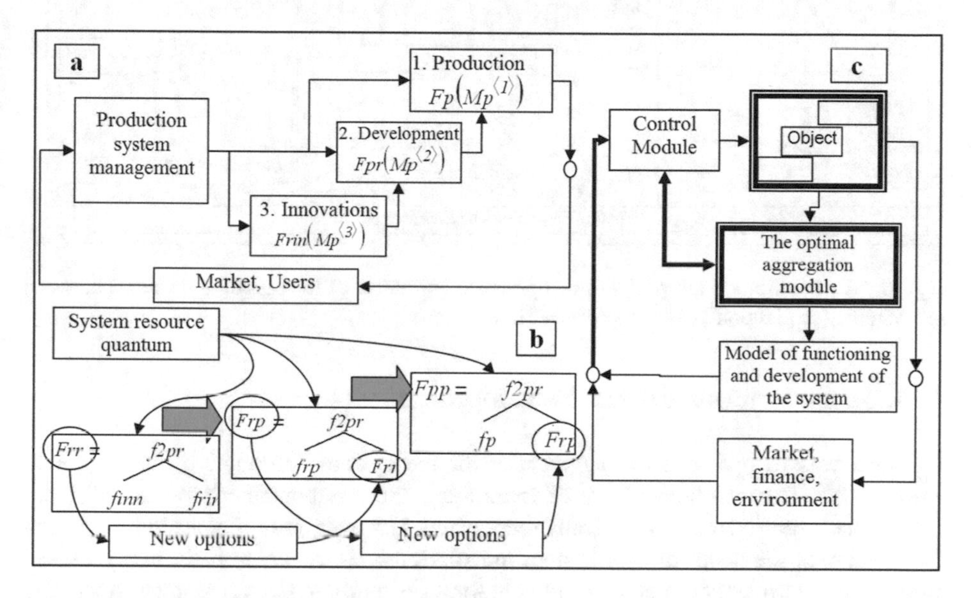

Fig. 7. Scheme for solving the problem of the system "production, development, innovation" "optimization"

Figure 8 shows the example of testing the module for optimal resource allocation in the system "production, development, innovation" for three values of the vector of starting rates of subsystem costs. The logic of variation of test data is the following: the dominance of production, equally, the dominance of innovation. It should be remembered that there are significant differences in the parameters and dynamics of changes in various areas of production: in the garment industry, automotive industry, residential construction and the aircraft industry. It should be remembered that we solved half of the practical problem – we obtained the optimal static dependencies "costs, output". When implementing the optimal control, it is necessary to take into account the dynamics of the subsystems. This is the subject of another article.

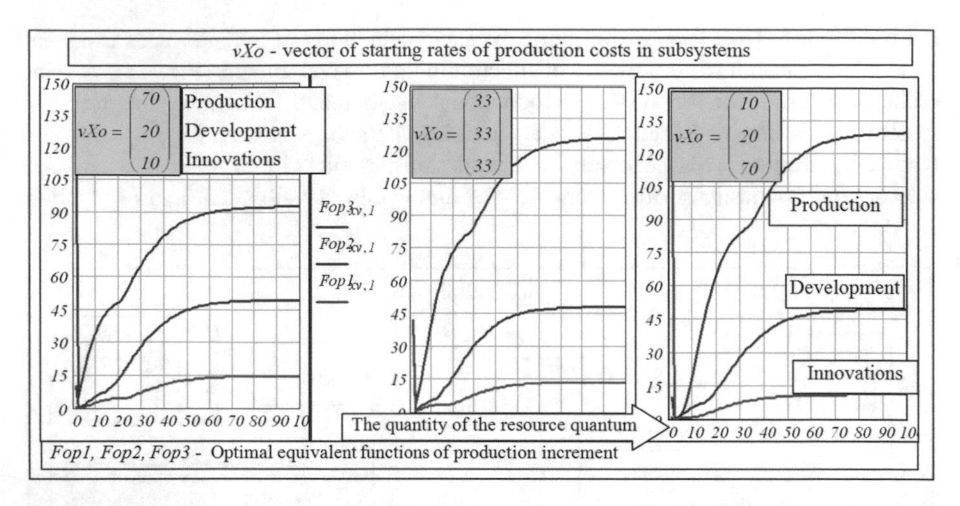

Fig. 8. The solution of the problem of optimal distribution of the resource quantum between subsystems "production", "development", "innovations"

4 Conclusion and Future Developments

The task was to optimize the aggregation of the system "production, development, innovation". The practical utility of integrating the components "production", "development", "innovation" is generally recognized, however, monotonous links between the functional elements of this system are obvious. However, holistic mathematical models and optimization methods, suitable for embedding in CAM are not known. The reasons for the small successes in this area of research are the difficulties in creating models of subsystems and the connections between them – significant nonlinearities, nonstationary, parametric connections between subsystems. Analogues of similar development are analyzed – a multi-level ecological system and two-dimensional optimization of the system "production, development, innovation". Unlike the analogues in this work, to achieve the goal, the optimal aggregation methodology based on the binary operator of optimal aggregation "production, development" was chosen.

The tasks of the research were solved:

- mathematical model of optimization of the system "production, development, innovation" was developed;
- analysis of the binary operator of optimal aggregation "production, development" is performed;
- the solution of the problem of optimal aggregation of the system "production, development, innovation" was obtained on the basis of the binary operator of optimal aggregation "production, development".

Three-dimensional non-linear programming problem is reduced to three one-dimensional optimization problems. Due to the parameterization of the solution, it is possible to search for critical modes of technologies change.

Prospects for further research – expanding the boundaries of the system: the construction of three-level models of regional systems, taking into account the demographic and environmental components. While at this level, constructive mathematical models are unknown.

Studies, based on the results of this article, allowed stating and solving a number of new tasks (appropriate papers are being published):

- the analysis and optimization of the efficiency and survivability of the enterprise, based on the methodology of optimal aggregation (the obtained integrated optimal function depends on two variables: production costs and cost of failures);
- the optimal credit strategies for modern production systems that operate in an active environment of competitors and consumers.

In the latest study, authors used the enterprise model as an integral system of "production, development, innovation", and proposed an approach to understand credit as a functional subsystem of the enterprise.

References

1. Borovska, T.N., Kolesnik, I.C., Severilov, V.A.: Optimal Aggregation Method in Optimization Problems. Universum Vinnitsa, Vinnitsa (2009)
2. Borovska, T.N.: Optimal aggregation of production systems with parametric connections. East. Eur. J. Enterprise Technol. **4**(11), 9–19 (2014)
3. Opoitsev, V.I.: Equilibrium and Stability in Models of Collective Behavior. World, Moscow (1977)
4. Forrester, J.W.: Basics of Cybernetics Enterprises (Industrial Dynamics). Progress, Moscow (1971)
5. Leontiev, V.: Theoretical assumptions and nonobservable facts. USA Econ. Ideol. Polit. **9**, 15 (1971)
6. Rüttimann, B.: Introduction to Modern Manufacturing Theory. Springer (2018)
7. Rüttimann, B., Stockli, M.: Going beyond triviality: The Toyota production system—lean manufacturing beyond Muda and Kaizen. J. Serv. Sci. Manag. **9**, 140–149 (2016)
8. Nersessian, N.J., Chandrasekharan, S.: Hybrid analogies in conceptual innovation in science. Cogn. Syst. Res. **10**, 178–188 (2009)
9. Borovska, T.N.: Mathematical models of functioning and development of production systems based on the optimal aggregation methodology. VNTU, Vinnitsa (2018)
10. Bellman, R.E., Kalaba, R.E.: Dynamic Programming and Modern Control Theory. NAYKA Publishing house, Moscow (1969)

The Information Model of Cloud Data Warehouses

Nataliya Shakhovska(✉), Nataliya Boyko, and Petro Pukach

Lviv Polytechnic National University, Lviv 79013, Ukraine
{nataliya.b.shakhovska,nataliya.i.boyko}@lpnu.ua

Abstract. In the article, the concept of cloud technologies, the benefits of the cloud approach are considered. There are provided advantages of cloud approach of data processing. Standards for data processing centers that are related to the objects with high requirements for reliability and resiliency of all the segments of the network and engineering infrastructure are reviewed. Levels concepts of data processing centers are analyzed. The paper has developed a model of cloud storage and data transfer methods. The model of hybrid protocols of data transmission through the cloud storage was introduced, the model of network traffic was substantiated, data warehouse load simulation was carried out.

Keywords: Cloud data warehouse · Information system
Information technology · Cloud technologies · Cloud computing
Cloud services · Remote processing · Data processing center

1 Introduction

Cloud technology is a specific approach that provides remote processing and data storing. This approach solves many problems that arise in the local data processing and data storing, where each customer has its local copy of the data set and software. However, there are also some disadvantages.

Cloud storage (cloud data warehouse) facilities can dramatically increase the availability of client data and the necessary network elements for their reliable transmission and storage. Today, the methods of local processing and storage have an extremely low level of consolidation of computing resources and memory (less than 18%). The geographical distribution of client applications, their mobility, while simultaneously maintaining the integrity of the data, gives rise to contradictions, which is the need to increase the bandwidth capacity of the existing telecommunication component of distributed data warehouses in a context of increased requirements for their availability, as well as regarding unauthorized access and protection against damage.

Along with this, there are a number of underdeveloped scientific tasks that prevent the efficient organization of cloud storage facilities and, accordingly, distributed computing systems on their basis, namely:

N. Shakhovska and M. O. Medykovskyy (Eds.): CSIT 2018, AISC 871, pp. 182–191, 2019.
https://doi.org/10.1007/978-3-030-01069-0_13

- Insufficiently developed theoretical basis, which would replace the classical theory of mass service in the design of modern telecommunication systems for the distribution of information with self-similar traffic;
- The issues of determining the quality indicators of the operation of transmission and distribution systems in distributed heterogeneous network environments are not sufficiently worked out;
- Insufficiently developed methods and algorithms that provide quality of service, in particular, throughput in heterogeneity of network platforms.

2 State of Arts

2.1 Advantages of Cloud Technologies

The main advantages of cloud-based data processing approach:

1. There is no need for custom equipment in a large computing power. The so-called cloud approach is able to reduce significantly the cost of the solution to a particular business problem, only in those cloud service, which has quite cheap terminal connected to the network. Another reason for the reduction is no need to purchase a license for each client - as computing machine is a cloud service so that the license is required only for it. However, these licenses are usually more expensive than conventional (single-user) and the profitability of such decision is achieved by a large number of users.
2. Flexible expansion of computing capacity - since in recent years the industry of cloud technologies has made great breakthrough towards virtualization and clustering, any cloud system can be expanding without any impact on it. We can add one more computing core or additional gigabyte of memory but this does not affect the operating system in the cloud, except – the increase of the computing capacity.
3. Placement of computer equipment in specialized infrastructure facilities – data processing center (DPC). The cloud user does not need to care about the physical equipment, this task is assumed by the company that manages the data center and the user pays only for giving him the computing resources. Modern data centers have been built in a way that failure of any unit will be immediately detected and faulty unit in the shortest period of time will be replaced by a reserve. A possibility of replacement this unit allows to replace the damaged one without stopping cloud service. But there is one nuance - namely, sometimes the company wants to place computer equipment in their buildings and that affects this advantage.
4. The big advantage is also a constant remote access to such resources. Data processing center provides at least two inclusions in trunk lines. That means if for some reason the route that links the user to the cloud service through one of the main channels is not be available, reserve communication route will be selected through another main channel, in most cases without users noticing [1, 2].

2.2 The Model of Cloud Data Warehouse

Based on the data warehouse model, it is advisable to use the data warehouse model proposed by Roberts [3] for cloud computing:

$$S = \langle F, D, G, C, L \rangle, \tag{1}$$

where $F = \{f_1, f_2, \ldots, f_n\}$ is the set of data elements, $f = \{p_1, p_2, \ldots, p_m\}$ – is the set of data packages, $D = \{d_1, d_2, \ldots, d_k\}$ the set of storage devices, $G : F \to D$ is the deployments od storage devices, $C : D \to Z$ is the capacity of storage devices, $L : D \to Z$ is the storage capacity of the devices.

For the needs of modeling cloud data storage, it is better to use its scalable version introduced by Petrov [4]:

$$S_m = \langle F, D(t), G, C, L \rangle. \tag{2}$$

Then the cloud data warehouse model is presented as:

$$S_{cloud} = \left\langle D, D_{free}, S_{ms} \right\rangle, \tag{3}$$

where $D_{free} \subseteq D$ subset of free data storages, $S_{ms} = \{S_{m1}, S_{m2}, \ldots, S_{ml}\}$ is the set of scalable data warehouses.

Scalable devices in this case are devices from a plurality of common devices that do not include a subset of free devices, $D_i(t) = D \backslash D_{free}$. Scalable devices do not have shared storage devices: $\forall t, i, j, i \neq j \Rightarrow D_i(t) \cap D_j(t) = \varnothing$.

Proposed models haven't operation under cloud data warehouses and don't allow data analysis.

3 The Improved Cloud Data Warehouse Model

The model of the cloud storage is improved as an algebraic system:

$$C_{dw} = \langle S_{cloud\_m}; Y; L \rangle,$$

$$S_{cloud\_m} = \left\langle D, D_{free}, S_{ms}, PR \right\rangle, \tag{4}$$

$$Y = \left\{ I_{cc,} I_{mpp}, I_{mpd} \right\},$$

where I_{cc} is the gateway selection method based on the complexity of the request, I_{mpp} is the method of multiprotocol transmission of stream data, I_{mpd} is the method of multiprotocol transmission of stream data, *PR* is the data transfer protocol, *L* load predicate S_{cloud}. Data element $f_i \in St \cup SemSt \cup UnSt$ can be structural, semistructural, unstructural.

The predicate of the load capacity of the cloud storage is given as the ratio of the load capacity of the cloud storage at the time t_1 and t_2. For its definition, data traffic in the clouds is studied, aggregated data flows are analyzed, the dependence of the level of fractality of the total current is established:

$$L(S_{cloud\_m_{t1}}, S_{cloud\_m_{t2}}) \rightarrow Z. \tag{5}$$

The parameters of cloud data warehouse S_{cloud_m} are:

- input and output traffic,
- count of working processors,
- load and unload processors,
- average load on the processor,
- amount of cache memory.

Let us present such a model of a cloud storage in the form of a graph model. Let the cloud data storage include N data servers $D = \{d_1, \ldots, d_N\}$ that are interconnected by communication channels. The peculiarity of organization of reliable data warehouse assumes that all servers of the repository have direct connection with any other. Thus, such a storage server network is a full-fledged graph $G_D = (D, L_D)$ representing arcs L_D of the graph, which represent the channels of communication between the data servers.

In addition to data servers, cloud storage also includes K satellites $St = \{St_1, \ldots, St_K\}$ that are also interconnected by communication channels. And, as in the case of servers, all satellites have direct connections to each satellite (L_{St}). In addition, all satellites are linked by separate links (L_Z) to data servers, forming a complete model of the data storage network, which also represents a full-fledged graph $G = (P, L)$ in which $P = St \cup D$, $L = L_D \cup L_{St} \cup L_Z$.

For a user, a cloud storage repository is represented as a satellite set $G_{St} = G \cap G_D$, through which it can access data warehouse.

This model of representation of a cloud storage is not limited to its type - private, public or hybrid. It allows you to abstract from the type of repository, but concentrate on the organization of access to it.

Using this model, you can evaluate the effective exchange time between users and data warehouse.

So to get data from a cloud storage, no matter which cloud server this data is in the form of a file f it takes time to spend

$$T_{down}(f) = \frac{F}{V_{down\_k}(f)}, \tag{6}$$

where k is user of data, T_{down} is loading time, F is size of data file f.

The speed of getting file f for user $V_{down\_k}(f)$ is given as:

$$V_{down\_k}(f) = \min(V_{down\_k}, V_{down\_St_j}(f)), \tag{7}$$

where V_{down_k} is speed of getting data to client, $V_{down\_St_j}(f)$ is file f transfer speed from data warehouse to satellite St_j.

And the speed of transferring a file f from the repository to satellite St_j is determined by the formula, given that the file can be placed on several servers and read in parallel with them

$$V_{down\_St_j}(f) = \min\{V_{upl\_St_j}, V_{down\_St_j}, \sum_{i=1}^{K} V_{down\_d_i}(St_j, f)\}, \tag{8}$$

where S_{t_j} is the satellite j, d_j is storage i, $V_{upl\_St_i}$ is transferring data speed from satellite, $V_{down\_St_i}$ is speed of data transferring to satellite.

Therefore, the total formula for retrieving a file from the repository will be determined:

$$T_{down}(f) = \frac{F}{\min\{V_{down\_k}, V_{upl\_St_j}, V_{down\_St_j}, \sum_{i=1}^{K} V_{down\_d_i}(St_j, f)\}}. \tag{9}$$

The estimation of the time of data loading from the client in the cloud storage can be carried out similarly, taking into account the fact that downloading is required not on several storage servers, but only on one:

$$T_{upl}(f) = \frac{F}{V_{upl\_k}(f)}. \tag{10}$$

The download speed of the k-th client:

$$V_{upl\_k}(f) = \min\{V_{upl\_k}, V_{upl\_St_j}(f)\}, \tag{11}$$

$$V_{upl\_St_j}(f) = \min\{V_{down\_St_j}, V_{upl\_St_j}, \max(V_{down\_d_i}(St_i, f))\}.$$

So, the boot time in the repository is determined by:

$$T_{upl}(f) = \frac{F}{\min\{V_{upl\_k}, V_{down\_St_j}, V_{upl\_St_j}, \max\{V_{down\_d_i}(St_j, f)\}\}}. \tag{12}$$

Consequently, the proposed model of representation of the organization of a cloud data warehouse has allowed to estimate the time of downloading and reading files in the data warehouse.

4 Modeling the Load of Cloud Data Warehouse

Research of the main characteristics of the cloud server, such as allocation of RAM, central processor download, the state of the processes of the operating system in the background of intense network traffic, is an urgent task.

The model should identify the predicate of the cloud storage capacity, which can only be done using a real data warehouse. To do this, you need to: investigate the data traffic in cloud storage, analyze the combined data stream, establish the dependence of the level of fractality of the total flow. The relevant parameters of the data store S_{cloud} for simulation are determined from the practical values of input/output traffic, the number of running processes, the load capacity and the simple processors, the average load on the processor and the amount of cache memory:

$$L(CH(S_{cloud_i}), CH(S_{cloud_k})) \rightarrow Z. \tag{13}$$

The predicate of the load capacity of the cloud storage is given as the ratio of the load factor of the cloud data warehouse at times t_1 and t_2. Loaded cloud data warehouse is function of the values of the parameters of the cloud data warehouse:

$$CH(S_{cloud}) = \frac{IT \cdot OT \cdot LCPU \cdot MC}{V \cdot PN \cdot (FCPU + LCPU) \cdot ALCPU}, \tag{14}$$

where IT is incoming traffic, OT is outbound traffic, V is channel, PN is number of running processes, $LCPU$ is load capacity, $FCPU$ is simple processors, $ALCPU$ is average load on the processor and MC is the amount of cache memory.

One of the most pressing problems of probabilistic-time characteristics of cloud data warehouses is the adequate consideration of network traffic features. It is advisable to consider different models of network traffic and analyze the most promising model for cloud data warehouses, which takes into account the self-similar properties of traffic as a time series.

On the example of a real cloud data warehouse, a dynamic model of input and output traffic characteristics, as well as a distribution of hardware capabilities of the cloud server, is built. Self-similarity has been established for all processes, which confirms the possibility of using fractal models to work with cloud storage, in particular, to predict the behavior of cloud storage servers.

Problems of the self-similarity of network traffic involved many scientists. In particular, in [1] the study of the properties of real traffic in networks with packet switching. Using the R/S analysis method, the self-similar nature of network traffic in the information networks is shown. Based on this approach, a traffic generator model was developed that implements multifractal behavior of data flows in real information systems, which allows to simulate traffic with given self-sufficiency indicators. In [4], it is shown that in the 802.16b networks, self-similar traffic properties are manifested both on the channel and on transport levels. Obtained values of the main indicators of the degree of fractal network traffic and proposed methods of aggregation of source statistics [5].

Work [6] is devoted to the experiment to remove network traffic of one of the major Intranet providers, as well as the results of analysis of the structural features of this traffic. The authors prove that self-similar properties manifest themselves both at the channel and transport levels. In [7], American scientists studied processes with long-term dependence. To generate such processes, authors suggest the use of a fractal model of an integrated variable mean.

Self-similarity is a property of an object whose parts are similar to the whole object as a whole. Many objects in nature have such properties, such as the coast, clouds, the circulatory system of man or animals.

Informally, the self-similar (fractal) process can be defined as an accidental process, the statistical characteristics of which exhibit the property of scaling. The eigenstone process does not significantly change the form when viewed on different scales on a time scale. In particular, unlike processes that do not have fractal properties, there is no rapid "smoothing" of the process when averaging over a time scale - the process retains a tendency to bursts.

If we consider the process of data transfer through a cloud data warehouse $\{X_k; k = 0, 1, 2, \ldots\}$ by a stationary random process and taking into account stationary and assumptions about the existence and finality of the first two moments, we can use:

- $m = E[X_t]$ is the mean value;
- $\sigma^2 = E[X_t - m^2]$ is dispersion;
- $R(k) = E[(X_{t+k} - m)(X_t - m)]_{\infty}$ is the correlation function
- $r(k) = R(k)/R(0) = R(k)/\sigma^2$ is the correlation coefficient.

The server used a cloud storage repository for the study. The physical server is divided into several virtual zones using the Solaris operating system, each of which is used to perform a number of tasks. Most of the traffic is transmitted over HTTP/HTTPS, FTP/FTPS and SFTP.

Zabbix application [8] is used for remote monitoring of real-time data store parameters. Zabbix is a client-server application that is used to collect, store and process information about network status, network load, as well as the status of a real-time data warehouse server operating system. For further processing the following data storage parameters were used:

- incoming/outgoing traffic;
- number of running processes;
- load and simple processors;
- average load on the processor;
- Cache size.

The data obtained was consolidated within a week, so we can assume that they represent the real picture of the use of a cloud storage. The time dependence of traffic volume is shown in Fig. 1 for input (a) and output (b).

Since the data were analyzed during the week, the graphs clearly show a certain periodicity. The graphical representation of the coefficient of auto-correlation allows visually to make sure that the traffic under study has a long-term relationship.

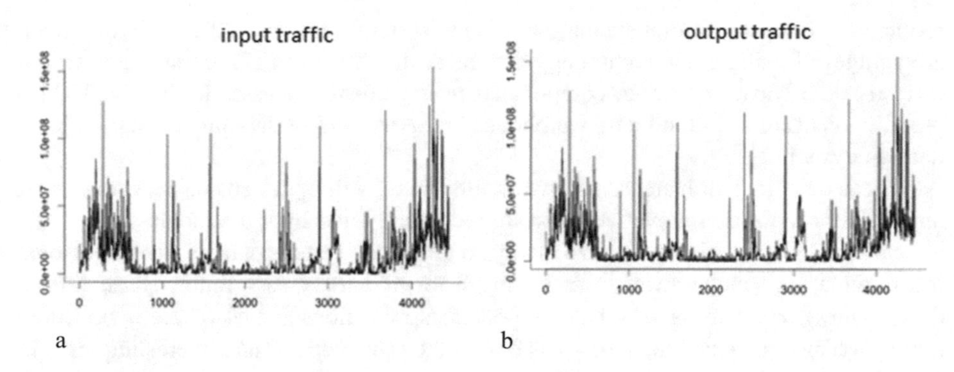

Fig. 1. Time dependencies of the incoming (a) and outgoing (b) traffic

5 Selection of the Suitable Technology

The above analysis allows to assume that an intranet concept is a solution that represents some of the immensely important aspects that can facilitate the knowledge-based information system in the information processing company.

Intranet supports the alignment, retrieval and exchange of information within a company, therefore the knowledge base behind it is invisible and inaccessible to unauthorized users. An intranet is based on the same technologies and services as the Internet and represents the Internet at the enterprise level.

Intranet is considered here as a tool for information management and not only as a collector of documents. That is, the focus or idea behind the use of the Intranet technology should be foremost motivated by the goals set and are not by the documents. The reason for this is obvious: people use information units to accomplish specific tasks.

Information technology companies strive to find a solution for their information system while their business processes are handled efficiently and profitably. Modern knowledge-based information systems should therefore be able to provide the right information at the right time in the right place available.

The technological possibilities of the Hyperwave Information Server (version 5.5 SP2) allow to establish the core architecture of the information system. Therefore, the intranet environment will be implemented on the Hyperwave Information Server [9].

6 Perspectives of Cloud Information Systems

The cloud environment is often used in IT: on global Internet and at enterprises or by target users in particular. The huge potential of this area allows not only to use the special purpose information products, but, also to expand the boundaries of the administration and user data service, systematization with own funds in the context of the cloud technology [10].

Using cloud technology will allow to use much more efficient the technical and economic potential of the country, enterprises and individuals, by significantly reducing the cost of implementation of IT system services. After all, the main advantages of using cloud technology are the scalability and agility, that allow to use infrastructure capacity without complicated management processes in the cloud. Also cloud computing and cloud data warehouse are used for Big data processing [11] and data analysis [12].

These benefits will increase user work efficiency, will optimize business processes, and will allow to use innovative software technology and hybrid structures.

So cloud - is the ability to have always a guaranteed and secure access to personal information, and move away from having a lot of unnecessary things (flash drives, disks, wires) or buying a new PC/components/applications/games. There is no doubt that nowadays, the cloud technology is one of the most popular and interesting topic in the IT-sector and there are more and more interesting solutions appearing in the world, that are associated with cloud.

Of course, for user it is difficult to fully appreciate (and reveal) the potential but, without a doubt, the future of the cloud technology is perspective, because such giants (Microsoft, Apple and Google) have visited the area and obviously not going to leave. A few years ago the concept of "cloud" seemed as a nice idea and a bold experiment, and today benefits are appreciated by people who are not actually connected with the development of software, web technology and other highly specialized things (Xbox Live, Windows Live, On Live, Google Docs - a vivid example).

7 Conclusion

The analysis of problems of implementation and functioning of cloud storage warehouses is carried out. Basis of analysis technological and functional problems the models of cloud data warehouses are described. The attention is focused on the unresolved problems of the construction of distributed data transmission systems.

The model of the cloud data warehouse is improved by presenting it as an algebraic system. It differs from the existing ones of the architecture of the related telecommunication network system of data processing methods on the basis of protocol means of the session level, which made it possible to more accurately and fully identify and use its bandwidth in accordance.

The cloud storage data repository for real storage has been simulated, which enabled the cloud storage data warehouse to be loaded as a self-similar process with long-term dependency. Based on existing approaches and their drawbacks, a model of cloud storage and data transmission methods has been developed. The model of hybrid protocols of data transmission through a cloud storage was introduced, the network traffic model was substantiated, data warehouse load simulation was carried out. The model of a cloud data warehouse as a system of algebraic equations is improved. The simulation of the cloud data warehouse has been carried out, which allowed to identify processes of self-similarity with the transmission of data through it.

References

1. Huo, Y., Wang, H., Hu, L., Yang, H.: A cloud storage architecture model for data-intensive applications. In: International Conference on Computer and Management (CAMAN), pp. 1–4 (2011)
2. Trenkic, V., Christopoulos, C., Benson, T.M.: Efficient computational algorithms for TLM. In: 1st International Workshop TLM, pp. 77–80. University of Victoria, Canada (1995)
3. Roberts, J.W.: Internet traffic, QoS, and pricing. Proc. IEEE **92**, 1389–1399 (2004)
4. Petrov, V.V., Platov, V.V.: Investigation of the seamless teletraffic structure of a wireless network. Electr. Inf. Syst. Syst. **3**, 38–49 (2004)
5. Belkov, D.V.: Investigation of network traffic. Sci. Pap. Donetsk Natl. Tech. Univ. **153**, 212–215 (2009)
6. Leland, W.E., Taqqu, M.S., Willinger, W., Wilson, D.V.: On the self-similar nature of Ethernet traffic (extended version). IEEE/ACM Trans. Netw. **1**, 1–15 (1994)
7. Harmantzis, F., Hatzinakos, D.: Heavy network traffic modeling and simulation using stable FARIMA processes. IEEE Trans. Sig. Process. **5**, 48–50 (2000)
8. Vacche, A.D., Lee, S.K.: Mastering Zabbix (2013)
9. Firestone, J.M.: Enterprise Information Portals and Knowledge Management, p. 230. A Volume in KMCI Press
10. Strubytskyi, R.P., Shakhovska, N.B.: Analysis of approaches to modeling of cloud data warehouses. Actual Probl. Econ. **149**(11), 263 (2013)
11. Shakhovska, N.: The method of Big Data processing for distance educational system. In: Conference on Computer Science and Information Technologies, pp. 461–473. Springer, Cham (2017)
12. Boyko, N.: A look trough methods of intellectual data analysis and their applying in informational systems. In: XIth International Scientific and Technical Conference Computer Sciences and Information Technologies (CSIT), pp. 183–185 (2016)

Information Systems for Processes Maintenance in Socio-communication and Resource Networks of the Smart Cities

Danylo Tabachyshyn[1,3], Nataliia Kunanets[1(✉)], Mykolay Karpinski[2], Oleksiy Duda[3], and Oleksandr Matsiuk[3]

[1] Information Systems and Networks Department, Lviv Polytechnic National University, St. Bandera street, 12, Lviv, Ukraine
tabachyshyn.danylo@gmail.com, nek.lviv@gmail.com

[2] Computer Science, University of Bielsko-Biala, Willowa Street, 2, Bielsko-Biala, Poland
mpkarpinski@gmail.com

[3] Computer Science, Ternopil Ivan Puluj National Technical University, Ruska street, 56, Ternopil, Ukraine
oleksij.duda@gmail.com, oleksandr.matsiuk@gmail.com

Abstract. Practical implementation of the Smart city concept is based on the heavy use of so-called "soft" domains characterizing the formation of comfortable urban socio-communicative environment and "hard" domains providing the operation of diverse community services, provision of appropriate services by means of modern information technologies. The analysis of numerous publications provides a strong evidence that at present there are no systematic information-technological developments concerning processes support in social communication and resource networks of the Smart cities.

The authors developed a number of information systems based particularly on consolidated socio-communicative resources and database containing information concerning accounting and cost analysis in resource networks of the Smart City. The context diagrams of the consolidated socio-communicative resource and the system of accounting and cost analysis in resource networks of the Smart City are constructed. Function sets of information systems possibilities are formed and corresponding diagrams of UseCases are designed. The database structures of the consolidated socio-communicative resources system and the information system of accounting and cost analysis in resource networks of the Smart City are formed. Relative data base sizing is carried out. The current state of practical implementation of software-algorithmic tools is presented. The obtained results are analyzed, the relevant conclusions are made and the promising trends of further researches are defined.

Keywords: Smart City · Databases · Information system · Application Resouce network · Socio-communication

N. Shakhovska and M. O. Medykovskyy (Eds.): CSIT 2018, AISC 871, pp. 192–205, 2019.
https://doi.org/10.1007/978-3-030-01069-0_14

1 Introduction

The concept of the Smart City is based on the formation of an effective innovative "social" capital with the large-scale use of modern information and communication technologies [1] aimed to form and support the municipal infrastructure necessary for stable economic, environmental and social development [2] and to improve life quality [3]. The concept of the Smart City includes two types of entities: so-called "soft" domains characterizing the processes of forming the comfortable urban socio-communicative environment [4, 5] and "hard" domains, which, on the basis of modern information technology platforms, form ecological safe environment [6], provide effective functioning of city services, the provision of quality services and services on the basis of municipal resource networks [7].

"Soft" domains of the Smart Cities are based particularly on the information resources of museums, archives, libraries, state institutions and organizations, territorial and regional mass media, information catalogues and portals, social networks, as those that form information bases of socio-communicative environment of the Smart City. "Hard" domains are based on information resources of municipal services and organizations providing relevant services using diverse sensors integrated into relevant urban engineering infrastructures, particularly resource networks for supplying water, gas, heat and electricity. Increase in the level of urbanization and steady growth in the proportion of urban population will undoubtedly update new researches and practical implementation of their results in socio-communicative and information-technological components of the Smart Cities. Information resources of the corresponding "soft" and "hard" domains require systematic management and integrated approach in order to improve and optimize processes for their efficient use [8].

2 State of Problem Investigation

Numerous scientific publications deal with the investigation of the problem of developing effective databases as informational models of various complexes and systems of the Smart City. Particularly in [9] integrated environmental information system for environmental monitoring developed by means of PHP and DBMS MySQL is presented. The papers [10, 11] describe databases of urban buildings integrated with geoinformation 3D platforms. Papers [12–14] deal with solution of transport problems of the Smart Cities using databases, cloud technologies and IoT-devices. In [15] the system of energy manager is considered using IoT-devices and NoSQL BD. Sta [16] analyzes the conceptual bases for processing various types of information sets received from different sources in the Smart Cities environment. Investigations of Souza et al. [17] deals with the problems of data integration in the Smart City. Zambon et al. [18] analyzed more than twenty software-algorithms developments for the "Smart cities" and identified the platforms using informational models based on traditional relational SQL databases, NoSQL BD and BigData technologies. Papers [19, 20] deal with monitoring systems organizations using BD for the smart health care field. The organization of the "Smart city" BD is described in paper [21]. The problems of preserving the cultural heritage in the Smart Cities with using of the BD are described in [22].

Pereira, João André Rosa highlighted the importance of database creation for museums [23], and [24] provided procedures for creating smart applications to create database for visual images of museum exhibits.

Having analyzed a wide range of systems and complexes on the basis of database technologies in the Smart Cities, we come to the conclusion that at present there is no sufficient systematic developments concerning complex formation of information resources, and, accordingly, systems in the socio-communicative environments of the Ssmart Cities which could systematically unite museums, archives, libraries, authorities, local media and regional information portals.

In a number of papers [25–31] considered the information systems of accounting and monitoring of one or several types of resources are considered at the same time insufficiently presenting the implementation of integrated accounting and cost analysis in resource networks of the Smart Cities designed specifically for water, heat, gas and electricity supply.

The objective of the work is to develop information systems to support the processes occurring in the consolidated socio-communication and resource networks of the Smart Cities.

3 Systems of Consolidated Information Socio-communicative Resources and Accounting and Cost Analysis in Resource Networks of the Smart City

3.1 Context Diagrams of the Systems

The development of consolidated informational socio-communicative resource of the Smart City started with the highest abstract level of description, such as the context diagram presented in terms of IDEF0 standard (see Fig. 1).

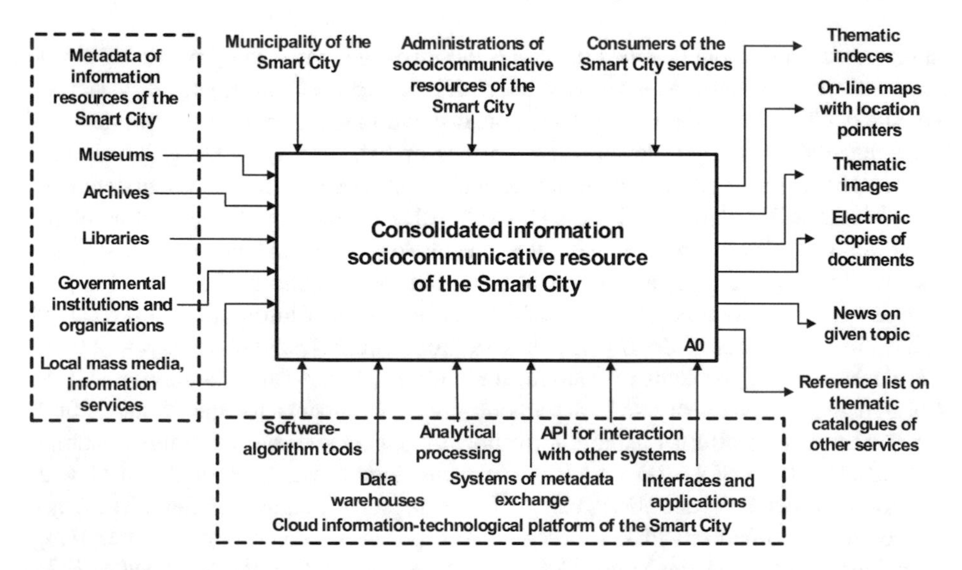

Fig. 1. Context diagram of consolidated informational socio-communicative resource of the Smart City

The metadata of information resources of museums, archives, libraries, state institutions and administrations, local mass media and information portals were recorded as the main inputs of consolidated information and socio-communicative resource of the Smart City. Cloud computing platform including software-algorithms means, data warehouses, analytical processing tools, metadata exchange systems, APIs for interacting with other systems, and visual interfaces is used as information technology platform. The management functions are carried out by the municipality, the administrations of socio-communicative resources and consumers of the relevant information services of the Smart City. It is specified that the functional block given by the context diagram provides the generation of thematic indexes and online maps with positioning of locations, search for thematic images and electronic copies of documents, formation of news list on predefined topics and lists of references to thematic catalogues of other services.

The formation of context diagram for "hard" domains of the Smart City is carried out on the example of information system of accounting and cost analysis in resource networks of the Smart City (see Fig. 2).

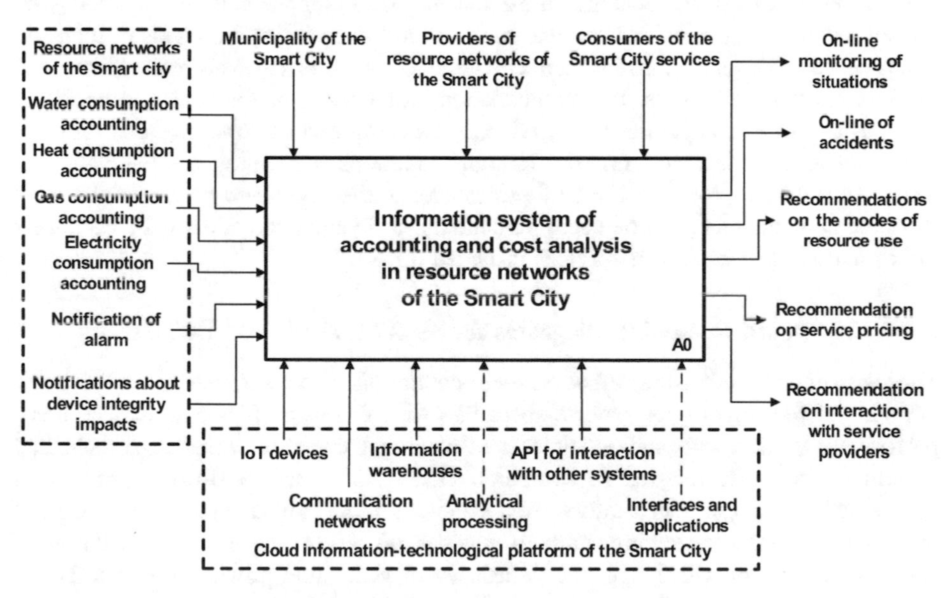

Fig. 2. The context diagram of the subsystem of accounting and cost analysis in resource networks of the Smart City

Let us fix such costs as water, heat, gas, electricity, emergency situations and events affecting the integrity of measuring devices integrated in the urban resource networks as the basic inputs of the information system of accounting and cost analysis in resource networks of the Smart City.

Cloud computing platform including IoT devices, ubiquitous municipal communications networks, data warehousing, analytical processing tools, APIs for interaction with other systems, visual interfaces and software-algorithmic applications are also used as the information-technological platform for accounting and cost analysis in resource networks of the Smart City. The management functions in this case are carried out by the municipality, providers of resource networks and consumers of the appropriate services of the Smart City.

The project specifies that the functional block represented by the context diagram provides the execution of operational monitoring of the situation and informing the engineering services about accidents, giving recommendations concerning optimal modes of resources consumption and services their price formation, recommendations concerning interaction with providers, etc. The execution of the mentioned function is focused first of all on the municipal administration, providers and users of services of resource networks of the Smart City. The results can be used to find sources and analyze losses in resource networks, to investigate dynamic characteristics for the processes of the relevant resources use.

Operational information on accidents is intended to reduce the time of emergencies localization and eliminate their consequences. IoT-devices with flow control functions in resource networks, which allow to turn on or off remotely the appropriate resources stream can be used while localizing the accident. The recommendations concerning the consumption of resources and services are intended to optimize the levels of loads on resource networks and transfer their consumption at time intervals with lower cost given by service providers. Recommendations for pricing of relevant resources and services are for municipalities and service providers in order to optimize consumption processes at minimum cost criteria. Recommendations concerning the interaction of users with service providers are for both municipalities and users of the "smart city" resource networks, and can be used, particularly, to identify providers with the lowest prices and highest quality of services provided (QoS).

3.2 Case-Based Reasoning Diagrams for "Soft" and "Hard" Domains

The users of consolidated information socio-communicative resource formed for "soft" domains of the Smart City, can be specialists of different profiles and subject areas, performing various information duties and playing the relative roles in complicated system - hereinafter referred to as actors. Taking into account diversity and polyfunctionality of consolidated information socio-communicative resource, the design of the UseCases-based reasoning diagram was carried out on the example of the actor "Administrator", provided with the widest list of rights and powers (see Fig. 3). To provide effective information services, each of the subordinate categories of user-actors is assigned the appropriate functionally-limited sets of rights and powers within the system.

Section "Analytical Processing" provides access to the tools, procedures and results of analytical processing of information entities. Section "System" allows you to change system settings, turn on or off debugging mode, perform actions with additional components, such as external APIs, metadata formats, or interfaces.

For "hard" domains of the Smart City, presented in the system of accounting and cost analysis in resource networks, the design of the UseCases-based reasoning

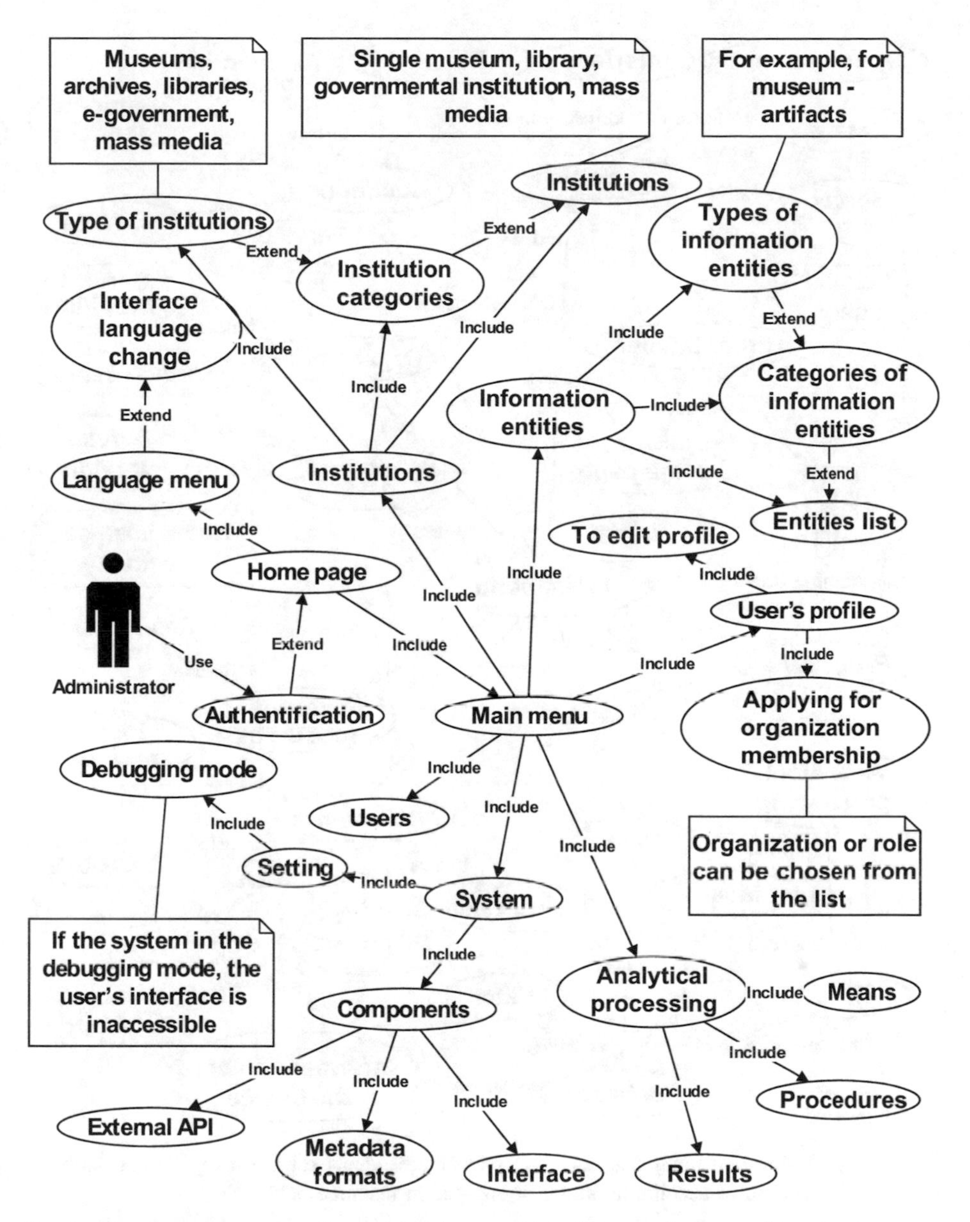

Fig. 3. Case-based reasoning diagram for the actor “Administrator” of consolidated information socio-communicative resource

diagram is carried out on the example of the actor “Super Administrator”, since he has the widest set of rights and powers (see Fig. 4).

To provide effective information services, each of the subordinate categories of user-actors is assigned the appropriate functionally-limited set of rights and powers. In order to authorize due to the personalized login and password in the system of accounting and

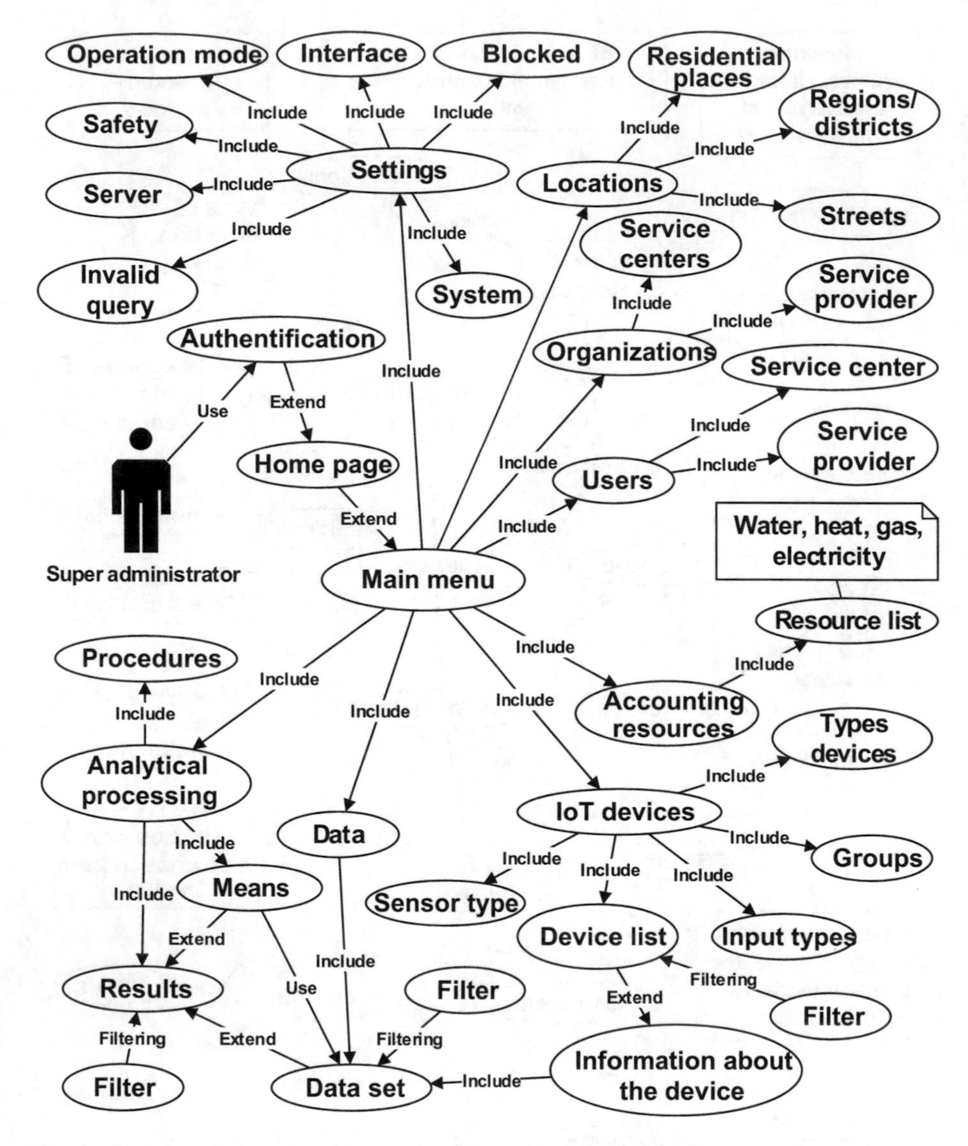

Fig. 4. Case-based reasoning diagram for Chart of precedents for the actor "Super Administrator" of the system of accounting and cost analysis in resource networks

cost analysis of the Smart City, the "Super Administrator" user can use the main navigation menu, which gives access to all functional sets. Section "Settings" includes the group of functions intended to service the system and modify its settings. Section "Locations" contains functions for processing parameters of geospatial placement of the system constituent elements. Section "Organization" allows you to manage accounts of service organizations and service providers. Section "Users" contains functions for servicing accounts registered in the user system. Section "Accounting resources" contains a list of recorded in resource system of the Smart City which at the moment includes water, heat, gas and electricity.

The "IoT device" section contains subcategories for processing input types, sensors, and devices with functionalities for grouping them. In the context of the IoT device, it is possible to view its parameters and sets of data generated by it. You can view and filter the data sets in the data menu on the main menu. The section "Analytical Processing" contains a list of appropriate procedures, means and results of data processing.

3.3 Database Structures for "Soft" and "Hard" Domains

The next step in creating systems for "soft" and "hard" domains of the Smart City is design of the relevant database structures. The general structure of database for consolidated information socio-communicative resource of "soft" domains of the Smart City is shown in Fig. 5. The diagram represents BD for storage of information about the corresponding categories of entities grouped according to the name of the database tables. The tables group "System" is intended to store information concerning the system parameters, its components, modules, interface templates, system panels and procedures (Actions), countries and geospatial locations. For certain types of entities, specialized parameters are available, information about which is stored in separate tables.

The table group "Users" is intended for storage of information about users registered in the consolidated information and social-communicational resource of the Smart City users, their functional rights and powers, belonging to groups, visits and actions performed in the system. The table group "Menu" is intended for storing information about available menu system, its types and elements. The table group "Panels" is designed to store information about available in the system panels with appropriate tool sets, and "Resources" for storing information concerning the types and list of information resources including museums, archives, libraries, government agencies, media, etc. The collection of tables "Data" is dynamically created while registering new resource and is intended to store information about the types and sets of information entities, their categories, user rights tuples and comments. The table group "Analytics" tables is designed to store information about the classes, methods, and means of analytical processing of data stored in the system.

The basic tuple of the database tables of the system of consolidated socio-communicative resources of the Smart City includes more than fifty entities for storing more than five hundred characteristics. In addition, for each information resource registered in the system, its own tuple with more than ten entities for storing information concerning more than sixty characteristics is dynamically generated.

Tables for storing information on the results of analytical processing of data stored in the system of data tuples are created dynamically and separately. While implementing consolidated informational socio-communicative resource of the Smart City of the average size with the population of 300–500 thousand people, where are museums, libraries, archives, government agencies and regional mass media, BD are measured by hundreds of gigabytes, and the volume of file storage for storing digitized photos, video and audio data will be measured by tens and hundreds of terabytes. BD and file

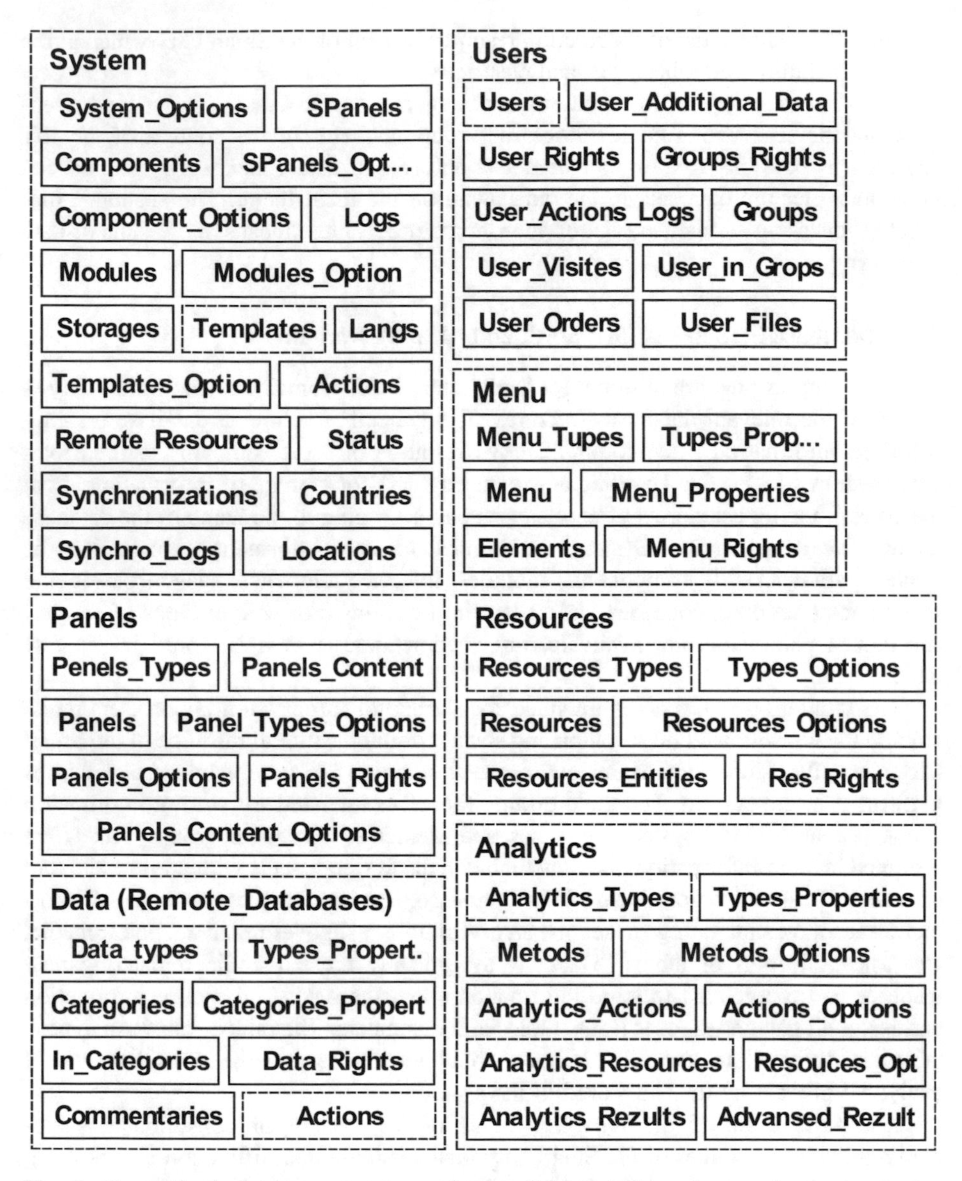

Fig. 5. Generalized database structure of consolidated information socio-communicative resource of the Smart City

storages will rapidly grow in the process of creating and storing of new collections of funds and portals.

While creating the database structure of BD "hard" domains of the Smart City it is necessary to take into account rapid growth of processes of corresponding data volumes. The general structure of BD of the information system for accounting and cost analysis in the resource networks of the Smart City is shown in Fig. 6.

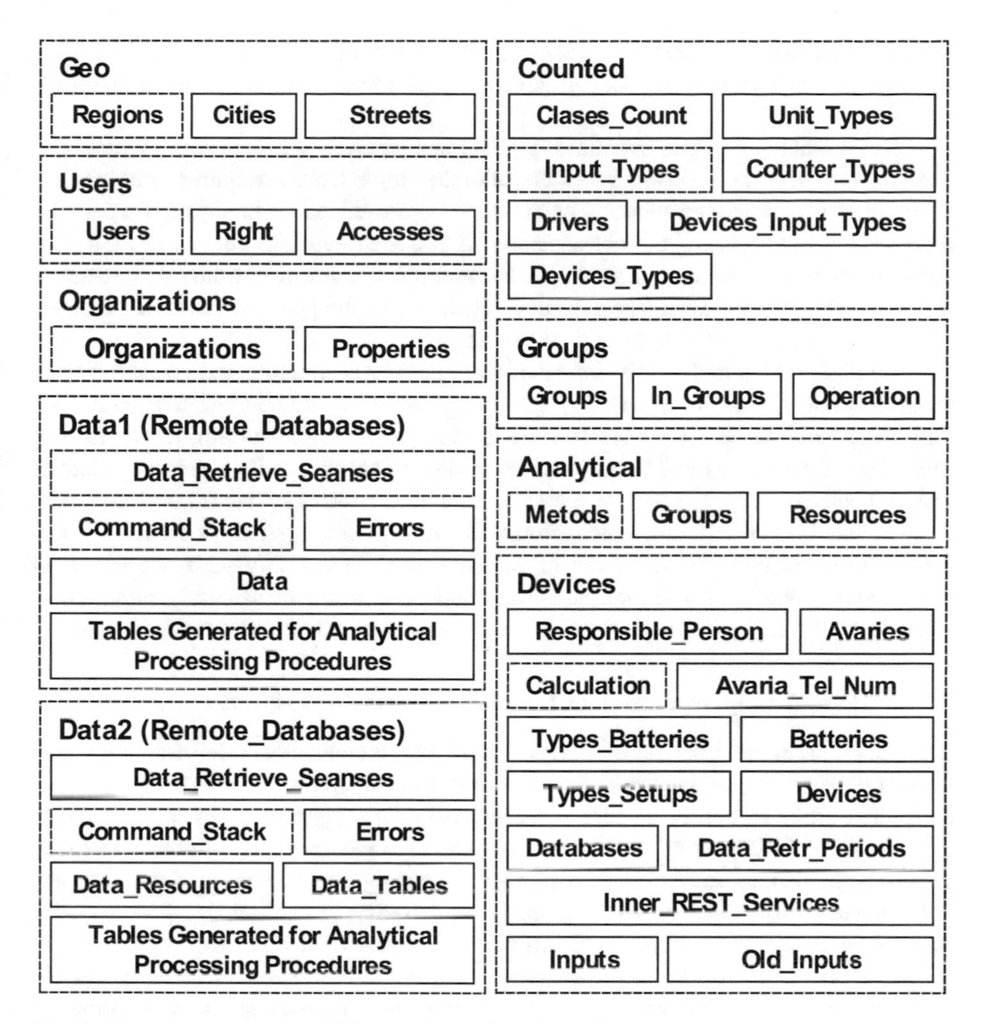

Fig. 6. The general structure of the database of the information system of accounting and cost analysis in resource networks of the Smart City

"Geo" tables group is designed for storage of information about geospatial locations of devices and sensors integrated into resource networks, and "Users" is designed for storage of information concerning the accounts registered in the system of users, their permissions and access rights. "Organizations" tables group is designed to store information regarding organizations registered in the system, including municipal and government institutions, resource providers, etc. "Counted" tables group is designed to store information concerning of accounting resource classes, including water, heat, gas and electric power, units of their measurements, types of inputs, sensors and types of IoT devices.

"Devices" tables are designed to store information concerning persons responsible for the device, accidents, calculations, batteries, installation types, data acquisition periods, databases, devices and their inputs, and "Groups" are designed to store information concerning device grouping information and corresponding group operations.

"Data1" tables group is dynamically generated when new IoT devices are added to the system where the received data storage in the single table is required, and "Data2" is dynamically generated when new IoT devices are added to the system where the received data storage in the tables generated for each separate collection of data is required. Both groups contain similar tables with information concerning data sessions, instruction stacks for transmission to devices, errors in the processes of data exchange and results of received data analytical processing.

The basic tables tuple of the information system of accounting and cost analysis in resource networks of the Smart City has more than thirty five entities and more than 125 characteristics. Besides for each IoT device added to the system, the tuple with more than five entities and twenty characteristics is automatically generated. Entities and characteristics for storing the results of analytical processing stored in data system are generated separately. Implementation of the information system for accounting and cost analysis in resource networks of medium-sized Smart City with population of 300–500 thousand with 150 thousand households, DB size is measured by hundreds of gigabytes, and will grow rapidly in the process of new data collections formation.

3.4 Practical Implementation of Information Systems Projects

The scaled partitioned database is implemented in the consolidated information socio-communicative resource of "soft" domains of the Smart City by means of MySQL. The structure of information catalogues is created. Software- algorithmic tools for dynamic generation of information resource tables and a wide range of system functional capabilities such as the subsystem of the processing of accounts registered in the consolidated information socio-communicative resource of institutions, the subsystem of classification and processing of information entities, subsystem of processing of accounts and user profiles are implemented by means of PHP. Web-interfaces are implemented for the software-algorithmic components of the consolidated information socio-communicative resource by means of HTML5 and CSS3. The subsystems of analytical data processing component servicing and integrated configuration of the consolidated informational socio-communicative resource of the Smart City are implemented.

The scaled partitioned database and structure of information catalogues are created for "hard" domains in the system of accounting and costs analysis in resource networks of the Smart City by means of MySQL. Software-algorithmic tools for dynamic generation of BD tables and declared functional capabilities of the information system by means of PHP are implemented, especially the subsystem of geospatial component location, processing of the organizations accounting means, users and resources are developed. The subsystem for processing of data received from IoT devices is designed. Particularly the software-algorithmic means of interaction with devices, import of data from text formats, calculation with data series, graphical representation of data, device grouping and group calculations performing are implemented. Web interfaces are implemented for the developed subsystems by HTML5 and CSS3. Tools for system complex configuration, APIs for interaction with other systems, procedures for integrating software-algorithmic means of analytical data processing are implemented.

4 Conclusions and Further Investigations

The formation of innovative high-efficient "social" capital due to the large-scale use of modern information and communication technologies becomes a strategic approach for the Smart Cities development. Modern cities implement the concept of the comfortable urban socio-communicative environment creation and implement the latest information and technology platforms for the efficient functioning of various city services, providing quality services and services on the basis of municipal resource networks in compliance with high environmental standards. The social component is combined in the environment of so-called "soft" domains, and the municipal engineering infrastructure is in the environment of "hard" domains of the Smart Cities. Comprehensive implementation of innovative information, communication and socio-communication technologies will improve the environment quality and the living standard in modern cities, tending to be promoted as Smart Cities.

The general objective of our research is to analyze the processes of database formation for the information systems of consolidated socio-communicative resources and accounting and cost analysis in resource networks of the Smart City. While developing the context diagrams of the specified information systems, their basic input entities are defined, the components of information-technological platforms are fixed, the influence of management and the results of the systems functioning of are defined. The functionality sets of information systems and scenarios for their use are developed. The generalized database structures of the consolidated information socio-communicative resource and the information system of the accounting and cost analysis in resource networks of the Smart City are designed. The first database contains more than fifty static and more than ten dynamic entities containing more than five hundred static and more than sixty dynamic characteristics. The second database has more than thirty five static and about ten dynamic entities with more than 125 static and more than 20 dynamic characteristics. The evaluation of the stored volumes in database sets, ranging within hundreds of gigabytes and characterized by high growth rates is carried out. In this turn it allows us to conclude that further implementation of information systems is reasonable to implement using BigData technology. It is planned to use the mechanisms of in-line analytical processing of Big Data sets.

References

1. Garau, C., Pavan, V.M.: Evaluating urban quality: indicators and assessment tools for smart sustainable cities. Sustainability **10**(3), 575 (2018)
2. European Parliament. Mapping smart cities in the EU; European Parliament, Directorate general for internal policies: Brussels, Belgium (2014)

3. Vázquez, J.L., Lanero, A., Gutiérrez, P., Sahelices, C.: The contribution of smart cities to quality of life from the view of citizens. In: Entrepreneurial, Innovative and Sustainable Ecosystems, pp. 55–66. Springer, Cham (2018)
4. Monfaredzadeh, T., Berardi, U.: Beneath the smart city: dichotomy between sustainability and competitiveness. Int. J. Sustain. Build. Technol. Urban Dev. **6**(3), 140–156 (2015)
5. Berardi, U.: Sustainability assessment of urban communities through rating systems. Environ. Dev. Sustain. **15**(6), 1573–1591 (2013)
6. Cugurullo, F.: Exposing smart cities and eco-cities: Frankenstein urbanism and the sustainability challenges of the experimental city. Environ. Plan. A (2017). https://doi.org/10.1177/0308518X17738535
7. Höjer, M., Wangel, J.: Smart sustainable cities: definition and challenges. In: ICT Innovations for Sustainability, pp. 333–349. Springer, Cham (2015)
8. Calvillo, C.F., Sánchez-Miralles, A., Villar, J.: Energy management and planning in smart cities. Renew. Sustain. Energy Rev. **55**, 273–287 (2016)
9. Wong, M.S., Wang, T., Ho, H.C., Kwok, C.Y., Lu, K., Abbas, S.: Towards a smart city: development and application of an improved integrated environmental monitoring system. Sustainability **10**(3), 623 (2018)
10. Monteiro, C.S., Costa, C., Pina, A., Santos, M.Y., Ferrão, P.: An urban building database (UBD) supporting a smart city information system. Energy Build. **158**, 244–260 (2018)
11. Yamamura, S., Fan, L., Suzuki, Y.: Assessment of urban energy performance through integration of BIM and GIS for smart city planning. Procedia Eng. **180**, 1462–1472 (2017)
12. Liu, Y., Weng, X., Wan, J., Yue, X., Song, H., Vasilakos, A.V.: Exploring data validity in transportation systems for smart cities. IEEE Commun. Mag. **55**(5), 26–33 (2017)
13. Xiao, Z., Lim, H.B., Ponnambalam, L.: Participatory sensing for smart cities: a case study on transport trip quality measurement. IEEE Trans. Ind. Inform. **13**(2), 759–770 (2017)
14. El-Wakeel, A.S., Li, J., Rahman, M.T., Noureldin, A., Hassanein, H.S.: Monitoring road surface anomalies towards dynamic road mapping for future smart cities. In: Signal and Information Processing (GlobalSIP), pp. 828–832 (2017)
15. Kamienski, C., Borelli, F., Biondi, G., Pinheiro, I., Zyrianoff, I., Jentsch, M.: Context design and tracking for IoT-based energy management in smart cities. IEEE Internet Things J. **5**(2), 687–695 (2017)
16. Sta, H.B.: Quality and the efficiency of data in “Smart-Cities”. Future Gener. Comput. Syst. **74**, 409–416 (2017)
17. Souza, A., Pereira, J., Oliveira, J., Trindade, C., Cavalcante, E., Cacho, N., Lopes, F.: A data integration approach for smart cities: the case of natal. In: Smart Cities Conference (ISC2), pp. 1–6 (2017)
18. Santana, E.F.Z., Chaves, A.P., Gerosa, M.A., Kon, F., Milojicic, D.S.: Software platforms for smart cities: concepts, requirements, challenges, and a unified reference architecture. ACM Comput. Surv. (CSUR) **50**(6), 78 (2017)
19. Muhammad, G., Alsulaiman, M., Amin, S.U., Ghoneim, A., Alhamid, M.F.: A facial-expression monitoring system for improved healthcare in smart cities. IEEE Access **5**, 10871–10881 (2017)
20. Ali, Z., Muhammad, G., Alhamid, M.F.: An automatic health monitoring system for patients suffering from voice complications in smart cities. IEEE Access **5**, 3900–3908 (2017)
21. Zotano, M.A.G., Bersini, H.: A data-driven approach to assess the potential of Smart Cities: the case of open data for Brussels Capital Region. Energy Procedia **111**, 750–758 (2017)
22. Koukopoulos, D., Koukoulis, K.: A trustworthy system with mobile services facilitating the everyday life of a museum. Int. J. Ambient. Comput. Intell. (IJACI) **9**(1), 1–18 (2018)
23. Pereira, J.A.R.: Smart augmented reality application for enhanced museum experience (2017)

24. Mulfari, D., Minnolo, A.L., Puliafito, A.: Building tensor flow applications in smart city scenarios. In: Smart Computing (SMARTCOMP), pp. 1–5 (2017)
25. Galvão, J.R., Moreira, L., Gaspar, G., Vindeirinho, S., Leitão, S.: Energy system retrofit in a public services building. Manag. Environ. Qual. **28**(3), 302–314 (2017)
26. Wang, X., Guo, M., Koppelaar, R.H., van Dam, K.H., Triantafyllidis, C.P., Shah, N.: A nexus approach for sustainable urban Energy-Water-Waste systems planning and operation. Environ. Sci. Technol. **52**(5), 3257–3266 (2018)
27. Spinsante, S., Squartini, S., Russo, P., De Santis, A., Severini, M., Fagiani, M., Minerva, R.: 14 IoT-enabled smart gas and water Grids. Internet Things Chall. Adv. Appl. **273** (2017)
28. Zhang, K., Ni, J., Yang, K., Liang, X., Ren, J., Shen, X.S.: Security and privacy in smart city applications: challenges and solutions. IEEE Commun. Mag. **55**(1), 122–129 (2017)
29. Sayegh, M.A., Danielewicz, J., Nannou, T., Miniewicz, M., Jadwiszczak, P., Piekarska, K., Jouhara, H.: Trends of European research and development in district heating technologies. Renew. Sustain. Energy Rev. **68**, 1183–1192 (2017)
30. Mosannenzadeh, F., Bisello, A., Vaccaro, R., D'Alonzo, V., Hunter, G.W., Vettorato, D.: Smart energy city development: a story told by urban planners. Cities **64**, 54–65 (2017)
31. Rismanchi, B.: District energy network (DEN), current global status and future development. Renew. Sustain. Energy Rev. **75**, 571–579 (2017)

Web Resources Management Method Based on Intelligent Technologies

Aleksandr Gozhyj[1(✉)], Victoria Vysotska[2], Iryna Yevseyeva[3], Iryna Kalinina[1], and Victor Gozhyj[1]

[1] Department of Computer Engineering, Petro Mohyla Black Sea National University, Nikolaev, Ukraine
alex.gozhyj@gmail.com, irina.kalinina1612@gmail.com, gozhyi.v@gmail.com

[2] Information Systems and Networks Department, Lviv Polytechnic National University, Lviv, Ukraine
Victoria.A.Vysotska@lpnu.ua

[3] Computing Science at the Faculty of Technology at de Montfort University, Leicester, UK

Abstract. This study explores a method for managing web content based on intelligent technologies, as a stage of the content life cycle in e-commerce systems. The method of content management describes the processes of creating web resources and simplifies the technology of managing web resources in conditions of uncertainty. Also, the main problems of e-commerce and functional content processing services are analyzed. The proposed method makes it possible to develop tools for managing web resources and commercial content.

Keywords: Intelligent technologies · Content management · Web resource Content · Content analysis · Content monitoring · Content search Content e-commerce system

1 Introduction

Rapid development of the Internet contributes to the growing need for fast data acquisition and the development of new forms of information services through the modern IT [1–3]. The entire spectrum of commercial content is sold through numerous web resources on the Internet, i.e. scientific and journalistic articles, music, books, movies, photos, software, etc. Famous corporations that develop content e-commerce are Google with its Play Market, Apple through Apple Store, I-Tunes, Amazon with Amazon.com and others [4–6]. Commercial content management is one of the main tasks that are solved during its lifecycle [7–12]. The management process involves monitoring, controlling, establishing and managing a range of parameters that determine the technological, consumer-related, contextual, commercial and other content qualities [12–19]. These parameters include, in particular, its significance, aging, completeness, accuracy, relevancy, authenticity, reliability, etc. [20–33]. The peculiarity of the e-commerce content management parameters is the complexity of their exact values determination. In this case, it is advisable to apply the intelligent methods

N. Shakhovska and M. O. Medykovskyy (Eds.): CSIT 2018, AISC 871, pp. 206–221, 2019.
https://doi.org/10.1007/978-3-030-01069-0_15

and management means which are based on the situational management [7] and fuzzy logic [8–10] principles. Nowadays, management systems and technologies which are using fuzzy values have become extremely widespread in various spheres, from home appliances control to the complex management of technological, industrial, medical, biological, social and other processes [8–11]. Therefore, the gained experience in this area allows applying the principles of fuzzy logic for the tasks of commercial content management [12–33].

2 Literature Review

The tasks which are solved within the commercial content management process have certain peculiarities that allow describing this process by the model classified by Golota [7] as a situational management model. According to the model, the management of a particular object is organized under conditions of its uniqueness, when formal descriptions and clear optimality criterion are missing or incomplete, in case of constant change of the object and its parameters [7–11]. Commercial content, as an object of management, fully meets these conditions. Nowadays, complex processes of IT management, decision support, analysis, forecasting, etc., often use fuzzy linguistic analogs instead of exact values. This approach is gaining rapid popularity due to its versatility, simplicity and active development of its methods together with the growth of IT artificial intelligence. This method of decision-making is used when obtaining exact values of a particular value is complicated, lengthy, time-consuming, expensive or impossible; when the value is incorrect or inaccurate; when the accuracy of the values is insignificant; within the management and decision-making processes when not the exact value, but the qualification, general estimate, category or class of the values are significant; when not the exact value, but its relation to other values are used. The above-mentioned factors are also implied in IT sphere and web content management in particular. The most popular tools that are used to develop intelligent technologies based on the use of fuzzy values are the classical fuzzy logic proposed by Osgood [10], the anthony logic developed by Grinyaev [8], semiotic models of Golota [7] and fuzzy situational nets and their varieties and Rashkevych et al. semantic differential technique [11]. The choice of fuzzy data presentation means and techniques largely depends on the tasks, subject area, value determination ways, their interpretation and application, etc. [12–14].

3 Statement of the Problem

The purpose of the study is as follows: (1) analysis of the management of commercial content; (2) development of a methodology for managing commercial content based on intelligent technologies; (3) study of the results of applying the methodology.

4 Analysis of the Commercial Content Management Process

The content management subsystem objectives are the following: formation and rotation of the databases and creating access to them; individualization of the users work, storing personal user and source requests, work statistics maintenance; search implementation within the databases; output forms generation; informational interaction with the databases; data source configuration. The commercial content management process is presented with the following scheme of connections:

$$User(q_d) \to q_d \to Q \to H(c_r, q_d) \to \beta(q_d, c_r, h_k, t_p) \to z_w \to User(z_w), \quad (1)$$

where $User(q_d)$ forms the user request q_d; $User(z_w)$ represents the response a user gets for the request q_d. Commercial content management operator $\beta : C \to Z$ is presented with the following superposition

$$\beta = \beta_4 \circ \beta_3 \circ \beta_2 \circ \beta_1, \quad (2)$$

where β_1 is the operator of commercial content editing and modification, β_2 is the operator determining the authority of a commercial content block; β_4 is the operator of the information resource page formation and presentation.

The value of the operators $\beta = \{\beta_1, \beta_2, \beta_3, \beta_4\}$ is adequate in the process of commercial content management, which is described by the following operator

$$\beta = \langle C, Q, H, U, T, Z, \beta_1, \beta_2, \beta_3, \beta_4 \rangle. \quad (3)$$

1. Commercial content editing and modification operator

$$\beta_1 \ : (c_r, h_k, u_l, t_p) \to c'_r. \quad (4)$$

2. Content search query formation and block authority evaluation operator

$$\beta_2 \ (c'_r, y_j, u_l, t_p) \to c''_r. \quad (5)$$

3. Core management values formation and representation operator

$$\beta_3 \ : (c''_r, h_k, u_l, t_p) \to h'_k. \quad (6)$$

4. Informational resource page configuration and representation operator

$$\beta_4 \ (c''_r, h'_k, z_w, q_d, t_p) \to z_{w+1}, \quad (7)$$

where $h_k \in H$, $h_k = \{h_{1k}, h_{2k}, \ldots, h_{mk}\}$ is the value of the commercial content management parameters (h_{1k} represents the pertinence, h_{2k} - relevance, h_{3k} - completeness, h_{4k} - authenticity, h_{5k} - reliability of the commercial content); $u_l \in U$, $u_l = \{u_{1l}, u_{2l}, \ldots, u_{nl}\}$ is the value of the commercial content management process criteria

(u_{1l} represents the commercial content block placement rate, u_{2l} is the block's keywords ratio, u_{3l} is the keyword statistical significance rate, u_{4l} is the value which reflects the presence of keywords used in a user's request, u_{5l} is the search query keywords volume ratio). Information resource pages formation is presented as follows.

$$z_w = \left\{ \bigcup_{r=1}^{n_C} c_r \,\middle|\, \begin{array}{l} \forall c_r \in C_{q_d}, \exists q_d \in Q, \exists h_k \in H_{c_r}, c_r \notin C_{\overline{q_d}}, h_k \notin H_{\overline{c_r}}, \\ C = C_{q_d} \cup C_{\overline{q_d}}, H = H_{c_r} \cup H_{\overline{c_r}}, d = \overline{1, n_Q}, k = \overline{1, n_H} \end{array} \right\}. \quad (8)$$

The commercial content pertinence rate h_{1k} is estimated employing the profitability indicator I_p and the growth of demand I_g for commercial content. Value of commercial content authenticity is a subjective value within [0, 5; 1] which is determined by the e-commerce content system moderator. Commercial content management method is a set of measures aimed at maintaining the commercial content $h_k \to h'_k$ management parameters (pertinence, completeness, relevance, authenticity, reliability) in accordance with the specified requirements based on the commercial content management standards.

$$z_{w+1}(h'_k, c''_r, t_{p+1}) = \beta(q_d, z_w, c_r, h_k, u_M, t_p). \quad (9)$$

Given that A is an overall volume of pages, N is the value defining the formed pages volume of the information resource, M is the amount of the relevant commercial content array processed to form the given volume of pages, U is the overall amount of commercial content used to form the volume of pages ($U = M + L$), and L is the array of irrelevant content. The function $Z = \beta_{A4}(C, M)$ presents the maximal number of the necessary selection sorting and filtering operations which must be performed by the algorithm A estimating the value M.

Linear search is the method used to estimate the values $\bigcup_{j=1}^{m} c_j(q_i, t_r)$ of true-value functions by a set of comparison criteria $\exists h_k \in H_c$ to form a number of pages Z of the information resource.

All commercial content management methods are classified according to the information resource page generation type. Those are a generation of pages on user request, a generation of pages set by a moderator, and a mixed type.

The generation of pages on user request is implemented in the following scheme which shows the connections between the commercial content management objects and subjects.

$$\textit{Moderator} \to \textit{Commercial Content Editing} \to \textit{Commercial Content Database} \\ \to \textit{Commercial Content Submission} \to \textit{Information Resource}.$$

Commercial content management with the generation of pages on user request within an e-commerce content system can be presented as follows.

$$\beta_Q = \langle C, Q, H, U, T, Z, \beta_1, \beta_2, \beta_3, \beta_4 \rangle. \tag{10}$$

The stage of commercial content editing and modification is presented by the operator $c_j(t_{r+1}) = \beta_1(c_j, t_r, h_k, u_l)$ considering that $c_j(t_{r+1}) \in C$. The stage of pages formation is defined by the operator $Z(t_r) = \beta_4(q_i, C, \beta_3(\beta_2(C), t_r))$ where

$$z_i = \left\{ \bigcup_{j=1}^{m} c_j(q_i, t_r) \middle| \begin{array}{l} \forall c_j \in C_q, c_j \notin C_{\overline{q}}, C_q = \beta_3(\beta_2(C_q)), \exists q_i \in Q_c, \exists h_k \in H_c, h_k \notin H_{\overline{c}}, \\ C = C_q \cup C_{\overline{q}}, Q_c \subset Q, H = H_c \cup H_{\overline{c}}, k = \overline{1, n_H}, i = \overline{1, n}, r = \overline{1, w} \end{array} \right\}.$$

The block's authority is defined as a sum of the commercial content authority coefficients.

$$\omega = \|C\| = \beta_2(C, \omega_1, \omega_2, \omega_3, \omega_4, \omega_5), \tag{11}$$

where $\omega_1(c_j)$ is the constituent of the block allocation within the commercial content, $\omega_2(c_j)$ is the block's keywords ratio, $\omega_3(c_j)$ is the terms statistical significance coefficient, $\omega_4(c_j)$ is the additional terms presence coefficient, and $\omega_5(c_j)$ is the coefficient defining the presence and quantity of terms from the user's request.

The generation of pages during the information resource editing is conducted through creating a set of static pages when changes to the commercial content or the information resource are done. Therefore, the following scheme of commercial content management objects and subjects interconnections and interactions is accomplished.

Moderator → Commercial Content or Information Resource Editing → Commercial → Content Database → Information Resource.

Commercial content management with the generation of pages set by a moderator is performed as follows

$$\beta_E = \langle C, H, T, Z, \beta_1, \beta_2, \beta_3 \rangle, \tag{12}$$

The pages formation stage is determined by the operator $Z(t_r) = \beta_3(C, H, t_r, \beta_1, \beta_2)$. This method does not take into consideration the interactivity between the visitor and the resource.

The mixed type of information resource page generation combines the advantages of the first two types. It has the following scheme of commercial content management objects and subjects interconnections.

Moderator → Commercial Content Editing → Content Database → Content Blocks Creation → Content Caching → Content Presentation → Information Resource.

Mixed type commercial content management is presented as follows

$$\beta_M = \langle C, Q, H, T, Z, W, \beta_1, \beta_2, \beta_3, \beta_4, \beta_5 \rangle, \tag{13}$$

where W is the cached commercial content volume, β_5 is the operator defining the formation of the cached commercial content or its information blocks considering that

$$W = \beta_5(C, \beta_3(\beta_2(\beta_1(C, t_r, H, U)), t_{r+1}) \text{ or } W = \beta_5(Z, \beta_3(\beta_2(\beta_1(C, t_r, H, U)), t_{r+1}),$$

$$w_l = \left\{ \bigcup_{i=1}^{n} c_i \middle| \forall c_i \in C_Q, C_Q \subset C, C_Q = \beta_3(\beta_2(C)) \right\},$$

$$w_l = \left\{ \bigcup_{j=1}^{m} z_j \middle| \begin{array}{l} z_j \in Z_c, \forall c_j \in C_z, \exists c_j \in Z_c, \forall c_j \in z_j, C_z = \beta_3(\beta_2(C)), \\ C_z \subset C, Z_c \subseteq Z, i = \overline{1,n} \end{array} \right\}.$$

The commercial content management subsystem is executed via caching. I.e., the subsystem generates a page once, and for the future requests, it is loaded several times faster from the cache, which is updated automatically within the defined period of time or when changes to the certain information resource sections are implemented, or manually upon the administrator's request. The commercial content management subsystem can be also implemented though the information blocks creation. Thus, the blocks are saved at the stage of information resource editing, and the displayed pages are constructed from these blocks in accordance with the related user request.

5 Method of Commercial Content Operations Management

The main task of the content management process is a database creation and accessibility; user experience personalization; a collection of individual queries and references; statistics maintenance; database search implementation; output forms generation; informational interaction with the databases of other subsystems. The main stages of the content management method, such as processing, analysis, and presentation of the commercial content through an information resource to the e-commerce content system are presented in Table 1.

Commercial content management operator β transmits the commercial content c_r to the new state c'_r which differs from the previous one by the values of the determinative parameters $h_k \rightarrow h'_k$ (pertinence, completeness, relevance, authenticity, reliability) that satisfy the predefined requirements and are used in the content search.

$$\beta : (q_d, z_w, c_r, h_k, u_M, t_p) \rightarrow (c''_r, h'_k, z_{w+1}, t_{p+1}).$$

The content management operator $\beta : C \rightarrow Z$ is represented by a superposition of functions $\beta = \beta_4 \circ \beta_3 \circ \beta_2 \circ \beta_1$, where β_1 is the operator of content editing and modification; β_2 is the operator which determines the content block authority; β_3 is the operator which forms core content management parameter values; β_4 is the operator which forms and presents information resource page using the content search.

Table 1. Stages of the content management process in an e-commerce content system

Name	Name of the step	Peculiarities
Processing	Content subject definition	The purpose of content creation, its meaning and structure
	Determination of the content presentation form	Graphic information; text (article, press release, job descriptions); HTML templates, back-end code etc.
	Selection of the processing means	HTML/text editors and object creation tools
Analysis	Access rights determination	Full or limited access to the content
	Process outline	Standard processes of content creation and publication
	Content saving	In the database or repository
	Processes logging	Creation, transmission, and storing processes
	Auto notification setting	Informing about the content created by other users
	Performance audit and examination	Saving the content previous versions and revisions
	Text content analysis	Quantitative or qualitative
	Revisions access	Content revisions support with the backup opportunity
	Business process analysis	Defining the goals, roles, and assignments; creating tasks for groups of users with different roles; business processes development
Presentation	Static	Disregarding any user behavior logic
	Dynamic (rules/filters)	Personalization; globalization; localization

Content search is a set of content analysis operations which is used to form the volume of content that matches the user request. Content search compares the content with the user search query. (Algorithm 1.)

Algorithm 1. Content Search

Stage 1. Formation of the SP $h_k \in H$ for commercial content c_r using the management criteria $u_l \in U$, where $h_k,= \{h_{1k}, h_{2k},..., h_{mk}\}$ is a set commercial content management process parameters (commercial content pertinence h_{1k}, relevance h_{2k}, completeness h_{3k}, authenticity h_{4k}, reliability h_{5k}); $u_l = \{u_{1l}, u_{2l},..., u_{nl}\}$ is a set of commercial content management process criteria (u_{1l} is the commercial content block placement coefficient, u_{2l} is the block's keywords coefficient, u_{3l} is the keyword statistical significance coefficient, u_{4l} is the coefficient which reflects the presence of keywords used in a user's request, u_{5l} is the search query keywords volume coefficient).

Step 1. Creation of the SP $h_k \in H$ for commercial content c'_r.

Step 2. Saving the SP $h_k \in H$ for commercial content c''_r in the database.

Stage 2. Formation of the SC $h'_k \in H'$ based on the user request.

Step 1. User SQ creation.

Stage 3. Search of the commercial content c'_r.

Step 1. Comparison of the SQ request $h'_k \in H'$ with the SP $h_k \in H$ of the content c'_r.

Step 2. Formation of the content value C'_r where the SP matches the SQ request.

Stage 4. Content analysis application.

Step 1. Quantitative analysis of the textual content c'_r.

Step 2. Qualitative analysis of the textual content c'_r.

Stage 5. Making a decision, i. e. presentation of the relevant content in accordance with the content analysis results within the range (0,7; 1] or (0,5; 1].

Stage 6. Presentation of the content c'_r which corresponds to the user request.

The presence/absence and frequency of the occurrence of a linguistic unit within the content has a great significance for the content search. A quantitative content analysis draws the conclusions concerning the content focus basing on the number of the used linguistic units, such as keywords or positive/negative reviews (Fig. 1). Therefore, a qualitative content analysis forms conclusions regarding the presence of the searched linguistic unit and its context.

The Author registers/authorizes in the system, comments/posts content.

The User registers/authorizes in the system, evaluates, investigates, searches, and comments on the content.

The user's content view results in the change of the content views number, which leads to the display of an updated popular content list. Moreover, the views and parameter searches result in the change of the revised content details, which involves the display of an updated similar content list.

The Moderator forms/modifies the content search parameters.

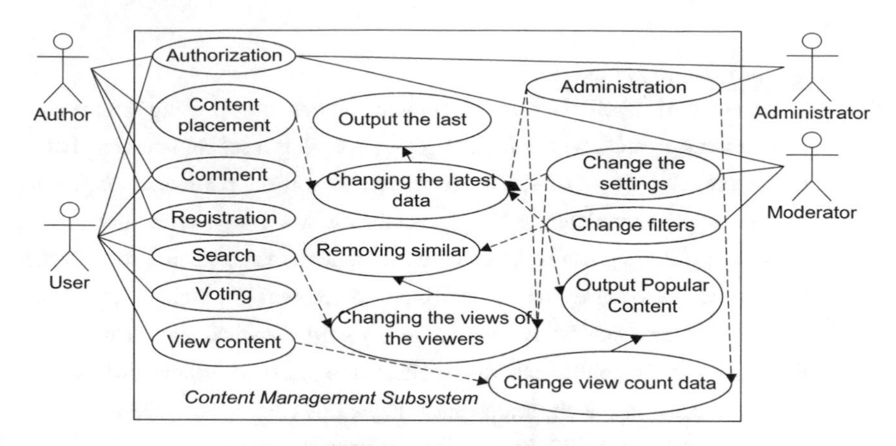

Fig. 1. Content management process outline by user role

Inbound non-processed content is submitted to the content processing block. After that, the revised content, e.g. an article, is assigned to the categories (popular, similar, selected by search preferences, or recently viewed) according to the search parameters, dates, and keywords of the revised content, and is stored in the database.

The processed content is arranged into blocks to search for popular, recently viewed, parametric, and similar content and to display the results. After authorizing, the author/user sends the content for processing. Systematically, there is a data exchange with the database to form a list of new, popular, relevant to the search parameters and similar to the recently viewed by user's content. These lists are sent to the user. The author's content is processed and stored in the database. The user search parameters are submitted to the content search block by the parameters, which generate the database query and the list of content that matches the search parameters, and transfers them to the user and to the block which forms a list of the viewed content. The viewed content block sends the generated list to the similar content forming block. The blocks which create the list of new and/or popular content generate queries in the database and display the generated lists for viewing. On the basis of the received lists, the user makes a new request and the process repeats. Having conducted these operations, the unprocessed author content is transformed into processed and saved in database. It is used for searches by size, keywords, author, and popularity. The system generates and displays the list of viewed content. The processed content correlates with the search matches block by the keywords previously viewed by the user. The formed content list is sorted for further selection by the number of matches, which are the most similar to those viewed by the user. The first units of content, the number of which is set in the system settings are selected. The author's works which are currently being viewed are moved to the top of the list and displayed to the user.

The most recent/popular content lists with more views are presented to the user. The user selects the content for viewing or searches by the parameters within the selected/similar/found content. An information resource consists of a strictly defined set of components (nodes) which are partially organized (some of the components come at a well-defined order, while their location in the graph is not necessarily defined).

The number of components of an information resource and the corresponding volume of component sets are precisely defined/estimated.

The navigational graph construction is done through the information resource page formation $Z(t_r) = \beta_4(q_i, C, \beta_3(\beta_2(C), t_r))$ and presentation operator on the basis of the given logical implication within the information resource components value.

The navigation is performed without interruption and the transition to a new navigation node of the graph is strictly logical. Each arc of the graph reflects an elementary connection of the information resource and has an obligatory orientation. Information resource creation involves several stages (Table 2).

Table 2. The stages of resource creation in e-commerce content systems

Stage	Definition
Technical task (TT) formulation	Subdivision of the TT in order to allocate separate tasks: specification of the graphic design, design and structure of the information resource, structure of the blocks and units; navigation connections; functional modules; programming technologies
Design creation	The result of creating a graphic design concept is implemented in the information resource homepage and internal pages design
Content management	It is a key factor in the informational resource creation process because visitors need an easily navigated operational information
Implementation	Software development and software modules creation, implementation of the design with a modular grid, HTML page markup
Production and testing	Filling a resource with the content; an input of the HTML documents; loading speed optimization, troubleshooting, pages interconnection testing
Transfer to hosting	Domain name choice and registration; server and mailbox settings; placing an information resource on a fast hosting

The template manager generates form and structure templates needed for the information resource functionality. The intellectual part of the process is the generation of a flexible web template. The content manager is a content management service used to create publications and creating relevant content categories. Content management service provides an effective way to manage commercial content and integrate it into an information resource. Commercial content has various structured/unstructured forms: instant messages, documents, multimedia assets, application data, integrated workflows, emails, integrated records, transaction content, reports, rights, images, digital information and software archives, which are used for e-commerce automation. Regulatory requirements for content retention require the access to it on a request. That is done in order to run its audit and management in accordance with the established policy. The processes of authentication and access rights separation are required for virtually all the components and are included in the core of the system. A flexible system of an information resource which uses regular values is a simple hierarchical structure, where the pst pages are assigned to one level and are descendant in relation to the posts archives page. In order to guarantee a more effective content management flow, it is necessary to implement web analytics and digital marketing methods to form and analyze the web resource key performance indicators, analyze information about its visitors, and promote it.

6 The Main Features of the Study of the Process of Managing Web Resources

For the systems development and IT content management, the situational management model is used [7]. This model is used if the system has a number of specific features in processes management. These particular features, according to [7], are the following.

1. The uniqueness, object and its management means are identified with the maximum consideration of its specific indicators, aimed at achieving the system-wide objective.
2. The absence of a formalized goal of the system functioning, which includes the management object, that arises as a consequence of the reducing clarity, multiplicity, complexity of the specification or non-determination of the final result.
3. The absence of a clear optimality criterion, when the best result is characterized by an ambiguous, interval or non-numerical value.
4. Dynamism - the ability of the system and it's management object to change its attributes and behavior.
5. Incompleteness of the description is a circumstance under which the value of some management parameters is missing, unknown, inaccurate or unreliable.
6. Free choice of the decision making options and object management actions. It involves absent or minimal restrictions which are applied to the object behavior and its management means.

Web content as the object of management within the system fully meets these requirements. Consequently, it is reasonable to build its IT management process basing on a situational model [7]. Situational management model implies incompleteness and inaccuracy of some parameters. That's why it is necessary to use special IT to process them. A characteristic feature of all problem-solving approaches through the use of linguistic estimates instead of exact values [7–11, 21–27] is the replacement of the exact value v_i, used for a certain task, with the linguistic estimate $f_j, j = \overline{1, m}$, which is a generalization or the result of the initial value qualification. For instance, the quantity of anything that exceeds 80% of the total amount can be defined as "much", at the same time less than 20% can be defined as "little", etc. [3]. The basis of such transformation is the value $\mu: v_i \rightarrow f_j$, $i = \overline{1, n}, j = \overline{1, m}$, which establishes the correspondence between the exact (numeric) and fuzzy (linguistic) values. In different methodologies, this display has different names and means of implementation. Specifically, it can be a semantic differentiation produced with a special scale [11]; an independent function [10], which not only establishes the correspondence but also specifies a certain measure, which characterizes the obtained value; specialized logical and linguistic means [7] as well as the antonyms logics [8] which involves the use of opposition pairs of values in conjunction with the function of measure. They have common characteristics, such as the replacement of numerical values by linguistic estimates; the use of semantically correct values, which are perceptible and have a clear interpretation; the simplification of the problem-solving procedures though reducing the number of options for values use [7].

The analysis of the mentioned indicators shows that the estimation of their exact values is quite difficult, and sometimes impossible to implement. The fact that commercial web resource content meets the requirements of the situational management [7] by its entire means makes it possible to take management decisions basing on incomplete, inaccurate or fuzzy values without any loss in quality and efficiency. This allows replacing the content management parameters with some generalized values that make it possible to make management decisions and estimate the outcome of the implemented changes. Another peculiarity of the parameters which determine the content properties is the lack of formal techniques and procedures to estimate their values. For example, the content relevance indicator h_1 is determined on the basis of its profitability value h_6 and popularity value h_7, taking into account the content character. Content validity index h_1 is a subjective value determined by the web resource moderator or administrator through the expert evaluation. This approach creates significant possibilities to apply the principles of fuzzy logic [10], due to which the estimation results are presented in a verbal-linguistic form, and the management is based not on the actual values, but on their fuzzy counterparts. Therefore, in IT content management, fuzzy equivalents h_1^*–h_7^* are used instead of the exact values h_1–h_7.

7 Results of the Web Recourse Management Investigations

The use of fuzzy logic in processes management involves the following scheme: exact value → independent function determination → fuzzification (transition to fuzzy values) → fuzzy calculations → defuzzification (transition to the exact values). Content peculiarities and the use of situational management model for IT sphere require a different way of fuzzy parameter values generation and application. The first step is to form the verbal expert estimate h_i^*, $i = \overline{1, n}$ of the ith management value without determining its exact value and independent function. It should be mentioned that according to the form and interpretation, fuzzy estimates differ. For their common use in content management processes and techniques, they are standardized. It means that various fuzzy values of the parameters h_1^*–h_7^* are brought to the same syntax and interpretation. The standardization is carried out by means of semantic differentiation [11] with the use of a special scale. Such action involves replacement of the verbal estimate h_i^*, $i = \overline{1, n}$ with the numerical value d_i^*, $i = \overline{1, n}$ taking into account the content and a mutual ratio of the linguistic values. In this case, numerical values do not specify any quantitative concepts; they solely formalize the corresponding fuzzy verbal estimates and their correlations. It is sufficiently convenient to use the semantic differentiation scale in the range of numerical values [0; 1] (Table 3, Fig. 2).

Table 3. Interdependence of the content management parameters

Parameter	Name	Explanation	A	B	C
h_1	Relevance	An indicator which defines the relevance and appropriateness of the content data to the currently available real values	+	±	±
h_2	Completeness	A measure which defines that web content data meets the needs of all user categories	+	±	±
h_3	Relevance	A measure which defines the correlation of the total web resource content with the amount of data demanded by users	+	±	±
h_4	Authenticity	A measure that evaluates the web resource content authorship, independence and its connection to the information source	±	+	±
h_5	Certainty	An indicator of the content relevance to real values and the reliability of the information source	±	+	±
h_6	Popularity	A demand characteristic that determines the quantity of effective web resource visits	±	+	±
h_7	Profitability	A measure which defines the amount of financial revenue generated from the commercial content	±	±	+
h_8	Uniqueness	A measure of content quality, which reflects the presence of original exclusive data in the web resource content	±	±	+

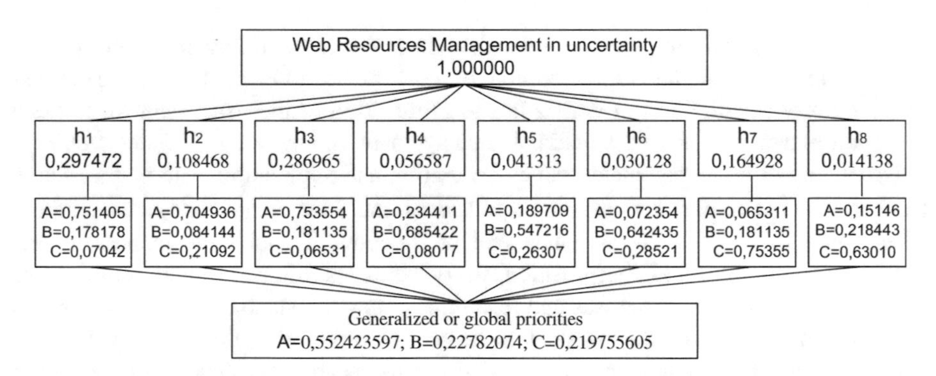

Fig. 2. Hierarchical tree of the web resource management alternatives.

According to the priority vector, the case A (superiority of h_1–h_3 over all other parameters) has a significant advantage over the cases B (superiority of the parameters h_4–h_6) and C (superiority of the parameters h_7–h_8).

Basing on standardized fuzzy estimates, the integrated index of commercial content marketing status $D(C_T)$ is formed. It characterizes the overall correspondence of its properties to the commercial and technological management requirements.

$$D(C_T) = k_1 d_1^* + k_2 d_2^* + k_3 d_3^* + k_4 d_4^* + k_5 d_5^* + k_6 d_7^* + k_7 d_7^*.$$

In this case, the main criterion for content management is:

1. $D_{min} \leq D(C_T) \leq 1$ – content marketing state value entry into the range of the valid values;
2. $D(C_T) \to Max$ – reaching local maximum, which reflects the best content marketing status at the time *T*.

8 Discussion

The results of the investigation are practically implemented in Internet projects "Vgolos" (vgolos.com.ua), "Victana" (victana.lviv.ua), "Tatjana" (tatjana.in.ua), "Presstime" (presstime.com.ua), "AutoChip" (autochip.vn.ua), "Vysotsky Photo-gallery" (fotoghalereja-vysocjkykh.com), "Exchange rates" (kursyvalyut.com), "Good morning" (dobryjranok.com), "Information for business" (goodmorningua.com), "Lviv secondary school № 3" (www.zsh3liviv.in.ua).

Figure 3 shows the results of the developed systems in operation on the charts, which imply that during all stages of the content lifecycle, the overall amount of sessions and the number of unique users significantly increases.

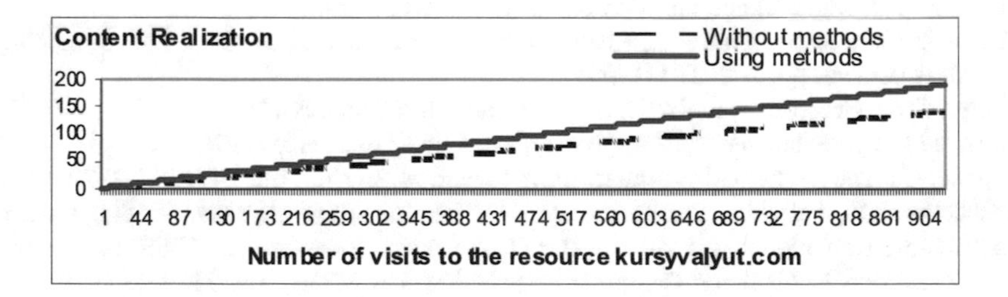

Fig. 3. Analysis of the commercial content sales increase

9 Conclusion

The maintenance of the informational resource traffic statistics allows estimating how the content sales volume increase changes in proportional dependence on the visits number increase, regular user quantity, and marketing campaigns. Subsystems aimed at commercial content creation, management, and maintenance within the content e-commerce systems guarantee a 9% increase in the number of commercial content sales made by regular customers; involvement of unique and potential visitors; increase of the target and regional audience by 11%, of the revised pages by 12%, and of the average amount of time spent on the informational resources by 7%. Analysis of the content maintenance results allows determining the target audience formation principles based on the set of characteristics of the similar systems functioning. By regulating the set of content, its uniqueness, the speed of its generation, and its adequate management according to the individual needs of a regular user, it is possible to model the boundaries of the target social audience and the number of unique visitors from the search engines.

References

1. Mykich, K., Burov, Y.: Uncertainty in situational awareness systems. In: Modern Problems of Radio Engineering, Telecommunications and Computer Science, TCSET, pp. 729–732 (2016)
2. Mykich, K., Burov, Y.: Algebraic framework for knowledge processing in systems with situational awareness. In: Advances in Intelligent Systems and Computing, pp. 217–228 (2016)
3. Vysotska, V., Hasko, R., Kuchkovskiy, V.: Process analysis in electronic content commerce system. In: Proceedings of the International Conference on Computer Sciences and Information Technologies, CSIT 2015, pp. 120–123 (2015)
4. Vysotska, V., Chyrun, L., Chyrun, L.: Information technology of processing information resources in electronic content commerce systems. Computer Science and Information Technologies, 212–222 (2016)
5. Vysotska, V., Rishnyak, I., Chyrun, L.: Analysis and evaluation of risks in electronic commerce. In: 9th International Conference CAD Systems in Microelectronics, pp. 332–333 (2007)
6. Pospelov, D.: Situatsionnoe upravlenie: teoriya i praktika (1986)
7. Golota, Y.: Logika antonimov i nechetkaya logika: skhodstva i razlichiya. Soft Computing and Measurement, pp. 208–210 (1998)
8. Grinyaev, S.: Nechetkaya logika. www.computerra.ru/offline/2001/4
9. Zade, L.: Ponyatie lingvisticheskoy peremennoy i ego primenenie (1976)
10. Osgood, C.: The nature and measurement of meaning. Psychol. Bull. **49**, 197–237 (1952)
11. Rashkevych, Y., Peleshko, D., Vynokurova, O., Izonin, I., Lotoshynska, N.: Single-frame image super-resolution based on singular square matrix operator. In: IEEE 1st Ukraine Conference on Electrical and Computer Engineering (UKRCON), pp. 944–948 (2017)
12. Rusyn, B., Lutsyk, O., Lysak, O., Lukeniuk, A., Pohreliuk, L.: Lossless image compression in the remote sensing applications In: Proceedings of the IEEE First International Conference on Data Stream Mining & Processing (DSMP), pp. 195–198 (2016)
13. Kowalik, D.: Polish vocational competence standards for the needs of adult education and the European labour market. In: International Conference on Advanced Information Engineering and Education Science, ICAEES, pp. 95–98 (2013)
14. Davydov, M., Lozynska, O.: Mathematical method of translation into Ukrainian sign language based on ontologies. In: Advances in Intelligent Systems and Computing, vol. 689, pp. 89–100 (2018)
15. Davydov, M., Lozynska, O.: Information system for translation into Ukrainian sign language on mobile devices. In: 12th International Scientific and Technical Conference on Computer Sciences and Information Technologies (CSIT), Lviv, pp. 48–51 (2017)

16. Davydov, M., Lozynska, O.: Linguistic models of assistive computer technologies for cognition and communication. In: 12th International Scientific and Technical Conference on Computer Sciences and Information Technologies (CSIT), Lviv, pp. 171–175 (2017)
17. Kravets, P.: The control agent with fuzzy logic. In: Perspective Technologies and Methods in MEMS Design, pp. 40–41 (2010)
18. Basyuk, T.: The main reasons of attendance falling of internet resource, In: Xth International Scientific and Technical Conference Computer Sciences and Information Technologies (CSIT), Lviv, pp. 91–93 (2015)
19. Mobasher, B.: Data mining for web personalization. In: The Adaptive Web, pp. 90–135. Springer (2007)
20. Dinucă, C.E., Ciobanu, D.: Web content mining. In: University of Petroşani, Economics, 85 (2012)
21. Vysotska, V.: Linguistic analysis of textual commercial content for information resources processing. In: Modern Problems of Radio Engineering, Telecommunications and Computer Science, TCSET 2016, pp. 709–713 (2016)
22. Khomytska, I., Teslyuk, V.: The method of statistical analysis of the scientific, colloquial, belles-lettres and newspaper styles on the phonological level. In: Advances in Intelligent Systems and Computing, vol. 512, pp. 149–163 (2017)
23. Khomytska, I., Teslyuk, V.: Specifics of phonostatistical structure of the scientific style in English style system. In: XIth International Scientific and Technical Conference Computer Sciences and Information Technologies (CSIT), Lviv, pp. 129–131 (2016)
24. Xu, G., Zhang, Y., Li, L.: Web content mining. In: Web Mining and Social Networking, pp. 71–87. Springer (2011)
25. Jivani Anjali Ganesh: A comparative study of stemming algorithms. Int. J. Comp. Tech. Appl. **6**, 1930–1938 (2011)
26. Shakhovska, N., Vovk, O., Hasko, R., Kryvenchuk, Y.: The method of big data processing for distance educational system. In: 12th International Scientific and Technical Conference on Computer Sciences and Information Technologies (CSIT), Lviv, pp. 461–473 (2017)
27. Shakhovska, N., Vysotska, V., Chyrun, L.: Features of E-learning realization using virtual research laboratory. In: XIth International Scientific and Technical Conference Computer Sciences and Information Technologies (CSIT), Lviv, pp. 143–148 (2016)
28. Shakhovska, N., Medykovsky, M., Stakhiv, P.: Application of algorithms of classification for uncertainty reduction. Przeglad Elektrotechniczny **89**(4), 284–286 (2013)
29. Schahovska, N., Syerov, Y.: Web-community ontological representation using intelligent dataspace analyzing agent. In: 10th International Conference the Experience of Designing and Application CAD Systems in Microelectronics, CADSM 2009, pp. 479–480 (2009)
30. Shakhovska, N.: Consolidated processing for differential information products. In: 2011 Proceedings of VIIth International Conference on Perspective Technologies and Methods in MEMS Design (MEMSTECH), pp. 176–177. IEEE, May 2011
31. Zhezhnych, P., Markiv, O.: Linguistic comparison quality evaluation of web-site content with tourism documentation objects. In: Advances in Intelligent Systems and Computing, vol. 689, pp. 656–667 (2018)
32. Gozhyj, A.: Development of fuzzy situational networks with time constraints for modeling dynamic systems. Naukovi visti KPI **5**, 15–22 (2015)
33. Wanget, P.P.: Fuzzy logic: Theoretical and Practical Issues. Springer, Berlin (2007)

Tourist Processes Modelling Based on Petri Nets

Valeriia Savchuk(✉) and Volodymyr Pasichnyk

Information Systems and Networks Department,
Lviv Polytechnic National University, Lviv, Ukraine
{Valeriia.V.Savchuk,Volodymyr.V.Pasichnyk}@lpnu.ua

Abstract. The paper is devoted to decision-making processes modelling in the field of tourism. Before developing models a wide range of information sources was analysed and its result is described in the paper. To develop models Petri nets method was chosen. The following models are described in the paper: Tourist travelling process model, Tourist preferences defining process model, Trip planning process model, Trip planning process model, Excursion process model, Excursion content formation process model. The main features that unite these models that distinguish them among others are considering individual characteristics of tourists and dangerous factors of tourist destinations. Modelling of processes in the field of individual tourism can lead to developing new better information technologies that can provide personalized information support to the user. As a result of research new tasks are distinguished and described in the conclusion part.

Keywords: Modelling · Tourism · Individual tourism · Decision-making Petri nets · Decision support systems · Tourism information technologies

1 Introduction

Individual tourism is gaining popularity according to the development of the tourism industry and the availability of international trips. The peculiarities of this type of tourism are personalized approach to the tourist and an individual tour program. A tourist can plan his own trip on his own or apply for assistance to tourist organizations. It should be noted that the support of tourism agencies at all stages of trip (planning and implementation) costs extra charges. The disadvantage of self-planning tours is the need to make a lot of different decisions. Modern tourist information technologies are aimed at supporting decision-making by a tourist independently [1].

The personal safety is also very important aspect as inexperienced traveller by making wrong choices can get into situations that will make a trip a disaster [2]. Main dangers during the trip are social (terrorism, cultural diversity, etc.) and natural (extreme climate, dangerous plants and animals). The tourist can mainly avoid them when planning the trip.

According to the analyses of tourist information technologies presented in the previous articles [3] only few researches take into account individual preferences of tourist and safety of the tourist while travelling.

N. Shakhovska and M. O. Medykovskyy (Eds.): CSIT 2018, AISC 871, pp. 222–234, 2019.
https://doi.org/10.1007/978-3-030-01069-0_16

The development of high-quality tourist information systems requires a clear structuring and identification of the main aspects of the tourist processes that accompany an individual trip, including planning, tour guidance, tourist preferences identification, etc.

2 Aspects of Modelling Decision-Making Processes

2.1 Basic Definitions

Individual tourism – a trip of one person (or a small group of people: family, friends) according to an individually developed program [4].

International tourism – tourism, which considers crossing a state border [5].

Cognitive tourism – a trip that is conducted in order to gain new knowledge about popular or unique objects, monuments in social and natural field [4].

Tourist process – a consistent change of stages of a tourist trip and its subprocesses.

Decision-making is the process of choosing the most suitable alternative from an existing set according to the task [6].

Decision support system is a software for analyzing data, modeling processes, making prognoses and, as a result, decisions [7]. So, in simple words, this is kind of information system that gives recommendations on what decision to make according to the situation or preferences.

2.2 Making Decision in the Field of Individual Tourism

The decision-making process of planning and implementing a trip consists of the following stages [6]:

- defining goals;
- obtaining the necessary information;
- consideration of possible alternative solutions;
- decision-making;
- assessment of the decision.

Defining goals – the stage that is based on the process of forming a task and determining the results that should be obtained [7]. The main purpose of the cognitive journey is to acquire new knowledge about the locality, nature, culture, population of the region (regions), and so on. At the stage of setting goals, the following main aspects of travel are selected: a type of holiday, a type of tourist destination, available budget and period of the tour, etc.

The problem identification is to clearly distinguish the obstacles that are on the way to the goal [6. The main problem when planning and realizing a trip that faces a tourist is the lack of information, as knowledge itself is the basis of qualitative planning of all aspects of the trip.

The stage of obtaining the necessary information is to collect data on tourist destinations, the level of prices for accommodation and food, transportation solutions, sources of danger, interesting tourist objects, etc.

The stage of considering possible alternative solutions is devoted to development of options for solving the issue that is based on the information received and takes into account all aspects of trip, their advantages and disadvantages [6].

Making the right decision about further trips is one of the most difficult tasks, since it is necessary to take into account the wishes of all members of the group, possess and operate a wide range of data on the tourist destinations, different accommodation and their cost, correctly set priorities and take into account the positive and negative aspects of each alternative decision.

The decision-making process is completed after the analysis of the results of the decision that is made. Such result is the realization of the trip [7]. Qualitative analysis of the results of the tour will take into account the disadvantages and advantages of the decision when planning and conducting the next trip.

2.3 Aspects of Modelling Processes and Objects

Development of innovative information technologies is a very costly process, both from an economic and from a technical point of view. In order to design and develop a qualitative conclusive product that will support and implement various processes, it is necessary to analyze and simulate them [8].

Modelling process is the method of various phenomena and processes analysis in the way of replacing real objects with their conditional analogues [9].

When constructing models the following aspects should be taken into account [10]:

- nature laws, including physical capabilities of people;
- variety of final solutions;
- parallel paths;
- hierarchy of the stages of the model;

The process of creating a model can be divided into the following interconnected steps [10]:

Verbal Description of the Object of Modelling: At this stage, a primary analysis is performed. As a result the main criteria of the model, a description of the desired result, are given etc.

Numerical Representation of the Object Characteristics: The stage is designed to accurately assess the object of the simulation and desired results of modelling.

Determination and Selection of the Model: At the stage, an analysis of analogues of developed models is processed. As a result one model is chosen. Then it should be formalized.

Expansion of the Model: At this stage, new details should be added to the model in order to make it more similar to the real object. But it should be noted, that only the details that are valuable to the goal of modelling should be added so that there are no overload of the model.

Model Analysis: At the end of the modelling process, it is necessary to determine the extent to which the original object is similar to the model.

Based on set theory, logic, linguistic or graphic methods of modeling give the opportunity to take into account the wide range of different by type and importance factors, that have influence on decision making [10].

3 Mathematical Modelling of Tourist Processes Based on Petri Nets

A tourist trip is naturally divided into several stages [1]: planning, implementation, analysis of the results. Each of them requires of certain decisions to be made and a number of specific tasks to be solved.

At the stage of planning of a trip, a tourist has to solve a number of issues [1]: where to go, which transport to choose, which tourist attractions to visit, etc. At this stage, the tourist constantly changes in trip plans. When travelling, the tourist still solves a number of tourist issues, and makes changes to future plans.

The analysis of the results of the trip that has passed affects the next one, because, in the case of high-quality analysis, it is possible to improve the quality of the next trip.

In order to formalise and precisely analyse in general the whole tourist trip as a multicomponent process the model was developed based on Petri nets method (Fig. 1). The model is presented in the form of system:

$$G = (P, T, I, O), \tag{1}$$

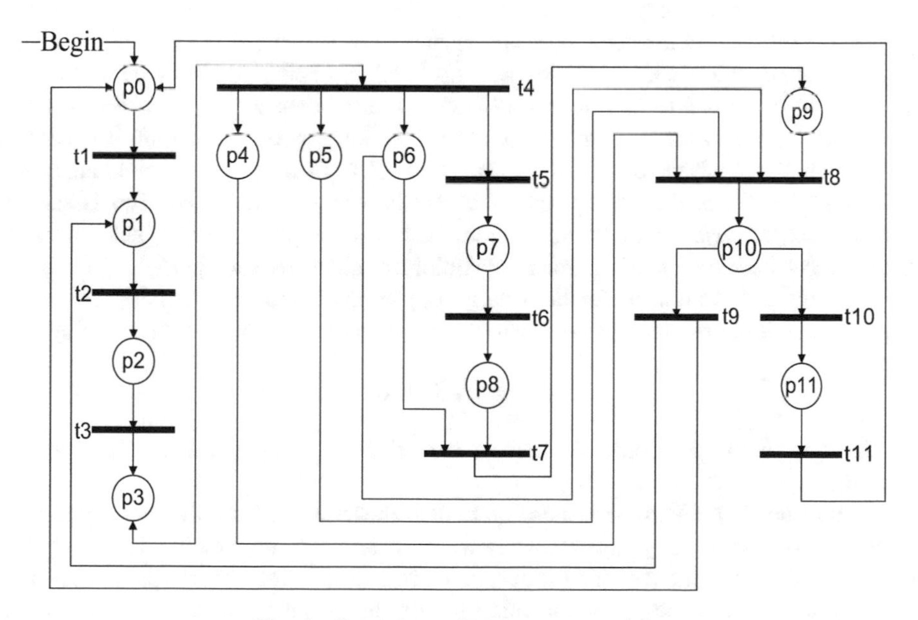

Fig. 1. Tourist travelling process model

where P is set of positions, T – set of transitions, I – input function, O – output function.

Transitions in the Petri net according to their meaning are the following actions: t_1 – defining basic trip aspects, such as type and time, t_2 – choosing tourist destinations, t_3 – planning a tourist route of a future trip, t_4 – choosing main aspects of the trip, such as accommodation, nutrition, transport and so on, t_5 – planning points of interest for the excursions, t_6 – excursion routes formation, t_7 – forming excursion content, t_8 – travelling, t_9 – changing plans, t_{10} – analysing trip results, t_{11} – making decisions on the next trip according to the results of the past.

The positions in the model are variables, that get there value and are used when the concrete transition is activated. They are: p_0 – the idea of trip, p_1 – the set of basic aspects of trip, p_2 – chosen tourist destination, p_3 – trip route, p_4 – accommodation, p_5 – transport decision, p_6 – trip information content, p_7 – the set of points of interests, p_8 – the set of excursion routes, p_9 – excursion content of the trip, p_{10} – the trip, p_{11} – conclusions on the past trip.

It should be mentioned that the activation of the next transition is impossible before activating all previous. The first initial marking μ_0 is – one point in the first position p_0. This assertion is right for all following models in the paper.

The originality of the presented model consists in a systematic approach to the tourist cognitive trip. Moreover, it takes into account that the analysis of the past trips as it can influence future decisions. This is important because modern tourist information systems should give proper support according to personal preferences to both new tourists and people with very big travelling experience.

A personalized approach to tourists is a component of a tourist trip, which is currently provided by tour operators and agencies. As a result of the analysis of a wide array of information sources, there was no available software that would take into account all personality characteristics of the traveler when giving him recommendations. According to the conclusion mentioned in the previous paragraph it should be noted that the analysis of previous trip can't give clear idea of individual tourist preferences of the user of the system when he/she has never travelled. This is also the reason why the user can't be sure about his/her own preferences. That's why we decided to take into account human psychological characteristics [11].

The model of tourist preferences defining process is shown on the fig. 2.

The model is presented in the form of Petri net and formalized as the next system:

$$Pp = (P, T, I, O), \tag{2}$$

where P is set of positions, T – set of transitions, I – input function, O – output function.

Transitions in the Petri net according to their meaning are the following actions: t_1 – collecting archive data, t_2 – interviewing the user, t_3 – filtrating archive data, according to the level of importance (long stops, recurring features of tourist spots, etc.), t4 – defining user preferences, t5 – analysing the results of interview, t6 – forming tourist profile.

The positions in the model are unique or sets of variables, they are: p0 – a query to form tourist profile, p1 – information on past interviews and cognitive tourist trips,

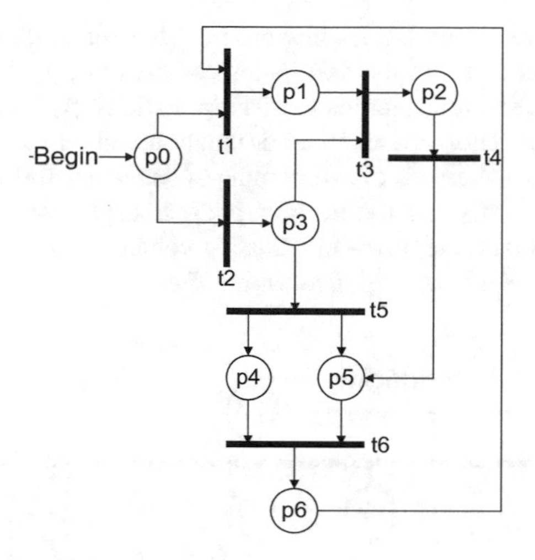

Fig. 2. Tourist preferences defining process model

p2 – filtrated data, p3 – interview results, p4 – psychological tourist profile, p5 – tourist preferences, p6 – user tourist profile.

Planning a tourist trip is the most important and most responsible step for a tourist trip. A well-planned tourist tour will require further efforts to overcome various problems.

One of the main problems that should be taken into account by a tourist during his tourist trip is safety. In the researches that are currently available insufficient attention is paid to the safety of tourist trip [3]. As the analysis of the vast amount of information on the dangerous and catastrophic events of recent years, neglecting of the safety factor in the planning and implementation of tourist trips leads to numerous human deaths and injuries, not to mention the severe psychological trauma among large groups of tourists. Therefore, at the planning stage of the trip it is necessary to prevent possible dangers and save yourself from traveling to potentially dangerous tourist destinations.

The model of trip planning process is presented in the form of Petri net and, as others, formalized as the same system:

$$Tp = (P, T, I, O), \tag{3}$$

Transitions in the Petri net according to their meaning are the following actions: t_1 – defining basic trip aspects, t_2 – choosing tourist destinations, t_3 – analysing danger level in every destination that is likely to be chosen, t_4 – classification of tourist destinations according to the defined dangers, t_5 – choosing dangerous destination, t_6 – choosing safe destination, t_7 – defining methods of reducing the danger level, t_8 – forming the set of points of interests, t_9 – choosing such aspects of trip as accommodation, nutrition, transport, information support etc., t_{10} – forming the route, information content, and set of things to take on a trip, t_{11} – combining all made decisions, t_{11} – changing decisions.

The positions in the model are variables, that get there value and are used when the concrete transition is activated. They are: p_0 – the idea of trip, p_1 – budget, p_2 – duration,

p_3 – period, p_4 – type of tourist destination, p_5 – the set of destinations of the type, p_6 – the set of danger levels in the destinations of the type, p_7 – the set of safe tourist destinations, p_8 – the set of dangerous tourist destinations, p_9 – chosen safe destination and its specifics, p_{10} – chosen dangerous destination and its specifics, p_{11} – the set of points of interests, p_{12} – methods of overcoming of dangers defined in the destination p_8, p_{13} – accommodation, p_{14} – nutrition type, p_{15} – transport decision, p_{16} – tickets on attractions, p_{17} – tourist route, p_{18} – information content, p_{19} – luggage, p_{20} – trip.

The following Petri net is presented on the Fig. 3.

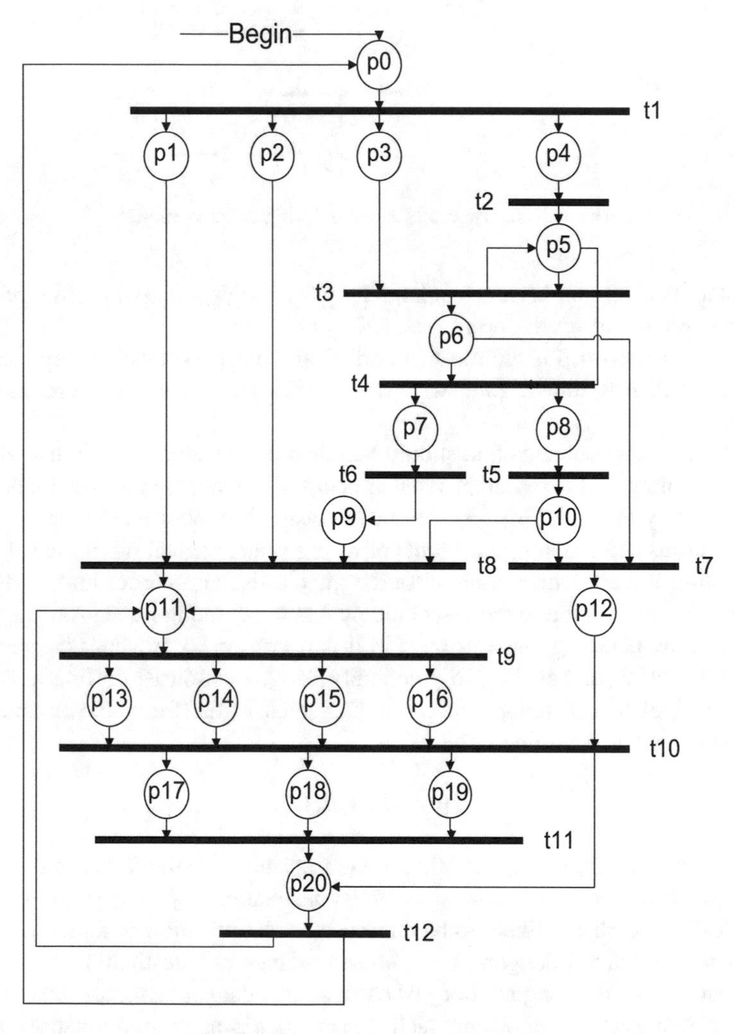

Fig. 3. Trip planning process model

As a result of the research that was conducted [12] the authors developed the excursion process model according to the tourist. The model is based on Petri net, that is performed as the next system:

$$E = (P, T, I, O), \tag{4}$$

where P is set of positions, T – set of transitions, I – input function, O – output function (Fig. 4).

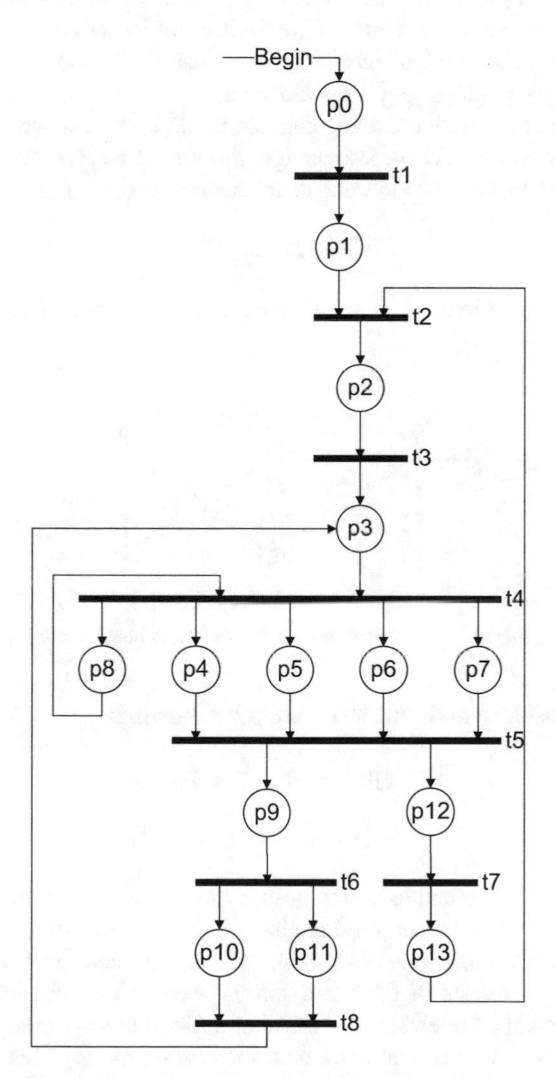

Fig. 4. Excursion process model based on Petri net

Transitions in the Petri net according to their meaning are the following actions: t_1 – defining free time periods, t_2 – choosing excursion from the proposed set, t_3 – coming to the start point, t_4 – studying point of interests, t_5 – taking pictures, t_6 – studying information between points of interests, t_7 – finishing excursion and analysing the results, t_8 – visiting next point of interests.

The positions in the model are variables, that get there value and are used when the concrete transition is activated. They are: p_0 – the idea of having excursion, p_1 – period and duration, p_2 – chosen excursion, p_3 – point of interest, p_4 – interior and exterior of the point of interests, p_5 – historical value, p_6 – cultural value, p_7 – the point of interests peculiarities, p_8 – the set of exhibits on the territory of the point of interests, p_9 – time needed on studying the point of interests, p_{10} – historical facts, p_{11} – information on cultural value, p_{12} – pictures, p_{13} – conclusions.

But the important issue when choosing excursion is its content. Excursion content is that set of information and so, knowledge, the tourist receive. The authors propose the following model of excursion content formation process (Fig. 5):

$$C = (P, T, I, O), \tag{5}$$

where P is set of positions, T – set of transitions, I – input function, O – output function.

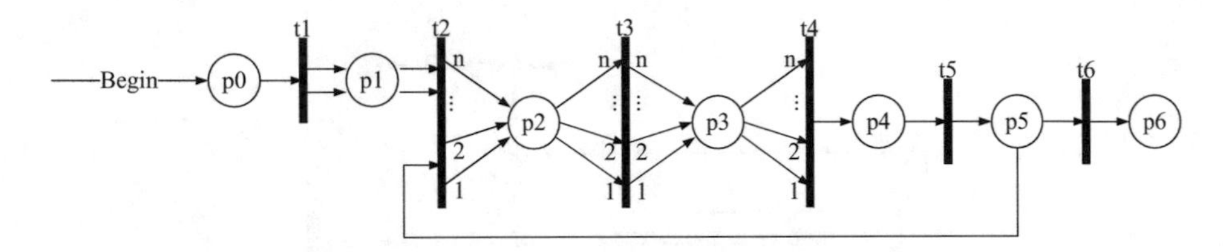

Fig. 5. Excursion content formation process model

The sets of positions and transitions are the following:

$$P = \{p_1, p_2, p_3, p_4, p_5, p_6\}, \tag{6}$$

$$T = \{t_1, t_2, t_3, t_4, t_5, t_6\}. \tag{7}$$

Transitions t_1–t_6 are operations, that a tour-guide, tourist organisation or information system should fulfil to form good quality excursion content: t_1 – defining the type of excursion (general-cognitive, historical, art-concentrated) and its duration, t_2 – choosing points of interests of the excursion, t_3 – choosing information content separately for every point of interests, t_4 – forming optimal tourist route according to type and duration of excursion, t_5 – making decisions on doing any changes in plans, t_6 – forming the resulting excursion content.

Positions p_1– p_6 are the results of the fulfilled operations (transitions) and so, the following data: p_0 – the idea to organise excursion, p_1 – type and duration, p_2 – points of interests of tourist route, p_3 – the information on point of interest, p_4 – tourist route, p_5 – data to change, p_6 – excursion content.

The first initial marking μ_0 is – one point in the first position p_0. When launched transition t_1, position p_1 gets two points that are symbols of data on type and duration of excursion. After second transition is launched, the p_2 gets n points (Fig. 5). Each of them is a point of interest and n – their general number. After launching third transition – p_3 gets n points, but this time they are information content sets on each chosen point of interests.

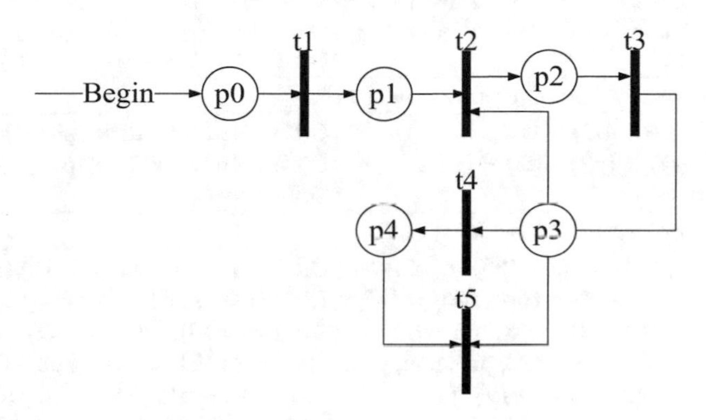

Fig. 6. Navigation process model

For systematization of navigation process [13] when developing tourist information systems the authors developed the model using Petri nets method (Fig. 6).

The model can be described by the following system:

$$N = (P, T, I, O), \tag{8}$$

where P is set of positions, T – set of transitions, I – input function, O – output function.

The transitions in navigation process model are: t1 – loading the system (any information system with navigation function), t2 – collecting data about environment (images of things around (streets, buildings etc.), satellite data, wireless nets finger-prints (Wi-Fi, GSM, etc.)), t3 – defining device location and movement directions (analysing data), t4 – forming optimal tourist route, t5 – visualisation of route location and movement directions.

Positions – are the following: p_0 – navigation query, p_1 – testing (data taken when teaching the navigation network or system) and archive data, p_2 – up-to-data data, p_3 – coordinates of device location, p_4 – optimal route.

The main purpose of proposed models of excursion processes is to generalise and systematize the processes, define their main characteristics, stages and needed information.

Input and output functions of all described models are shown in the Table 1.

Table 1. Input and output functions of the models

№	Model	
	Input function (I)	Output function (O)
1	*Tourist travelling process model (G)*	
	I(t1) = {p0}, I(t2) = {p1}, I(t3) = {p2}, I(t4) = {p3}, I(t5) = {p3}, I(t6) = {p7}, (t7) = {p6, p8}, I(t8) = {p4, p5, p6, p9}, I(t9) = {p10}, I(t10) = {p10}, I(t11) = {p11}	O(t1) = {p1}, O(t2) = {p2}, O(t3) = {p3}, O(t4) = {p3, p4, p5, p6, p7}, O(t5) = {p7}, O(t6) = {p8}, O(t7) = {p9}, O(t8) = {p10}, O(t9) = {p0, p1}, O(t10) = {p11}, O(t11) = {p0}
2	*Tourist preferences defining process model (Pp)*	
	I(t1) = {p0, p6}, I(t2) = {p0}, I(t3) = {p1, p3}, I(t4) = {p2}, I(t5) = {p3}, I(t6) = {p4, p5}	O(t1) = {p1}, O(t2) = {p3}, O(t3) = {p2}, O(t4) = {p5}, O(t5) = {p4, p5}, O(t6) = {p6}
3	*Trip planning process model (Tp)*	
	I(t1) = {p0}, I(t2) = {p4}, I(t3) = {p3, p5}, I(t4) = {p5, p6}, I(t5) = {p8}, I(t6) = {p7}, I(t7) = {p6, p10}, I(t8) = {p1, p2, p9, p10}, I(t9) = {p11}, I(t10) = {p12, p13, p14, p15, p16}, I(t11) = {p17, p18, p19}, I (t12) = {p20}	O(t1) = {p1, p2, p3, p4}, O(t2) = {p5}, O(t3) = {p5, p6}, O(t4) = {p7, p6}, O(t5) = {p10}, O(t6) = {p9}, O(t7) = {p12}, O(t8) = {p8}, O (t9) = {p11, p13, p14, p15, p16}, O(t10) = {p17, p18, p19}, O(t11) = {p20}, O(t12) = {p0, p11}
4	*Excursion process model based on Petri net (E)*	
	I(t1) = {p0}, I(t2) = {p1, p13}, I(t3) = {p2}, I(t4) = {p3, p8}, I(t5) = {p4, p4, p6, p7}, I(t6) = {p9}, I(t7) = {p12}, I(t8) = {p10, p11}	O(t1) = {p1}, O(t2) = {p2}, O(t3) = {p3}, O(t4) = {p4, p4, p6, p7, p8}, O(t5) = {p9, p12}, O(t6) = {p10, p11}, O(t7) = {p13}, O(t8) = {p3}
5	*Excursion content formation process model (C)*	
	I(t1) = {p0}, I(t2) = {p1, p1, p5}, I(t3) = {p2, p2, …, p2}, I(t4) = {p3, p3, …, p3}, I(t5) = {p4}, I(t6) = {p5}	O(t1) = {p1, p1}, O(t2) = {p2, p2, …, p2}, O(t3) = {p3, p3, …, p3}, O(t4) = {p4}, O(t5) = {p5}, O(t6) = {p6}
6	*Navigation process model (N)*	
	I(t1) = {p0}, I(t2) = {p1, p3}, I(t3) = {p2}, I(t4) = {p3}, I(t5) = {p3, p4}	O(t1) = {p1}, O(t2) = {p2}, O(t3) = {p3}, O(t4) = {p4}, O(t5) = {}

The described models of processes in the field of tourist trips take into account such issues as personal characteristics of traveller and sources of danger which are very important especially when planning the trip or choosing where to go. Every decision-making process starts with the idea of trip and choosing its main aspects (time, duration, transport and budget). According to these aspects the variants of trips are chosen and then compared. When comparing variants of trips such aspect as safety, points of interests and personal preferences are taken into account. All processes are modelled as cycles because: there is a need to make changes in plans or decisions, previous decisions influence the next ones.

Developed models give a clear view on what actions should the expert or an information system do to make the right decision

4 Conclusions

The paper provides a detailed analysis of basic decision-making processes during the tourist trip, which can highly influence on the quality of the results of the trip.

Petri's network is chosen to simulate the processes in the tourism industry, as it is a powerful tool for analyzing complex decision-making processes. Its advantage is the ability to describe both the subprocesses or simple actions and their results. Moreover, it can simulate parallel subprocesses that are extremely important to study in tourist trip processes. So, the use of this method of mathematical modelling allows you to analyze in detail the ways to achieve the desired result at each stage of the process.

During the research, a set of models in the field of individual tourist trips has been developed, namely: Tourist travelling process model, Tourist preferences defining process model, Trip planning process model, Trip planning process model, Excursion process model, Excursion content formation process model. The main feature of these models is taking into account the following aspects: the danger level on the territory of tourist destinations and individual preferences of tourists. These gives an opportunity to create high quality information technologies to support the tourist while preparing and doing cognitive tourist trips.

According to the result of deep analysis of information sources the main advantage of described models of decision-making processes is that they take into aspects as personal characteristics of tourists and the level of danger in traveling area [14]. Moreover, the models were developed directly with the purpose to be used when developing methods and information technologies of personalized tourist support on all stages of the trip, so they do not have redundant data, but they do not take into account transport tasks.

Developing models of processes in the field of individual trips is the first step to develop good quality tourist information systems with personalised attitude to the user.

For the future researches the following tasks are planed: detailing developed models processes and their subprocesses, improve developed by authors system "MIAT" according to the developed models.

As a result the authors concluded that the systematization of the processes in the field of tourism by developing the following models positively influence to the developing of modern tourist information technologies.

References

1. Chalmers, M.: Tourism and mobile technology. http://www.dcs.gla.ac.uk/~matthew/papers/ECSCW2003.pdf
2. Kennedy-Eden, H.: A taxonomy of mobile applications in tourism. University of Wollongong (2012)

3. Savchuk, V.V., Kunanec, N.E., Pasichnyk, V.V., Popiel, P. Weryńska-Bieniasz, R., Kashaganova G., Kalizhanova, A.: Safety recommendation component of mobile information assistant of the tourist. In: Proceedings of SPIE 10445, Photonics Applications in Astronomy, Communications, Industry, and High Energy Physics Experiments, 104455Z (2017)
4. Types of tourism. Groupe and individual tours. https://www.kursy-tourism.com/2014/08/08/
5. Kuzyk, S.P.: Tourism geography schoolbook. Knowledge, Kiev (2011)
6. Demyanenko, S.I.: Management of agrarian factories schoolbook. KNEU, Kiev (2005)
7. Power, D.J., Sharda, R., Burstein, F.: Decision support systems. In: Cooper, C.L. (ed.) Wiley Encyclopedia of Management. Wiley, Chichester (2015)
8. Melnykov, V.V.: Modelling process of supporting decision-making in innovational clasters. Bus. Inform. **2**, 172–177 (2016)
9. Tretiak, A.M., Dorosh, O.C.: Land management schoolbook. New book, Vinnytsia (2006)
10. Formal methods of building models. http://wiki.kspu.kr.ua/index.php/
11. Venugopal, K.R., Srinivasa, K.G. Patnaik, L.M.: Algorithms for web personalization. In: Soft Computing for Data Mining Applications, pp. 217–230. Springer, Heidelberg (2009)
12. Let the city talk to you. Location based audio tours. Pocket guide. http://pocketguideapp.com
13. Giaglis, G.M., Pateli, A., Fouskas, K., Kourouthanassis, P., Tsamakos, A.: On the potential use of mobile positioning technologies in indoor environments. In: 15th Bled Electronic Commerce Conference eReality: Constructing the eEconom. Bled, Slovenia (2002)
14. Smallman, C., Moore, K.: Process studies of tourists' decision-making. Annal. Tourism Res. **37**(2), 397–422 (2010)

Selective Dissemination of Information – Technology of Information Support of Scientific Research

Antonii Rzheuskyi[1(✉)], Halyna Matsuik[3], Nataliia Veretennikova[1], and Roman Vaskiv[2]

[1] Information Systems and Networks Department, Lviv Polytechnic National University, Lviv, Ukraine
antonii.v.rzheuskyi@lpnu.ua, nataverl9@gmail.com
[2] Tech StartUp School, Lviv Polytechnic National University, Lviv, Ukraine
vaskivroman@gmail.com
[3] Department of Ukrainian and foreign languages, Ternopil Ivan Puluj National Technical University, Ternopil, Ukraine
galuna.matsiuk@gmail.com

Abstract. A highly organized form of information support for interdisciplinary research is the selective dissemination of information (SDI) technology. The selection of information is a process of extracting the most valuable documents from the information flow, their individual parts or factual information. That is why the successful strategy for information service of scientists is the introduction of a system of selective dissemination of information.

The authors of the article improved the procedure of processing, providing information requests and giving answers in the system of selective dissemination of information. The Information Support of the Research service proposed by the authors is based on SDI. The technological cycle of user servicing in the system of SDI is represented by a few algorithms in the article. An algorithm for the implementation of SDI at the modern stage is presented. The means of individual and group informing of users and source search algorithm in library information systems are introduced.

For the first time, it is proposed to use an ontological approach to refine the information search to expand the ability to search relevant documents using keywords. As the use of a set of keywords is based on any information retrieval, an ontology is a new intellectual means for obtaining information, a contemporary method for representing and processing knowledge and queries.

Keywords: Selective dissemination of information · Information support
Scientific research · e-Science · Ontology · Information retrieval

1 Introduction

In the world of information flows, the implementation of effective scientific activity is impossible without continuous information support and information provision. The definition of the concept of information support, given in various scientific sources, highlights the purpose of information support, its object, means of implementation, process and activity as well as information and resource components. We consider that

N. Shakhovska and M. O. Medykovskyy (Eds.): CSIT 2018, AISC 871, pp. 235–245, 2019.
https://doi.org/10.1007/978-3-030-01069-0_17

information support is a set of processes for the preparation and provision of specially prepared information for solving management, scientific, technical and other tasks in accordance with the stages of their solution. The efficiency of any information support system depends on the quality of the used information. The required level of quality is ensured by searching and filtering the incoming stream of information. Information selection is the process of extracting the most valuable documents, their parts or factual information from an information flow according to the defined criteria. That is why a successful strategy of information service of scientists is the introduction of selective dissemination of information system (SDI). The selective dissemination of information system facilitates the search for information and its filtering in accordance with the information needs of users. With the rise of the Internet as a source of information, the amount of available information within the scientific interests has rapidly increased and this requires a more thorough review and selection of information.

From the beginning of the information resource creation and directly the search in automated systems was carried out by specialists who had skills in working with computing machines. Later, this work began to be performed by information workers. In late 1990 – early 2000, scientific or scientific and technical libraries started switching to the service in SDI mode on the basis of their own information resources. Implementing SDI using such technology, the system functioned in a semi-automated mode, where the selection of information was carried out manually, all other processes, including the delivery of notifications were done with the help of a computer. This marked the development of a new aspect of the implementation of information support in the mode of selective dissemination of information. According to today's realities, SDI system as a form of information support has undergone significant changes, both in resource and in technological aspects. The technology of SDI implementation varies with the development of tools for processing information, accumulating information resources and using the Internet. It becomes clear that with the help of databases of own generation, library institutions will not be able to implement SDI system fully. Providing the information needs of scholars with relevant and up-to-date information, librarians should use electronic journals of powerful foreign publishers, both open and closed full-text databases of electronic publications of various formats, scientific abstract and reference databases.

In recent years, there has been a tendency for the creation of geographically distributed virtual scientific teams, the need for scientists to conduct the research on the electronic science platform is generated, but the issue of identifying and developing information and communication tools for research in virtual creative teams on the e-Science platform remains relevant.

2 Background

Despite the recent popularization of the term of electronic research [1] as a substitute for the concept of e-science, new approaches to data management and curation help to focus on the information support of scientific research conducted on the E-science platform in all scientific fields [2] (Fig. 1).

Selective dissemination of information is a type of personalized current information service that includes document checking, the selection of information that accurately

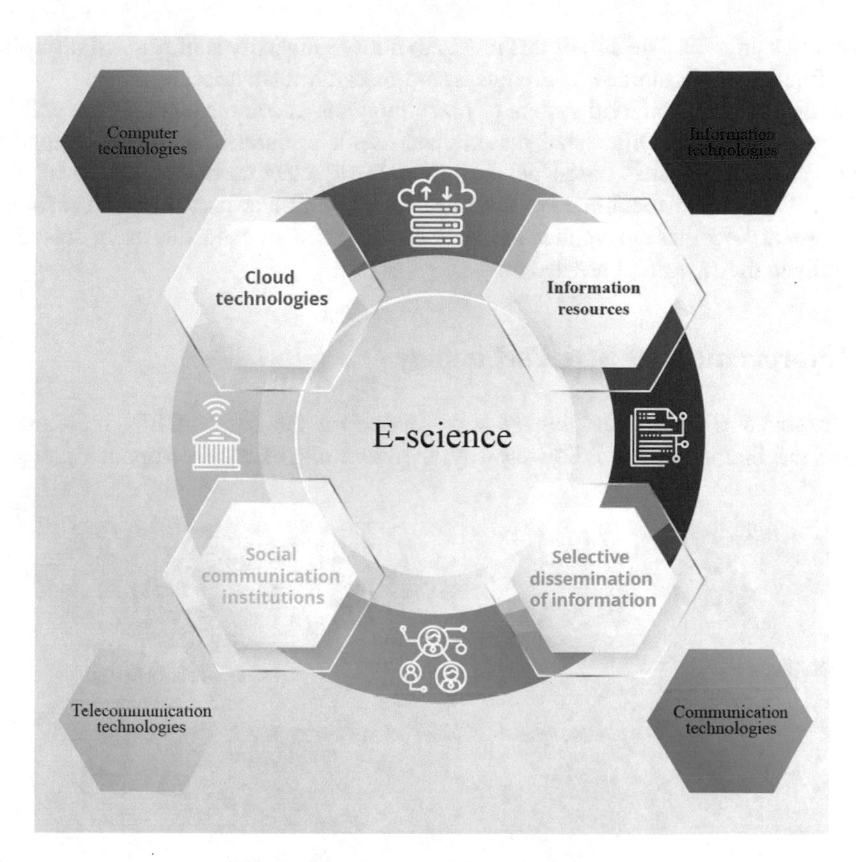

Fig. 1. The decomposition of e-Science.

responds to the specific research needs of each user or group of users [3], and the provision of information directly to each individual or group so that the user can independently monitor the latest developments in their area of specialization.

In the opinion of Chatterjee [4], SDI service is a type of alerting service designed for individual users, mainly professionals, which is exclusively limited to the area of their interests. This is a user-centric service.

Uzohue and Yaya [5] consider selective dissemination of information as a specialised form of current awareness services designed to disseminate information. It provides educational resources [6] for information support for healthcare professionals and researchers. Thus, they are aware of new developments in the field of research and patient care. Keeping up to date with latest information is competitive advantages in decision making. For medical libraries, SDI is a priority service for medical staff supporting with authoritative, reliable, relevant, accurate, up-to-date, and timely published information or sources.

Another example of using the system of selective dissemination of information was implementation of electronic message system in Research Authority at the Bar-Ilan university's unit [7]. The aim of this unit is to notify university staff from all disciplines about funding opportunities, conferences and research related issues.

In the paper [8], a hybrid system of fuzzy linguistic recommendations that will help Technology Transfer Office staff disseminate useful resources for research support is proposed. The system is based on the principle of selective dissemination of information. The system recommends both specialized and auxiliary research resources. This system becomes an application that can be used to help the users selectively disseminate the research knowledge.

3 Improvement of SDI Technology

The Information support of the research service on the basis of SDI presented by authors can be modeled in following way, by presenting its main components (Fig. 2).

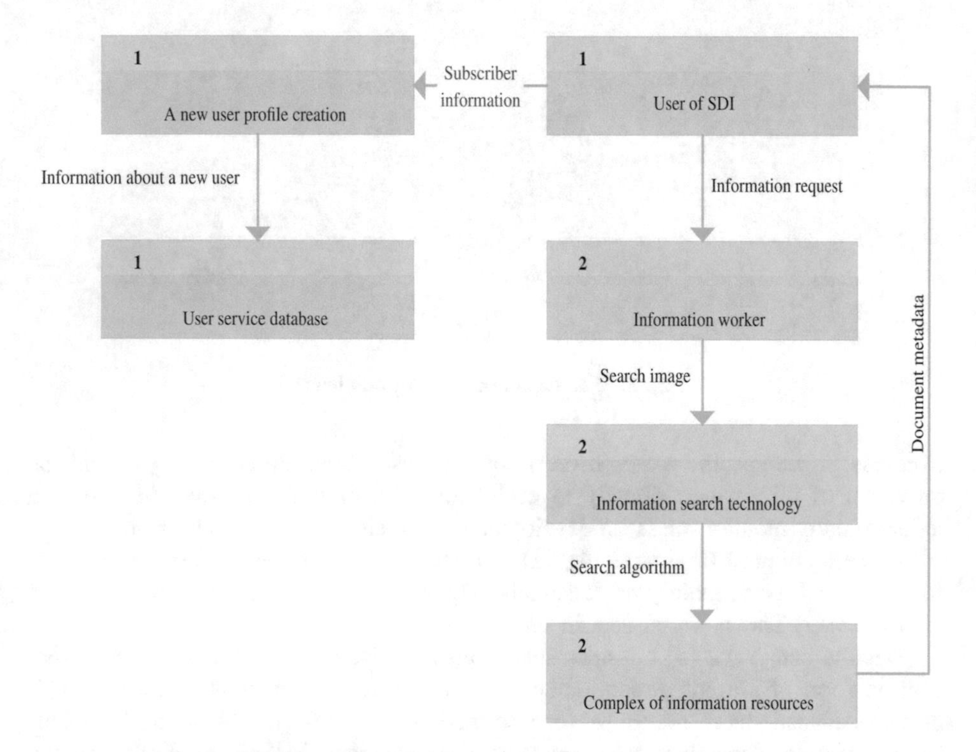

Fig. 2. The interaction of the main components in the SDI system.

The technological cycle of user service in SDI system can be represented by a number of the following algorithms (Fig. 3).

The proposed algorithm of the SDI system (Fig. 4) consists of series of consecutive operations.

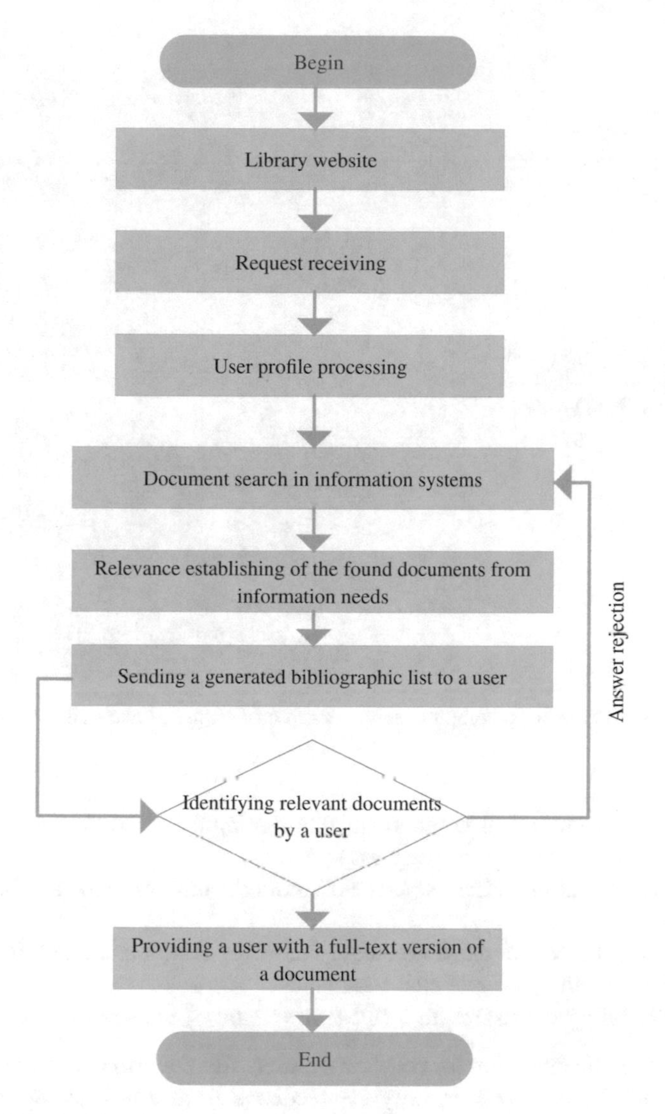

Fig. 3. The algorithm of SDI implementation at the present stage.

Step 1. User access to the SDI service via the web form on the library website and user profile creation.
Step 2. Processing of information from a profile, saving the user's personal data for constant information. The user provides information about the field of scientific interests, keywords or subject topics, according to which the information worker will generate a response to their request and generate metadata.
Step 3. Searching documents in information systems is the most labor-intensive, as a large staff of information professionals is engaged. It is a presence of a human

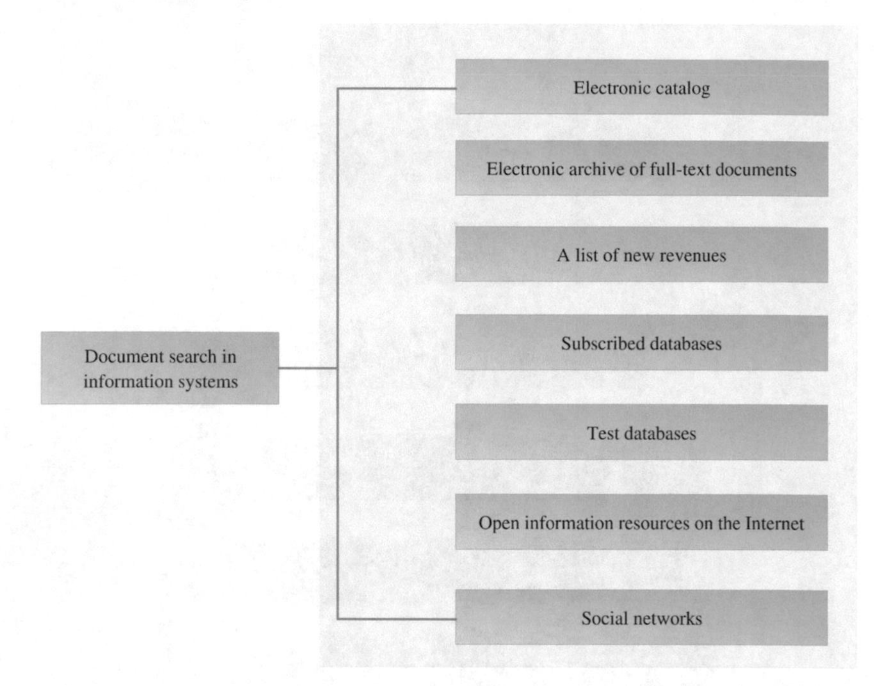

Fig. 4. A set of sources for a comprehensive search of documents according to the information needs of the user.

factor, as a technological component, that distinguishes our project from foreign ones, which are based exclusively on software.
Step 4. Determination of the relevance of found documents to the user's information needs.
Step 5. Send the consolidated metadata list to the recipient's email box.
Step 6. User defining of relevant documents.
Step 7. Providing the user with a full-text version of the documents.

At this stage, the operation for providing a user with the primary information can be considered completed. However, only the feedback from a user along with a library specialist can determine the relevance of the received bibliographic sources. In this case, if the answer satisfies a user, a user can generate a request for obtaining full-text versions of documents. Otherwise, the user sends the received source list to the information worker for revision. The cycle repeats until the information needs of the user are fully met.

Today, there are a number of technical tools that can be used in the process of informing users about new documents (Fig. 5), however, they function discretely and can't provide all operations described in the algorithm above.

Dobko T. points out that the selection of sources, containing potentially necessary information, requires their preliminary studying, evaluation and analysis. Undoubtedly, these processes can be carried out qualitatively by a specialist in the field of information search, who has certain practical skills.

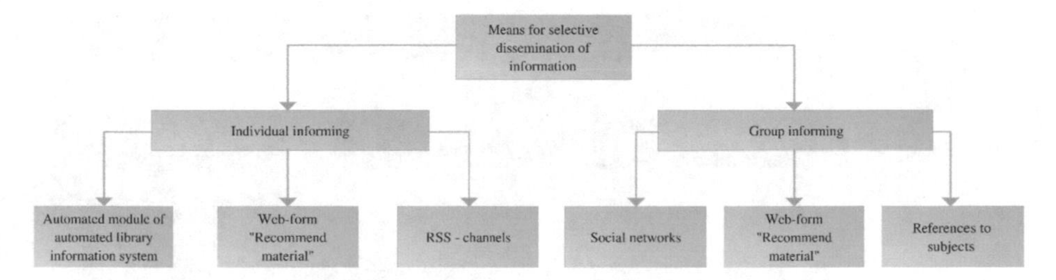

Fig. 5. Means of individual and group users informing.

In our opinion, modern librarians should be included in the group of information analysts. First of all, it concerns librarians being involved in the selective dissemination of information system. Their practical skills contribute to more effective satisfaction of information users' needs, active information support of scientific research, and the provision of needs of management decision-making.

At present, a library specialist has a large array of information resources for the organization of information work. It is up to them to decide which complex of information resources is more appropriate to operate for more clear consideration of the nature of information users' needs.

The information worker needs to understand the information needs of the users to meet them. If the information worker does not understand the needs of a user, they will not be able to perform effectively their professional duties.

At the initial stage, the information worker plays a role of a consultant and develops a special information search program. He or she provides the following list of works:

- formation of databases and data warehouses;
- information services;
- information analysis and compilation of information reports.

At the same time, the search for relevant documents on the user's request should be carried out in a complex with the involvement of all possible information retrieval systems and electronic resources (Fig. 6).

Among the main databases, in which information search on the requests of SDI system subscribers is carried out, are electronic catalogues, electronic (institutional) archives, science-metric databases such as Scopus and Web of Science.

Thus, in the proposed technology of organization of SDI modern system for information search, much more resources are used, not limited to information support of users based on own library document collections. Each information worker is assigned a search subject area for information search. That is why this approach makes the proposed implementation of SDI unique among other technological approaches, providing the team work of information workers.

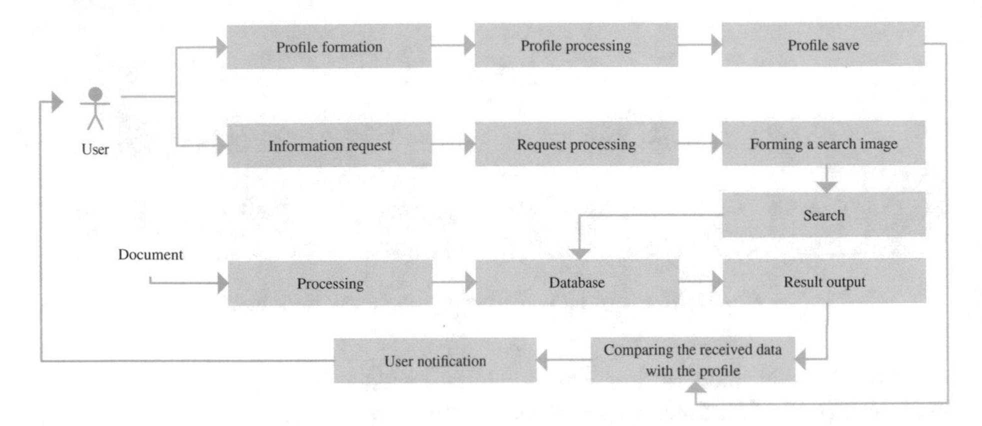

Fig. 6. The scheme of source searching in information systems of the library.

4 Applying an Ontological Approach to Optimize Information Retrieval

We have proposed an ontological approach in order to expand an ability to search for relevant documents using key words.

Ontologies outline the terms using which the subject area is described and structured, therefore, seeking and combining information from different sources and environments it is effectively to use ontologies. The ontology language is used to represent clearly defined values and is a generic set of terms for describing and presenting the subject area being studied. Ontologies are served as a means of communication between a library and a user.

Little attention is paid to the ontologies of libraries that provide the design process. To build such an ontology, you need to use knowledge of the subject area, especially a dictionary and rules for designing a specific tool that is useful for a user of the library. Creating such an ontology is a complex task, which depends on the means used to describe the subject area.

Ontology is a new intellectual means for obtaining information on the Internet, a new method of presentation and processing of knowledge and queries. It is suitable for accurate and effective description of the semantics of the subject area data. Ontology has its own logical conclusions (means of processing) in accordance with the tasks of semantic processing of information. When referring to the search engine, the user has an opportunity to receive relevant resources, semantically relevant to a request, due to ontology.

Ontologies can be used to compile library catalogs. The presentation of library catalogs in the form of ontologies allows to automate their processing and perform semantic search in the ontological space of libraries.

However, the lack of standardized requirements for collecting and processing knowledge in the library area in Ukraine greatly complicates the exchange of information in electronic form and leads to difficulties in its processing and use. In order to ensure the possibility of exchanging information in the library area, it is necessary to create a single information space, that is, the creation of common standards for the information dissemination, transmission systems and knowledge updating. The term ontology was firstly used in philosophy to refer to the system of knowledge related to the surrounding world. So, ontology is a science about the nature of things and the connections between them. Later, this term appeared in the context of information technology of knowledge presentation as a mechanism or method used to describe a particular area. Ontologies in information systems are created for solving complex problems that require intellectual data processing. In engineering knowledge an ontology is understood as a detailed description of the subject area in which the study of a specific problem is done, which has a large amount of data and it needs to be processed.

Ontology is a detailed formalization of the knowledge area, presented using a conceptual scheme [10]. Such scheme consists of a hierarchical structure of concepts (concepts, classes, entities), properties of concepts (slots, attributes, roles), relationships between concepts (relationships, dependencies, functions), additional constraints (defined by axioms, facets), and instances (individuals). A concept can be a description of a task, function, action, strategy, process of reasoning, etc. They are usually organized in taxonomy. Relationships represent the type of interaction between the concepts of the subject area. "Part-of" and "connected-to" serve as examples of binary relations. Axioms are used to simulate allegations that are always true. They can perform the following roles: limit the information contained in the ontology, verify the validity of the information or display new information. Copies are used to represent elements in a subject area, that is, an element of the given concept. Copies may be physical and abstract objects. An ontology can be constructed without specific objects, but as one of the main goals of ontology is the classification of such objects, therefore they are also included. Ontology, along with a plurality of individual copies, forms the basis of knowledge.

There are three levels of ontology representation:

- meta-ontology (ontology of the upper level) describes the most general concepts, for example, place, time, substance, object, event, action, etc., that do not depend on subject areas;
- ontology of a subject area is a formal description of the subject area, is usually used to clarify the concept, definition in meta-ontology (if it is used) and/or determine the general terminology database of the subject area;
- ontology of a specific task is an ontology that defines a general terminology base that relates to a characteristic area or characteristic task or activity by means of the specialization of the terms represented in the upper level ontology.

The main tasks that must be solved by ontologies are to:

- use the overall structure by humans and program agents;
- accumulate and reuse knowledge in the subject area;
- have a possibility to make explicit assumptions in the subject area;
- separate knowledge of the subject area from operational knowledge;
- analyze knowledge of the subject area.

The development of ontologies involves several stages:

- defining the industry and the scale of ontology;
- studying options for re-using existing ontologies;
- establishment of important terms for ontology;
- defining classes and hierarchy of classes;
- clarification of class properties - slots;
- defining facets of properties;
- creation of copies.

Several ontologies [11] are widely used in the world, based on which information systems of knowledge database management in various subject areas are made. Most works on the creation and use of ontologies are carried out abroad.

5 Conclusion

This paper analyzes the effectiveness of the information support of scientific research with the help of the Information support of research service, which is based on a highly organized system of selective dissemination of information. SDI is considered by authors as one of the communication tools for supporting information needs of virtual teams on the platform of E-science [12]. The success of SDI online service especially depends on the ability to identify related documents to the topic, selecting broader and more auxiliary or related subject fields and appropriate keywords that will lead the document to its core content.

The modern approach is characterized by the complexity of application of the latest information technologies and a comprehensive range of sources that are involved in the informing process (databases of abstract and full-text type, electronic archives of foreign and domestic higher educational institutions, electronic journals of foreign publishing houses). It should be noted that a staff of specialists can perform this process who specialize in a particular subject area and have heuristic skills of working with information resources. The entire technological cycle should be carried out in an automated mode, with continuously active feedback by means of information and communication technologies.

The main elements of SDI system are:

- subscribers of SDI system such as scientists and specialists (individuals or groups);
- formulation of information requirements as a list of topics (ongoing requests);
- regular provision the users with bibliographic information;
- use of modern technical means of mechanization and automation;
- permanent feedback;
- providing subscribers with primary documents or their copies.

The users of the Information support of the research service receive not only the metadata about the new documents from a certain field of knowledge, but also at the second stage, they get acquainted with full-text content. Thus, the two-contour system is provided. Improving the efficiency of SDI system is ensured using web technologies and automation of technological processes. The modern approach to user information support is a regular sending of messages in accordance with the information user needs as the library collections and databases are enriched.

Consequently, it can be argued that the ontology question is relevant and actual today because it is an effective way of presenting and systemizing knowledge. To solve the issue of libraries entering a competitive web environment, the development and implementation of ontologies in this subject area is one of the main conditions for a qualitative presentation of its own information product. This multidimensional and complex process is necessary for a modern user, as it contributes to the productivity of its information retrieval.

References

1. Rzheuskiy, A., Veretennikova, N., Kunanets, N., Kut, V.: The information support of virtual research teams by means of cloud managers. Int. J. Intell. Syst. Appl. **10**(2), 37–46 (2018)
2. Wang, M.: Academic library, e-Science/e-Research, and data services in a broader context. http://www.ala.org/acrl/sites/ala.org.acrl/files/content/conferences/confsandpreconfs/2013/papers/Wang_AcademicLibrary.pdf
3. Veretennikova, N., Kunanets, N.: Recommendation systems as an information and technology tool for virtual research teams. In: Advances in Intelligent Systems and Computing, vol. 689, 577–587 (2017)
4. Chatterjee, A.: Elements of Information Organization and Dissemination. Chandos Publishing, Cambridge (2017)
5. Uzohue, C.E., Yaya, J.A.: Provision of current awareness services and selective dissemination of information by medical librarians in technological era. Am. J. Inf. Sci. Comput. Eng. **2**(2), 8–14 (2016)
6. Rzheuskyi, A., Kunanets, N., Kut, V.: Methodology of research the library information services: the case of USA University libraries. In: Advances in Intelligent Systems and Computing, vol. 689, pp. 450–460 (2017)
7. Zimmerman, E., Kedar, R., Mackler, Y.: Select dissemination of information on research funding opportunities to university scientists. http://citeseerx.ist.psu.edu/viewdoc/download?doi=10.1.1.594.9227&rep=rep1&type=pdf
8. Porcel, C., Tejeda-Lorente, A., Martínez, M.A., Herrera-Viedma, E.: A hybrid recommender system for the selective dissemination of research resources in a technology transfer office. Inf. Sci. **184**(1), 1–19 (2012)
9. Dobko, T.: Reference and bibliographic activity of scientific libraries of the National Academy of Sciences of Ukraine: formation and development (XX century – the first decade of XXI century), Kyiv (2013)
10. Pfaff, M.: Ontology-based semantic data integration in the domain of IT benchmarking. http://mediatum.ub.tum.de/node?id=1392030
11. Wullinger, P.: Supporting format migration with ontology model comparison. University of Bamberg, Germany (2018)
12. Shakhovska, N., Vysotska, V., Chyrun, L.: Features of E-learning realization using virtual research laboratory. In: XIth International Scientific and Technical Conference Computer Sciences and Information Technologies (CSIT), Lviv, pp. 143–148 (2016)

Models of Decisions Support Systems in the Employment Industry

Iryna Zavushchak(✉), Iryna Shvorob, and Zoriana Rybchak

Lviv Polytechnic National University, Lviv, Ukraine
iryska2009@ukr.net

Abstract. The article deals with the problem of building decision support systems that are planned to be used in the field of employment. The features of the implementation of decisions are analyzed and the requirements to the decision support system are formulated. A stratified scheme of constructing a decision support system model is proposed, which makes it possible to simplify the formalization of the decision-making process. The process of modeling the context in the decision making process is considered. The article presents a formal representation of contextual models for business operations of the employment sector. The Context Graphic Model has been improved to address the problem of analyzing the context of business operations.

Keywords: Decision support system · Context · Employment Ontology

1 Introduction

The problem of decision-making arises in many areas of human activity. Moreover, each branch has specific requirements that determine the existence of different approaches to building a decision support system. Decision-making is not a random choice of one option with at least two elemental set of possible actions. The choice of this action is carried out in such a way that some definite goal is met that will satisfy the decision maker. Among the large number of solutions, it is possible to separate the so-called managerial decisions that relate to human actions aimed at achieving the full goal of management [1].

Management actions are differentiated depending on the subject area, importance, time horizon, degree of uncertainty of the situation on which the decision is made, and the degree of repeatability. Decisions taken in practice have different degrees of repeatability - from one-time solutions (unique) to decisions with high repeatability. In this regard, there is a classification of decision-making problems that takes into account knowledge of their structure [2]. According to this classification, the following problems are distinguished:

- those with a precisely defined and well-known structure;
- those with a partially known structure;
- those that do not have or have only the least-known structure.

N. Shakhovska and M. O. Medykovskyy (Eds.): CSIT 2018, AISC 871, pp. 246–255, 2019.
https://doi.org/10.1007/978-3-030-01069-0_18

It is necessary to pay special attention to the relationship between the degree of inaccuracy and the degree of knowledge of the structure of the problem, as well as the correlation between the degree of inaccuracy of the solution and the horizon of time. The less well-known is the structure, the more precise the decision-making situation is. In turn, the longer the implementation period, the more difficult it is to determine the conditions that can have an impact on success, and all the possible consequences of the decision.

The complex and obscure structure of the problem situation, the decisions taken, the remote time horizon and the high degree of inaccuracy are the hallmarks of so-called strategic decisions. In this case, inaccuracy may relate to knowledge of the consequences that may lead to the choice of a particular option or the knowledge of a plurality of options among which it is necessary to make choices. Structured solutions cover routine repetitive problems, for which the relevant decision-making procedures were previously developed. Therefore, there is no need to create such procedures again.

New unique problems with specific characteristics or structure which are extremely complex represent the branch of unstructured solutions. In making such decisions, as a rule, there are not even auxiliary procedures, since they were not the subject of research. In this case, when making decisions, it is necessary to address the abilities of the decision maker, his intelligence, associations, etc. [1].

The purpose of this article is to justify the approach to building decision support systems, which should ensure the solution of poorly structured tasks that arise during the process of employment.

2 Characteristics of Decision Making Tasks

Analyzing existing approaches to solving decision-making problems, one can single out the characteristics of these tasks, which are the most significant in terms of constructing methods for their solution.

The first such characteristic is the presence (or absence) of an objective model that binds most of the main task parameters. There is a class of decision-making tasks for which it is possible to build a reliable model (similar to models in the study of operations), in which the quality of the solution is estimated by many criteria.

The second characteristic combines the requirements put forward in the form of a final decision. The most common among these requirements are:

1. The selection of one best solution;
2. The distribution of the options under consideration into several classes of solutions;
3. Ordering solutions for quality.

The third important characteristic is related to how new to the decision maker the problem under consideration is. On the basis of human awareness, it is possible to distribute decision-making problems to two essentially different classes - a problem where the decision maker can themselves be an expert (she can evaluate the decision options as a whole and according to certain criteria), and the problems where the role of this person and experts differ significantly.

For problems of the first class is characterized by the presence of a person who makes a decision on the idea of an alternative variant of the holistic image - the gestalt. Often this gestalt is much wider and deeper than its formal image in the form of a set of ratings according to many criteria. If the problem is familiar to a person, she confidently uses a set of gestalt in its solution [3]. This class of problems is called the problem of holistic choice. Problems of the second class are typical for cases when the decision maker does not have sufficient information to have an idea of possible alternatives. To obtain such information, it is necessary to help experts who have special knowledge.

The last important feature of the problem of decision-making is its dimension. Under the dimension understand the number of criteria - the number of alternative solutions. It is clear that the dimension of the problem affects the choice of the method of its solution. In the case of decision-making in complex employment problems, which are characterized by a large dimensionality and a wide range of external factors influencing decision-making, use decision support systems that are implemented on the basis of modern computer technologies.

3 Requirements for Decision Support Systems

Decision support systems can be classified differently. If they are considered an information tool consisting of the appropriate combination of computer and software, as well as databases (databases) and models, then one can distinguish the following construction approaches:

- Organizational
- Managerial
- Instrumental

Representatives of the organizational approach emphasize the construction of systems that are compatible with the actions of decision makers, real processes, an appropriate strategy and tactics of enterprise management. The managerial approach emphasizes the nature of the tasks that are solved in the decision-making process (according to the level of use - strategic, tactical, operational, according to the degree of complexity - weakly structured, partially structured and unstructured). The instrumental approach focuses on the attention of computer scientists who should ensure the rapid development of the main tools for building a decision support system: system generators, examples of the specific application of the system, etc.

Each of the approaches described is based on the quality requirements of the decision support system, which are summarized by such models: management structure, management hierarchy, management process phase, independence model of the decision maker.

The model of governance structure focuses mainly on unstructured and partially structured issues.

The management hierarchy model provides support for decision makers at all levels of management, and helps coordinate these levels wherever possible. Such a model is associated with the implementation of the task of supporting structured problems. Its main task is the integration and coordination of decisions for a person dealing with the synthesis of common problems.

The model of the phases of the management process is the implementation of the idea of supporting a consistent decision-making process. If you divide the decision-making process into several phases (for example, recognition, design, selection, evaluation), then you can get additional information regarding the general problem. The latter may be unstructured only in relation to some phases within the overall decision-making task. At the same time, the structured problem must be structured at all its phases. This means that it is necessary to build algorithms and define principles that allow not only fixing the problem, but also to design alternative solutions and choose the best ones.

The personal model embodies the idea of universal support for heterogeneous decision-making processes. Operational level information systems are less dependent on an individual's or an expert's intuition and, therefore, can largely automate actions performed by a person. This is characterized by well-structured tasks. At the same time, with the shift of the direction to unstructured tasks and systems of strategic planning, the role of intuitive knowledge increases. Thus, intuitive human skills are beginning to play a dominant role in the decision-making process when solving complex problems.

The model of human independence in decision-making should emphasize the degree of its participation in management processes. The decision-support system must be capable of solving the problems that determine the dependent or independent level of participation of the person in the decision-making process. That is, when decisions are taken by a team or partially by several individuals in turn, or by one person in particular.

In terms of the design process, three levels of decision support systems can be identified as given in [4] or [17]. They characterize the types of hardware and software, as well as the specifics of the tasks for which the system is built. The first level - these are systems that now improve work. Such systems, in contrast to the typical applications of data processing methods, allow a particular person or team to solve a specific set of problems. An example of such a system is the system of investment financing in Western banks. The second level is the decision support system generator. It represents a software package that allows you to quickly and easily build a specific system. The third level, the system tools, covers the core technology that is used to build a decision support system. The largest number of developments is precisely this category of information tools, which include, first of all, new special-purpose languages, advanced operating systems, tools for the projection of color graphics, etc.

4 Approaches to Building Decision Support Systems

Taking into account the characteristics of tasks and the corresponding requirements to decision-support systems, it is possible to determine a number of approaches to their construction based on decision analysis, decision-making, computational solutions and their implementation.

The decision analysis is based on a modern theory of decision making with many goals in terms of uncertainty or risk. This approach is primarily aimed at developing recommendations and guidelines on how to make decisions. It provides a methodology for both structuring decision-making situations and for determining rational choices. The essence of the decision analysis is to break down complex problems into simpler components that are manageable. The decision is depicted as a sequence of elections, due to the effects of previous elections and events that can not be managed. The distribution of functions between decision makers and means of support is that the person provides all the input information, and the system integrates this information to carry out the ranking of different alternatives.

At the heart of the approach to research decisions is the choice of the way the person acts as a rationalist to a limited extent. The decision research approach is a procedure for creating a decision support system as an opportunity to improve the decision-making process and increase its efficiency. When constructing a system, on the basis of the decision research approach, complex models are used together with empirical methods for their application, which are based on the study of human behavior when making a decision in order to reflect the current decision-making mode. Thus, although in the approach to studying system solutions are created for the existing decision-making process, their application should guide human behavior within the desired process development.

The decision-making approach is based on the application of methods for investigating operations in the process of creating a system. Such systems use a set of procedures that implement a model approach to data processing and judgments. A model is an element of an organization that is created to support a person with consideration of her thoughts and experiences. The procedure first reveals the verbally expressed intuitive version of the model of the decision maker. Subsequently, this model can be transformed into a formalized version in a mathematical form. At the next stage, all available data that combines estimates based on real data and human considerations are used to refine the model parameters.

Particular attention is paid to the implementation of the decision-making support system. The main idea of this phase is that the implementation of the system should take place not at the final stage of its creation, but in parallel with the very beginning of the system. Existing recommendations for the implementation of the system insist on starting the development from a simple version, quickly implementing it in practice, and gradually improving and distributing the system based on the experience gained through the interaction between the user, the system and those who design it.

Processes that implement decision support systems cannot always be described with only one approach. For example, when using an approach based on decision analysis, the person who accepts these solutions provides all the input information, and the system processes it. But in processing, to some extent, an approach based on computational solutions is used. That is, when creating a system for solving specific problems, different combinations of the approaches described above should be used.

5 Modeling the Context in Decision-Making Processes

A common feature of the approaches to building decision support systems is that they are oriented towards the creation of systems designed to address a wide range of problems. One of the decisive trends in the development of enterprise information systems today is the growing demand for quality management solutions. This problem is solved by the intellectualization of information systems [1], which was expressed in the transition to the concept of a cognitive enterprise. One of the main requirements of [2.3] to take account of such an enterprise is contextual information in decision-making, i.e. understanding and definition of these contextual elements as syntax and semantics of information, time, location, and features subject area, user profile information about business processes, current task and goals.

The concept of context is used and researched in many branches of science. The first attempts to understand and use this concept were made in linguistics to solve the tasks of automated translation. Later, with the advent of intelligent systems based on knowledge, it became clear that all knowledge is context-dependent, i.e. relevant only within a certain, often implicitly specified conditions. The lack of consideration of the contextual dependence of knowledge was the cause of failure in many intellectual systems projects [4].

The paper [12] is devoted to modeling the context for decision support systems. It describes and compares contextual models for solving contextual model management problems. Article [14] considers context models based on the use of ontologies for the Internet of things.

At the same time, existing comparisons of context models take into account only the features of the subject area. We believe that is useful for comparing a large number of parameters, which takes into account the mathematical apparatus universality/specificity of models, options for using the context as a generic group tasks.

The basis of our approach to comparing the models of the context is to take into account a greater number of factors. As a result of the analysis of the tasks of the domain [15], the following requirements for the context models were defined (Table 1):

Table 1. Requirements for contextual models.

Requirement	Justification
The focal object is a client - a person who is search employment	The main purpose of the employment service is to employ a client
Support as static and dynamic context	Determined by the variability of the situation in the labor market
Take into evaluation and experience of other people	Allows you to discard false and subjective information
Inaccurate and incomplete information	The information provided by the search provider or the employer is often incomplete and inaccurate and requires additional testing
Consideration and resolution of conflicts	Possible conflict of interest between the unemployed and the employer
Variety of sources of information and data formats	Employers provide information about vacancies in various formats
Take into account the history of changes	The need to take into account the history of the applicant's recruitment and employment processes
Ability to offer and choose alternative solutions	Job seeker satisfaction satisfies several vacancies

6 Graph-Oriented Models of Working Out the Context for the Employment Sector

In [15], the method of constructing ontology of the subject industry on the basis of the analysis of the context of business operations is proposed, and an example of the creation of ontology for the employment industry is given.

The solution of the ontology construction task based on the analysis of operational contexts begins with the creation of a business process model using one of the process modeling languages. As a result, the structure of the process, its components of business operations and the connections between them are determined.

The second step is to analyze each business transaction. Determine:

- the entities and their attributes that participate in the transaction. If possible, refer them to types that are already defined in the general ontology on. If objects of these types are not present on ontology, they add to the ontology new types of objects or attributes to existing ones.
- the relation and their attributes are relevant to the operation. Similarly, (a) refers them to known types of relationships, or creates new types of relationships.

As a result of the analysis of business operations, they receive the ontology of this operation,

$$On_{op} \in On \tag{1}$$

In [10], the presentation of business processes by context graphs is proposed. The authors note that the structure of such a graph, which corresponds to the structure of decision-making in the business process, is well structured and is often standardized in normative documents. But in practice, the adoption of a specific solution depends on the context of the business operation, must take into account the nuances of the situation and occurs as a result of an analysis of this context. At the same time, [10] does not specify how to formalize the process of analyzing the context of a business operation and choosing specific solutions.

The ontological model of business operations containing ontology operation On_{op} and set SSt situations, which are defined in the context of this operation. Each situation is defined by the conditions specified using the values of the objects included in the ontological model of the operation:

$$Md_{op} = (On_{op}, SSt) \tag{2}$$

The set of situations is determined by the expert and determined in the form of an ontological model of the situation as a triple:

$$Sit = (Sig_{st}, \ SAc_{st}) \tag{3}$$

where Sig_{st}, - the signature of the situation, SAc_{st} - a set of actions. The signature of a situation is a condition given to the values of the ontology entities of the operation. If this condition is fulfilled, then it is assumed that the situation is taking place, and actions from SAc_{st} must be executed.

The process of analyzing the context of a business operation is as follows (Fig. 1). First, the current business operation and the corresponding ontological model are determined. The essence and ratio of the model are filled with current data. An analysis of the conditions included in the signature of model situations is underway. Depending on which situation is detected, appropriate actions are initiated.

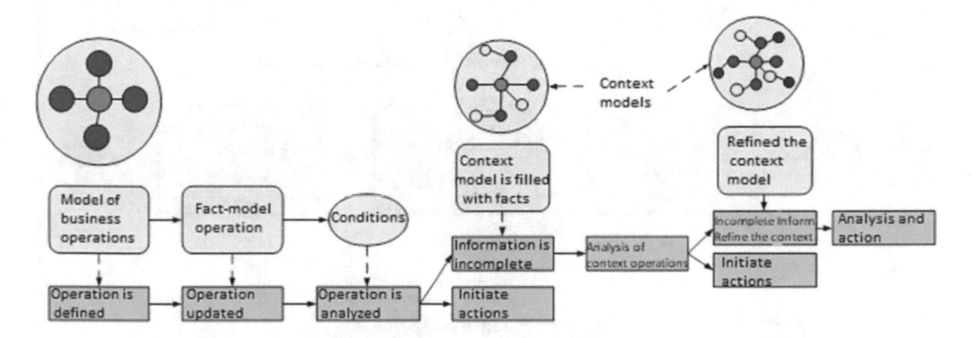

Fig. 1. The process of analyzing the context of a business operation

In practice, for some sets of input information values for decision making is not enough and it is necessary to obtain additional information from the context of the objects included in the ontological model of the business operation. The contextual model in which the business operation model is supplemented with entities and relationships from the context of existing objects is considered. In [16, 18, 19], it is suggested to use the chain of ontology elements for the formal representation of such contextual objects. This allows starting with the particular element of the ontology operation to determine the value of the context element, taking into account the values of all intermediate chain elements. The context model is filled in by the facts of the subject area. The analysis of this model determines the necessary actions. One possible action is to further refine the context for obtaining information necessary for decision-making.

7 An Example of a Context Analysis for the Employment Process

One of the business processes of the employment industry is the processing of the client's request (Fig. 2). The request is analyzed and entered into the Uniform Information and Analytical System (EIAS), appointed by the consultant and the date of admission. When analyzing a client request, it is expedient to investigate the client's context, for example, or for the first time he asks for a request. If the request is not submitted for the first time, then it is expedient to expand the search context, in particular, to consider the history of previous appeals and the results of the implementation of job placement recommendations. As a result, an excerpt from the history of previous appeals for a consultant will be formed. The system can also direct the client to a consultant with a wealth of experience that can analyze the causes of past job havoc.

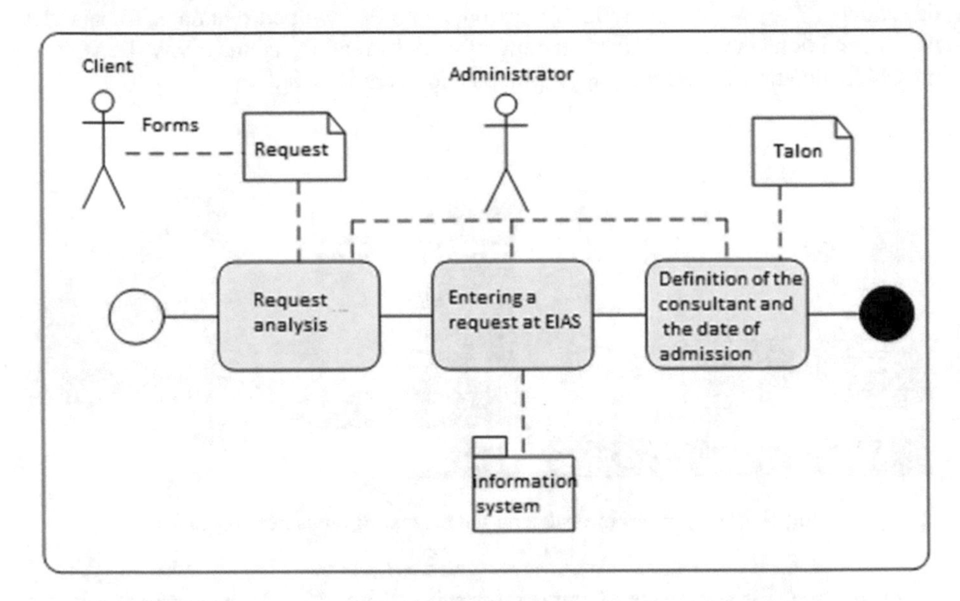

Fig. 2. Business operation model submission of employment application

Similarly, the system verifies the client's context in order to identify the client's level of qualification, his health and other information that can be used to improve the quality of service.

8 Summary

The analysis of the peculiarities of decision-making processes made it possible to formulate requirements for decision support systems that can be applied in the field of employment. One of the main requirements is the need to develop a model of the decision-making process, which should take into account the complexity and different degree of structuring of the primary decision-making problem.

Using ontological modeling to present and analyze the context of business operations in the field of employment allows you to provide knowledge about the nature and dependence of the subject area of employment and use them to build different types of decision support models. One of these models is a business process model in the context of a context graph, in which the vertices of the graph correspond to the operations of the analysis of the context of the business operations of the process, and the arcs determine the main directions of in-depth analysis of the context and the transitions between business operations.

The introduction of components of the analysis of the context of business operations as part of the information system will enable formalizing expert knowledge of decision-making in typical situations and automatically detecting such situations, conduct a detailed analysis of the context of the client, employer and job offerings in order to avoid mistakes and improve the quality of service.

References

1. Quinn, J.B.: The intelligent enterprise a new paradigm. Exec. **6**(4), 48–63 (1992)
2. Lewis, B., Lee, S.: The Cognitive Enterprise. Meghan-Kiffer Press, Tampa (2015)
3. Slovic, P., Lichtenstein, S.: Comparison of Bayesian and regression approaches to study of information in judgement. Organ. Behav. Hum. Perfom. **6**, 649–744 (1971)
4. Shakhovska, N., Vysotska, V., Chyrun, L.: Features of e-learning realization using virtual research laboratory. In: XIth International Scientific and Technical Conference Computer Sciences and Information Technologies (CSIT), Lviv, Ukraine, pp. 143–148 (2016)
5. Context - definition from merriam-webster dictionary. http://www.merriam-webster.com/dictionary/context
6. Dey, A.K.: Understanding and using context. Pers. Ubiquitous Comput. **5**, 4–7 (2001)
7. Kokar, M.M., Matheusb, C.J., Baclawski, K.: Ontology-based situation awareness. Int. J. Inf. Fusion **10**(1), 83–98 (2009)
8. Schmidt, A.: There is more to context than location. Computers **23**, 893–901 (1999)
9. Raz, D., Juhola, A.T., Serrat-Fernandez, J., Galis, A.: Fast and Efficient Context-Aware Services. Wiley, Chichester (2006)
10. Brézillon, P.: Task-realization models in contextual graphs. In: Modeling and Using Context, pp. 1–8 (2005)
11. Veres, O.M.: Aspects of manifestation of uncertainty in the processes of developing decision support systems. Bulletin of the National University "Lviv Polytechnic". Information systems and networks, vol. 829, pp. 58–75 (2015)
12. Smirnov, A., Levashova, T., Pashkin, M.: Models of context-managed decision support systems in dynamic structured fields. Pervasive Mob. Comput. **6**(2), 161–180 (2010)
13. Juan, Ye., Coyle, L., Dobson, S., Nixon, P.: Ontology-based models in pervasive computing systems. Knowl. Eng. Rev. **22**(4), 315–347 (2007)
14. Zavuschak, I., Burov, Ye.: The context of operations as the basis for the construction of ontologies of employment processes. Int. J. Modern Educ. Comput. Sci. **11**, 13–24 (2017)
15. Zavuschak, I.: Methods of processing context in intelligent systems. Int. J. Modern Educ. Comput. Sci. **3**, 1–8 (2018)
16. Burov, Ye.: Working out the context in the cognitive information system of managed models. East Eur. J. Adv. Technol. **1/7**(43), 40–47 (2010)
17. Melnykova, N., Shakhovska, N., Sviridova, T.: The personalized approach in a medical decentralized diagnostic and treatment. In: 14th International Conference The Experience of Designing and Application of CAD Systems in Microelectronics (CADSM), Lviv, Ukraine, pp. 295–297 (2017)
18. Shakhovska, N., Shvorob, I.: The method for detecting plagiarism in a collection of documents. In: 2015 Xth International Scientific and Technical Conference Computer Sciences and Information Technologies (CSIT), pp. 142–145. IEEE, September 2015
19. Shakhovska, N., Kaminskyy, R., Zasoba, E., Tsiutsiura, M.: Association rules mining in big data. Int. J. Comput. **17**(1), 25–32 (2018)

On Restricted Set of DML Operations in an ERP System's Database

Pavlo Zhezhnych(✉) and Dmytro Tarasov

Department of Social Communication and Information Activities,
Lviv Polytechnic National University, Lviv, Ukraine
{pavlo.i.zhezhnych, dmytro.o.tarasov}@lpnu.ua

Abstract. Information security is very important and critical indicator of reliability and efficiency of modern information systems. Violation of information integrity and availability usually causes to financial and reputational losses and incorrect decision making for owners of information. This paper proposes some approaches to avoid these information threads with the restricted set of DML operations that are available to users of an ERP system. These approaches are based on an analysis of semantics of data modification operations in terms of ERP-system developers and ERP security system violators that results special rules of applying certain DML operations during data processing. The analysis allowed identifying potential losses that may be caused by unauthorized usage of DML operations like inserting incorrect and redundant information, erasing necessary information, information faking, erasing the traces of previous interventions into the ERP system, blocking database data objects etc. The proposed approach to adapting the database schema to store the whole history of data records processing as regular data provides elimination of these losses because of disallowing the UPDATE operation and controlling the ability to use the DELETE operation for different types of ERP-system users.

Keywords: Information system · Database · DML · Database schema
Information security · ERP system · SQL

1 Introduction

Information security is very important and critical indicator of reliability and efficiency of modern information systems. Criteria of information protection are conveniently defined as ensuring integrity, availability and confidentiality [2, 18]. Violation of information confidentiality and availability usually causes to financial and reputational losses for owners of information or information systems. Violation of integrity additionally leads to threats of faking data and documents. Finding reliable and effective means of data threats blocking from different categories of security violators is a challenge for many years.

Modern technologies of corporate data storing, especially storing of large amount of structured information, involves usage of databases. Database technologies allow efficient and fast analyzing of information, formalizing the process of designing

N. Shakhovska and M. O. Medykovskyy (Eds.): CSIT 2018, AISC 871, pp. 256–266, 2019.
https://doi.org/10.1007/978-3-030-01069-0_19

modern corporate Enterprise Resource Planning (ERP) system, quick processing large amount of information [1, 10, 19].

Taking into account wide usage of relational databases, large amount of information accumulated in existing databases from various data domains, and existence of standardized means for database management systems (DBMS), this paper is focused on methods and technologies of information protection just in relational databases.

Data protection in ERP systems is carried out with a whole set of technical means at different levels (network rules, OS authentication, firewalls, database management system authentication, VPN authentication, etc.) [6, 16, 20, 21]. But bypassing one of network security levels, and compromised user's password or unreliable authentication means lead to unauthorized access to a database. Accordingly, ensuring data integrity, confidentiality and availability at the DBMS level is the last barrier in the way of blocking information threats in ERP systems [4, 11].

This paper is mainly focused on blocking threats of data integrity and availability caused by authorized users of ERP systems built with the use of relational databases. The peculiarity of authorized users' working with databases is that the users have successfully passed most levels of data protection associated with networks and ERP system components, and accessed protocols of interaction with a database with the authority defined for the a legal ERP-system user. This means that basic approaches of data protection for authorized users are setting appropriate database rights and privileges and data access control at the logical level of a database [22]. Correct implementation of the approaches requires correct determining of rules of data processing operations usage.

2 State-of-the-Art

Traditionally, there are distinguished two kinds of data processing operations in databases: selections and data modifications. These operations are implemented with commands of Data Manipulation Language (DML) [7]. Rights and privileges to perform these operations must be granted to most ERP-system users registered at a DBMS.

Selection operations include projection, filtering, join and union of tables, etc. In SQL, they are implemented with the following query.

- SELECT < expressions > FROM < from_item > …

Unauthorized usage of selection operations leads to violation of data confidentiality. So, there is no immediate impact on an ERP-system working caused by unauthorized usage of selection operations. The main problems are unauthorized usage of computing resources and reduction of database performance.

Data modification operations include adding records, removing records and changing values of record attributes in tables [5, 8]. In SQL, they are implemented with the following queries.

- INSERT INTO < table_name > (< columns_names >) VALUES (< expressions >) …

- DELETE FROM < table_name > …
- UPDATE < table_name > SET (< column_name >) = (< expression >) …

Unauthorized usage of data modification operations leads to violation of data integrity, information authenticity in an ERP-system, decision making efficiency, data confidentiality [13].

Consequently, unauthorized or wrong applying of data modification operations may invoke partial or total termination of ERP-system working with possible suspension of ERP-based enterprise functions caused by wrong or incorrect data stored in the database.

To reduce a level of information threats in ERP-systems traditional mechanisms of database user access restriction are used. User privileges on DML commands running are formalized with methods of access control and are distinguished on CRUD (Create Read Update Delete) operations. Therefore, data access is granted to a user with SQL commands at the level of DBMS.

In order to reduce a possibility of performing incorrect data modifications, CU’D’ and CU’ methods of restricted data processing were proposed [24]. These special methods are based on a concept of data processing stage and reduce a possibility of misapplication of user authorities when general CRUD data processing restriction is used.

Alternative approaches to avoiding data mistakes in an ERP system or data errors correction without losing existing information in a database provide preservation of historical information in the database.

The first approach of storing historical information is based on temporal databases and data versioning [12, 14, 22, 23]. Temporal databases naturally allow storing previous data values of attributes or data records but the corresponding DBMS are not widespread.

The second approach of storing historical information is information archiving with alternative databases and/or data structures, replacing old values of attributes with new ones. The approach has the following disadvantages:

- Double duplication of data values and/or data structures;
- Separation of actual and historical information complicates data analysis procedures;
- Additional special program units are needed.

3 Semantics of DML Operations

3.1 Semantics of Data Modification Operations in Terms of ERP System Developers

When designing a database schema, ERP system developers take into account DML capabilities and rules for working with specific tables, e.g. the rules may allow only data inserting, value substituting, batch processing of tuples, etc. (Fig. 1).

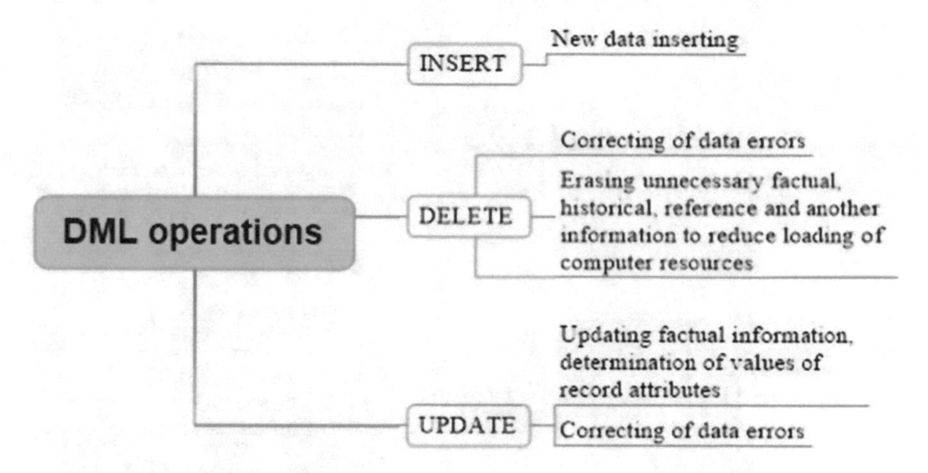

Fig. 1. Semantics of data modification operations in terms of ERP system developers.

Figure 1 shows that the purpose of data modification operations is inserting new information, errors correcting, updating factual information, erasing unnecessary information etc.

3.2 Semantics of Data Modification Operations in Terms of an ERP System Security Violator

A potential violator of ERP security system often accesses a database as a legal user (e.g., with a stolen password). Another way may include the usage of defects in security policies, authentication tools [3, 15], and obtaining additional administrator rights on database structures. Often, enterprise employees and legal users of an ERP system are conscious or unconscious security violators.

In all these cases, the violator has an ability to perform operations provided by the ERP system developers and the current security policy on data processing available for that purpose, in particular, to create and delete records, correct data errors etc. But the violator uses existing opportunities for another purposes (Fig. 2).

Legal ERP users often become unconscious data security violators due to misuse of data modification operations (Fig. 3).

So, the main task is losses reducing with the full preservation of the intended purpose of data modification operations while a security violator runs data modification operations.

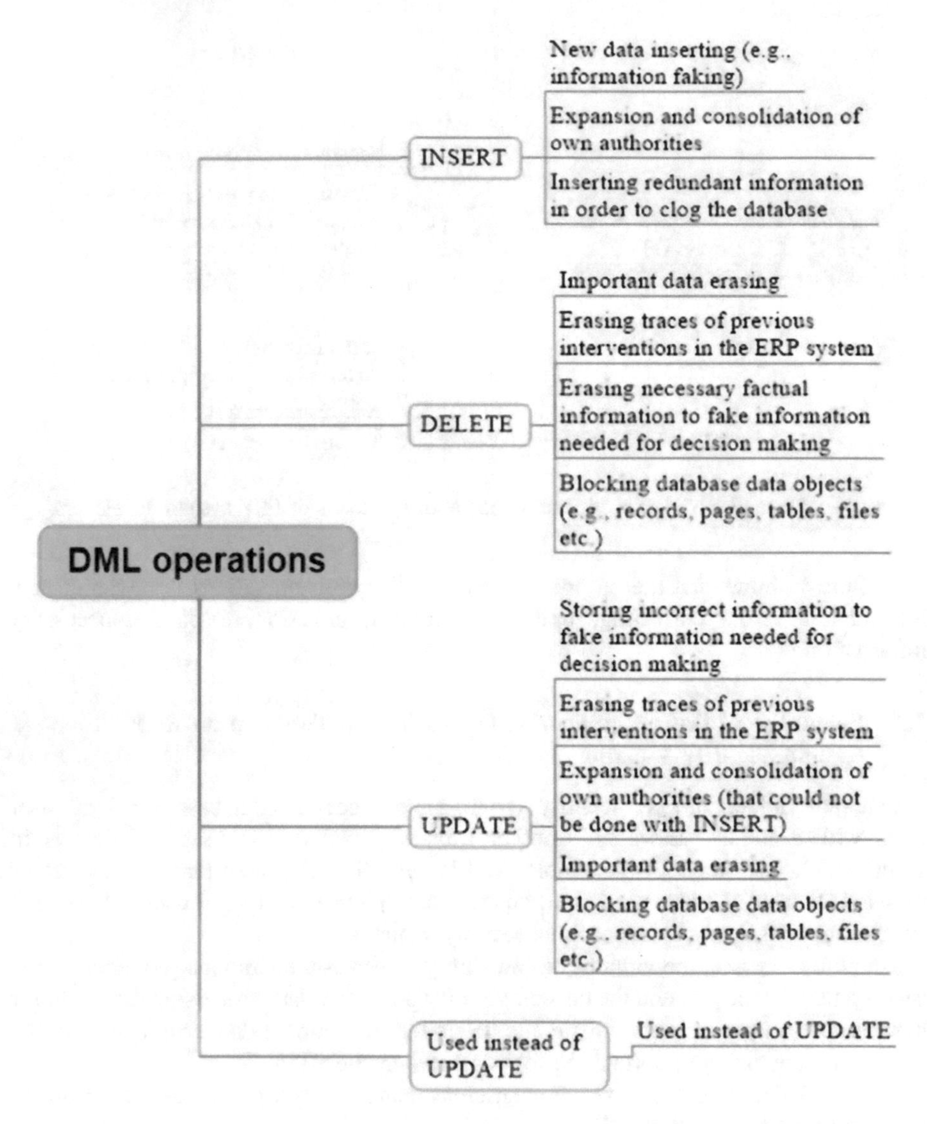

Fig. 2. Semantics of data modification operations in terms of ERP security system violators.

4 Approaches to Providing ERP Functionality with a Limited Set of Data Manipulation Operations

4.1 Database Schema Modifying

If users can correct erroneous data in an ERP system with replacing particular data (UPDATE) or removal of incorrect information then the correspondence between the database information and real information is violated. And this violates data integrity requirements.

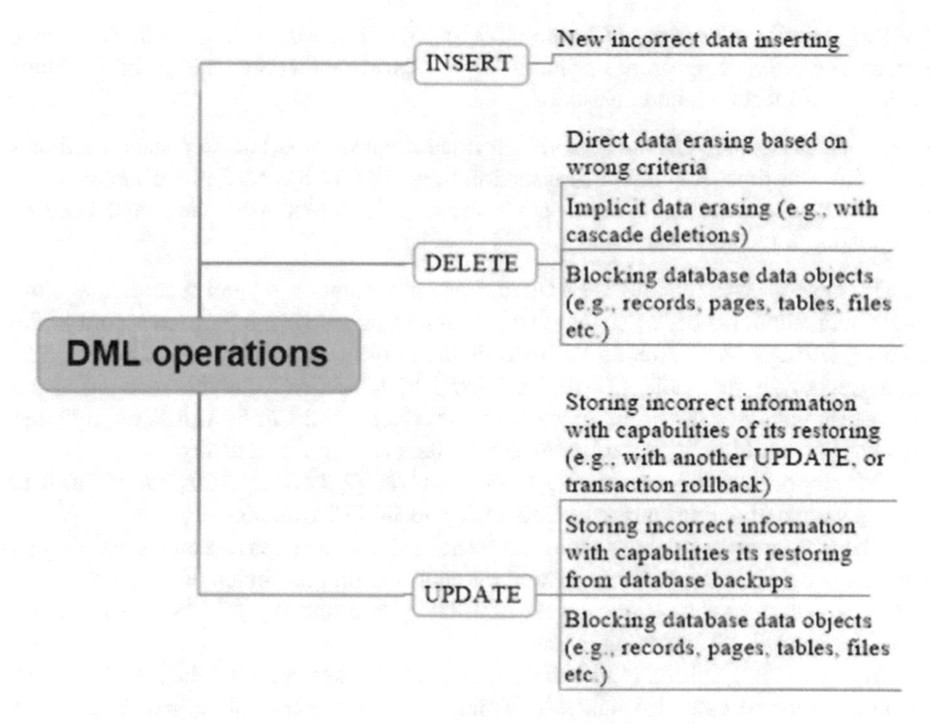

Fig. 3. Possible mistakes in using data modification operations.

A convenient method of storing historical information is designing of a database schema that ensures further accumulation of historical information in conjunction with actual latest values of attributes.

In [24], it is proposed to add additional attributes to each database table schema to separate 5 data processing states and additional attributes for data creation audit [9, 17]. The corresponding table schema *Q_TABLENAME* is represented by (1), where *id* is a primary key, *data_attr* is a set of user data attributes, *processing* is a data processing stage, *audit_attr* is a set of audit attributes.

$$\text{Q\_TABLENAME}\,(\text{id}, \text{data\_attr}, \text{processing}, \text{audit\_attr}) \tag{1}$$

For the purpose of more detailed storing information about data processing stages and audit information, and to avoid UPDATE operation needed to save changes in data processing stages, we propose to decompose the table *Q_TABLENAME* into two Tables (2) and (3).

$$\text{Q\_TABLENAME}\,(\text{id}, \text{data\_attr}, \text{audit\_attr}) \tag{2}$$

$$\text{Q\_TABLENAME\_PROCESSING}\,(\text{id}, \text{tablename\_id}, \text{processing}, \text{audit\_attr}) \tag{3}$$

The attribute *tablename_id* in the table *Q_TABLENAME_ PROCESSING* denotes foreign key referencing on the primary key *Q_TABLENAME.id*. The set of attributes *audit_attr* is identical and consists of:

- Attributes *crtuser, crtdate* – store information about creation user name (author's name), and time of creation (transaction time [23]) of a record in the table;
- Attributes *upduser, upddate* – store information about user name and time of updating a record in the table.

The table *Q_TABLENAME_PROCESSING* may contain several records that store information about history of changes of processing stages values for each record in the table *Q_TABLENAME*. Among the records the only one is considered as actual. The actual record in the table *Q_TABLENAME_PROCESSING* can be recognized by consequent values of the primary key *id*, or by creation time, or by values of any other attribute that could be added to *audit_attr* to denote record's actuality.

The proposed data structures *Q_TABLENAME, Q_TABLENAME_PROCESSING* allow data manipulating only with SELECT and INSERT queries.

When designing a database schema, it is advisable to take into account relationships between audit data *audit_attr* and standard audit logging attributes of a DBMS. Such integration of audit data at the levels of ERP system and the DBMS allows saving resources needed for analyzing events.

Thus, providing storing of historical data in an ERP system gives the possibilities of detailed audit and extended analysis of data in the context of time, and in addition allows the following:

- Eliminating the needs of data removing (DELETE);
- Correcting wrong data with only the insertion operation (INSERT);
- Omitting the update operation both for fixing errors and for defining values of individual attributes of stored records (UPDATE);
- Containing information about the whole state of the ERP-system at a given time;
- Integrating historical information with audit data of the ERP-system;
- Carrying out an analysis of personnel efficiency, the workflow speed, etc.

4.2 Restricted Set of DML Operations

In order to minimize losses when a security violator runs modification operations, it is needed to restrict usage of data modification operations with full disallowing of some operations and partial disallowing of other ones.

It is clear that the resulting set of operations should be complete in the sense that users can perform all their functions with operations of the set.

Therefore we propose:

1. Deny the operation UPDATE for changing attribute values.
2. Limit record removing permissions for regular ERP-system users to make it impossible:

- Replacing UPDATE with DELETE and INSERT operations;
- Erasing historical information and audit data;
- Cascade deleting;
- Erasing traces of unauthorized interventions.

Consequently, a restricted set of manipulation operations for database users contains operations INSERT, SELECT, DELETE. Usage of DELETE is allowed only for database administration, and it must be disallowed for regular ERP system users.

In order to maintain functionality of an ERP-system without usage of UPDATE and DELETE operations, it is necessary to provide storing of historical information in the database that needs proper data classifying and database schema designing, taking into account information security requirements and the current DBMS audit system.

Let us consider peculiarities of usage of DELETE operation and its combinations with INSERT by ERP system users for the purpose of data security violation.

Firstly, the DELETE operation allows the violator to directly erase records causing violation of correspondence between the database information and real data in documents and on other media. Secondly, the user has an ability to erase traces of interventions in the ERP system.

Especially the possibility of cascade data deletion is undesirable. In the case of record deletion in one database table, the cascade deletion erases (often without any warning) records of several tables that have a references to the first one.

The combination of DELETE and INSERT operations allows performing data updating. We suggest to restrict usage of data deletion operation (it is possible to completely disallow the usage of the operation). In particular, it is prohibited to remove records containing historical information related to protection data and audit data, and records referenced by foreign keys.

Of course, the database administrator is allowed to use all operations, including UPDATE and DELETE, because the operations sometimes significantly simplify the database administering tasks. E.g., erasing unnecessary information is a database administration task. Accordingly, rights to delete data are required only by database administrators.

5 Implementation of the Restricted Set of DML Operations

To implement the restricted set of DML operations, the following components should be used:

- Adapted database schema (according to the paragraph 4.1);
- Standard access control means of DBMS (like database objects VIEW, ROLE, SQL-commands GRANT, REVOKE etc.).

The considered approaches to database design and limiting users' data modification operations to INSERT and SELECT commands were used in designing and implementing a University Management Information System (UMIS) in Lviv Polytechnic National University that automates most of business processes of the university. The corresponding corporate database consists of 245 tables, and 21,63% of them are designed to support the restricted set of DML operations (Fig. 4).

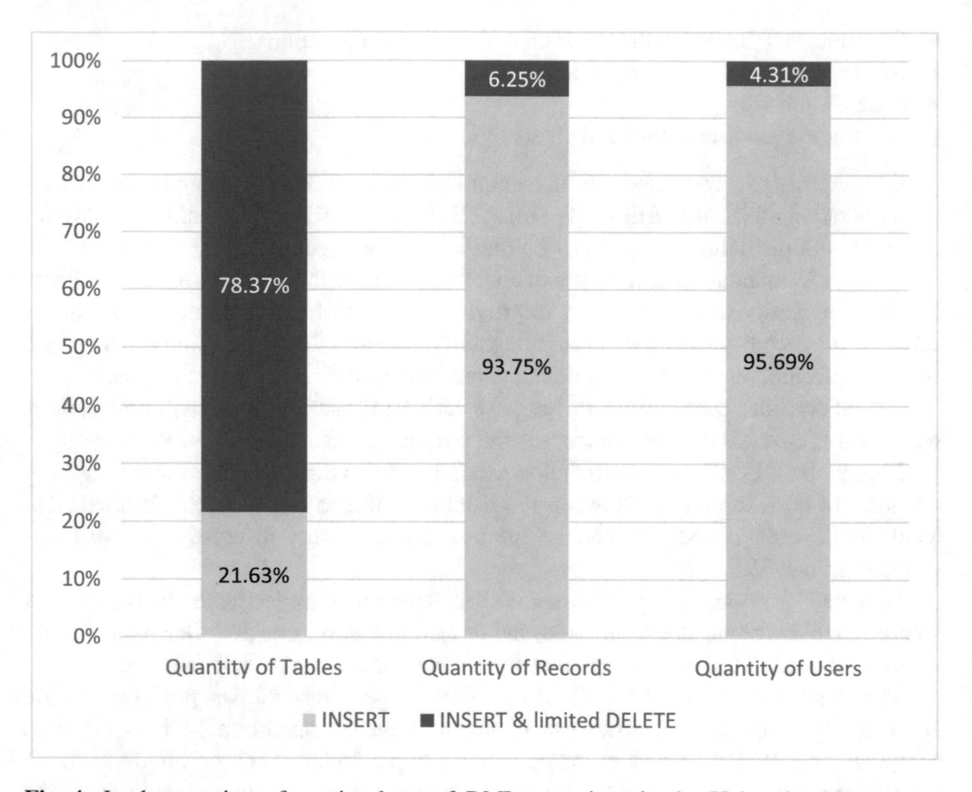

Fig. 4. Implementation of restricted set of DML operations in the University Management Information System.

As shown on Fig. 4 the restricted set of DML operations covers 93,75% of data records processed during a year by 95,69% of users (total quantity of users is about 3 thousand).

Database schema optimizing provided all needed operations that UMIS users perform to read and write data. But most of users are no allowed to run DML commands UPDATE, DELETE, so they are not able to violate information integrity and availability. and are limited to the INSERT, SELECT commands.

6 Conclusions

This paper considers approaches to increasing the level of integrity and availability of information stored in a database with the restricted set of DML operations that are available to users of an ERP system. These approaches are based on the expediency of applying certain DML operations during data processing that may provoke certain data threats.

An analysis of semantics of data modification operations in terms of ERP-system developers and ERP security system violators has allowed identifying potential losses that may be caused by unauthorized usage of DML operations. Unlike the "correct" data manipulation provided by an ERP-system developer, in terms of ERP security system violator, the main purpose of DML operations is to insert incorrect and redundant information, erase necessary information and make information fakes, erase the traces of previous interventions into the ERP system, block database data objects (e.g. records, pages, tables, files etc.).

In order to eliminate these losses, it is proposed to adapt the database schema to store the whole history of data records processing in the database as regular data. At the same time, this history of data records processing is isolated in separate tables related to the main tables of user data. This approach allows complete eliminating usage of the UPDATE operation and controlling the ability to use the DELETE operation for different types of ERP-system users.

To verify the effectiveness of the proposed approaches, an implementation of the restricted set of DML operations in the University Management Information System has been analyzed. With the help of adapted corporate database schema, 21.63% of the database tables allow avoiding of the above-mentioned violations of information integrity and availability for 93.75% of data records and for 95.69% of users during a year.

References

1. Alagic, S.: Relational Database Technology. Springer Science & Business Media (2012)
2. Bagiński, J., Rostański, M.: The modeling of business impact analysis for the loss of integrity, confidentiality and availability in business processes and data. Theor. Appl. Inform. **23**, 73–82 (2011)
3. Banyal, R., Jain P., Jain V.: Multi-factor authentication framework for cloud computing. In: Fifth International Conference on Computational Intelligence, Modelling and Simulation (CIMSim), pp. 105–110 (2013)
4. Basharat, I., Azam, F., Muzaffar, A.W.: Database security and encryption: a survey study. Int. J. Comput. Appl. **47**(12), 28–34 (2012)
5. Chaudhuri, S., Kaushik R., Ramamurthy R.: Database access control and privacy: is there a common ground? In: CIDR, pp. 96–103 (2011)
6. Daya, B.: Network security: History, importance, and future, vol. 4. Department of Electrical and Computer Engineering, University of Florida (2013)
7. Deutsch, D.R.: The SQL standard: how it happened. IEEE Ann. Hist. Comput. **35**(2), 72–75 (2013)
8. Ferretti, L., Colajanni, M., Marchetti, M.: Supporting security and consistency for cloud database. In: Cyberspace Safety and Security, pp. 179–193. Springer, Heidelberg (2012)
9. Govinda, K., Nelge, P., Malwade, M.: Database audit over cloud environment using forensic analysis algorithm. Int. J. Eng. Technol. **5**, 696–699 (2013)
10. Grabski, S.V., Leech, S.A., Schmidt, P.J.: A review of ERP research: a future agenda for accounting information systems. J. Inf. Syst. **25**, 37–78 (2011)
11. Jain, S., Ingle, M.: Software security requirements gathering instrument. Int. J. Adv. Comput. Sci. Appl. (IJACSA) **2**(7) (2011)
12. Künzner, F., Petković, D.: A comparison of different forms of temporal data management. In: International Conference: Beyond Databases, Architectures and Structures, pp. 92–106. Springer, Cham (2015)

13. Pascu, C.: Security principles in ERP systems. J. Mob. Embed. Distrib. Syst. **5**(1), 36–44 (2013)
14. Radhakrishna, V., Kumar, P.V., Janaki, V.: A survey on temporal databases and data mining. In: Proceedings of the International Conference on Engineering & MIS 2015, p. 52. ACM (2015)
15. Shoewu, O., Idowu, O.: A: development of attendance management system using biometrics. Pac. J. Sci. Technol. **13**(1), 300–307 (2012)
16. Spears, J.L., Barki, H.: User participation in information systems security risk management. MIS Q. 503–522 (2010)
17. Tarasov, D., Andrukhiv, A.: Algorithms of the corporate information system's protection analyses. In: Proceedings of the International Conference on Computer Science and Information Technologies (CSIT 2006), pp. 178–183 (2006)
18. Teixeira, A.: Attack models and scenarios for networked control systems. In: Proceedings of the 1st International Conference on High Confidence Networked Systems, pp. 55–64. ACM (2012)
19. Wang, M.T.: The design and implementation of enterprise management system based on ERP. Appl. Mech. Mater. **644**, 6221–6224 (2014)
20. Wang, X.: Network database security detection and the realized management program design. Netinfo Secur. **2**, 009 (2012)
21. Whitman, M.E., Mattord, H.J.: Principles of information security. Cengage Learning, Boston (2011)
22. Zhezhnych, P., Burak, T., Chyrka, O.: On the temporal access control implementation at the logical level of relational databases. In: XIth International Scientific and Technical Conference Computer Sciences and Information Technologies (CSIT), Lviv, Ukraine, pp. 84–87 (2016)
23. Zhezhnych, P., Peleschychyn, A.: Time aspects of information systems. In: Proceedings of the 9th International Conference on The Experience of Designing and Application of CAD Systems in Microelectronics (CADSM), pp. 530–533 (2007)
24. Zhezhnych, P., Tarasov, D.: Methods of data processing restriction in ERP systems. In: Proceedings of the 13th International Scientific and Technical Conference Computer Science and Information Technologies (CSIT 2018), Lviv, Ukraine (2018)

Vertically-Parallel Method and VLSI-Structures for Sorting of Arrays of Numbers

Ivan Tsmots[1], Skorokhoda Oleksa[1](✉), Vasyl Rabyk[2], and Antoniv Volodymyr[1]

[1] Lviv Polytechnic National University, Lviv 79013, Ukraine
oleksa.v.skorokhoda@lpnu.ua
[2] Ivan Franko National University of Lviv, Lviv 79000, Ukraine

Abstract. The vertically-parallel method for sorting one-dimensional arrays of numbers has been developed. The graph of the algorithm for vertically-parallel sorting of arrays has been built. The structure of the hardware for vertically-parallel sorting of one-dimensional arrays of large numbers has been designed. Components of the device for vertically-parallel sorting of arrays of numbers using FPGA have been implemented.

Keywords: Sorting algorithms · Vertically-Parallel calculations
VLSI · FPGA

1 Formulation of the Problem

The current stage in the development of information technology is characterized by the accumulation of large amounts of data. When processing such data arrays, it is often necessary to use data sorting and searching operations. The purpose of data arrays sorting is to accelerate the search for the necessary information. The main ways of increasing the speed of sorting operations are the development of new sorting algorithms, their adaptation to the architecture of modern mass-parallel computer tools (software implementation) and their implementation in the form of a very large scale integrated circuit (VLSI), whose hardware architecture reflects the structure of the sorting algorithm.

Hardware implementation of data sorting requires the development of new parallel algorithms and structures focused on VLSI-implementation. Hardware-oriented parallel sorting algorithms should be:

- well-structured with deterministic data movement;
- based on the same type of operations with regular and local connections;
- use conveyor and spatial parallelism;
- have a minimum number of interface outputs.

The orientation of hardware structures on VLSI-implementation requires reducing the number of interface outputs and implementation of sorting algorithms based on the same type of processor units (PUs) with regular and local connections.

N. Shakhovska and M. O. Medykovskyy (Eds.): CSIT 2018, AISC 871, pp. 267–284, 2019.
https://doi.org/10.1007/978-3-030-01069-0_20

Therefore, the problem of developing new parallel algorithms oriented on VLSI implementation becomes especially relevant.

2 Analysis of Recent Research and Publications

An analysis of known methods for data sorting has shown that they are based on a basic operation, which reduces to pairwise comparison and rearrangement of numbers [1–3]. Sorting methods based on such a basic operation differ from each other by selecting different pairs of values for comparison. These sorting methods can be grouped into the following groups: insertion sort, exchange sort, merge sort and displacement sort. Each of these sorting group is oriented toward sequential implementation.

In insertion sort algorithms, the process of pairwise comparison of numbers is combined with their permutation. The peculiarity of insertion sort algorithms is that they define not "numbers for places", but "places for numbers". From the family of these algorithms, the most suitable for VLSI implementation is direct insertion algorithm, since it is well structured with deterministic data movement. An increase in the number of simultaneously executed basic operations of pairwise comparison and permutation of numbers reduces the time of sorting the array of numbers and complicates the hardware implementation [1, 2].

Algorithms for merge sort, in comparison with algorithms that implement other methods, are more structured, homogeneous, and oriented both to sorting one-dimensional and two-dimensional data arrays. The basis of the algorithms for merge sort is the basic operation of combining two or more ordered arrays into one ordered array. In most cases, when sorting data arrays, the basic operation of combining two ordered arrays into one ordered array, i.e., a two-way merger, is used. The disadvantage of existing algorithms for the implementation of two-way merging is low speed since all of them are based on operations of pairwise comparison of data elements [1, 2]. One of the ways to improve the performance of a two-way merger is to implement it on the basis of the basic operation of multichannel merging and sending of data groups. The number of simultaneously executed basic operations determines the speed of data sorting tools, which may be different due to the symmetry of most algorithms for sorting the numbers.

An analysis of the methods for parallel sorting of one-dimensional data arrays [1–3] shows that the methods of sorting by counting and merging are the most oriented on hardware implementation. The disadvantage of parallel algorithms for merge sorting of one-dimensional data arrays is the low speed compared with the counting sorting algorithms, and their hardware implementation requires a large number of interface outputs.

The method of parallel sorting by counting involves comparing each number in an array with all other numbers. The parallel counting sort algorithm is performed in two stages [1]. In the first stage, using a simultaneous pairwise comparison of each number with all other numbers of the array, the amount of numbers greater, smaller and equal to such number is determined. In the second stage, based on the results of pairwise comparisons permutation of data is performed. The parallel algorithm for counting sort is distinguished by high speed. The disadvantage of such an algorithm is heterogeneity, a large number of interface outputs and significant hardware costs that are needed for its implementation.

The paper [4.5] considers the method of vertically-parallel search of maximal and minimal numbers in both one-dimensional and two-dimensional arrays. The disadvantage of usage of vertically-parallel maximum and minimum numbers search for data sorting is increasing of the sorting time compared to the methods under consideration. The development of the method of vertically-parallel search of the maximum and minimum numbers is a method of vertically-parallel sorting of one-dimensional arrays of numbers [6], which must be oriented to hardware and software implementation on mass-parallel computer tools [7].

The development of devices for vertically-parallel sorting of data arrays is advisable to perform using reconfigurable Systems-on-a-Chip [8, 9]. Field-programmable Gate Array (FPGA) is used as the main design base. Unlike conventional digital microcircuits, the logic of FPGA operation is not determined during manufacturing but is determined by programmable design. Special hardware programmers and debugging environments are used to program the chip and to set the desired structure of a digital device in the form of a principal circuit diagram or program in special languages for the description of the apparatus: Verilog, VHDL, AHDL, and others [10].

From the analysis of publications [1–10], it follows that the reduction of the sorting time of the array of numbers, the number of interface outputs and hardware costs can be achieved by developing of new methods, parallel hardware-oriented algorithms and VLSI structures for sorting data arrays.

3 The Purpose and Objectives of the Research

The purpose of the work is to develop a vertically-parallel method and means for sorting one-dimensional arrays of large numbers.

To achieve this goal, it is necessary to solve the following problems:

- develop a vertically-parallel method for sorting one-dimensional arrays of numbers;
- develop a graph of the algorithm for vertically-parallel sorting of arrays;
- develop a structure of the hardware for vertically-parallel sorting of one-dimensional arrays of large numbers;
- implement components of the device for vertically-parallel sorting of arrays of numbers using FPGA.

4 Vertically-Parallel Method of Sorting One-Dimensional Arrays of Numbers

This method of sorting involves the parallel receipt of N numbers by bit cuts with higher bits forward and the parallel formation of bit cuts of sorted numbers. Sorting of a one-dimensional array of numbers $\{D_k\}_{k=1}^{N}$ involves performing of $N \times n$ basic operations. When performing each of the i-th (i = 1,…, n) stage of sorting the array of numbers N basic operations are performed at the same time, which ensures the formation of i-th bits of N sorted numbers. At the first stage of sorting, the formation of higher bits of all numbers is performed. This stage is reduced to the following operations:

- counting the number of "1" in a bit cut:

$$S_1 = \sum_{k=1}^{N} D_{1k} \wedge y_{1k}, \tag{1}$$

where y_{1k} – the value of k-th bit of the first control word, which is equals $y_{1k} = 1$;

- formation for the $D_1^*, \ldots, D_{s_1}^*$ outputs the value of the 1st bit cut P_1 by the formula:

$$P_1 = \bigvee_{k=1}^{N} D_{k1} \wedge y_{1k}, \tag{2}$$

where D_{k1} – the value of the 1st bit of k-th number;

- formation for the $D_{S_1+1}^*, \ldots, D_N^*$ outputs the value of the 1st bit cut P_1 by the formula:

$$P_1 = \bigvee_{k=1}^{N} D_{k1} \wedge \bar{y}_{1k}, \tag{3}$$

where $\bar{y}_{1k}$ – the inverse value of the k-th bit of the 1st control word;

- calculation of 1st (high) bits for outputs $D_{11}^*, \ldots, D_{N1}^*$ by the following expression:

$$D_{k1}^* = \begin{cases} 0, & when \quad P_1 = 0 \\ 1, & when \quad P_1 = 1 \end{cases}; \tag{4}$$

- calculation of the second control word for outputs $D_1^*, \ldots, D_{s_1}^*$ by the following expression:

$$y_{21}, \ldots, y_{2S_1} = \begin{cases} 0, & when \quad P_1 = 1, \quad D_{k1} = 0 \\ 1, & when \quad P_1 = 1, \quad D_{k1} = 1 \end{cases} \tag{5}$$

- calculation of the second control word for outputs $D_{S_1+1}^*, \ldots, D_N^*$ by the following expression:

$$y_{2(S_1+1)}, \ldots, y_{2N} = \begin{cases} 0, & when \quad P_1 = 0, \quad D_{k1} = 1 \\ 1, & when \quad P_1 = 1, \quad D_{k1} = 0 \end{cases}. \tag{6}$$

The method of parallel sorting by counting involves comparing each number in an array with all other numbers. The parallel counting sort algorithm is performed in two stages. In the first stage, using a simultaneous pairwise comparison of each number with all other numbers of the array, the amount of numbers greater, smaller and equal to such number is determined. In the second stage, based on the results of pairwise comparisons permutation of data is performed. The parallel algorithm for counting sort is distinguished by high speed. The disadvantage of such an algorithm is heterogeneity, a large number of interface outputs and significant hardware costs that are needed for its implementation.

The following sorting steps for each output group (numbers) $\left(D_1^*, \ldots, D_{S_1}^*\right)$ and $\left(D_{S_1+1}^*, \ldots, D_N^*\right)$ are performed independently and in the same way as the first step.

5 Graph of the Algorithm for Vertically-Parallel Sorting of One-Dimensional Arrays of Numbers

Such a graph of the algorithm should provide a spatio-temporal representation of the process of vertical-parallel sorting of a one-dimensional array of numbers $\{D_k\}_{k=1}^N$. A feature of the vertical-parallel sorting of a one-dimensional array of numbers is the parallel bitwise incoming of N input numbers (bit cutoff) by high bits forward and parallel bitwise formation of N sorted numbers. The flow graph of the algorithm for vertically-parallel sorting of a one-dimensional array of numbers is shown in Fig. 1, where F_1 and F_c are respectively functional and control operators, and PU – processor units.

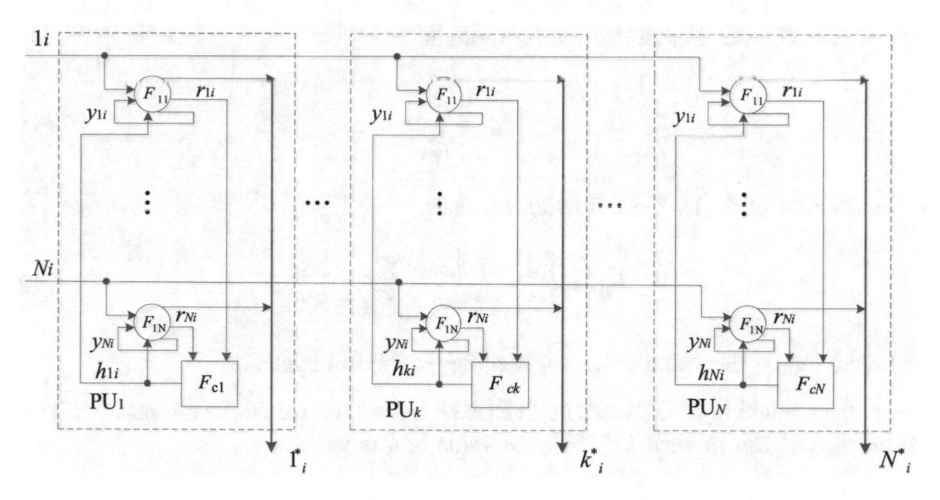

Fig. 1. Flow graph of the algorithm for vertically-parallel sorting of a one-dimensional array of numbers

The peculiarity of the flow graph of the algorithm for vertically-parallel sorting of a one-dimensional array of numbers is that data from each k input simultaneously enters all PUs. The number of PUs is determined by the dimension of the array that is sorted. In each i-th cycle of the PU work an i-th bit of the sorted k-th number is formed. The largest sorted number is formed at the output of the PU_1, and the smallest – at the output of the PU_N. Each PU is implemented using N functional operators F_1 and one functional operator F_C. In PU_k, the functional operator F_{1k} in each i-th cycle of work provides the following operations:

- calculating the value of the i-th bit for the k-th input as follows:

$$P_{ki} = \overline{D_{ki} \wedge y_{ki}}; \tag{7}$$

- forming a signal to calculate the number of "1" by the formula:

$$r_{ki} = (\bar{D}^{*}_{ki} \vee D_{ki}) \wedge y_i; \tag{8}$$

- calculating the k-th bit of the (i + 1)-th control word by the formula:

$$y_{k(i+1)} = r_{ki} \oplus h_{ki}, \tag{9}$$

where h_{ki} – transmission control signal for the (i + 1)-th control word; $\bar{D}^{*}_{ki}$ – inverse value of the i-th bit of the k-th sorted number, which is calculated in PUk by the formula:

$$\bar{D}^{*}_{ki} = \bigvee_{k=1}^{N} P_{ki}. \tag{10}$$

In PU_k the functional control operator F_{Ck} provides the following operations:

- counting the number of "1" by the formula:

$$R_{ki} = \sum_{k=1}^{N} r_{ki}; \tag{11}$$

- calculating of h_{ki} by the following formula:

$$h_{ki} = \begin{cases} 0, & when \quad \sum_k \geq Rg_{ki} \\ 1, & when \quad \sum_k < Rg_{ki} \end{cases}, \tag{12}$$

where Rg_{ki} – the value in the register Rg_k in the i-th cycle.

At the beginning of the work in all PUs all bits of control words are set in "1", and in the register Rg_k of each k-th PU_k the value of k is written.

6 Designing of Hardware for Vertically-Parallel Sorting of a One-Dimensional Array of Numbers

Hardware for vertically-parallel sorting of a one-dimensional array of numbers is advisable to implement using VLSI. The cost of VLSI, which implements vertical-parallel sorting of a one-dimensional array of numbers, depends mainly on the area of the crystal. The crystal area is mainly determined by the number of transistors needed for implementation, and the number of external outputs, the number of which is limited by the level of technology and the size of the crystal.

The development of highly effective VLSI for implementation of vertically-parallel sorting of a one-dimensional array of numbers can be ensured with an integrated approach that covers:

- development of new parallel methods and algorithms for sorting large data arrays;
- development of new structures oriented to VLSI-implementation;
- use of a new element base and automated design tools for VLSI.

For the development of VLSI for vertically-parallel sorting of a one-dimensional array of numbers, it is proposed to use the same type of PUs, which are connected by regular links.

Vertically-parallel sorting of a one-dimensional array of numbers $\{D_k\}_{k=1}^{N}$ assumes the entry in each cycle of a bit cut of N numbers and its sorting. Receipt of such bit cuts begins with the higher bits. To synthesize a structure of vertically-parallel sorting device of a one-dimensional array of numbers, we develop a PU that implements formulas (7), (8) and (9). The structure of the device for vertically-parallel sorting of a one-dimensional array of numbers synthesized based on implemented *PU* is shown in Fig. 2, where *Clk* – the input of synchronization, *Reset* – input of initial reset, D^{*}_{ki} – the output of the i-th bit of the k-th sorted number, *CU* – a control unit, *TgD* – a data trigger, *TgC* – a control trigger.

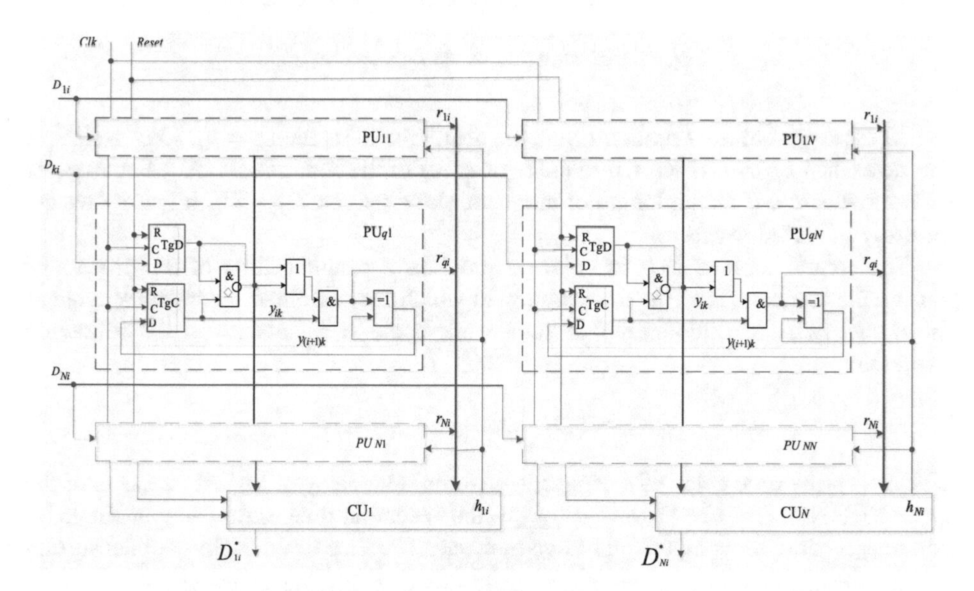

Fig. 2. The VLSI structure of the device for vertically-parallel sorting of a one-dimensional array of numbers

The structure of the device for vertically-parallel sorting of a one-dimensional array of numbers is matrix and consists of $N \times N$ PU and N CU. The formation of the i-th bit of the k-th sorted number is carried out with the help of N PUs, which are vertically joined together by a common tire and form the p-th column, where $p = 1, \ldots, N$. The control of each p-th column of PUs is carried out by CU_p, the structure of which is shown in Fig. 3, where MAdd is a multi-input adder, Rg is a register, Sub is a subtractor, CS is a comparison scheme, Tg is a trigger.

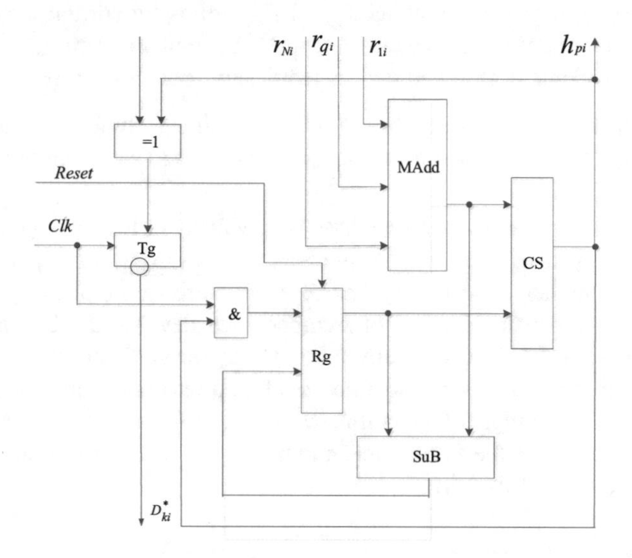

Fig. 3. The structure of the k-th control unit

At the output of the subtractor Sub, we obtain the difference $L = p_i - Z_{pi}$, which, in the case when $h_{pi} = 1$, is written to the register Rg of the control unit CU_p. For n cycles of work, where n is the bit length of numbers, at the outputs $D^*_{1i} - D^*_{Ni}$ bitwisely we get an array of sorted numbers.

The combination of PUs in columns provides a parallelization of the process of sorting the numbers, the time of execution of which is determined by the work cycle of the device. The execution time of such work cycle is calculated by the following expression:

$$T_S = t_{Tg} + 4t_{AND} + t_{MAdd} + t_{CS}, \tag{13}$$

where t_{Tg} – the time delay of a trigger; t_{AND} – the time delay of logical elements of the type OR, AND, AND-NOT, XOR; t_{MAdd} – the execution time of the N-input single bit adder; t_{CS} – time for comparison of two numbers. The time of vertically-parallel sorting of one-dimensional array of numbers $\{D_k\}_{k=1}^{N}$ is calculated as follows:

$$t_S = T_S n. \tag{14}$$

The cost of equipment for hardware implementation of the device for vertically-parallel sorting of a one-dimensional array of numbers (Fig. 1) is equal to:

$$W_S = N^2(2W_{Tg} + 4W_{AND}) + N(2W_{Tg} + W_{Rg} + W_{Sub} + W_{MAdd} + W_{CS} + 2W_{AND}), \quad (15)$$

where W_{Tg}, W_{AND}, W_{Rg}, W_{MAdd}, W_{CS} та W_{Sub}, – the cost of equipment for the implementation of the trigger, logical elements of type AND, register, N-input single-bit adder, comparison scheme and subtractor correspondingly.

7 Implementation of VLSI-Device for Vertical-Parallel Sorting of One-Dimensional Arrays of Numbers

The VLSI device for vertically-parallel sorting of a one-dimensional array of numbers is implemented in the integrated environment Quartus II for FPGA EP3C16F484 of the Altera Cyclone III family on the hardware description language VHDL using libraries from the integrated environment. The algorithm expects the receipt in each step of the synchronization of the bit cut Dki of N numbers, starting with the higher bits. When implementing a VLSI-device for vertically-parallel sorting of a one-dimensional array of numbers, it must be possible to work with both a different number of elements of the array N and their bit length n.

The main components of the device for sorting a one-dimensional array of numbers are registers, the former of vertical bit cuts, one-bit multi-input adders, comparators, converters of vertical bit sections in the sequence of data of the bit length n. The simulation of their work is developed and executed. For the synthesis of the VLSI sorting device, its main components were developed, and their work was modeled.

The appearance of the symbol of the vertical bit cuts former with the dimension of the

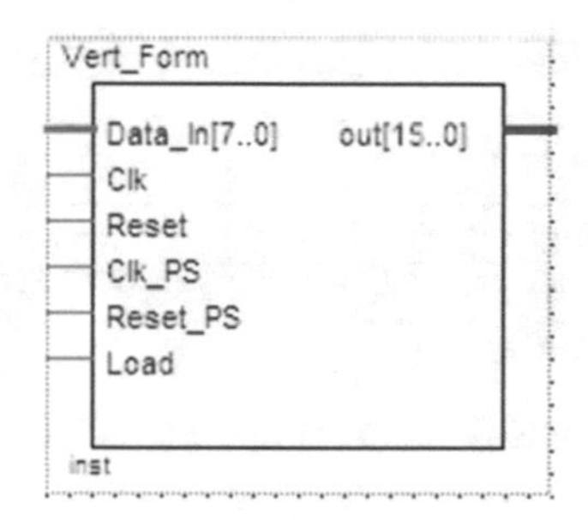

Fig. 4. The appearance of the symbol of the vertical bit cuts former

array N = 16 and the bit length of numbers n = 8 (Vert_Form_16_8) is depicted in Fig. 4.

The vertical bit cuts former consists of *N* registers of the parallel-parallel type and *N* registers of the parallel-serial type, the symbols of which are depicted in Fig. 5a and b correspondingly.

The inputs of the vertical bit cuts former are as follows: Data_In[7..0] – D_k, (k = 1, …, N) length of the one-dimension array; Clk – the input of synchronization of loading of numbers of an array; Reset (active signal level “0”) – input of initial reset to “0” of

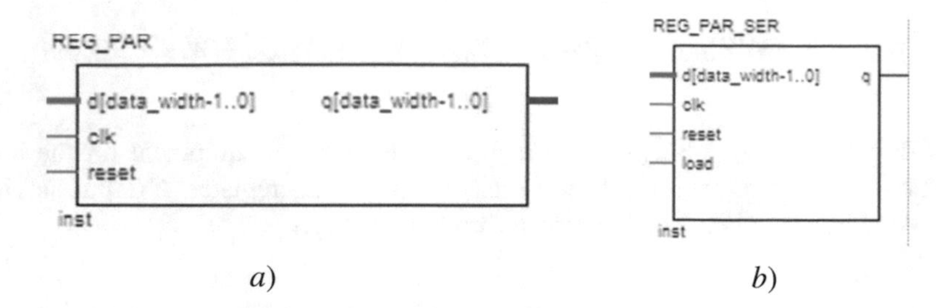

Fig. 5. The appearance of the register symbols: (a) parallel-parallel type; (b) parallel-serial type

the REG_PAR register output; Clk_PS – data synchronization input for register REG_PAR_SER; Reset_PS (active signal level "1") – input of initial reset to "0" of the REG_PAR_SER register output; Load (active level of signal "1") – the signal of permission to load data in the register REG_PAR_SER. The output of the former Vert_Form_16_8 is the vertical parallel cut Out [15..0]. During the first N pulses of Clk synchronization data is loaded into registers REG_PAR. During each pulse of Clk_PS synchronization, a vertical parallel cut Dki is formed, starting from the higher bits. The synchronization is implemented along the front edge of Clk and Clk_PS pulses.

Figure 6 shows the timing diagram of the former of vertical bit cuts Vert_-Form_16_8, which generates vertical parallel cuts for 16 eight-bits numbers (N = 16; n = 8): 0x2C, 0x35, 0xA0, 0x0F, 0xB7, 0x49, 0x64, 0x72, 0xF3, 0xE8, 0xA8, 0x83, 0x63, 0x2F, 0x7A, 0x2C. Bit cuts for an array of these numbers (from the higher bits): 0x0F14, 0x53E0, 0xF7D7, 0x4192, 0xE629, 0xA059, 0x7998, 0x393A.

The appearance of the symbol of the output data sequence former with dimension

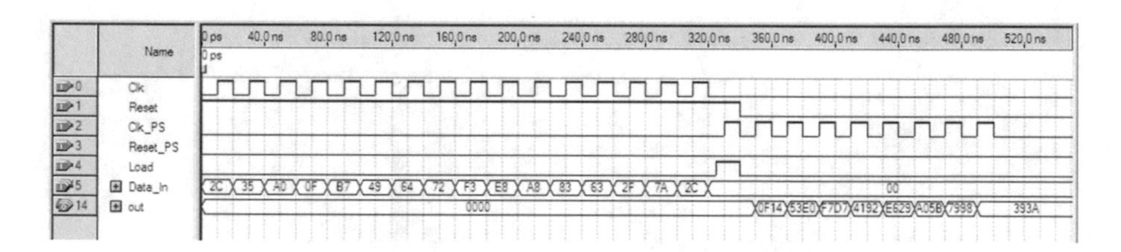

Fig. 6. Timing diagrams of the vertical bit cuts former

N = 16 and the bit length n = 8 (Out_Data_16_8) is shown in Fig. 7.

The structure of the output data sequence former is similar to the structure of the vertical bit cuts former. The output data sequence former consists of N registers of the parallel-parallel type and N registers of the parallel-serial type.

Data_In [15..0] input of the output data sequence former – it is Dki, vertical parallel cuts, which are coming from the input data sorting module. Their number (n) is determined by the bit length of the one-dimensional array. The inputs Clk, Reset, Clk_PS, Reset_PS, Load have the same purpose and active signal levels as in the vertical parallel bit cuts former. The output of former Out_Data_16_8 (Out [7..0]) is a

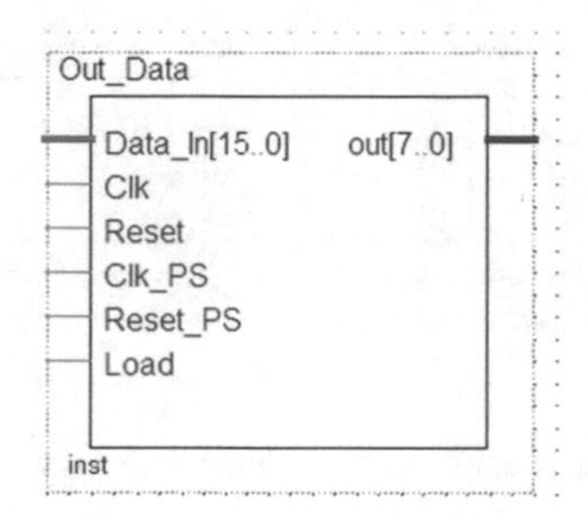

Fig. 7. The appearance of the symbol of the output data sequence former

sequence of N numbers with bit length n. For the former Out_Data_16_8, during the first n Clk synchronization pulses, data is loaded into registers REG_PAR. Then, at the time of each synchronization pulse Clk_PS, output data D*k is generated at REG_PAR_SER outputs. The number of such synchronization pulses is N.

Figure 8 shows the timing diagram of the output data sequence former Out_-Data_16_8, which generates a sequence of 16 8-bit numbers for 8-input vertical parallel bit cuts (N = 16; n = 8): 0x003F, 0x07C3, 0x7BDF, 0x08C5, 0xF44A, 0xF904, 0x92E5, 0x9E25. The data sequence on the output of the former: 0xF3, 0xE8, 0xB7,

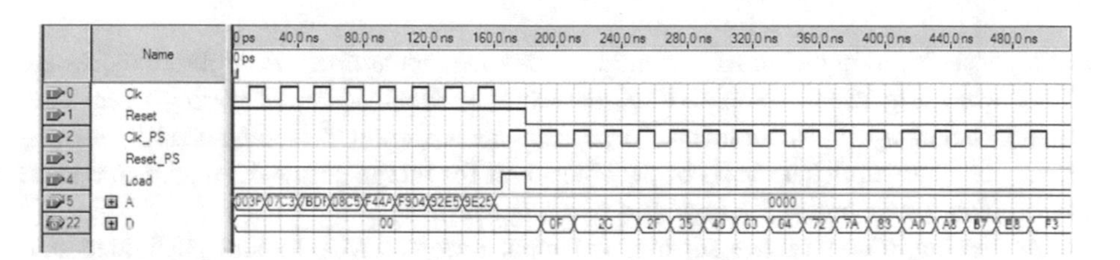

Fig. 8. The timing diagram of the output data sequence former

0xA8, 0xA0, 0x83, 0x7A, 0x72, 0x64, 0x63, 0x49, 0x35, 0x2F, 0x2C, 0x2C, 0x0F.

The devices are implemented and the simulation of the work of one-bit 3, 7, 15 and 31 input adders is executed. The symbols of the 7-input (ADD_7_3_Sym) and the 15-input (ADD_15_4_Sym) adders are shown in Fig. 9, where Clk – the input of synchronization; Reset (active signal level “1”) – input of initial reset of the adder, C_IN – data inputs combined into a data bus, S – output of multi-input adders.

The sum in multi-input adders is calculated on the front of the sync pulses. Figure 10 shows the timing diagram of the 15-input adder ADD_15_4.

At the inputs of the C_IN [15..1] 15-input adder at each step arrives data 0x3FC4, 0x250D, 0x3703, 0x006D, 2AFD corresponding to the 9th, 6th, 7th, 5th, and 10th logical units at the inputs. At the output S of the adder an amount equal to the number of units entering the inputs of the 15-input adder, respectively, 9, 6, 7, 5, and 10, is formed in each tact.

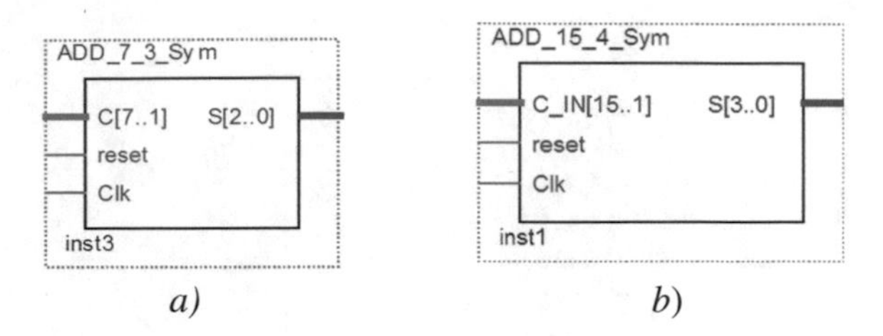

Fig. 9. The appearance of the symbol of: (a) 7-input one-bit adder; (b) 15-input one-bit adder

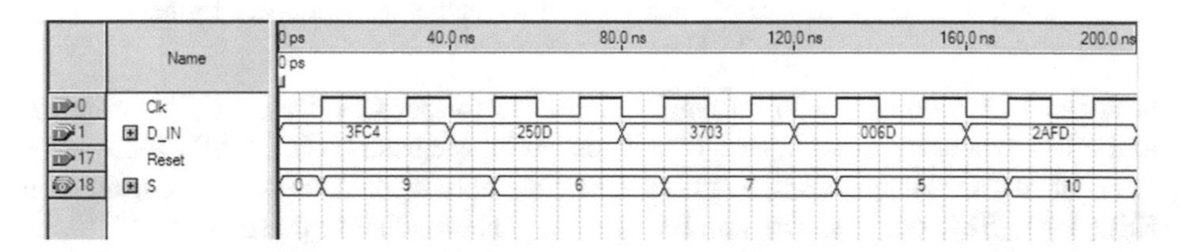

Fig. 10. Timing diagram of the 15-input adder

8 Sorting Large-Sized One-Dimensional Arrays of Numbers Using the Developed VLSI Device

Let's consider the question of sorting an one-dimensional array of M numbers, where $M = N \times b$, using the developed VLSI-device of vertically-parallel sorting of an one-dimensional array of N numbers. To perform such sorting it is suggested to use the merge sort algorithm. The basis of the algorithms for merge sort is the macro-operation of combining two ordered arrays into one ordered array.

At the beginning of the sorting, the input array of M numbers is divided into b = M/N arrays of length N, which are sorted using the developed device. The macro-operation of the first type (combining two ordered arrays of N numbers into one ordered array of 2 N numbers) is performed over the sorted arrays of N numbers. As a result of performing macro-operations of the first type, $b/2$ arranged arrays of length 2 N are formed. The number of types of macro-operations for sorting an array of M numbers with the use of a basic operation of sorting N numbers is determined by the following formula:

$$K = |\log_2 b|. \tag{16}$$

Macro-operations of the first type are implemented on three VLSI-devices for vertically-parallel sorting of N numbers, which are combined into a sorting unit of the first type according to the scheme shown in Fig. 11, where SU – a sorting unit of the first type, D_{1i} ,…, D_{Ni} i $D_{(N+1)i}$,…, D_{2N-} – inputs which receive two arranged arrays of the size N each, D^*_{1i} ,…, D^*_{2Ni} – outputs of the sorted array of $2N$ numbers.

From the inputs D_{1i} ,…, $D_{N/2i}$ on $VLSI_1$ arrives $N/2$ larger numbers of the first array, and from the inputs $D_{(N+1)i}$,…, $D_{(N+N/2)i}$ – $N/2$ larger numbers of the second array.

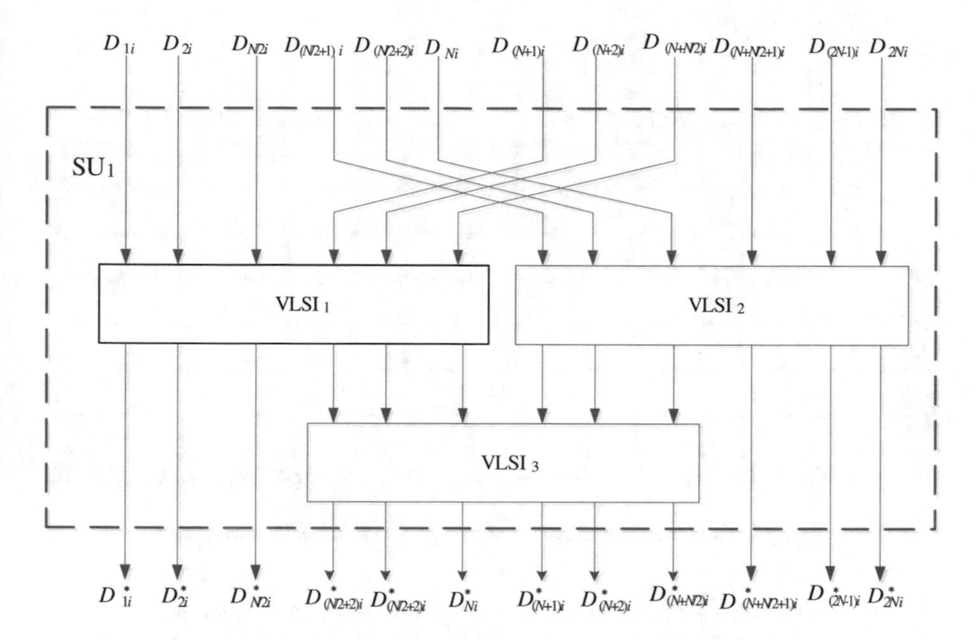

Fig. 11. Scheme of sorting block of the first type

On the inputs of $VLSI_2$ from the inputs $D_{(N/2+1)i}$,..., D_{Ni} arrives $N/2$ smaller numbers of the first array, and from the inputs $D_{(N+N/2+1)i}$,..., D_{2Ni} – $N/2$ smaller numbers of the second array. At 1 ,..., $N/2$ outputs of the $VLSI_1$ we obtain $N/2$ larger numbers of a sorted array of 2 N numbers, and smaller $N/2$ numbers from the outputs $N/2$ + 1,..., N of the device arrive on the $VLSI_3$. At the outputs ($N/2$ + 1),..., N of $VLSI_2$ we get $N/2$ lower numbers of the sorted array of 2N numbers. The numbers, sorted by means of $VLSI_3$, come to the outputs $D^*_{(N/2+1)i}$,..., $D^*_{(N+N/2+1)i}$ of the sorting unit SU1.

The macro-operation of the s-th type (combining two ordered arrays of N2s–1 numbers into a single ordered array of $N2^s$ numbers), where s = 1 ,..., K, are implemented by the sorting unit SU_s, which is obtained by combining of three units SU_{s-1}. A scheme for sorting an array of 8 N numbers using merge sort with the use of VLSI-devices for vertically-parallel sorting of N numbers is shown in Fig. 12, where SU_2 and SU_3 are sorting units of the second and third types respectively.

To sort an array of 8N numbers, three types of sorting units are used, which are implemented on the basis of VLSI-devices for vertically-parallel sorting of arrays of N numbers. The number of VLSI that is required to sort an array of 8 N numbers is 65. Sorting an array of 8 N numbers using a conveyor principle is performed with clock tick $T_{S1} = t_{Tg} + 4t_{AND} + t_{MAdd} + t_{CS}$. The sorting time of an array of 8 N numbers is determined by the formula:

$$t_S = T_{S1}(n+6). \tag{17}$$

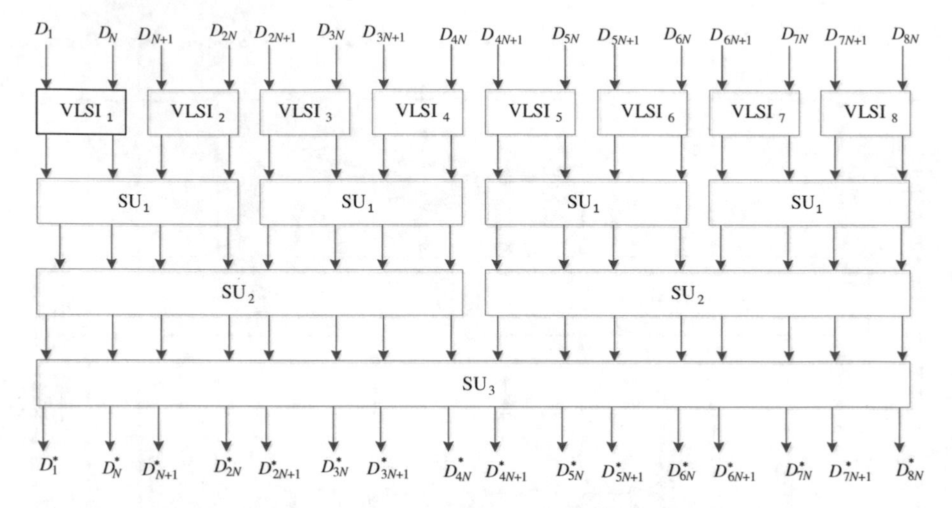

Fig. 12. Scheme for sorting an array of 8 N numbers by merge sort

Vertically-parallel sorting of large arrays of numbers using developed VLSI provides a reduction in the sorting time due to the conveyor organization of sorting and forming of sorted bit cuts in each cycle.

9 Sorting of Two-Dimensional Arrays of Numbers Using the Developed VLSI-Device of Vertical-Parallel Sorting of One-Dimensional Array of Numbers

Sorting a two-dimensional array of numbers $\{D_{hj}\}_{h=1;j=1}^{N/2;M}$ using a VLSI-device of vertically-parallel sorting of one-dimensional array of numbers assumes that the input data arrives in parallel from $N/2$ channels. It is expedient to perform such sorting by the method of displacement, the basic macro-operation of which is vertical-parallel sorting of a one-dimensional array of N numbers. In each macro-tact of the implementation of this sort, the sorting of N numbers is performed, of which $N/2$ are input numbers, and N/2 are larger numbers from the previous macro-tact of work. Sorted $N/2$ larger numbers remain for the next macro-operation with new input data, and $N/2$ smaller ones are written in a memory. The number of macro-tacts needed to sort a two-dimensional array of numbers $\{D_{hj}\}_{h=1;j=1}^{N/2;M}$ based on one VLSI-device of vertically-parallel sorting of one-dimensional arrays of N numbers is determined by the formula:

$$g = \sum_{j=1}^{M} (M - j). \tag{18}$$

One of the ways to increase the speed of sorting two-dimensional arrays of numbers is to increase the number of basic macro-operations that are executed in parallel by developing parallel-flow structures. The flow graph of the algorithm of parallel-stream sorting of a two-dimensional array of numbers $\{D_{hj}\}_{h=1;j=1}^{N/2;M}$ by the displacement method is shown in Fig. 13, where C – control input, 1–$N/2$ – data inputs, PU_j – j-th processor unit, F_{Cj1}, F_{Cj2} – the first and second control operators for PU_j, F_{Cmj1}, F_{Cmj2} – the first and second commutation operators for PU_j, F_{Mj1}, F_{Mj2} – the first and second memory operators for PU_j, F_S – the operator of sorting of N numbers.

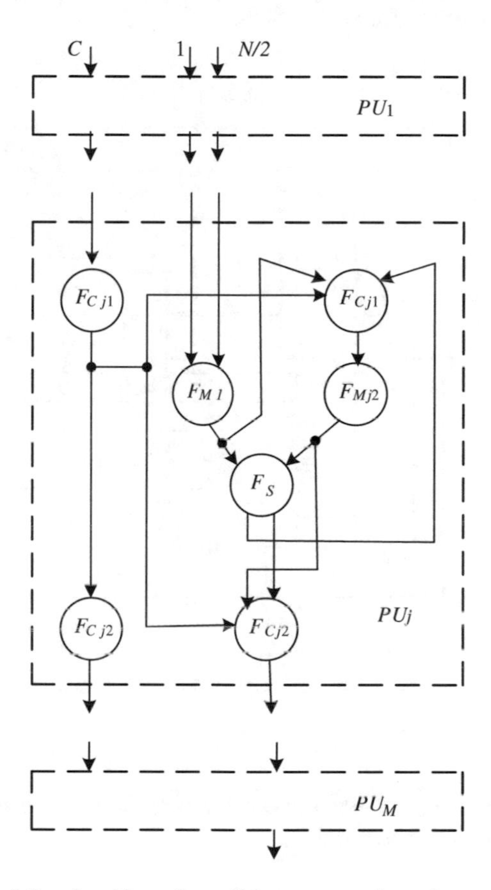

Fig. 13. Flow graph of the algorithm of parallel-stream sorting of a two-dimensional array of numbers using the displacement method

In the parallel-stream sorting of a two-dimensional array of numbers $\{D_{hj}\}_{h=1;j=1}^{N/2;M}$ using the displacement method, M PU are used. Each PU_j is implemented based on the first and second control operators Φ_{Cj1} and Φ_{Cj2}, the first and second commutation operators Φ_{Cmj1} and Φ_{Cmj2}, the first and second memory operators Φ_{Mj1} and Φ_{Mj2}, and the operator of sorting of N numbers Φ_S. The feature of the parallel-stream sorting is the use of a VLSI-device of vertical-parallel sorting of one-dimensional arrays of numbers for the implementation of the sorting operator Φ_S. This operator is performed for n clock ticks, that is, one macro-tact. The developed flow graph of the algorithm of parallel-stream sorting of a two-dimensional array of numbers by the method of displacement is oriented on the sorting of continuous streams of data in real time.

The structure of a parallel-streaming device for sorting a two-dimensional array of numbers by the method of displacement is shown in Fig. 14, where *Clk* – the input of synchronization, *C* – the control input, In_1–$In_{N/2}$ – data inputs, *PU* – processor unit, *Tg* – trigger, *M* – memory, *Cm* – commutator, $VLSI_{sort}$ – VLSI for sorting of *N* numbers, Out_1–$Out_{N/2}$ – outputs for sorted data.

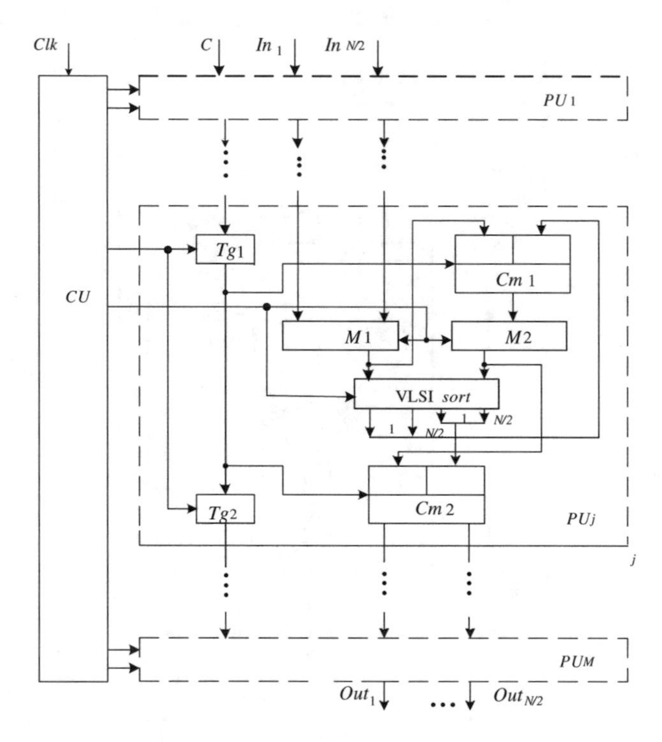

Fig. 14. The structure of a parallel-streaming device for sorting a two-dimensional array of numbers by the method of displacement

The work of the parallel-streaming device of sorting a two-dimensional array of numbers by the method of displacement begins with a signal log.1 on the C input, which indicates the beginning of the arrival of the two-dimensional array. In each step of the work, the bit cuts of the input data from the inputs In_1–$In_{N/2}$ are stored in memory M_1. After n clock cycles in memory M_1 we get the first one-dimensional array $\{D_{h1}\}_{h=1}^{N/2}$. After this, the first macro-pulse in the trigger Tg_1 of the first PU_1 writes log.1 in it, which sets the Cm_1 and Cm_2 to transmit information from the first inputs. In the next macro-tact into M_1 the second one-dimensional array is written $\{D_{h2}\}_{h=1}^{N/2}$, and in M_2 the first one is written $\{D_{h1}\}_{h=1}^{N/2}$. The next macro-pulse sets Tg_1 in log.0 and Tg_2 in log.1. In each subsequent cycle, the data from the outputs M_1 and M_2 arrives by bit cuts on the *VLSIsort*. At outputs of *VLSIsort* we get bit cuts of the sorted array of N numbers. Larger $N/2$ numbers through Cm_1 go to the inputs of memory M_2, whereas $N/2$ smaller ones through Cm_2 go to the outputs of the next PU_2. In the following macro-tacts the parallel-stream sorting device works the same way.

Immediately after loading M-th one-dimensional array of numbers $\{D_{hM}\}_{h=1}^{N/2}$ in M1 of the first PU1, the next two-dimensional array can be loaded and sorted. Simultaneous sorting of two two-dimensional arrays reduces the sorting time and ensures high efficiency of the equipment use.

Sorting a two-dimensional array in the parallel-stream sorting device is performed at such a time

$$t_s = 2Mn(t_M + t_S + t_{Cm}), \tag{19}$$

where M – the number of one-dimensional arrays, n – bit lengths of input data, t_M – memory latency, t_S – latency of the vertically-parallel sorting device, t_{Cm} – latency on the switch.

10 Conclusion

1. A vertical-parallel method and a VLSI-device of data sorting have been developed, which due to the parallel processing of the *i*-th bit cut of the array of numbers and the parallel formation of the *i*-th bit cut of the sorted array of numbers reduces the sorting time.
2. The method of parallel merge sort for large one-dimensional arrays of numbers has been improved, which due to the conveyance, the use of the basic operation of vertically-parallel sorting of N numbers and the formation of sorted bit cuts arrays in each cycle reduces the sorting time.
3. The method of insertion sort of two-dimensional arrays of numbers has been improved, which due to the time-alignment of the two-array sorting processes reduces the sorting time and increases the efficiency of the equipment use.
4. A parallel structure for sorting one-dimensional large-scale arrays of numbers and a parallel-flow structure for sorting two-dimensional arrays of numbers have been developed using a VLSI vertical-parallel sorting device for a one-dimensional array of N numbers, that ensure high performance, the efficiency of the equipment use and work in the real time.

References

1. Knuth, D.: The Art of Computer Programming: Sorting and Searching, 844 p. (1978)
2. Melnychuk, A., Lutsenko, S., Gromov, D., Trofimova, K.: Analysis of methods for sorting the array of numbers. Technol. Audit Prod. Reserves **4/1**(12), 37–40 (2013)
3. Gryga, V., Nykolaychuk, Y.: Methods and hardware for sorting binary arrays. In: ASIT 2017, Ternopil, 19–20 May 2017, pp. 58–61 (2017)

4. Tsmots, I.G., Skorokhoda, O.V., Medykovskyy, M.O., Antoniv, V.Y.: Device for determining the maximum number from a group of numbers. Patent of Ukraine for invention No. 110187, 25 November 2015, Bul. No. 22 (2015)
5. Tsmots, I., Rabyk, V., Skorokhoda, O., Antoniv, V.: FPGA implementation of vertically parallel minimum and maximum values determination in array of numbers. In: 2017 14th International Conference on the Experience of Designing and Application of CAD Systems in Microelectronics (CADSM) Proceedings, Polyana, 21–25 February 2017, pp. 234–236 (2017)
6. Tsmots, I., Skorokhoda, O., Antoniv, V.: Parallel algorithms and structures for implementation of merge sort. Int. J. Adv. Res. Comput. Eng. Technol. (IJARCET) **5**(3), 798–807 (2016)
7. Gergel, V.: High-Performance Computing for Multi-processor Multinuclear Systems, 544 p. Moscow University Press (2010)
8. Palagin, A.V., Yakovlev, Y.S.: Features of designing computer systems on the FPGA/AV crystal. Math. Mach. Syst. **2**, 3–14 (2017)
9. Palagin, A., Opanasenko, V.: Reconfigurable Computing Systems, 295 p. Prosvita, Kyiv (2006)
10. Erkin, A.: Review of modern CAD for FPGA. ChipNews **10-11**(134–135), 17–29 (2008)

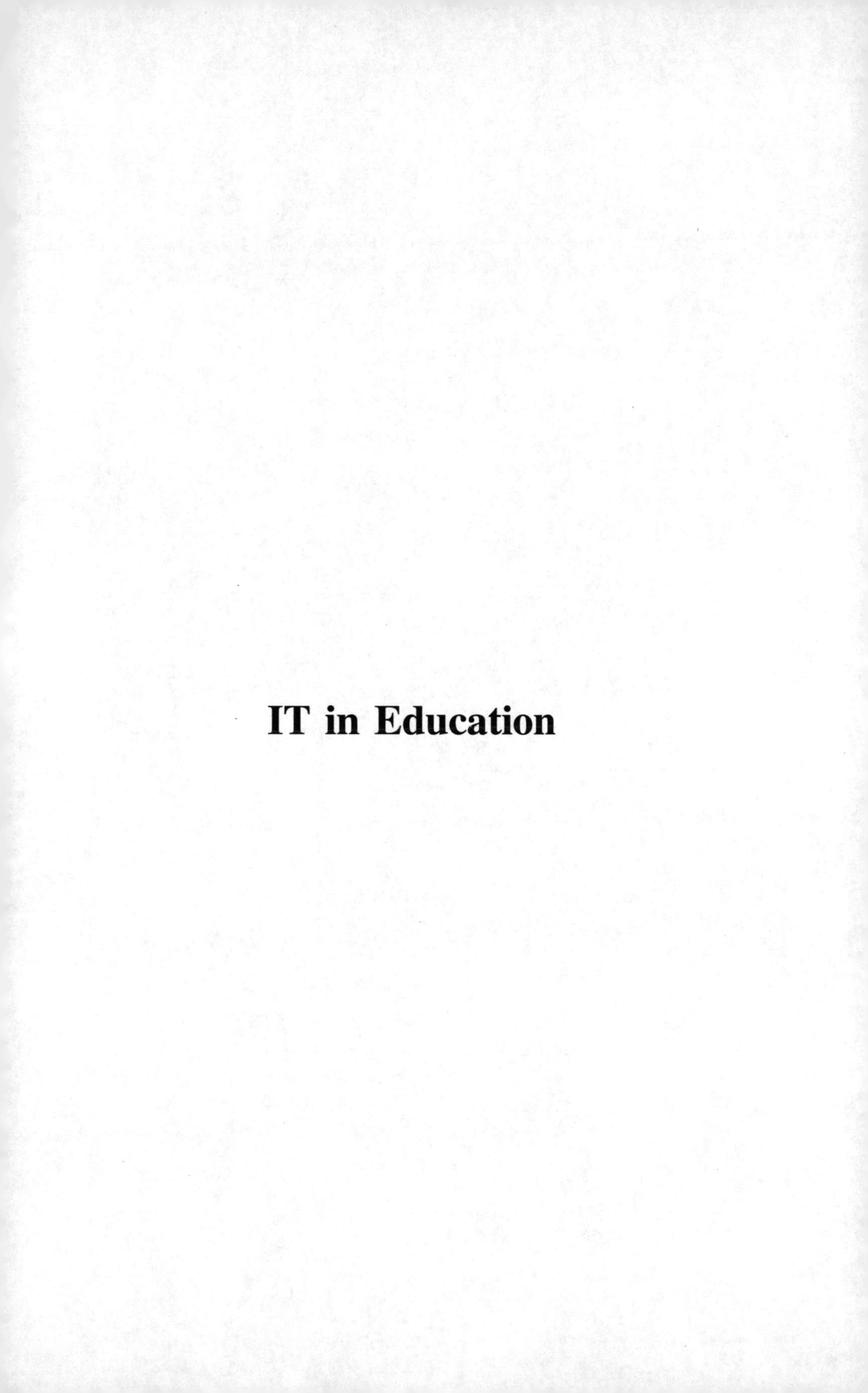

IT in Education

Education and Research in Informatics and Automation at Faculty of Mining and Geology

Roman Danel(✉), Michal Řepka, Jan Valíček, Milena Kušnerová, and Marta Harničárová

VSB-Technical University of Ostrava, Ostrava, Czech Republic
roman.danel@vsb.cz

Abstract. In this article we describe the teaching of automation and informatics at the Faculty of Mining and Geology that primarily educates professionals in the field of geology and resource industries. Even in mining, the need for specializations in automation and IT has arisen in the past and, for this reason, these fields of study have been created. We also operate our branch office in Most, where part of the lessons is realized via videoconference, which has its plus and minus. We describe the problems we have with the lessons, deficiencies in study plans, and suggestions for further development. At the end of the paper, we present some of our research projects and areas of action - information systems for coal preparation plants, brownfields revitalization projects database, landscape modeling and visualization system, application of RFID technology, methane flow modeling or fault tolerant systems designing.

Keywords: Automation and IT education
Teaching using videoconference system · Preparing IT specialist

1 Introduction

The Faculty of Mining and Geology ("FMG") of VŠB - Technical University of Ostrava is the oldest faculty and provides teaching of numerous areas, from historically specified fields in the field of mining, geology and geodetics, through automation and economics in the raw material industry, to modern fields related to geo-sciences and environment engineering. The faculty is active primarily in the Ostrava region, where mining of black coal is concentrated.

The automation study at the FMG was historically based on the needs of introducing automation and control in the extraction of minerals. At the FMG, therefore, the study course "Automated Control Systems in the Mining Industry" started to be taught in 1962. In the 1990s, the Czech Republic started to reduce mining, which led to the necessity of adjusting the study courses. "System Engineering" and "Information and System Management" courses were gradually designed and accredited, focusing on the application of information technology in the raw material industry (Fig. 1).

A separate Institute of Geoinformatics has also been created to deal with the application of GIS in the raw material industry and the remote earth sensing. Due to the

N. Shakhovska and M. O. Medykovskyy (Eds.): CSIT 2018, AISC 871, pp. 287–300, 2019.
https://doi.org/10.1007/978-3-030-01069-0_21

low level of interest from students, the course dealing with automation in the raw material industry was closed. It is ironic that the low level of interest from students in this field does not correspond with the increase in requests from industry, where companies are increasingly looking for experts in automation [12]. This creates a situation where students tend to seek out less technical fields whereas companies experience a distinct lack of technically oriented professionals. While the field of automation at the FMG was closed, the steelmaker Mittal Steel in Ostrava will have to employ automation specialists from abroad (e.g. Ukraine) in 2016, because the required positions cannot be filled by experts from the region [3].

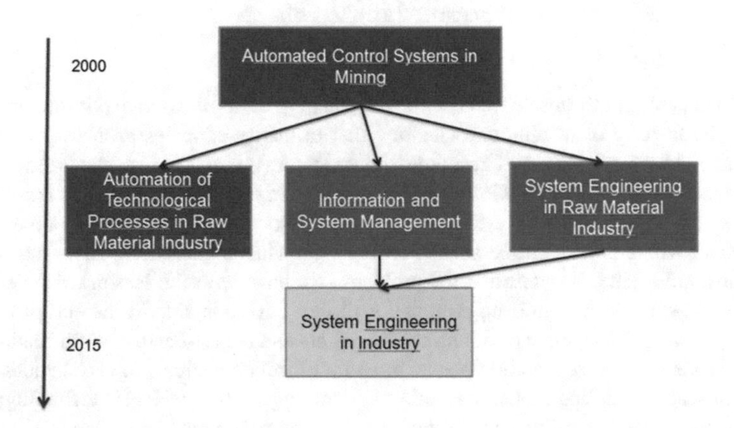

Fig. 1. Changes in study fields at FMG from the year 2000 to 2018

2 Teaching Using Videoconferencing System

Extensive mining activity takes place in North Bohemia (mining brown coal); therefore, VŠB opened a branch office in North Bohemia – Institute of Combined Studies in Most. The form of remote teaching is used here for teaching numerous fields taught in Ostrava; the Institute also provides for attendance and also postgraduate re-qualification. The interest in studying at the Institute in Most is very high, which can be seen from the numbers of registered students [11].

Throughout the existence of the Institute, fields related to automation and IT/ICT in the raw material industry are also taught here – bachelor′s courses "Information and systems management" (until 2014) and "Systems engineering in industry" (bachelor's degree and also follow-on).

Due to the distance of the Institute from Ostrava (approx. 500 km, Fig. 2), it was necessary to make use of frequent business trips by teachers to ensure teaching, which represents a substantial cost. Therefore, FMG purchased a videoconference system (see Figs. 3 and 4) which enables the provision of some of the teaching from Ostrava (parallel teaching of combined fields in Ostrava and Most), or conversely teaching from Most (for Most and Ostrava simultaneously).

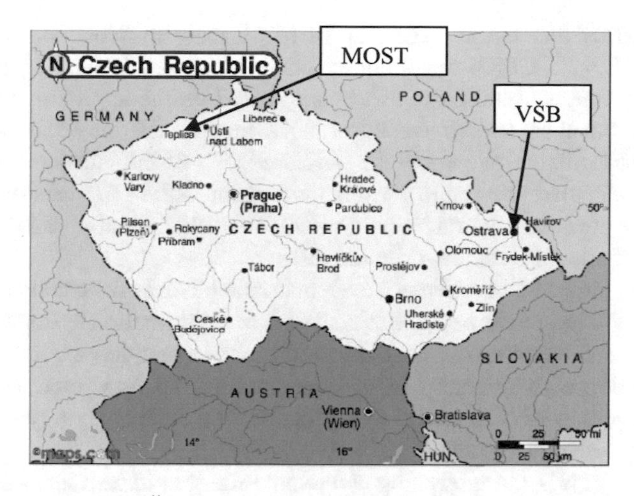

Fig. 2. Location of VŠB university and Institute of Combined Studies in Most

Fig. 3. The view how students see the teacher in Most [Photo: M. Řepka]

Fig. 4. The Classroom with videoconferencing system in Ostrava with two cameras, data projector and four TV screen [Photo: R. Danel]

If we consider that the average stay of one teacher in Most constitutes 2 nights (CZK 1,600, 1€ = 25 CZK), travelling expenses (approx. CZK 1,500) and board, we can easily calculate the total costs of teaching based on the approximate cost of CZK 3,500 for one business trip. If we know what income the university receives from teaching one student, and we know the purchase costs of the videoconference system, we can easily calculate the return on this investment. After ten years of operation we can state that the videoconference system has significantly affected the cost-effectiveness of such a remote workplace [9].

Teaching using a videoconference system is suitable only for certain subjects which are more descriptive in nature (e.g. subjects such as "Information systems"). In the case of subjects where experiments must be performed, examples explained and which require interaction with students, direct contact with students is more suitable.

Remote access to laboratories in this field is only a complementary tool and cannot replace the presence of a teacher.

The method of providing teaching at a remote site should be a sensible compromise between cost-cutting using teaching via videoconference for selected subjects, and the direct presence of teachers in classes which require a personal approach. Excessive pressure on changing the majority of teaching to videoconferencing can paradoxically lead to a reverse economic effect, where a school will receive lower income based on a decrease in the numbers of students due to dissatisfaction with the quality of education.

3 Remote Access to Laboratory

To support teaching of automation subjects remote access to the laboratory was proposed where, through an internet application, students have access to physical equipment in the laboratory. Equipment is monitored by a webcam and students can remotely perform simple experiments using robotic laboratory equipment Bioloid from the Robotis Company [18].

Basic concept here is that applicants can watch the whole process via camera, which is installed in the place of our experiment. It was necessary to select a proper type of such camera. At the very beginning we have used only USB cameras. After several experiments it proved, however, that those types of cameras at our disposal are not suitable, as they showed delay of several seconds. This delay appeared, even though the student was using the computer to which this camera was connected.

Because of this delay, and also because of its simpler configuration to our web application, we have decided to us the IP cameras instead. With this type, however, the experiments showed similar delays as the USB cameras did. As it was not possible to detect, what was the cause of these delays in both cases, we have conducted experiments with IP cameras of various producers, to find out that the shortest delay can be reached using cameras by D-Link company, namely of the DCS-9xx line [2].

4 Trends in Education of Automation and Informatics

4.1 Courses of Informatics

At universities in Czechia, there are currently two opinions of conceiving the bachelor's degree study and continuing education in information fields.

The first concept is based on the notion that the bachelor's degree is to provide students with a basic overview of existing technologies, approaches and directions, while the specific tools should be deeply presented only in the subsequent stage. This concept stems from systemic thinking, based on the process from abstract, general and global perspectives to details and specific knowledge.

The second concept is based on the assumption that the bachelor's degree study should lead to a mastery of specific tools and practices, i.e. "the job". Follow-up studies should then deal with more complex and abstract approaches, the integration of knowledge and penetration into deeper knowledge, and should be aimed at more abstract concepts [3].

At present, it is necessary to adapt the concept of teaching to the development of society. In the age of the Internet and online mobile devices, when previously unavailable information is available immediately, the educational structure "lecture - exercise" ceases to make sense [http://www.vet.utk.edu/enhancement/pdf/feb11-2.pdf]. Paper [14] points out the practical use of Facebook for universities, where the multiplication effects of this social network were activated by cooperation between teachers and students, when solving scientific and innovative projects or in the actual learning process. Students have no reason to attend lectures where the teacher repeats what is downloadable from the Internet; if there is no added value, it is a waste of time for students.

A long-term problem in IT fields is also reflected as a declining interest of students in studying programming, while the industry mostly demands programmers [15]. One of the reasons is the wrong way of teaching programming, which is too focused on mastering techniques and algorithm design. Students often do not understand what the respective procedures are good for, how they could be used in practice. We believe that the programming education at bachelor level should take place using specific tasks and start from visual aspects of the application. The most common programming task is a form for data input or extract (whether on screen or as a report). Therefore, it makes no sense to devote a semester to programming console applications and studying pointer arithmetic if this type of task is almost never met by students in practice.

Another problematic element of the study of IT fields, except non-informatics faculties, is the underestimation of the importance of databases. Relational databases and SQL language are one of the most commonly used technologies. For a number of academics, databases are of a secondary importance, something which does not have sufficient "scientific" potential. This opinion is especially held by people who do not have practical experience. After graduating from university and entering a job, students find with surprise that database applications are the most common types of applications which they encounter, and that data processing is a key activity [3].

Students often ask what tools, programming languages or techniques they should learn to be competitive in the labour market. Discussions with representatives of the corporate sector show that companies especially prefer a desire to learn and work on oneself to the specification of particular technologies [16]. Firms expect active workers who will take the initiative. In this area, we see the largest reserve in the current university education.

4.2 Project INOHGF

In 2012–2015, the Faculty of Mining and Geology solved the project "Innovation of Bachelor's and Master's Degree Courses at the Faculty of Mining and Geology, VSB-TUO – INOHGF" financed by Ministry of Education. The main objective of the project was defined as an "innovation and restructuring of the existing system and the content of bachelor's and following master's degree courses at the Faculty of Mining and Geology, VSB-TUO, in accordance with the requirements of the knowledge economy and labour market needs while respecting demographic trends and priorities of the Strategic Plan of Activities of the VSB-TUO for 2011–2015" [3].

The outputs of the project are as follows:

- Restructured system of innovated bachelor's and follow-up master's degree courses with an overall lower number of fields of study.
- Creation of new and innovated existing teaching materials.
- Enhanced internationalization of education at the FMG.
- Strengthening of the cooperation between the FMG VSB-TUO and practice, updating of the system of study fields in relation to practical requirements.
- Better language and professional competencies of academic staff at the FMG.
- Upgraded technical, software and laboratory equipment at the HGF.
- Processed accreditation materials for restructured fields.

In further analysis, we will discuss the INOGF project in its part concerning the study areas of automation and informatics. Although the project objectives as defined have been met, we think that the opportunities to improve given by the project were not fully utilized, especially in the area of strengthening cooperation with practice. The project resulted in upgrading study programs – this specifically relates to the newly created bachelor's course "System Engineering in the Industry", which replaced the existing course "Information and System Management" in 2014. Several new subjects (courses) were created in the curriculum (study plan), and many subjects and study materials were updated. Unfortunately, the project did not include any deeper analysis of requirements for graduates on the part of companies. The new course therefore continues in teaching according to ingrained practices that already do not always comply with the current requirements for education and skills of graduates. The project paid considerable attention to the internationalization of education (especially cooperation with Japanese and Korean universities), which is accessible only to a limited number of students. Notions of local companies regarding the graduates were not addressed [3].

4.3 Proposal for Changes

- Strengthen the teaching of specific skills and the knowledge of techniques and tools in the bachelor's degree program (SQL, programming skills, soft skills, projects administration…).
- Stronger integrate the involvement of companies in the study (teaching subjects by external specialist – practitioners) [12].
- Enable students in the bachelor's degree program to choose subjects for specialization and arrange the study plan.
- Strengthen the "soft skills" – presentations, teamwork, analytical and communication skills.
- Companies in the raw material industry prefer specialists who have a wider range of knowledge, who are not "only" programmers.
- Firms expect active workers who will take the initiative.
- Involving the corporate sphere in the formulation of semestral work themes instead of "search" on Internet topics.

4.4 Using Bioloid Robots for Teaching of Programming

Since 2011 we have started to teach programming languages using the robotic kit Bioloid (from Korean manufacturer Robotis [1]).

From our own experience with teaching subjects such as automation, programming, mathematics, and controlled systems, we have learned that some full-time students see application of mathematics or programming procedures as purely theoretical subjects. Those students have no idea what to do with this information and how their knowledge can be later utilized. That is why we have tried to make our lessons more attractive and enliven. Robots created from a modular construction kit are great equipment for teaching basic programming structures and how to integrate the hardware and software. Students can see real results of their work - moving robots - which makes teaching attractive for those of them who have not yet met with programming. Bioloid Robots in education can also be successfully used for a clarification of terms of higher mathematics, such as integral or derivative and to show how the explained theory is to apply in practice. This approach, when students are able to see how their theoretical knowledge can be applied into the field practice, proved to be very beneficial.

As an example of using robots to explain some ideas from theory control, we can describe a task with two vehicles that demonstrate system with integral behavior (Fig. 5). The aim is to control the set value of distance between these two vehicles (we demand constant distance as a result of control). If we want to show the integration character of this controlled system, it must be emphasized that the type of controlled system behavior is possible to provide by eliminating the disturbance. In this example, the disturbance is change the distance between two moving vehicles against each other. While the barrier is not moving (distance to the vehicle standing), it's possible to find the behavior of system. Of course, the system has more disturbance values e.g. temperature (caused by changes in the parameters of electronic devices). But this disturbance you have neglected, because in this case are not so essential. We can measure the step response of the system. Step response gives us information that the controlled system has integral character. Using P controller doesn't cause the permanent control deviation. This example of P controller behavior is possible to show using the Bioloid robots [19].

Fig. 5. Example from theory of control - P controller ensures the same distance between two vehicles [Photo: M. Řepka]

5 Areas of Research and Our Projects

5.1 Automation and Information Systems in the Raw Material Industry

The automation of coal preparation plants is an area of research in which we have been engaged for a long time. Coal preparation aims to achieve quality parameters of the coal according to the customers' demands. In particular, technological processes in the preparation plant can influence the content of ash and content of water. In the past, we have cooperated on the development of information systems for the control of preparation plants in the OKD Company (main producer of black coal in Czechia) and the development of systems for sales management (1997–2014). The highest benefit of such systems is real-time monitoring of coal quality parameters, utilizing continual sensors (ash meters, moisture meters). The system then calculates the trends of the individual parameters and predicts the possibility of insufficient quality. Thus the executive personnel of the preparation plant have the possibility to intervene in the production process still during the production. Without this information support, insufficient quality was not often ascertained until the coal had already been loaded on wagons, ready for shipment to customers, causing economic losses [6].

5.2 Fault-Tolerant Solution of Information and Control Systems

Characteristically, the raw material industry often involves uninterrupted continual production. This poses increased claims on the reliability and sturdiness of information and control systems. In the case of critical systems (the failure of which may endanger or stop the technological process), a so-called fault-tolerant solution is created, characterized by the system being able to provide information regardless of any failure. This can only be achieved by duplication of the systems (production and back-up) and, if the production system fails, immediate changeover to the back-up system, enabling continuation without interruption. Moreover, if the back-up system is separated geographically, one speaks of a disaster- tolerant system. In 2006, I participated in the development of a fault-tolerant control system for the coal preparation plant of the Darkov Mine. This system was in operation in the preparation plant from 2007 to 2011.

The critical part of a fault-tolerant solution is the database. If an information system is to satisfy the fault-tolerant requirements, it is necessary to ensure that all data from the production system database is available in the back-up system. Manufacturers of database systems offer a variety of technologies enabling on-line data replication between two or more database systems (replication, mirroring, clustering). In 2014, we implemented an internal grant at our faculty with students' participation, within which we tested technologies for the fault-tolerant security of databases from leading manufacturers. The tests focused on the length of data unavailability in the case of a database failure, and ability of automatic failover; moreover, crash tests were executed to analyse the sturdiness and resistance of database systems in the case of failure. With regard to the output-price ratio in the area of standard database system editions, Microsoft SQL Server systems and Caché (InterSystems) databases were evaluated as the most suitable. In the Enterprise software category, the most sophisticated solution is provided by Oracle; however, higher costs should be considered in this case [4].

5.3 Information Support for Reclamation Projects

Damage to the landscape is a negative effect of the underground mining of minerals – subsidence and deformation of the surface occurs; in addition, sludge lagoons (waste from the treatment process which could no further be processed by the technologies of the time) and heaps were created in the past. In old heaps containing coal remains, endogenous fires occurred in the past, which have lasted for decades (for instance, the Ema heap in Ostrava). At our faculty, we are concerned with computer modelling of landscape – we have developed a methodology which enables the creation of 3D models of the landscape from the supplied background materials by means of ArcGIS, 3Dmax and Autocad software (Fig. 6). By means of such models, it is possible to depict the influence of undermining on the appearance of the landscape over the course of time. Furthermore, it is possible to present consequences of the individual reclamation processes, thus enabling reclamation designers to select the most suitable solution.

Our visualizations use the environment of 3ds Max which generates the final landscape model; this software is also used for drawing the individual objects which are subsequently inserted into the model. The output is a realistic 3D model which is suitable for viewing, making movie clips of walk-throughs or flybys, or, in the final result, for converting into Virtual Reality world (see example at the Fig. 7).

From 2013 to 2015 we worked on an EU project called "SPOLCZECH", within the Czech-Polish territorial cooperation, the main subject of which was the training of specialists in the reclamation and rehabilitation of landscapes affected by mining activities in the Czech Republic and Poland. Within the project, methods of computer landscape visualisation and work with GIS systems were also deepened. Public and commercial data sources usable for GIS systems in the area of landscape modelling were analysed together with their availability, formats and conditions under which they

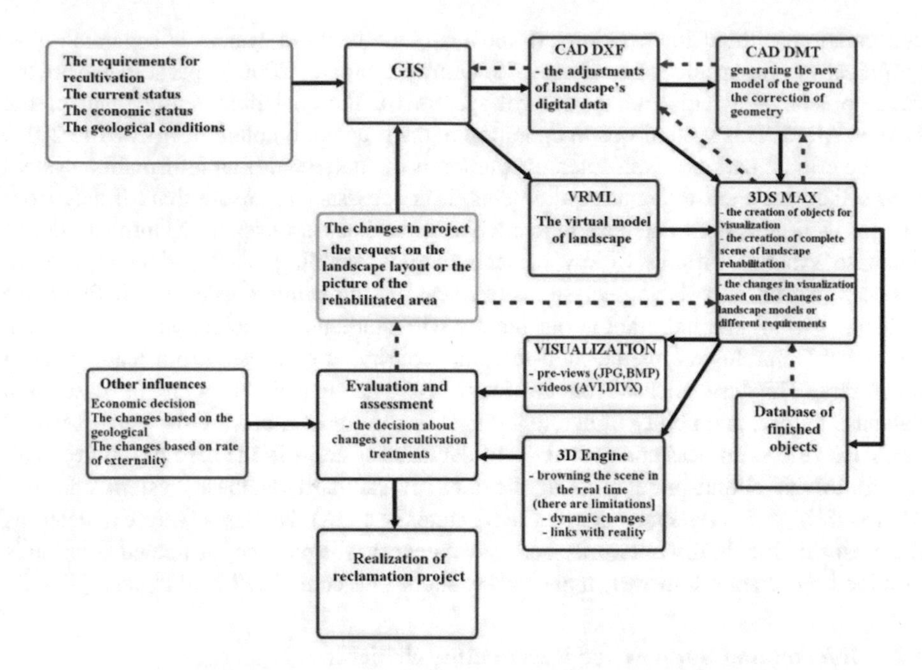

Fig. 6. Methodology of landscape modelling by Z. Neustupa – block diagram of the process [7]

Fig. 7. Model of "Louky church" in Karviná in a landscape affected by mining [Z. Neustupa]

are provided. An overview of data sources for the Czech Republic was published in [7]. Moreover, data sources available in Poland and Slovakia were analysed. One of the project output is a best practices guide with the methodology of landscape model creation and its 3D visualisation.

5.4 Revitalisation of Brownfields

Another area with which we have been concerned for a long time is the revitalisation of brownfields (localities and areas affected by industrial or agricultural activities which have already been terminated; these areas contain abandoned buildings, can suffer from contamination or are not utilized in any way). In the Ostrava-Karviná region, this particularly involves brownfields of the mining, metallurgical and chemical industries. From 2009 to 2013, we participated as co-solvers in the international COBRAMAN project (CENTRAL EUROPE Project 1CE014P4, 2009–2013), which dealt with the issue of brownfield revitalisation, including several solutions of pilot revitalisation projects in Central Europe.

Within the project COBRAMAN the study program "European School for Brownfield Management" (master course) was also created. The course was accredited in 2014. The study program is a multidisciplinary study that is based on the combination of natural, economic, construction and technical sciences and was created as a result of discussions with managers and investors of the brownfields revitalization or municipalities' needs.

Among others, our part of the project consisted in the creation of a database of brownfield revitalisation projects categorized according to the type of brownfield so that this database could be used as the so-called best practices. On the basis of analogy, project solvers in other localities can find in the database a similar project already resolved in the past, thus gaining an insight into the project intensity and possible recommendations concerning what to avoid in the solution. The database includes both examples of brownfield revitalisation projects implemented in Central Europe and, in the second part, detailed information on brownfields in localities around Ostrava, as compiled by students of the mentioned study program as their diploma thesis. It is planned to develop an expert system acting as an "advisor" for managers and employees responsible for the solution of re-use of these localities [5].

5.5 Monitoring of Methane Escape from Old Mines

When mining operations stopped and forced ventilation of mine areas ended in Ostrava in the mid-1990s, a problem arose – the uncontrollable escape of mine gases. Their most hazardous ingredient is explosive methane. In the city territory, the most affected cadastral areas involve Slezská Ostrava, Muglinov, Petřkovice at Ostrava, Hrušov, Koblov, Hošťálkovice and Lhotka at Ostrava. Methane is exhausted to the atmosphere in a controlled way by means of exhaust chimneys. At present, in connection with the solution of exhaust of mine gases, the project of "Complex Solution of the Issue of Methane in Relation to Old Mines in the Moravian-Silesian Region" is implemented in the region under the aegis of the Diamo company. The project is implemented in the years of 2014–2019. It also includes computer models of methane flow and escape as well as models of methane dissipation in the atmosphere. The Fluent program is used for the modelling [13].

5.6 RFID Technologies

Since 2009, a laboratory of RFID technologies has been used at FMG, primarily concerned with the application of RFID in practical technical tasks. The laboratory closely cooperates with the universities of Dongguk (Korea) and Kyushu (Japan). Furthermore, it closely cooperates also with industry – on both the supply sphere (Gaben) and the development of commercial applications (such as Hyundai) [17].

In underground coal mines, it is necessary to solve the safety-related problem of the explosion risk which the control technology means can cause in an explosive atmosphere (because the explosive properties of coal dust or occurrence of methane). Thus all equipment used in mines must be of such a design so that safety, health and lives of workers are not endangered. Therefore, the intrinsic safety requirements given by pertinent safety regulations also apply to RFID tags, readers and gateways [21].

In terms of the price intensity of such a solution, a common RFID tag can be acquired for a few crowns. Special tags resistant to the given specific conditions will certainly be more expensive; the price of a special RFID tag suitable for a potentially explosive atmosphere will be about CZK 100 (for example, the Metso RFID tag which can be placed directly in the raw material transported on the belt [20]). For its operation, the reader needs an antenna emitting waves by means of which the reader communicates with the RFID tag. The price of a reader suitable for these conditions ranges from about CZK 30,000 to 40,000; a single antenna is then worth of approx. CZK 10,000.

The RFID technology has also its operating limits. Reading in the presence of metals is very problematic. The greatest obstacle for the RFID technology systems, however, is water, which disables reading totally [4].

Within the project financed from the EUREKA programme solving the use of RFID tags for the identification of materials at sub suppliers of the Hyundai Company in Nošovice, we developed in 2015 a certified methodology of transmission of information obtained by means of RFID technology to the ERP system SAP.

5.7 Security of Industrial Information Systems

Ensuring the security of information and control systems in industry is an important task. Therefore, high priority is also given to the security of information systems in the study fields we teach. Students are acquainted with the issue of safeguarding technology in the form of IP covers, explosion-proof design and with requirements on equipment employed in environments with increased humidity, temperature or dustiness. Detailed attention is also paid to risk analysis and the proposal of safety measures, methodology and standards to ensure security in IT/ICT [10].

Although the penetration of new technologies into the information systems for production control is fairly slow, the system creators definitely cannot avoid it. It is necessary to take the long sustainability (often about 10 years) of these systems into account; for this reason, it is suitable to use to the maximum extent standard means which enable simple administration and sustainability.

In the area of security, we also cooperate with the commercial sphere. An example of successful cooperation is the project of employment of biometric fingerprint scanners for identification of employees in the Ardagh Company in Teplice. Implementation of the project and the problems which we encountered have been published in the paper [8] at the conference of "IT for Practice 2016".

6 Conclusion

Teaching automation and informatics at the faculty that focuses on mining, geology and the environment at first glance may seem unnecessary, especially considering the fact that VŠB has a Faculty of Electrical Engineering and Informatics. However, the practice shows that companies in the raw material industry are looking for specialists in this field, who also have a basic overview of controlled technologies. That's why the FMG has been teaching automation for 50 years. On the other hand, this education cannot be rigid and it is necessary to constantly monitor the current trends and requirements of the practice to flexibly adapt the teaching and preparation of the graduates.

References

1. BIOLOID - Robotis educational kits. http://en.robotis.com/index/product.php. Accessed 12 Nov 2017
2. Černín, J.: Usage of Web Application for Remote Control of Four Wheel Car Model [In Czech: Využití webové aplikace pro vzdálené řízení laboratorního modelu čtyřkolového vozidla], bachelor thesis, supervisor: M. Řepka, VŠB-Technical University of Ostrava (2014)
3. Danel, R.: Adapting IT/ICT education to current requirements from practice. In: IDIMT - 2016 Information Technology and Society Interaction and Interdependence 24th Interdisciplinary Information Management Talks, pp. 63–68. Trauner Verlag Universitat, Linz, Poděbrady, Czech Republic (2016)
4. Danel, R.: Trends in information systems for production control in the raw industry. In: Liberec Informatics Forum, LIF 2016, Liberec, pp. 19–26 (2016)
5. Danel, R., Neustupa, Z., Stalmachová, B.: Best practices in design of database of Brownfield revitalization projects. In: 12th International Multidisciplinary Scientific GeoConference and EXPO - Modern Management of Mine Producing, Geology and Environmental Protection - SGEM 2012, Albena, pp. 49–56 (2012)
6. Danel, R., Otte, L., Vančura, V., Neustupa, Z.: Software support for quality control in coal and coke production in OKD a.s. In: 14th International Carpathian Control Conference, ICCC 2013, Rytro, pp. 33–37 (2013)
7. Danel, R., Neustupa, Z.: Information support for brownfield revitalization projects. In: XIth International Scientific and Technical Conference Computer Sciences and Information Technologies (CSIT), Lvov, pp. 111–115 (2016)
8. Danel, R., Růžička, V.: Biometric employee identification system implementation at Ardagh Metal Packaging. In: 19th international conference Information Technology for Practice, pp. 139–149. VŠB-Technical University of Ostrava, Ostrava (2016)
9. Danel, R., Řepka, M.: Experience with teaching using videoconferencing system. In: IT for practice, pp. 191–198. VŠB-Technical University of Ostrava, Ostrava (2017)

10. Hons, L.: Bezpečnost v oblasti MES systémů – kde začít [in Czech]. In: Workshop IT v průmyslu 2016. MES Centrum & Trade International, Brno (2016)
11. Institute of Combined Studies in Most [Institut kombinovaného studia Most], VŠB–Technical University of Ostrava. https://www.hgf.vsb.cz/instituty-a-pracoviste/cs/512/. Accessed 22 Jan 2018
12. Ministr, J., Pitner, T.: Academic-industrial cooperation in ICT in a transition economy – two cases from the Czech Republic. Inf. Technol. Dev. **21**(3), 480–491 (2015)
13. Kohut, V., Staša, P., Kodym, O.: Modeling of flow using CFD and virtual reality. In: Proceedings of the 13th International Multidisciplinary Scientific GeoConference SGEM, Albena (2013)
14. Kozel, R., Chuchrová, K.: Creation of system support for decision-making processes of managers. In: IDIMT 2015: Information Technology and Society - Interaction and Interdependence: 23rd Interdisciplinary Information Management Talks, pp. 163–170. Poděbrady, Czech Republic. Universitatsverlag Rudolf Trauner, Linz (2015)
15. Kozel, R., Vilamová, Š., Baránek, P., Friedrich, V., Hajduová, Z., Behún, M.: Optimizing of the balanced scorecard method for management of mining companies with the use of factor analysis. Acta Montanistica Slovaca **22**(5), 439–447 (2017)
16. Mikoláš, M., Kozel, R., Vilamová, Š., Paus, D., Király, A., Kolman, P., Piecha, M., Mikoláš, M.: The new national energy concept 2015 - the future of brown coal in the Czech Republic. Acta Montanistica Slovaca **20**(4), 298–310 (2015)
17. RFID Laboratory. http://ilabrfid.cz/?lang=en. Accessed 10 Mar 2018
18. Řepka, M., Danel, R.: Remote control of laboratory models. In: Proceedings of CIAAF 2015 - 1st Ibero-American Conference of Future Learning Environments, Porto, pp. 59–63 (2015)
19. Řepka, M., Danel, R., Neustupa, Z.: Use of the Bioloid robotic kit in the teaching of automation and programming. In: SGEM 2012 - 12th International Multidisciplinary Scientific Geo-Conference, Albena, vol. III, pp. 1229–1236 (2012)
20. Tracking mineral materials with RFID tags and detectors. http://www.aggbusiness.com/categories/quarry-products/features/tracking-mineral-materials-with-rfid-tags-and-detectors/. Accessed 01 Sept 2016
21. Vestenický, P., Mravec, T., Vestenický, M.: Analysis of inductively coupled RFID marker localization methods. In: Proceedings of the Federated Conference on Computer Science and Information Systems – FedCSIS, Łodź, pp. 1291–1295 (2015)

Information Support of Scientific Researches of Virtual Communities on the Platform of Cloud Services

Kazarian Artem(✉), Roman Holoshchuk, Nataliia Kunanets, Tetiana Shestakevysh, and Antonii Rzheuskyi

Lviv Polytechnic National University, Lviv 79013, Ukraine
artem.kazarian@gmail.com, nek.lviv@gmail.com, holoshchuk@vlp.com.ua, artem.kazarian@gmail.com, nek.lviv@gmail.com

Abstract. The development of modern information technology generates the need in scientific research on an innovation platform called the e-science. Implementation of projects aimed at informational and technological support for research activities requires effective communication among members of virtual teams and information workers. The authors have developed a project aimed at creating a complex system of informational and technological support for scientific research conducted by virtual research teams on the platform of electronic science using cloud computing technologies. The information system provides the establishment of effective scientific communication when conducting research on the platform of electronic science. This approach increases the effectiveness of communication processes of participants in the virtual scientific team for interdisciplinary research; contributes to solving the informational aspects of scientific communication and the exchange of scientific data, information and knowledge for the convenient interaction of researchers, it also facilitates the effective search, consolidation, preservation and dissemination of research results among the scientific community.

Keywords: E-science · Information system · Digital science · Cloud services Virtual communities · Communication

1 Introduction

In the information society there are systemic changes in the processes of scientific information exchange, presented both in text and in multimedia formats. In the system of scientific communications during research on the platform of electronic science circulating information flows, presented in the form of text, data, images, videos, blogs, etc. This requires the implementation of information and technology projects aimed at information support of research work.

The emergence of the latest cloud-based technologies has changed the approaches to the formation of the infrastructure of information provision of scientific research. Dimitrios and Dimitrios [1] in this context substantiates the need to ensure the integrity and confidentiality of communications and information messages. A rapid transition to

N. Shakhovska and M. O. Medykovskyy (Eds.): CSIT 2018, AISC 871, pp. 301–311, 2019.
https://doi.org/10.1007/978-3-030-01069-0_22

the clouds raised concerns about information systems, communication and information security. As a result, they evaluate cloud security and present a solution that eliminates these potential threats. Badii, Bellini, Cenni, Difino, Nesi, Paolucci [2] have proposed decisions made about intelligent architectures of information systems for aggregation of data using ontologies and knowledge bases in the production of intelligent services. This substantiates the possibility of re-storing data. David De Route and Carole Goble [3] introduce the 'Semantic Grid' based on Semantic Web, which forms a unified infrastructure for data storage, computing, collaboration and automation of scientific activity. Hock Beng Lim has developed a smart e-Science cyber-infrastructure for cross-disciplinary scientific collaborations. The proposed infrastructure is available resource pooling systems for sharing resources of each scientific community, providing access to their computing and intellectual resources. Amitava Biswas explores semantic technologies for searching in e-Science grids [4]. Kitowski, Wiatr, Dutka, Twardy, Szepieniec, Sterzel, Slota and Pajak [5] suggest to use the distributed computing infrastructure of PLGrid to provide communication interoperability. It is a flexible, large-scale electronic infrastructure that offers homogeneous, easy-to-use access to organizationally distributed, heterogeneous hardware and software resources. This electronic infrastructure [6–8] is focused on the requirements and needs of users conducting interdisciplinary research, equipped with specific environments, solutions and services. The authors analyze the necessity of establishing interaction between scientists in social networks, providing communication with the organizers of conferences in real time, developing the processes of filtering and sorting the comments text of the users of the Twitter [9–11] network, since this network is widely used by the scientific community during scientific events. Thus, in many studies the necessity for development of effective tools for scientific and technical support of scientific activities, in particular on the platform of electronic science, and the creation of innovative information and technological infrastructure is substantiated. The purpose of the article is to introduce an analysis of the project aimed at creating an integrated system of information-technology support for research work conducted on the platform of electronic science, using cloud computing technologies.

2 The Use of SaaS Cloud Computing Model in Research

The development of information technology aimed at the formation of effective scientific communications for conducting research on the platform of electronic science is at its initial stage. A comprehensive information system project that facilitates the formation and processing of information resources using the Big Data technology, taking into account the factor of their weak structuring, is suggested in the article. The suggested system of information and technological support for scientific research on the platform of electronic science solves a number of key problems for creating effective scientific communication when conducting research on the e-science platform. Innovative technology of scientific research involves:

- formation of a virtual creative team for conducting multidisciplinary research;
- solving informational and technological aspects of scientific communication;

- exchange of scientific data, information, and knowledge;
- providing a convenient interaction of researchers;
- Search, consolidation, storage, and dissemination of scientific knowledge in a scientific community.

3 "Client-server" System Architecture of Research Support by the Electronic Science Platform

The architecture of the information provision system for research on the electronic science platform is based on client-server technology built on the basis of the SaaS cloud service model. The specified architecture contains the following main components: the client part, the server part of the vendor, the communication between the client and the server part. The client part of the system is compatible with any web browser. This allows members of the virtual research team to access the system's functions through a user interface, formed as a web application. Data transmission is carried out over the Internet and Intranet networks. Functionality of the system is heterogeneous, and the specifics of each individual module of the system requires the use of various technologies for its implementation. This has forced developers to predict the use of technology for optimization of the load on the system. The developed system is structured as a coordinated project. At the same time, our project needs to be created for a monolithic information system, despite the fact that within the framework of an integrated information system such architecture may have certain disadvantages. In the process of project implementation, they will be eliminated. An important aspect of the project (Fig. 1) implementation is the creation of the architecture of the information and technology support system for scientific research on the platform of electronic science, based on different approaches for its development. In this case, each complex function of the system is developed as a separate project. A certain disadvantage of this approach is the reduction of flexibility of the system in the monitoring process.

Communication between the server and the client is achieved on the basis of an HTTP protocol, which provides the identification of the resource of the global URI, without preserving the intermediate state of the connection between the "request-response" pairs. Queries are generated using GET methods, which include invitations to the contents of the specified resource and obtaining the same results for each iteration of the query. Retaining information about the state of the connection associated with the latest requests and answers, if necessary, is performed by the components themselves. Data about a specific resource is transmitted using the POST method, integrating them into the body of the query itself. For members of the virtual scientific community, the system of information and technological support of research on an electronic scientific platform based on the client-server architecture provides a large set of programs and tools focused on maximizing the utilization of computing capabilities of client machines. Server resources are used mainly for temporary storage and data exchange, as well as for access of external researchers-partners.

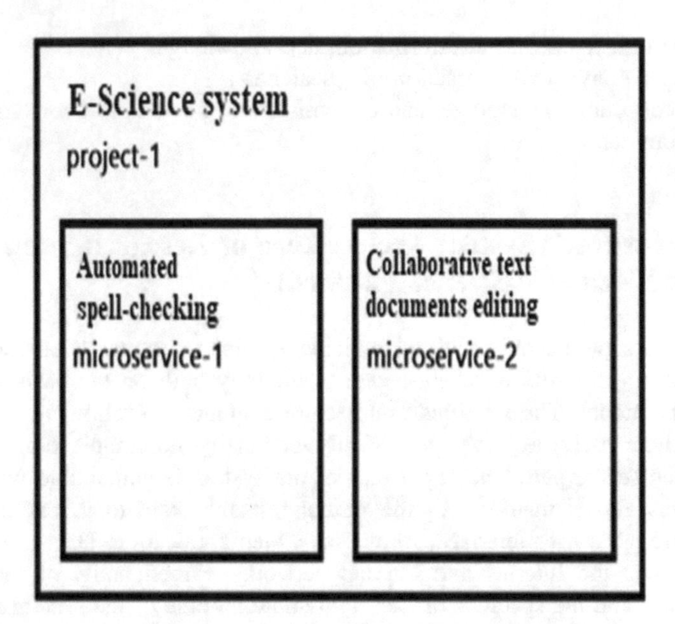

Fig. 1. System microservice structure

4 Information System Based on the Development of Microservices

With monolithic system architecture, the processes of query processing are performed simultaneously, it is provided by the advantages of the chosen programming language and the ability to split applications into classes, functions and namespaces. At the same time, the ability to run and test the application on the user's computer is implemented and the standard deployment process is used to check changes before publishing them for end users. Scaling of monolithic applications is achieved by running multiple physical servers with a load balancing between them. Architecture of information system (Fig. 2) based on the development of microservices (Fig. 3) and implementation of applications in form of services set is used to prevent the inconvenience. In addition to the possibility of independent deployment and scaling, each service gets a clear physical boundary that allows to implement services in different programming languages. Given the complexity of the task, most of the system's functions are separate and complete projects. The architecture of the information system contributes to the implementation of its declared functions and the following objectives:

- Define clear rules for interactions between different services.
- Apply independent cycles of the deployment process.
- Allow simultaneous test running for each of the subsystems.
- Minimize the cost of test automation and quality assurance.
- Improve logging and monitoring quality.
- Increase the overall scalability and reliability of the system.

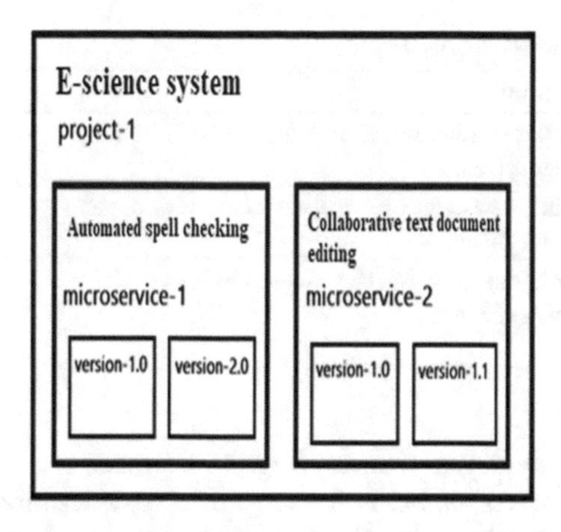

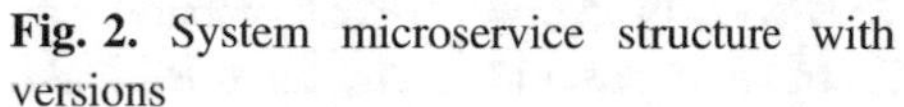
Fig. 2. System microservice structure with versions

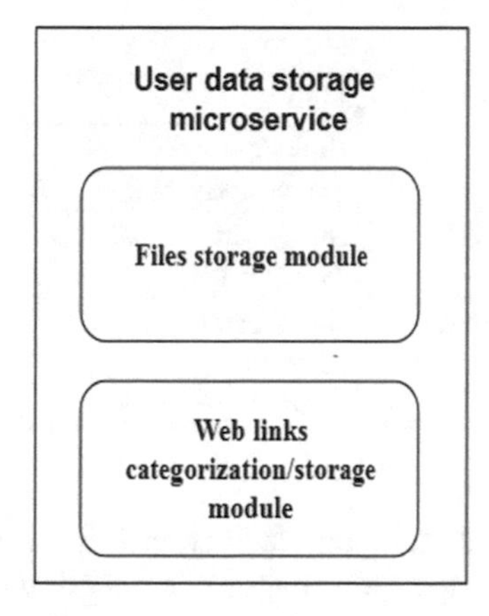

Fig. 3. User data storage microservice

The developed system deploys several modules. These services have the complete isolation of the program code. One way to execute the code in these services allowed by HTTP calls, such as a user request or RESTful API call. The program code of one service can not directly call the code of another service. The software code is deployed for each service independently, and various services are written in different languages, such as Python, Java, Go, NodeJS and PHP. Auto-scale and load balancing processes are driven by independent services, presented in Table 1.

Table 1. User data storage microservice endpoints

Method	Path	Description
GET	/files/{userId}/list	Get the list of files, that are available for the user with unique identifier userId
GET	/files/show/{fileId}	Download file with unique identifier fileId
POST	/file/add	Upload file to the server
POST	/file/edit/{fileId}	Rewrite file on the server with unique identifier fileId
DELETE	/files/{fileId}	Delete file from the server with unique identifier fileId
GET	/links/{userId}/list	Get the list of references, that are available to user with unique identifier userId
GET	/links/show/{linkId}	Get detailed information about reference (name, path to hyperlink, description and category) with unique identifier linkId
POST	/links/add	Add new reference
POST	/links/edit/{linkId}	Edit information about reference (name, path to hyperlink, description and category) with unique identifier linkId
DELETE	/links/{linkId}	Delete reference with unique identifier linkId

(continued)

Table 1. (*continued*)

Method	Path	Description
GET	/links/category/{categoryName}	Get list of references of category with name categoryName
POST	/links/category/add/{categoryName}	Create new category of references with name categoryName
DELETE	/links/category/remove/{categoryName}	Delete category of references with name categoryName

In addition, each service has several deployed versions (Fig. 4). For each service, one of these versions is a default working version, but at the same time provides direct access to any deployed version of the service (Fig. 5), since each version of each service has its own separate address. This structure provides many features, including testing a new version, testing interdependencies between different versions and simplifying the revert operations to the previous version. System functions of information and technological support of scientific research on the electronic science platform are divided into four microservices, which solve related problems. Each service is designed according to the separate task:

User data storage microservice functions:

- Files storage module
- Web links categorization/storage module

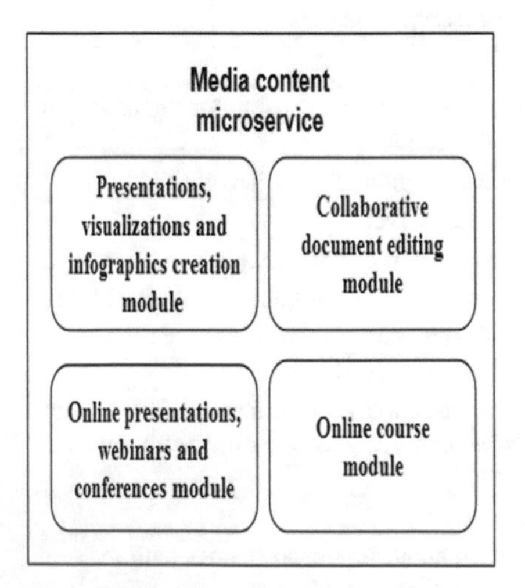

Fig. 4. Media content microservice

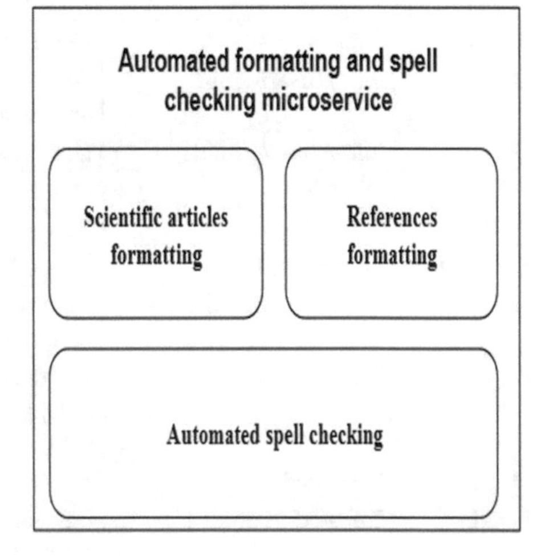

Fig. 5. Automated formatting and spell checking microservice

Interfaces of this microservice presented in the next table.
Media content microservice functions:

- Presentations, visualizations and infographics creation conferences module
- Collaborative document editing module
- Online course module

Interfaces of this microservice presented in Table 2.

Table 2. Media content microservice endpoints

Method	Path	Description
GET	/presentation/{userId}/list	Get list of presentations, that available for user with unique identifier userId
GET	/presentation/show/{presentationId}	Download presentation with unique identifier presentationId
POST	/presentation/add	Upload presentation on the server
POST	/presentation/edit/{presentationId}	Rewrite presentation on the server with unique identifier presentationId
DELETE	/presentation/{presentationId}	Delete presentation from the server with unique identifier presentationId
GET	/document/{userId}/list	Get list of documents, that are available for user with unique identifier userId
GET	/document/show/{documentId}	Show the content of document on server with unique identifier documentId
POST	/document/add	Create new document on the server
POST	/document/edit/{documentId}	Edit the document on server with unique identifier documentId
DELETE	/document/{documentId}	Delete document from the server with unique identifier documentId
GET	/course/{userId}/list	Get list of study courses, that are available for user with unique identifier userId
GET	/course/show/{courseId}	Show the content of study course on server with unique identifier courseId
GET	/course/{courseId}/video/list	Get list of video materials, that are available for specific course with unique identifier courseId
GET	/course/{courseId}/video/{videoId}	Show video material with unique identifier videoId
GET	/course/{courseId}/documents/list	Get list of text materials, that are available for specific course with unique identifier courseId
GET	/course/{courseId}/documents/{documentId}	Show text material with unique identifier documentId

Automated formatting and spell checking microservice functions:

- Scientific articles formatting.
- References formatting.
- Automated spell checking (Fig. 6).

Interfaces of this microservice presented in the next Table 3.

Scientific works processing microservice functions (Fig. 7):

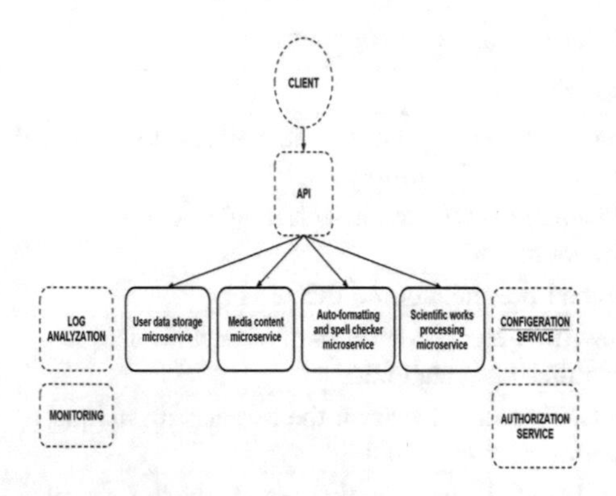

Fig. 6. System components structure

Scientific works processing microservice

Scientific works, journals, books database

Plagiarism search

Publication

Article reviews

Articles ratings

Fig. 7. Scientific works processing microservice

Table 3. Automated formatting and spell checking microservice endpoints

Method	Path	Description
POST	/articles/format	Send the text of article (title, authors, content, references) for automated formatting according to the standards
POST	/references/format/{formatName}	Send data of reference (title, authors, content, reference, pages, year) for automated formatting of references according to the standard with name formatName
POST	/spellcheck/{languageName}	Send the text for spell checking for language languageName

- Scientific works, journals, books database
- Article reviews
- Plagiarism search
- Publication
- Articles ratings

Interfaces of this microservice presented in Table 4.

Table 4. Scientific works processing microservice endpoints

Method	Path	Description
GET	/db/article/search/{articleName}	Search scientific article in database by name articleName
GET	/db/book/search/{bookName}	Search book in database by name bookName
GET	/db/journal/search/{journalName}	Search journal in database by name journalName
GET	/reviews/{articleName}/search	Search reviews in database by name of article articleName
GET	/reviews/{articleId}/show	Get full version of reviews for article with unique identifier articleId
POST	/plagiat/check	Send text for plagiarism checking
POST	/publications/request	Send file of scientific work for publication in the system
POST	/publications/rating/categories	Get list of categories for article ratings
POST	/publications/rating/{categoryName}	Get article ratings of category with name categoryName

The system has additional modules for system's infrastructure support: log analysis, monitoring, configuration service, and authorization service. *Configuration service.* Configuration service performs scalable horizontal storage of a distributed system for file versions control in various formats. *Authorization service.* The authorization functions are delegated to the microservice, which return OAuth2 tokens for access to the system server part resources. The authorization service allow to authorize members of the virtual research team. Its function also involves the protection of "service-service" communication. *Gateway API.* All main services of software application interfaces are provided for a virtual research team members. In information systems with microservice architecture there is a rapid increase in the number of components. A virtual team member have access to each of the services using a single entry point for external requests and routing.

5 Distributed Systems in Electronic Science Projects

In order to provide the information and communication needs of members of virtual scientific teams it is expedient to use the proposed information system based on the principles of information integration, using the principle of distributed information systems. When implementing a new information system, you can certainly use existing ready-made components that can provide a certain direction of the system in its functions and extend its functionality. In this case, it is necessary to take into account compatibility issues that can not be eliminated during implementation, but only at the stage of system development. If you do not comply with the requirements of compatibility, some components do not understand each other, they can not work together.

For such situations, there is a mechanism or set of mechanisms for adapting independent developed information and computing resources to the common functionality. During the development of the information system for research support on the platform of electronic science, the operating systems and network protocols were identified to be used.

6 Conclusions

The developed system of information and technological support for scientific research on the platform of electronic science is presented in the form of a distributed information system, which is based on the use of client-server architecture. Each server, integrated into the network system, is a stand-alone element of the information system. Any member of the virtual research team can perform data operations on their computer connected to the distributed network. Distributed system provides communication between individual local databases, microservices, located on separate local servers, adhering to the basic principle of creating distributed databases, providing users of the system with ease of use in the same way as not distributed. The proposed system solves a number of key problems for establishing effective scientific communication in conducting research on the platform of electronic science, including information provision of the virtual team for research; solution of informational aspects of scientific communication; exchange of scientific data, information and knowledge; ensuring interaction of researchers; search, consolidation, storage and dissemination of scientific knowledge in the scientific community. Distributed databases, implemented in the information system developed by the authors, can be characterized as a set of logically interconnected databases that are freely distributed on the Internet.

References

1. Dimitrios, Z., Dimitrios, L.: Addressing cloud computing security issues. Futur. Gener. Comput. Syst. **28**(3), 583–592 (2012)
2. Badii, C., Bellini, P., Cenni, D., Difino, A., Nesi, P., Paolucci, M.: Analysis and assessment of a knowledge based smart city architecture providing service APIs. Futur. Gener. Comput. Syst. **75**, 14–29 (2017)
3. De Roure, D., Hendler, J.A.: E-science: the grid and the semantic web. Intell. Syst. **19**(1), 65–71 (2004)
4. Chen, H., Wang, Y., Cheung, K.-H.: Semantic e-Science. Springer, New York (2010)
5. Kitowski, J., Wiatr, K., Dutka, L., Twardy, M., Szepieniec, T., Sterzel, M., Slota R., Pajak, R.: Distributed computing infrastructure as a tool for e-Science. In: Parallel Processing and Applied Mathematics, vol. 9573, pp. 271–280 (2016)
6. Cheptsov, A., Koller, B., Adami, D., Davoli, F., Mueller, S., Meyer, N., Lazzari, P., Salon, S., Watzl, J., Schiffers, M., Kranzlmueller, D.: E-infrastructure for remote instrumentation. Comput. Stand. Interfaces **34**(6), 476–484 (2012)
7. Fernández-del-Castillo, E., Scardaci, D.,, García, Á.L.: The EGI federated cloud E-infrastructure. Procedia Comput. Sci. **68**, 196–205 (2015)

8. Dash, S., Pani, S.K.: E-governance paradigm using cloud infrastructure: benefits and challenges. Procedia Comput. Sci. **85**, 843–855 (2016)
9. Lee, N.Y., Kim, Y., Sang, Y.: How do journalists leverage Twitter? Expressive and consumptive use of Twitter. Soc. Sci. J. **54**(2), 139–147 (2017)
10. Flores, C.C., Rezende, D.A.: Twitter information for contributing to the strategic digital city: towards citizens as co-managers. Telemat. Inform. **35**(5), 1082–1096 (2018)
11. Wu, T., Wen, S., Xiang, Y., Zhou, W.: Twitter spam detection: survey of new approaches and comparative study. Comput. Secur. **76**, 265–284 (2018)

Organization of the Content of Academic Discipline in the Field of Information Technologies Using Ontological Approach

Serhii Lupenko[1(✉)], Volodymyr Pasichnyk[2], and Nataliya Kunanets[2]

[1] Computer Systems and Networks Department, Ternopil Ivan Pul'uj National Technical University, Ternopil, Ukraine
lupenko.san@gmail.com

[2] Information Systems and Networks Department, Lviv Polytechnic National University, Stepan Bandera Street, 32a, Lviv 79013, Ukraine
Volodymyr.V.Pasichnyk@lpnu.ua, nek.lviv@gmail.com

Abstract. The paper presents strategy for organizing the content of the academic discipline using an ontological approach, which enables an effective system solution of a range of important methodological, methodical and technological tasks for the development of intellectualized systems of electronic education in the field of information technologies. The mathematical structures that describe and detail the abstract logic-semantic core of the discipline in the form of a set of axiomatic systems are developed. It is shown that the ontological approach of the organization of educational content ensures the presence of a clear, compact, ordered structure of the knowledge organization about the subject area of the academic discipline and is well consistent with the theoretical formal basis for the development of modern ontologies - a family of descriptive logics. As an example of the application of the proposed approach, the elements of the glossary and taxonomies of the concepts of the discipline "Computer Logic" are developed.

Keywords: Ontology · Ontological approach · E-learning system
Axiomatic-deductive system

1 Introduction

The training of a modern highly qualified specialist in the field of computer sciences and information technologies should take into account the main statements of the Bologna process, be oriented on the latest model of credit-modular organization of the educational process, rely on the international legal framework and scientifically substantiated standards of training, which requires the development and application of the latest educational concepts, models and technologies, in particular, the active use of systems of individually-oriented e-learning. An important component of the quality of the educational process, which realized using e-learning systems, is the quality of the content of the disciplines, which are covered by the relevant training program. Known approaches to the organization of the contents of academic disciplines in the field of information technology, implemented in a number of existing e-learning systems, are

N. Shakhovska and M. O. Medykovskyy (Eds.): CSIT 2018, AISC 871, pp. 312–327, 2019.
https://doi.org/10.1007/978-3-030-01069-0_23

mainly based on intuitive, heuristic paradigm, rather than on a clear formalized strategy using an adequate mathematical apparatus, which often leads to low quality content of electronic courses.

The application of the ontological approach to the presentation and organization of knowledge greatly enhances the quality of content organization in e-learning systems as it enables an effective systemic solution to a range of important methodological, methodical and technological tasks, in particular: (1) unification, standardization of the technology of presenting information (data and knowledge) in the subject area, which enables to overcome the problem of semantic heterogeneity of weakly structured and highly formalized knowledge; (2) creation of a qualitative dictionary (glossary) and knowledge base (thesaurus) in the subject area of the discipline with the properties of completeness, consistency, interpretation, unification, integration with other academic disciplines; (3) multiple reuses of knowledge, which greatly simplifies and intensifies the development of intellectualized systems in the field of e-learning; (4) realization of effective search of the information on the Internet on the basis of technologies WEB 2.0, which will provide high relevance of searched information [1–7].

An ontological approach involves the development of an ontology of a subject area, which is studied by the corresponding academic discipline. According to the generally accepted definition of ontology, an explicit machine-interpretive specification of conceptualization is understood [8–12]. Conceptualization is the process of constructing a conceptual model of a subject area, which is studied by a certain discipline, which in the form of a set of concepts (classes), their properties (attributes, slots, roles), limitations (facets), which are imposed on the properties, and the relations between concepts, reflects objects of the subject area, their properties, attitudes and regularities. The conceptual model of the subject area is the core of the content of the academic discipline. The success of all the subsequent stages of constructing the ontology of the subject area of the academic discipline and onto-oriented (onto-based) intellectualized systems of e-learning depends exactly on the correct, qualitative conceptualization.

To a large extent, the methodology of projecting the ontology of the subject area of the academic discipline uses an axiomatic-deductive strategy, which, in its most complete form, manifested itself in the field of mathematics and mathematical logic as a formal axiomatic system. In the formal axiomatic system, the complete abstraction from the semantics of the words of the natural language is carried out, and the rules for manipulating the characters in explicit form are given in the form of axioms and the rules of derivation from the axioms of theorems. The axioms of the formal system are presented as sequences of symbols of a certain alphabet, and the methods of proofing the theorems as formal methods of obtaining one formulas from the other, by applying formal operations over symbols. This approach guarantees the preciseness of the initial assertions and univocity of conclusions. In this regard, despite its purely formal construction of the theory, there is always a possibility of meaningful interpretation of the created abstract objects.

Despite the onto-orientation of the modern systems of electronic learning, the existence of high-quality examples of the organization of mathematical knowledge in an axiomatic-deductive form and the good consistency of the axiomatic-deductive approach with the theoretical formal basis for the development of modern ontologies - a family of descriptive logics, the development of a coherent axiomatic-deductive strategy of organizing of academic content in e-learning systems, is insufficiently represented in scientific papers. Therefore, it is advisable to develop an ontological approach which correlated with the mathematical and deductive strategy of organizing the contents of the academic discipline in e-learning systems. The axiomatic-deductive strategy of organizing academic content ensures the presence of a clear, compact, ordered structure of knowledge organization about the subject area of the discipline, which gives it significant advantages over non-axiomatic strategies.

In fact, the purpose of this paper is to develop an axiomatically-deductive strategy of organizing the content of the academic discipline in the field of information technology using the ontological approach. To achieve the goal, it is necessary to solve such problems:

1. To formulate clear quality requirements for the content of the discipline.
2. To develop a general axiomatic-deductive strategy for organizing academic content that would satisfy the requirements of content quality.
3. To develop a mathematical apparatus that formalizes and specifies the main stages of the strategy of organizing the contents of the discipline.
4. To realize the elements of the proposed axiomatic-deductive strategy for organizing the context of academic discipline “Computer Logic” in the Protégé environment using the OWL ontology description language.

2 Problem Statement

The studies aimed at using an ontological approach began at the end of the twentieth century. The theoretical foundations for the development of ontologies are proposed by Gruber [9], in works by Guarino [10] approaches to the formation of ontologies are suggested. In the further researches of scientists, the concepts of the conceptual graph Sowa [13], and the features of its application in the construction of ontologies (Montes-y-Gómez) [14] are introduced. The usage of ontological approach for the first class of intelligent systems is considered by Dosyn [15]. The researchers [16] Guarino analyze the possibilities of automating the process of constructing ontologies with the help of the genetic and automated programming method, which facilitate the generation of approaches to solve the problem in automatic mode, based on their developed converters. T.J. Watson Research Center (New York, USA) IBM proposed a method of constructing ontologies with the help of scientific queries based on the use of text analysis technologies. Researchers have suggested to use knowledge representation and formal reasoning in ontologies with Coq [17].

3 Conceptual Foundations of the Axiomatic-Deductive Strategy of the Academic Discipline Content

Content of the academic discipline is in a certain way a structured text that covers the semantics (semantic space) of the academic discipline. This semantic space is a sophisticated heterogeneous system whose fundamental components are the set of terms-concepts that define the terminology-conceptual apparatus (glossary) of the academic discipline, the set of relations between these concepts, a set of true statements and reasoning (inferences, proofs) of the academic discipline. Given that the assertion of a academic discipline can be considered as certain functions on a set of its actual concepts, which actualize, reflect explicit and implicit relations (relationships) between terms-concepts of academic discipline, and inferences are certain relations between true statements, it can be argued, that the content (semantic space) of the discipline consists of its terminological-conceptual apparatus and the system of the true statements that determine the subject of the study of the academic discipline (see Fig. 1).

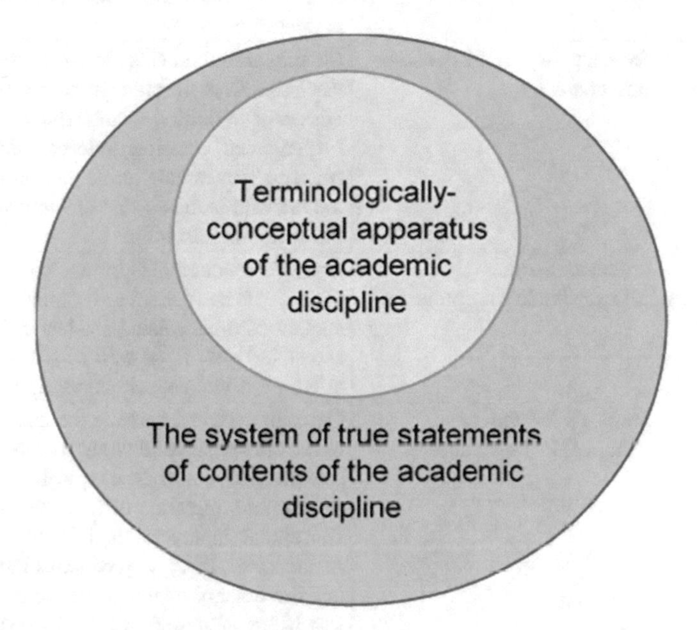

Fig. 1. Terminologically-conceptual apparatus and the system of true statements of contents of the academic discipline

The quality of the content of the academic discipline is given by a certain set of requirements to it. Table 1 provides information about the main groups of requirements to the content of the academic discipline, which determines its quality.

Given the above requirements for the content of the academic discipline and the expediency of using the ontological approach, it is appropriate to organize the academic discipline content in accordance with the axiomatic-deductive strategy as this strategy provides satisfaction of all the above groups of requirements. In the general case, the axiomatic-deductive strategy of organizing academic content consists in the sequential implementation of the next steps:

1. Formation of a metadisciplinary logical-semantic core of the academic discipline, consisting in outlining the set of metadisciplinary (general scientific) concepts, relations between concepts and true statements that underlie the logical-semantic core of the academic discipline.

Table 1. Main groups of requirements for the organization of the content of academic discipline

Requirements for the organization of the content	Description of the group requirement to the content organization
A group of requirements of logicality of the academic discipline content	Contains requirements to the content of the academic discipline from the standpoint of satisfaction of the principles of logical rigor (satisfaction of the laws of identity, consistency, completeness, and compactness) presentation of knowledge (concepts, statements, models and methods) of content presentation
A group of requirements of visibility of the academic discipline content	Contains requirements for the content of the discipline from the standpoint of satisfying the criteria of visibility, experimental (experienced, practical) interpretation of concepts, statements, principles, concepts, models and methods that are considered by the academic discipline
A group of requirements for the compatibility of the academic discipline content	Contains requirements for compatibility of the content of the academic discipline with the content of other related academic disciplines associated with it, as well as with general scientific principles, concepts, and theories
A group of requirements for the convenience of using of the academic discipline content	Contains requirements for the characteristics of the convenience of using the content of the discipline in terms of the logical, technical and educational operation of its terminology-conceptual apparatus, models, and methods. In particular, this group contains requirements that the content of the academic discipline should be submitted in three forms: a contestable verbal form, as a formalized (formal) system and in the form of computer ontology

2. Formation of a set of basic (atomic) general (abstract) concepts of the discipline that are characterized by the highest level of abstraction and the maximum possible scope of coverage of the entire semantic space of the academic discipline.
3. A derivation from a set of basic general concepts of the set of new derivatives of general concepts of the academic discipline, by applying logical operations (operations of combining, intersections, complements, definitions of concepts) to basic general concepts. Basic and derivative general concepts in their aggregate form the terminological and conceptual apparatus of the abstract content core of the academic discipline.

4. Formation of a set of relations between the above general concepts (both basic and derivative) that capture the logical-semantic interconnections between the fundamental concepts of the highest level of abstraction in the subject area of the academic discipline.
5. Formation of a set of mutually not contradictory and mutually independent axioms - true statements (judgments), the truth of which is taken without proof in the framework of this academic discipline. From the formal point of view, axioms are functions (predicates) from general (basic and derivative) concepts and clearly reflect (postulate, actualize) the logical-semantic relationships (relationships) between them.
6. Formation of the set of logical rules of derivation from the set of axiomatic statements of derivatives of true statements (theorems), which together with axiomatic statements form a set of true statements of the abstract content core of the academic discipline.
7. Formation of the set of taxonomies of the concepts of the academic discipline by repeatedly applying the operation of the division of the general (abstract) concepts by highlighted fundamentals of division in advance, which provides a derivation from the most general concepts of the discipline of its derivative concepts of a lower level of universality and abstraction. The set of all general concepts and newly formed notions of a lower level of universality, which are elements of the above taxonomy, form a terminology-conceptual apparatus of the academic discipline that covers all its hierarchical levels of abstraction (including specific concepts).
8. Formation of a set of truthful statements of a lower level of abstraction (including specific statements) of academic discipline as predicates given on elements of taxonomy concepts, which provides a strictly logical transition from abstract statements of a academic discipline to concepts of a lower level of universality and abstraction, including specific statements of a academic discipline.

The result of the axiomatic-deductive strategy of organizing the contents of the academic discipline is its logical-semantic core, on which “are nailed down” all other elements of this content. The core is a highly structured component of academic content. The content periphery is a derivative additional complement to its logical-semantic core, which fills the content with examples, explanations of the components of this core. In this case, the content (semantic space) of the academic discipline can be presented as the union of its logical semantic core and periphery (see Fig. 2).

Given the axiomatic-deductive strategy of organizing the contents of the academic discipline described above, the structure of the semantic space of the academic discipline, which is conventionally depicted in Fig. 2, can be detailed by explicitly allocating the structure of its logical-semantic core, organized in accordance with the axiomatic-deductive strategy, and depicting in the form of a diagram, which is given in Fig. 3.

According to the axiomatic-deductive strategy of content organization as a heterogeneous system of knowledge that contains the concepts of a subject area, propositions (judgments, statements) attributed to true values and considerations (inferences, proofs), all knowledge is divided into two large groups. The first group includes knowledge in the form of a set of basic primary concepts of the subject area

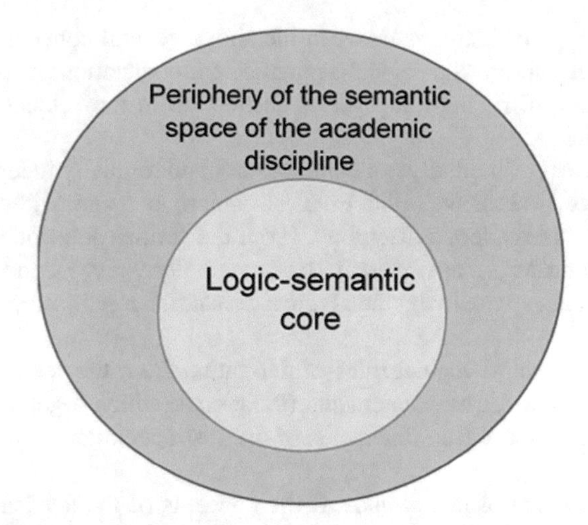

Fig. 2. Logic-semantic core and periphery of the semantic space of the academic discipline

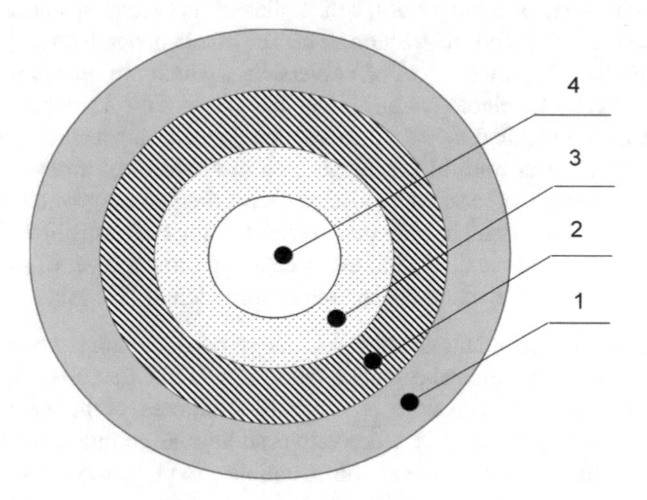

Fig. 3. Structural constituents of the semantic space of the academic discipline: (1) the periphery of the semantic space; (2) logical-semantic core of the content of the academic discipline, organized by the axiomatic-deductive strategy; (3) abstract logic-semantic core of the content of the academic discipline, organized by the axiomatic-deductive strategy; (4) metadisciplinary logic-semantic core of the semantic space of the academic discipline.

and a system of axiomatic statements (system of axioms) that are not demanding of its proof and are evident within this the discipline. As a rule, it is put forward the requirements to an axiom system about their consistency, completeness and independence. The second group of knowledge includes derivative concepts of the subject area, derived from its fundamental primary concepts, as well as derivative statements, which are a direct logical consequence of axiomatic statements. In view of the existence of highly abstract knowledge (concepts, statements) and their derivative knowledge of a lesser level of abstraction (in particular, specific knowledge), in the axiomatic-deductive strategy of organizing the contents of the academic discipline one can distinguish three of its subtrategies, namely, the axiomatic-deductive substrategy of the organization of the abstract conceptual apparatus of academic discipline, axiomatic-deductive substrategy of the organization of fundamental abstract statements of academic discipline and taxonomically oriented substrategy of deployments of discipline content (see Table 2).

Table 2. Axiomatic-deductive substrategies for organizing the contents of the academic discipline

Axiomatic-deductive substrategy of the organization of the conceptual apparatus of the academic discipline	1. Selection of the set of fundamental concepts of the academic discipline 2. The definition of derivative concepts of the discipline, based on its fundamental concepts
Axiomatic-deductive substrategy of organizing the statements of the academic discipline	1. Selection of the plural of true axioms, the truth of which is accepted without proof 2. Setting the set of rules for the logical deduction (proof) of all statements (theorems) of the discipline from the set of its axioms
Taxonomically oriented substrategy of deployment of discipline content	1. Formation of the set of taxonomies of the concepts of the academic discipline by the multiple use of the operation of the division of abstract concepts in the predefined basis of division, which provides a derivation from the most general concepts of the discipline of its derivatives of concepts of a lower level of universality and abstraction 2. Formation of the set of true statements of a lower level of abstract academic discipline of predicates, given on elements of taxonomy concepts

Given that the content of the academic discipline in e-learning systems should have a clear logical structure with the necessity of implement in the environment of the corresponding software, it is natural to allocate three forms of presentation of content of the academic discipline (see Table 3).

Thus, taking into account the axiomatic-deductive nature of the organization of the content core of the discipline content, as well as the three forms of its presentation, we consider the symbolic mathematical structures that describe both the terminological and conceptual apparatus and the set of true assertion and conclusions of the discipline content.

Table 3. Forms of presentation of content of academic discipline

Forms of content presentation	Description of the form of content presentation of academic discipline
Content form of presentation of the content of the academic discipline	Presentation of the contents of the academic discipline in a form of verbal-conceptual by means of natural language
Content of a discipline as a formalized (formalized) system	Presentation of content of academic disciplines using artificial languages of mathematics and mathematical logic (descriptive logic), which allow to get accurate, consistent and compact description of it
Content of the academic discipline as a computer ontology	Presentation of the content of the academic discipline in a form of the computer knowledge base by the language of ontology development (for example, OWL), which makes it possible to use modern artificial intelligence systems in e-learning systems

4 Axiomatic-Deductive Substrategy of the Organization of the Fundamental Terminological-Conceptual Apparatus of the Academic Discipline in the Systems of Electronic Learning

4.1 The General Terminological and Conceptual Apparatus of the Academic Discipline in a Content Form

The first stage of the organization of the terminology-conceptual apparatus of the academic discipline is the formation (selection) of the set $\mathbf{BC}_V = \{C_{V_1}, C_{V_2}, \ldots, C_{V_N}\}$ basic (atomic) general concepts in natural language with a finite alphabet $\mathbf{AlC}_V$, which are the fundamental abstract concepts of the subject area, which studies the corresponding academic discipline. These atomic concepts should be characterized by the highest level of abstraction (the scope of these concepts is the largest) within the framework of academic discipline with aim to cover all of its semantic space. In addition, it is important that the atomic concepts of the academic discipline are independent of each other, namely that any atomic concept can not be defined through the totality of other concepts.

The second stage of the organization of the abstract conceptual apparatus of the academic discipline is the formation of the set $\mathbf{RC}_V = \{R_{V_1}, R_{V_2}, \ldots, R_{V_K}\}$ basic relations between atomic concepts $\mathbf{BC}_V = \{C_{V_1}, C_{V_2}, \ldots, C_{V_N}\}$, which fix the logical-semantic interconnections between fundamental concepts in a given subject area. Among the basic relations between atomic concepts, which are mutually independent, there can be no relation of generic-form type, namely, the ratio of generic-form subordination (denoted by abbreviation AKO “A Kind Of” or term “SubsetOf”), which connects a set (class) and a subset (subclass) among themselves.

The third stage of the organization of the abstract conceptual apparatus of the academic discipline is the formation of the set $\mathbf{Ruls\_for\_C}_V = \{RC_{V_1}, RC_{V_2}, \ldots, RC_{V_K}\}$ of logical rules (operations) generation from basic general concepts $\mathbf{BC}_V = \{C_{V_1}, C_{V_2}, \ldots, C_{V_N}\}$ new derivatives of common concepts $\mathbf{DC}_V = \{C_{V_{N+1}}, C_{V_{N+2}}, \ldots, C_{V_{N+L}}\}$. Preferably, such rules have certain logical operations over concepts, such as join, crossing, additions and definitions of concepts.

The fourth stage of the organization of the terminology-conceptual apparatus of the academic discipline is the derivation of derivative concepts $\mathbf{DC}_V = \{C_{V_{N+1}}, C_{V_{N+2}}, \ldots, C_{V_{N+L}}\}$, that along with the basic concepts $\mathbf{BC}_V = \{C_{V_1}, C_{V_2}, \ldots, C_{V_N}\}$ form the fundamental general terminology of the academic discipline $\mathbf{TC}_V = \mathbf{BC}_V \cup \mathbf{DC}_V = \{C_{V_1}, C_{V_2}, \ldots, C_{V_{N+L}}\}$, as a collection of words-terms in the natural (national) language, which is formed from the elements of its alphabet $\mathbf{AlC}_V$ in accordance with the accepted rules of grammar.

Thus, the fundamental general terminology-conceptual apparatus of academic discipline in verbal form by means of natural language can be described as an axiomatic informal system that can be represented in the form of such structure:

$$\mathbf{AS}_{CV} = \{\mathbf{AlC}_V, \mathbf{TC}_V, \mathbf{BC}_V, \mathbf{Ruls\_for\_C}_V\}$$

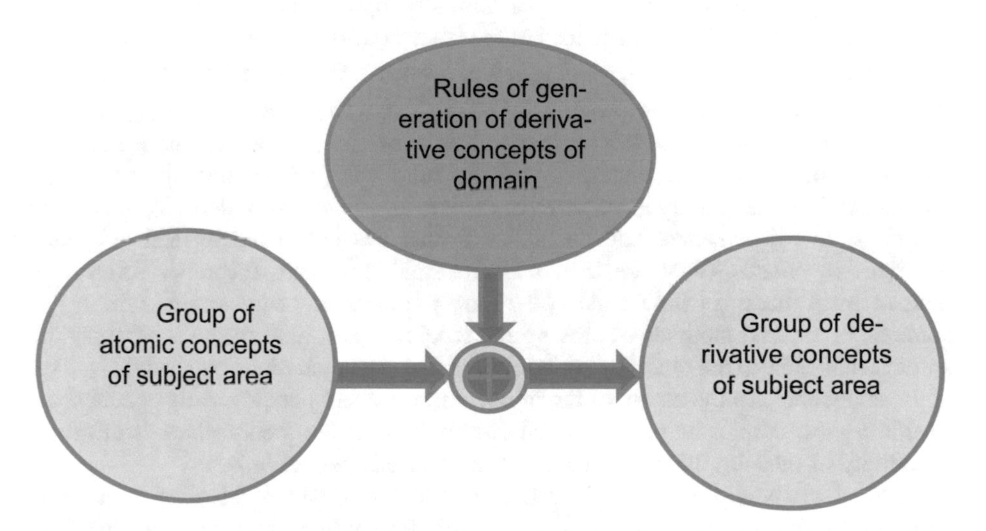

Fig. 4. Conditional scheme of axiomatic-deductive substrategy organization of the fundamental general terminological-conceptual apparatus of the academic discipline

The schematic diagram of the axiomatic-deductive substrategy of organizing the fundamental general terminology-conceptual apparatus of the discipline is shown in Fig. 4.

4.2 The General Terminological and Conceptual Apparatus of the Academic Discipline as a Formal Axiomatic System

The transition from the content form of representation of the general terminological-conceptual apparatus of the academic discipline to its presentation by means of artificial languages of mathematical logic, namely, descriptive logic, is carried out through its formalization - the procedures for mapping the axiomatic informal system into the formal axiomatic system, which is given as such a four:

$$\mathbf{AS}_{CF} = \{\mathbf{AlC}_F,\ \mathbf{TC}_F,\ \mathbf{BC}_F,\ \mathbf{Ruls\_for\_C}_F\}$$

where $\mathbf{AlC}_F$ is the alphabet of the formal language of a certain type of descriptive logic (for example, logics) - a finite set of characters from which finite sequences are formed-correctly created formulas (formulas) of the artificial language of the corresponding logic system; $\mathbf{TC}_F = \{C_{F_1},\ C_{F_{N+2}},\ \ldots, C_{F_{N+L}}\}$ is the set of all correctly created formulas in the alphabet $\mathbf{AlC}_F$, which mutually unequivocally correspond to the terms of the academic discipline in the natural language $\mathbf{TC}_V = \{C_{V_1},\ C_{V_{N+2}},\ \ldots, C_{V_{N+L}}\}$; $\mathbf{BC}_F = \{C_{F_1},\ C_{F_2},\ \ldots, C_{F_N}\}$ is the set of names of basic (atomic) concepts that are mutually unequivocally consistent with the basic concepts of $\mathbf{BC}_V = \{C_{V_1},\ C_{V_2},\ \ldots, C_{V_N}\}$ and is a subset $\mathbf{TC}_F(\mathbf{BC}_F \subset \mathbf{TC}_F)$; $\mathbf{Ruls\_for\_C}_F = \{RC_{F_1},\ RC_{F_2},\ \ldots, RC_{F_K}\}$ is the set of formally-logical rules (operations) of the derivative from the set $\mathbf{BC}_F$ the names of the basic concepts of new derivative concept names $\mathbf{DC}_F = \{C_{F_{N+1}},\ C_{F_{N+2}},\ \ldots, C_{F_{N+L}}\}$ of academic discipline. From the foregoing follows the following relation: $\mathbf{TC}_F = \mathbf{BC}_F \cup \mathbf{DC}_F = \{C_{F_1},\ C_{F_{N+2}},\ \ldots, C_{F_{N+L}}\}$.

In formal systems, there is a complete abstraction from semantics, the meaning of the words of the natural language, and the rules for manipulating the characters completely in the explicitly given form of axioms and the rules for deriving them from theorems. In mathematical knowledge, as formal systems, mathematical axiomatic theories are created (they are then called formal axiomatic theories). Axioms of mathematical theory in the formal system are presented as sequences of symbols of some alphabet, and methods of proving the theorems as formal methods of obtaining some formulas from the other, by applying mathematical operations over symbols. This approach guarantees the clarity of the initial assertions and unequivocally conclusions. In this regard, despite its purely formal construction of the theory, there is always a possibility of meaningful interpretation of the created abstract objects.

Formal systems are created to reflect (simulate) certain regularities of a given subject area, and therefore regardless of its purely formal (abstract, syntactic) character, they should have means of interpretation of their formulas and output procedures in this subject area. Therefore, the inverse procedure to the procedure of formalization is an interpretation procedure. Being within the formal system, there is no need to pay attention to nature, semantic meaning of the components of the formal system. However, in practice it is important to correctly interpret, to interpret correctly created formulas and the rules of derivation in terms and concepts of the corresponding subject area. To solve this problem, an interpretation procedure is used - the attribution of meanings (sense, meaning) to the primary concept of a formal system.

From the mathematical point of view, the formalization procedure will be presented as such a two $\mathbf{Formal}_C = \left(\mathbf{TC}_F, f_C^{Form}(\cdot)\right)$, which consisting of a set of correctly created formulas $\mathbf{TC}_F$ and formalization functions $f_C^{Form}(\cdot)$, which is given on the set $\mathbf{TC}_V$, with a the range of values $\mathbf{TC}_F$. That is, the procedure of formalizing a certain concept C_{V_n} from $\mathbf{TC}_V$ is determined by comparing it with the function of formalization $f_C^{Form}(\cdot)$, a certain correctly created formula C_{F_n} from $\mathbf{TC}_F$, namely:

$$\forall C_{V_n} \in \mathbf{TC}_V,\ \exists (C_{F_n} \in \mathbf{TC}_F),\ \text{that}\ C_{F_n} = f_C^{Form}(C_{V_n})$$

Being within the axiomatic formal system $\mathbf{AS}_{CF} = \{\mathbf{AlC}_F, \mathbf{TC}_F, \mathbf{BC}_F, \mathbf{Ruls\_for\_C}_F\}$, there is no need to pay attention to the meaning of verbal expressions, which makes it possible to reduce the logical conclusion to simple rules of symbolic transformations, similar to arithmetic operations. That is, the operation with semantic, ideal conceptual structures that make up the terminology-conceptual apparatus of the academic discipline, the means of formalization can be reduced to trivial operations over material systems and processes that represent certain sign systems and can be automated using modern languages for the description of knowledge and development environments ontologies.

On the other hand, the transition from the formal axiomatic system $\mathbf{AS}_{CF} = \{\mathbf{AlC}_F, \mathbf{TC}_F, \mathbf{BC}_F, \mathbf{Ruls\_for\_C}_F\}$ to the axiomatic informal system $\mathbf{AS}_{CV} = \{\mathbf{AlC}_V, \mathbf{TC}_V, \mathbf{BC}_V, \mathbf{Ruls\_for\_C}_V\}$ is accomplished through the interpretation procedure, which is given as a two $\mathbf{I}_C = \left(\mathbf{TC}_V, f_C^I(\cdot)\right)$ and consists of a set $\mathbf{TC}_V$ totality of words-terms of the academic discipline (called the subject area of interpretation), and interpretation functions $f_C^I(\cdot)$, which is given on the set $\mathbf{TC}_F$ with an range of values $\mathbf{TC}_V$. That is, the procedure for interpreting some correctly created formula C_{F_n} is $\mathbf{TC}_F$ is given by comparing it with the function of interpretation $f_C^I(\cdot)$, of a certain concept C_{V_n} from $\mathbf{TC}_V$, namely:

$$\forall C_{F_n} \in \mathbf{TC}_F,\ \exists (C_{V_n} \in \mathbf{TC}_V),\ \text{that}\ C_{V_n} - f_C^I(C_{F_n}).$$

4.3 General Terminological-Conceptual Apparatus of Academic Discipline in Machine-Interpretive Form

The transition from the representation of the general terminological-conceptual apparatus of the academic discipline in the form of a formal axiomatic theory to its presentation by means of development of ontology in the environment Protégé using the language of describing the ontology OWL, is carried out by applying a procedure of coding, which reflects the axiomatic formal theory in the formal machine-interpreted ontology, which can be filed as such four:

$$\mathbf{AS}_{CO} = \{\mathbf{AlC}_O, \mathbf{TC}_O, \mathbf{BC}_O, \mathbf{Ruls\_for\_C}_O\},$$

where $\mathbf{AlC}_O$ is the alphabet of machine-interpreted language for description of ontology, for example, OWL; $\mathbf{TC}_o = \left\{C_{O_1}, C_{O_{N+2}}, \ldots, C_{O_{N+L}}\right\}$ is the set of all correctly created formulas in the alphabet $\mathbf{AlC}_O$, which mutually unequivocally correspond to the terms of the academic discipline in the natural language $\mathbf{TC}_v = \{C_{V_1}, C_{V_{N+2}}, \ldots, C_{V_{N+L}}\}$; $\mathbf{BC}_o = \{C_{O_1}, C_{O_2}, \ldots, C_{O_N}\}$ is the set of names of the basic (atomic) concepts in language of the description of ontology, which mutually unequivocally correspond to the basic concepts of $\mathbf{BC}_v = \{C_{V_1}, C_{V_2}, \ldots, C_{V_N}\}$ and is a subset $\mathbf{TC}_o(\mathbf{BC}_o \subset \mathbf{TC}_o)$; $\mathbf{Ruls\_for\_C}_O = \{RC_{O_1}, RC_{O_2}, \ldots, RC_{O_K}\}$ is the set of formal-logic rules (operations) of the derivative from the set $\mathbf{BC}_o$ the names of the basic concepts of new derivative concept names $\mathbf{DC}_O = \{C_{O_{N+1}}, C_{O_{N+2}}, \ldots, C_{O_{N+L}}\}$ the academic discipline in the syntax of the language of the description of ontologies. From the foregoing follows the following expression: $\mathbf{TC}_o = \mathbf{BC}_o \cup \mathbf{DC}_o = \{C_{O_1}, C_{O_{N+2}}, \ldots, C_{O_{N+L}}\}$.

From a mathematical point of view, the coding procedure we will present as such a two $\mathbf{OntCod}_C = \left(\mathbf{TC}_O, f_C^{Cod}(\cdot)\right)$, which consist of a set of correctly created formulas $\mathbf{TC}_O$ in the syntax of the language of describing ontology and coding functions $f_C^{Cod}(\cdot)$, which is given on the set $\mathbf{TC}_F$, with an range of values $\mathbf{TC}_O$. That is, the procedure of ontological coding of some correctly created formula C_{F_n} is $\mathbf{TC}_V$ is given by comparing it with the function of coding $f_C^{Cod}(\cdot)$, a certain expression C_{O_n} in the language of describing ontology with $\mathbf{TC}_O$, that is:

$$\forall C_{F_n} \in \mathbf{TC}_F,\ \exists (C_{O_n} \in \mathbf{TC}_O),\ \text{what}\, C_{O_n} = f_C^{Cod}(C_{F_n})$$

In the environment of the developed ontology of the subject area of the academic discipline, it is possible to use the built-in mechanisms of logical derivations and to maintain the correctness of taxonomic connections between concepts, which makes it possible to creation complex concepts from simpler ones and organize them into taxonomy.

Figure 5 gives a generalized structure of processes of interactions of the verbal level of the description of the subject area of the academic discipline, of the formal level of description of the subject area and the description of the subject area at the level of computer ontology, which implemented through the formalization, interpretation, encoding and decoding in the computer-ontology development environment.

Transition from the ontological system $\mathbf{AS}_{CO} = \{\mathbf{AlC}_O, \mathbf{TC}_O, \mathbf{BC}_O, \mathbf{Ruls\_for\_C}_O\}$ to the axiomatic formal system $\mathbf{AS}_{CF} = \{\mathbf{AlC}_F, \mathbf{TC}_F, \mathbf{BC}_F, \mathbf{Ruls\_for\_C}_F\}$ is accomplished through the decoding procedure, which is given as $\mathbf{OntDecod}_C = \left(\mathbf{TC}_F, f_C^{Decod}(\cdot)\right)$. Decoding function $f_C^{Decod}(\cdot)$ given on set $\mathbf{TC}_O$ with an range of values $\mathbf{TC}_F$. That is, the procedure of decoding of a certain expression C_{O_n} in the language of describing ontology with $\mathbf{TC}_O$ determined by comparing it with the decoding function $f_C^{Decod}(\cdot)$ a certain correctly created formula C_{F_n} is $\mathbf{TC}_F$, that is:

$$\forall C_{O_n} \in \mathbf{TC}_O,\ \exists (C_{F_n} \in \mathbf{TC}_F),\ \text{what}\, C_{F_n} = f_C^{Decod}(C_{O_n}).$$

We note that all the above axiomatic systems representing the general of terminological-conceptual apparatus of the core of the content of the academic

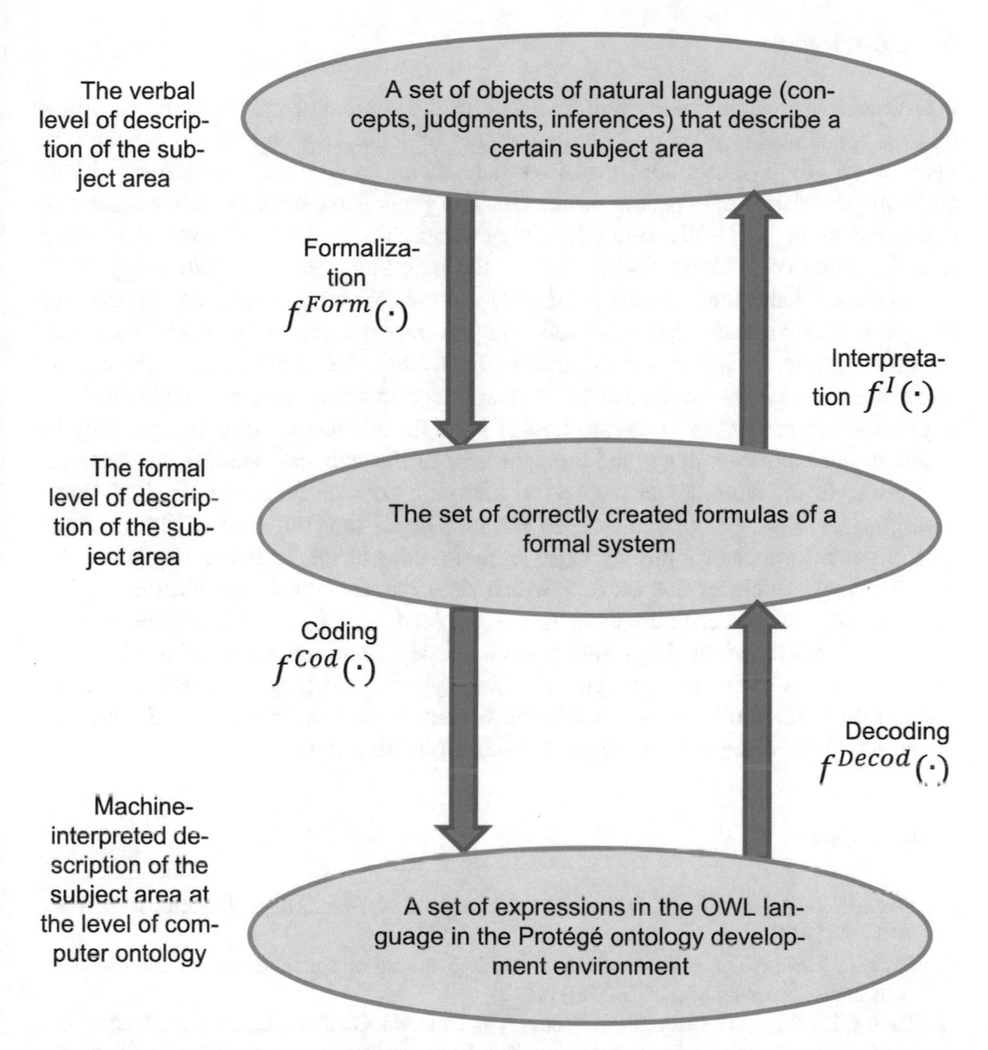

Fig. 5. General structure of processes of interaction of the verbal level of description of the subject area, the formal level of description of the subject area and machine-interpreted description of the subject area at the level of computer ontology in the systems of e-learning.

discipline are isomorphic to each other, since there is a mutually unequivocally correspondence between their elements and the preservation of the corresponding structures, which is provided by the bijective functions of interpretation $f_C^I(\cdot)$, formalization $f_C^{Form}(\cdot)$, coding $f_C^{Cod}(\cdot)$ and decoding $f_C^{Decod}(\cdot)$.

5 Conclusion

It has been formulated quality requirements for the content of the academic discipline, that is, are allocated a group of requirements of logic, a group of requirements for visibility, a group of requirements of consistency and a group of requirements for the convenience of the use of academic content. The generalized structure of the axiomatic-deductive strategy of the organization of academic content is developed, which includes three of its subtrategies, that is, the axiomatic-deductive subtrategy of the organization of the fundamental terminological-conceptual apparatus of the academic discipline, the axiomatic-deductive subtrategy of the organization of general statements of the academic discipline and the taxonomically oriented substrategy of the deployment of the content academic discipline. The semantic space of the academic discipline is proposed to present as an connection of its logical-semantic core, organized by the axiomatic-deductive strategy, and the periphery of the semantic space of the academic discipline. In the structure of the logical-semantic core of the academic discipline is identified its basic components as an abstract logic-semantic core of the academic discipline and a metadisciplinary logic-semantic core of the academic discipline. The mathematical structures are created which describe and detail the abstract logical-semantic core of the academic discipline in the form of a group of axiomatic systems, that is: axiomatic informal systems general concepts and assertions of the academic discipline in a verbal form, formal axiomatic systems of general concepts and assertions of the academic discipline, axiomatic systems of general concepts, and statements of the academic discipline in the machine-interpretative form.

References

1. Shute, V.J., Zapata-Rivera, D.: Adaptive educational systems. Adapt. Technol. Train. Educ. **1**, 5–27 (2011)
2. Rui, L., Maode, D.: A Research on E-learning resources construction based on semantic web. Phys. Procedia **25**, 1715–1719 (2012)
3. Barros, H., Silva, A., Costa, E., Bittencourt, I., Holanda, O., Sales, L.: Steps, techniques, and technologies for the development of intelligent applications based on semantic web services: a case study in E-learning systems. Eng. Appl. Artif. Intell. **24**, 1355–1367 (2011)
4. Vega-Gorgojo, G., Bote-Lorenzo, M.L., Asensio-Pérez, J.I., Gómez-Sánchez, E., Dimitriadis, Y.A., Jorrín-Abellán, I.M.: Semantic search of tools for collaborative learning with the ontological search system. Comput. Educ. **54**(4), 835–848 (2010)
5. Isotani, S., Mizoguchi, R., Isotani, S., Capeli, O.M., Isotani, N., de Albuquerque, A.R.P.L., Bittencourt, I.I., Jaques, P.: A semantic web-based authoring tool to facilitate the planning of collaborative learning scenarios compliant with learning theories. Comput. Educ. **63**, 267–284 (2013)
6. Lytvyn, V., Vysotska, V., Burov, Y., Veres, O., Rishnyak, I.: The contextual search method based on domain thesaurus. In: Advances in Intelligent Systems and Computing, vol. 689, pp. 310–319 (2017)
7. Junuz, E.: Preparation of the learning content for semantic E-learning environment. Procedia Soc. Behav. Sci. **1**(1), 824–828 (2009)

8. World Wide Web Consortium (W3C): W3C Semantic Web Activity. http://www.w3.org/2001/sw/
9. Gruber, T.: A translation approach to portable ontologies. Knowl. Acquis. **5**(2), 199–220 (1993)
10. Guarino, N.: Formal ontology, conceptual analysis and knowledge representation. Int. J. Hum Comput Stud. **43**(5–6), 625–640 (1995)
11. Kut, V., Kunanets, N., Pasichnik, V., Tomashevskyi, V.: The procedures for the selection of knowledge representation methods in the "virtual university" distance learning system. In: First International Conference on Computer Science, Engineering and Education Applications (ICCSEEA2018), Kiev, pp. 713–723 (2018)
12. Bomba, A., Kunanets, N., Nazaruk, M., Pasichnyk, V., Veretennikova, N.: Information technologies of modeling processes for preparation of professionals in smart cities. In: First International Conference on Computer Science, Engineering and Education Applications (ICCSEEA2018), Kiev, pp. 702–712 (2018)
13. Sowa, J.: Conceptual graphs as a universal knowledge representation. semantic networks in artificial intelligence. Special Issue Int. J. Comput. Math. Appl. **23**(2–5), 75–95 (1992)
14. Montes-y-Gómez, M., Gelbukh, A., LópezLópez, A.: Comparison of Conceptual Graphs. http://ccc.inaoep.mx/ ~ mmontesg/publicaciones/2000/ComparisonCG
15. Dosyn, D.G., Lytvyn, V.V., Nikolskyi, Y.V., Pasichnyk, V.V.: Intelligent Systems Based on Ontologies. Civilization, Lviv (2009)
16. Guarino, N.: Formal ontology and information systems. In: FOIS 1998, pp. 3–15. IOS Press, Amsterdam (1998)
17. Lenko, V., Pasichnyk, V., Kunanets, N., Shcherbyna, Y.: Knowledge representation and formal reasoning in ontologies with Coq. In: Advances in Intelligent Systems and Computing, vol. 689, pp. 759–770 (2018)

The Virtual Library System Design and Development

Bohdan Rusyn[1,5(✉)], Vasyl Lytvyn[2], Victoria Vysotska[2,3], Michael Emmerich[4], and Liubomyr Pohreliuk[5]

[1] University of Technology and Humanities of Radom, Radom, Poland
rusyn@ipm.lviv.ua
[2] Silesian University of Technology, Gliwice, Poland
[3] Lviv Polytechnic National University, Lviv, Ukraine
{Vasyl.V.Lytvyn, Victoria.A.Vysotska}@lpnu.ua
[4] Leiden Institute of Advanced Computer Science, Leiden University, Leiden, The Netherlands
[5] Karpenko Physico-Mechanical Institute of the NAS of Ukraine, Lviv, Ukraine
liubomyr@inoxoft.com

Abstract. **Annotation.** An overview of the design and development features of the Virtual Library information system was conducted. A new approach is proposed for designing and developing the Virtual Library information system for saving and development of e-books in the MARC 21 format. The model of information system Virtual Library is proposed.

Keywords: Virtual Library · E-library · Cloud computing Information system · MARC 21 · E-book

1 Introduction

Nowadays, with the exponential growth in the demand for operational information for a modern civilized person, it becomes relevant to digitize a book fund of different directions and provide access to it at any time from any part of the globe [1]. Attending libraries does not have enough time or this process becomes rather uncomfortable (the necessary information is distributed in several libraries, as well as in several geographically located cities/countries) [2]. In addition, the list of services and their quality will not always satisfy the average consumer of information [3]. Therefore, such libraries does not longer satisfy in some ways the requirements for informatization of modern society [4, 5].

2 General Formulation of the Problem

Firstly, the Internet can solve problems in finding relevant information [6–9]. But here we have a bunch of problems [10–20]:

N. Shakhovska and M. O. Medykovskyy (Eds.): CSIT 2018, AISC 871, pp. 328–349, 2019.
https://doi.org/10.1007/978-3-030-01069-0_24

- loss of time to search for relevant consumer-specific information, but not "popular" among most Internet users;
- the lack of unique information on the Internet due to the fact that it is necessary only for a narrow circle of consumers;
- confusion among the large number of search results obtained on a consumer request;
- rapid change in the dynamics of access to certain sites with the necessary information;
- lack of option as recommended literature for highly specialized topics;
- lack of confirmation of the reliability of the information received (anonymity of publications, modification of credible information, fraud, etc.);
- there is a high probability of inaccurate and incomplete information.

The traditional library should be responsible for the information contained on its territory [1–5]. This puts IT professionals with a variety of tasks in operational identification, processing, searching, storing and providing relevant access to large databases of this book fund, e-library [21–32]. According to [1–5], the digital library is a distributed information system (IS) for storing heterogeneous collections of electronic documents (text, graphics, audio, video, etc.) and providing them with access through the Internet in a user-friendly form. This is too inaccurate definition of this concept. It defines which standards of preservation must be respected and which information technologies (IT) are desirable to use.

Therefore, the purpose of the creation of an electronic (virtual) library is to provide operational information to regular users of access to information relevant to their limited access data (rare and manuscripts, photo albums, dissertations, archives that are not available in most libraries) or to such data that exist only in electronic form. An additional bonus of the electronic library is to provide consumers with better information services for working with electronic data (documents, books, manuscripts, etc.) of large volumes. Therefore, an electronic library is an IS that provides high-quality and timely access to relevant information in online mode with the effect of visiting a traditional library (the presence of shelves with books, the ability to view and select books on shelves, view a scanned book).

3 Researches and Publications Analysis

Virtual Library is an environment of such an IS that specializes in informational objects (books, manuscripts, documents, manuscripts, etc.) that are stored and processed only in electronic-digital form, to which the consumer receives information through Internet search engines or through specialized IC [1–5]. In the latter case, the user must often be a registered user of this IS. That is, there are two types of electronic library [1–5]:

- distributed in the information space of the public network, focused on the exchange of data between libraries through search engines or specialized IS;
- specialized IS, which stores and processes data objects in data warehouses, and provides them with access to the consumer through their services.

In the first case, the environment of such Virtual Library may consist of one to several electronic libraries that are geographically separate from one another [1–5]. In turn, the IS of such a virtual library acts as an intermediary between them and supports the processes of communication, integration and obtaining data about relevant library resources [1–5].

In the second case, the information objects of such Virtual Library are stored in a data warehouse or in a cloud, regardless of which library belongs to this information object [1–5]. In this case, such an IS serves as an integrated environment for accessing the information objects of a particular library that provides access through its particular IS to its resources. In any case, there are a number of Virtual Library benefits over traditional libraries, including [1–5]:

- Access to the Virtual Library information object at any time and anywhere in the planet, where the Internet is available, for the user of this IS;
- an expanded range of opportunities for the search and development of descriptive information about the desired information object;
- support for the ability to compose search terms for search efficiency, where part of the expression can be any word, phrase or phrase of the desired information object;
- Information objects in the data warehouse are available 24 h a day and everywhere;
- the possibility of choosing a consumer's presentation of information objects in various formats convenient for the latter;
- Possibility to present informational materials in various formats (text, database, diagram);
- support the sharing of a certain amount of information to avoid duplication, for example, for little-used materials;
- access to a unique information object, access to which previously required physical presence and obtaining appropriate levels of access;
- increasing the range of library users and expanding access to their own resource funds by digitizing them;
- the possibility of updating the electronic version of the information object;
- Enabling libraries to constantly maintain their own information resources funds in their current state, in accordance with the analysis of consumer information requests of the virtual objects of this Virtual Library.

According to [1–5], today there are certain well-known e-libraries targeting a certain number of consumers of information (Table 1). These electronic libraries have their advantages and disadvantages.

Table 1. Well known multilingual e-libraries

Name	Feature	Advantage	Disadvantages
Europeana	European e-library	Over 2 million digitized objects, fast repository filling of information objects	Due to the large number of hits (10 million per hour), the project is closed
Open Library	Internet Archive and Open Content Alliance project	Rapid filling of the repository with information objects, the possibility of free digitization at the request of any book from the list of the Boston Public Library	Requests must be done in Latin alphabet
Gallica	Funds of the National Library of France	One of the largest e-libraries in the world (increased by 100,000 titles per year)	Many free works are mostly French
Google	Funds of Michigan, Oxford, University, Madrid, Ghent Universities, Complutense, National Library of Catalonia and Lausanne, University of Keio in Japan	A significant number of digitized publications in free access (including Ukrainian literature)	Due to legal restrictions, objects are provided in limited access (users outside the U.S. are defined for IP)
Gutenberg	The oldest electronic library	A significant number of digitized editions	The works are mostly German and English
WDL	World Digital Library	Submitted objects of world culture	Western Cultural Objects are not Submitted
MDZ, GDZ	German regional centers of digitization	A significant number of digitized editions	The works are mostly German and English
AON	Austrian Newspaper Online	More than 4 million digitized pages	Archive of the historic Austrian periodicals
arxiv.org	Library of Cornell University	The largest collection of free scientific papers and preprints in the world	Narrowly oriented
International Music Score Library Project	Library of music	The largest music collection	Narrowly oriented

(*continued*)

Table 1. (*continued*)

Name	Feature	Advantage	Disadvantages
The Great Online Fiction Library	Over 100 thousand works of art in fb2, txt and html formats	Easy to use, powerful, relevant search. The personal list of recommended literature is created on the basis of comparison of individual book reviews. Discussions, comments, reviews, and impressions about read books. Personal book shelf. Ability to download all books from the bookshelf in one archive	Narrowly oriented

4 Problem Emphasis

The main purpose of the IS virtual library (ISVL) is to integrate information resources for high-quality, operational and efficient navigation: $S_{DL} = <B, R, Q, P, f_{itg}, f_{shr}>$, where B is a set of integrated documents from a plurality of information resources R, Q is a set of requests for information consumers, P is a set of relevant content as a result of a search on a consumer request; f_{itg} is the operator in the integration of information resources and f_{shr} is the operator of navigation in full-text databases. In case, the integration of the set of information resources R consists in combining them in a certain set of conditions U_{ing} in order to use different information with preservation of its properties, features of presentation and possibilities to process it: $B = f_{itg}(R, U_{ing})$. Such integration should provide the consumer with information to take the necessary information for him as a single information space, that is, ISVL must provide work with large data warehouses, for example, through cloud computing, and due speed, quality and effectiveness of finding relevant content according to the search terms $U_{shr} : P = f_{shr}(Q, B, U_{shr})$. Efficient and high-quality navigation in the electronic library is to enable the consumer to find relevant and relevant information in the entire accessible Virtual Library information space with the greatest completeness and accuracy at the least cost of effort on his part.

5 Goal Formulation

ISVL should provide a certain set of functionalities for working with certain information objects, and not with their content (Table 2), in particular:

- entering/removing Virtual Library information objects in a certain format;
- integration/restructuring of information objects according to certain requirements;
- automatic/semi-automatic formation of the information space of the Virtual Library, accessible to the consumer of information;
- ensuring the cataloging of information objects and their various associations in accordance with predetermined goals (the formation of courses in education, the formation of a personal library, the formation of thematic discussions, etc.) created Virtual Library information space.

It is necessary to develop the IS of integration of the technological processes of the library processing of the input scanned stream and the technologies of forming the funds of electronic resources of information objects.

The formation of resources of information objects consists of two tasks, which are devoted to the following research:

1. Determination of the optimal structure of the electronic resources of information objects, which determines the technology of the location of electronic information objects and is related to the effectiveness of access to the latest on the demand of information consumers;
2. The development of information technology that will ensure the prompt processing of electronic information flows and the effective formation of resources of information objects Virtual Library.

6 Received Scientific Results Analysis

To implement the basic processes of the Virtual Space Information Space, the main modules of such an IS must be technological, server and client (Fig. 1).

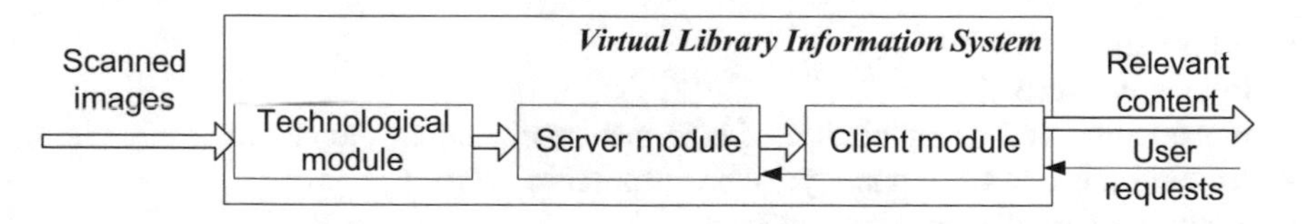

Fig. 1. The conceptual structure of the Virtual Library Information System

The following main processes of Virtual Library work are separated:

- technological processing of scanned information flows:
  - recognition of information objects;
  - storing images of information objects in the cloud;

  - Preservation of the location information in the traditional library of these scanned information objects in the database;
  - Marking of information objects in a specific pattern, for example, MARC 21 and storing this information in a database;
  - forming a descriptive file for a marked information object;
- formation of electronic funds of information resources of a certain traditional library in the cloud as a result of the technological process;
- processing of consumer demand streams for forming the correct information search expression;
- support for the implementation of effective information search for information objects in accordance with consumer information requests, in particular implementation: lexical search; symbolic search; attributive search; linguistic search;
- formation of a set of operative relevant responses to specific requests of information consumers;
- support of technology of interaction of information resources components in IS;
- analysis of frequent requests from users to form a cache for standard responses to these queries;
- support for cloud computing for the speed of access to electronic resources of information resources Virtual Library, namely: the fund of electronic documents; abstract databases; bibliographic databases; semantic means (linguistic support), that is, the knowledge base used for information retrieval, namely:
  - Dictionaries of the rules of linguistic (morphological and semantic) search (used for the nature of the language search in the content of information objects);
  - dictionaries of lexical search rules;
  - dictionaries of the rules of character search;
  - dictionaries for attribution search rules;
- search vocabularies in the format MARC 21.

Typically, the repository of electronic funds for information resources Virtual Library is built on a two-tier scheme:

- a file storage for storing and storing information objects;
- Knowledge base for information search of these information objects.

As a result, the Virtual Library module consists of subsystems (Fig. 2):

- information retrieval;
- interactive access;
- preservation and accumulation of information objects in the cloud;
- forming a cache of frequently asked information objects;
- analysis of user queries for caching.

The reference database of Virtual Library IS is the basis of the search engine subsystem of the electronic library. Effective search for the context of an information object provides a high relevancy of consumer-relevant content, presented in various

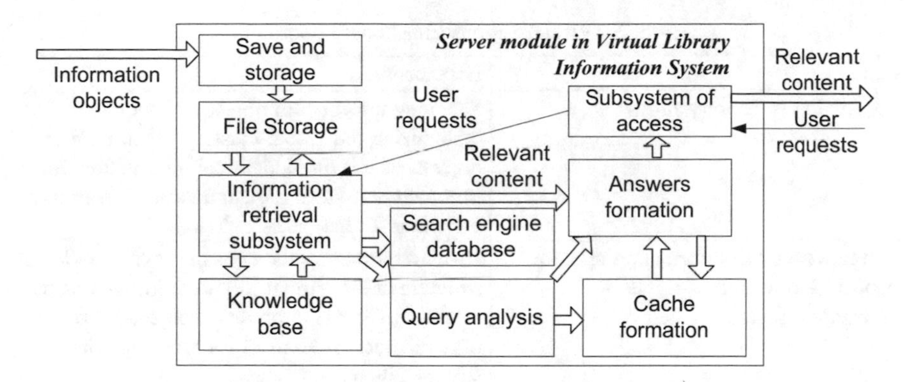

Fig. 2. The conceptual structure of the server module of Virtual Library IS

formats. The use of abstract-bibliographic databases in the information retrieval subsystem also allows for the real-time search of actual content without informational noise (Fig. 3). Virtual Library IS must provide:

- formation of information resources;
- preservation and support in the current state;
- providing access to information resources.

Table 2 provides us with the main functionality of Virtual Library.

On Fig. 4 the structure of the proposed Virtual Library information system is presented, taking into account its main functionalities and work with clouds.

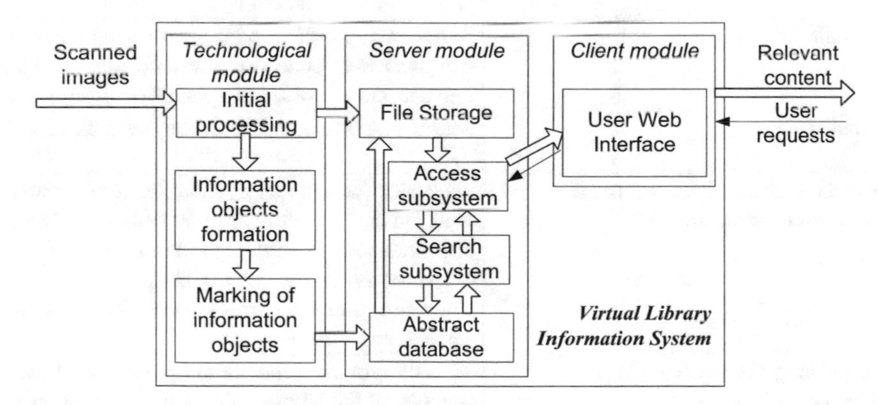

Fig. 3. Virtual Library Information System model

Table 2. Virtual Library functional opportunities

Name	Explanation
Digitization of Library funds	Scanning information object repositories, recognizing individual clusters of information objects, identifying individual information objects in the cluster, storing them and describing in a particular format, such as MARC 21
Navigation on the information space according to its access rights Information search	Visually providing the consumer with the logical structure of the Virtual Library information space, catalogs of ISVL information objects, and providing access to tools for working with the Virtual Library
Informational search	Formation of a set of information objects, the value of characteristics of which satisfies the conditions of the search query of the consumer of this information. Search results can be sorted by the values of any field defined as the key. You must allow the use of AND, OR, NO logical operators, as well as the ability to search for values >1 characteristics at a time
Lexical search	Searching for the free vocabulary of the national language and languages using the Latin alphabet
Symbol search	Search by lexical unit, which is a certain sequence of admissible characters
Attributes search	Search for information objects according to their characteristics (author, title, place of publication, date of publication, etc.)
Linguistic search	Search taking into account grammatical features with the use of context-distance operators, taking into account the order of application of operands
Format search	Search in a bibliographic format, such as MARC 21
View the content of the information object and its structure	Sequential (page by page) and selective (transition to any given page or to any element of an object). The structure and context of the information object are synchronized: any change in the structure causes the corresponding change in the context, and vice versa
Manipulating the structure of the information object	Visually providing the consumer with the logical structure of the Virtual Library information object and providing access to the tools for working with them
Support for the set of hypertext and hypermedia links	Providing the consumer of information objects an operational transition from him or some of his element to another interconnected information object or element

(*continued*)

Table 2. (*continued*)

Name	Explanation
Recording user's session	Support for the possibility of transition to each of the previously existing states of the system
Customizing the consumer system	Providing the consumer with the information to adjust the interface and corresponding to its level of access to the functionality of the ISVB according to their preferences and clutches
Install bookmarks in the text information objects	Possibility of marking the text of the information object with the support of the operative transition of these markers within this information object
Information export	Providing the possibility to export from the ISIF an information object (if it has the property as free to access) or part thereof (if the eligible consumer is eligible, or this part can be exported) with indication of the source
Formation of a cache	Ability to fill a cache by a set of frequently asked information objects
Consumer information requests analysis	Periodic automated analysis of consumer requests for: – further formation of the cache and its updating – updating of information resources in case of new demand from ISVL's permanent users – Update information search rules to improve the relevance of information objects
Comments and feedback from consumers analysis	Similar to the analysis of consumer information requests to support the effectiveness of ISVL
Formation and support of the informative portrait of the constant consumer	Collecting and analyzing sessions of the constant user of ISVL to formulate a list of recommendations for him in relation to other similar to his requests, information objects
Formation and support of the professor's virtual cabinet	The type of consumer, the professor can compile their lists of recommended information objects for a particular course or direction of research and recommend their subscribers or other user of ISVL
Professor's virtual cabinet work analysis	Collection and analysis of sessions of Professor ISVL to form a list of recommendations for him in relation to other, similar to his requests, information objects

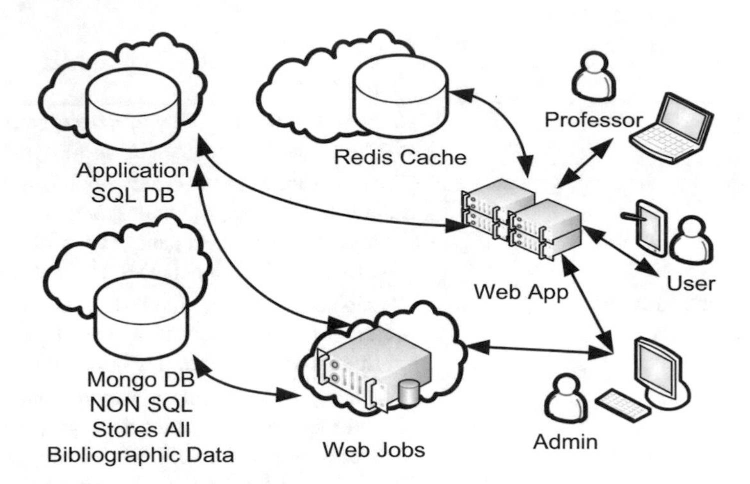

Fig. 4. Virtual Library IS structure

Virtual Library is a specialized system for informational objects (books, manuscripts, documents, manuscripts, etc.), which are stored and processed only in electronic-digital form, to which the consumer accesses information through search engines, or through specialized systems [1–5]. Virtual Library Information System Model S_{VL} (ISVL) will be presented as a tuple

$$S_{VL} = <A, S_\alpha, S_\beta, S_\chi, S_\delta, S_\varepsilon, S_\varphi, S_\gamma, S_\eta, W, \alpha, \beta, \chi, \delta, \varepsilon, \varphi, \gamma, \eta>, \quad (1)$$

where A is input data into ISVL in the form of a description of the information object, including the format of the MRK21, and the very information object; S_α is a subsystem of working with users to generate the results of their queries; S_β is subsystem of work with professors for the formation of the results of their inquiries; S_χ is subsystem of input/modification of the rules of operation of other subsystems from the administrator of ISVL (for example, linguistic search rules, cache updates, etc.); S_δ is subsystem of formation of unstructured database based on MARC 21; S_ε is subsystem of the formation of a structured database based on MARC 21; S_φ is cache processing subsystem for generating reports on popular consumer queries; S_γ is subsystem of cache formation; S_η is subsystem of formation of results of work of subsystems of generation of reports; W is Output from ISVL in the form of reports of relevant content; α is operator to work with users to generate the results of their queries; β is the operator of work with professors for the formation of the results of their inquiries; χ is the operator of the input/modification of the rules of operation of other subsystems from the administrator ISVL; δ is Operator of the formation of unstructured database based on MARC 21; ε is operator for structuring the database based on MARC 21; φ is cache processing operator for generating reports on popular consumer queries; γ is update cache operator; η is the operator of the formation of the results of the work of subsystems for generating reports.

Non SQL database contains all bibliographic data from books. There's a lot of data and therefore you need to keep them in Non SQL. The SQL database contains only what the cloud solution works for. Redis Cache is a specific database used for caching, that is, in order to store some queries or data that users are looking for to optimize then use and not go all the time into a heavy SQL database. Web jobs are cloud services that update data in a SQL database by processing data in a Non SQL database. They also update the cache accordingly, when needed, and the other, for example, how to search various information about books, their covers, and so on.

If $W = \Phi(A)$, then for ISVL according to the different roles of ISVL users (visitors, regular consumers, professors and administrators) we will receive $A_\iota \cap A_\alpha = \varnothing$, $A = A_\kappa \cup A_\iota \cup A_\alpha \cup A_\beta \cup A_\chi$, $A_\beta \subset A_\alpha$, $A_\beta \subset A_\eta$, $A_\iota \cap A_\beta = \varnothing$, $W = f(A_\eta \cup L_\beta)$, $A'_\eta \cap L_\beta = \varnothing$, where A_κ is the set of input data from libraries about information objects and their scanned covers; A_ι is a set of information objects in which the visitor is interested (he has access to a much smaller set of such objects than to other users), A_α – a set of information objects in which the consumer is interested, A_β is a set of information objects in which the professor is interested, A_χ is a set of information objects, and which the administrator is interested in, A_η is a set of information objects that meet the criteria for information retrieval, L_β is a set of recommended literature references for consumers from the professor. Generating the results of subsystems to generate reports of relevant content in accordance with (1) will provide a superposition of functions

$$W = \eta \circ \varphi \circ (\eta', \varepsilon \circ \delta), \tag{2}$$

where η' is the operator of the formation of the previous results of the subsystems of the generation of reports when $W = W_\alpha \cup W_\beta \cup W_\iota \cup W_\chi$, where W_α is a set of generated reports of relevant content according to user queries; W_β is the set of generated reports of relevant content according to the professor's requests; W_ι is the set of generated reports of relevant content according to ISVL visitors' requests; W_χ is a set of generated reports of relevant content according to the ISVL administrator's requests. In the general subsystem of the formation of results in the form of the generation of reports (Fig. 5), taking into account (1) and (2), will be presented as

$$S_\eta = < W_\eta, Q_\alpha, U_\eta, U_\mu, U_\alpha, \lambda, \mu, \eta > , W_\eta = \eta(U_\eta, \lambda(U_\mu, \mu(A, U_\alpha, Q_\alpha))), W_\eta = \eta \circ \lambda \circ \mu, \tag{3}$$

where Q_α is a set of requests from ISVL users; U_α is conditions set for analyzing user queries; U_η is conditions set for reporting; λ is operator for finding information objects in the cloud; μ is operator for finding information objects in cache.

In fact, this is just an interface, it's not a service part. All that cloud, Web jobs and

Fig. 5. User interface – search by keyword or reviewing a professor's course

the rest is the whole cloud and it's the server part. The subsystem of work with professors for the formation of the results of their inquiries will be presented as

$$S_\beta = <A_\alpha, U_\alpha, Q_\alpha, A_\beta, U_\beta, Q_\beta, B_\varepsilon, C_v, W_\beta, U_\eta, L_\beta, \alpha, \beta, \lambda, \mu, \eta>, L_\beta = \beta \circ \eta \circ \lambda \circ (\mu, \beta' \circ \alpha),$$
$$L_\beta = \beta(W_\beta, U_\beta, \eta(U_\eta, Q_\beta, A_\beta(\lambda(B_\varepsilon, f_{qc}(C_v, \beta'(A_\alpha, U_\alpha, Q_\alpha, \alpha)))))), \quad (4)$$

where U_β is a set of conditions for working with the profile of the professor to complete the training courses; Q_β is set of requests from a professor; β is operator with professor's profile.

Libraries should prepare the information in the MARC 21 format and submit it to the ISVL. Moderator ISVL corrects the function of importing data and they fall into the system. An additional software is created, a small program that works with predetermined MARC 21 data formats. It is provided with data/archives and it automatically fetches them in the system. The subsystem of the formation of an unstructured database based on MARC 21 will be presented as

$$S_\delta = <B_o, A_\kappa, U_\pi, U_\theta, U_\rho, U_o, \delta, \rho, \theta, \pi>,$$
$$B_o = \delta \circ \rho \circ f_{rp} \circ \pi, B_o = \delta(U_o, \rho(U_\rho, \theta(U_{rp}, \pi(A_\kappa, U_\pi)))), \quad (5)$$

where B_o is a set of unstructured data in a database, A_κ is a set of information objects that need to be digitized, U_π is a set of conditions for scanning the location of objects and objects themselves, U_θ is set of recognition conditions from scanned images, U_ρ is set of conditions and rules for marking recognized bindings in MARC 21 format, U_o is set of conditions for the preservation of new formed descriptions for recognizable information objects, δ is the operator of the storage or updating of the processed information object, ρ is operator of labeling information object, θ is operator of identification and identification of the information object, π is operator to scan the location of the information object and its covers. Scanning is done using planetary or robotic scanners, which are usually equipped with high-resolution digital cameras. The result of the scan is a bitmap image (Fig. 6). Then the shelves with books are recognized, and then the covers on each shelf (Fig. 7).

Recognized cover is marked by tags about this particular book (author, title,

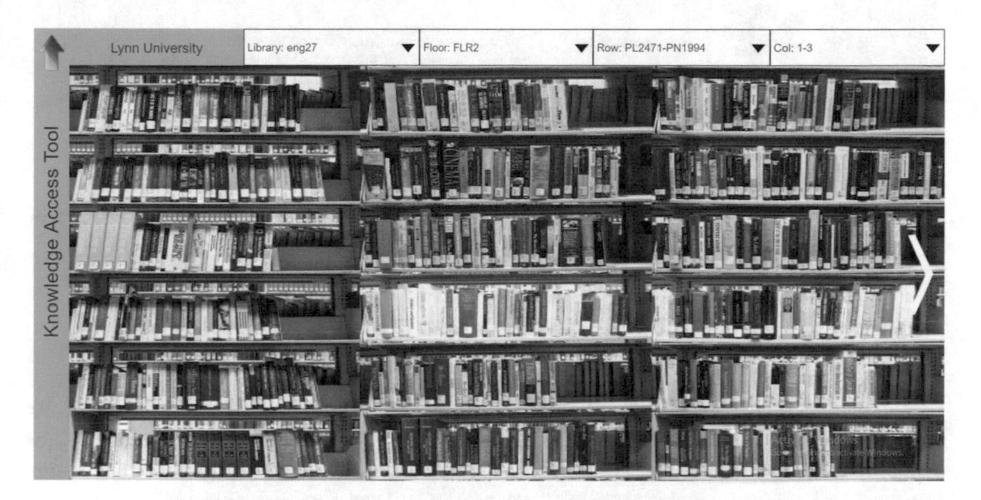

Fig. 6. The result of recording with wide-format scanner

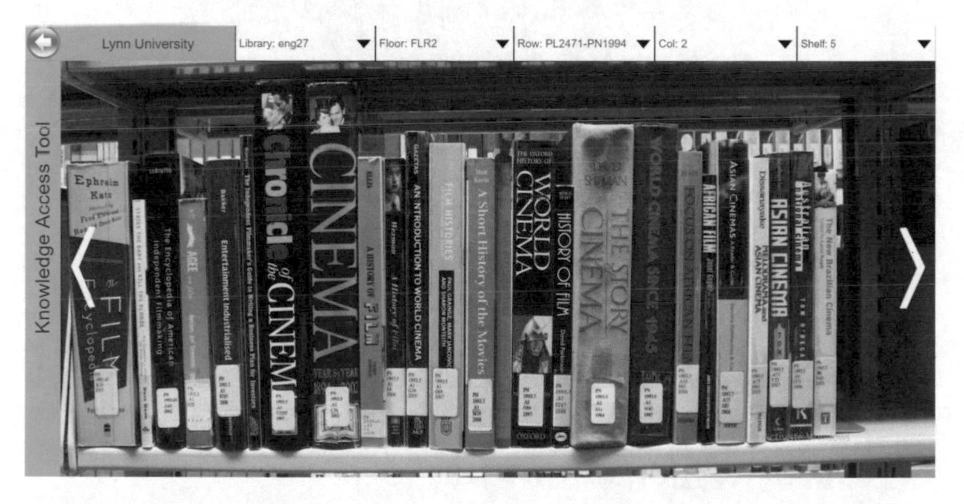

Fig. 7. A distinguished shelf is recognized for recognizing a certain book library shelf

publishing house, year of publication, photo titles, annotation, cipher in MARC21 format, etc.). So the files with the pictures of the cover on a specific shelf on a specific rack in a certain library are combined with the text, allowing the possibility of a text search along with the stored original appearance of the information object (Fig. 8).

Fig. 8. Book description in e-library

Formation of a structured database from an unstructured database is made by joboons automatically on a schedule (from time to time they start by themselves) or when they are called by the admin from the admin page. The subsystem of the formation of a structured database based on MARC 21 will be presented as

$$S_{\varepsilon} = <B_{\varepsilon}, B_{o}, U_{\varepsilon}, \varepsilon> , B_{\varepsilon} = \varepsilon(B_{o}, U_{\varepsilon}), \tag{6}$$

where B_{ε} is a set of descriptions of information objects in a structured database, U_{ε} is a set of conditions for the formation of structured descriptions of information objects, ε is the operator of recording or updating the description of the information object in a structured database.

The subsystem of cache formation when $B_{\varepsilon} \supset C_{v}$ lets present as

$$S_{\gamma} = <B_{\varepsilon}, C_{v}, Q_{\alpha}, Q_{\beta}, U_{v}, \gamma> , C_{v} = \gamma(B_{\varepsilon}, Q_{\alpha}, Q_{\beta}, U_{v}), \tag{7}$$

where C_{v} is a set of popular information objects in the cache, γ is the operator updating and filling the cache according to the analysis of requests from users and professors, U_{v} is the set of conditions for updating the cache. Then, based on the analysis of formulas (4)–(7), the subsystem of work with the users to generate the results of their queries will be submitted by a superposition $W_{\alpha} = \eta \circ \lambda \circ \mu \circ \alpha$ and by respective tuple $S_{\alpha} = <A_{\alpha}, U_{\alpha}, Q_{\alpha}, B_{\varepsilon}, C_{v}, W_{\alpha}, U_{\eta}, \alpha, \lambda, \mu, \eta>$, i.e.

$$W_{\alpha} = \eta(U_{\eta}, \lambda(B_{\varepsilon}, \mu(C_{v}, (\alpha(A_{\alpha}, U_{\alpha}, Q_{\alpha}))))), \tag{8}$$

where Q_{α} is a set of requests from ISVL users; U_{α} is a set of conditions for analyzing user queries; B_{ε} is a set of information objects in the database; C_{v} is a set of information objects in the cache; U_{η} is a set of conditions for reporting; λ is operator of search of information objects in a cloud; μ is operator for finding information objects in the cache. Taking into account the formulas (3)–(8), we replace the formula (1) and describe the ISVL model as a tuple:

$$S_{VL} = \langle A_{\iota}, U_{\pi}, B_{o}, U_{o}, B_{\varepsilon}, U_{\varepsilon}, Q_{\alpha}, U_{\alpha}, L_{\beta}, U_{\beta}, C_{v}, U_{v}, W_{\eta}, U_{\eta}, \delta, \beta, \eta, \pi, \varepsilon, \gamma, \theta, \rho \rangle . \tag{9}$$

The set of user queries for ISVLv (Fig. 9) looks like $Q_{\alpha} = \alpha(A_{\alpha}, U_{\alpha})$.

Fig. 9. Recommended by professor literature for studying the course

Accordingly, the formation of a professor's list in his own virtual cabinet $L_{\beta} = \beta(A_{\beta}, Q_{\beta}, B_{\varepsilon}, C_{v}, U_{\beta})$, and cache update $C_{v} = \gamma(Q_{\alpha}, Q_{\beta}, U_{v})$. Then the results of the queries will be presented as a superposition $W_{\eta} = \mu \circ \eta$ or $W_{\eta} = \lambda \circ \mu \circ \eta$, i.e.

$$W_{\eta} = \mu(\eta(Q_{\alpha}, U_{\eta}), U_{\mu}) \text{ or } W_{\eta} = \lambda(\mu(f_{\eta}(Q_{\alpha}, U_{\eta}), U_{\mu}), U_{\lambda}). \tag{10}$$

The quality of obtaining relevant content for user queries directly depends on the quality of the description of the scanned information objects. But the efficiency of forming a qualitative set of relevant content depends on the quality of the cache update and the effectiveness of the implemented algorithms of informational search.

Administrator: Manages everything in the system. The user can view books on shelves, search for books, watch their pdf files. A professor can build his own recommendation lists for a particular course in the system (Figs. 10, 11, 12, 13 14 and 15). When users search for a course - they will see a list of all professors who are professors for this course. There is a database containing all the basic data. There is a database that serves as a cache for faster searching. The data in them is filled with jobs or an admin.

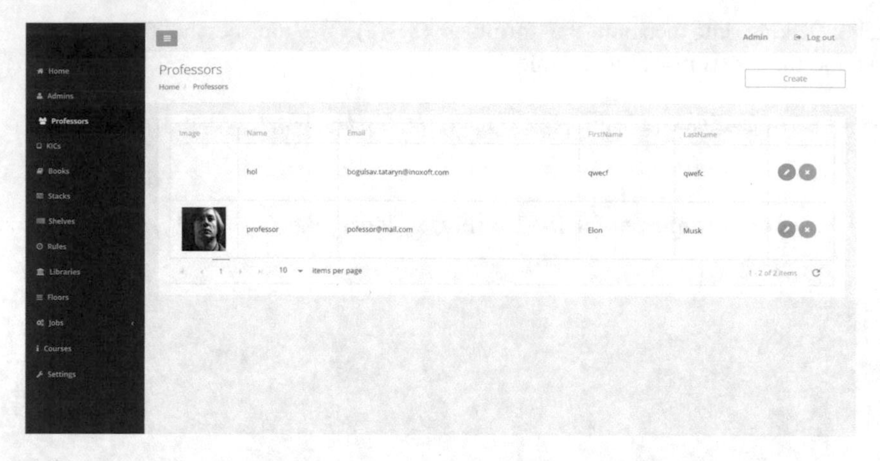

Fig. 10. Course professor's informational page

Fig. 11. Course professor correction (edit) page

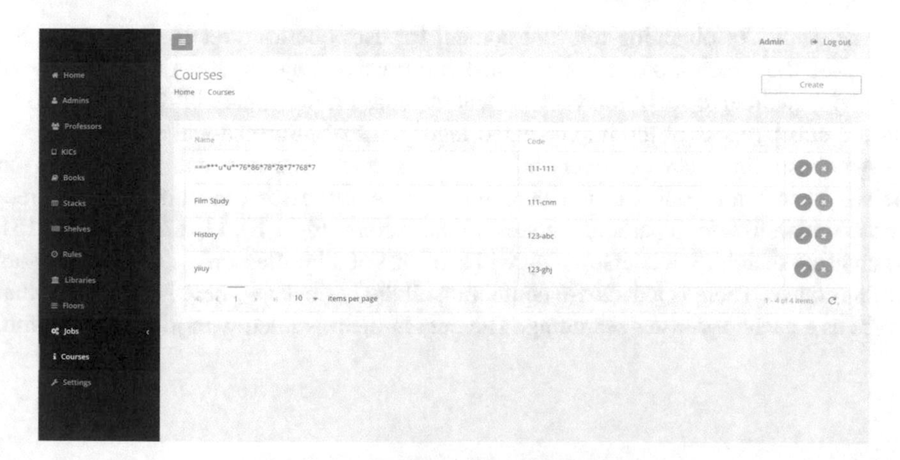

Fig. 12. Informational page about courses

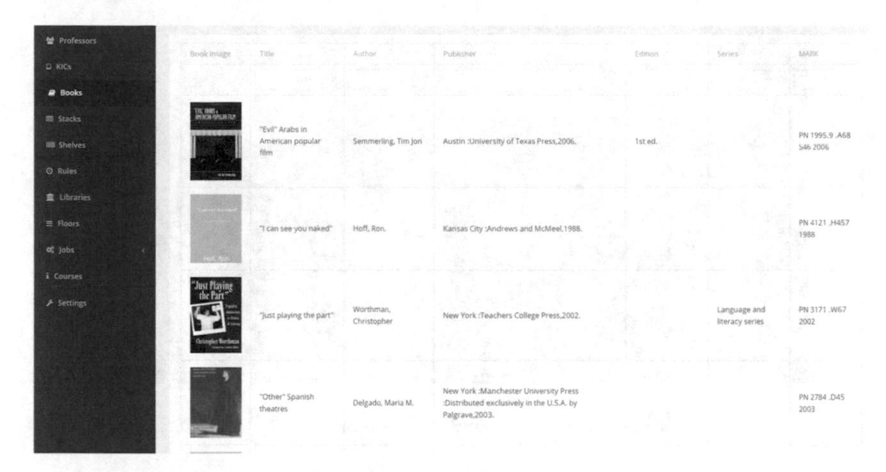

Fig. 13. Professor's book list

Admin can manage WebJobs, but in general they are automatic and work on the chart for themselves for the specific libraries (Fig. 16). Main system functions (Fig. 17): user management in the system, book management, shelves management, shelf stack management, library management; the system supports many libraries at the same time and switch between them, the floors in the library to navigate them, courses (you can create courses for which professors will create advisory lists of books). Developed with WebJobs actually supports such processes (Fig. 17c) as:

- Job, which surfs the internet and looks for book covers that do not have it.
- 'Builder' Job. It collects books, knowing where they should stand and builds a shelf first, and then a stack of shelves, and then it is available for users. It's all a CGI work, we automatically create all the pictures.

Fig. 14. Page for correction course professor's book list

Fig. 15. Formed shelf with course book list

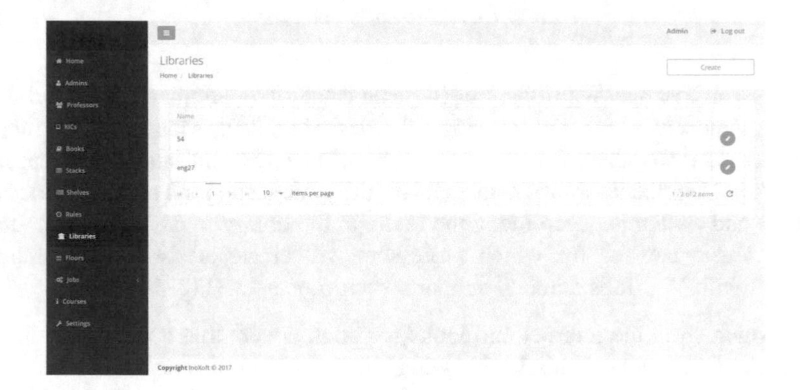

Fig. 16. Library in ISVL

- Job, which expands the shelves if book appeared that had not previously stood there. CGI expands the shelf and adds a new book there (Fig. 17).
- Job, which builds links between MARC's books and their graphic data, namely, covers and books' side pictures.
- Job, which re-indexes all in the system, creating a spreadsheet index and clustered indexes for a fast search.

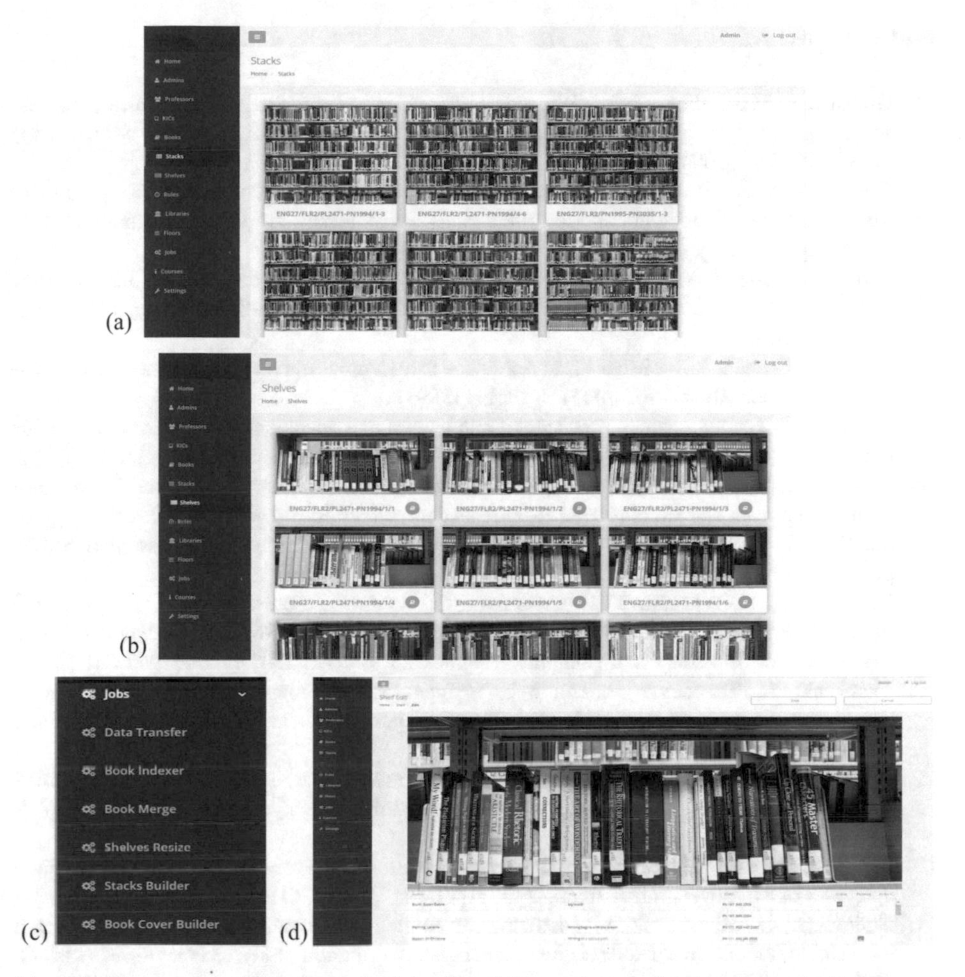

Fig. 17. The process of forming a shelf for the addition of a new book: (a) search of racks and identification of the rack; (b) search and identification of shelf on the rack; (c) choosing and running an exact job; (d) filled shelf with a new book.

7 Conclusions

An overview of the features of the design and development of the virtual library information system is carried out. A new approach is proposed for the design and development of the Virtual Library Information System for saving and development of e-books in the MARC 21 format. The structure of the Virtual Library information system, the model of the Virtual Library Information System and a detailed description of the system decomposition have been proposed.

Particular virtual library solution was designed for US Florida state. Right now we are working on integration and testing of the solution in 2 Florida State.

References

1. Gonçalves, M.A., Fox, E.A., Watson, L.T., Kipp, N.A.: Streams, structures, spaces, scenarios, societies (5S): a formal model for digital libraries. ACM Trans. Inf. Syst. (TOIS) **22**(2), 270–312 (2004)
2. Gonçalves, M.A.: Streams, structures, spaces, scenarios, and societies (5S): a formal model for digital library framework and its applications. Ph.D. thesis, Virginia Polytechnic Institute and State University (2004)
3. Pérez, A., Enrech, M.: Defining library services for a virtual community. In: Libraries Without Walls Conference, Lesvos, Grecia, Centre for research in Library and Information Management (1999)
4. Pérez, A., Enrech, M.: Virtual library services for a virtual university: user-oriented virtual sites in an open library. In: EADTU, Paris (1999)
5. Stoffle, C.J.: The emergence of education and knowledge management as major functions of the digital library. http://www.ukoln.ac.uk/services/papers/follett/stoffle/paper.html
6. Lytvyn, V., Vysotska, V., Chyrun, L., Chyrun, L.: Distance learning method for modern youth promotion and involvement in independent scientific researches. In: Proceedings of the IEEE First International. Conference on Data Stream Mining & Processing (DSMP), pp. 269–274 (2016)
7. Rashkevych, Y., Peleshko, D., Vynokurova, O., Izonin, I., Lotoshynska, N.: Single-frame image super-resolution based on singular square matrix operator. In: IEEE 1st Ukraine Conference on Electrical and Computer Engineering (UKRCON), pp. 944–948 (2017)
8. Tkachenko, R., Tkachenko, P., Izonin, I., Tsymbal, Y.: Learning-based image scaling using neural-like structure of geometric transformation paradigm. In: Studies in Computational Intelligence, vol. 730, pp. 537–565 (2018)
9. Maksymiv, O., Rak, T., Peleshko, D.: Video-based flame detection using LBP-based descriptor: influences of classifiers variety on detection efficiency. Int. J. Intell. Syst. Appl. **9** (2), 42–48 (2017)
10. Peleshko, D., Rak, T., Izonin, I.: Image superresolution via divergence matrix and automatic detection of crossover. Int. J. Intell. Syst. Appl. **8**(12), 1–8 (2016)
11. Chernukha, O., Bilushchak, Y.: Mathematical modeling of random concentration field and its second moments in a semispace with Erlangian distribution of layered inclusions. Task Q. **20**(3), 295–334 (2016)
12. Rusyn, B., Lutsyk, O., Lysak, Y., Lukeniuk, A., Pohreliuk, L.: Lossless image compression in the remote sensing applications. In: Proceedings of the IEEE First International Conference on Data Stream Mining & Processing (DSMP), pp. 195–198 (2016)
13. Kowalik, D.: Polish vocational competence standards for the needs of adult education and the European labour market. In: International Conference on Advanced Information Engineering and Education Science ICAEES, pp. 95–98 (2013)
14. Zhezhnych, P., Markiv, O.: Linguistic comparison quality evaluation of web-site content with tourism documentation objects. In: Advances in Intelligent Systems and Computing, vol. 689, pp. 656–667 (2018)
15. Mobasher, B.: Data mining for web personalization. In: The Adaptive Web, pp. 90–135 (2007)
16. Dinucă, C.E., Ciobanu, D.: Web content mining. University of Petroşani, Economics 85 (2012)
17. Xu, G., Zhang, Y., Li, L.: Web content mining. In: Web Mining and Social Networking, pp. 71–87 (2011)

18. Jivani, A.G.: A comparative study of stemming algorithms. Int. J. Comput. Technol. Appl. **6**, 1930–1938 (2011)
19. Vysotska, V.: Linguistic analysis of textual commercial content for information resources processing. In: Modern Problems of Radio Engineering, Telecommunications and Computer Science TCSET 2016, pp. 709–713 (2016)
20. Lytvyn, V., Vysotska, V., Pukach, P., Brodyak, O., Ugryn, D.: Development of a method for determining the keywords in the slavic language texts based on the technology of web mining. East. Eur. J. Enterp. Technol. **2/2**(86), 4–12 (2017)
21. Davydov, M., Lozynska, O.: Linguistic models of assistive computer technologies for cognition and communication. In: 12th International Scientific and Technical Conference on Computer Sciences and Information Technologies (CSIT), Lviv, pp. 171–175 (2017)
22. Khomytska, I., Teslyuk, V.: The method of statistical analysis of the scientific, colloquial, belles-lettres and newspaper styles on the phonological level. In: Advances in Intelligent Systems and Computing, vol. 512, pp. 149–163 (2017)
23. Khomytska, I., Teslyuk, V.: Specifics of phonostatistical structure of the scientific style in English style system. In: XIth International Scientific and Technical Conference Computer Sciences and Information Technologies (CSIT), Lviv, pp. 129–131 (2016)
24. Shakhovska, N., Vysotska, V., Chyrun, L.: Features of E-learning realization using virtual research laboratory. In: XIth International Scientific and Technical Conference Computer Sciences and Information Technologies (CSIT), Lviv, pp. 143–148 (2016)
25. Shakhovska, N., Medykovsky, M., Stakhiv, P.: Application of algorithms of classification for uncertainty reduction. Przeglad Elektrotechniczny **89**(4), 284–286 (2013)
26. Schahovska, N., Syerov, Y.: Web-community ontological representation using intelligent dataspace analyzing agent. In: 10th International Conference the Experience of Designing and Application of CAD Systems in Microelectronics, CADSM 2009, pp. 479–480 (2009)
27. Shakhovska, N., Vovk, O., Hasko, R., Kryvenchuk, Y.: The method of Big Data processing for distance educational system. In: Conference on Computer Science and Information Technologies, pp. 461–473. Springer, Cham (2017)
28. Shakhovska, N., Vovk, O., Kryvenchuk, Y.: Uncertainty reduction in Big Data catalogue for information product quality evaluation. East. Eur. J. Enterp. Technol. **1**(2), 12–20 (2018)
29. Davydov, M., Lozynska, O.: Mathematical method of translation into Ukrainian sign language based on ontologies. In: Advances in Intelligent Systems and Computing, vol. 689, pp. 89–100 (2018)
30. Davydov, M., Lozynska, O.: Information system for translation into Ukrainian sign language on mobile devices. In: 12th International Scientific and Technical Conference on Computer Sciences and Information Technologies (CSIT), Lviv, pp. 48–51 (2017)
31. Kut, V., Kunanets, N., Pasichnik, V., Tomashevskyi, V.: The procedures for the selection of knowledge representation methods in the "virtual university" distance learning system. In: Advances in Intelligent Systems and Computing, pp. 713–723 (2018)
32. Veretennikova, N., Kunanets, N.: Recommendation systems as an information and technology tool for virtual research teams. In: Advances in Intelligent Systems and Computing, vol. 689, pp. 577–587 (2018)

Web-Products, Actual for Inclusive School Graduates: Evaluating the Accessibility

Tetiana Shestakevych[1(✉)], Volodymyr Pasichnyk[1], Maria Nazaruk[1,2], Mykola Medykovskiy[1,2], and Natalya Antonyuk[2]

[1] Lviv Polytechnic National University, Lviv 79013, Ukraine
tetiana.V.Shestakevych@lpnu.ua
[2] Ivan Franko Lviv National University, Lviv 79000, Ukraine
nantonyk@yahoo.com

Abstract. Over the years, the understanding of a person with special needs evolved into the worldwide tendency of support of groups of people in need of social protection. The structure of social and educational inclusion became a complex system, with numerous members of different kinds and hierarchical organization. Modern society promotes a concept of "design for all" when products and environments are created and managed in such a way that they could be used by the widest range of people without the need for adaptation or special design. Such concept of universal design is also applicable to information and communication technologies. Developed web-accessibility manuals (WCAG 2.0 Guidelines, for example) and tools, discussed in this paper, were used to evaluate web-products, available for graduating schoolchildren in Ukraine. The formal conceptual model of the WCAG 2.0 Guideline was used to describe the current compliance state of appropriate web products. The websites characteristics, that do not meet WCAG 2.0 success criteria, should be emphasized and taken into account not only by IT professionals while developing ICTs, but also by teachers, lecturers, educators, and tutors, who are the subject of teaching disciplines, related to programming and information technologies.

Keywords: Inclusive education · Complex system · Universal design in ICT WCAG 2.0

1 Introduction

Ukraine is on the way to improving the state support of people in need of additional social protection, taking into account international standards. Such people include children, refugees and persons in need of additional protection, as well as foreigners and stateless persons who are legally resident in Ukraine, members of the anti-terrorist operation (ATO), internally displaced persons, hostages, and people with disabilities. A similar understanding of the needs of socially vulnerable individuals was announced in January 2018 by experts from the European Commission [1] and in the program of European scientific and innovative research Horizons 2020 [2]. The documents

N. Shakhovska and M. O. Medykovskyy (Eds.): CSIT 2018, AISC 871, pp. 350–363, 2019.
https://doi.org/10.1007/978-3-030-01069-0_25

mentioned above have a similar understanding of groups of people in need of social protection (Fig. 1).

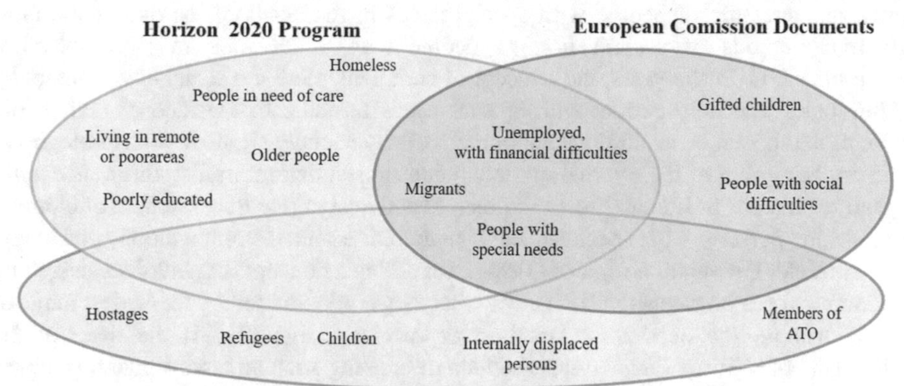

Fig. 1. Groups of people in need of social protection and support

An urgent task is the thorough study and research of a complex system of inclusion as a multicomponent complex system, designed to ensure lifelong socialization of socially vulnerable individuals.

2 A Complex System of Social and Educational Inclusion

The system of inclusive education is multifactorial, with a large number of components, and, besides people with special needs, involve diverse professionals [3]. But inclusive education is only one of many aspects of the functioning of a complex system of inclusion of socially vulnerable individuals.

A system of social and educational inclusion is characterized as a complex system according to its features [4–6]. Such features are a large number of components, variability, a large number of diverse resources are involved in the processes of the system, self-organization, diversity, dynamic and viable, adjustment to the environment, interaction, nonlinearity, selectivity, feedback, lack of central control, the hierarchy of the organization, emergence, and evolution.

An important aspect of a complex system of educational and social inclusion is its IT component. As the first stage of the analysis of a complex system of educational and social inclusion, as a member of informational society, authors suggest the study of Web products, required by participants of the inclusion. In this paper, authors will explore web products that are accessible to people with special needs at the critical stage of their life, meaning, after the graduation of an inclusive school. The selected web products will be tested for compliance with web accessibility requirements and the needs of users with special needs.

3 Universal Design as an Aspect of Inclusion

The development of the inclusion system has begun in 1970th when the specialists stated an education of people with special needs in mass schools as one of the most effective methods of socialization of people with a deviation in psychophysical development. Over the years, the concept of complementing the social environment by technologies that help people with special needs turned into the idea of taking into account the special needs in the development of any socially significant technologies at the very beginning of their verbal modeling and substantiation. In fact, this concept has spread to all areas of human life – architecture and design (the development of clothing suitable for persons with special needs), automotive and instrumentation, pedagogy, medicine, etc. Universal design or “design for all” is a concept that involves designing products and environments in such a way that they could be used by the widest range of people without the need for adaptation or special design [7]. At the heart of the philosophy of universal design lies the idea of creating such an environment, products and services that would be useful and convenient for everyone, and not just people with disabilities. In the simplest sense, universal design is the design of all things, the focus of which is a person and which takes into account the needs of everyone [8]. Access to such living conditions in [7] is called an important aspect of social sustainability. Authors will consider the system of inclusion in Ukraine from the positions declared in the European Commission’s social initiatives. Such initiatives, on the one hand, call a lifelong learning as a driving force of socialization, and on the other, promotes man-centered universal design.

The concept of universal design is based on seven basic principles proposed in 1997 by a group of designers and researchers [9].

- Equal use: design neither neglect, nor exclude, or classify any group of users.
- Flexibility in an application: design should offer a wide range of personal solutions and capabilities.
- Simple, intuitive way of use: design should be easy to understand, regardless of user experience and skills, language skills, current ability to focus.
- Perceptive information: design effectively provides information regardless of the current level of user experience.
- Tolerance to mistakes: design minimizes the risk and loss from unforeseen or unintentional actions.
- Low physical effort: the design should be used skillfully and with comfort, causing the least fatigue.
- Size and space for accessibility and use: design is implemented in such a way that users, regardless of body size, posture or mobility, have space enough to access and exploitation.

Such concepts concerned, first of all, the architectural accessibility of buildings, but eventually acquired a much broader understanding, spreading to all components of life, becoming a philosophical system, and even a way of thinking.

4 Universal Design in ICT: Web Accessibility Requirements

The development of the information society has caused conscious and systematic implementation of principles, methods, and means of universal design also in the field of computer sciences. According to [10], the implementation of the principles of universal design in information and communication technologies is one of the aspects of sustainable social development. With the development of information technology and the Internet, compliance with the requirements of convenience and accessibility has been standardized in documents of international organizations ISO. In this way, the increase in the level of socialization of people with special needs is affected by the opportunity to study throughout their lives, and the support of such training, which is based on information technologies, can be improved through the implementation of the principles of universal design of ICT (Fig. 2).

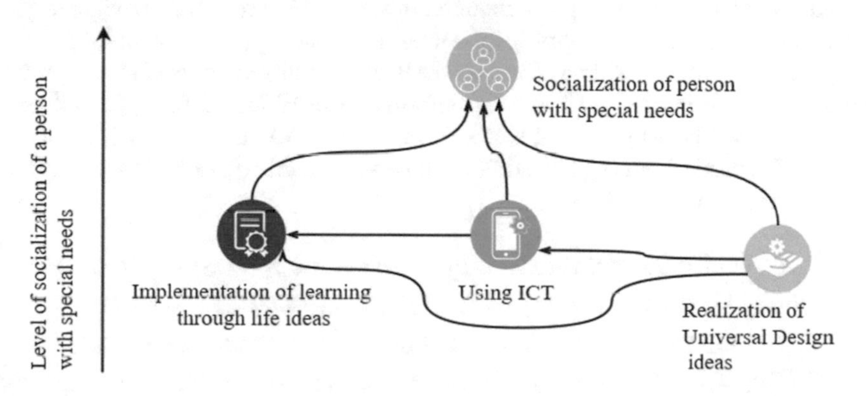

Fig. 2. Conceptual scheme of ways to deepen the socialization of a person with special needs

In the late 1990s, the World Wide Web Consortium (W3C), an international organization that develops and implements technology standards for the global network, launched the Web Accessibility Initiative (WAI). This initiative was intended to improve the accessibility of the World Wide Web to people with special needs. As a result of the W3C work, the Web-accessibility manuals were developed. One of them is the Web Content Accessibility Guidelines (WCAG). The current version of this Guide (WCAG 2.0) was published in December 2008, and it which became an ISO/IEC 40500:2012 standard in October 2012 [11].

Criteria for implementing the WCAG 2.0 recommendations are presented in the form of verifiable statements and are not tied to the technology. These recommendations are aimed at ensuring wider availability of web content for people with diverse special needs, e.g. visual impairment, hearing impairment, peculiarities of mental development, and their combinations. In addition, the implementation of these recommendations will make the web content of the site more accessible to users, regardless of the presence or absence of any restrictions. The WCAG 2.0 recommendations should be used by web design professionals, designers, software developers, executives, teachers, students, etc.

5 Model of WCAG 2.0 Guidelines

The model of WCAG 2.0 guidance is given as a tuple M:

$$M = <P, G, C, R>,$$

where P is a set of principles of web content accessibility, G is a set of guidelines of web content accessibility, C is a set of success criteria of web content accessibility for each guideline, R is a set of sufficient and advisory techniques for each of the guidelines and success criteria.

The set of principles of web content accessibility P = $\{p_1, p_2, p_3, p_4\}$ consists of p_1 as perceivability, p_2 as operability, p_3 as understandability, and p_4 as robustness. Elements of the set of guidelines of web content accessibility $G = \{g_1, \ldots g_{12}\}$ are: $g_{1.1}$ – text alternatives, $g_{1.2}$ – time-based media, $g_{1.3}$ – adaptability, $g_{1.4}$ – distinguishability, $g_{2.1}$ – keyboard accessibility, $g_{2.2}$ – enough time, $g_{2.3}$ – seizures, $g_{2.4}$ – navigability, $g_{3.1}$ – readability, $g_{3.2}$ – predictability, $g_{3.3}$ – input assistance, $g_{4.1}$ – compatibility.

The set C of success criteria of web content accessibility consists of four subsets: C_1 has the criteria of lowest (A) level of conformance in WCAG 2.0, C_2 has all criteria from C_1, and additionally a set of a criteria of middle (AA) level of conformance, and set C has all criteria from C_2, and additionally a set of a criteria of highest (AAA) level of conformance in WCAG 2.0:

$$C_1 = \{c_{1.1.1}, c_{1.2.1}, c_{1.2.2}, c_{1.2.3}, c_{1.3.1}, c_{1.3,2}, c_{1.3.3}, c_{1.4.1}, c_{1.4.2}, c_{2.1.1}, c_{2.1.2}, c_{2.2.1}, c_{2.2.2}, c_{2.3.1}, c_{2.4.1}, c_{2.4.2}, c_{2.4.3}, c_{2.4.4}, c_{2.4.5}, c_{3.1.1}, c_{3.2.1}, c_{3.2.2}, c_{3.3.1}, c_{3.3,2}, c_{3.3.4}, c_{4.1.1}, c_{4.1.2}\};$$
$$C_2 = C_1 U\{c_{1.2.4}, c_{1.2.5}, c_{1.4.3}, c_{1.4.4}, c_{1.4.5}, c_{2.4.6}, c_{2.4.7}, c_{3.1.2}, c_{3.2.3}, c_{3.2.4}, c_{3.3.3}\};$$
$$C = C_2 U\{c_{1.2.6}, c_{1.2.7}, c_{1.2.8}, c_{1.2.9}, c_{1.4,6}, c_{1.4.7}, c_{1.4.8}, c_{1.4.9}, c_{2.1.3}, c_{2.2.3}, c_{2.2.4}, c_{2.2.5}, c_{2.3.2}, c_{2.4.8}, c_{2.4.9}, c_{2.4.10}, c_{3.1.3}, c_{3.1.4}, c_{3.1.5}, c_{3.1.6}, c_{3.2,5}, c_{3.3.5}, c_{3.3.6}\}.$$

Such model can be conceptually presented as a tree (Fig. 3).

6 Web-Products for Inclusion

In the inclusive system, inclusive education covers pre-school, school and out-of-school, vocational and higher education, as well as self-education and, accordingly, all institutions and organizations that provide and accompany such training – kindergartens, schools, institutes of higher education of different levels of accreditation, libraries and so on. Teaching a person with special needs during a lifetime also involves vocational training, advanced training, internship, etc. An important component of social adaptation is the completion of the school education and the planning of a future. To analyze the current state of compliance of web products, relevant to people with special needs (with different needs [12, 13]) that are finishing school, authors selected a set of web products related to future education or the beginning of a professional career in Ukraine.

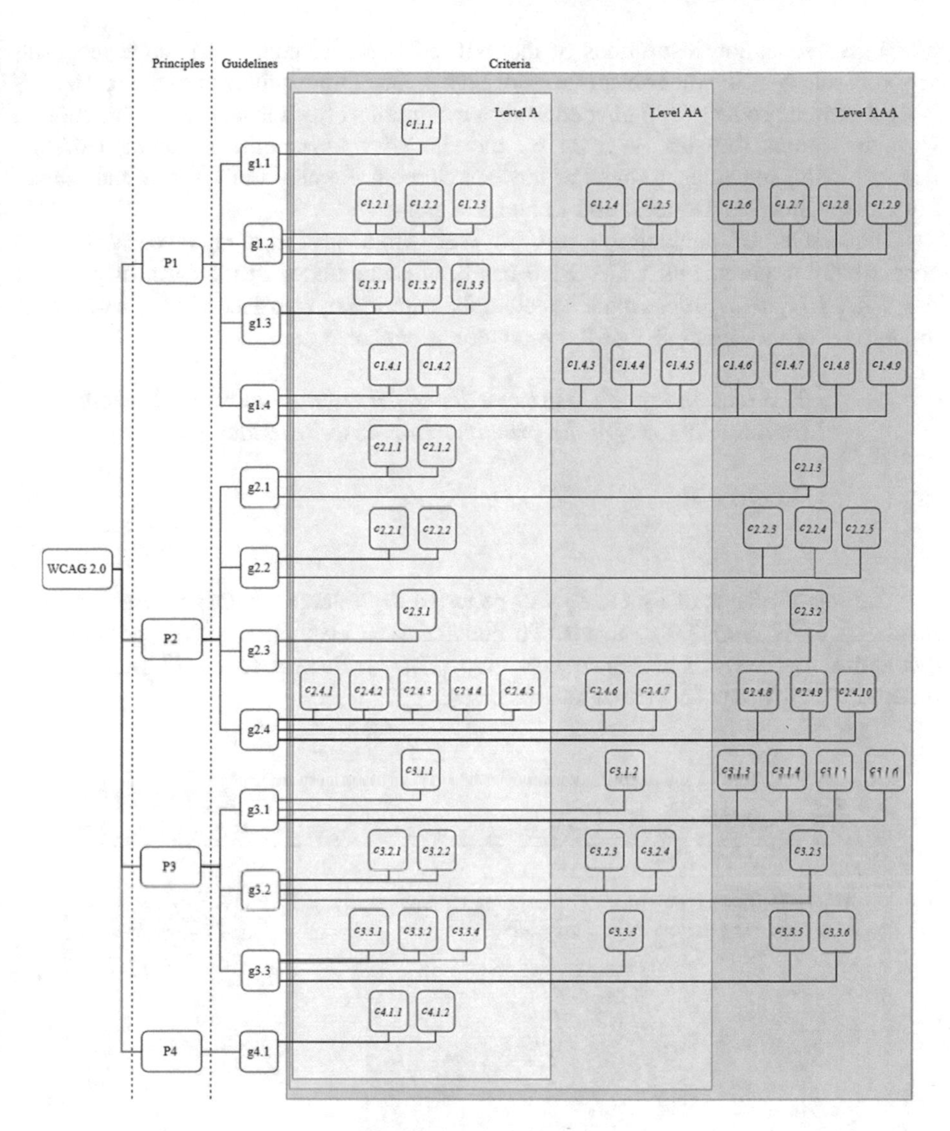

Fig. 3. The Conceptual Model of the WCAG 2.0 Guide

Let us consider a range of web products, connected to the future education or starting a professional career. To set S_1 belongs websites for a job search, as well as a State Employment Service. The S_1 contains ten unique web products proposed by searching system Google for the "find a job" request («знайти роботу») (www.work.ua/, www.rabota.ua/ua, www.ria.com/uk/work/, www.olx.ua/uk/, www.ua.jooble.org/uk, www.rabota-tut.ua, www.hh.ua/, vashmagazin.ua/robota-ta-navchannya/, www.jobs.ua/ukr, www.talent.ua/uk). Together with State Employment Service website (https://www.dcz.gov.ua/), the power of S_1 is eleven.

Higher education institutions of the I–II levels of accreditation include technical schools, colleges, and other institutions of higher education equivalent to them. Then S_2 is a set of web products of higher education institutions of I–II levels of accreditation in Ukraine regions (the list is given on the site www.osvita.ua), excluding (for this research only) the web products of the Autonomous Republic of Crimea and educational institutions of Donetsk and Luhansk region.

The results of analyzing a set of Web products for compliance with web accessibility requirements WCAG 2.0 has been accumulated in the form of a matrix $A = [{}_za_{i,j,k}]$. Here ${}_za_{i,j,k}$ determines whether the criterion of web accessibility k of for the guideline j of the principle i of the product z is executed (1).

$$
{}_za_{i,j,k} = \begin{cases} 1, \text{ if there is a problem in execution of the criterion of web accessibility } k \\ \text{for the guideline } j \textit{ of the principle } i \textit{ for the web product } z \\ \\ 0, \textit{otherwise} \end{cases} \tag{1}
$$

The web products of S_1 and S_2 will be tested to evaluate whether they conform to level AA of WCAG 2.0 demands. To conduct such evaluation a Web accessibility evaluation tool www.achecker.ca was used (Fig. 4). Such a tool [14] was used in research [15, 16] and in Virtual labs [18, 19].

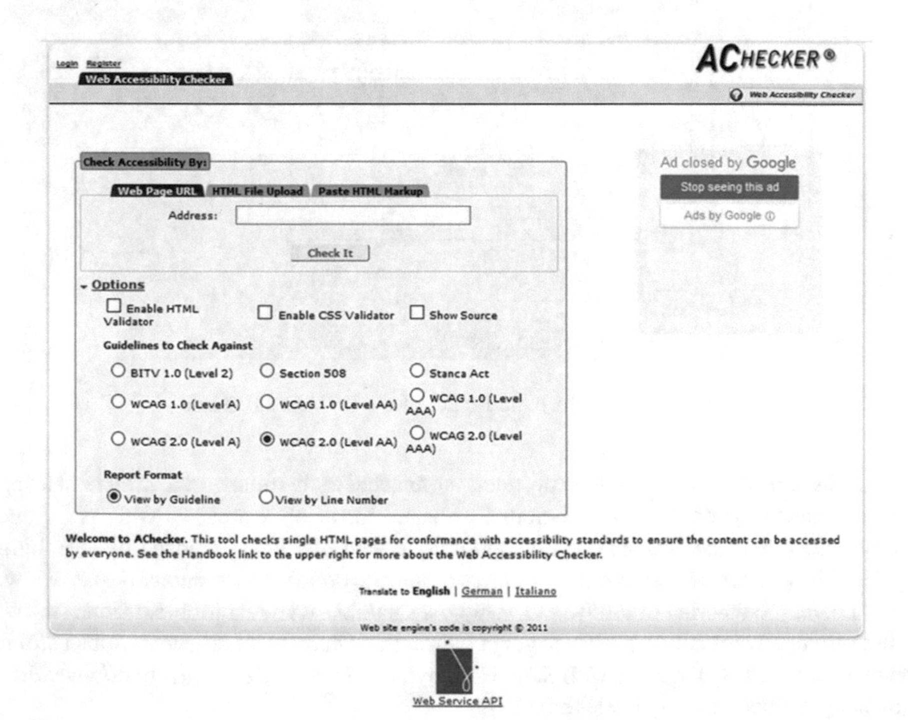

Fig. 4. A tool for Web accessibility evaluation

7 Conformance with Accessibility Standards for a Set of Web Products S_1

Each web product from the S_1 set was analyzed using www.achecker.ca tool and found the number of known, likely, and potential problems. Of 61 criteria in WCAG 2.0, 38 are conforming Web accessibility of the AA level (not to forget that conforming AA level means also an A level). Out of 38 accessibility criteria, thirty were fully realized in all Web products of S_1 set. The criteria with problems are in Table 1.

Table 2 contains a matrix $A = [_z a_{i.j.k}]$, here columns captions are $c_{i.j.k,}$ which denotes criteria of Web accessibility k for the guideline j of principle i. Here z is a number of an appropriate Web products from S_1 set, and in matrix as «1» we denoted an existence of at least one problem, and «0» means that there were no problems detected. We shall denote Web products as follows: r_1 (address https://www.dcz.gov.ua/), r_2 (https://www.work.ua/ua/), r_3 (https://rabota.ua/ua), r_4 (https://ua.jooble.org/uk), r_5 (https://www.ria.com/uk/work/), r_6 (https://www.olx.ua/uk/), r_7 (https://vashmagazin.ua/robota-ta-navchannya/), r_8 (https://jobs.ua/ukr), r_9 (https://talent.ua/uk), r_{10} (https://hh.ua/), r_{11} (http://rabota-tut.ua/uk).

Table 1. Criteria [17] of S_1 web products that do not meet success

Criterion denotation	Web access criterion	Explanation
$c_{1.1.1}$	Non-text Content	All non-text content that is presented to the user has a text alternative that serves the equivalent purpose, except for the situations listed below
$c_{1.3.1}$	Info and relationships	Information, structure, and relationships conveyed through presentation can be programmatically determined or are available in text
$c_{1.4.3}$	Contrast (Minimum)	The visual presentation of text and images of text has a contrast ratio of at least 4.5:1
$c_{1.4.4}$	Resize text	Except for captions and images of text, text can be resized without assistive technology up to 200 percent without loss of content or functionality
$c_{2.4.4}$	Link purpose (In Context)	The purpose of each link can be determined from the link text alone or from the link text together with its programmatically determined link context, except where the purpose of the link would be ambiguous to users in general
$c_{3.1.1}$	Language of page	The default human language of each Web page can be programmatically determined
$c_{3.3.2}$	Labels or Instructions	Labels or instructions are provided when content requires user input
$c_{4.1.1}$	Parsing	In content implemented using markup languages, elements have complete start and end tags, elements are nested according to their specifications, elements do not contain duplicate attributes, and any IDs are unique

Table 2. Summary results of S_1 Web product research for meeting WCAG 2.0 AA level criteria

Web product	$C_{i.j.k}$							
	$c_{1.1.1}$	$c_{1.3.1}$	$c_{1.4.3}$	$c_{1.4.4}$	$c_{2.4.4}$	$c_{3.1.1}$	$c_{3.3.2}$	$c_{4.1.1}$
r_1	0	0	0	0	0	0	0	0
r_2	1	1	0	1	1	0	1	0
r_3	1	1	0	1	1	1	0	0
r_4	0	1	0	0	0	0	1	0
r_5	1	1	0	1	1	0	1	0
r_6	1	0	0	0	0	0	0	1
r_7	1	1	1	1	0	1	1	0
r_8	0	1	0	1	1	0	1	1
r_9	1	1	0	1	0	0	1	1
r_{10}	1	1	0	0	1	1	1	1
r_{11}	1	1	0	1	1	0	1	0

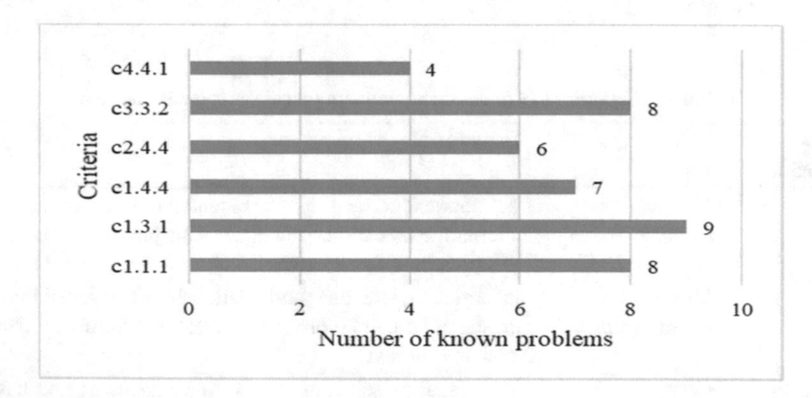

Fig. 5. The top-6 problems according to the success criteria for a set of web products S_1

8 Conformance with Accessibility Standards for a Set of Web Products S_2

As mentioned above, to the set S_2 are classified web products of 613 educational institutions of I–II levels of accreditation. The addresses of each site were placed in the www.achecker.ca tool, as a result, the address of several web products were excluded from S_2 set because the system reported on the impossibility of verification. Thus, 578 website addresses were analyzed, that is 94% of the available website addresses. 418 of these sites had problems according to WCAG 2.0. 160 websites conformed to the AA level, which is 28% of the web products studied. Detailed analysis results are given in the Table 3.

Figure 5 has a graphical interpretation of the results from the Table 2 (six most numerous errors).

Table 3. Results of analysis of sites using www.achecker.ca tool

№	Regin	Number of educational institutions	Websites checked	Websites with problems	Average number of problems per site
1	Vinnytsia	29	29	20	26
2	Volyn	23	22	13	75
3	Dnipropetrovsk	65	60	42	17
4	Zhytomyr	22	20	14	35
5	Zakarpattia	19	18	14	23
6	Zaporizhzhia	30	30	25	20
7	Ivano-Frankivsk	25	24	17	28
8	Kyiv, city	53	50	29	16
9	Kyiv	21	20	14	44
10	Kirovograd	18	15	12	32
11	Lviv	38	36	24	16
12	Mykolaiv	17	16	11	125
13	Odesa	32	32	26	9
14	Poltava	25	22	21	14
15	Rivne	19	19	17	25
16	Sumy	23	23	15	37
17	Ternopil	18	16	10	57
18	Kharkiv	43	42	32	9
19	Kherson	17	16	10	36
20	Khmelnytskyy	16	16	14	22
21	Cherkasy	21	19	16	19
22	Chernivtsi	19	13	8	113
23	Chernihiv	20	20	14	30

The results of this comparison are also given in Fig. 6.

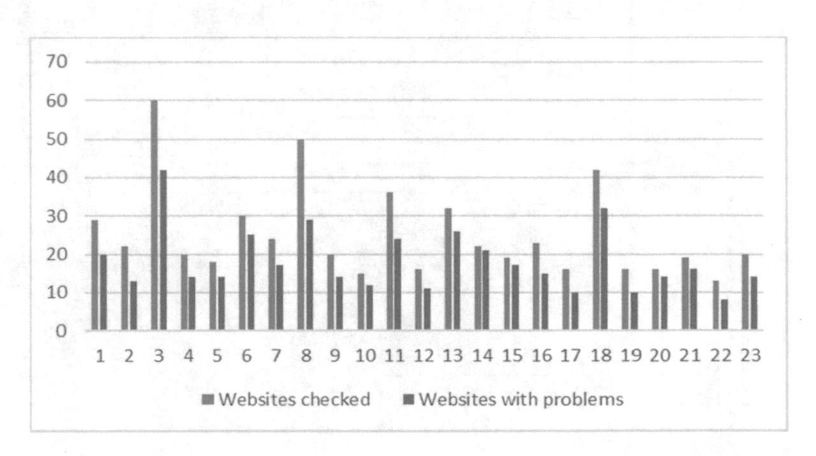

Fig. 6. Criteria that have not been met by one or more of the investigated web products

The largest average number of problems is on the website of Kyiv city and Chernivtsi region.

Of the 38 criteria (level AA) 17 did not meet the WCAG 2.0 criteria (in Fig. 7, the relevant principles, guidelines and criteria are marked in color).

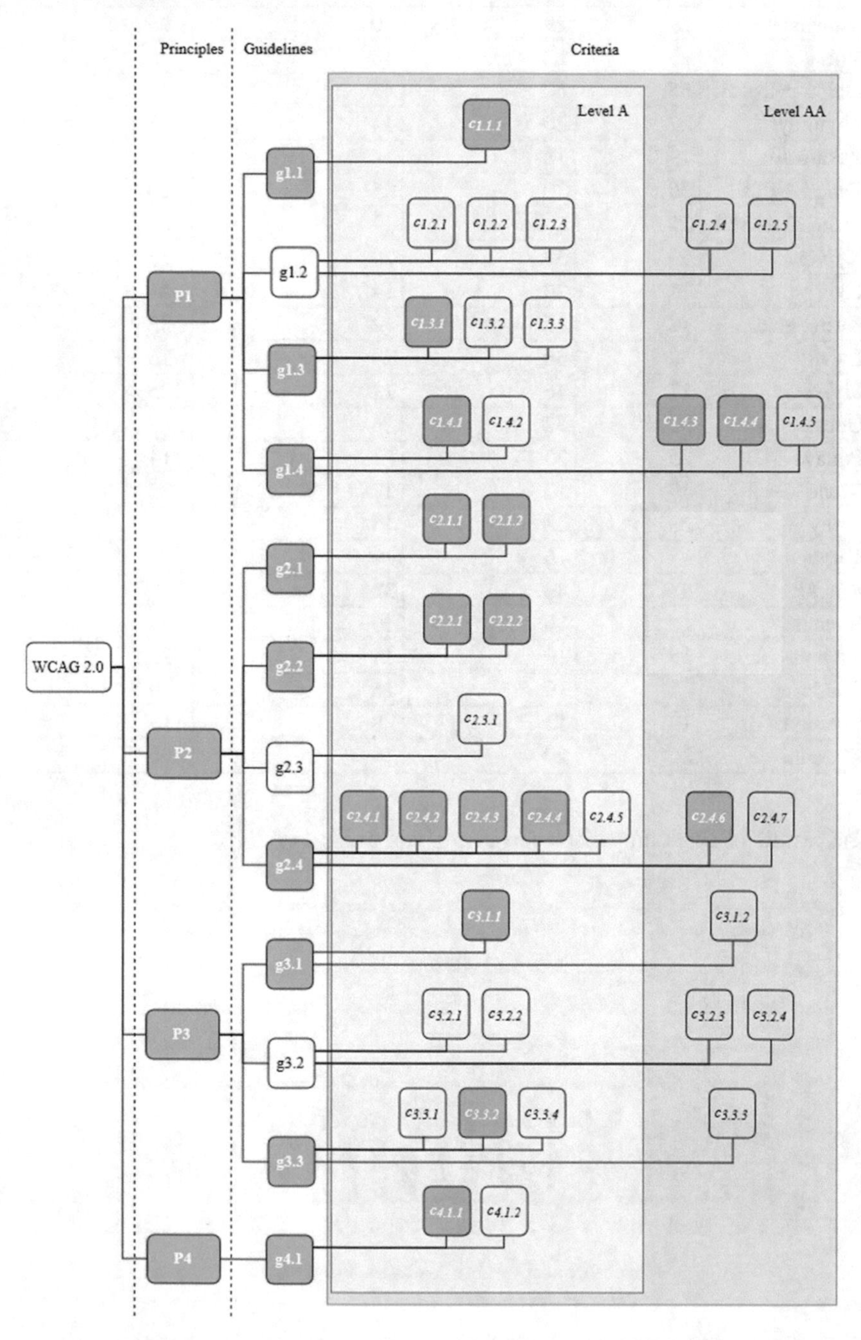

Fig. 7. Results of the comparison of the number of checked sites from the set S_2

The top-6 problems according to the success criteria for a set of web products S_2 are presented at Fig. 8.

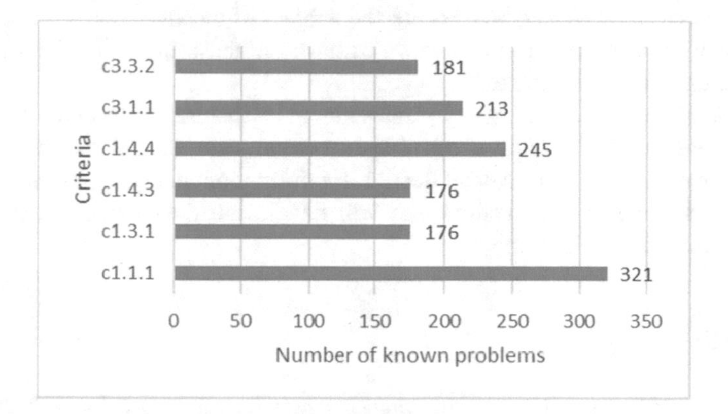

Fig. 8. The number of top-6 problems criteria for a set S2 of web products (according to www.achecker.ca).

There are four common problems for Web products of both sets: $c_{1.1.1}$, $c_{1.3.1}$, $c_{1.4.4}$, $c_{3.3.2}$. The presence of these problems evidences the disadvantages of developing sites, namely the absence of accompaniment of text representation by visual means, lack of connectivity of information on the site, mismatch font size requirements, as well as an insufficient number of tips for errors.

9 Conclusions

Improving the information and technology support of a complex system of educational and social inclusion is an urgent task. While developing new or improving existing technologies, it should be taken into account the requirements of the universal design. The most complete requirements of universal design in ICT are set in Web Content Accessibility Guideline WCAG 2.0, which underlies four principles of web accessibility: perceivability, Operability, Understandability, and Robustness. These principles are implemented in 12 guidelines, each having certain criteria. Developers of WCAG 2.0 also offered a variety of techniques for the implementation of each guideline and criterion.

Web products, available to people with special needs in Ukraine, were checked on compliance with web accessibility criteria, and most of such products had a number of problems. Such criteria should be taken into account by not only the developers of such web products but also by those who order the development of such products (educational institutions in this study). More accessible web products will be helpful in attracting additional customers – students with special needs. Even more, presenting their institution using web-accessible website might improve the reputation of the site's developer and owner.

Such conclusions are important not only for IT professionals, which can take into account these shortcomings in their future developments. It is important that teachers, lecturers, educators, and tutors, who are the subject of teaching disciplines, related to programming and information technologies have emphasized on, at least, such inadmissible programming mistakes.

The foundation of the culture of programming so that the information technologies developed are convenient and accessible to people with special needs will help to establish the principles of universal design in information technologies and education, and thus contribute to socialization and will have a positive impact on the complex system of inclusion in Ukraine.

References

1. Council recommendation promoting common values, inclusive education, and the European dimension of teaching. https://ec.europa.eu/education/sites/education/files/recommendation-common-values-inclusive-education-european-dimension-of-teaching.pdf. Accessed 21 May 2018
2. Horizon 2020 work programme 2014–2015. http://ec.europa.eu/research/participants/data/ref/h2020/wp/2014_2015/main/h2020-wp1415-societies_en.pdf. Accessed 21 May 2018
3. Pasichnyk, V., Shestakevych, T.: The application of multivariate data analysis technology to support inclusive education. In: Xth International Scientific and Technical Conference "Computer Sciences and Information Technologies" (CSIT), Lviv, Ukraine, pp. 88–90 (2015)
4. Kożuch, B., Sienkiewicz-Małyjurek, K.: Information sharing in complex systems: a case study on public safety management. In: Procedia – Social and Behavioral Sciences, vol. 213, pp. 722–727 (2015). http://www.sciencedirect.com/science/article/pii/S1877042815058486. Accessed 27 Apr 2018
5. Campean, F., Yildirim, U.: Enhanced sequence diagram for function modelling of complex systems. http://www.sciencedirect.com/science/article/pii/S2212827117300549. Accessed 27 Apr 2018
6. Rocha, L.E.C.: Dynamics of air transport networks: a review from a complex systems perspective. Chin. J. Aeronaut. **30**(2), 469–478 (2017). http://www.journal-aero.com/EN/10.1016/j.cja.2016.12.029. Accessed 11 May 2018
7. BS 7000-6:2005 Design management systems: Part 6: Managing inclusive design guide. https://www.researchgate.net/publication/240989498_BS_7000-62005_Design_management_systems_Part_6_Managing_inclusive_design_guide_Book_review. Accessed 21 June 2018
8. Universal design. http://ud.org.ua/. Accessed 21 June 2018
9. Center for universal design at North Carolina State University. https://projects.ncsu.edu/design/cud/. Accessed 17 Mar 2018
10. Universal design as a significant component for sustainable life and social development. http://www.sciencedirect.com/science/article/pii/S1877042813024762. Accessed 12 Dec 2017
11. Information technology - W3C Web content accessibility guidelines (WCAG) 2.0. https://www.iso.org/standard/58625.html. Accessed 21 June 2018

12. Davydov, M., Lozynska, O.: Information system for translation into Ukrainian sign language on mobile devices. In: 12th International Scientific and Technical Conference on Computer Sciences and Information Technologies (CSIT), Lviv, Ukraine, vol. 1, pp. 48–51 (2017)
13. Davydov, M., Lozynska, O.: Mathematical method of translation into Ukrainian sign language based on ontologies. In: Advances in Intelligent Systems and Computing II, vol. 689, pp. 89–100 (2018)
14. Gay, G., Li, C.Q.: Achecker: Open, interactive, customizable, web accessibility checking. http://doi.acm.org/10.1145/1805986.1806019. Accessed 01 May 2018
15. Ismail, A., Kuppusamy, K.S., Kumar, A., Ojha, P.K.: Connect the dots: accessibility, readability and site ranking – An investigation with reference to top ranked websites of Government of India. http://www.sciencedirect.com/science/article/pii/S1319157816301550. Accessed 27 Apr 2018
16. Youngblood, S.A., Youngblood, N.E.: Usability, content, and connections: How county-level Alabama emergency management agencies communicate with their online public. http://www.sciencedirect.com/science/article/pii/S0740624X17300266. Accessed 02 Apr 2018
17. Web content accessibility guidelines (WCAG) 2.0. https://www.w3.org/. Accessed 27 Apr 2018
18. Shakhovska, N., Vysotska, V., Chyrun, L.: Features of e-learning realization using virtual research laboratory. In: XIth International Scientific and Technical Conference Computer Sciences and Information Technologies (CSIT), Lviv, Ukraine, pp. 143–148 (2016)
19. Shestakevych, T.: The method of education format ascertaining in program system of inclusive education support. In: 12th International Scientific and Technical Conference on Computer Sciences and Information Technologies (CSIT), Lviv, Ukraine, vol. 1, pp. 279–283 (2017)

Information Analysis of Procedures for Choosing a Future Specialty

Oleksandr Matsyuk[3], Mariia Nazaruk[2(✉)], Yurii Turbal[4], Nataliia Veretennikova[1], and Ruslan Nebesnyi[3]

[1] Information Systems and Networks Department, Lviv Polytechnic National University, Lviv, Ukraine
nataver19@gmail.com

[2] Informatics and Applied Mathematics Department, Rivne State Humanitarian University, Rivne, Ukraine
marinazaruk@gmail.com

[3] Computer Sciences Department, Ternopil Ivan Puluj National Technical University, Ternopil, Ukraine
oleksandr.matsiuk@gmail.com, nebesnyi@gmail.com

[4] Department of Applied Mathematics, National University of Water and Environmental Engineering, Rivne, Ukraine
turbaly@gmail.com

Abstract. The authors analyze the decision-making process for choosing a future specialty by an entrant in a large city, which is associated with solving a number of tasks of multicriteria choice. The information analysis of the procedures for choosing a future specialty by school graduates using cognitive cards is carried out. The analysis takes into account those factors whose impact is considered to be the most significant. A model of the data analysis procedure is developed to determine the person's professional inclinations and abilities on the basis of the results of vocational guidance tests, which enabled to estimate comprehensively the professional personality traits.

Keywords: Smart city · Social and communication environment
Profession choice · Cognitive card · Data analysis

1 Introduction

The city as a constructive public social component is a producer and a main consumer of a wide range of resources of diverse nature, one of the types of which is educational information resources. A characteristic modern vision of the recent years on the concept of a city is its interpretation from the standpoint of a smart city, as a modern model of urban transformation, where using information technology the most complex problems of qualitative changes in the management system and the creation of conditions for the development of each inhabitant and the community as a whole are solved.

Modelling the processes of development of the social and communication environment of a modern city is an important tool for the formation of a holistic innovative educational system in the city, which provides a wide range of information,

N. Shakhovska and M. O. Medykovskyy (Eds.): CSIT 2018, AISC 871, pp. 364–375, 2019.
https://doi.org/10.1007/978-3-030-01069-0_26

telecommunication and technological services, which promote increasing the efficiency of processes of obtaining new and consolidating previously acquired knowledge; acquiring necessary professions for the city; improving information exchange processes; spatial approximation and social and psychological adaptation of informative and cognitive educational materials to the end user (Fig. 1).

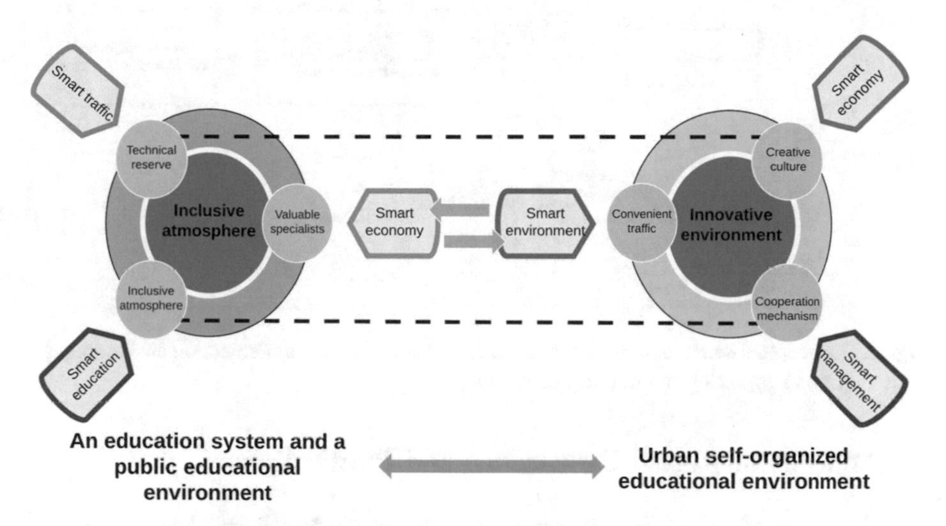

Fig. 1. Education environment of a smart city

For today, the issues of modeling the urban social and communication environment remain unresolved, such as development of information technologies for training specialists considering interests, abilities, personality traits, features of individuals and employers' needs, which would link the development of an education system in a regional dimension with the transformations that take place in the economy of cities and territorial communities (Fig. 2).

There is an imbalance between the education system, the vocational guidance work and the labor market and all this has a number of reasons:

- an uncertainty (a lack of knowledge of the profession specifics, its requirements);
- an existence of external influence (the profession choice is formed based on other opinions, "prestige" of the future specialty, etc.);
- the task difficulty (the choice of a profession is a multicriteria question);
- limited time (the entrant must choose a profession for a short period of time).

The described situation forms the purpose of this article and it is the development of models and methods of data analysis to determine the professional inclinations and abilities of a person for their further professional orientation and choice of a specialty.

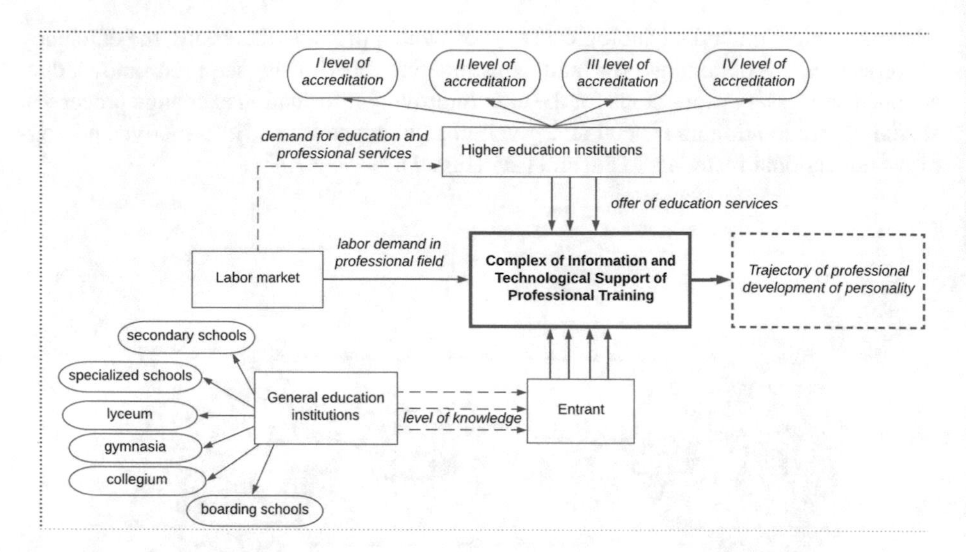

Fig. 2. Conceptual scheme of modeling educational social and communication environment in a large city and supporting the training of specialists

2 Analysis of Recent Researches and Publications

The problems of choosing a future profession, professional self-determination and becoming a specialist are described in the works by Buser et al. [1], Fouad [2], Nota et al. [3], Eesley and Wang [4], Meijers [5], van Aalderen-Smeets and Walma van der Molen [6], Mann [7], Ceschi [8], van der Gaag and van den Berg [9], Holland [10], Super [11], Fukuyama [12], Ukke [13], Dimitrakopoulos and Kostas [14], etc. The researchers reasonably argue that the right choice of a profession affects the success and productivity of professional activity in the future, the realization of personal potentials and, as a result, the person's satisfaction with their life. A few researchers are proposing to use information technologies and information systems to accompany the processes of choosing a profession by residents in smart cities. At the same time, existing profile information systems in this regard are not sufficiently effective yet. Certain technological developments are in the context of choosing a particular specialty when entering an institution of higher education. In several works it is proposed to use the modular principle during the development of a recommender system that facilitates the implementation of the process of selecting a specialty in a particular institution [15].

In modern systems, the complex possibility of analyzing data about a person as an object of vocational guidance and educational work and obtaining comprehensive analytical data on the regional labor market and opportunities for obtaining relevant educational services is practically unrealized.

The analysis shows that the availability of information and technological means of the processes of choosing a profession is usually shred and very uneven. There is practically no comprehensive information technology supporting the personalized choice of a profession, which would integrate the main stages of the choice of specialty, preparation of specialists, taking into account their personal needs and inclinations, and also considering the level of economic and social development and labor market requirements in the city, territorial community, region or country in general.

Making decisions about the future professional direction and the specialty choice is related to the tasks of multicriteria selection, such as the choice of a profession, the choice of an educational institution, the choice of a perspective place of the future employment, etc. (Fig. 3).

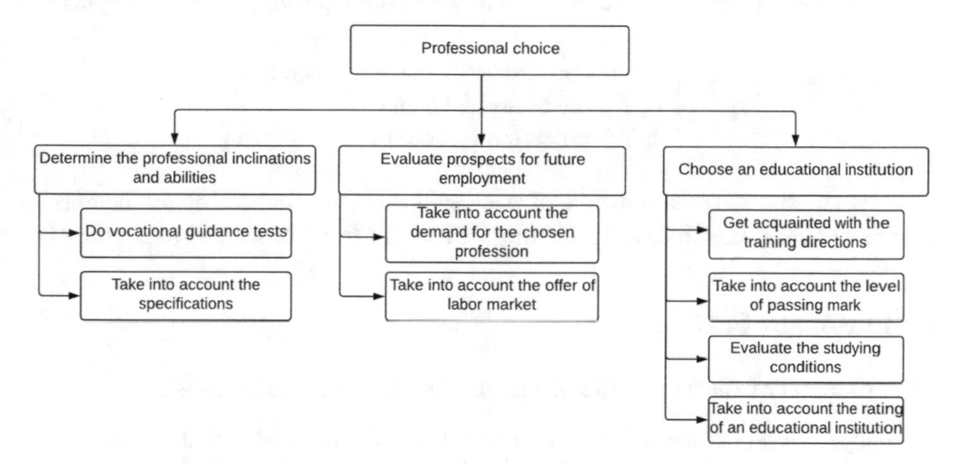

Fig. 3. Scheme of the process of choosing a future profession

The following tasks of decision-making are considered the most common [3, 4], namely:

1. Classification of the space of alternatives (tasks of breaking down alternatives to classes, for example, educational institutions are classified by ownership, accreditation level, etc.).
2. Arranging alternatives (modeling the rating problem).
3. Choice of the best alternative (choice of an educational institution, profession, place of work).

In general, analytical and non-analytical methods are used when deciding on the choice of a future profession [5, 6].

Analytical methods are:

- factor comparison, which involves comparing the profession using factor by factor, considering a monetary value scale that is directly related to the chosen specialty;
- analytical comparison based on the analysis of a number of certain factors, as a result, we obtain profiles of degrees and levels that determine the characteristics of occupations for each class of graded structure in relation to these factors;
- batch-factor schemes provide the decomposition of a profession into factors or key elements (professional inclinations, necessary competencies).

Non-analytical methods:

- the classification of occupations, which consists in comparing professions with established degrees, which are defined, for example, by occupational classifications and based on job descriptions;
- pair comparison involves comparing the professions in pairs with one another and ultimately ranking them using statistical methods;
- ranking of professions is a comparison of occupations with each other and determining their position in the hierarchy, depending on the professional abilities and preferences of a person. The professions are described using the following matrix:

$$r_{ij} = \begin{cases} -1, \text{if } i \text{ profession is less important than } j \\ 0, \text{if professions are equal} \\ 1, \text{if } i \text{ profession is more important than } j \end{cases}$$

r_{ij} is an element of the matrix of occupations, i, j are indices of the number of occupations considered by a person during the analysis from the set of N professions [7].

3 The Main Part

3.1 Analysis of the Motivational Factors for the Choice of a Specialty

We analyzed a set of factors influencing an entrant who is faced with the choice of the future specialty, which will be obtained after the school graduation. The analysis of the "Family" factor clearly illustrates the presence of its significant interaction with other factors. The entrant is consulting with relatives in which school he or she is to enter, which profession to choose and how much he can rely on the family support. At the same time, the existing cumulative family experience, which significantly influences the determination of the decision on the selection of the same profession, whose representatives are close relatives, plays an important role. Moreover, there is a situation when the future entrant has a preconceived conviction and an established view of choosing a future profession. This may particularly be intended to enter a technical university and after graduation the applicant places himself in close association with the IT industry [22–25]. Experimentally, this factor is assigned with two identical marks of 5 points.

The graph formed in such way can have loops (Fig. 4). In our case, all elements except the factor of External Independent evaluation have a long-term impact, the level of which can change over time. The specifics of the factor of External Independent evaluation are the following: firstly, the time during which it substantially affects is relatively short, and secondly, it can be measured realistically, and the estimates obtained in this case will play a prominent role. The impact of this factor is determined by experts as high and therefore it received a weight of 10 points. It is clear that the entry is influenced not only by the results of external independent evaluation, but also by the average mark of the graduate certificate, the prestige of an educational institution, the faculty, the competitive situation, the conditions of an entry and a number of other factors.

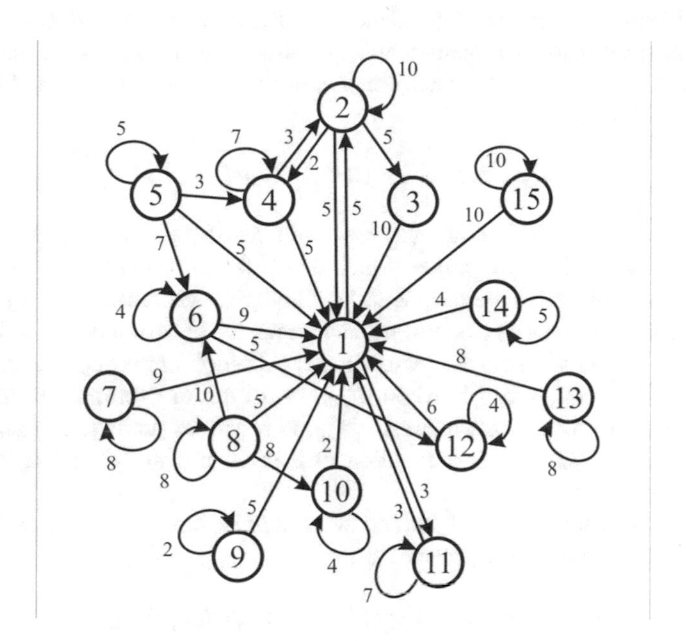

Fig. 4. The graph of the factor influence on the choice of a specialty by an entrant.

The graph apexes represent short-term and long-term goals that influence the decision-making process. Let's say the factor U15 "Employment" is considered to be important by entrants, since the general objective of selecting a specialty by an entrant is to receive a prestigious high-paying job in the future that is why the influence of this factor is estimated at 10 points, as having a significant influence on the decision-making.

The U8 factor "Place of residence" is closely linked to the factors "Access to the Internet", "Health" and has also a long-term impact, provided that the applicant lives in urban areas for a long time. Applicants living in rural areas do not usually have full Internet access, which significantly reduces the impact of this factor.

The "Family", "School" and "Internet Access" vertices have three links for each, and the "External Independent Evaluation", "Employment", "Internet Access" and "Place of Residence" vertices are the most important.

Factors: 1 – Entrant, 2 – Family, 3 – External independent evaluation, 4 – School, 5 – Abundant, 6 – Internet access, 7 – Self-education, 8 – Place of residence, 9 – Club, 10 – Health, 11 – Friends, 12 – Social networks, 13 – College, 14 – Street, 15 – Employment.

3.2 Analysis of Data on the Determination of Professional Inclinations and Abilities of a Person

In order to identify the general dependencies on the basis of which decisions are taken regarding the professional direction and the choice of a specialty, a model of the process of data analysis for the determination of the person's professional inclinations and abilities is proposed:

$$M = (A, V, R_{test}, EscC(v), T, ClasR(v), Evl(v)),$$

where: A is a number of persons (agents) who participated in vocational guidance testing; V is a set of their properties, which is divided into subsets: $V = \{V_1, V_2, V_3\}$, where V_1 is informative properties, V_2 is psychological characteristics, V_3 is personal characteristics; R_{test} is results of testing according to the methods by J. Holland, L. Yovashi, E. Klimov and A. Golomstok, divided into equivalence classes $R_{test} = \{R_{test_1}, R_{test_2}, R_{test_3}, R_{test_4}\}$, where R_{test_1} is a type of professional environment, R_{test_2} is a circle of professional interests, R_{test_3} is a type of profession, R_{test_4} is professional inclinations; $EscC(v)$ is a function that eliminates non-essential attributes by constructing redoubts.

The decision-making table T, which is created in the subprocess of forming the description of the subject area, takes the form of:

$$T = (A, \{V_1, V_2, V_3\} \cup \{R_{test_1}, R_{test_2}, R_{test_3}, R_{test_4}\}).$$

The proposed data structure for attribute sets assumes that the data is uncertain and redundant. To eliminate them, reduce the data size, and reduce the time to execute procedures from the detection of dependencies, it is introduced the EscC (v) function, which eliminates the non-essential attributes by constructing redundancies (the attributes on which professional decision-making depends). Reducers were determined using the well-known Johnson algorithm [8].

Based on the attributes included in the reducer, the function $ClasR(v)$ builds a classifier in the form of a set of classification rules that represent dependence between a set of values of the conditional attributes and attributes of the decision-making of the T-table and a decision is made as to the person's belonging to one of the 6 professional types (Fig. 5 and Table 1):

$$P\_type = (p\_type_1, p\_type_2, p\_type_3, p\_type_4, p\_type_5, p\_type_6),$$

where p_type_1 = «realistic», p_type_2 = «intellectual», p_type_3 = «social», p_type_4 = «conventional», p_type_5 = «entrepreneurial», p_type_6 = «artistic».

The quality of the rules $Evl(v)$ was evaluated by the following numerical characteristics:

(1) support is the number of study examples for which both the rule condition and its consequence are fulfilled;
(2) accuracy is the ratio of the number of study examples for which the rule is followed up to the number of study examples for which the rule condition is met;

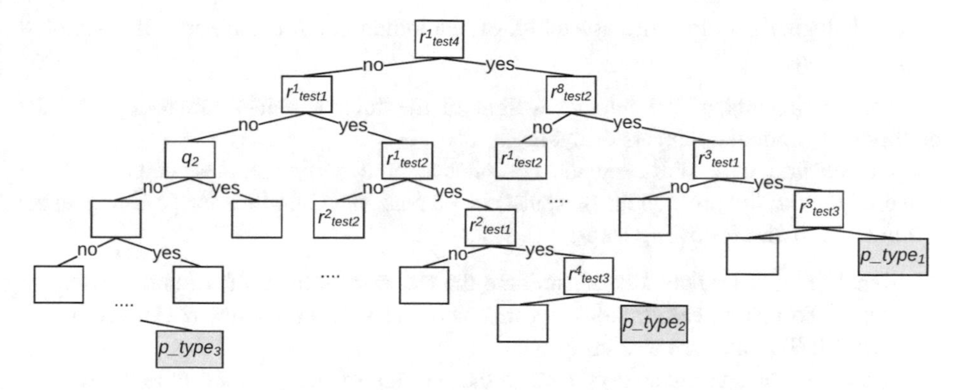

Fig. 5. A fragment of the decision tree to search a professional type

Table 1. A fragment of the rule base of the decision-making table

IF	Professional inclinations (R_{test_4}) = *work with people* ($r^1_{test_4}$))
I	professional interests (R_{test_2}) = *pedagogy and medicine* ($r^9_{test_2}$)
I	professional environment (R_{test_1}) = *social* ($r^3_{test_1}$)
I	type of a profession (R_{test_3}) = *person-person* ($r^3_{test_3}$)
THAN	professional type (P_type) = social (p_type_1)

(3) coverage is the ratio of the number of study examples, for which the rule is followed, to the number of study examples for which the result of the rule is followed.

To establish the correspondence of a certain professional type of a personality *P_type* to the professions *P_prof*, described in the National Classification of Professions (CP), the basic characteristics of professional work (professions, positions) were formed given in 9 chapters of the CP [9], especially the meaning of these characteristics, the criteria by which these values are estimated and the coefficients of the relative importance of the values of the signs.

Thus, the professions are described by the three-dimensional tuple of the values of the following features:

1. *Educational and qualification level* (according to the educational and professional programs): (qualified worker, junior specialist, bachelor, master).
2. *Activity branch*: education (libraries, kindergartens, schools, technical schools and universities); medicine (hospitals, pharmacies, dentistry); production (factories); finances (banks, insurance companies); transport (freight and route transportation); service sector (grocery stores, shopping centers, pizzerias and hairdressers).
3. *Qualification level of work* (groups of professions): legislators, senior civil servants, managers; professionals; specialists; technical staff; workers in the sphere of trade and services; skilled workers in agriculture and forestry, fish breeding and fishing; skilled workers with a tool; maintenance, operation and control staff for the work of

technological equipment, assembly of equipment and machinery; the simplest professions.

The coefficients of the relative weight of the feature values are formed by the methods of immediate expert evaluation.

The method of establishing the compliance of a particular professional type of personality with the professions presented in the National Classification of Occupations is made up of the following steps:

Step 1. To accumulate and consolidate the test results for professional guidance.
Step 2. To pre-process the data (at this step, the user's response is analyzed).
Step 2.1. To structure and unify data.
Structuring and unification of data is the process of bringing attribute values to a single structure, provided different scales for evaluating the values of certain attributes.
Step 2.2. To sample data.
Sampling is a reduction in the number of values of a continuous variable by dividing the range of values into a finite number of non-intersecting intervals, which are referred to as certain symbols, usually, by the order of numbers of these intervals.
The algorithm of the sampling process consists of the following steps: sorting examples by the value of the investigated continuous attribute, which should be discretized; interval setting; implementation of the value evaluation of the studied continuous attribute to one of the intervals, going to the next example.
Step 3. To conduct an evaluation and interpretation of the results of the previous data processing and to establish the relevance of the consolidated results of the professional testing using the set of attributes of the decision-making.
Step 4. To put a set by experts $S = \{s_{\psi p}\},\ \psi = (1, N), p = (1, q)$ of coefficients of similarity of ψ-tuple of the values of features with p-th profession, which is described in the classifier.
Step 5. To determine the set $W = \{w_{tp}\}$, $\mathrm{t} = (1, 6)$ of weighting factors of the determined, because of vocational guidance of *t-th* type of personality in relation to the *p-th* profession.
Step 6. To determine the degree of independence $Deg_{\psi t}$ of the *t-th* type of personality to the profession, which is given by ψ tuple of sign values: $Deg_{\psi g} = \sum_{p=1}^{q} s_{\psi p} \sum_{t=1}^{6} w_{tp}$.
Step 7. To add recommendations to the database on the choice of profession in accordance with the State Classification of Professions.

The block diagram of the algorithm for determining the correspondence of the profession to the National Classifier of professions is shown in Fig. 6.

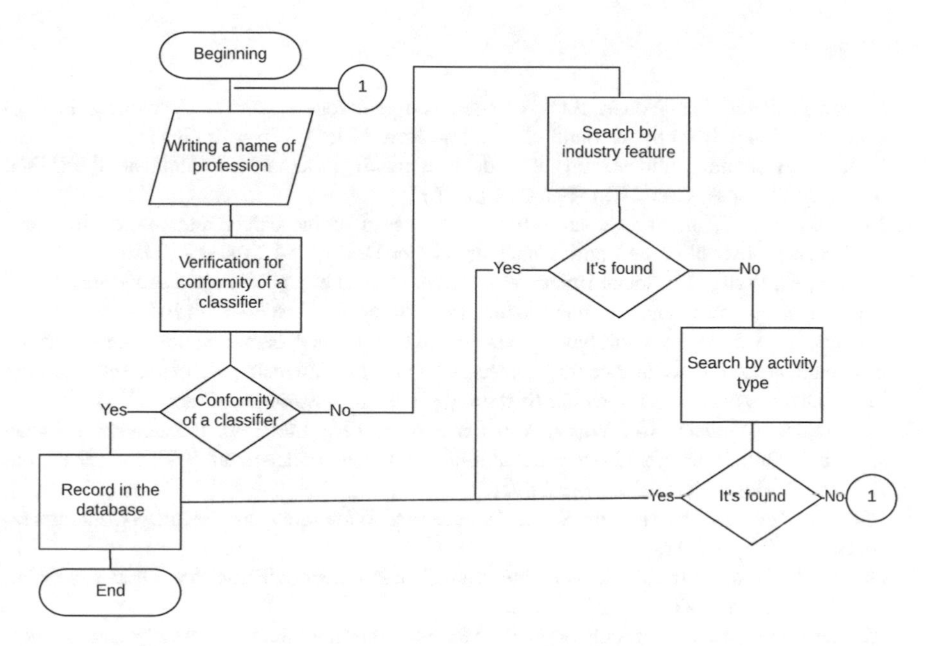

Fig. 6. Algorithm for determining the correspondence of a professional type to a classifier of professions

4 Conclusion

The main goal of the formation of a modern high-tech educational social and communication environment of a smart city is to maximize the satisfaction of educational needs of the youth in cities as a future generation of specialists in a full range of levels and forms of education, diverse educational institutions and informational and educational resources, regardless of the location of these resources, or educational services that they may need, using the most advanced information and telecommunication technologies.

In this article, the authors analyzed the problem of choosing a professional orientation and considered the main social and economic factors that influence the decision making on the choice of a profession by an entrant. A model of the process of data analysis for the determination of professional inclinations and abilities of a person was proposed based on the results of vocational guidance testing in the part of the complex assessment of an individual, which made it possible to optimize the process of identifying professional personality peculiarities and to formulate recommendations on choosing a profession in accordance with the State Classifier of professions.

References

1. Buser, T., Peter, N., Wolter, S.C.: Gender, competitiveness, and study choices in high school: evidence from Switzerland. Am. Econ. Rev. **107**(5), 125–130 (2017)
2. Fouad, N.A.: Family influence on career decision making: validation in India and the United States. J. Career Assess. **24**(1), 197–212 (2016)
3. Nota, L., Santilli, S., Soresi, S.: A life design based online career intervention for early adolescents: description and initial analysis. Career Dev. Q. **64**(1), 4–19 (2016)
4. Charles, E., Wang, Y.: Social influence in career choice: evidence from a randomized field experiment on entrepreneurial mentorship. Res. Policy **46**(3), 636–650 (2017)
5. Meijers, F.: A dialogue worth having: vocational competence, career identity and a learning environment for twenty-first century success at work. In: Enhancing Teaching and Learning in the Dutch Vocational Education System, pp. 139–155. Springer, Cham (2017)
6. van Aalderen-Smeets, S.I., Walma van der Molen, J.H.: Modeling the relation between students' implicit beliefs about their abilities and their educational STEM choices. Int. J. Technol. Des. Educ. **28**(1), 1–27 (2018)
7. Mann, I.: Hacking the Human: Social Engineering Techniques and Security Countermeasures. Routledge (2017)
8. Ceschi, A.: The career decision-making competence: a new construct for the career realm. Eur. J. Train. Dev. **41**(1), 8–27 (2017)
9. van der Gaag, M.A.E, van den Berg, P.: Modeling the individual process of career choice. In: Advances in Social Simulation, pp. 435–444. Springer, Cham (2017)
10. Holland, J.L., Gottfredson, G.D., Bacer, H.G.: Validity of vocational aspiration and interests of inventories: extended replicated and reinterpreted. J. Consult. Psychol. **37**, 337–342 (1990)
11. Super, D.E., Montross, D.H., Shinkman, S.L.: Toward a comprehensive theory of career development. Career development: theory and practice, pp. 35–64. Charles C. Thomas, Springfild IL (1992)
12. Fukuyama, S.: Theoretical foundations of vocational guidance (1989)
13. Ukke, Y.V.: Test of S. Fukuyama to determine the awareness of the choice of profession by Japanese students. Pedagogy **5**(6) (1992)
14. Dimitrakopoulos, I., Kostas, K.: Decision making using multicriteria analysis: a case study of decision modeling career in education. Case Stud. Bus. Manag. **4**(2), 24 (2017)
15. Ilyasov, B.G., Startseva, E.B., Yangurazova, N.R.: Approach to building the knowledge base of the expert system for choosing the best specialty for an applicant at admission to a university. Bull. USATU **7**(2), 102–106 (2006)
16. Ertelt, B.J.: Nowe trendy w poradnictwie zawodowym. Pedagogika pracy. Doradztwo zawodowe (2005)
17. Lotov, A.V., Pospelova, I.I.: Multi-criteria problems of decision making, Moscow (2008)
18. Royzenson, G.V.: Multicriteria choice of computing clusters. Methods Decis. Support Collect. Work. Inst. Syst. Anal. Russ. Acad. Sci. **12**, 68–94 (2005)
19. Zaritskyi, O.V.: Theoretical and methodological bases of development of intellectual information technologies of analytical estimation of professional activity. Dissertation of Doctor of Technical Sciences 05.13.06, Kyiv (2018)
20. Zavaliy, T.I., Nikolsky, Yu.V.: Analysis of data and decision-making on the basis of the theory of approximate sets. Bulletin of the National University "Lviv Polytechnic" 610: "Information systems and networks", pp. 126–136 (2008)
21. National classification of occupations DK 003 (2010). http://www.dk003.com/

22. Hasko, R., Shakhovska, N.: Tripled learning: conception and first steps. In: 14th International Conference on ICT in Education, Research and Industrial Applications, Integration, Harmonization and Knowledge Transfer, pp. 481–484 (2018)
23. Bobalo, Y., Stakhiv, P., Shakhovska, N.: Features of an eLearning software for teaching and self-studying of electrical engineering. In: 16th International Conference Computational Problems of Electrical Engineering (CPEE), Lviv, Ukraine, pp. 7–9 (2015)
24. Rzheuskyi, A., Kunanets, N., Kut, V.: Methodology of research the library information services: the case of USA university libraries. Adv. Intell. Syst. Comput. **689**, 450–460 (2017)
25. Kut, V., Kunanets, N., Pasichnik, V., Tomashevskyi, V.: The procedures for the selection of knowledge representation methods in the "virtual university" distance learning system. Adv. Intell. Syst. Comput. 713–723 (2018)

Methods and Technologies of Inductive Modeling

Opinion Mining on Small and Noisy Samples of Health-Related Texts

Liliya Akhtyamova[1](✉), Mikhail Alexandrov[2,3], John Cardiff[1], and Oleksiy Koshulko[4]

[1] Institute of Technology Tallaght, Dublin, Ireland
liliya.akhtyamova@postgrad.ittdublin.ie, john.cardiff@it-tallaght.ie
[2] Autonomous University of Barcelona, Barcelona, Spain
malexandrov.uab@gmail.com
[3] Russian Presidential Academy of National Economy and Public Administration, Moscow, Russia
[4] Glushkov Institute of Cybernetics, Kyiv, Ukraine
koshulko@gmail.com

Abstract. The topic of people's health has always attracted the attention of public and private structures, the patients themselves and, therefore, researchers. Social networks provide an immense amount of data for analysis of health-related issues; however it is not always the case that researchers have enough data to build sophisticated models. In the paper, we artificially create this limitation to test performance and stability of different popular algorithms on small samples of texts. There are two specificities in this research apart from the size of a sample: (a) here, instead of usual 5-star classification, we use combined classes reflecting a more practical view on medicines and treatments; (b) we consider both original and noisy data. The experiments were carried out using data extracted from the popular forum AskaPatient. For tuning parameters, GridSearchCV technique was used. The results show that in dealing with small and noisy data samples, GMDH Shell is superior to other methods. The work has a practical orientation.

Keywords: Classification · Health social networks · Unbalanced data
Noise immunity · GMDH

1 Introduction[1]

1.1 Motivation

Social media is a modern phenomenon that has opened new possibilities for analysis of various aspects of the human society life in total or some group of peoples [1]. The medical domain is presented in various forums, where users discuss both general topics as the state of the healthcare system or the specific questions concerning medicine,

[1] Akhtyamova, L., Alexandrov, M., Cardiff, J., Koshulko, O.: Building Classifiers with GMDH for Health Social Networks (DB AskaPatient). In: Proc. of the Intern. Workshop on Inductive Modelling (IWIM-2018), IEEE, 5 pp (2018) [To be published].

N. Shakhovska and M. O. Medykovskyy (Eds.): CSIT 2018, AISC 871, pp. 379–390, 2019.
https://doi.org/10.1007/978-3-030-01069-0_27

treatment etc. Such information is of interest to various governmental and private institutions. The former has an opportunity to evaluate the reaction of community on the laws and acts concerning healthcare as well as monitor the health condition of citizens and the latter can see a market to produce medicines [2].

On the other hand, social media has provoked new developments in the Natural Language Processing (NLP) field, namely new models, methods and program systems. The medical domain presented in social media uses traditional approaches of NLP related to (1) retrieval of given cases in data, and (2) opinion mining concerning these cases. Speaking of cases, we mean specific medicines or treatments.

First, we should mention here adverse drug reactions (ADRs) which are proved to be the reason of serious injury and death of more than 700,000 people in the USA [3]. So, most of the methods developed for the analysis of health social networks are related to these ADRs. Other topics are utilizing smoking cessation patterns on Facebook [4] as well as organizing different anti-smoking and other campaigns revealing drug abuse [5] and monitoring malpractice on Twitter [6].

1.2 Problem Setting

The motivation behind this research is the consideration of limitation and noisiness of information, in this case regarding drug use. Indeed, there are often just certain users who write about problems with their health and often provide irrelevant information, pointing out their initial condition or possible the side effects of drugs, rather than their own experience. By adding noise to their reports, we reflect this issue.

In this paper we consider possibilities of GMDH-based algorithms to build useful noise-immunity classifiers for processing texts from health social networks. It is our contribution to the problem of analysis of health social media. By the term "useful classifiers", we mean classifiers which allow detecting negative or extreme cases in social media. As we mentioned above, the traditional 5-star classification includes classes = {very negative, negative, satisfactory, positive, very positive}. We denote them as {1*, 2*, 3*, 4*, 5*} respectively. Our 2-class scale includes the negative class = (1*, 2*) and the class 'others' = (3*, 4*, 5*). The 3-class scale includes the very negative class = (1*), the satisfactory class = (2*, 3*, 4*), and the very positive class = (5*). These classifications were introduced in [7]. Noise in data reduces the discriminatory between classes, therefore decreasing model accuracy. At the same time, GMDH simplifies a model to make it more stable [8]. When we speak about noise-immunity algorithms we mean here algorithms whose results are worsened less than noise grows. This worsening and growth are considered in relative units.

We intend to study various ways of text parameterization that is a transformation of the dataset to its vector form by putting attention on using not only one-word terms but also n-grams of terms and n-grams of characters.

It should be mentioned here about convolutional neural networks (CNN) having successful applications in opinion mining health social networks, see e.g. [9, 10]. However, they dealt with the traditional 5-star ratings rather than the combined classes used in this paper. They did not consider the stability of results with respect to data, and we cannot directly compare our results with those related to CNNs.

The content of this paper is as follows: Sect. 1 is the introduction, Sect. 2 describes dataset AskaPatient. In Sect. 3 we give a short description of GMDH and GMDH Shell. Section 4 presents the results of experiments with the original data. Section 5 shows the results of experiments with noisy and shortened data. We discuss our research in Sects. 6 and 7 concludes the paper.

2 Dataset

2.1 General Description of AskaPatient

The dataset AskaPatient consists of 8 fields, which are rating, the reason for taking the medication, side effects, comments, gender, age, duration and date added (AskaPatient, n.d.). As the comments usually reflect patient opinions about the drug, we left only this field for rating prediction purposes. The dataset we retrieved consists of 48,088 comments, among them there are 32,437 comments without duplicates (67%). The 5-star rating distribution among comments is presented in Table 1.

Table 1. Rating distribution.

1*	2*	3*	4*	5*
12823 (27%)	5713 (12%)	8202 (17%)	9152 (19%)	12093 (25%)

With the new class distribution, the class imbalance essentially increased, as can be seen in Table 2.

Table 2. Distribution of documents on combined classes.

Contents	Class 1	Class 2	Class 3
2 classes	18536 (39%)	29449 (61%)	
3 classes	12823 (27%)	23069 (48%)	12093 (25%)

The distribution of the lengths of reviews is presented in Table 3. It could be observed that there are only 10% of reviews with a word count exceeding 200 words.

Table 3. Descriptive statistics on term count in each review.

Min number	Max number	Aver. number	90% percentile
1	823	54.4	200

For calculation simplicity, we have chosen 1,000 texts for both classification tasks, preserving the class distribution among texts. This leads to a small loss in accuracy of models, however, we were able to carry out more experiments trying different modes of the GMDH Shell platform.

2.2 Parameterization and Normalization for ML and GMDH-Based Methods

We have chosen bag-of-words (BoW) as our primary parametrization technique due to its simplicity. By choosing the best parameters for preprocessing as well as tuning models we used the GridSearchCV technique. This technique allows us to conduct an exhaustive search over specified parameter values for a classifier.

The vocabulary size varied between 100 and an unlimited number of terms. Here 'term' means word or character n-grams, where n varied between interval of [1–6].

We filtered terms which were encountered in more than 50%–75% texts. Such a limit corresponded to the first transition point with respect to the number of terms, providing discriminative power while preserving information value of obtained vectors. These vectors then were normalized to the interval [0, 1] using L2-norm. Table 4 presents all parameters that were used for tuning the BoW model.

Table 4. Word representation tuning in a grid search.

Character/word parametrization
n-gram range
tokenization
tf-idf rate
size of dictionary

While dealing with ML methods from the scikit-learn library [11], we also tried to add a range of model-specific parameters but it did not give noticeable results. We conclude that the correct preprocessing of text data itself is more important than model specification tuning.

However, this is not the same for GMDH-based algorithms where with further model tuning in GMDH Shell platform it was possible to get the significant model improvements.

Overall, character n-grams are always superior to word ones and character n-gram in the range from 1 to 7 gives the best results. The term 'range' means here that in the process of parameterization, n-grams of different sizes are used simultaneously. Maximum dictionary size is the best option for methods in scikit-learn. Due to the computer limitations, the vocabulary size of 150 is the best option in GMDH Shell. Therefore, in all the experiments described in Sects. 4 and 5, we deal with documents presented with maximum dictionary size for ML methods in scikit-learn and in the space of dimensionality 150 for GMDH-based methods. Throughout, for ML methods character n-grams are used.

In our experiments, we used 5-fold cross-validation and weighted F-score averaged among all folds to correctly measure the model quality with unbalanced data.

2.3 Noisy Data

To form noisy data, we added an independent Gaussian noise to parameterized and normalized data with the mean = 0 and the standard deviation $s = 0.1; 0.2$.

As for the neural networks in comparison to feature-based methods they do not presume to have normalized word vectors. Indeed, as it was stated in [12] that words that are used in a similar context have longer vectors than words that are used in different contexts. Thus, the usage of raw vectors makes a model more accurate increasing its performance. Moreover, we fed to the neural networks word an embedding matrix rather than bag-of-words vectors. This is due to the fact that word2vec format embeddings give in all cases better results for neural networks than one-dimension parameterization. That is why we do not perform noise immunity analysis with neural network algorithms.

3 Methods and Tools

3.1 GMDH-Based Classifiers with Applications

Group Method of Data Handling (GMDH) is a technology of machine learning (ML) for creating noise immunity models. The ideas and applications of GMDH are presented in many publications, see for example [13–15]. Theoretical bases of GMDH are described in the well-known paper [8]. GMDH does not orient on certain class functions, but the most popular GMDH-based tools use polynomial functions of many variables [16, 17]. This fact has the simple explanation: any continuous function of many variables on hypercube can be presented in the form of uniformly-convergent polynomial series.

GMDH itself has many applications in NLP. For example, the paper [18] demonstrates the GMDH based technique for building empirical formulae to evaluate politeness, satisfaction, and competence reflecting in dialogs between passengers and Directory Enquires at a railway station in Barcelona. The formulae contain the sets of linguistic indicators preliminary assigned by experts separately for each mentioned problem (politeness, satisfaction, and competence).

The paper [19] shows the possibility of building a classifier of primary medical records using GMDH Shell. The linguistic indicators are extracted from the training dataset related to six stomach diseases. The accuracy of results on a real corpus of medical documents proved to be close to 100%. Such a result essentially exceeded the results of other methods which had been used on the same dataset.

In another paper [20], the authors present opinion classifiers for Peruvian Facebook, where users discuss the quality of various products and services. These classifiers use linguistic indicators prepared by qualified experts. The indicators form two variables reflecting the contribution of positive and negative units and then GMDH-based algorithms build polynomial models with these two variables. The total accuracy reached in the experiments significantly improved the results obtained by other researchers.

In this paper, GMDH algorithms are implemented on the platform called GMDH Shell. All algorithms related to classification realizes the One-Vs-All approach [21] which reduces multi-class classification to the binary one. The variety of preprocessing options for this instrument could be learned from [15].

At the moment, we do not know any publications in which GMDH has been used for opinion analysis of health social networks. For this reason, it would be useful to study the possibilities of GMDH-based algorithms to classify any typical network. It would be also interesting to test the stability of results having in view the well-known property of noise immunity of GMDH-based algorithms. This paper continues our applied research presented in [22].

3.2 Standard ML Classifiers

In our experiments, we have tested several ML techniques: Random Forest, Logistic Regression, Extremely Randomized Trees, Support Vector Machine classifiers from Python scikit-learn package [11].

These tools have a long history with successful application in many research fields, e.g. Sentiment Analysis tasks. For pharmacovigilance, it was applied for example to the tasks of ADR detection [23] and monitoring prescription medical abuse on Twitter [5]. Usually, these algorithms are enriched with huge set of additional features to get better results.

3.3 Neural Networks

For comparison purposes, we included here the results of deep learning methods, however, the advantage of them is more pronounced while dealing with large data.

In this work, we construct a LSTM-CNN model for dealing with user posts. It was shown that such combined methods often achieve better results in a variety of text classification tasks [24, 25]. The intuition behind this type of networks is that output tokens from the LSTM layer store an information about not only the current token but also any previous tokens. This output of the LSTM layer is then fed to a convolutional layer which is now get enhanced information, thus making better predictions.

For preprocessing health, word embeddings were used [26]. It turned out results to be better on data with any modifications (normalization, stemming).

4 Experiments on Original Data

4.1 Experiments in GMDH Shell

Here, by original data we mean noise-free data reflected in 1,000 documents. Overall, the investigated parameters of GMDH Shell are presented in Table 5. Here *lin* denotes linear members and *sq/div.* denotes squares/divisions. The latter means the model includes linear, pairwise and square members. Complexity or rank of model means the number of features to consider which keeps some number of the most important variables according to the selected ranking algorithm. This number dramatically increases the running time of an algorithm if pairwise and square members were included. The number of final parameters could be reduced by selecting a model complexity value.

Table 5. Options for GMDH shell tuning.

Balance	Ensemble	Form	Complexity	Rank
yes/no	yes/no	lin/sq/div.	20–200	20–300

GMDH Shell is presented in four algorithms: the combinatorial, neural network type, forward and mixed selections. The first two ones are the classical GMDH-based algorithms [14, 15]. The last two ones are the well-known algorithms of stepwise regression [27] where GMDH is used for generation of variants.

The preliminary experiments showed the following results which we considered while testing different methods of classifications:

- balancing impairs results quality;
- data transformation to different forms lead to model accuracy increase;
- ensembling, in general, leads to slightly better results;
- the model complexity, i.e. number of coefficients in a model of about size of vocabulary is always the best adjustment.
- ranking boosts model accuracy.

On the original, noise-free data mixed selection algorithm showed the best results and was chosen for the further analysis on the noisy and reduced data (Table 6). It can be observed from this table that the mixed selection algorithm exceeded the baseline by 29% for 2 classes and 26% for 3 classes (32% and 35% in relative units). The baseline here is equal to the proportion of the biggest class in a classification problem, Table 2.

Table 6. F-score for different algorithms from GMDH Shell platform, original data.

Methods	2 classes	3 classes
Combi	0.66	0.61
Forward	0.82	0.47
Mixed	0.90	0.74
NN	0.61	0.47

4.2 Building Classifiers with Other Methods

The results with the best parameters are presented in Table 7. Here the SVM algorithm is superior to other methods, which can be explained by the fact that it is a less sophisticated algorithm, thereby less prone to overfitting. In the case of small data size that quality is essential. SVM exceeded the baseline by 15% for 2 classes and by 8% for 3 classes (20% and 14% in relative units). Other methods gave worse results.

Table 7. F-score for different algorithms from the scikit-learn package and neural network algorithm, original data.

Methods	2 classes	3 classes
Random forest	0.63	0.41
Extra trees	0.62	0.43
SVM	0.76	0.56
Logistic regression	0.62	0.42
RCNN	0.66	0.54

For 2 classes all other ML methods gave slightly higher than baseline results. On the 3-class problem other methods did not exceed the baseline; RCNN exceeded it by 6% (12% in relative units). However, as stated before, significant advantages of RCNN can be shown only when the sample size is tens and hundreds of thousands of documents.

5 Experiments with Noisy and Reduced Data

5.1 Building Classifiers for Noisy Data

In this sub-section, we test the noise-immunity of the best algorithm from GMDH Shell and methods from scikit-learn library. The results of the analysis in terms of rates to original means are presented in Table 8 for 2 and 3 classes accordingly.

Table 8. F-score rate for noisy data and different level of noise.

Methods	2 classes			3 classes		
	s = 0	s = 0.1	s = 0.2	s = 0	s = 0.1	s = 0.2
(GMDH) mixed	1.00	0.82	0.81	1.00	0.91	0.90
SVM	1.00	0.74	0.71	1.00	0.63	0.64
Tree-based	1.00	0.84	0.83	1.00	0.91	0.90

Tree-based and GMDH-based mixed selection methods are more stable to the noise. SVM algorithm is less prone to the noise increase stability, although outperforming tree-based methods in terms of weighted F-score.

5.2 Building Classifiers with Reduced Data

In this section, we test model performance on very small samples of data: 500 and 250 samples. This allows us to check the stability of models on the extremely small text samples. The results of the experiments for two and three classes are presented in Tables 9 and 10 accordingly.

Table 9. F-score for different ML algorithms, reduced data (500 samples).

Methods	2 classes	3 classes
(GMDH) mixed	0.92	0.79
Random forest	0.63	0.42
Extra trees	0.63	0.47
SVM	0.67	0.52
Logistic regression	0.55	0.38

Table 10. F-score for different ML algorithms, reduced data (250 samples).

Methods	2 classes	3 classes
(GMDH) mixed	0.98	0.96
Random forest	0.53	0.42
Extra trees	0.57	0.44
SVM	0.63	0.47
Logistic regression	0.46	0.34

It is noticeable that GMDH-based mixed selection algorithm is more efficient when dealing with very small data samples. Amusingly, the results for GMDH turned out to be even better with sample size reduction. The reason for this lies in the flexibility of GMDH-based algorithms which are well adjusted to the variability in the data. It is not the same for scikit-learn methods.

6 Discussion

In our paper [22], we began to study the possibilities of GMDH-based algorithms on opinion mining of typical texts related to health social networks. In the paper we conducted our experiments on the same dataset AskaPatient used in this paper. Our interest in GMDH as a technology of text mining was provoked by the following circumstances: GMDH can successfully deal with small amount of experimental data; moreover, it works well even when the dataset size is less than the number of parameters used; GMDH builds models of optimal complexity that provide their high noise immunity. With these circumstances, our study of GMDH-based algorithms was quite limited: we did not consider the sensibility of models to size of experimental data and we did not consider the noise-immunity of models built.

GMDH-based classifiers are not the only ones that can be used for opinion mining. Last year the great popularity came to program language Python and tools based on it. This fact provokes comparison of classifiers built on GMDH technology [13–17] and classifiers included in the well-known Python library scikit-learn [11].

In the current research, we tried to explore all mentioned problems by putting special attention to parameter tuning, in particular, experiments with different type of parametrization: character or word n-grams. In the paper [22] we used only one-word terms. To select these terms, we used the criterion of term specificity which considers term frequency in a given document corpus and any basic corpus [28, 29].

In that research, we used word frequency list related to British National Corpus as this basic corpus. In the current research, we carefully studied different combinations of n-grams of terms and n-grams of characters to select the best parameters. Such a process is described in the Sect. 2.2. Table 11 shows results of classification for n-grams of terms and n-grams of characters. We studied also results of classification related to different number of posts where a given term occurs. The results are presented in Table 12.

Table 11. Study of different sizes of vocabularies.

Options	Sizes	Results
n-grams of characters	50, 150, 250, 400	150–250 give the best and close results
n-grams of terms	150, 250, 400, 800	150–250 give the best and close results

Table 12. Study of different number of posts.

Option	Number of posts	Results
Posts with a given term	>25%, >50%, >75%	75% gives the best results

The results presented above defined options which we used in this paper.

7 Conclusions

In the paper, we investigated the noise-immunity and data size sensitivity of different algorithms on health-related texts. It was stated that user reports on drugs are good examples of very noisy data where it is often that the information is quite limited on some drugs and especially their side effects. Thus, while dealing with imbalance it is needed to deal with small samples of text and noise in data. For these purposes, we built different machine learning classifiers including standard machine learning classifiers as well as GMDH-based algorithms and neural networks.

We tested different preprocessing options and found out that character n-grams with absent lemmatization and stemming work the best in all cases. Overall, GMDH-based mixed selection algorithm performs better on small and extremely small text samples. Moreover, it is more stable to adding a noise in comparison to the standard ML methods. This might be explained by the fact of more simplicity and flexibility of the GMDH-based algorithms in comparison to tree-based and SVM algorithms. The results have clear practical implications and can be used in further research.

References

1. Kaplan, A.M., Haenlein, M.: Users of the world, unite! The challenges and opportunities of Social Media (2007). https://doi.org/10.1016/j.bushor.2009.09.003
2. Ventola, C.L.: Social media and health care professionals: benefits, risks, and best practices. P T **39**, 491–520 (2014)
3. Lehne, R.A., Rosenthal, L.D.: Pharmacology for Nursing Care. Elsevier Health Sciences (2013)
4. Struik, L.L., Baskerville, N.B.: The role of Facebook in crush the crave, a mobile- and social media-based smoking cessation intervention: qualitative framework analysis of posts. J. Med Int. Res. **16**(7), e170 (2014). https://doi.org/10.2196/jmir.3189
5. Sarker, A., O'Connor, K., Ginn, R., Scotch, M., Smith, K., Malone, D., Gonzalez, G.: Social media mining for toxicovigilance: automatic monitoring of prescription medication abuse from Twitter. Drug Saf. **39**, 231–240 (2016)
6. Nakhasi, A., Passarella, R.J., Bell, S.J., Paul, M.J., Dredze, M., Pronovost P.J.: Malpractice and Malcontent: analyzing medical complaints in Twitter. In: AAAI Technical Report FS-12-05, Information Retrieval and Knowledge Discovery in Biomedical Text, pp. 84–85 (2012)
7. Alexandrov, M., Skitalinskaya, G., Cardiff, J., Koshulko, O., Shushkevich, E.: Classifiers for Yelp-reviews based on GMDH-algorithms. In: Proceedings of the Conference in Intelligent Text Processing and Comput. Linguistics (CICLing-2018). LNCS, pp. 1–18. Springer (2018)
8. Stepashko, V.S.: Method of critical variances as analytical tool of theory of inductive modeling. J. Autom. Inf. Sci. **40**, 4–22 (2008). https://doi.org/10.1615/J.AutomatInfScien.v40.i3.20
9. Huynh, T., He, Y., Willis, A., Uger, S.: Adverse drug reaction classification with deep neural networks. In: Proceedings of 26-th International Conference on Computational Linguistics (COLING-2016), pp. 877–887 (2016)
10. Akhtyamova, L., Ignatov, A., Cardiff, J.: A Large-scale CNN ensemble for medication safety analysis. In: Proceedings of 22th International Conference on Applications of Natural Language to Information Systems (NLDB 2017). LNCS, pp. 1–6. Springer (2017)
11. Pedregosa, F., Varoquaux, G., Gramfort, A., et al.: Scikit-learn: machine learning in python. J. Mach. Learn. Res. **12**, 2825–2830 (2011)
12. Schakel, A.M.J., Wilson, B.J.: Measuring word significance using distributed representations of words, CoRR, abs/1508.02297 (2015)
13. Madala, H.R., Ivakhnenko, A.G.: Inductive Learning Algorithms for Complex Systems Modelling. CRC Press, New York (1994)
14. Farlow, S.J.: Self-Organizing methods in modeling: GMDH type algorithms. In: Statistics: A Series of Textbooks and Monographs, Book 54, 1-st edn. Marcel Decker Inc., New York, Basel (1984)
15. Stepashko, V.: Developments and prospects of GMDH-based inductive modeling. In: Shakhovska, N., Stepashko, V. (eds.) Advances in Intelligent Systems and Computing II / AISC book series, vol. 689, pp. 346–360. Springer, Cham (2017)
16. Platform GMDH Shell. www.gmdhshell.com
17. Resource GMDH in IRTC ITS NAS of Ukraine. mgua.irtc.org.ua/
18. Alexandrov, M., Blanco, X., Catena, A., Ponomareva, N.: Inductive modeling in subjectivity/sentiment analysis (case study: dialog processing). In: Proceedings of 3-rd International Workshop on Inductive Modeling (IWIM-2009), pp. 40–43 (2009)

19. Kaurova, O., Alexandrov, M., Koshulko, O.: Classifiers of medical records presented in free text form (GMDH shell application). In: Proceedings of 4-th International Conference on Inductive Modeling (ICIM-2013), pp. 273–278 (2013)
20. Alexandrov, M., Danilova, V., Koshulko, A., Tejada, J.: Models for opinion classification of blogs taken from Peruvian Facebook. In: Proceedings of 4-th International Conference on Inductive Modeling, pp. 241–246 (2013)
21. Tax, D.M.J., Duin, R.P.W.: Using two-class classifiers for multiclass classification. In: Proceedings of 16-th International Conference on Pattern Recognition, pp. 1051–1054. IEEE (2002)
22. Akhtyamova, L., Alexandrov, M., Cardiff, J., Koshulko, O.: Building classifiers with GMDH for health social networks (DB AskaPatient). In: Proceedings of the International Workshop on Inductive Modelling (IWIM-2018). IEEE (2018). [to be published]
23. Sarker, A., Gonzalez, G.: Portable automatic text classification for adverse drug reaction detection via multi-corpus training. J. Biomed. Inform. **53**, 196–207 (2015). https://doi.org/10.1016/j.jbi.2014.11.002
24. Lai, S., Xu, L., Liu, K., Zhao, J.: Recurrent convolutional neural networks for text classification. In: Proceedings of 16th International Conference on Artificial Intelligence, pp. 2266–2273 (2015)
25. Stojanovski, D., Strezoski, G., Madjarov, G., Dimitrovski, I.: Finki at SemEval-2016 Task 4: deep learning architecture for Twitter sentiment analysis. In: Proceedings of SemEval-2016, pp. 149–154 (2016)
26. Miftahutdinov, Z., Tutubalina, E., Tropsha, A.: Identifying disease-related expressions in reviews using conditional random fields. In: Proceedings of International Conference on Computational Linguistics and Intellectual Technologies (Dialog-2017), pp. 155–166 (2017)
27. Draper, N., Smith, H.: Applied Regression Analysis. Wiley, New York (1981)
28. Gelbukh, A., Sidorov, G., Lavin-Villa E., Chanova-Hernandez, L.: Automatic term extraction using Log-likelihood based comparison with General Reference Corpus. In: Proceedings of 15-th International Conference on Applications of Natural Language to Information Systems (NLDB-2010). LNCS, vol. 6177, pp. 248–255. Springer (2010)
29. Lopez, R., Alexandrov, M., Barreda, D., Tejada, J.: LexisTerm – the program for term selection by the criterion of specificity. In: Artificial Intelligence Application to Business and Engineering Domain, vol. 24, pp. 8–15. ITHEA Publ., Rzeszov-Sofia (2011)

Formation and Identification of a Model for Recurrent Laryngeal Nerve Localization During the Surgery on Neck Organs

Mykola Dyvak(✉) and Natalia Porplytsya

Department of Computer Science, Ternopil National Economic University, Ternopil, Ukraine
mdy@tneu.edu.ua, ocheretnyuk.n@gmail.com

Abstract. In the article, a structural identification method for models of objects with distributed parameters is considered. The method is based on the artificial bee colony behavioral model as well as the interval data analysis. The artificial model imitates the foraging behavior of a honey bee colony. The proposed method of structural identification makes it possible to build models of objects with distributed parameters in the form of interval discrete difference scheme. This method is applied when solving the problem of recurrent laryngeal nerve (RLN) monitoring during the surgery on neck organs. The principles of building RLN localization systems based on the electrophysiological method of surgical wound tissue stimulation are considered. Based on the results of previous researches, an actual task of the model building of the main spectral component amplitudes (signal of reaction on surgical wound tissues stimulation) spatial distribution on the surface of surgical wound is solved. Using the method of structural identification and based on the results of electrophysiological researches of surgical wound tissues during the surgery, such a model for RLN localization is built. The model with the appropriate adjustments for each patient makes it possible to identify the RLN location and to reduce the risk of its damage during the neck organs surgery.

Keywords: Neck surgery · Recurrent laryngeal nerve
Structural identification of models · Interval discrete model
Interval data analysis · Discrete difference scheme

1 Introduction

Recurrent laryngeal nerve (RLN) monitoring during the neck organs surgery is a very important procedure [1, 2]. Special neuro monitors are used for this purpose. Their working principles consist in stimulation of surgical wound tissue and estimation of results of such stimulation [2–8]. However, these methods are intended only for the monitoring of RLN.

In [9, 10], the methods of RLN localization are described. In particular, the problem of visualization of RLN location based on the estimation of amplitude of reacting signal to its stimulation by alternating current is considered in [9]. However, this method does not provide high sensitivity and the accuracy of the model is low. As a result, the risk

N. Shakhovska and M. O. Medykovskyy (Eds.): CSIT 2018, AISC 871, pp. 391–404, 2019.
https://doi.org/10.1007/978-3-030-01069-0_28

of RLN damage is high. In [10], the method of building a difference scheme as a model for RLN localization based on the results of the interval analysis of the reaction to stimulation of surgical wound tissues by alternating current is considered. However, the method requires forming a uniform grid for stimulation of surgical wound tissues. It is difficult to adjust such grid to a particular patient. It should be noted that the informative parameter in both methods is the maximal amplitude of the signal of reaction to the surgical wound tissues stimulation. At the same time, in [11], the amplitude of main spectral component (the highest amplitude) is chosen as the informative parameter of the reacting signal to the stimulation of the surgical wound tissues. This method is characterized by higher sensitivity. Thus, the building of model of the main spectral component amplitudes (signal of reaction on surgical wound tissues stimulation) spatial distribution on the surface of surgical wound is an actual task. Its solving will ensure the visualization of RLN location and reduce the risk of its damage during the neck organs surgery. It is advisable to present such a model in the form of a difference scheme in order to simplify its setting for a particular patient. The structure of the difference scheme is unknown and the data for building of this model are represented with errors in the interval form.

For the synthesis of the difference scheme, we develop in this paper a method of structural identification based on the artificial bee colony behavioral model and the interval data analysis [12].

2 Statement of the Problem

The stimulation of surgical wound tissues during the neck organs surgery using electrophysiological method gives the possibility to identify the tissue type with the purpose of RLN localization. The requirements for building of the system for RLN localization are based on neurochronaxic theory of voice production introduced by Raul Husson, the French scientist, in 1952 [13].

Muscles and other tissues in the surgical wound have low sensitivity to the alternating current with strength ranging from 0.5 to 2 mA [11]. The second, technical aspect of the method consists in substantiating and developing a technology of obtaining the information about change of the vocal cords position during the electrical stimulation of larynx nerves by the current with given parameters. It can be executed based on the analysis of the neurochronaxic theory of voice production [13].

The theory establishes the "central genesis" (brain nature) of vocal cords vibration. Its essence consists in the following: vocal cords vibrate not passively under the influence of exhaled air stream but move actively because of the impulses of biocurrents that are transmitted from the central nervous system to the relevant larynx muscles [11].

Husson has established that the frequency of vocal cords vibration and the frequency of impulses received by the nerve from the center are the same. Confirmation of this theory was obtained during experiments with the electrical stimulation of the lower recurrent nerve. It is stroboscopically proved that the series of electric impulses to nerve with the frequency of 100 and 600 per second caused the vocal cords vibration with the same frequency [13].

So, the vocal cords vibrate under the influence of larynx muscles that are contracted because of the rhythmic impulses transmitted from the brain, with the sound frequency.

Based on the conducted analysis, the main requirements to the method of localization among tissues in surgical wound were established. The scheme of this method is represented in Fig. 1.

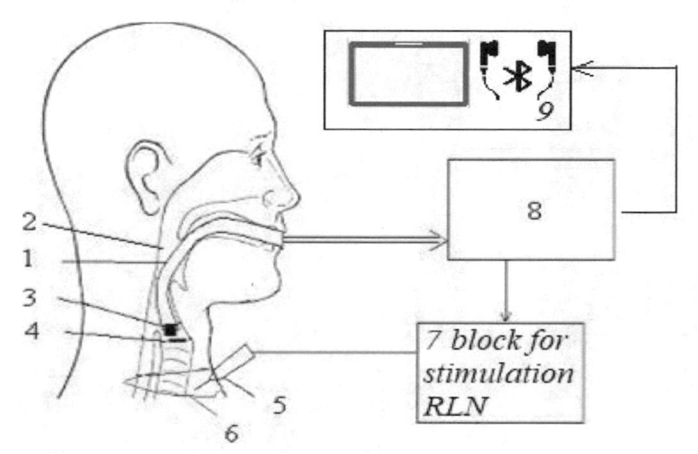

1 is respiratory tube, 2 is larynx, 3 is sound sensor, 4 are vocal cords, 5 is probe, 6 is surgical wound, 7 is block for RLN stimulation, 8 is single-board computer, 9 is output part

Fig. 1. Method of RLN localization among tissues in surgical wound.

In respiratory tube 1 inserted into larynx 2, the sound sensor 3 is implemented and positioned above vocal cords 4.

Probe 5 is connected to the stimulation block functioning as a current generator controlled by the single-board computer 8. Surgical wound tissues are stimulated by the block 7 via probe. As a result, vocal cords 4 are stretched.

Air flow passing through patient's larynx is modulated by the stretched vocal cords. The result is registered by the voice sensor 3 and the obtained signal amplified by the amplifier 8 is processed by the single-board computer.

To process the obtained signal, special software was installed on a single-board computer. The main functions of the software are:

- segmentation of information signal based on the analysis of its amplitude;
- analysis of the amplitude spectrum using the Fourier-transformation;
- calculation of the main spectral component (with maximum amplitude);
- classification of tissues of surgical environment at the points of stimulation using threshold method; and
- calculation distance between stimulation points and RLN [14, 15].

The last function of software was realized based on the previously conducted researches by the authors team. In Fig. 2, the fragments of amplified information signal obtained from the sound sensor and fragments of their spectral characteristics are shown.

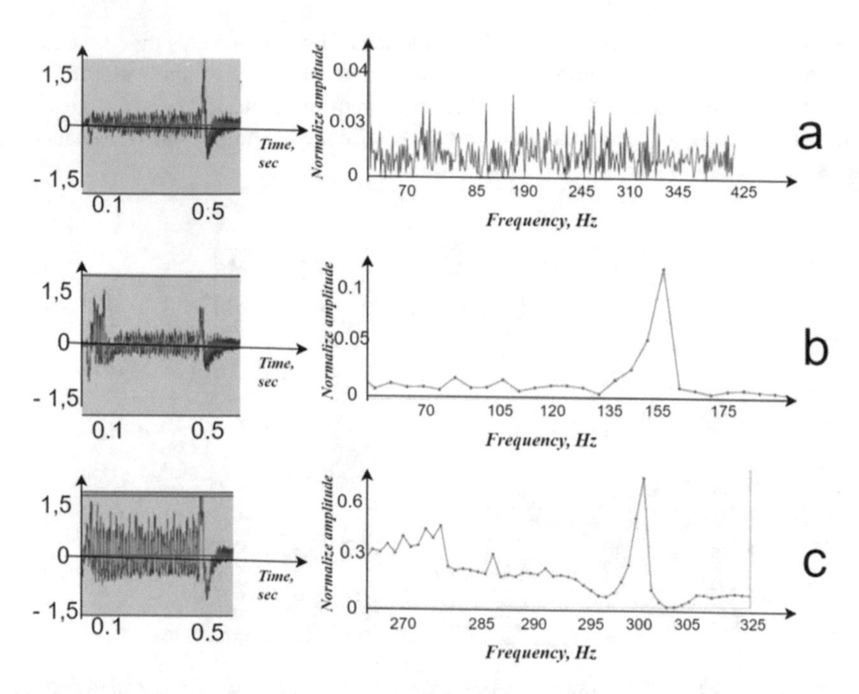

Fig. 2. Result of stimulation of RLN by alternating current with frequency 300 Hz.

We see in Fig. 2a the result of stimulation of the muscle tissue at a distance of more than 1 cm to RLN with a specific blurred spectrum, without a clearly distinguished main spectral component.

Figure 2(b) reflects the result of stimulation of the muscle tissue at a distance of no more than 3 mm, with a specific distinguished main spectral component with a small amplitude value. Finally, the result of RLN stimulation with a specific main spectral component with sufficiently high normalized amplitude (6 times higher than in the previous case) is illustrated in Fig. 2(c).

These results allow to affirm that this characteristic can be used for the RLN localization.

Let us represent the obtained set of points in such form:

$$[z_{i,j}] = [z_{i,j}^{-}, z_{i,j}^{+}], i = 1, \ldots, I, j = 1, \ldots, J, \tag{1}$$

where $[z_{i,j}]$ is an interval estimation of the normalized amplitude of main spectral component; *i, j* are indices of discrete increments of coordinate values on *X* and *Y* axes relatively to some initially given point. Interval estimation of the amplitude $[z_{i,j}]$ is caused by the fact that different values of main spectral component amplitude $z_{i,j}$ may be obtained for equal values of *i* and *j*. In addition, there is some error of detecting a point with coordinates *i, j*. Let us denote indices of points from neighborhood of point with coordinates i, j by $i_o, j_o, o = 1,\ldots,O$. Lower and upper values of estimation intervals of main spectral component amplitude are obtained from the equations:

$$z_{i,j}^{-} = \min_{i_o j_o}\{z_{i_o,j_o}, o = 1,\ldots,O\};$$
$$z_{i,j}^{+} = \max_{i_o j_o}\{z_{i_o,j_o}, o = 1,\ldots,O\}.$$

A mathematical model for RLN localization is considered as a discrete difference model (DDM), that is, as the difference scheme in such form [1, 9, 10]:

$$[\hat{v}_{i+1,j+1}] = [\hat{v}_{i+1,j+1}^{-}; \hat{v}_{i+1,j+1}^{+}] = \vec{f}^{T}([\hat{v}_{0,0}],\ldots,[\hat{v}_{0,j}],\ldots,[\hat{v}_{0,j}],\ldots,[\hat{v}_{i,j}])\cdot\hat{\vec{g}}, \quad (2)$$
$$i = d+1,\ldots,I,\ j = d+1,\ldots,J,$$

Where $\vec{f}^{T}(\bullet)$ is the vector of unknown basis functions defining the structure of DDM; $\hat{v}_{i,j}$ is the predicted value of main spectral component amplitude in the point with discrete specified spatial coordinates i, j; $\vec{g}$ is the vector of unknown parameters of DDM; d is the DDM order. Further, the model (2) will be called an interval discrete difference model (IDDM).

Based on the requirements of ensuring the accuracy of the model within the accuracy of the experiment, the setting of IDDM (2) will be realized using such criterion [9, 10, 12]:

$$[\hat{v}_{i,j}^{-}; \hat{v}_{i,j}^{+}] \subset [z_{i,j}^{-}; z_{i,j}^{+}], \forall i = 1,\ldots,I, \forall j = 1,\ldots,J. \quad (3)$$

After substituting the recurrent formula (2) in the expression (3) instead of the interval estimations $[\hat{v}_{i,j}^{-}; \hat{v}_{i,j}^{+}]$, together with the defined initial interval values, we obtain the following interval system of non-linear algebraic equations (ISNAE):

$$\begin{cases}
\left[\hat{v}_{0,0}^{-}; \hat{v}_{0,0}^{+}\right] \subseteq \left[z_{0,0}^{-}; z_{0,0}^{+}\right]; \ \ldots \ \left[\hat{v}_{d,d}^{-}; \hat{v}_{d,d}^{+}\right] \subseteq \left[z_{d,d}^{-}; z_{d,d}^{+}\right]; \\
\left[\hat{v}_{d+1,d+1}^{-}; \hat{v}_{d+1,d+1}^{+}\right] = \vec{f}^{T}([\hat{v}_{0,0}],\ldots,[\hat{v}_{0,j}],\ldots,[\hat{v}_{i,0}]\ldots,[\hat{v}_{d,d}],\vec{u}_0,\ldots\vec{u}_k)\cdot\hat{\vec{g}}; \\
\vdots \\
\left[\hat{v}_{I,J}^{-}; \hat{v}_{I,J}^{+}\right] = \vec{f}^{T}([\hat{v}_{0,0}],\ldots,[\hat{v}_{0,j}],\ldots,[\hat{v}_{i,0}]\ldots,[\hat{v}_{I,J-1}],\vec{u}_0,\ldots\vec{u}_k)\cdot\hat{\vec{g}}; \\
z_{d+1,d+1}^{-} \le \vec{f}^{T}([\hat{v}_{0,0}],\ldots,[\hat{v}_{0,j}],\ldots,[\hat{v}_{i,0}]\ldots,[\hat{v}_{d,d}],\vec{u}_0,\ldots\vec{u}_k)\cdot\hat{\vec{g}} \le z_{d+1,d+1}^{-}; \\
\vdots \\
z_{I,J}^{-} \le \vec{f}^{T}([\hat{v}_{0,0}],\ldots,[\hat{v}_{0,j}],\ldots,[\hat{v}_{i,0}]\ldots,[\hat{v}_{I,J-1}],\vec{u}_0,\ldots\vec{u}_k)\cdot\hat{\vec{g}} \le z_{I,J}^{-}; \\
i = d+1\ldots I, \quad j = d+1\ldots J, \quad k = 0\ldots K.
\end{cases} \quad (4)$$

3 Method of Structural Identification for Interval Discrete Difference Model

To build the IDDM (2) for RLN localization, we use the known method of combination of structural identification of interval discrete models based on the behavioral model of artificial bee colony [12] which imitates the foraging behavior of honeybees [16–20].

The application of this method of IDDM structural identification involves the implementation of activity phases of all functional groups of honey bees in the colony: employed bees (they execute food search in the neighborhood of known sources and inform onlooker bees about result), onlooker bees (they process information obtained from employed bees and make decision on what known food source they must fly to) and scout bees (randomly search new food sources) [12, 19].

Let us consider procedures and modules of behavioral model of artificial bee colony in more details (Fig. 3). Procedure of detecting of exhausted food sources realizes mechanism of decision-making by employed bee on whether researched food source is exhausted. In the case if known food source is still not exhausted, the procedure of research of the source neighborhood is called. This procedure realizes flight of bee to the neighborhood of the source for researching with further call of procedure of it's quality identification and procedure of remembering of its coordinates and quality.

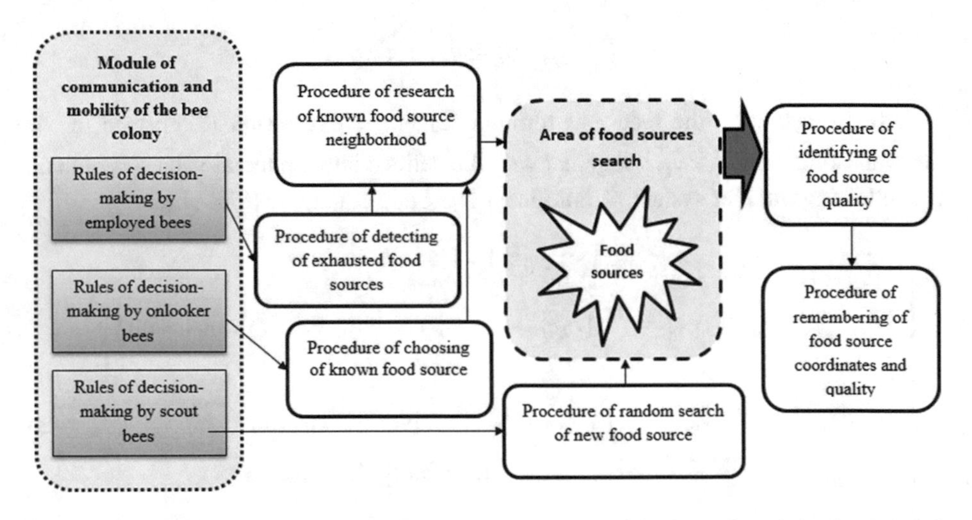

Fig. 3. The scheme of interconnection between components of the bee colony behavioral model.

Procedure of choosing of known food source by onlooker bees means the following: based on information obtained from employed bees dances, onlooker bees chose known food source to the neighborhood of which they will fly. Let us note that more onlooker bees fly to the neighborhood of "better" food source. To the neighborhood of "worse" food source, may not fly even single one.

Procedure of random search of new food source realizes flights of scout bees in random directions to search new food sources with further call of procedures of its quality identification and remembering its coordinates and quality. The procedure of remembering coordinates and quality of food source for employed bees means mechanism of decision-making on location of which food source must be remembered: found in the neighborhood one or known one. Module of communication and mobility of bees in the colony ensures information transfer among the bees of the colony and mobility between different groups. Module of mobility of bees in the colony ensures transitions between different functional groups, in particular: scout bee –> employed bee, onlooker bee –> employed bee, employed bee –> scout bee [17].

Rules of decision-making by different bee groups are defined by quantitative criteria based on which bees choose which next procedure must be executed.

Let us establish main analogies between the bee colony behavioral model and main procedures of structural identification method for mathematical models of distributed parameters objects. In particular, in context of IDDM structural identification task, bee behavior while choosing of food source, directly realizes the algorithm of current IDDM structure synthesis; the area of food search is the set of all possible IDDM structures with known estimates of components of parameters vector $\hat{\vec{g}}$; the neighborhood of food source is the set of IDDM structures that can be generated based on the current one in the way of partial replacement of its structural elements; coordinates of food source are the current IDDM structure λ_s; food source quality is identified by the value of objective function $\delta(\lambda_s)$ for the current IDDM structure λ_s, that quantitatively presents the approximation of mathematical model built based on current structure λ_s to the "optimal" one.

Let us consider realization of all stages of activity of honeybee colony functional groups in the context of IDDM structural identification task in more details.

Initialization. At this stage, in order to identify the IDDM, the researcher must set the values of initial parameters of the algorithm. These are: MCN is the maximal number of iterations of method implementation (technical parameter of the method which is used to avoid looping while applying the method), *LIMIT* is the maximal value of counter of the IDDM structure "exhaustion" (method parameter, imitates the process of "exhaustion" of food source in behavioral model of honeybee colony), S is the initial number of IDDM structures, $[I_{\min}; I_{\max}]$ are the minimal and maximal values of IDDM structural elements number, F is the set of structural elements.

Further, at the initialization stage, it is necessary to randomly generate the initial set of structures of discrete equations Λ_{mcn} (at the initialization stage, the value of iteration counter mcn is set to zero: $mcn = 0$) with cardinality S from the set of structural elements F. Let us note that generating of the sets is executed by software in such a way as to avoid repeating of elements from set F in IDDM structures λ_s, $s = 1\ldots S$ from Λ_{mcn}. Herewith, the number of structural elements m_s in each structure λ_s is defined in such a way: $m_s = rand([I_{\min}; I_{\max}])$, $s = 1...S$, and, for each of structures, the value of "exhaustion" counter is initialized: $Limit_s = 0$, $s = 1...S$.

Stage of employed bees activity. At this stage, the quality of current nectar sources is researched. From point of view of the structural identification task, this means application of operator $P(\Lambda_{mcn}, F)$ for synthesis of the set of current interval discrete models structures Λ'_{mcn}. Let us note that operator $P(\Lambda_{mcn}, F)$ executes transformation of structures set Λ_{mcn} (mcn is the number of a current iteration) into a set of structures Λ'_{mcn} randomly replacing n_s elements of each structure from set Λ_{mcn} by elements from the set of structural elements F.

Indicator n_s defines the number of structural elements that must be replaced in the current IDDM structure based on the following principle: the "worse" is the quality of IDDM structure, the more structural elements of the structure must be replaced. Let us note that the value of variable n_s is calculated based on only the quality of current structure λ_s (i.e., the value of the objective function $\delta(\lambda_s)$ calculated for it) within the current set Λ_{mcn}.

To calculate the n_s value, the following equation is used:

$$n_s = \begin{cases} \operatorname{int}\left(\left(1 - \frac{\min\{\delta(\lambda_s)|s=1...S\}}{\delta(\lambda_s)}\right) \cdot m_s\right), \\ \qquad \delta(\lambda_s) \neq \min\{\delta(\lambda_s)|s = 1...S\} \text{ and } n_s \neq 0; \\ \\ 1, \quad \delta(\lambda_s) = \min\{\delta(\lambda_s)|s = 1...S\} \text{ or } n_s = 0. \end{cases} \tag{5}$$

After this, it is necessary to conduct a pairwise selection of best IDDM structures using operator $D_1(\lambda_s, \lambda'_s)$ and obtain the set of "best" structures $\lambda^1_s \in \Lambda^1_{mcn}$, $s = 1...S$, namely such ones, for which the objective function values are smaller:

$$\lambda^1_s = \begin{cases} \lambda_s, & \text{if } \delta(\lambda_s) \leq \delta(\lambda'_s); \\ \lambda'_s, & \text{if } \delta(\lambda_s) > \delta(\lambda'_s). \end{cases} \tag{6}$$

where $\lambda_s \in \Lambda_{mcn}$, $\lambda'_s \in \Lambda'_{mcn}$, $\lambda^1_s \in \Lambda^1_{mcn}$, $s = 1...S$.

Herewith, in the case if the first condition from (6) is met, the counter is incremented: $Limit_s = Limit_s + 1$, where $s = 1...S$; in the case if second condition from (6) is met, the value of the counter is set to zero: $Limit_s = 0$.

Stage of onlooker bees activity. At this stage, the nectar is collected, and the higher is the quality of nectar source, the more bees fly there. From point of view of the structural identification task, this means application of operator $P_\delta(\Lambda_{mcn}, F)$. It executes the transformation of each structure λ^1_s from the set of structures Λ^1_{mcn} into the set of structures Λ'_s (where $s = 1...S$) in the way of random replacing n_s elements of each structure λ^1_s by elements from the set F. Unlike $P(\Lambda_{mcn}, F)$, operator $P_\delta(\Lambda_{mcn}, F)$ executes replacing only for that structures of IDDM $\lambda^1_s \in \Lambda^1_{mcn}$ for which $R_s > 0$. Let us note that R_s means number of structures that will be generated based on s-th structure from the set Λ^1_{mcn}. Elements of Λ^1_{mcn} set must be ordered in accordance with corresponding decreasing values of objective function $\delta(\lambda^1_s)$.

$$R_s = ToInt\left(\frac{S \cdot (2 \cdot \max\{\delta^2(\lambda_1^1)\ |s = 1 \ldots S\} - \delta^2(\lambda_s^1) - \delta^2(\lambda_{s-1}^1))}{\sum\limits_{s=1}^{S}(\max\{\delta^2(\lambda_1^1)\ |s = 1 \ldots S\} - \delta^2(\lambda_s^1))} - R_{s-1}\right), \ s = 2 \ldots S. \quad (7)$$

Value of the indicator R_s in (7) is calculated based on the following principle: the number of onlooker bees that fly to the neighborhood of food source informed by specific employed bee depends on its quality. This dependence was researched in [21]. It was demonstrated that the optimal one, from point of view of the minimization of computational complexity of application of the IDDM structural identification method, is the quadratic kind of this dependence.

After this, group selection of current IDDM structures is conducted using the operator $D_2(\lambda_s, \Lambda_s')$. In such a way, the set of "best" structures $\lambda_s^2 \in \Lambda_{mcn}^2$, $s = 1 \ldots S$, is obtained from current sets Λ_{mcn}^1 and Λ_{mcn}'', with using of the next equation:

$$\lambda_s^2 = \begin{cases} \lambda_s^1, & if\ (R_s = 0); \\ \lambda_s^1, & if\ ((\delta(\lambda_s^1) \leq \delta(\lambda_r)) \wedge (R_s \neq 0)), \\ & \forall \lambda_r \in \Lambda_s', r = 1 \ldots R_s; \\ \lambda_r^s, & if\ ((\delta(\lambda_s^1) > \delta(\lambda_r)) \wedge (R_s \neq 0)), \\ & \exists \lambda_r \in \Lambda_s', r = 1 \ldots R_s. \end{cases} \quad (8)$$

If the first or second condition from (8) is met, the counter is incremented: $Limit_s = Limit_s + 1$. In the case if the third condition from (8) is met, the value of the counter is set to zero: $Limit_s = 0$.

Thus, the set of IDDM structures Λ_{mcn}^2 of second stage of forming at *mcn*-th algorithm iteration is obtained.

Stage of scout bees activity. At this stage, bees leave the exhausted nectar sources and fly to search new qualitative nectar sources. From point of view of the structural identification task, this means checking the condition $Limit_s \geq LIMIT$ for all structures $\lambda_s^2 \in \Lambda_{mcn}^2$. Meeting condition $Limit_s \geq LIMIT$ for a concrete structure of the mathematical model means that it is "exhausted" and there is no need to take it into account in further iterations. Instead of "exhausted" structures, generating "new" ones is executed using operator $P_N(F, I_{\min}, I_{\max})$.

If at least one structure for which $\delta(\lambda_s^2) = 0$ was found at this stage, the procedure of structural identification is completed. Otherwise, it is necessary to go back to onlooker bees activity stage and also to increment the value of the counter: $mcn = mcn + 1$.

4 Experimental Research

Let us consider an example of building the model of distribution on the surface of surgical wound of amplitudes of main spectral component as a signal of reaction on the stimulation of the surgical wound tissues. The model is built in the form of interval discrete difference scheme using the above method of structural identification.

The fragment of data obtained during the thyroid gland surgery is presented in the Table 1 where four values of the main spectral component amplitude measured in the neighborhood of each from 25 grid nodes ($i = 0\ldots4,\quad j = 0\ldots4$) are shown; also, interval representation of the main spectral component amplitude $[z_{i,j}]$ for each of 25 grid nodes is represented.

Table 1. Fragment of the set of normalized values of the maximal amplitude of the main spectral components.

№	Coordinates			Normalized amplitude values	Interval values of the amplitude
	i	j	o	z_{i_o,j_o}	$[z_{i,j}]$
1	0	0	1	0,917233	[0,589233; 0,917233]
1	0	0	2	0,768333	
1	0	0	3	0,713517	
1	0	0	4	0,589233	
2	0	1	1	0,540567	[0,456483; 0,540567]
2	0	1	2	0,50005	
2	0	1	3	0,481283	
2	0	1	4	0,456483	
3	0	2	1	0,443833	[0,389517; 0,443833]
3	0	2	2	0,417733	
3	0	2	3	0,402083	
3	0	2	4	0,389517	
…	…	…	…	…	…
23	4	2	1	0,043367	[0,0346; 0,043367]
23	4	2	2	0,038567	
23	4	2	3	0,0346	
23	4	2	4	0,040817	
24	4	3	1	0,034233	[0,025817; 0,034233]
24	4	3	2	0,030117	
24	4	3	3	0,027683	
24	4	3	4	0,025817	
25	4	4	1	0,02495	[0,016667; 0,02495]
25	4	4	2	0,021683	
25	4	4	3	0,019433	
25	4	4	4	0,016667	

After forming interval values of the main spectral component amplitude, the method of IDDM structural identification is realized. At first, the set of structural elements F with cardinality $L = 44$ was generated. The structural elements from the set F are not higher than the third degree and not higher than the second order and are given in the Table 2.

Table 2. The set of structural elements.

№	Structural element	№	Structural element
1	$v_{i,j-1}$	23	$v_{i-1,j-1} \cdot v_{i,j-1} \cdot v_{i,j-2}$
2	$v_{i,j-2}$	24	$v_{i-1,j-2} \cdot v_{i,j-1} \cdot v_{i-1,j}$
3	$v_{i-1,j}$	25	$v_{i-1,j} \cdot v_{i,j-2} \cdot v_{i-1,j-1}$
4	$v_{i-1,j-1}$	26	$v_{i-1,j-1} \cdot v_{i,j-2} \cdot v_{i-1,j-2}$
5	$v_{i-1,j-2}$	27	$v_{i-1,j-2} \cdot v_{i,j-2} \cdot v_{i,j-1}$
6	$v_{i,j-1} \cdot v_{i,j-1}$	28	$v_{i-1,j-1} \cdot v_{i-1,j} \cdot v_{i,j-2}$
7	$v_{i,j-2} \cdot v_{i,j-1}$	29	$v_{i-1,j-2} \cdot v_{i-1,j} \cdot v_{i-1,j-1}$
8	$v_{i-1,j} \cdot v_{i,j-1}$	30	$v_{i-1,j-2} \cdot v_{i-1,j-1} \cdot v_{i,j-1}$
9	$v_{i-1,j-1} \cdot v_{i,j-1}$	31	$v_{i,j-1} \cdot v_{i,j-1} \cdot v_{i,j-1}$
10	$v_{i-1,j-2} \cdot v_{i,j-1}$	32	$v_{i-1,j} \cdot v_{i,j-1} \cdot v_{i,j-1}$
11	$v_{i,j-2} \cdot v_{i,j-2}$	33	$v_{i-1,j-1} \cdot v_{i,j-1} \cdot v_{i,j-1}$
12	$v_{i-1,j} \cdot v_{i,j-2}$	34	$v_{i-1,j-2} \cdot v_{i,j-1} \cdot v_{i,j-1}$
13	$v_{i-1,j-1} \cdot v_{i,j-2}$	35	$v_{i,j-2} \cdot v_{i,j-1} \cdot v_{i,j-1}$
14	$v_{i-1,j-2} \cdot v_{i,j-2}$	36	$v_{i-1,j} \cdot v_{i-1,j} \cdot v_{i-1,j}$
15	$v_{i-1,j} \cdot v_{i-1,j}$	37	$v_{i,j-1} \cdot v_{i-1,j} \cdot v_{i-1,j}$
16	$v_{i-1,j-1} \cdot v_{i-1,j}$	38	$v_{i-1,j-1} \cdot v_{i-1,j} \cdot v_{i-1,j}$
17	$v_{i-1,j-2} \cdot v_{i-1,j}$	39	$v_{i,j-2} \cdot v_{i-1,j} \cdot v_{i-1,j}$
18	$v_{i-1,j-1} \cdot v_{i-1,j-1}$	40	$v_{i-1,j-2} \cdot v_{i-1,j} \cdot v_{i-1,j}$
19	$v_{i-1,j-2} \cdot v_{i-1,j-1}$	41	$v_{i,j-1} \cdot v_{i-1,j-1} \cdot v_{i-1,j-1}$
20	$v_{i-1,j-2} \cdot v_{i-1,j-2}$	42	$v_{i-1,j} \cdot v_{i-1,j-1} \cdot v_{i-1,j-1}$
21	$v_{i,j-2} \cdot v_{i,j-1} \cdot v_{i-1,j}$	43	$v_{i,j-2} \cdot v_{i-1,j-1} \cdot v_{i-1,j-1}$
22	$v_{i-1,j} \cdot v_{i,j-1} \cdot v_{i-1,j-1}$	44	$v_{i-1,j-2} \cdot v_{i-1,j-1} \cdot v_{i-1,j-1}$

Also, the values of the following parameters of the structural identification method implementation algorithm are set: $MCN = 60$, $LIMIT = 4$, $S = 12$, $[I_{\min}; I_{\max}] = [4, 7]$.

Then, the randomly generating the initial set of interval discrete difference model structures with the cardinality $S = 12$ is executed, the results are given in the Table 3.

Each of IDDM structures presented in the Table 3 is given as ordered set of decimal numbers of structural elements from the Table 2. Also, for each structure from set $\Lambda_{mcn=0}$, the parametric identification was executed and the values of the objective function were estimated. Results of this stage are represented in the Table 3 as well.

Table 3. The initial set of IDDM structures $\Lambda_{mcn=0}$.

№	Numbers of structural elements from set F that preset structures λ_s, $s = 1\ldots S$.	$\delta(\lambda_s)$	m_s
1	12, 17, 27, 36, 38	0,811	5
2	4, 22, 28, 30, 39	0,733	5
3	6, 10, 36, 42	0,654	4
4	13, 18, 20, 24, 28, 43	0,92	6
5	2, 5, 39, 44	0,401	4
6	1, 12, 17, 20, 26, 27, 35	0,688	7
7	16, 24, 33, 40	0,543	4
8	6, 9, 14, 30, 33	0,602	5
9	8, 10, 19, 28, 37, 42	0,367	6
10	5, 29, 31, 34	0,411	4
11	3, 8, 22, 29, 32, 33	0,92	6
12	17, 23, 28, 34, 38	0,456	5

Next, the stages of activity of all bees groups are repeated until a model structure for which the value of objective function equals zero will be found. Such a structure of the model for RLN localization by predicting the amplitude of the main spectral component was obtained on the ninth iteration of the method implementation:

$$\begin{gathered}
[\hat{v}^-_{i,j}; \hat{v}^+_{i,j}] = -0.0161 + 0.503 \cdot [\hat{v}^-_{i,j-2}; \hat{v}^+_{i,j-2}] \\
+ 0.2145 \cdot [\hat{v}^-_{i-1,j}; \hat{v}^+_{i-1,j}] + 0.7969 \cdot [\hat{v}^-_{i,j-1}; \hat{v}^+_{i,j-1}] \cdot [\hat{v}^-_{i,j-1}; \hat{v}^+_{i,j-1}] \\
+ 0.6344 \cdot [\hat{v}^-_{i-1,j-1}; \hat{v}^+_{i-1,j-1}] \cdot [\hat{v}^-_{i,j-1}; \hat{v}^+_{i,j-1}] \cdot [\hat{v}^-_{i,j-1}; \hat{v}^+_{i,j-1}], \\
i = 1\ldots 4, j = 2\ldots 4,
\end{gathered} \tag{9}$$

where $[\hat{v}^-_{i,j}; \hat{v}^+_{i,j}] \subset [z^-_{i,j}; z^+_{i,j}] = [z_{i,j} - z_{i,j} \cdot 0,02; z_{i,j} + z_{i,j} \cdot 0,02]$ and also, $\{i = 0, j = 0, \ldots, 4\} \vee \{i = 0, \ldots, 4, j = 0, 1\}$ are the given initial conditions.

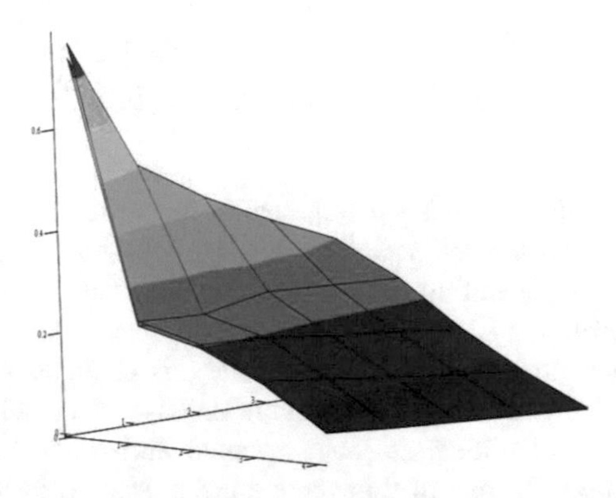

Fig. 4. The corridor of interval models for distribution of the main spectral component amplitude.

The corridor of interval models for distribution of the main spectral component amplitude obtained from (9) is shown in the Fig. 4.

Using model (9), it is possible to obtain a guaranteed estimation of RLN location area in the surgical wound. It is estimated as a projection of the "crest" of corridor of the maximal amplitude of the main spectral component distribution onto the area of surgical environment.

5 Conclusion

In this paper, the principles of building RLN localization systems based on the electrophysiological method of surgical wound tissue stimulation have been considered. Based on the results of previous researches, it was substantiated to implement the model of the main spectral component amplitudes spatial distribution on the surface of surgical wound into the block of surgical wound tissues classification of the existing system of RLN localization. For its building, the method of structural identification for models of distributed parameters objects is proposed. Using this method and based on the results of electrophysiological researches of surgical wound tissues during the neck organs surgeries, the example of building the model for the RLN localization has been represented. The obtained model, with the appropriate adjustments for each patient, makes it possible to identify the RLN location and reduce the risk of its damage during the neck organs surgery. The advantage of the proposed model is the simplicity of setting for a particular patient. For this, the surgeon must stimulate the surgical wound tissues in 13 points.

It is expedient to conduct further researches in the direct of improving the predicting properties of the obtained model.

References

1. Anuwong, A.: Transoral endoscopic thyroidectomy vestibular approach: a series of the first 60 human cases. World J. Surg. **40**(3), 491–497 (2016)
2. Abstract book of First World Congress of Neural Monitoring in Thyroid and Parathyroid Surgery, Krakow, Poland (2015)
3. Poveda, M.C.D., Dionigi, G., Sitges-Serra, A., Barczynski, M., Angelos, P., Dralle, H., Randolph, G.: Intraoperative monitoring of the recurrent laryngeal nerve during thyroidectomy: a standardized approach part 2. World J. Endocr. Surg. **4**(1), 33–40 (2012)
4. Dhillon, V.K., Tufano, R.P.: The pros and cons to real-time nerve monitoring during recurrent laryngeal nerve dissection: an analysis of the data from a series of thyroidectomy patients. Gland. Surg. **6**(6), 608–610 (2017)
5. Kim, H.Y., Liu, X., Wu, C.W., Chai, Y.J., Dionigi, G.: Future directions of neural monitoring in thyroid surgery. J. Endocr. Surg. **17**(3), 96–103 (2017)
6. Davis, W.E., Lee Rea, J., Templer, J.: Recurrent laryngeal nerve localization using a microlaryngeal electrode. Otolaryngol. Head Neck Surg. **87**(3), 330–333 (1979)
7. Varaldo, E., Ansaldo, G.L., Mascherini, M., Cafiero, F., Minuto, M.N.: Neurological complications in thyroid surgery: a surgical point of view on laryngeal nerves. http://dx.doi.org/10.3389/fendo.2014.00108. Last Accessed 10 Apr 2018

8. Genther, D.J., Kandil, E.H., Noureldine, S.I., Tufano, R.P.: Correlation of final evoked potential amplitudes on intraoperative electromyography of the recurrent laryngeal nerve with immediate postoperative vocal fold function after thyroid and parathyroid surgery. JAMA Otolaryngol. Head Neck Surg. **140**(2), 124–128 (2014)
9. Dyvak, M., Kozak, O., Pukas, A.: Interval model for identification of laryngeal nerves. Przegląd Elektrotechniczny **86**(1), 139–140 (2010)
10. Porplytsya, N., Dyvak, M.: Interval difference operator for the task of identification recurrent laryngeal nerve. In: Proceedings of the 16th International Conference on Computational Problems of Electrical Engineering (CPEE 2015), pp. 156–158 (2015)
11. Dyvak, M., Kasatkina, N., Pukas, A., Padletska, N.: Spectral analysis the information signal in the task of identification the recurrent laryngeal nerve in thyroid surgery. Przegląd Elektrotechniczny **89**(6), 275–277 (2013)
12. Porplytsya, N., Dyvak, M., Dyvak, T., Voytyuk, I.: Structure identification of interval difference operator for control the production process of drywall. In: Proceedings of 12th International Conference on the Experience of Designing and Application of CAD Systems in Microelectronics, (CADSM 2013), pp. 262–264 (2013)
13. Husson, R.: Etude des phénomènes phisiologiqes et acoustiqes fondamentaux «de îa voix chantée» . Thése Fac Sciences, Paris (1952)
14. Cantelon, M., Harter, M., Holowaychuk, T.J., Rajlich, N.: Node.js in Action. Manning Publications, Shelter Island (2013)
15. Teixeira, P.: Professional Node.js: Building Javascript Based Scalable Software. Wiley, Indianapolis (2012)
16. Karaboga, D., Gorkemli, B., Ozturk, C., Karaboga, N.: A comprehensive survey: artificial bee colony (ABC) algorithm and applications. Artif. Intell. Rev. **42**(1), 21–57 (2014)
17. Karaboga, D., Basturk, B.: A powerful and efficient algorithm for numerical function optimization: artificial bee colony (ABC) algorithm. J. Glob. Optim. **39**(3), 459–471 (2007)
18. Karaboga, D., Basturk, B.: A survey: algorithms simulating bee swarm intelligence. Artif. Intell. Rev. **31**, 68–85 (2009)
19. de Vries, H., Biesmeijer, J.C.: Modelling collective foraging by means of individual behaviour rules in honey-bees. Behav. Ecol. Sociobiol. **44**(2), 109–124 (1998)
20. Sean, L.: Essentials of Metaheuristics, 2nd edn. Lulu, Raleigh (2013)
21. Dyvak, M., Porplytsya, N., Maslyiak, Y., Kasatkina, N.: Modified artificial bee colony algorithm for structure identification of models of objects with distributed parameters and control. In: Proceedings of the 14th International Conference on Experience of Designing and Application of CAD Systems in Microelectronics (CADSM 2017), pp. 50–54 (2017)

Probabilistic Energy Forecasting Based on Self-organizing Inductive Modeling

Frank Lemke(✉)

KnowledgeMiner Software, 13187 Berlin, Germany
frank@knowledgeminer.com

Abstract. Self-organizing inductive modeling represented by the Group Method of Data Handling (GMDH) as an early implementation of Deep Learning is a proven and powerful data-driven modeling technology for solving ill-posed modeling problems as found in energy forecasting and other complex systems. It develops optimal complex predictive models, systematically, from sets of high-dimensional noisy input data. The paper describes the implementation of a rolling twelve weeks self-organizing modeling and probabilistic ex ante forecasting, exemplarily, for the Global Energy Forecasting Competition 2014 (GEFCom 2014) electricity price and wind power generation forecasting tracks using the KnowledgeMiner INSIGHTS inductive modeling tool out-of-the-box. The self-organized non-linear models are available analytically in explicit notation and can be exported to Excel, Python, or Objective-C source code for further analysis or model deployment. Based on the pinball loss function they show an overall performance gain of 67.3% for electricity price forecasting and 47.5% for wind power generation forecasting relative to corresponding benchmark measures.

Keywords: Inductive modeling · Deep Learning · Energy forecasting GMDH · Machine learning

1 Introduction

Energy forecasting is the foundation for utility planning and is a fundamental business problem in the power industry. With the transition towards a regionalized, secure, affordable, and 100% renewable, low carbon energy supply, every single percent increase of renewable energy on the energy mix, has been introducing new major problems on energy companies concerning secure and cost-effective energy supply due to the uncontrollable and volatile nature of renewables.

The power industry today is facing high volatility in electricity and heat demand as well as in renewable energy generation within a single day resulting in highly fluctuating electricity market prices (Weron 2014), which has brought significant challenges to power systems planning and operations. In such an uncertain environment, the companies have to rely on data and analytics in addition to their experience to make informed decisions. Accurate energy forecasting, i.e., load, renewable power generation, and price forecasting, is a crucial and fundamental step in this analytics workflow for energy companies.

N. Shakhovska and M. O. Medykovskyy (Eds.): CSIT 2018, AISC 871, pp. 405–420, 2019.
https://doi.org/10.1007/978-3-030-01069-0_29

Many factors influence energy forecasting accuracy, such as geographic diversity, data quality, forecast horizon, customer segmentation, and forecasting method. A model that works well in one region may not be the best model for another. Similarly, a model that provides good forecasts in one customer segment may not be appropriate for another. Also, within the same utility, a model that forecasts well in one year may not generate a good forecast for another year.

To improve energy independence of Europe and to strengthen Europe's leading and innovative role in making the transition towards a 100% renewable and low carbon energy supply, on regional scales, a self-adapting forecasting solution for energy firms covering uncertain future energy consumption, renewable power generation, and associated costs and benefits is required. In this context, availability of accurate, timely, regional energy forecasts for a specific customer segment is a key factor for energy supply.

In addition, the Energy Union has highlighted the importance of smart heating systems and storage solutions (including district heating) (European Union 2015) to lower the carbon content of energy in Europe and as a means to bring additional flexibility to balance the electric grid by combined use of heat pumps, district heating and decentralized combined heat and power generation (CHP), thus enhancing the system integration of gas, electric and heat grids. The use of advanced forecasting approaches can enhance the operational performance of heat storage at various time scales.

Many statistical and machine learning methods have been applied for energy forecasting (Hong et al. 2016). In this paper results following a self-organizing inductive modeling approach are presented.

2 Self-organizing Inductive Modeling

The inductive approach of self-organizing modeling, which is represented by the Group Method of Data Handling (GMDH), was developed by A.G. Ivakhnenko (Ivakhnenko 1968) and has been further developed and improved since then by several other authors (Farlow 1984; Madala and Ivakhnenko 1994; Müller and Lemke 2000; Kordik 2006; Kondo and Ueno 2007; Stepashko, 2008). Today, it is seen as an early implementation of the concept of Deep Learning (Schmidhuber 2015). In result of intense research a dedicated noise immunity theory has been developed (Stepashko 1983; Ivakhnenko and Stepashko 1985) and implemented as a central part of this modeling technology.

The basic principle of GMDH that makes it different from other well-known machine learning and data mining methods is that of *induction*. The concept of induction is composed of three ideas:

- The principle of self-organization for adaptively evolving a model from noisy observation data without subjective points given;
- The principle of external information to allow objective selection of a model of optimal complexity (noise immunity), and
- The principle of regularization of ill-posed tasks.

GMDH inductive modeling is based on complete (combinatorial GMDH) or incomplete induction (multi-layered GMDH networks with active neurons) approaches. For incomplete induction, self-organization is considered in identifying connections between the network units by a learning mechanism to represent discrete items. For this approach, the objective is to identify networks of sufficient size by a successively evolving structure controlled by the learning algorithm. A process is said to undergo self-organization if identification emerges through the system's environment.

To realize an inductive self-organization of models from a given number of input-output data the following conditions must exist to be fulfilled (Müller and Lemke 2000):

First condition: There is a very simple initial organization that enables the description of a large class of systems through the organization's evolution.

A common class often used is that of dynamic systems, which can be described by Volterra functional series. Discrete analogues of the Volterra functional series describing systems with a finite memory are higher-order polynomials of the Kolmogorov-Gabor form. For one input variable x it is:

$$y_t^M = k_{0,t} + \sum_{s=0}^{g} a_s x_{t-s} + \sum_{s_1} \sum_{s_2} a_{s_1} a_{s_2} x_{t-s_1} x_{t-s_2} + \ldots \tag{1}$$

where $k_{0,t}$ is some trend function, g memory depth, and a_s are coefficients of memory s. It is possible to develop any sub-model of the general model y_t^M by evolution of networks of initial simple elementary models like:

$$f_1(v_i, v_j) = a_0 + a_1 v_i + a_2 v_j, \text{ or} \tag{2}$$

$$f_2(v_i, v_j) = f_1(v_i, v_j) + a_3 v_i v_j + a_4 v_i^2 + a_5 v_j^2. \tag{3}$$

Second condition: There is an algorithm for development of the initial or already evolved organizations (intermediate models).

In inductive modeling, a gradual increase of model complexity is used as a key principle. The successive combination of many variants of mathematical models with increasing complexity has proven to be a universal solution in the theory of self-organization presenting variability of models in a way like that in biological selection processes. In most self-organizing inductive modeling algorithms, a pairwise combination of M inputs is used to develop model candidates (intermediate models) of growing complexity.

Third condition: There is an external selection criterion for validating the usefulness of a model relative to the intended task of modeling.

The principle of selection in inductive modeling is closely linked to the principle of self-organization in biological evolution. It is applied if a complete induction of models becomes inefficient, i.e., when the number of all possible model candidates is going to become too large due to exponential complexity. Using a threshold value, all model candidates are selected which satisfy a given quality function (survival-of-the-fittest) that embeds noise immunity to avoid overfitting of the design data. As in biological evolution selected models are used as inputs for the development of a next generation of model candidates.

The overall process of model evolution and selection stops automatically when a new generation of model candidates provides no further improvement of model quality as expressed by the external selection criterion. Then, a final *optimal complex analytical model* composed of self-selected relevant inputs is obtained.

Overfitting a model on the training data results in bad generalization of the obtained model and this always have been a problem in experimental systems analysis. At a certain point in model induction approximation power and prediction power of the model start to diverge. The model must have appropriate structure and complexity to be powerful enough to approximate the known data (training data), but also constrained enough to generalize successfully, that is, to do well on new data (testing data) not yet seen during modeling. There are always many models with a similar closeness of fit on the training data. Generally, simpler models generalize better on testing data than more complex ones with same or higher accuracy. According to this heuristic principle (Occam's Razor), we have to optimize the trade-off between model complexity and the model's accuracy on training *and* testing data.

The idea of systematically building *optimal complex models* from noisy observational data has been developed and introduced into GMDH by Ivakhnenko and Stepashko (Stepashko 1983; Ivakhnenko and Stepashko 1985). The multidimensional problem of model optimization can be solved by an inductive sorting-out procedure:

$$m^* = argmin_{m \in M} Q(m), Q(m) = \Phi(P, C, \sigma, T, V), \tag{4}$$

where M is a set of considered models, Q is an external criterion that measures the quality of model m from set M, P a set of variables, C model complexity, σ the noise variance, T the type of data sample transformation, and V the type of neuron reference function. For a definite reference function (e.g. linear polynomial), each set of variables (number of variables) corresponds to a definite model structure (max. number of model terms) $P = C$. The optimization problem then transforms to a much simpler one-dimensional problem $Q(m) = f^*(C)$ if σ, T and V are constant(s). Ivakhnenko (Ivakhnenko and Stepashko 1985) has shown that in the case of linear models with complexity C, the criterion $Q(m)$ depends on C and σ as a unimodal function (Fig. 1):

1. $Q(C, \sigma)$ is a unimodal function. The minimum C_i exists and is unique: $Q(C_i, \sigma) = \min_C Q(C, \sigma)$, where C_i is the complexity of the selected model (model of optimal complexity),
2. In absence of noise, the optimal complexity C_0 is equal to the unbiased model (the complete physical model that contains all significant variables),
3. With increasing noise, the complexity of the model decreases, leading to an optimal model with simpler structure and higher error value compared to the unbiased model,
4. In case of completely random input data the optimal model is equal to the mean value of the output variable with $Q = 1$, $Q \in [0, 1]$.

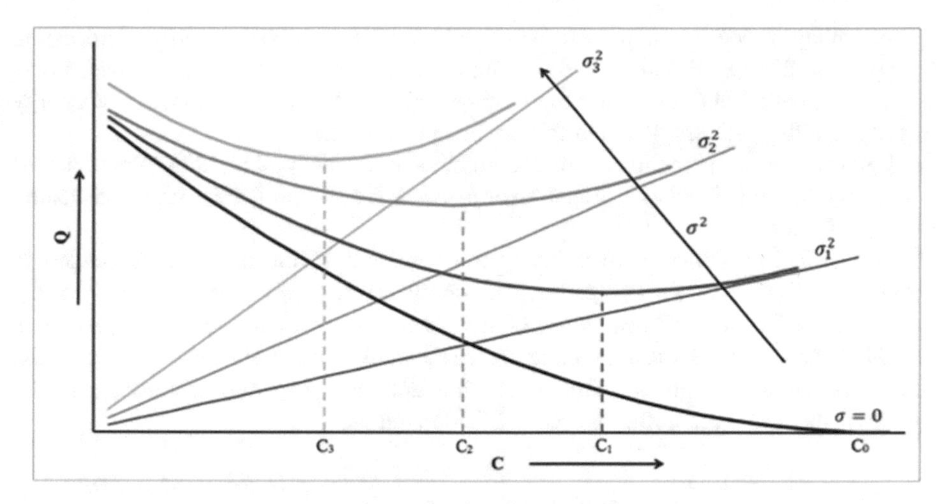

Fig. 1. Selection of models of optimal complexity (Q - average value of external error criterion; C - complexity of model structure; σ^2 noise variance)

3 Self-organizing Inductive Modeling Applied to Probabilistic Energy Forecasting

The GMDH approach has been proven to be very efficient in data-driven modeling of complex systems in economics, climate forecasting, toxicology and environmental engineering (KnowledgeMiner 2018A) with several advantages over conventional neural networks, from which these are key for energy forecasting problems:

- Inductive self-organization of the forecasting model of interest from short and noisy data. This includes both model structure identification (the model formulation step in theory-driven modeling approaches) and coefficients estimation.
- Systematically and automatically builds validated predictive models of optimal complexity that do not overfit the design data.
- Generates sets of alternative models (composites), autonomously, that reflect forecasting uncertainty by design.
- Analytical expression of the generated models (linear/non-linear, dynamic regression model), which makes models interpretable, implementable, and transparent;
- Developed theory of robust modeling (noise immunity) as condition for high-dimensional modeling also on small number of samples.
- Self-selection of a small set of relevant variables from a given high-dimensional vector of potential inputs and self-detection when modeling stops (built-in feature selection).

These key features do not only allow employing rolling forecasting but also implementing rolling modeling easily and reliably using most recent data. Modeling is based on the assumption that the functional relations between variables are constant over the evaluated period of time. Therefore, to satisfy this requirement, rather short time series of most recent data can be used for electricity forecasting.

A rolling modeling approach was applied to the Global Energy Forecasting Competition 2014 (GEFCom 2014) (Hong et al. 2016) in all four tracks (load, price, wind, and solar power generation forecasting) organized by the IEEE Power & Energy Society and the University of North Carolina at Charlotte.

For ease of implementation and communication the pinball loss function was chosen by the organizers as a proper error measure for probabilistic energy forecasting in all four tracks.

For each time period τ over the forecast horizon T, the participants needed to provide the 1st, 2nd,…, 99th percentiles, calling these $q_1, \ldots q_{99}$, with $q_0 = -\infty$, or the natural lower bound, and $q_{100} = \infty$, or the natural upper bound. The full predictive densities composed by these quantile forecasts were to be evaluated by the quantile score calculated through the pinball loss function. For a quantile forecast q_a, with $a/100$ as the target quantile, this score L is defined as:

$$L_\tau(y_\tau) = \frac{1}{99}\sum\nolimits_{a=1}^{99} L(q_a, y_\tau),\ with\ L(q_a, y_\tau) = \begin{cases} \left(1 - \frac{a}{100}\right)(q_a - y_\tau), if\ y_\tau < q_a \\ \frac{a}{100}(y_\tau - q_a), if\ y_\tau \geq q_a \end{cases} \quad (5)$$

$$L = \frac{1}{T}\sum_{\tau=1}^{T} L_\tau(y_\tau),$$

where y_τ is the observation at forecast step τ.

4 Results of Probabilistic Electricity Price Forecasting

For each of the 12 competition weeks of probabilistic electricity price forecasting track these basic steps of a rolling modeling were accomplished:

- Data updating and self-organization of a model composite
- Extraction of hourly error distributions
- Forecasting and calculation of percentiles

4.1 Data Updating and Self-organization of a Model Composite

After adding the provided weekly data update to the historical hourly data set consisting of variables $\boldsymbol{x}_1$ (Forecasted Total Load), $\boldsymbol{x}_2$ (Forecasted Zonal Load), and $\boldsymbol{x}_3 = \boldsymbol{y}$ (Zonal Price, Fig. 2), $N = 4008$ most recent observations for all $\boldsymbol{x}_i, i = 1, 2, 3,$ were selected as the information matrix to self-organize a set of analytical models (model composite) to forecast the Zonal Price 24 h ahead.

From the information matrix $\mathbf{X}$, $m = 54$ potential input variables v_k were synthesized in the pre-processing step of model self-organization:

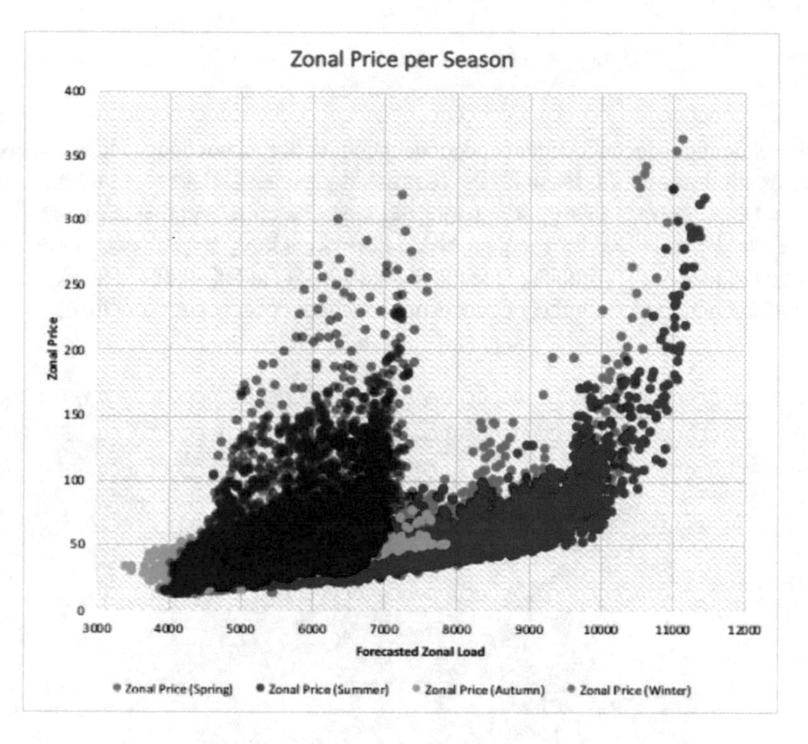

Fig. 2. Historical hourly zonal price data (x_3) per season as provided by ISO New England (USA) relative to forecasted zonal load (x_2).

$$v_k = \sum_{i=1}^{2} \sum_{j=0}^{g} x_{i,t-j},$$

$$v_{53} = x_{3,t-24}, v_{54} = x_{3,t-25}, g = 25, k = (g+1)(i-1) + j + 1 \tag{6}$$

Note that these are *potential* inputs for final model composite development. The actual set of *relevant* inputs used in the models are self-selected during the process of self-organization of non-linear, dynamic regression models. For example, for the competition weeks 10, 11, and 12 these non-linear dynamic models with different sets of relevant inputs have been self-organized:

$$y_t^{(10)} = f_1^{(10)} \begin{pmatrix} x_{1,t}, x_{1,t-1}, x_{1,t-18}, x_{1,t-25}, x_{2,t-2}, \\ x_{2,t-11}, x_{2,t-24}, x_{2,t-25}, y_{t-24} \end{pmatrix} \tag{7}$$

$$y_t^{(11)} = f_1^{(11)} \begin{pmatrix} x_{1,t}, x_{1,t-22}, x_{1,t-23}, x_{2,t-10}, x_{2,t-12}, \\ x_{2,t-25}, y_{t-24} \end{pmatrix} \tag{8}$$

$$y_t^{(12)} = f_1^{(12)}\begin{pmatrix} x_{1,t}, x_{1,t-1}, x_{1,t-21}, x_{2,t-1}, x_{2,t-9}, \\ x_{2,t-24}, x_{2,t-25}, y_{t-24} \end{pmatrix} \quad (9)$$

Model self-organization was accomplished using KnowledgeMiner Software's inductive modeling tool INSIGHTS (KnowledgeMiner 2018A) out-of-the-box that incorporates a number of important internal steps, such as synthesis of variables v_k, variables transformation (normalization/de-normalization), hypothesis (model candidates) generation and hypothesis testing (validation), generation of the model equation, and model evaluation including evaluation of model robustness (stability).

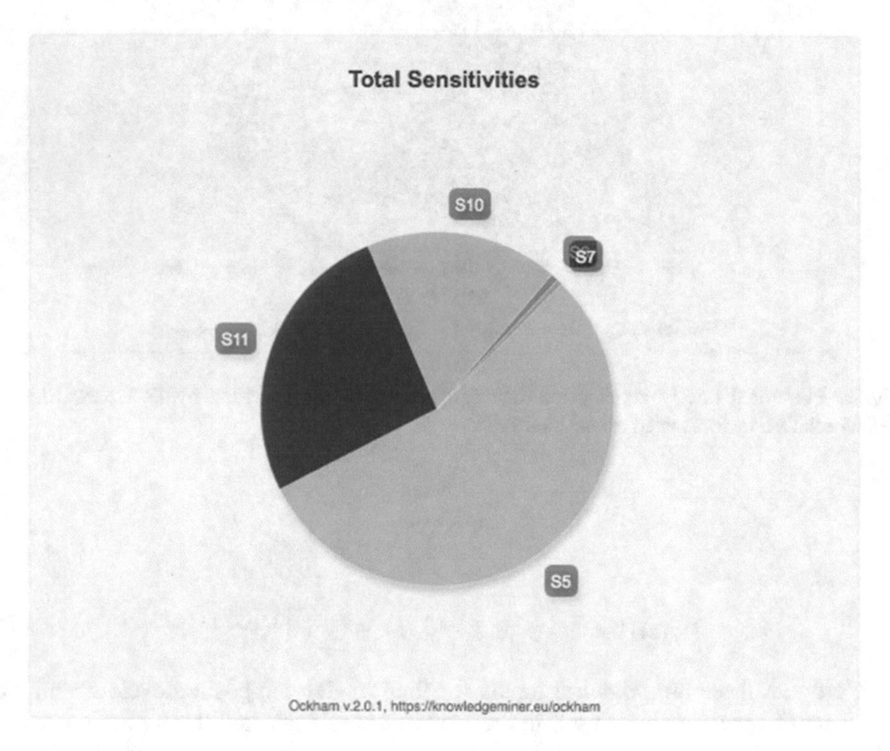

Fig. 3. Total sensitivities of the self-organized non-linear model $f_1^{(11)}$. S5: $x_{1,t}$ (relevance: 55%), S11: y_{t-24} (26%), S10: $x_{2,t-25}$ (18%). The identified max. order of interactions is 5.

As shown in (7) to (9) the self-organized models may also contain auto-regressive inputs. Running a Global Sensitivity Analysis (Lambert et al. 2016; KnowledgeMiner 2018B) turns out that the obtained models are not mainly auto-regressive but that other input variables, including their second and higher order interactions, have a major impact on the identified model behavior (Fig. 3).

The result of model self-organization is an ensemble of up to $I_{max} = 10$ alternative individual models of comparable quality and accuracy forming a model composite to express forecasting uncertainty by simultaneous application of the model set.

For ill-posed modeling problems there always exists a number of models which show comparable overall performance on the design data but which predict differently on a per-sample basis building a more or less wide forecasting interval along with a most likely forecast as shown in Fig. 4.

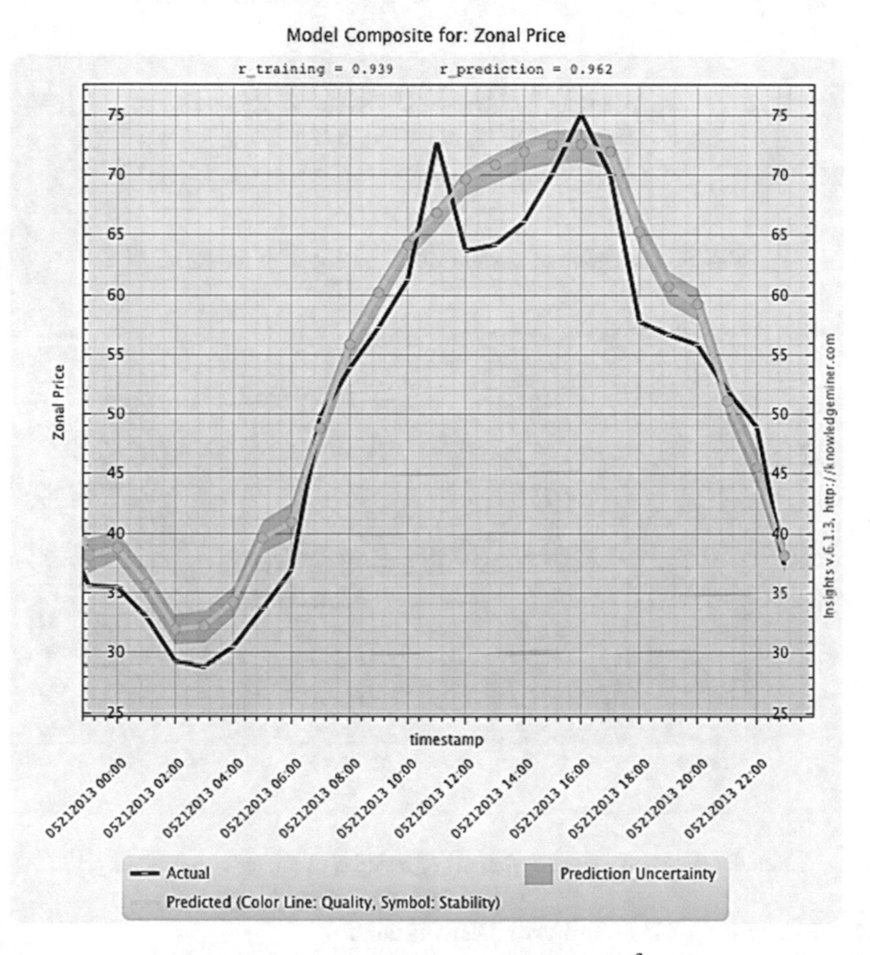

Fig. 4. 24 h ex-ante forecast of a self-organized model composite (R^2 = 0.93; green dots: most likely forecast or 50th percentile; gray area: per-sample forecasting interval (prediction uncertainty) obtained from the individual models of the composite).

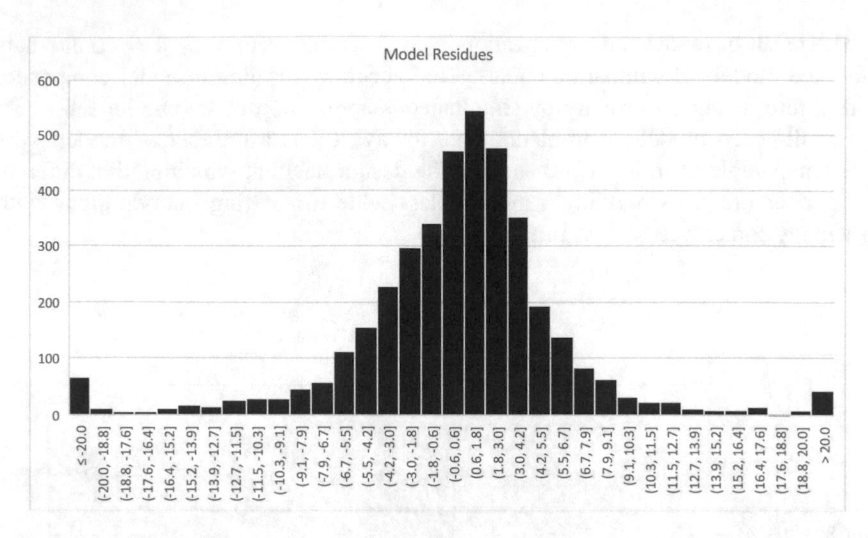

Fig. 5. Histogram of model residues exemplarily for model $f_1^{(12)}$

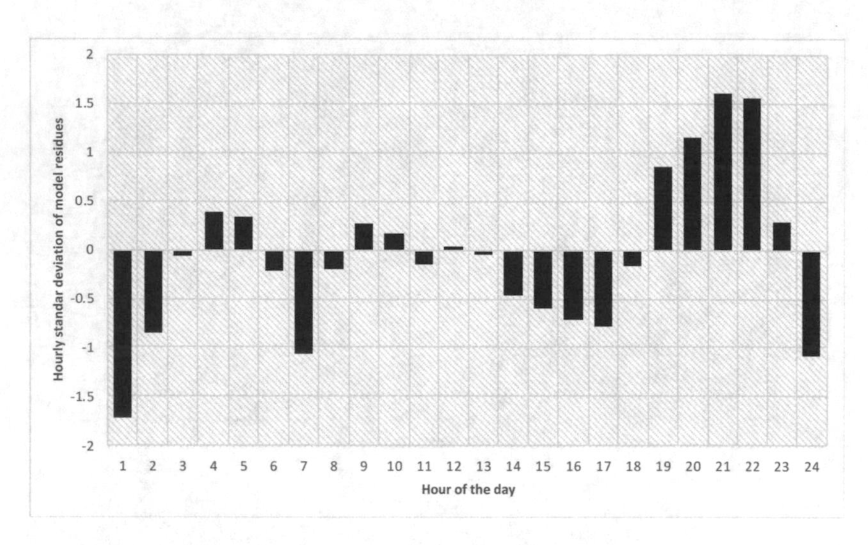

Fig. 6. Hourly standard deviations of residues exemplarily for model $f_1^{(12)}$

4.2 Extraction of Hourly Error Distributions

From the residues e_t (Fig. 5)

$$e_t = \sum\nolimits_{t=g+1}^{N} y_t - \hat{y}_t \tag{10}$$

24 discrete hourly error distributions are extracted then, where the obtained standard deviations $\sigma_h, h = 0, 1, \ldots, 23$, and N observations will be used in step 4.3 for percentiles calculation (Fig. 6).

4.3 Forecasting and Calculation of Percentiles

From the ex-ante forecasts of the composite models self-organized in step 4.1 a most likely forecast is obtained:

$$\hat{\hat{y}}_{t+p} = \frac{1}{I}\sum\nolimits_{i=1}^{I} \hat{y}_{i,t+p}, p = 1, 2, \ldots, P, \tag{11}$$

with I as the actual number of individual models in the composite, $I \leq I_{max}$, forecast horizon $T = 24$, and $\hat{y}_i$ the forecast of the i-th model. The most likely forecast is assigned to the 50th percentile (P_{50}). Using the hourly standard deviations σ_h obtained in step B, the lower forecast interval bounds

$$\hat{y}_{min,t+p} = P_{1,t+p} = \hat{\hat{y}}_{t+p} - a\sigma_h, a = 1, h = p - 1 \tag{12}$$

and the upper interval bounds

$$\hat{y}_{max,t+p} = P_{99,t+p} = \hat{\hat{y}}_{t+p} + b\sigma_h, b = 2, h = p - 1 \tag{13}$$

are calculated. The remaining percentiles, for each forecasting step p, are calculated by dividing the differences $d_1 = P_{50} - P_1$ and $d_2 = P_{99} - P_{50}$ into equidistant intervals $s_1 = \frac{d_1}{49}$ and $s_2 = \frac{d_2}{49}$, accordingly:

$$P_i = \sum\nolimits_{i=2}^{49} P_1 + (i - 1)s_1, \textit{ and} \tag{14}$$

$$P_j = \sum\nolimits_{j=51}^{98} P_{50} + (j - 50)s_2. \tag{15}$$

The average ex ante forecasting accuracy obtained for the P_{50} percentile over the twelve competition weeks is $r = 0.966$, $MAPE = 10.5\%$, and a pinball loss $L = 3.33$. Table 1 lists the pinball loss over the twelve competition weeks along with corresponding benchmark values. The average performance gain of the probabilistic forecast of the inductive models over the competition benchmark is 67.3% (Table 1).

Table 1. Comparison of the pinball loss function between inductive models and benchmark over the twelve competition weeks. As benchmark the corresponding price value of one week before was given.

Week	1	2	3	4	5	6	7	8	9	10	11	12
Benchmark	4.03	7.97	5.70	12.1	38.3	44.2	18.2	31.6	42.9	2.86	3.20	22.4
Inductive models	2.37	1.99	1.07	2.79	4.23	4.71	8.41	1.25	2.24	3.68	1.06	6.28
Performance gain	0.41	0.75	0.81	0.77	0.89	0.89	0.54	0.96	0.95	-0.3	0.67	0.72

5 Results of Probabilistic Wind Power Generation Forecasting

The task of the wind power generation forecasting track for each of the 12 competition weeks was to forecast wind power generation one month ahead (i.e., 720 respectively 744 h ahead) for each of ten wind park sites $\boldsymbol{y}_i, i = 1, 2, .., 10$, located in the state of New England, USA. Given are hourly ex post and ex ante forecasts of wind speed components u and v at 10 m $(\boldsymbol{x}_1, \boldsymbol{x}_2)$ and 100 m height $(\boldsymbol{x}_3, \boldsymbol{x}_4)$ for each wind park site $\boldsymbol{y}_i$ (Fig. 7).

Additional input variables were synthesized for model self-organization, $x_{i+4,j} = \sqrt{x_{i,j}}, i = 1, 2, 3, 4; j = 1, 2, .., N,$ to extend freedom of choice for model synthesis. Time lags of up to 2 h were applied to all eight input variables, which sums up to 24 inputs used for modeling, finally.

The forecasting process followed the procedure described in Sects. 4.1 to 4.3, correspondingly.

For the target site 1 of week 11 the following non-linear analytical model, which is composed of four self-selected relevant inputs, only, $\boldsymbol{x}_{1,t}(relevance : 53.4\%)$, $\boldsymbol{x}_{1,t-2}(<1\%)$, $\boldsymbol{x}_{2,t}(42.3\%)$, $\boldsymbol{x}_{5,t-2}(3.5\%)$, was provided on the fly by the used inductive modeling tool (KnowledgeMiner 2018A) and which can be exported to Excel or other formats such as Python or Objective-C source code for further analysis and use:

$$\begin{aligned}
y_{1,t}^{(11)} = {} & 0.0102133 * x_{1,t} + 0.00433531 * x_{1,t-2} - 0.0396636 * x_{5,t-2} \\
& + 0.00252824 * x_{1,t} * x_{2,t} + 0.000272581 * x_{1,t} * x_{5,t-2} \\
& + 0.00872022 * x_{1,t}^2 + 0.000375902 * x_{1,t-2}^2 + 0.00881341 \\
& * x_{2,t}^2 + 0.00952582 * x_{5,t-2}^2 + 6.74758 * 10^{-5} * x_{1,t} * x_{2,t} \\
& * x_{5,t-2} + 0.000233561 * x_{1,t}^2 * x_{5,t-2} - 5.35799 * 10^{-5} * x_{1,t} \\
& * x_{2,t}^2 - 1.53701 * 10^{-5} * x_{1,t}^2 * x_{2,t} + 0.000235219 * x_{2,t}^2 \\
& * x_{5,t-2} - 5.32021 * 10^{-5} * x_{1,t}^3 - 4.78123 * 10^{-5} * x_{1,t}^2 * x_{2,t}^2 \\
& - 1.31699 * 10^{-5} * x_{1,t}^3 * x_{2,t} - 1.32634 * 10^{-5} * x_{1,t} * x_{2,t}^3 \\
& - 2.27931 * 10^{-5} * x_{1,t}^4 - 2.3118 * 10^{-5} * x_{2,t}^4 - 0.0764179
\end{aligned}$$

This model shows an accuracy $r = 0.81$ of the most likely forecast (50th percentile P_{50}) for the 720 h forecasting period November 1, 2013, 1:00 to December 1, 2013, 0:00 (Fig. 8) and a pinball loss $L = 0.0369$. Table 2 lists the average pinball loss of over the ten locations for the twelve competition weeks along with corresponding benchmark values. The average performance gain of the probabilistic ex ante forecast of the inductive models relative the competition benchmark is 47.5% (Table 2).

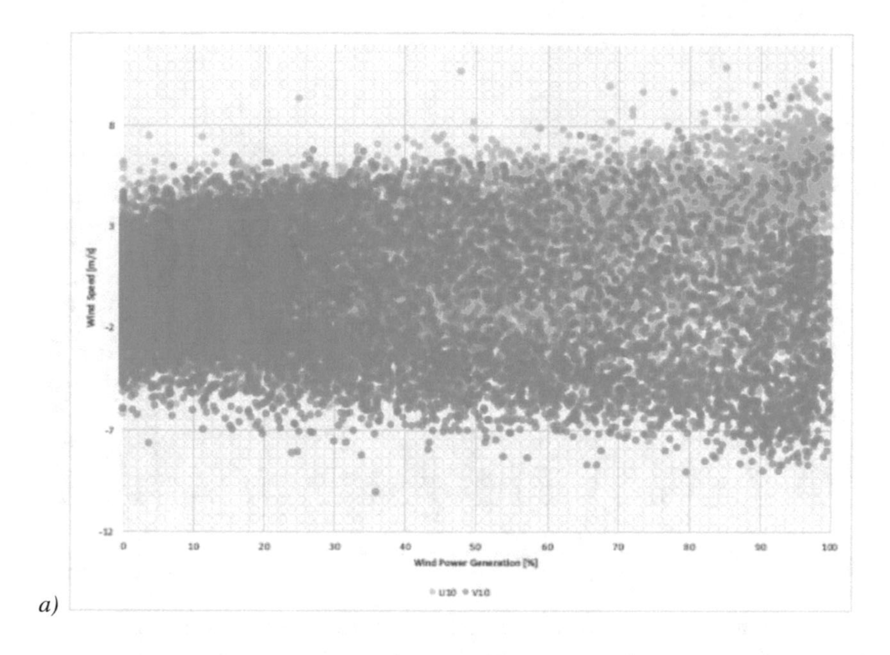

a)

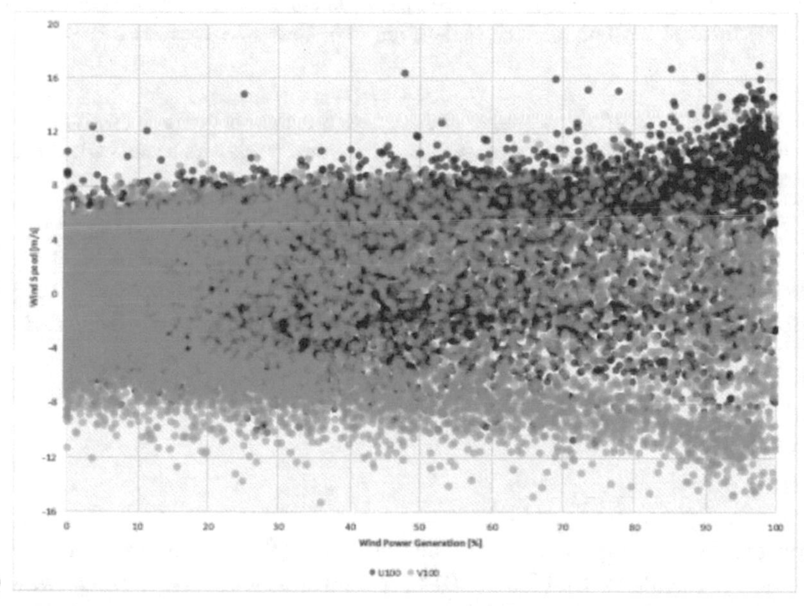

b)

Fig. 7. Wind power generation vs wind speed forecasts at 10 m *(a)* (x_1, x_2) and 100 m height *(b)* (x_3, x_4) for target location 1. The data shows no significant correlation between the variables (r = 0.19 for x_2, for example). More advanced modeling approaches are required for predicting wind power generation sufficiently accurate.

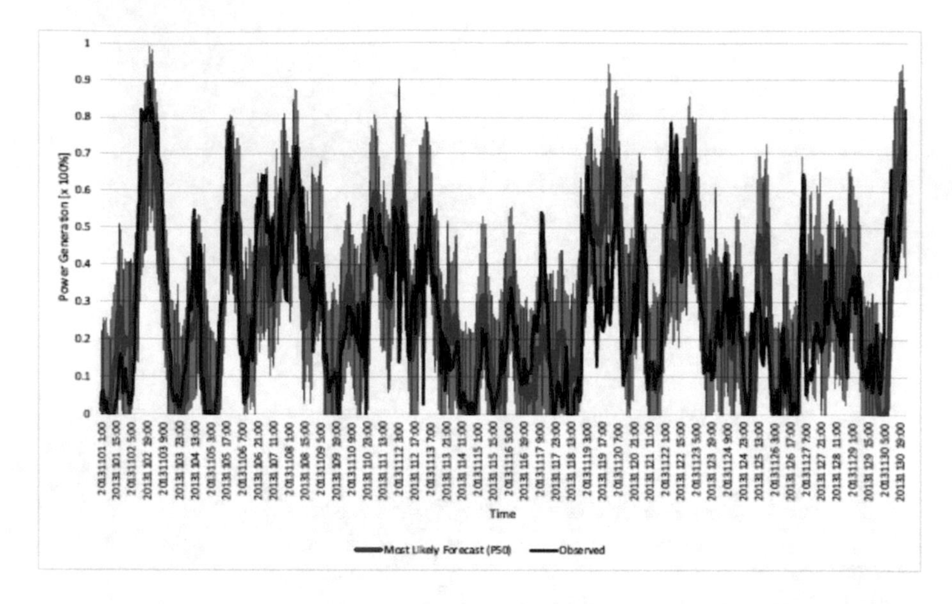

Fig. 8. Ex ante probabilistic forecast of model $y_{1,t}^{(11)}$ for the period November 1, 2013, 1:00 to December 1, 2013, 0:00 (720 h). The percentiles P_1 to P_{99} are displayed in light red color. The most likely forecast (P_{50}) highlighted in dark red color shows an accuracy of $r = 0.81$.

Table 2. Comparison of the average pinball loss function between inductive models and benchmark of all ten locations over the twelve competition weeks. As benchmark climatological values were used.

Week	1	2	3	4	5	6	7	8	9	10	11	12
Benchmark	0.08	0.07	0.08	0.08	0.08	0.08	0.10	0.12	0.10	0.10	0.08	0.08
Inductive models	0.04	0.04	0.04	0.04	0.05	0.04	0.04	0.05	0.05	0.05	0.06	0.04
Performance gain	0.48	0.40	0.47	0.50	0.45	0.51	0.57	0.62	0.51	0.49	0.25	0.45

6 Summary

A rolling self-organizing, inductive modeling on a number of most recent observational data was implemented for GEFCom2014 probabilistic electricity price and wind power generation forecasting, exemplarily. Other complex energy forecasting problems such as load and solar power forecasting have also been accomplished successfully by this highly automated and proven approach, which is an early implementation of Deep Learning. Key features of self-organizing modeling for energy forecasting are:

- *Knowledge extraction from data and model transparency (i.e., no "black box")* Inductive self-organization of the forecasting model of interest from short and noisy observation data. This includes both model structure identification and coefficients estimation as well as delivery of an analytical model in explicit notation.

- *Objectivity and reliability*
  Systematically builds validated predictive models of optimal complexity that do not overfit the design data.
- *Feature extraction*
  Self-selection of a small set of relevant inputs from a given high-dimensional vector of potential inputs.
- *Accuracy improvement*
  Generates ensembles of alternative models that reflect forecasting uncertainty by design.

The described rolling modeling approach generated in average 67.3% more accurate probabilistic forecasts of zonal electricity price 24 h ahead based on the pinball loss function than the competition benchmark over the entire competition period of twelve weeks. Similar performance improvements with 47.5% over all ten target locations were obtained for 744 respectively 720 h ahead probabilistic wind power forecasting, solar power forecasting (52%), and load forecasting (33%).

Out of more than 500 teams from over 40 countries who registered for the competition only four teams successfully finished all four challenging GEFCom2014 forecasting tracks, which underlines the high predictive power and productivity of the presented self-organizing inductive modeling approach.

References

European Union: A Framework Strategy for a Resilient Energy Union with a Forward-Looking Climate Change Policy (2015). http://eur-lex.europa.eu/legal-content/EN/TXT/?uri=COM:2015:80:FIN

Farlow, S.J., (ed.) Self-organizing Methods in Modeling. GMDH Type Algorithm. Marcel Dekker, New York (1984). ISBN 0-8247-7161-3

Hong, T., Pinson, P., Fan, S., Zareipour, H., Troccoli, A., Hyndman, R.J.: Probabilistic energy forecasting: global energy forecasting competition 2014 and beyond. Int. J. Forecast. **32**, 896–913 (2016)

Ivakhnenko, A.G.: Group method of data handling as a rival of stochastic approximation method. Sov. Autom. Control. **3**, 58–72 (1968)

Ivakhnenko, A.G., Stepashko, V.S.: Pomechoustojcivost' modelirovanija (Noise-immunity of modeling). Naukova dumka, Kiev (1985). (In Russian)

KnowledgeMiner Software: INSIGHTS - Self-organizing modeling and forecasting tool, v6.1.3 (2018 A). https://www.knowledgeminer.eu. Last Accessed 05 May 2018

KnowledgeMiner Software: OCKHAM – Global Sensitivity Analysis tool, v2.0.1 (2018 B). https://www.knowledgeminer.eu/ockham. Last Accessed 02 May 2018

Kondo, T., Ueno, J.: Feedback GMDH-type neural network self-selecting optimum neural network architecture and its application to 3-dimensional medical image recognition of the lungs. In: Proceedings of II International Workshop on Inductive Modelling, Czech Technical University, Prague, pp. 63–70 (2007)

Kordik, P.: Fully automated knowledge extraction using group of adaptive model evolution. Ph. D. thesis, Department of Computer Science and Computers, FEE, CTU in Prague (2006)

Lambert, R., Lemke, F., Kucherenko, S., Song, S., Shah, N.: Global sensitivity analysis using sparse high dimensional model representations generated by the group method of data handling technique. J. Math. Comput. Simul. **128**, 42–54 (2016)

Madala, H.R., Ivakhnenko, A.G.: Inductive Learning Algorithms for Complex Systems Modelling. CRC Press Inc., Boca Raton, Ann Arbor, London, Tokyo (1994). ISBN 0-8493-4438-7

Müller, J.-A., Lemke, F.: Self-organizing Data Mining. Libri, Hamburg (2000). ISBN: 3-89811-861-4

Schmidhuber, J.: Deep learning in neural networks: an overview. Neural Netw. **61**, 85–117 (2015)

Stepashko, V.S.: Potential noise immunity of modelling using a combinatorial GMDH algorithm without information regarding the noise. Sov. Autom. Control. **16**(3), 15–25 (1983)

Stepashko, V.S.: Method of critical variances as analytical tool of theory of inductive modeling. J. Autom. Inf. Sci. **40**(3), 4–22 (2008)

Weron, R.: Electricity price forecasting: a review of the state-of-the-art with a look into the future. Int. J. Forecast. **30**, 1030–1081 (2014)

A Method for Reconstruction of Unmeasured Data on Seasonal Changes of Microorganisms Quantity in Heavy Metal Polluted Soil

Olha Moroz(✉) and Volodymyr Stepashko

Department for Information Technologies of Inductive Modelling, International Research and Training Centre for Information Technologies and Systems of the NAS and MES of Ukraine, Akademik Glushkov Avenue 40, Kyiv 03680, Ukraine
olhahryhmoroz@gmail.com, stepashko@irtc.org.ua

Abstract. The article presents results of application of the hybrid combinatorial-genetic algorithm COMBI-GA to building models simulating the dependence of quantity of microorganisms in soil on the meteorological conditions and concentration of a heavy metal in an experimental plot. The models built on the rarely measured data during the vegetation seasons are used then for reconstructing the unmeasured decade data on seasonal changes of microorganisms quantity in the soil of a polluted plot during the whole season taking into account the complete support series of the decade meteorological data. This method is demonstrated on the results of modelling amylolytic microorganisms quantity dependence on measured weather factors and concentration of copper in the soil of experimental plots. Meteorological data included the humidity and temperature of air of the current and previous decades. Linear and nonlinear models of changing the microorganisms quantity in control and polluted plots are build based on the rarely measured data during the vegetation seasons. Nonlinear models are used for reconstructing the unmeasured decade data taking into account the complete support series of the decade weather data. Such a methodology can reduce in the future the cost of expensive and time-consuming experiments. A generalized model of amylolytic microorganisms quantity dependence on copper concentration and weather factors is created for predicting critical ecological situations.

Keywords: Inductive modelling · GMDH · Combinatorial algorithm COMBI
Genetic algorithm GA · Hybrid algorithm COMBI-GA
Amylolytic microorganisms · Soil · Heavy metals · Data reconstruction
Generalized model

1 Introduction

One of the main components of most ecosystems is soil in which microorganisms play an important role in the evolution and formation of the fertility. Anthropogenic pollution of the biosphere impacts all living components of biogeocenoses including soil microorganisms [1].

N. Shakhovska and M. O. Medykovskyy (Eds.): CSIT 2018, AISC 871, pp. 421–432, 2019.
https://doi.org/10.1007/978-3-030-01069-0_30

Present agroecosystems are subject to the considerable anthropogenic influence resulting frequently in pollution of arable soils. Pollutants influence negatively on soil mikrobiotics which causes the necessity to carry out long-term observations after its current status. In parallel with monitoring, there is the task of determination of critical deviations and prediction of the mikrobiotum state dependently on pollutants concentration in soil. Solving this task is possible based on formalization of monitoring data in the form of mathematical models [2].

Investigation of the effect of specific anthropogenic factors such as heavy metals on microbial community functioning is very important. Negative influence of heavy metals on microbial kenosis, soil and biological activity is well-known. In this connection, the organization and carrying out of regular control of soil state for the purpose of critical situation detecting and forecasting are actual.

But such kind of researches is conducted on irregular base in view of technological difficulties of permanent microbiological analysis. It causes a necessity to apply mathematical modeling to describing the microbial processes on the bases of small measured data and restoring unmeasured data for obtaining even series of ecological observations with the purpose of further detailed analysis. Such a modeling from observation data is the necessary condition of the ecological monitoring as it allows operative estimating current ecological situations and forecasting their evolution.

To construct such models, it is reasonable to use the Group Method of Data Handling (GMDH) [3] as an effective method for the analysis, modeling and forecasting of complex processes from experimental data under conditions of incompleteness of a priori information and short data samples.

For the analysis of organotrophic microorganisms functioning in dark gray podzol soil (Kyiv region) polluted with heavy metals (copper and mercury), the GMDH-based interactive modeling system ASTRID [4] was initially used.

In [5], the results are presented on the application of the hybrid combinatorial-genetic algorithm COMBI-GA [6, 7] for building optimal linear models from small data samples of microbiological observations on the change in the quantity of amylolytic microorganisms in the soil contaminated by copper. In [8], COMBI-GA was used for building nonlinear models under conditions given in [5].

In this research, for the same data we build more accurate nonlinear models of dependence of the amylolytic microorganisms quantity on weather conditions and the pollutant concentration and use them for reconstruction of the unmeasured decade data of the microbiological process on the basis of known weather factors during a vegetation season.

Section 2 of this paper main presents characteristics of the experimental data being used. In Sect. 3, the method of unavailable data reconstruction is suggested based on inductive modeling. Section 4 considers hybrid combinatorial-genetic GMDH algorithm COMBI-GA and their features for solving this task. Section 5 describes results of modelling, data reconstruction and generalized model of amylolytic microorganisms quantity dependence on copper concentration. Section 6 presents concluding remarks.

2 Characteristics of the Experimental Data

Experiments regarding functioning of amylolytic microorganisms under copper contamination were carried out on small plots in deep-gray podzolic soil (Kyiv region). The traditional chart of experiments [9] was used: several plots of the same soil type were selected for experiments and one plot was used as non-contaminated control soil. The contamination of soil was carried out by the annual one time applying solutions of Cu^{2+} salts in the soil at the beginning of a vegetation season. The amount of the applied metal (content of the ions) corresponds to contamination doses of 2 maximum permissible concentrations (MPC).

The soil specimens for the analyses was taken from the arable layer depth (0–20 cm) three times during vegetation seasons in period 1993 to 1996, approximately in 2nd, 30th and 100th days (these days vary from year to year) after applying the metal salt. It was hence received three measuring time points during every of four years, or 12 measuring points together. The amount of amylolytic microorganisms in the control and polluted soils was determined using conventional microbiological technique [9].

Figure 1(a)–(d), visualizes initial data of measured results in control and experimental plots. The ordinates correspond to the quantities of microorganisms and days are indicated on the abscissas.

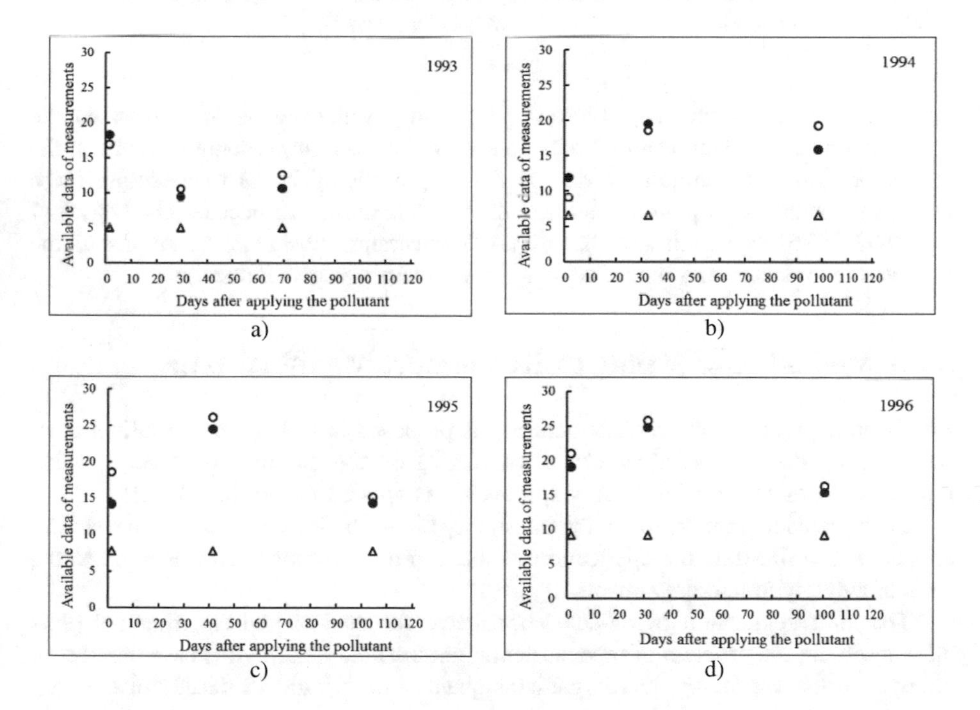

Fig. 1. Measured initial data for control "O" and experimental "●" plots with copper concentration "Δ".

Figure 2 shows generalized initial data of measurements all together in control and experimental plots during 1993–1996 years. The ordinate indicates the quantity of microorganisms and the abscissa corresponds to the whole number of 12 measuring points during these four years.

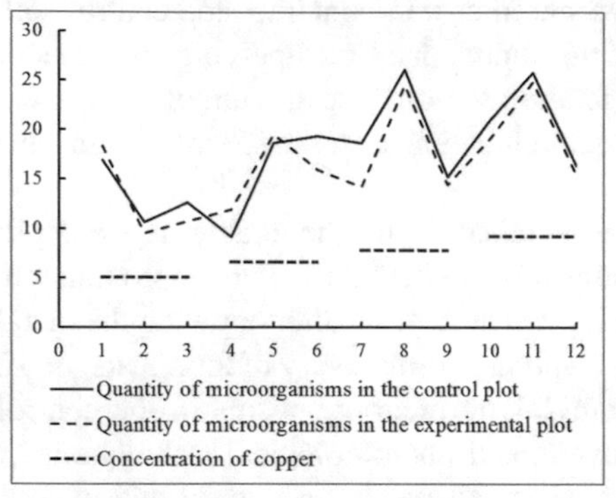

Fig. 2. Copper concentration, quantity of microorganisms in the control and experimental plots during four years (3 measurement points every year).

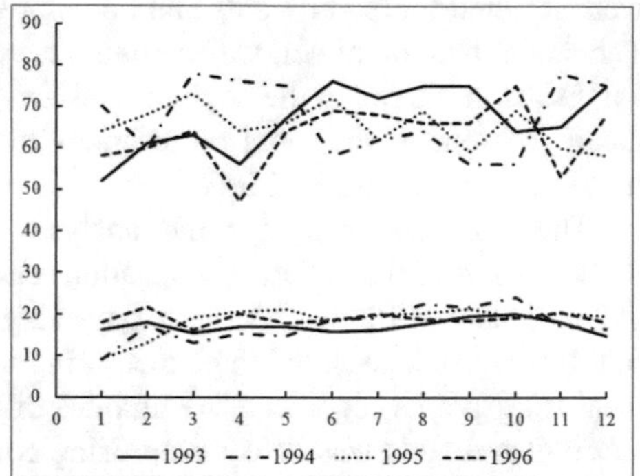

Fig. 3. Decade data charts of the humidity (above, %) and the temperature (below, °C) changes during 12 decades of the 1993–1996 vegetation seasons.

Since the functioning any microorganisms in a soil depends first of all on the weather conditions, there is a necessity to take in this task into account the data on the air temperature and humidity. Like the first approaches [10–12] to modeling these tasks, we use the corresponding average decade data during 12 decades (120 days) of the 1993–1996 vegetation seasons. Figure 3 illustrates substantial variations of the recorded temperature and humidity data during these 4 vegetation seasons.

3 A Method of Modeling to Reconstruct Microbial Data

Firstly an approach to the reconstruction was proposed in [10] where an influence of some heavy metals (including Cu^{2+}, and Hg^{2+}) on the amounts of *organotrophic* microorganisms was modeled. This approach was applied further in [11, 12].

Let us consider in more detail the possibility to further develop and improve similar approach. We illustrate the application of the improved reconstruction method to the case of amylolytic microorganisms.

The joint application of models constructed for control and experimental plots allows solving the problem of reconstructing unmeasured data, in this case the decade changes of the *amylolytic* microorganisms quantity during the 12 decade of the vegetation period of a year. This is possible if, first, these models are accurate enough in observation time points, and second, for intermediate points (decades) there are “supporting” data series of the regularly registered meteorological indicators which substantially determine the vital activity of any microorganisms in a soil.

As input variables for construction of models for the experimental plots, the following indicators were used: Q_{contr} – quantity of amylolytic microorganisms in the control plot (millions in 1 g of dry soil); D – number of days from the date of applying the pollutant; concentration of copper CC in mobile forms of Cu^{2+} (mg/kg of soil); average air temperature of a current T and previous T_p decades (°C); average air humidity of a current H and previous H_p decades (%). The quantity of amylolytic microorganisms in soils of experimental plots polluted by the copper salt Q_{exp} was the output (dependent) variable. The weather data for previous decades were used to take into account possible inertia of ecological processes. Based on this data, the models of microorganisms amount in soil were built. Unlike [10–12], we do not use data on other pollutants because any kind of them was applied on different experimental plots hence there was no relationships between them.

The inductive modelling problem in this task may be defined as follows. We are given: the data set of the type $W = [X \vdots y]$, $\dim W = 12 \times 8, \dim X = 12 \times 7$, of 12 observations after 7 inputs $x = (Q_{contr}, CC, D, T, T_p, H, H_p)^T$ and one output $y = Q_{exp}$ variables. Traditionally for GMDH, this set W was divided into three subsets: A (training), B (checking), and C (validation):

$$W = \begin{bmatrix} A \\ B \\ C \end{bmatrix} = \begin{bmatrix} X_A \; y_A \\ X_B \; y_B \\ X_C \; y_C \end{bmatrix}.$$

The GMDH task is to find a model $Q_{exp} = f(Q_{contr}, CC, D, T, T_p, H, H_p, \theta)$ with minimum value of a given model quality criterion $CR(f)$, where θ is unknown vector of model parameters. The optimal model $f^* = \arg\min_{\Phi} C(f)$ is to be built, where Φ is a set of models of various complexity, $f \in \Phi$.

Taking this into account, the data reconstruction method proposed in this paper has the following main stages:

1. Building optimal model using the measured data (12 points of measurements) for the control plot, namely quantity of amylolytic microorganisms Q_{contr} dependence only on the weather factors (unlike to [10–12]): $Q_{contr} = g(T, T_p, H, H_p, \theta)$.
2. Building optimal model using 12 points of the measured data for the experimental plot: $Q_{exp} = f(Q_{contr}, CC, D, T, T_p, H, H_p, \theta)$.
3. Reconstruction of unmeasured, unavailable data of the changing the decade quantity Q_{contr} concentrations of amylolytic microorganisms in the control plot during vegetation periods of all 4 years using the corresponding model for Q_{contr} and known average decade values of the weather factors.
4. Reconstruction of unmeasured, unavailable data of the changing the decade concentrations Q_{exp} in the experimental plot during vegetation periods of all 4 years using the corresponding model for Q_{exp} and known decade values of the quantity Q_{contr}, number of days D after applying the copper of concentration CC as well as the weather factors.
5. Visualization and analysis of the obtained results from the viewpoint of explaining dynamics of the amylolytic microorganisms reaction on the copper contamination.

Application of this method to the case of reconstructing unmeasured data on the quantity change of amylolytic microorganisms in soil polluted by copper is illustrated in the Sect. 5.

4 Algorithm COMBI-GA as the Modelıng Tool

The genetic algorithm [13] is one of the meta-heuristic procedures of global optimization constructed as a result of simulation in artificial systems and application of such properties of living nature as natural selection of species, adaptability to changing environmental conditions, inheritance by offspring of vital properties from parents and others.

Since GA is based on the principles of biological evolution and genetics, some biological terms are used to describe them: *individual* is a potential solution of a problem; *population* is a set of individuals; *offspring* is usually an improved copy of potential solution (father); *fitness* is usually a quality characteristic of the solution; *chromosome* is some encoded data structure of an individual in the form of an array/string of a fixed length and *gene* is an element of this array. With each GA step, the mean fitness value of the current population is improved and the evolution procedure converges to the solution of an optimization problem.

Any GA algorithm efficiency largely depends on its characteristics and genetic operators: the *selection* operator which stores a certain amount of chromosomes at any iteration with the best values of fitness function of GA, as well as the *crossover* and *mutation* operators for the creation of new offspring-chromosomes. A crossover operator creates offspring by partly exchanging genetic material between the parent chromosomes, and a mutation operator does that by changing one chromosome in accordance with certain rules.

The combinatorial-genetic algorithm [6, 7] as a hybrid architecture of COMBI [14] and GA performs the following main operations:

(1) generating a random set of partial model structures of a given size as an initial population of the COMBI-GA;
(2) computing coefficients of every partial model using least squares method (LSM);
(3) calculating an external criterion value (as the fitness function of the GA) for each model, for example, the regularity criterion being typical for GMDH;
(4) current selecting the best partial models (elite selection in GA) or reduction/rejection of worst individuals from the parent and offspring populations, and then formation of new population of the same size;
(5) checking a stop criterion, for example, achieving a given accuracy or number of iterations; stop if it is fulfilled, otherwise go to the next step;
(6) the use of genetic operators (crossover and mutation) with a given probability to selected individuals of the population and forming a set of partial model structures (new population) for the next generation; go to step 2.

5 Modelıng Results

5.1 Linear and Nonlinear Models for Measured Data Simulation

Based on the available experimental data, models of quantity change of amylolytic microorganisms were built for the control plot as well as for the copper polluted soils. In all cases we use the COMBI-GA algorithm with the following division of all data sample (12 points, 3 observations during 4 years 1993–1996): 6 points (2 years) as the training set *A*, 3 points (1 year) as the checking set *B*, and 3 points (1 year) as the validation set *C*. Probabilities of crossover and mutation operators were 0, 7 and 0, 1 respectively.

According to the above introduced method of reconstruction, the model was built at the first stage for the dependence of quantity Q_{contr} of microorganisms in the control plot on all measured weather factors, and at the second stage the model of dependence of the value Q_{exp} for polluted experimental plots on all 7 measured factors. We have built and compare linear and nonlinear models for both cases.

Linear Modelling. In this case, the quantity of amylolytic microorganisms in the control soil Q_{contr} is described by the model built using COMBI-GA [5]:

$$Q_{contr} = 0.2136T - 0.7149T_p + 0.5412H_p. \tag{1}$$

The linear model for the quantity of amylolytic microorganisms in the copper polluted soil was built taking into account the quantity Q_{contr} in the control soil [5]:

$$Q_{exp} = 0.743Q_{contr} + 1.6516CC - 0.9182T_p - 0.2845H_p. \tag{2}$$

Proper graphs for control and experimental plots are given on Fig. 4.

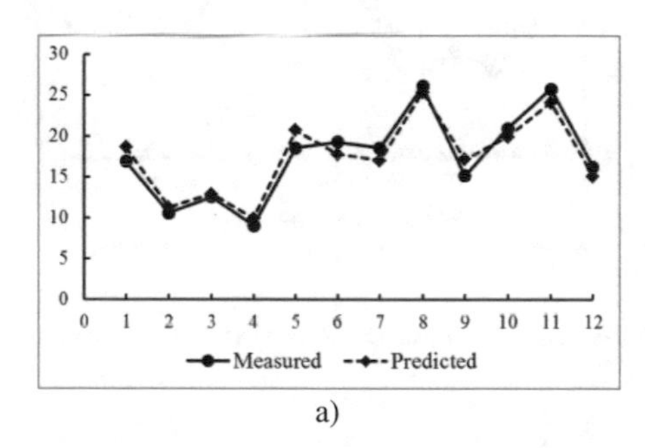

a)

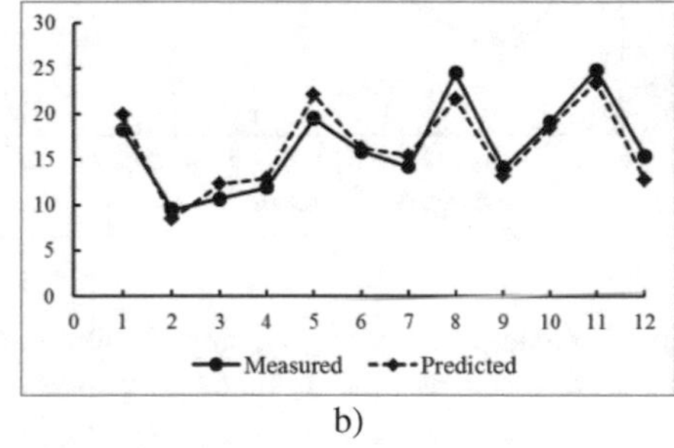

b)

Fig. 4. Change of quantity of amylolytic microorganisms (linear models) in (a) control plot and (b) experimental plot polluted by copper.

The accuracy measures MSE for models (1) and (2) are given in the Table 1. This accuracy level is insufficient for quality monitoring needs and it was the reason to build more complex nonlinear (polynomial) models. To do that, we use COMBI-GA for finding the model of optimal complexity on the basis of monomials of the second order polynomial of 5 (for Q_{contr}) and 7 (for Q_{exp}) arguments. This means that the complexity of the task for building the polynomial models increased substantially.

Table 1. Characteristics of accuracy for all obtained models.

	Linear case		Nonlinear case	
	Control	Experimental	Control	Experimental
AR	0,138	0,214	0,054	0,058
MSE_{AB}	0, 738	0,814	0,124	0,138
VE_C	0, 120	0, 157	0, 092	0, 107

Nonlinear Modelling. The quantity of amylolytics in a control soil is described by the following model:

$$Q_{contr} = 1.438T - 0.839H + 0.249T \cdot H. \tag{3}$$

The model of changing the amylolytic microorganisms in soil polluted by copper:

$$Q_{exp} = 1.315Q_{contr} + 0.9176Q_{contr} \cdot CC - 0.415T \cdot H. \tag{4}$$

Measured and predicted data of amount of amylolytic microorganisms for control and experimental plots are presented on Fig. 5.

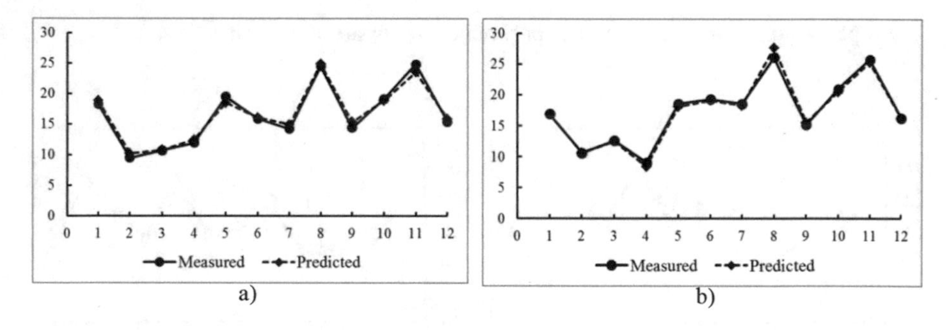

Fig. 5. Change of quantity of amylolytic microorganisms (nonlinear models) in (a) control plot and (b) experimental plot polluted by copper.

These graphs show that in most points the data measured and predicted by the model fit well, that is the models adequately represents the change of microorganisms quantity. Last three validation points on the graphs testify good results of models verification in the forecasting mode.

Table 1 presents the accuracy of models (1)–(4): values of the regularity criterion AR_B, MSE on subsamples A and B, and validation error VE_C on subsample C:

$$AR_B = \left\| y_B - X_B \widehat{\theta}_A \right\|^2, \ \widehat{\theta}_A = (X_A^T X_A) X_A^T y, \tag{5}$$

$$\mathrm{MSE}_{AB} = \sqrt{\frac{1}{9}\sum_{i=1}^{9}(y_i - \bar{y})^2}, \ \bar{y} = \frac{1}{9}\sum_{i=1}^{9} y_i, \tag{6}$$

$$VE_C = \left\| y_C - X_C \widehat{\theta}_A \right\|^2, \tag{7}$$

As it is evident from the models, the functioning of amylolytic bacteria in soil is substantially influenced first of all by the temperature and humidity of air.

In the nonlinear case we obtain much more accurate results as compared to the linear one which can help to solve different ecological tasks based on microbial monitoring. In particular, we can now solve the task of reconstruction of unmeasured data for control and polluted soil as it was indicated above at the 3rd and 4th stages of the complex method, see Sect. 3.

5.2 Reconstruction of Unavailable Decade Data

To reconstruct the unavailable decade data Q_{contr} for the control plot, we use the "supporting series" of the air temperature and humidity, see Fig. 3. Then, substituting this data from (3) into (4), we reconstruct decade data Q_{exp} for the experimental plot.

Figure 6 gives reconstructed graphs for (a) control and (b) polluted soils which shows that the pollution by copper inhibits the development of the microorganisms during practically all the vegetation season but later they are able to recover.

Figure 6 shows that 1993 and 1995 are extreme years with min and max quantity of microorganisms. For the years, the comparative graphs for the two plots are shown in Fig. 7 where marks ♦ indicate the measured quantities of amylolytic microorganisms. The positions of these points confirm the correctness of the reconstructed graphs.

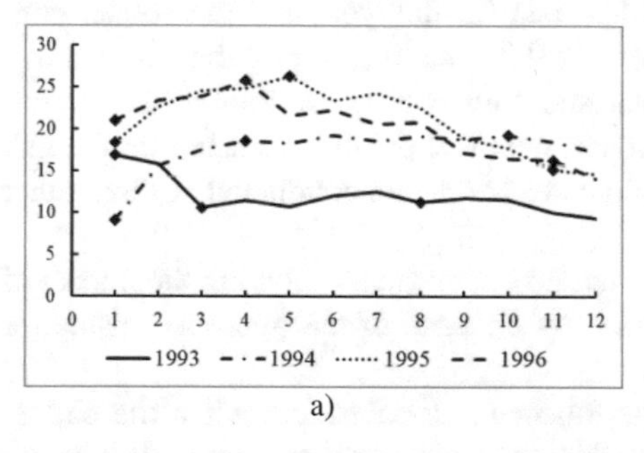

a)

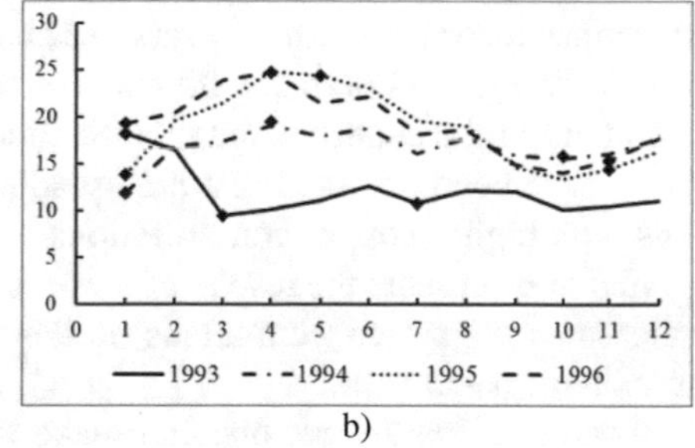

b)

Fig. 6. Reconstructed decade data of the amylolytic microorganisms amount during the vegetation periods of 1993–1996 years for (a) control plot and (b) experimental plot

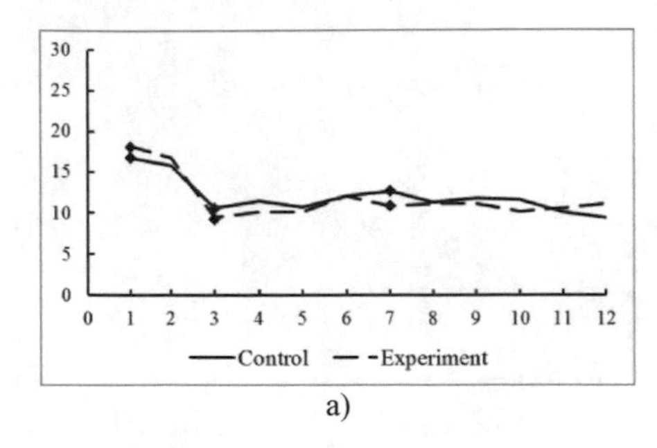

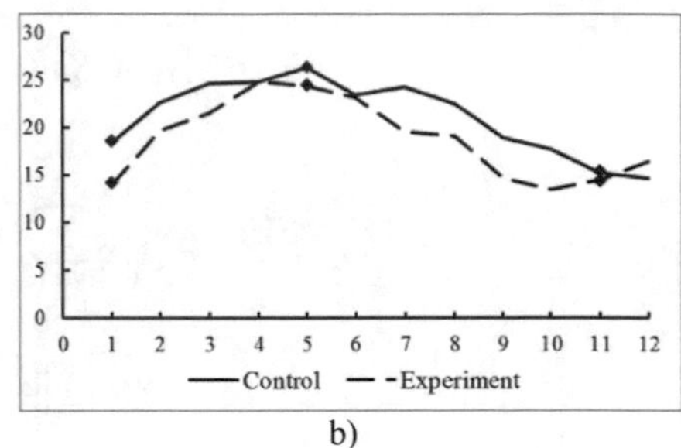

Fig. 7. Comparison of the reconstructed decade data for control and experimental plots for extreme years (a) 1993 and (b) 1995.

5.3 Generalized Model of Amylolytic Microorganisms Quantity Dependence on Copper

Consider the model obtained after substituting the model (3) for the control plot into the model (4) for the experimental plot:

$$Q_{exp} = 1.438T - 0.839H + 1.319CC \cdot T - 0.769CC \cdot H + 0.228T \cdot H \cdot CC - 0.166T \cdot H. \quad (8)$$

This model describes dependency of amylolytic microorganisms quantity only on the weather conditions during all 4 years together. As we demonstrate above, it makes it possible to reconstruct unmeasured decade data for the experimental plot.

As it is evident from (8) this model can be used to study the functioning of the amylolytic microorganisms under other weather conditions and other concentrations of copper. The appropriate results are given below for the 1993 extreme year.

Let we use the weather conditions of the year 1993 with the smallest quantity of microorganisms. For these conditions, we can study the virtual behavior of microorganisms at higher copper concentrations by substituting in the generalized formula (8) the concentrations that were observed in the next three years.

Thus, we calculate the decade quantity changes of the microorganisms during the season taking into account the decade weather data for this year and the copper concentrations recorded during 4 years: 5.1; 6.6; 7.8; 9.2. Finally, we plot the comparative chart showing the curve for the control plot and four curves for four different concentrations at the experimental plot. Such a chart would show the predicted results that would have been obtained in 1993 by additional experiments conducted at three other plots with higher copper concentrations.

Figure 8 presents the results of these computing experiments showing the impact of 4 measured copper concentrations on the quantity changes of the microorganisms for the extreme 1993 year.

Analyzing this figure, one can make the following conclusions: when the copper concentration is CC = 5.1 and CC = 6.6, the microorganisms can restore their quantity during the vegetation season; under concentration CC = 7.8, the microorganisms lose their ability to recover and live on a weakened vital level; upon concentration 9.2 they die out during six decade. Hence, this computing experiment shows that the generalized model can be effectively used to predict the effect of various copper concentrations on the quantity of microorganisms in soil and their ability to survive.

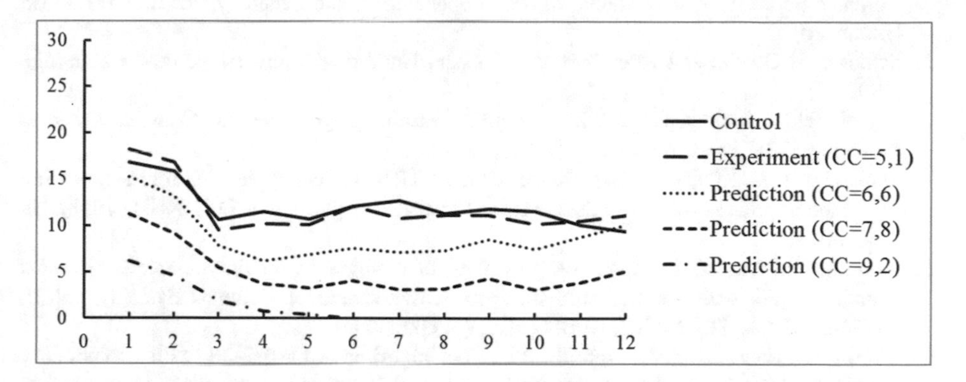

Fig. 8. Reconstructed decade data for control and experimental plots for extremal 1993 year and four copper concentration levels.

6 Conclusion

This research manifests effectiveness of the GMDH-based inductive approach, particularly the combinatorial-genetic algorithm COMBI-GA, to research behavior of microorganisms in polluted soil.

The presented modeling results for the case of amylolytic microorganisms demonstrate that among meteorological factors the temperature and humidity of the current decade make dominative effect on the activity of the microorganisms during vegetation seasons of 1993–1996 years. In the soil polluted by cooper their quantity is generally decreased but they begin to restore at the end of the each vegetation season. Thus, contamination of soils by copper is not critically dangerous for the vital activity of this kind of microorganisms under recorded concentrations.

The built nonlinear models highly fit experimental data and can be used for reconstruction of unmeasured data with the purpose of obtaining the uniform series of ecological observations and operative predicting the dynamics of microorganisms functioning under various ecological conditions.

The constructed generalized model of the dependence of amylolytic microorganisms activity on copper concentration makes it possible to determine critical ecological situations, particularly to simulate the effect of various copper concentrations on the change of quantity of microorganisms and to predict the conditions being dangerous for their survival. The results are especially important for a polluted environment impacting natural ecosystems.

References

1. Andreyuk, K.I., Iutynska, H.O., Antypchuk, A.F.: The Functioning of Soil Microbial Communities Under Conditions of Anthropogenic Load. Oberehy, Kyiv (2001). (in Ukrainian)
2. Schlegel, H.G.: General Microbiology. 7th edn. Cambridge University Press, Cambridge (1993)
3. Madala, H.R., Ivakhnenko, A.G.: Inductive Learning Algorithms for Complex Systems Modeling. CRC Press, New York (1994)
4. Stepashko, V.S., Koppa, Y.: Experience of the ASTRID system application for the modeling of economic processes from statistical data. Cybern. Comput. Tech. **117**, 24–31 (1998). (in Russian)
5. Iutynska, G., Moroz, O.: Inductive modeling of changes of amylolytic microorganisms quantity in plot with polluted soil. In: Inductive Modeling of Complex Systems, vol. 9, pp. 85–91. IRTC ITS NASU, Kyiv (2017). (in Ukrainian)
6. Moroz, O., Stepashko, V.: Hybrid sorting-out algorithm COMBI-GA with evolutionary growth of model complexity. In: Shakhovska, N., Stepashko, V. (eds.) Advances in Intelligent Systems and Computing II. AISC book series, vol. 689, pp. 346–360. Springer, Cham (2017)
7. Moroz, O.H.: Sorting-Out GMDH algorithm with genetic search of optimal model. Control. Syst. Mach. **6**, 73–79 (2016). (in Russian)
8. Moroz, O., Stepashko, V.: Inductive modeling of amylolytic microorganisms quantity in copper polluted soils. In: International Conference Advanced Computer Information Technologies, ACIT 2018, Ceske Budejovice, pp. 71–74 (2018)
9. Andreyuk, E.I., Iutynska, G.A., Petrusha, Z.V.: Homeostasis of microbial communities of soils contaminated with heavy metals. Mikrobiol. J. **61**(6), 15–21 (1999). (in Russian)
10. Stepashko, V.S., Koppa, Y.V., Iutynska, G.O.: Method for the missed data recovery in ecological tasks based on GMDH. In: Proceedings of the International Conference on Inductive Modeling, vol. 1, Part 2, pp. 113–117. DNDIII, Lviv (2002). (in Ukrainian)
11. Iutynska, G., Koppa, Y.: Modeling of the dependence of the number of microorganisms on the concentration of heavy metals in the soil. Control. Syst. Mach. **2**, 121–127 (2003). (in Russian)
12. Iutynska, G., Stepashko, V.: Mathematical modeling in the microbial monitoring of heavy metals polluted soils. In: Book of Proceedings of IX ESA Congress. Institute of Soil Science and Plant Cultivation, Warsaw, Poland, Warsaw-Pulavy, Part 2, pp. 659–660 (2006)
13. Holland, J.: Adaptation in Natural and Artificial Systems. An Introductory Analysis with Application to Biology, Control, and Artificial Intelligence. University of Michigan, Ann Arbor (1975)
14. Stepashko, V.S.: A combinatorial algorithm of the group method of data handling with optimal model scanning scheme. Sov. Autom. Control. **14**(3), 24–28 (1981)

On the Self-organizing Induction-Based Intelligent Modeling

Volodymyr Stepashko(✉)

International Research and Training Centre for Information Technologies and Systems of the NAS and MES of Ukraine, Kyiv 03680, Ukraine
stepashko@irtc.org.ua

Abstract. The article considers the issues of intellectualization of data-driven means for modeling of complex processes and systems. Some relevant terms of the modeling subject area are analyzed for an adequate explanation of the difference between theory-driven and data-driven approaches. The results are presented of the Internet retrieval for journal and book sources containing the term "intelligent modeling" and its variations in their titles and texts. Analysis of these sources made it possible to suggest an advanced conception of the intelligent modeling. It introduces three main levels of intellectualization of such means: offline intelligent modeling for constructing models of objects from available data; online intelligent modeling in an operating system of control or decision making; comprehensive intelligent modeling of work modes of a complex system. The original features of GMDH-based self-organizing inductive modeling are characterized showing that GMDH is one of the most powerful methods of data mining and computational intelligence for tasks being solved under conditions of uncertain and incomplete prior information. The inductive modeling algorithms can be the reasonable basis for creating advanced intelligent modeling tools.

Keywords: Complex system · Data-driven modeling · Inductive modeling Intelligent modeling · Self-organization · GMDH · Intelligent interface

1 Introduction

Mathematical modeling of complex systems is a base for effective solution of control and decision-making tasks. The construction of adequate predictive models is intended to avoid undesirable development of processes in such systems. There are many methods and tools to construct them. In recent decades, computer systems for control and decision making support have been widely developed and applied, and raising their level of intelligence, including modeling tools, is a topical task.

In this paper, the issues are considered regarding intellectualization of the tools for modeling of complex processes and systems from statistical or experimental data under uncertainty conditions. A general process of modeling is analyzed; some relevant terms of the modeling subject area are discussed to adequately explaining the difference between the known theory-driven and data-driven approaches.

N. Shakhovska and M. O. Medykovskyy (Eds.): CSIT 2018, AISC 871, pp. 433–448, 2019.
https://doi.org/10.1007/978-3-030-01069-0_31

There are presented some results of the Internet retrieval for journal and book sources containing the term "intelligent modeling" and its variations in their titles and texts. Based on the analysis of these available approaches, an advanced concept of intelligent modeling is proposed to deepen the existing viewpoint. It is suggested to implement various GMDH algorithms of inductive modeling as the base for constructing efficient intelligent modeling tools.

The term *inductive modeling* can be defined as a self-organizing process of evolutional transition from initial data to mathematical models reflecting some functioning patterns of the modeled objects and systems implicitly contained in the available experimental, trial or statistical information under the uncertainty conditions. The task of inductive modeling consists in an automated construction of a mathematical model approaching an unknown regularity of functioning the simulated object or process.

The paper is divided into four sections: Sect. 2 contains a discussion of the relevant terminology, Sect. 3 briefly analyzes some publications containing the term "intelligent modeling", and Sect. 4 introduces and describes a knowledge-based concept of such kind of modeling.

2 Discussion of the Subject Area Terminology

For an adequate explanation of the proposed concept of the intelligent modeling, it is expedient to analyze the content of some relevant terms.

2.1 On the Main Stages of a Modeling Process

Modeling in the broadest sense is a process of studying objects on their models. It assumes the replacement of the object-original with its conditional object-model for the research of the original properties over its model. A model can be an abstract, physical or other object whose properties are in some sense similar to the properties of the object under study. However, this conventional understanding can be considered insufficiently complete. Indeed, in modern conditions, in most cases not only existing "objects of cognition" are modeled, but also virtual, planned, designed, invented ones etc.

Among all the variety of existing and possible types of modeling – mental, figurative, verbal, physical, natural, abstract, schematic, and others – in this article we are interested primarily in mathematical and computer modeling as the most relevant to the tasks of control and decision making.

Mathematical modeling is a process of constructing and studying explicit mathematical models of objects. Moreover, this term covers a large range of tasks with different specialized names: designing models, approximation of dependencies, identification (structural and parametric) of models, regression analysis, recognition, classification, clustering, forecasting, etc.

Computer modeling in the narrow sense is a numerical study of mathematical models, while in the broadest sense it is simulative modeling, that is, the construction of aggregated models reflecting the structure of a complex system and the carrying out computational experiments to study possible modes of the system functioning under various conditions. Simulation modeling assumes the preliminary application of methods of system analysis for a correct and sufficiently complete computer representation of the studied system and its environment.

From the above we can conclude that the term "modeling" may cover three generally different but interrelated processes:

(1) the process of *constructing* a model;
(2) the process of model *analysis*, theoretically or numerically;
(3) the process of *computer-based study* of a complex system.

Generally, a modeling process can be represented in the form of the following successive stages: (1) the object observation; (2) construction of his model; (3) studying this model; (4) verification of its adequacy; (5) application of the model.

With this in mind, a generalized "life cycle" of the process of modeling of an arbitrary object or system can be represented in the form of Fig. 1.

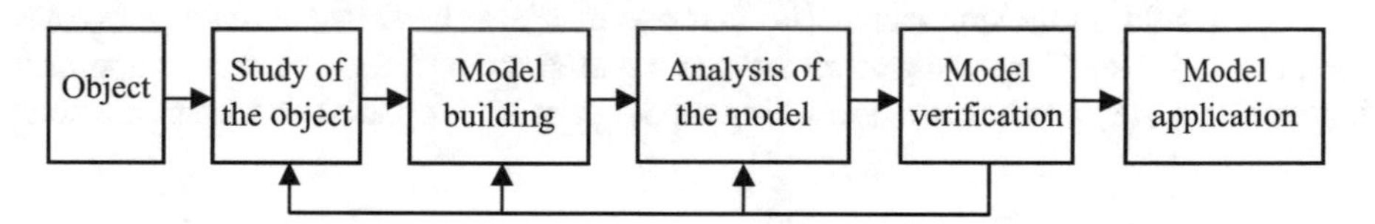

Fig. 1. General representation of main stages of the modeling process.

After unsatisfactory verification of the model's adequacy by given criteria one should return to some of the previous stages. The term "application" of the constructed model can be treated as the realization of its intended purpose, in particular: studying the object regularities, recognition of its status, testing its possible reaction on external impacts, prediction of its behavior, and also for supporting actions of a decision maker.

Note that a model may be called *adequate* when it helps to achieve a given modeling goal, e.g. increasing the efficiency of decision-making. A model being adequate in this sense does not necessarily have to be a "physical" one. It may not reflect at all the internal structure of the object and the laws of its functioning. It is enough for model to be plausible and not contradict the available data and a priori information [1].

2.2 Two Basic Approaches to Building Models

The process of mathematical modeling (Fig. 1) involves the mandatory presence of at least the following four main actors:

(a) the target modeling *object*, real or virtual;
(b) the *subject* that builds the model (designer, constructor, or modeler);
(c) the *goal* of the model building;
(d) the *model* being created according to the goal.

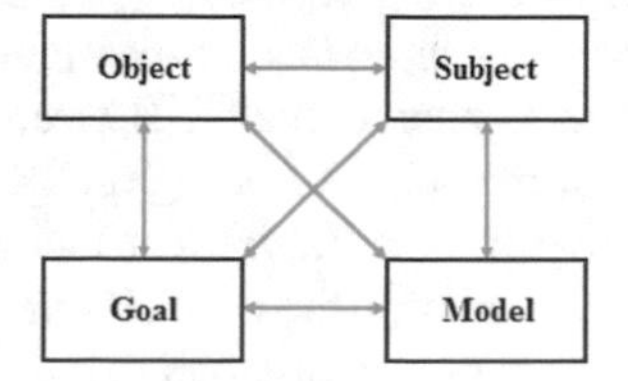

Naturally, the leading role in this process is played by the subject who defines the modeling goal, performs the model synthesis, checks its adequacy and makes decision regarding its application.

Obviously, the stage of model construction is the main, most labor-intensive and demanding intellectual efforts of the model designer in the whole modeling process; all the other stages may consist of rather routine operations. It is well known that to build models, a modeler may use two main approaches that can be treated as opposite: (1) model creation based on the study of laws of the object's functioning; (2) model synthesis based on analysis and generalization of experimental, statistical, or trial data on the object functioning.

In modern literature, these two approaches are called *theory-driven* and *data-driven* ones, a qualified comparative analysis of them is done in [2]. The fundamental difference between these approaches can be visualized in the form of Fig. 2. It is clear from the figure why the first of them is characterized as a "top-down approach" and the latter as a "bottom-up approach". The first one may also be called *deductive*, because the process is run "from the general laws to a partial model", and the second approach is then *inductive*, given that the resulting model generalizes partial observation results.

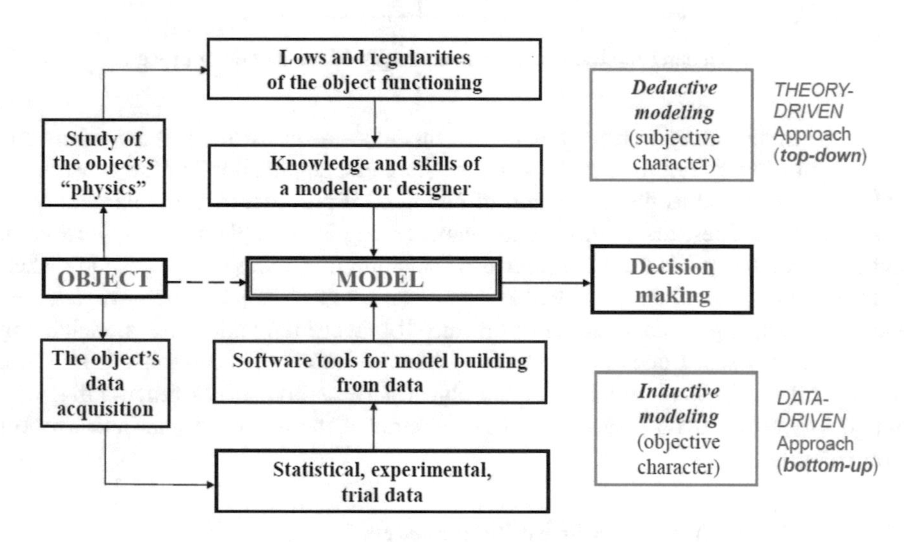

Fig. 2. Two basic approaches to the model construction

Consequently, the *deductive modeling* is the process of transition from the general laws and regularities of the object functioning to its specific (partial) model, while *inductive modeling*, respectively, is the process of transition from specific (partial) data to the general model. Hence, any constructed model can act as a partial or general phenomenon depending on the applied approach. These two approaches are sometimes called also as *theoretical* and *empirical* ones.

In spite of the explicit contrast between the approaches of deductive and inductive modeling, it is obvious that they complement each other as scientific methods and the best way for solving modeling tasks is the balanced combining both approaches. The attention to this fact is paid in [2] where the two approaches, top-down and bottom-up, are attributed to the field of artificial intelligence (AI) on the grounds that the first one comes to the development of expert systems with the use of inference, while the second provides now a wide arsenal of computational intelligence tools as an integral part of AI. The same paper discusses significant intersection of tasks, methods and means of scientific directions of the inductive type like Data Mining, Computational Intelligence, Soft Computing and Machine Learning.

However, it should be noted that the distinction between these deductive and inductive approaches has another peculiarity: the first of them has an explicitly *subjective* nature since the quality of the built model is largely determined by the knowledge and skills of a particular modeler or researcher; whereas the second approach is more *objective* since it uses widely recognized and well-tested computing instruments and tools.

Therefore, it is obvious that two models constructed for the same object according to these two approaches will, as a rule, be different. It is clear that a designer or a group of researchers who are able to harmoniously and skillfully combine both approaches will have the higher productivity in solving such problems.

With this in mind, we can give the following most general description of the intelligence of computer modeling systems (similarly to the well-known Turing test): intelligent modeling is the process of constructing models of objects with the use of knowledge and tools that ensure the quality of models at the level of a qualified modeler (user, designer).

Such a functional (albeit unconstructive) definition presupposes the implementation in the modeling system the knowledge of a modeler or a group of researchers both on tools for supporting the modeling process and on the subject domain of the object, as well as on ways to organize the interface with the system. Due to that we can in fact to emphasize these three elements as main features of an intelligent modeling system.

Thus, taking the above into consideration, one can say that *intelligent modeling* is a process of constructing models based on the

(a) computer system of supporting modeling tools,
(b) knowledge base of the subject area and the modeling experience, and
(c) intelligent interface with the system capable working both in automatic and interactive modes.

Obviously, the intelligence level of any computer modeling system may vary depending not only on computational intelligence tools but on the depth of knowledge and skills of an expert (or their group) in the given subject area, the qualification of the system designers and the interface flexibility.

3 Publications Regarding Intelligent Modeling Issues

The results of the Internet retrieval for journal and book sources containing the term "intelligent modeling" and its variations in their names and texts showed that there are quite a few such publications. As a result, only several articles from Ukraine [3–8] and the neighboring countries [9–15] were found on this subject, whereas in the English-language literature, the corresponding term is more widely spread [16–23]. It indicates that this topic has already stated its existence and relevance, and, on the other hand, that the concept of intelligent modeling is still not sustainable, commonly used and universally recognized.

As a result of the analysis of approaches to the development of intelligent methods and tools for modeling of complex processes and systems available in these sources, it is possible to make a completely unambiguous conclusion: absolute majority of the existing publications operating with the term “intelligent modeling” justify its application simply by using artificial neural networks, evolutionary algorithms, fuzzy logic and other methods and means of computational intelligence.

It should be noted that in the last decade this is already a fully developed and even dominant terminological trend. Based simply on the fact that the specialized groups of methods and means of Data Mining and its generalization Computational Intelligence have been formed in the field of artificial intelligence, many authors call “intelligent” any modeling, control and decision-making systems only because of the use, for example, neural networks (in the majority of cases), genetic algorithms or fuzzy logic.

This point of view, even stereotypical but formally acceptable, should be taken into account, but at the same time the depth of its validity may be questioned. For example, neural networks are specific means for constructing nonlinear input-output models of the “black box” type, although this significantly reduces the requirements to the level of “physical” knowledge on the modeling object. At the same time, any trained network is only an evaluator of the model output reaction to the input signals but it cannot disclose the internal operation mechanism of the object and, accordingly, does not increase the amount of knowledge to enhance the level of efficiency and intelligence any control or decision-making system.

On the other hand, among the sources under consideration [3–23] there are several publications that go beyond this stereotypic trend. Namely, in articles of early 1990th [9, 16] the emphasis is on the fact that intellectual resources of the interface with the user are needed to be used in modeling and control systems. In [7, 11], the intelligence of modeling tools is proposed to provide using knowledge of experts, experienced operators and decision makers (DM). Finally, in [4, 8, 17] it is argued that in order to increase the intellectual level of modeling, it is expedient to build ontological models of the subject field.

Consequently, as a result of a concise analysis of these publications, one can conclude that the simplified, rather formal concept of the intelligence of modern computer systems of modeling, control and decision-making can be substantially deepened by suggesting the use of the following basic elements: (1) methods and means of computational intelligence; (2) knowledge base of the subject field; (3) means of an intelligent interface. Thus, we reached again the three main elements of the intelligent modeling system indicated above in the Sect. 2.

This conclusion confirms the expediency of proposing a deeper concept of the intelligent modeling described below.

4 On the Advanced Concept of Intelligent Modeling

4.1 On the Intellectualization Problem

When discussing possible approaches to the intellectualization of the overall modeling process (Fig. 1), the question often arises: can the modeling process in principle be non-intelligent? One can answer it with two statements. First, as noted in the Sect. 2, the term "modeling" is understood to mean three fairly different processes: building a model, its analysis, and computer implementation, and only the first one of them, namely the model construction, directly requires knowledge, skills and intellectual efforts from the modeler. Consequently, it is the first process or stage of the model construction that is always intelligent for a researcher or model designer while the other two ones may not correspond to this characteristic since they have a certain independence from the first one and are often performed as independent and even routine stages of modeling.

Second, in the modern sense, the problem of intellectualization of the modeling process implies the intelligent behavior not a designer/modeler but an appropriate computer system with the main focus on the stage of model building. Therefore, in order to construct such a system, we must firstly define the concept of intelligent modeling, then analyze in general the subject area of model building in order to structure the knowledge about the main stages of the modeling process, the methods used and the conditions of the model efficiency and then formulate the task for the design of an appropriate computer system. We note at once that the questions of designing and implementing such a system require special consideration and are not touched upon in this paper.

When forming the intelligent modeling concept, it is necessary to take into account the general conditions under which the appropriate methods and tools can be used. It is advisable to distinguish the three main variants of the conditions: (1) stand-alone application for constructing models outside the control circuit; (2) built-in use in the operating control system; (3) comprehensive simulation modeling. With this in mind, one can specify three main levels of intellectualization of such systems which are defined and discussed below: intelligent offline modeling for constructing models of objects from available data base; intelligent online modeling as part of the operating system of control or decision-making; comprehensive intelligent modeling of operation modes of a complex system.

4.2 Three Main Levels of the Intelligent Modeling

1. *Separate intelligent modeling* (SIM) or IntM-offline is a static problem of intelligent support for the process of constructing models outside the control system, that is, on a fixed base or sample of data. The corresponding system reflected in Fig. 3 should be based on instrumental means of inductive modeling or, in more broad sense, computational intelligence, contain the data and knowledge bases and tools for intelligent interface between such system and a user.

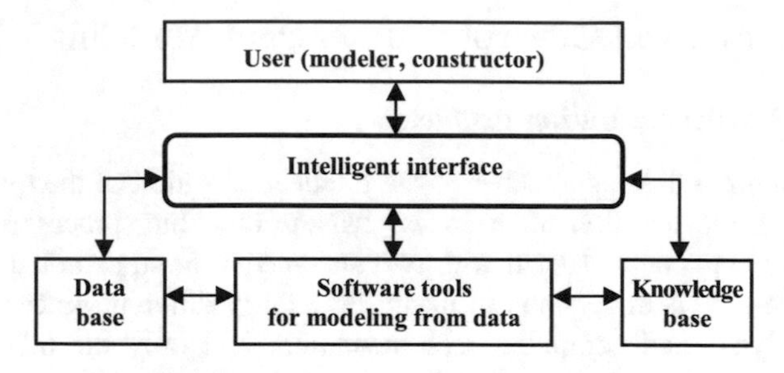

Fig. 3. Main structural components of the SIM system.

Intellectual resources are concentrated here not only in the knowledge base but also in the interface providing both interactive supports for the designer's decisions at all stages of the model constructing process and a fully automated solution of modeling tasks if needed, as well as all possible intermediate modes of the interactions.

2. *Embedded intelligent modeling* (EIM) or IntM-*online* is a dynamic task of automatic or automated construction, adjustment and modification of models that plausibly describe the behavior of objects under conditions of incomplete and uncertain a priori information on the properties of both objects and the environment in which they function, with the sufficient accuracy for effective decision making under conditions of changing situation. Such system, the main functions of which are shown in Fig. 4, should operate in the process of the object performing (with dynamic database) and based on accumulated and supplementing knowledge about the object and the environment of its operation. It should include all the elements of the previous system performing the "modeling" function as IntM-offline block, as well as means of supporting the process of constructing and applying models in the online mode.
3. *Comprehensive intelligent modeling* (CIM), or IntM-*complex* is the task of constructing and using a software for simulative modeling of a complex system in which there are tools that provide intelligent support for modeling processes of decision making in the simulated system with the goal of automatic detecting both optimal operating modes and possible adverse or critical scenarios. This software system should contain the following main subsystems (Fig. 5): (1) *information* subsystem for data collection, accumulation and retrieval; (2) *monitoring* subsystem executing functions of tracking, estimating and analyzing processes as well as online modeling and predicting for making informational support of current decision maker solutions – these in fact are the Int-online functions; (3) *DSS* subsystem, in which the feasible variants of possible solutions are formed and their effectiveness is evaluated according to certain criteria.

Such IntM-complex includes the two previous levels of intelligent modeling. It must necessarily have the function of accumulation of knowledge about the modeled object and the environment, as well as on the appropriate options for decision making in different changing situations.

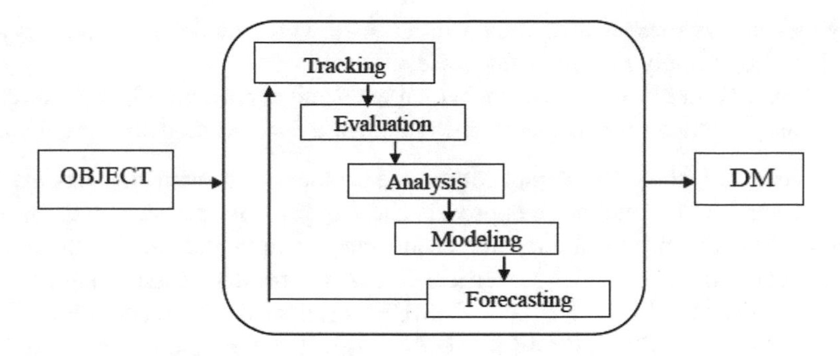

Fig. 4. Main functions of the EIM system.

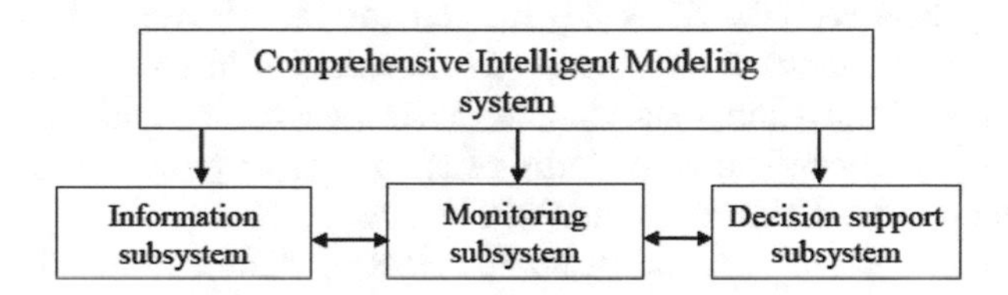

Fig. 5. Main subsystems of the CIM system.

4.3 GMDH as a Basis for Developing Intelligent Modeling Tools

Conditions of a Priory Information Uncertainty and Incompleteness. The complexity of the first two tasks, the separate and embedded modeling solved from statistical, experimental or observation data, is determined by the fundamental property of this class of problems: in practice, they are solved under conditions of uncertainty and incompleteness of information which significantly affects the quality of their solution. All varieties of such conditions can be attributed to two main groups:

(1) uncertainties ***related to data***, i.e. to a priori knowledge on a modeled object:

- *structural* meaning incomplete knowledge of the input-output relationships and not allowing to uniquely specify the structure of the model;
- *informational*: data is often of small volume, incomplete and inaccurate, and usually does not characterize all the variety of factors affecting the output;
- *stochastic* in the form of the unknown noise type and level in a data set;

(2) uncertainties ***related to data handling***, i.e. to the applied modeling techniques:

- *functional* consisting in the choice of an adequate basic set of functions or operators in which the model is constructed;
- *parametric* related to the choice of the method of solving the problem of parametric identification;
- *criterion-related* regarding the choice of the criterion for solving the main task of structural identification;

- *method-related*: it is unknown in advance which modeling method is appropriate to apply in a particular case;
- *technological*: it is unknown which adequate software tool to develop or to choose among the available ones to solve a specific modeling problem.

These aspects reflect the real complexity of the problem of constructing models within the framework of the inductive approach and suggest the intellectualization of the process of its solution. Existing methods and means of computational intelligence have to some degree the intellectual properties. One of the most successful among them is the group method of data handling (GMDH) originated by Academician Oleksiy Ivakhnenko in 1968 [24, 25] which greatly personifies the essence of the inductive approach and is still actively developing today [26].

GMDH-Based Inductive Modeling from Perspective of Data Mining and Computational Intelligence. GMDH is the method for model synthesis with automatic selection (self-organization) of structure and parameters of linear, nonlinear, difference and other models from short data sample under uncertainty and incomplete initial information to identify unknown patterns of an object or process functioning information on which is implicitly contained in the data.

Main principles of the inductive search. GMDH differs from other methods of constructing models by active applying the following fundamental principles:

(1) automatic generation of inductively complicated model variants;
(2) the use of non-finale decisions (freedom of choice) during the modeling process;
(3) sequential selection of best solutions according to external criteria for constructing models of optimal complexity.

GMDH application as an evolutionary process. GMDH has an original multilayer procedure for automatic generation of model structures which simulates the process of evolutionary biological selection with pairwise account (crossover) of sequential features. For comparison and selection of the best models, external criteria are used based on the sample division into two or more parts: parameter estimation and quality evaluation of models are performed on different subsets. This also automatically solves the known problem of "overfitting" of the network: in this method, such effect is consequently avoided due to the sample division.

In other words, in all the GMDH algorithms the sample division implicitly (automatically) ensures compliance with the known *trade-off principle* between the model *complexity* and its *accuracy* when constructing an optimal complexity model.

GMDH as an original neural network. Typical GMDH structure is also called a neural network, and the classic multilayer iterative algorithm MIA GMDH is called as the Polynomial Neural Network (PNN). In this case, one of the main elements of the algorithms, namely the polynomial partial description of two arguments, is considered as the elementary neural node of the PNN. There are the following features of originality and efficiency of the neural network of such neurons:

(a) absence of a predefined structure of the network;
(b) speed of the process of local training of neuronal weights;

(c) automatic global optimization (self-organization) of the network structure (numbers of nodes and hidden layers);
(d) fundamental destination exactly for construction of forecasting models;
(e) possibility to reduce the adjusted network into an explicit mathematical model unlike conventional NNs being input-output computing elements.

This means that the problem of the so-called "deep learning of neural network" [27] has been solved by the GMDH author just when creating his method: the number of layers of the GMDH neural network increases until the external criterion decreases and stops at the beginning its growth. The fact that GMDH was the very first example of the deep neural network is clearly noted in [28].

GMDH algorithms based on the fuzzy logic. For real-world tasks with interval errors in data, a Fuzzy GMDH method was introduced and some FGMDH algorithms were constructed and investigated with triangular, Gaussian and bell-wise membership functions [29, 30].

Similarly, the application of GMDH for structure optimization of fuzzy polynomial neural networks FPNN were developed and investigated, e.g., in [31, 32]. They implemented new modeling architectures combining polynomial neural networks (PNNs) and fuzzy neural networks (FNNs).

The development and investigations of combined GMDH-fuzzy neural networks were performed in [33, 34] where GMDH-wavelet neuro-fuzzy systems was suggested and investigated using advantages of neuro-fuzzy networks and GMDH.

GMDH-based hybrid architectures. New and efficient architectures of neuronets are recently intensively developed on the basis of hybridization of GMDH procedures and various approaches of computational intelligence and nature-inspired solutions, e.g.: particle swarm optimization [35], genetic selection and cloning [36], immune systems [37] etc. A survey of several approaches to construction and implementation of various hybrid GMDH algorithms is presented in [38]. The combinatorial-genetic algorithm COMBI-GA of the sorting-out type [39] was developed as the hybrid architecture of COMBI GMDH and GA.

GMDH neuronets with active neurons. Typical GMDH neurons in the form of quadratic polynomials of two arguments have the same fixed structure and may be called as "passive" ones, i.e. the PNN GMDH is the homogeneous net. In the 1990s, Ivakhnenko proposed a new type of GMDH network with active neurons [40, 41] or a heterogeneous network in which any neuron is in turn also a GMDH algorithm, due to that the structure of the neuron is optimized. As a result, all neurons can get different structures increasing the flexibility of configuring the network to a specific task. Networks of such type are also called as "twice multilayered" ones [42]. New types of architectures of such kind are described in [43, 44].

All the stated above allows attributing GMDH to the most effective methods of data mining and computational intelligence and suggesting to use it as the basis for the development of instrumental tools of intelligent modeling. This method and corresponding software means demonstrated good performance when solving real-world modeling problems of different nature in fields of environment, economy, finance, hydrology, technology, robotics, sociology, biology, medicine, and others [25, 26, 42–44].

On the Concept of Intelligent Interface for Modeling Tools. The problem of constructing and implementing various types of intelligent interface between users and software means has long history, see e.g. [45–47]. In this year, already the 23rd International Conference on Intelligent User Interface (IUI) will be held, see the home page [48]. As it is noted in [49], an IUI involves the computer-side having sophisticated knowledge of the domain and/or a model of the user. These allow the interface to better understand the user's needs and personalize or guide the interaction. Another interpretation of the IUI: a program which has an intelligent interface uses intelligent techniques in working with the user; it might use user models, or it might be knowledgeable about system functionality, or it might help its user [50].

Generally, there is an idea in most sources that an intelligent user-computer interface might predict what users want to do and present information based on this prediction adequately to their current needs. Besides, the following IUI features are eligible: adaptation to the needs of different users and the ability to learn new concepts and techniques.

In our study, we plan to realize the following main characteristics of the intelligent interface when designing the intelligent modeling system: user-system interactive mode at all successive stages of the modeling process; extracting and active utilizing the user's knowledge when performing the process; permanent monitoring, testing and correcting all the accepted user's decisions; training the user during interaction with the system; and others.

Under the user-system interactive mode we mean the spectrum of possible mutual actions starting from the fully automated mode for a novice user to the possibility of planning the whole modeling process by a skilled user (expert). This is possible only when the intelligent interface will have means for the ascertaining the level of user's skills and automatic adapting to them.

Evidently that to design and implement such kind of the intelligent user-system interface with the suggested features, there is the need to construct a knowledge base describing the whole inductive modeling process and containing the expedient rules of decision making at all stages of this process. Such a construction envisages the formalized structuring of theoretical, applied and expert knowledge in the subject area of inductive modeling, which is possible to realize on the basis of designing the ontology of this domain [51]. Hence, one must conclude that these two tasks of constructing the knowledge base and designing the intelligent interface should be solved in common and in strict coordination between them.

5 Conclusion

The issues of intellectualization of modeling tools for complex processes and systems were considered and an advanced concept of the intelligent modeling was proposed.

As a result of the analysis of approaches to the development of intelligent methods and tools for modeling complex processes available in actual publications, it is concluded that the vast majority of existing sources employing the term "intelligent modeling" justifies its implementation simply by using neural networks, evolutionary methods and other means of computational intelligence.

This simplified concept of intelligence of a modeling system is significantly deepened above. In contrast, in this study a new concept of intelligent modeling of complex processes and systems is developed, according to which it is proposed to distinguish the three basic levels of such process: (1) separate (offline) intelligent modeling as a static task of intellectual support of the process of building models out of a control system (from static data set); (2) embedded (online) intelligent modeling as a dynamic task of construction, adjustment and restructuring models in a system operation process (from dynamic database); (3) comprehensive intelligent modeling providing an intellectual support of modeling processes in a complex system to automatically detect optimal operating modes of a real system as well as possible adverse or dangerous modes.

It is shown that for the basis of the development of appropriate tools it is expedient to use algorithms of GMDH as an effective method of computational intelligence. This method represents the original and efficient means for solving a wide spectrum of artificial intelligence problems including identification and forecast, pattern recognition and clusterization, data mining and search for regularities.

References

1. Ljung, L.: System Identification: Theory for the User. Prentice-Hall, Englewood Cliffs (1999)
2. Manhart, K.: Artificial intelligence modelling: data driven and theory driven approaches. In: Troitzsch, K.G., Mueller, U., Gilbert, G.N., Doran, J.E. (eds.) Social Science Micromodeling, pp. 416–431. Springer, Heidelberg (1996)
3. Burov, Y.V.: A system of modeling of the intellectual network of business processes. In: Information Systems and Networks, Bulletin of the Lviv National Polytechnic University, vol. 610, pp. 34–39 (2008). (In Ukrainian)
4. Lytvyn, V.V.: Modeling of intelligent decision support systems using the ontological approach. Radioelektron. Inform. Control. **2**(25), 93–101 (2011). (In Ukrainian)
5. Zachko, O.B.: Intelligent modeling of product parameters of the infrastructure project (on the example of Lviv airport). East. Eur. J. Adv. Technol. **10**(1), 92–94 (2013). (In Ukrainian)
6. Korolev, O., Krulikovsky, A.P.: Intelligent methods for modeling of project control processes, Scientific notes of the V.I. Vernadsky Taurida National University, Economics and management series, vol. 26(1), pp. 73–86 (2013). (In Ukrainian)
7. Timashova, L.A., Vitkovski, T.: Technology of Intellectual Manufacturing Modeling of Virtual Enterprises. Informatics and modeling. Bulletin of the National Technical University "Kharkov Polytechnic Institute", vol. 32, pp. 136–147 (2015). (In Russian)
8. Valkman, Yu.R., Stepashko, P.V.: On the way to constructing the ontology of intelligent modeling. In: Inductive Modeling of Complex Systems, vol. 7, pp. 101–115. IRTC ITS NASU, Kyiv (2015). (In Russian)
9. Merkuriev, YuA, Teilans, A.A., Merkuryeva, G.V.: Intelligent modeling of production processes. Softw. Prod. Syst. **3**, 43–49 (1991). (In Russian)
10. Gladkiy, S.L., Stepanov, N.A., Yasnitsky, L.N.: Intelligent Modeling of Physical Problems. Institute of Computer Studies, Moscow (2006). (In Russian)
11. Mikoni, S.V., Kiselev, I.S.: Intelligent modeling of expert preferences on matrices of paired comparisons. In: Collected papers of Vseros Conference on "Simulation Modeling. Theory and Practice" SIMMOD-2007, CSRIETS, SpB, vol. 1, pp. 182–186 (2007). (In Russian)

12. Novikova, E., Demidov, N.: Means of intelligent analysis and modeling of complex processes as a key tool of situational control. World Inf. Technol. **3**, 84–89 (2012). (In Russian)
13. Gorbatkov, S.A., Rashitova, O.B., Solntsev, A.M.: Intelligent modeling in the problem of decision making within the tax administration. Bulletin Ufa State Aviation Technical University, vol. 1(17), pp. 182–187 (2013). (In Russian)
14. Polupanov, D.V., Khayrullina, N.A.: Intelligent modeling of segmentation of shopping centers on the basis of Kohonen self-organizing maps. Internet J. Naukovedenie, **1**, 1–15 (2014). (In Russian)
15. Glushkov, S.V., Levchenko, N.G.: Intelligent modeling as a tool to improve the management of the transport and logistics process. In: Proceedings of the International Scient.-tekhn. Conference of the Eurasian Scientific Association, pp. 1–5 (2014). (In Russian)
16. Amarger, R., Biegler, J.L.T., Grossmann, I.E.: An Intelligent Modelling Interface for Design Optimization. Carnegie Mellon University, Pittsburgh (1990)
17. Bille, W., Pellens, B., Kleinermann, F., De Troyer, O.: Intelligent modelling of virtual worlds using domain ontologies. In: Proceedings of the Workshop of Intelligent Computing (WIC) held in Conjunction with the MICAI 2004 Conference, Mexico City, pp. 272–279 (2004)
18. Balic, J., Cus, F.: Intelligent modelling in manufacturing. J. Achiev. Mater. Manuf. Eng. **24** (1), 340–349 (2007)
19. Al-Shareef, A.J., Abbod, M.F.: Intelligent modelling techniques of power load forecasting for the western area of Saudi Arabia. J. King Abdulaziz Univ. Eng. Sci **21**(1), 3–18 (2010)
20. Ćojbašić, Ž.M., et al.: Computationally intelligent modelling and control of fluidized bed combustion process. Therm. Sci. **15**(2), 321–338 (2011)
21. Kołodziej, J., Khan, S.U., Burczyński, T. (eds.) Advances in Intelligent Modelling and Simulation: Artificial Intelligence-Based Models and Techniques in Scalable Computing. Springer, Heidelberg (2012)
22. Sharma, A., Yadava, V., Judal, K.B.: Intelligent modelling and multi-objective optimisation of laser beam cutting of nickel based superalloy sheet. Int. J. Manuf. Mater. Mech. Eng. (IJMMME) **3**(2), 1–16 (2013)
23. Simjanoska, M., Gusev, M., Madevska-Bogdanova, A.: Intelligent modelling for predicting students' final grades. In: Proceedings of 37th International Convention on Information and Communication Technology, Electronics and Microelectronics (MIPRO), Opatija, pp. 1216–1221. IEEE Publisher (2014)
24. Ivakhnenko, A.G.: The group method of data handling – a rival of the method of stochastic approximation. Sov. Autom. Control. **1**(3), 43–55 (1968)
25. Madala, H.R., Ivakhnenko, A.G.: Inductive Learning Algorithms for Complex Systems Modeling. CRC Press, New York (1994)
26. Stepashko, V.: Developments and prospects of GMDH-based inductive modeling. In: Shakhovska, N., Stepashko, V. (eds.) Advances in Intelligent Systems and Computing II. CSIT 2017. AISC Series, vol. 689, pp. 474–491. Springer, Cham (2018)
27. https://en.wikipedia.org/wiki/Deep_learning
28. Schmidhuber, J.: Deep learning in neural networks: an overview. Neural Netw. **61**, 85–117 (2015)
29. Zaychenko, Y.: The investigations of fuzzy group method of data handling with fuzzy inputs in the problem of forecasting in financial sphere. In: Proceedings of the II International Conference on Inductive Modelling ICIM-2008, IRTC ITS NASU, Kyiv, pp. 129–133 (2008)
30. Zgurogvsky, M., Zaychenko, Yu.: The fundamentals of computational intelligence: System approach. Springer, Cham (2016)

31. Huang, W., Oh, S.K., Pedrycz, W.: Fuzzy polynomial neural networks: hybrid architectures of fuzzy modeling. IEEE Trans. Fuzzy Syst. **10**(5), 607–621 (2002)
32. Oh, S.K., Park, B.J.: Self-organizing neuro-fuzzy networks in modeling software data. Neurocomputing **64**, 397–431 (2005)
33. Bodyanskiy, Y., Vynokurova, O., Dolotov, A., Kharchenko, O.: Wavelet-neuro-fuzzy network structure optimization using GMDH for the solving forecasting tasks. In: Proceedings of the 4th International Conference on Inductive Modeling ICIM 2013, Kyiv, pp. 61–67 (2013)
34. Bodyanskiy, Y.V., Vynokurova, O.A., Dolotov, A.I.: Self-learning cascade spiking neural network for fuzzy clustering based on Group Method of Data Handling. J. Autom. Inf. Sci. **45**(3), 23–33 (2013)
35. Voss, M.S., Feng, X.: A new methodology for emergent system identification using particle swarm optimization (PSO) and the group method of data handling (GMDH). In: Proceedings of the Genetic and Evolutionary Computation Conference, pp. 1227–1232. Morgan Kaufmann Publishers, New York (2002)
36. Jirina, M., Jirina, Jr., M.: Genetic selection and cloning in GMDH MIA method. In: Proceedings of the II International Workshop on Inductive Modelling, IWIM 2007, pp. 165–171. CTU, Prague (2007)
37. Lytvynenko, V.: Hybrid GMDH cooperative immune network for time series forecasting. In: Proceedings of the 4th International Conference on Inductive Modelling, pp. 179–187. IRTC ITS NASU, Kyiv (2013)
38. Moroz, O., Stepashko, V.: On the approaches to construction of hybrid GMDH algorithms. In: Proceedings of 6th International Workshop on Inductive Modelling IWIM-2013, pp. 26–30. IRTC ITS NASU, Kyiv (2015). ISBN 978-966-02-7648-2
39. Moroz, O., Stepashko, V.: Hybrid sorting-out algorithm COMBI-GA with evolutionary growth of model complexity. In: Shakhovska, N., Stepashko, V. (eds.) Advances in Intelligent Systems and Computing II. AISC series, vol. 689, pp. 346–360. Springer, Cham (2017)
40. Ivakhnenko, A.G., Ivakhnenko, G.A., Mueller, J.-A.: Self-organization of neuronets with active neurons. Pattern Recognit. Image Anal. **4**(4), 177–188 (1994)
41. Ivakhnenko, A.G., Wunsh, D., Ivakhnenko, G.A.: Inductive sorting-out GMDH algorithms with polynomial complexity for active neurons of neural networks. In: Proceedings of the International Joint Conference on Neural Networks, pp. 1169–1173. IEEE, Piscataway, New Jersey (1999)
42. Muller, J.-A., Lemke, F.: Self-Organizing Data Mining. An Intelligent Approach to Extract Knowledge from Data. Springer, Heidelberg (1999)
43. Tyryshkin, A.V., Andrakhanov, A.A., Orlov, A.A.: GMDH-based modified polynomial neural network algorithm. In: Onwubolu, G. (ed.) Book GMDH-Methodology and Implementation in C (With CD-ROM), Chap. 6, pp. 107–155. Imperial College Press, London (2015)
44. Stepashko, V., Bulgakova, O., Zosimov, V.: Construction and research of the generalized iterative GMDH algorithm with active neurons. In: Shakhovska, N., Stepashko, V. (eds.) Advances in Intelligent Systems and Computing II. AISC series, vol. 689, pp. 474–491. Springer, Cham (2018)
45. Hancock, P.A., Chignell, M.H. (eds.) Intelligent Interfaces Theory, Research, and Design. North Holland, New York (1989)
46. Kolski, C., Le Strugeon, E.: A review of "intelligent" human-machine interfaces in the light of the ARCH model. Int. J. Hum. Comput. Interact. **10**(3), 193–231 (1998)
47. Rogers, Y., Sharp, H., Preece, J.: Interaction Design: Beyond Human-Computer Interaction, 3rd edn. Wiley, Chichester (2011)

48. http://iui.acm.org/2018/. Last Accessed 22 July 2018
49. https://en.wikipedia.org/wiki/Intelligent_user_interface. Last Accessed 29 July 2018
50. https://web.cs.wpi.edu/Research/airg/IntInt/intint-outline.html. Last Accessed 29 July 2018
51. Pidnebesna, H., Stepashko, V.: On construction of inductive modeling ontology as a metamodel of the subject field. In: Proceedings of the International Conference on Advanced Computer Information Technologies, University of South Bohemia, Ceske Budejovice, pp. 71–74 (2018)

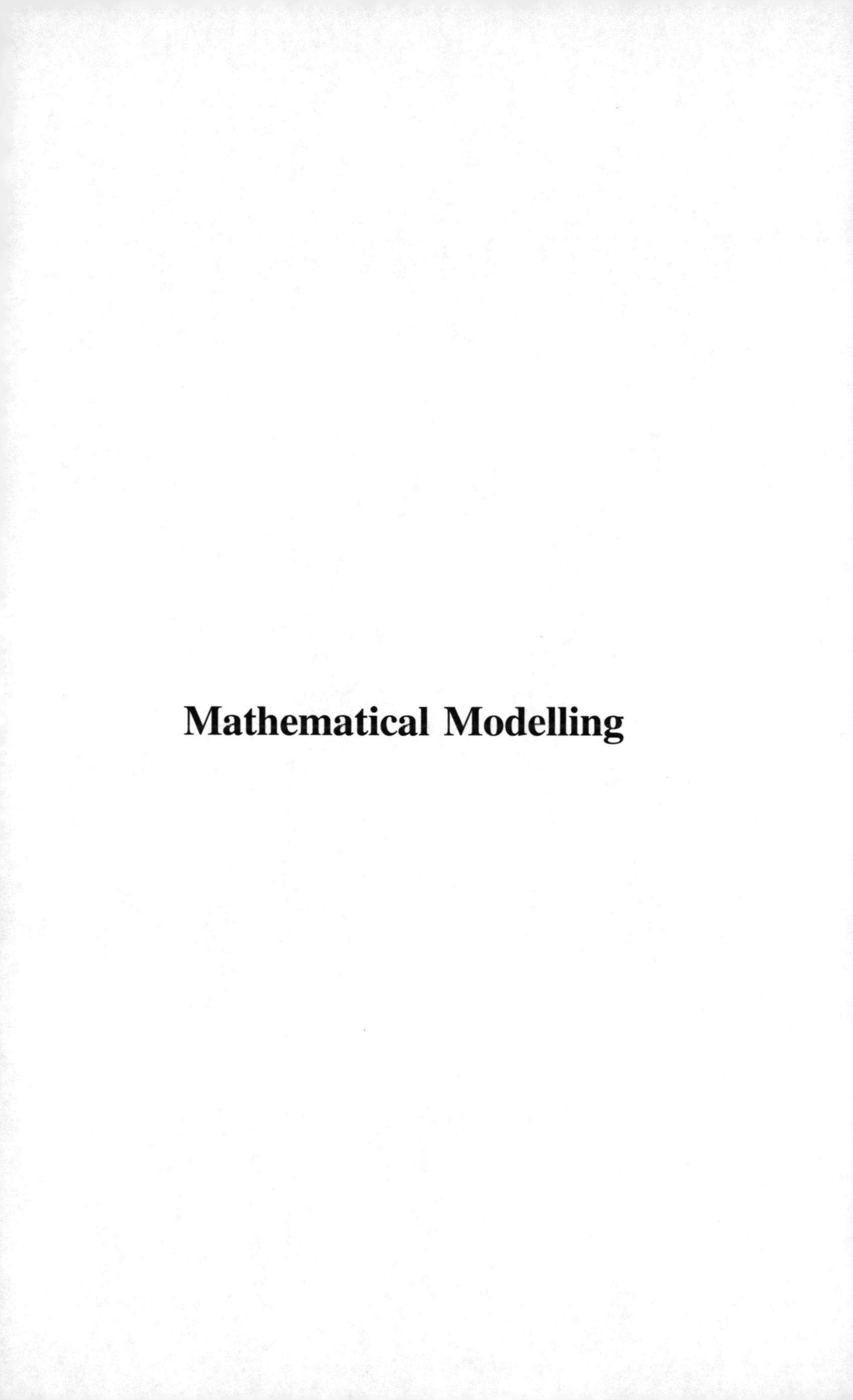

Mathematical Modelling

Modeling and Automation of the Electrocoagulation Process in Water Treatment

Andrii Safonyk[1(✉)], Andrii Bomba[2], and Ivan Tarhonii[1]

[1] Department of the Automation, Electrotechnical and Computer-Integrated Technologies, National University of Water and Environmental Engineering, Soborna Street, 11, Rivne 33028, Ukraine
safonik@ukr.net

[2] Department of Informatics and Applied Mathematics, Rivne State Humanitarian University, Ostafova Street, 29, Rivne 33000, Ukraine
abomba@ukr.net

Abstract. Electrocoagulation is successfully used for purifying various industrial wastewaters. In order to improve the quality of wastewater treatment and to reduce economic costs for achieving the specified pollution indicators, an automation system for controlling the wastewater treatment by an electrocoagulation method has been developed. A phenomenological mathematical model of the electrocoagulation process has been adapted for the investigation of the effect of current supply on the quality of wastewater treatment. The estimation of the dynamic characteristics of the electrocoagulation processes and the investigation of the influence of current supply on the concentration of the inlet pollution is made by means of simulation modeling. Based on experimental data a method of determination of regulator adjustment has been developed. A simulation model of the electrocoagulation process taking into account geometrical dimensions of the reactor, volume flow rate of the liquid and the applied current supply has been elaborated. An automation system for controlling concentration of nickel ions with an algorithm of operation for minimum power expenses has been proposed. The automation provides real-time control of the system based on SCADA-system WinCC Flexible. The automation control system with P-regulator can conserve up to 21.4% of power expenses.

Keywords: Mathematical model · Electrocoagulation · Current supply Wastewaters · Electrolyser · Volume flow rate

1 Introduction

In recent years, despite the slowdown in industrial production, problems of water pollution have become more severe. One of the ways of the pollution is the use of chemical methods, which are very common in technological processes. The chemical technology always releases waste that has a negative impact on the environment. For the treatment of such wastewaters end-of-life systems are often used. The application of worn out systems does not provide the allowable concentration of impurities which has

N. Shakhovska and M. O. Medykovskyy (Eds.): CSIT 2018, AISC 871, pp. 451–463, 2019.
https://doi.org/10.1007/978-3-030-01069-0_32

a negative impact on the ecosystems. Therefore, the development of modern water treatment systems is one of the priority tasks for the environmental protection.

Electrocoagulation is most efficient method of water treatment. Its advantages include: high productivity, low sensitivity to changes of composition of mixtures, the absence of the need for preliminary removal of dissolved organic substances, the availability of industrial-scale production of various electrocoagulation units, the absence of the need for coagulant addition. However, there are a number of shortcomings in these systems, among which are: the production of large amounts of sludge, high consumption of anode metal and electricity [1–5].

2 Formulation of the Problem

For the study of the effect of current supply on the quality of the wastewater treatment a phenomenological mathematical model of the electrocoagulation process has been adapted [6]:

$$\begin{cases} \frac{dC}{dt} = \frac{L}{V}(C_{in} - C) - (a_0 + a_1 T + a_2 C + a_3 I); \\ \tau \frac{dT}{dt} = \frac{(a_0 + a_1 T + a_2 C + a_3 I)VH}{KS} - (T - T_z) - \frac{c\rho L(T - T_{in})}{KS}, \end{cases} \quad (1)$$

$$C(0) = C_0, \; T(0) = T_0. \quad (2)$$

where C – concentration of the specified component in water; C_{in} – concentration of pollution at the inlet of the electrolyzer; T – water temperature in the reactor; T_z – ambient temperature; T_{in} – temperature of the inlet fluid; I – current supply; $a_0 \ldots a_3$ – empirical coefficients; S – surface area of the apparatus working zone; L – volume flow of liquid; V – volume of the working area of the electrolyzer; c – specific heat of a liquid; ρ – water density; K – coefficient of heat transfer; τ – time constant.

The model takes into account the processes occurring in the reactor as a combination of various factors depending on: the concentration of suspended substances in the water, the applied current supply, temperature of the liquid, temperature of the external environment, temperature of the water in the reactor, the design parameters of the coagulator.

3 Literature Review

In recent years, a great deal of scientific research has been devoted to simulation of electrocoagulation wastewater treatment:

- modeling of electrocoagulation processes for reducing concentration of nickel in galvanic wastewater, performed on experimental scale. Determination and evaluation of the following parameters of the electrocoagulation process: efficiency of removal, specific energy consumption and produced sediment at different direct current voltage (5 V, 7.5 V and 10 V) at different time scale (30, 60 and 90 min).

The results show that optimal electrocoagulation process was obtained at 5 V DC and 76.5 min [1];

- in work [2] the method of electrocoagulation is successfully applied for the treatment of various industrial waste waters;
- in [3] it was established that the efficiency depends mainly on the current density and the flow velocity in the reactor;
- work [4] is devoted to investigations of the effectiveness of the electrocoagulation process performed on the basis of artificial wastewater.
- this paper [5] presents a comprehensive review on its development and design. The most recent advances on EC reactor modeling are summarized with special emphasis on four major issues that still constitute the cornerstone of EC: the theoretical understanding of mechanisms governing pollution abatement, modeling approaches, CFD simulations, and techno-economic optimization.

At the same time, during the experimental studies of electrocoagulation units, the following problems take place:

- the choice of optimal operation modes of the electrocoagulator (due to the fact that the liquid is affected by electromagnetic and chemical processes in the liquid and physical and chemical processes within the electrode space);
- influence of the main parameters of the electrocoagulation process: the type of current (constant, alternating or reversible), the voltage applied, the electrodes material, the electrodes shape (flat, tubular, box, lamellar), the distance between the electrodes, the cleanliness of the electrodes surface and also hydraulic and thermodynamic processes and their change over time.

Usually for the description of the processes in the reactor mathematical models that do not take into account a number of electrocoagulation parameters are used. So in [7, 9] a mathematical model of the electrolyzer is proposed which takes into account the parameters that influence the speed and quality of the electrocoagulation treatment. However, for the completeness of the information it is necessary to study the influence of the change in the density of current on the concentration of suspended components which is the main controlling factor. In [10–12], the process of electrocoagulation is studied experimentally. The change in the concentration of pollution at various values of current density, fluid flow and acidity of the medium is investigated. But the proposed mathematical model describes a reactor of specified geometric sizes taking into account change of the basic perturbations within specified limits and does not take into account the effect of water temperature change on the quality of the purification.

Taking into account the above-mentioned, the purpose and tasks of this work are: to adapt the mathematical model of the electrocoagulation treatment which takes into account the processes occurring in the reactor and makes it possible to change the controlling factor (the current supply, the flow rate, the geometric dimensions of the reactor and the inlet concentration of suspended particles); to conduct simulation of the corresponding process; to investigate the influence of the current supply on the quality of wastewater treatment at the variable input pollution concentration; to develop an automation control system that minimizes power consumption for wastewater treatment in compliance with established environmental norms of pollution concentrations.

4 Materials and Methods

To simulate the problem (1)–(2) the Simulink application of the Matlab software is used, which allows to construct graphic block diagrams, to simulate dynamic systems, to examine the system performance and improve it [8, 13, 18].

The simulation model is presented in Fig. 1a. The input parameters of this model include: inlet electrolyzer concentration of pollution, volume flow rate of liquid, reactor working volume, applied current supply, effect of thermal processing of electrocoagulation, coefficient of heat transfer of the coagulator, surface area of the working zone, external temperature and inlet temperature of the liquid, heating time, the specific heat and density of the liquid. The output parameter is the concentration of the target component pollution at the reactor outlet.

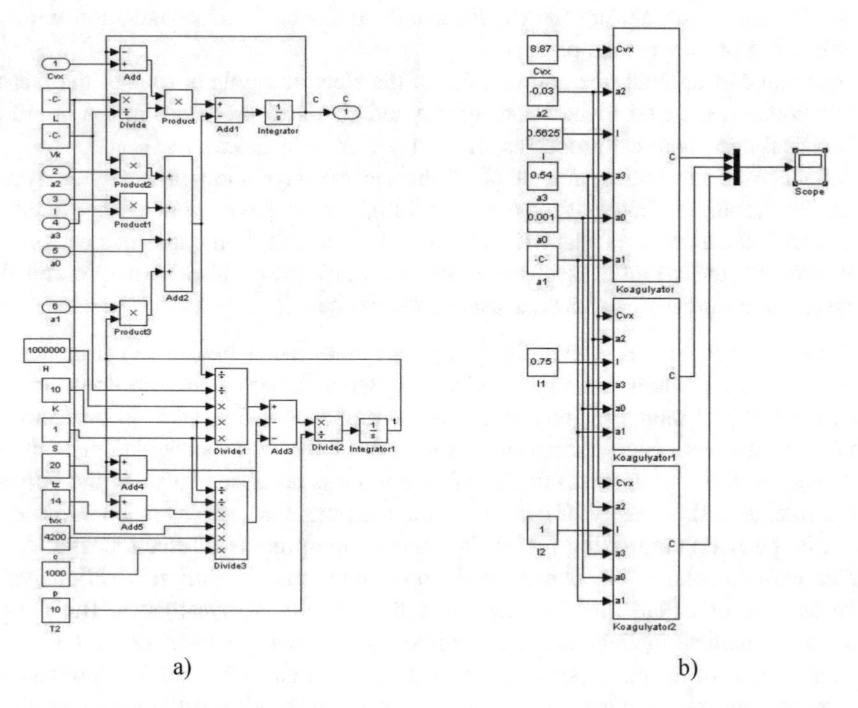

Fig. 1. Simulink model of the operation of the electrocoagulator (a) and Simulink model of the study of the effect of current supply on the concentration of the target component pollution (impurities) at the reactor outlet (b)

To study the effect of changes in the current supply value on the quality of wastewater treatment, a model 1, b has been developed, consisting of 3 reactors in which all input data is common in besides the current density. The simulation results will be displayed in the same coordinate plane.

To confirm the adequacy of the model, the experimental data from [14–16] are used and presented in Table 1.

Table 1. Output parameters of the experiment

Time, min	Current density I, A/m^2		
	9	12	16
	Current supply of the modeling electrolyzer, A		
	0,5625	0,75	1
	Concentration if the nickel ions, mg/l		
0	8,87	8,87	8,87
5	6,23	5,9	5,24
10	4,22	5,4	4,71
15	3,84	5,33	4,67
20	3,29	3,81	4,64
25	3,06	3,69	3,27
30	2,81	3,43	3,1
35	2,67	3,17	3,03

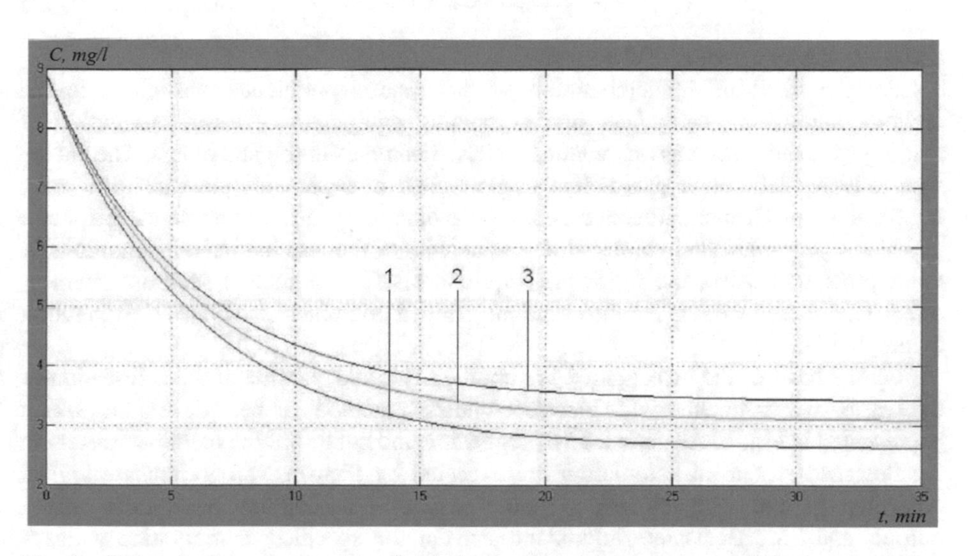

Fig. 2. Change in the concentration of nickel ion contamination at the coagulator outlet during time for current strength $I = 0.5625$ A – curve 3, $I_1 = 0.75$ A – curve 2, $I_2 = 1$ A – curve 1

The automated control provides energy efficient performance of the electrocoagulation process of treatment. The output data from the table allows to develop P-regulator for the proposed operating modes. To obtain the configuration of the systems approximation of the data from Table 2 is performed to obtain the coefficients of the polynomial series. The polyfit function has been applied and the appropriate code has been created.

Table 2. Dependence of current supply on the inlet concentration of nickel ions for automation control

№	Concentration if the nickel ions, mg/l	Current supply, A
1	8,87	0,75
2	4	0,132
3	5	0,258
4	6	0,385
5	7	0,511
6	8	0,638
7	10	0,891
8	15	1,524
9	20	2,155

The dependence of the change in the current supply on the input concentration of nickel ions has been obtained and represented by the Eq. (3)

$$I = 0.1265 \cdot C_{in} - 0.3738 \tag{3}$$

and acts as a P-regulator of the system.

For the hardware implementation of the system, preference was given to the devices, automation equipment and computing equipment of serial production of leading companies such as Schneider Electric, Siemens, Omron and others. The choice was influenced by such parameters of controlled environment (pressure, flow rate, temperature, mechanical influences, conditions of control and measurement), the nature and size of the controlled volume, the productivity of the installation, the conditions of labor protection. Also, the following requirements for automation such as accuracy, sensitivity, inertia were taken into account. Unified equipment facilitates the operation of the system.

On the basis of the investigated system the functional scheme of the automation of the wastewater treatment by the electrocoagulation method has been developed, which is presented in Fig. 6. All functions of regulation and control of the main parameters in the designed system of automation are executed by Siemens' programmable logical controller S-7300 with expansion modules, SM331 analog inputs, SM332 analog outputs and SM323 discrete inputs/outputs. For the specified configuration it has 2 analog inputs, 2 analog outputs, 8 discrete inputs and 8 discrete outputs which is enough to operate the system.

To provide allowable concentration of impurities in waste water, a system consisting of an electric coagulator, a sand separator, a primary sedimentation tank, a primary clarification chamber, an aeration chamber, 3 sensors of sediment level, a sensor concentration of hexavalent chromium, a programmable logic controller S7-300, 3 control valves, a frequency controlled compressor.

The system has 4 control loops which provide the following technological parameters:

(1) the concentration of nickel ions at the electrocoagulator inlet;
(2) the level of heavy mineral impurities (sand) in the sand separator;
(3) the level of sediment in the sedimentation tank;
(4) level of sediment in the electrocoagulator.

The automation control system is implemented in Simatic Step7 which has pre-configured hardware of the controller and developed character table that describes all inputs, outputs, labels with their symbolic, hardware denoting and comments.

In the developed system, the program of operational management of wastewater treatment is written by the language of ladder diagrams (Ladder Language - LD).

In order to provide an allowable concentration of nickel ions in wastewater, it is used a system which includes an analyzer with an analog output signal, a controller, a rectifier that controls the amount of voltage applied to the electrocoagulator. The signal from the unified sensor is supplied to the controller, which is processed in accordance to the limits of measurement of concentrator in OB1 network1. The obtained real value of the concentration of nickel ions in water is saved in the address MD14, which in the next block of the program is processed according to the received dependence of the change of the allowable concentration of impurities from the input concentration. Initially, the actual value of the parameter at the electrocoagulator inlet is scaled and the deflection rate is added. The optimal value of the applied voltage which minimizes the power consumption is determined and delivered to the analog output of the controller after time T1 during which the water flows from the electrocoagulator inlet to the working zone. The received signal from the controller enters the regulator, which amplifies the signal and controls the applied voltage for determining concentration of nickel ions. Such control algorithm is implemented in the controller and presented in Fig. 7. To monitor the operation of the system in real time a visualization window in the WinCCFlexible environment, shown in Fig. 8, has been developed. The system provides automatic and manual operation for quality maintenance of the installed equipment and the possibility of remote control of the system.

5 Results

On the basis of simulation the influence of current supply on the concentration of nickel ions has been carried out. State Standards of Ukraine was used as the baseline data, according to which the concentration of nickel ions in waste water should not exceed 3.5 mg/l. But due to the inertia of the regulator and the system an allowable concentration of impurities was on the level of 3.0 mg/l. Experimental data from Table 1 was also used for simulation according to which during (in) 35 min the system should provide allowable modes. So, the modeling was guided by the following principle: in 35 min the system should provide the concentration of nickel ions up to 3.0 mg/l. When the concentration of impurities at the inlet was changed the value of the current supply was obtained. The results of simulation are shown in Table 2.

During experimental studies the inlet concentration which varies within 8–10 mg/l was specified. According to environmental norms the outlet concentration should not exceed 3.5 mg/l. According to the technological requirements the rate flow of the coagulant should be 25 L per hour. In order to provide these conditions for the specified sizes of the reactor it is necessary to maintain at rectifier the current supply 0.9 A for 24 V using a linear regulator, which corresponds to 8.25 A of 220 V alternating current based on which electricity is paid. During the day of operation of this electrocoagulator without the automation system only rectifier will consume 44.9 kWh of electricity.

The developed system has very low inertia in the control channel and there are no perturbations that would change the concentration of nickel ions after measuring. So, it is not feasible to use a PI or PID a regulator which would only complicate the system. It is sufficient to use a P regulator.

According to the results of simulation the current supply between the electrodes in the proposed automation system with the P-regulator will be on the level of 0.7289 A. 6.68 A will be supported in the loop of the power supply respectively. Thus, the electricity consumption per day of the rectifier will be 35.3 kW-h. Therefore, the proposed automation system will save up to 21.4% of electricity costs. Taking into account the fact that the experimental installation supports the current supply not more than 1 A but in industrial systems the current supply between the electrodes is equal to 80 A, the application of this system will provide environmental standards with significant electricity cost savings.

As a result of simulation based on the input data $C_{in}|_{t=0} = 8.87$ mg/l, $T|_{t=0} = 18\ °C$, $L = 0.006$ m^3/s, $V = 0.03125$ m^3, $H = 1$ MJ/(g-s), $K = 10$ W/(m^2-K), $S = 1$ m^2 $T_z = 20$ °C, $T_{in} = 14$ °C, $c = 4200$ J/(kg-K), $\rho = 1000$ kg/m^3, $\tau = 10$, $a_0 = 0.001$, $a_1 = 0.000007$, $a_2 = -0.03$, $a_3 = 0.54$ the following results were obtained (Fig. 2). Using the experimental data from Table 1 and the results of modeling the system at a different current supply value using standard Matlab materials, the results were obtained on a grid of experimental data (Fig. 3).

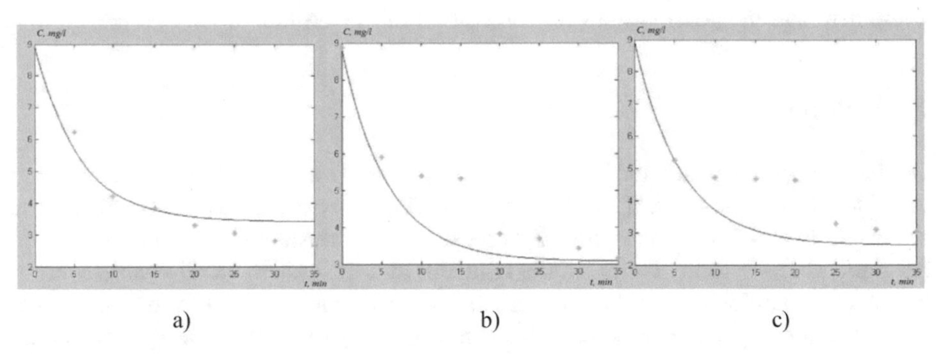

Fig. 3. Change of the concentration of pollution during time for current density $I = 9$ A/m^2 - (a), $I = 12$ A/m^2 - (b) and $I = 16$ A/m^2 - (c)

To check the operation of the regulator the simulation model has been developed and presented in Fig. 4. It consists of two subsystems: "Regulyator" (see Fig. 4(b) and "Koagulyator" (see Fig. 4(a).

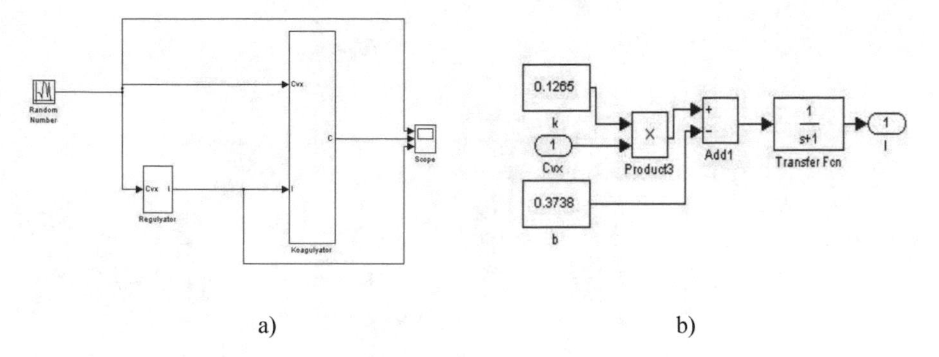

Fig. 4. Simulation model for testing the system (a) and simulation model of the regulator (b)

The functional dependency 3 which optimizes the current supply between the cathode and the anode and allows to save energy costs is implemented in the "Regulyator" subsystem. However, in real systems the current does not change instantaneously but in inertial manner. So an additional transfer function is introduced that describes the inertia of the system.

After simulation the results which are presented in Fig. 5 have been received. For inlet concentration of pollution the random signal generated by the "Random number" block has been used (see Fig. 5a).

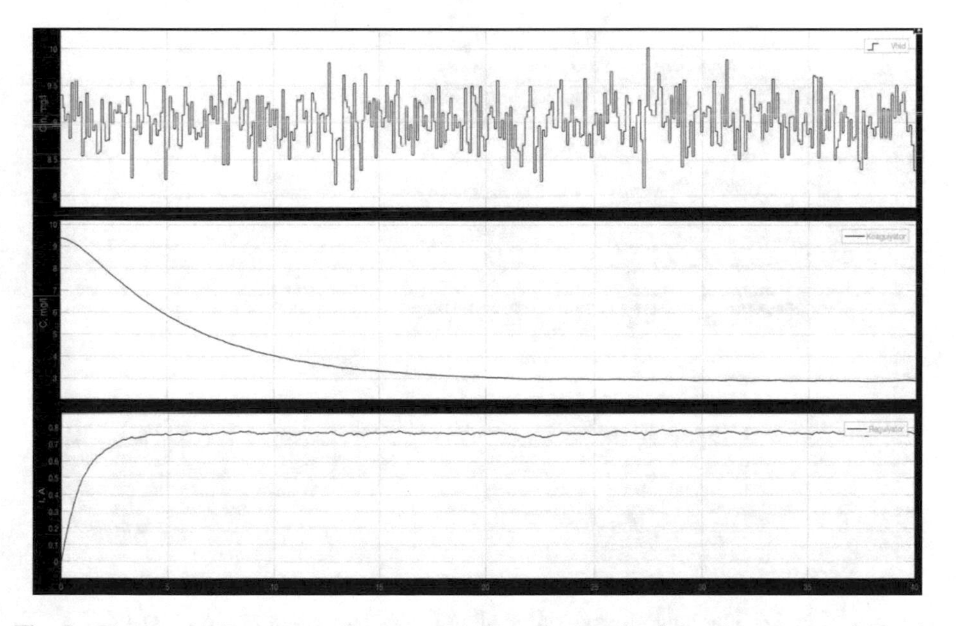

Fig. 5. Change of inlet concentration, concentration of contamination and current supply for the operation of the P-regulator

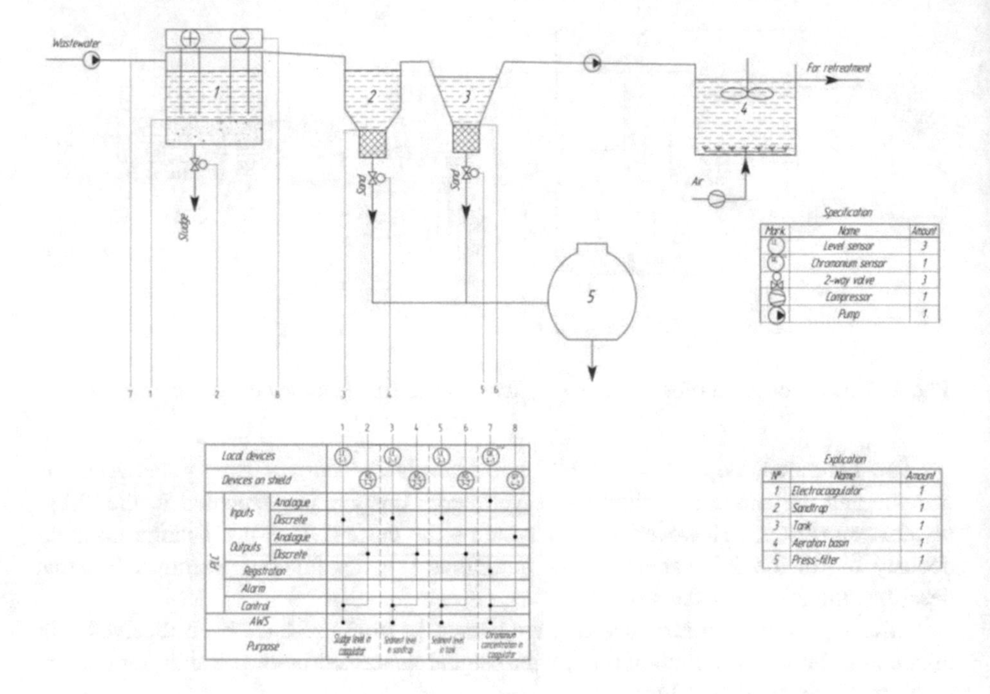

Fig. 6. Functional diagram of the electrocoagulator automation

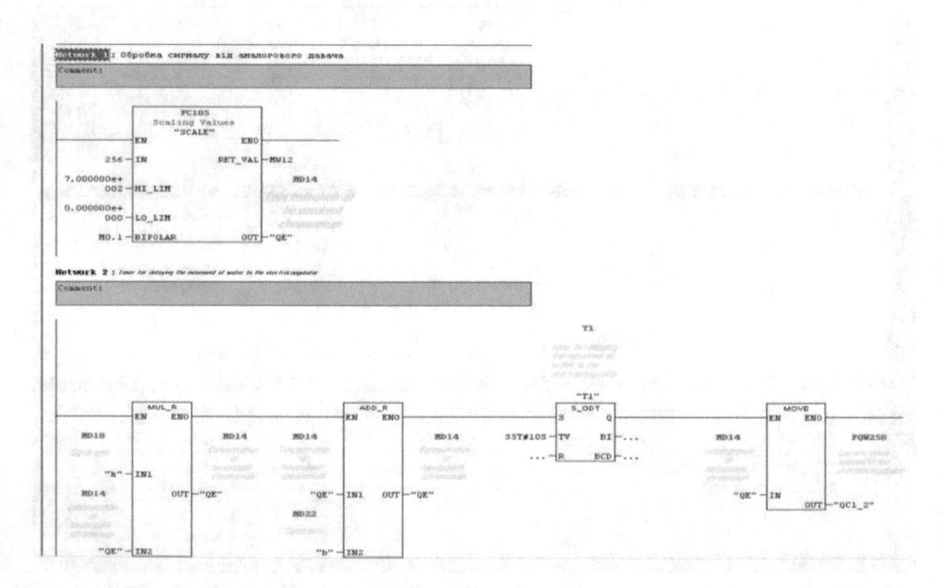

Fig. 7. Subroutine for providing allowable concentration of nickel ions

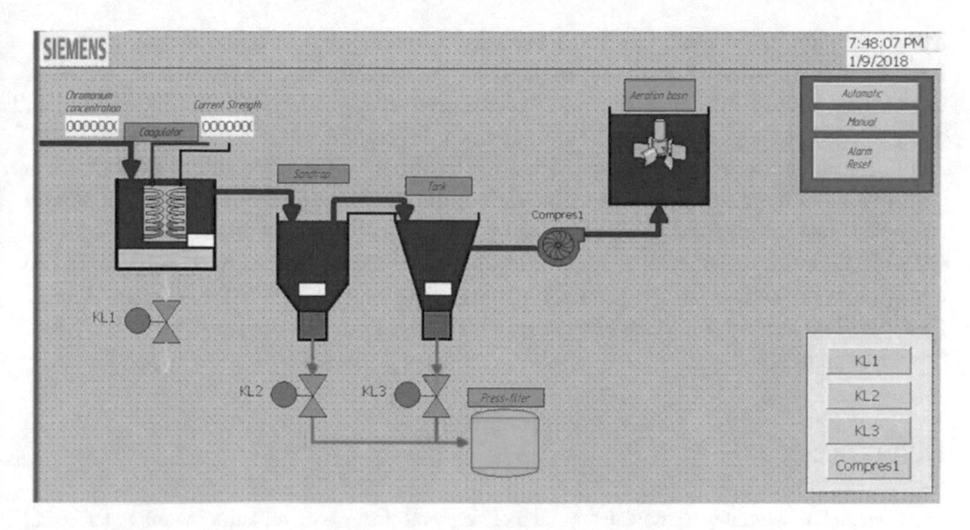

Fig. 8. Diagram of the process for operation control of the electrocoagulator

6 Discussion

Figure 2 deals with the case when $C_0 = C_{in} = 8.87$. However, during time, depending on the current density, it can be clearly seen that the quality of waste water treatment is better where there is higher current supply. The developed model allows to calculate the value of the applied current supply which provides the specified values of concentration of (impurities) target component for meeting environmental regulations. The obtained model makes it possible to investigate the dynamics of changes in the initial concentration of pollution not only with stable inlet parameters but also with the variables ones that takes place on real objects.

The presented results of experiments (see Table 1) and modeling (see Fig. 3) indicate that the model adequately describes the changes that take place in the reactor. The deviations that can be noticed in Fig. 3(c) are due to some inaccuracies of the experiment but on sufficient level demonstrate the nature and order of the output parameter change.

It is seen from Fig. 5 that the developed automation system responds to the change of the inlet concentration of pollution and influences the current supply for providing minimum power consumption. At the same time, as can be seen from Fig. 5, the concentration of nickel ions does not exceed the established norms.

7 Conclusions

The automation control system for waste water treatment by electrocoagulation method is developed. The system provides energy-saving mode of operation. A phenomenological mathematical model of the electrocoagulation process has been adapted for the investigation of the effect of current supply on the quality of wastewater treatment.

Means of simulation modeling have been used for the evaluation of dynamic characteristics of the processes occurring in the electrocoagulator and for the investigations of the current supply influence on the concentration of the inlet pollution. A method of determination of regulator adjustment has been developed based on experimental data. A simulation model of the electrocoagulation treatment taking into account geometrical dimensions of the reactor, volume flow rate of the liquid and the applied current supply has been elaborated. An automation system for controlling concentration of nickel ions with an algorithm of operation for minimum power expenses has been proposed. The automation provides real-time control of the system based on SCADA-system WinCC Flexible. The automation control system with P-regulator can conserve up to 21.4% of power expenses.

References

1. Djaenudin, Muchlis, Ardeniswan: Nickel removal from electroplating wastewater using electrocoagulation. IOP Conf. Ser. Earth Environ. Sci. **160**, 1–8 (2018)
2. Elnenay, A.E.M.H., Malash, G.F., Nassef, E., Magid, M.H.A.: Treatment of drilling fluids wastewater by electrocoagulation. Egypt. J. Pet. **26**(1), 203–208 (2016)
3. Pavón, T., del Río, G.M., Romero, H., Huacuz, J.M.: Photovoltaic energy-assisted electrocoagulation of a synthetic textile effluent. Int. J. Photoenergy **3**, 1–9 (2018)
4. Perren, W., Wojtasik, A., Cai, Q.: Removal of microbeads from wastewater using electrocoagulation. Am. Chem. Soc. Omega **3**, 3357–3364 (2018)
5. Nepo Hakizimana, J., Gourich, B., Chafi, M., Stiriba, Y., Vial, C., Drogui, P., Naja, J.: Electrocoagulation process in water treatment: a review of electrocoagulation modeling approaches. Desalination **404**, 1–21 (2017)
6. Kruglikov, S.S., Turaev, D.U., Borodulin, A.A.: Local electrochemical purification of washing waters of electroplating from heavy metal ions in a slit diaphragm electrocoagulator with an insoluble anode. Galvanotech. Surf. Treat. **12**(4), 35 (2004)
7. Philipchuk, V.L.: Cleaning of Multicomponent Metal-Containing Sewage from Industrial Enterprises. UDUVGP, Rivne (2004)
8. Bomba, A., Safonyk, A., Fursachik, E.: Identification of mass transfer distribution factor and its account for magnetic filtration process modeling. J. Autom. Inf. Sci. **45**(4), 16–22 (2013)
9. Yakovlev, S.V., Voronov, Y.: Water Disposal and Wastewater Treatment, 4th edn. Publishing of the DIA, Moscow (2006)
10. Ponkratova, S.A., Emelyanov, V.M., Sirotkin, A.S., Shulaev, M.V.: Mathematical modeling and management of sewage treatment quality. Vestn. Kazan Technol. Univ. **5–6**, 76–85 (2010)
11. Adetola, V., Lehrer, D., Guay, M.: Adaptive estimation in nonlinearly parameterized nonlinear dynamical systems. In: American Control Conference on O'Farrell Street, San Francisco, USA, pp. 31–36 (2011)
12. Filatova, E.G., Kudryavtseva, E.V., Soboleva, A.A.: Optimization of parameters of electrocoagulation process on the basis of mathematical modeling. Bull. Irkutsk. State Tech. Univ. **4**(75), 117–123 (2013)
13. Bomba, A.Y., Safonik, A.P.: Mathematical simulation of the process of aerobic treatment of wastewater under conditions of diffusion and mass transfer perturbations. J. Eng. Phys. Thermophys. **91**(2), 318–323 (2018)

14. Shantarin, V.D., Zavyalov, V.V.: Optimization of processes of electrocoagulation treatment of drinking water. Scientific and technical aspects of environmental protection. SAT Rev. Inf. **5**, 62–85 (2003)
15. Khalturaina, T.I., Rudenko, T.M., Churbakova, O.V.: Investigation of the technology of electrochemical treatment of sewage containing emulsified petroleum products. Izv. Univ. Constr. **8**, 56–60 (2008)
16. Nikiforova, E.Yu., Kilimnik, A.B.: The regularities of electrochemical behavior of metals under the influence of alternating current. Vestn. TSTU **15**(3), 604–614 (2009)
17. Bomba, A., Kunanets, N., Nazaruk, M., Pasichnyk, V., Veretennikova N.: Information technologies of modeling processes for preparation of professionals in smart cities. In: Advances in Intelligent Systems and Computing, pp. 702–712 (2018)
18. Shakhovska, N., Vysotska, V., Chyrun, L.: Features of e-learning realization using virtual research laboratory. In: XIth International Scientific and Technical Conference Computer Sciences and Information Technologies (CSIT), Lviv, pp. 143–148 (2016)

Application of Qualitative Methods for the Investigation and Numerical Analysis of Some Dissipative Nonlinear Physical Systems

Petro Pukach[1(✉)], Volodymyr Il'kiv[2], Zinovii Nytrebych[2], Myroslava Vovk[2], and Pavlo Pukach[3]

[1] Department of Computational Mathematics and Programming, Lviv Polytechnic National University, Lviv, Ukraine
ppukach@gmail.com
[2] Department of Mathematics, Lviv Polytechnic National University, Lviv, Ukraine
ilkivv@i.ua, znytrebych@gmail.com, mira.i.kopych@gmail.com
[3] Department of Applied Mathematics, Lviv Polytechnic National University, Lviv, Ukraine
pavlopukach@gmail.com

Abstract. The mathematical models of the oscillations for the important classes of the nonlinear physical systems with dissipation are considered in the paper. It is impossible to apply the asymptotic analytical methods to construct the solutions in the mathematical models of the dynamical processes in these systems. Consequently the qualitative approach is used, the solution existence and uniqueness is substantiated and estimated. The qualitative methods enable to use the corresponding specified numerical methods for the investigation of the mathematical model and solution construction. Basing on the numerical analysis and the fourth-order Runge-Kutta method there are analyzed some singularities of the dynamical processes in the considered systems classes. The effective combination of the theoretic and numerical approach allows to build the innovative procedure to analyze the mathematical models for the wide class of the nonlinear physical systems applied in the engineering.

Keywords: Mathematical model · Physical system · Dynamical process Dissipation · Nonlinear oscillations · Galerkin method

1 Introduction. Problem Actuality. Literature

The new trends development in the engineering and the modern scientific research reason the necessity of the nonlinear physical systems investigations, in particular, the well-posed soluble problems, in case, when the mathematical models cannot be analyzed by the analytical methods. Contrary to the linear models, the nonlinear models are characterized by the absence of the comfortable for the engineer applications apparatus enabling to define the parameters of the physical systems with the required

N. Shakhovska and M. O. Medykovskyy (Eds.): CSIT 2018, AISC 871, pp. 464–475, 2019.
https://doi.org/10.1007/978-3-030-01069-0_33

accuracy. For example, the analysis of the nonlinear mechanical oscillation system is mathematically complicated since the lack of the general analytical solving methods in the case of the nonlinear law of the material elasticity, and also in the case of the nonlinear dependence of the oscillation amplitude on the resistance force. This question in the general case is solved only for the bounded class. So the general procedure to define amplitude-accuracy characteristics of the oscillation process doesn't exist. At the same time such type physical systems and their mathematical models are used in the technical problems: the problems on oscillations of the elastic elements of the chain or strap transmissions, digital audiotapes systems, conveyer bands, cableways, the equipment for rolling up the paper, metal band, string, wire, the equipment for the oil drilling, pipelines.

For the majority of the applied problems in the nonlinear oscillation theory there is obvious action of the generalized forces of the internal dissipation in the oscillation system. In particular, the bending oscillations in the bar according to the Voight-Kelvin theory are described by the fifth-order linear equation, taking into account the influence of the dissipative forces on the dynamical process. The mathematical oscillations models in Voight-Kelvin bars [1] under the nonlinear resistant forces action are studied in the paper. The models are based on the next equation form

$$\frac{\partial^2 u}{\partial t^2} + a\frac{\partial^5 u}{\partial t \partial x^4} + b\frac{\partial^4 u}{\partial x^4} + g\left(x, t, \frac{\partial u}{\partial t}\right) = f(x, t). \tag{1}$$

It is occurred that for some types of the problems (1) and the similar equations the asymptotic methods of the nonlinear mechanics [2–5] are effective.

Let's notice, that the linear Eq. (1) is used in the mathematical oscillation model of the elastic isotropic environment, obtained on the basis of the integral variational Hamilton-Ostrogradsky principle [6]. This principle is generalized from the conservative systems characterizing the isotropic environment oscillations under the potential forces action to the nonconservative systems under the potential for and the internal dissipative forces action. This assumption needs the essential modification of Lagrange function inputing the internal dissipation of the mechanical energy. The dissipative properties of the elastic environment are under the influence of the hysteresis processes, explained by the nonlinear dependence behavior $\sigma = \sigma(\varepsilon)$, as well as under the friction loss between the conjugate elementary volumes of the elastic environment, related to the nonhomogenity material [7].

On the other hand, the qualitative methods of the general theory of the nonlinear boundary problems enable the solving results for the wide class of mentioned oscillation systems (namely about the existence, uniqueness and the continuous dependence on the initial data). The described procedure allows to justify the solution correctness in the model and to apply the different computational methods for the further investigations. That is why the qualitative methods issues of the nonlinear oscillation systems are actual. The qualitative methods to study the bounded and unbounded bodies under the resistance force action analyzed in the paper, is based on the general principles of the nonlinear boundary problems theory – Galerkin method and the monotonicity method [8]. The scientific newness is about the generalization of the investigative methods for the nonlinear problems to the new classes of the oscillation systems, the reasoning of the solution correctness for the mathematical models really applicated in the technical oscillation systems.

The investigation of the solution problems in the linear case and also some type of nonlinearity, modeling the processes in the physical systems is described in the papers [9–11]. The existence of the weak solutions for the mixed problems in the boundary domain for some linear partial equation system where one of the unknown functions describes the vertical displacement of the bar is studied in the [12]. The corresponding equation of such system is the linear case of the Eq. (1). In this paper, the substantially more complicated problem of studying a nonlinear mathematical model is considered. Such a model is often found in the description of technical oscillation systems The Eq. (1) generalizes the bar oscillation model in the resisting environment. The actuality of the analysis of the boundary problems for such type equations and systems is explained by the abrasion of the contact surfaces that is the main factor of the non-durable use of the equipment. The statement and investigation of the general mathematical models in the physical-mechanical processes, in particular, in the dynamical contacts in the elastic structures describing by the Eq. (1), is the actual and modern engineering question [13–24]. The different type boundary conditions describe the mechanical models being studied in the elasticity theory, and in the oscillation theory as well.

2 The Mathematical Model of the Nonlinear Oscillations in the Homogeneous Environment with Dissipation and the Nonlinear Resistant Forces

Let's introduce the procedure of the qualitative investigation of the solution for the nonlinear oscillation mathematical model, described by the mixed problem for the equation

$$\begin{aligned} &\frac{\partial^2 u}{\partial t^2} + \frac{\partial^2}{\partial x^2}\left(a(x,t)\frac{\partial^3 u}{\partial t \partial x^2}\right) + \frac{\partial^2}{\partial x^2}\left(b(x,t)\frac{\partial^2 u}{\partial x^2}\right) \\ &\quad + g(x)\left|\frac{\partial u}{\partial t}\right|^{p-2}\frac{\partial u}{\partial t} = f(x,t), \quad p > 2 \end{aligned} \tag{2}$$

with the initial

$$u(x,0) = u_0(x), \tag{3}$$

$$\frac{\partial u}{\partial t}(x,0) = u_0(x) \tag{4}$$

and the boundary conditions

$$u(0,t) = \frac{\partial^2 u}{\partial x^2}(0,t) = 0,\ u(l,t) = \frac{\partial^2 u}{\partial x^2}(l,t) = 0 \tag{5}$$

in the domain $Q_T = (0,l) \times (0,T)$, $l, T < +\infty$ In the relationships (2)–(5):

- $u(x,t)$ - lengthwise (lateral) environment movement with the coordinate x in the arbitrary time moment t;
- $a(x,t)$ - the function, characterizing the cross-sectional environment area, mass per unit length, the elastic characteristics of the environment, etc.;
- $b(x,t)$ - the function, characterizing the internal dissipative of the environment;
- $g(x) > 0$ - the function, taking into the account the mentioned characteristics and describing the nonlinearity of the resistant forces;
- $f(x,t)$ - the function, describing the distribution along the external forces environment;
- $u_0(x)$ and $u_1(x)$ - the functions, describing the environment initialization (the initial deviation – the form and the initial speed).

The *aim of this paper* is the investigation of the mixed problem (2)–(5) for the equation of the bending oscillations, modeling the influence of the internal dissipative forces and the nonlinear resistant forces on the dynamical process, and also the statement of the conditions on the solution correctness of the mathematical model – the sufficient conditions on the solution existence and uniqueness.

3 The Main Result Statement

Let suppose respect to the right side of the Eq. (2) and the initial data the next conditions are true

- **(a)** the function $a(x,t)$ is bounded in the domain Q_T, $a(x,t) \geq a_0 > 0$;
- **(b)** the functions $b(x,t)$, $\frac{\partial b(x,t)}{\partial t}$ are bounded in the domain Q_T, $b(x,t) \geq b_0 > 0$;
- **(g)** the function $g(x)$ is bounded on $(0,l)$, $g(x) \geq g_0 > 0$;
- **(f)** the right side $f(x,t)$ is function, integrable by Lebesgue with the power $p' = \frac{p}{p-1}$ in the domain Q_T;
- **(u)** the initial deviation $u_0(x)$ is squared integrable function by Lebesgue with the second order derivative on the interval $(0,l)$, satisfying the condition (3); the initial deviation speed $u_1(x)$ is the squared integrable function by Lebesgue on the interval $(0,l)$.

As the *generalized solution* of the problem (2)–(5) in the domain Q_T let's call the function $u(x,t)$, satisfying the conditions (3), (4), (5) and the integral identity

$$\int_{Q_\tau} \left[-\frac{\partial u}{\partial t}\frac{\partial v}{\partial t} + a(x,t)\frac{\partial^3 u}{\partial t \partial x^2}\frac{\partial^2 v}{\partial x^2} + b(x,t)\frac{\partial^2 u}{\partial x^2}\frac{\partial^2 v}{\partial x^2} \right] dxdt$$
$$+ \int_{Q_\tau} \left[-f(x,t)vg(x) + g(x)\left|\frac{\partial u}{\partial t}\right|^{p-2}\frac{\partial u}{\partial t} v \right] dxdt \tag{6}$$
$$+ \left[\int_0^l \frac{\partial u(x,\tau)}{\partial t} v(x,\tau) - u_1(x)v(x,0) \right] dx = 0$$

for the arbitrary $\tau \in [0;T]$ and for the arbitrary probe function v when the identity (6) is valid.

The qualitative characteristics of the solution are the next: the functions $u(x,t)$ and $\frac{\partial u(x,t)}{\partial t}$ are continuous on the variable t on the closed interval $[0;T]$, the function $\frac{\partial u(x,t)}{\partial t}$ is integrable on degree p by Lebesgue on $[0;T]$; on the variable x the function $u(x,t)$ with its second-order derivative is squared integrable on $(0,l)$ by Lebesgue; on the variable x the function $\frac{\partial u(x,t)}{\partial t}$ with its second-order derivative is squared integrable on $(0,l)$ by Lebesgue.

The main result: under the conditions (a), (b), (g), (f), (u) there exists the unique generalized solution $u(x,t)$ of the problem (2)–(5) in Q_T.

4 The General Scheme to Obtain the Main Result Using Galerkin Method

To substantiate the solution existence of the problem (2)–(5) one can use the scheme to get the approximate solution via Galerkin method [8–11]. Let's consider in the domain Q_T the sequence of approximations $u^N(x,t) = \sum_{k=1}^{N} C_k^N(t)\omega_k(x), N = 1,2,\ldots, \omega_k(x)$, is orthonormal in $L^2(0,l)$ system of the linear independent elements of the space $H^2(0,l) \cap L^p(0,l)$, and the linear combinations $\{\omega_k\}$ are dense in $H^2(0,l) \cap L^p(0,l)$. The functions C_k^N are defined as Cauchy problem solutions for the ordinary differential equations system

$$\int_0^l \left(\frac{\partial^2 u^N}{\partial t^2} + g(x)\left|\frac{\partial u^N}{\partial t}\right|^{p-2}\frac{\partial u^N}{\partial t} - f(x,t) \right)\omega_k dx$$
$$+ \int_0^l \left(a(x,t)\frac{\partial^3 u^N}{\partial t \partial x^2}\frac{\partial^2 \omega_k}{\partial x^2} + b(x,t)\frac{\partial^2 u^N}{\partial x^2}\frac{\partial^2 \omega_k}{\partial x^2} \right) dx = 0, \tag{7}$$

where $k = 1,2,\ldots,N$, with the initial conditions

$$C_k^N(0) = 0,\; u_0^N(x) = \sum_{k=1}^{N} u_{0,k}^N \omega_k(x),\; \frac{\partial C_k^N}{\partial t}(0) = u_{1,k}^N,$$

$$u_1^N(x) = \sum_{k=1}^{N} u_{1,k}^N \omega_k(x),\; \left\|u_1^N - u_1\right\|_{L^2(0,l)} \to 0,\; \left\|u_0^N - u_0\right\|_{H^2(0,l)} \to 0,\; N \to \infty.$$

Due to Carateodory theorem the continuous solution of such Cauchy problem exists with the absolutely continuous derivative on t on the $[0;T]$. Using the analysis analogical to the [10], from the equation system (7), one can get the next a priory approximation solution estimation:

$$\int_0^l \left(\left(\frac{\partial u^N(x,\tau)}{\partial t} \right)^2 + \left(\frac{\partial^2 u^N(x,\tau)}{\partial x^2} \right)^2 \right) dx + \int_{Q_\tau} \left(\left(\frac{\partial^3 u^N}{\partial x^2 \partial t} \right)^2 + \left| \frac{\partial u^N}{\partial t} \right|^p \right) dxdt \le C \quad (8)$$

for the arbitrary $\tau \in [0,T]$, where C – the positive constant independent on N. Let's make the conclusion from the inequality (8) about the existence of some subsequence $\{u^{N_k}\}$ of the sequence $\{u^N\}$ such, that $u^{N_k} \to u$ and $\frac{\partial u^{N_k}}{\partial t} \to \frac{\partial u}{\partial t}$ in some Sobolev functional spaces. Besides that, one can show $\left|\frac{\partial u^{N_k}}{\partial t}\right|^{p-2} \frac{\partial u^{N_k}}{\partial t} \to \left|\frac{\partial u}{\partial t}\right|^{p-2} \frac{\partial u}{\partial t}$ in the corresponding space. Also it is possible to show that, the function u satisfies the integral identity (6), the conditions (3), (4), possessing the described in the solution definition qualitative characteristics.

To reason the uniqueness let's note $w = u^1 - u^2$, where u^1, u^2 – two generalized solutions of the problem (2)–(5). Since $u^1(x,0) = u^2(x,0)$, $\frac{\partial u^1}{\partial t}(x,0) = \frac{\partial u^2}{\partial t}(x,0)$, then, analogically to (8), one can obtain

$$\int_0^l \left(\left(\frac{\partial w(x,\tau)}{\partial t} \right)^2 + \left(\frac{\partial^2 w(x,\tau)}{\partial x^2} \right)^2 \right) dx + \int_{Q_\tau} \left(\frac{\partial^3 w}{\partial x^2 \partial t} \right)^2 dxdt$$

$$+ \int_{Q_\tau} \left(\left| \frac{\partial u^1}{\partial t} \right|^{p-2} \frac{\partial u^1}{\partial t} - \left| \frac{\partial u^2}{\partial t} \right|^{p-2} \frac{\partial u^2}{\partial t} \right) \left(\frac{\partial u^1}{\partial t} - \frac{\partial u^2}{\partial t} \right) dxdt \le 0$$

From the last inequality $w = 0$, namely the solution uniqueness.

5 The Example of the Developed Procedure Application. Runge-Kutta Method to Integrate Numerically the Motion Equation in the Beam Oscillation Mathematical Model

The numerically simulation results for the problem considered in the Sect. 2 are obtained. Let's study the problem for nonlinear equation of the free oscillations of the bounded beam with the length l under the internal dissipative forces action and under the nonlinear vincler resistant force in the form

$$\frac{\partial^2 u}{\partial t^2} = a\frac{\partial^5 u}{\partial t\partial x^4} + b\frac{\partial^4 u}{\partial x^4} + c\frac{\partial^3 u}{\partial t\partial x^2} + d\frac{\partial^2 u}{\partial x^2} - g\left|\frac{\partial u}{\partial t}\right|^{p-2}\frac{\partial u}{\partial t},\ p\geq 2.$$

Let's notice that this equation can be treated as weakly nonlinear case of the Eq. (2). The constants b, d, characterizing the physical-mechanical features of the beam material, and constants a, c, g – the coefficients of the internal and the external dissipation of the environment (defining the character and the value of the nonlinear dissipative forces) in this equation are dimensionless. The model equation would be considered under the conditions the fixed hinged joints of the beam ends $u(0,t) = u(l,t) = 0$ and the initial deviation of the beam points $u(x,0) = u_0(x)$, where $u_0(x) = \begin{cases} \frac{2hx}{l}, 0\leq x\leq \frac{l}{2}, \\ 2h - \frac{2hx}{l}, \frac{l}{2} < x\leq l, \end{cases}$ h - some positive constant. Moreover, let's suppose that the initial speed of the beam points is absent, namely $\left.\frac{\partial u(x,t)}{\partial t}\right|_{t=0} = 0$. This problem is similar to the problem (2)–(5). There exists the unique generalized solution of this problem in the domain $Q_T = (0,l)\times(0,T)$ as was shown above. To realize numerically integration of the motion equations the considered equation would be transformed into the next two equation system

$$\begin{cases} \frac{\partial u(x,t)}{\partial t} = v(x,t), \\ \frac{\partial v(x,t)}{\partial t} = a\frac{\partial^4 v(x,t)}{\partial x^4} + b\frac{\partial^4 u(x,t)}{\partial x^4} + c\frac{\partial^2 v(x,t)}{\partial x^2} \\ \quad + d\frac{\partial^2 u(x,t)}{\partial x^2} - g_0|v(x,t)|^{p-2}v(x,t) \end{cases}$$

Let's make the partition of the closed interval $[0;l]$ by the discretization points $x_i = i\frac{l}{n}$ to the n parts with the length $\Delta = \frac{l}{n}$. Let's approximate the derivatives on the space variable by the finite differences

$$\frac{\partial^4 u(x,t)}{\partial x^4} = \frac{u(x_{i-2},t) - 4u(x_{i-1},t) + 6u(x_i,t)}{\Delta^4} - \frac{4u(x_{i+1},t) - u(x_{i+2},t)}{\Delta^4},$$
$$\frac{\partial^4 v(x,t)}{\partial x^4} = \frac{v(x_{i-2},t) - 4v(x_{i-1},t) + 6v(x_i,t)}{\Delta^4} - \frac{4v(x_{i+1},t) - v(x_{i+2},t)}{\Delta^4}.$$

The numerical simulation of the differential equation system

$$\begin{cases} u' = v(t), \\ v' = L(t, u, v) \end{cases},$$

where

$$\begin{aligned} L(t,u,v) &= a\frac{\partial^4 v(x,t)}{\partial x^4} + b\frac{\partial^4 u(x,t)}{\partial x^4} + c\frac{\partial^2 v(x,t)}{\partial x^2} \\ &+ d\frac{\partial^2 u(x,t)}{\partial x^2} - g_0|v(x,t)|^{p-2}v(x,t) \end{aligned}$$

is realized by the fourth-order Runge-Kutta method

$$\begin{cases} u_{k+1} = u_k + v_k\Delta t + \frac{1}{6}\Delta t(k_1 + k_2 + k_3), \\ v_{k+1} = v_k + v_k\Delta t + \frac{1}{6}(k_1 + 2k_2 + 2k_3 + k_4) \end{cases},$$

besides $t_k = k\Delta t$, $u_k = u(t_k)$, $v_k = v(t_k)$, the values k_1, k_2, k_3, k_4, are choosing as:

$$\begin{aligned} k_1 &= L(t_k, \varphi_k, v_k)\Delta t, \\ k_2 &= L\left(t_k + \tfrac{\Delta t}{2}, u_k + v_k\tfrac{\Delta t}{2}, v_k + \tfrac{k_1}{2}\right)\Delta t, \\ k_3 &= L\left(t_k + \tfrac{\Delta t}{2}, u_k + v_k\tfrac{\Delta t}{2} + \tfrac{k_1}{4}\Delta t, v_k + \tfrac{k_2}{2}\right)\Delta t, \\ k_4 &= L\left(t_k + \Delta t, u_k + v_k\Delta t + \tfrac{k_2}{2}\Delta t, v_k + k_3\right)\Delta t. \end{aligned}$$

The graphic dependences of the oscillation amplitude on time for the point being the middle of the beam under the conditions of the different initial deviation from equilibrium are presented on the figures (curve 1-$h = 0,1$; curve 2-$h = 0,5$ curve 3-$h = 1$). On the Fig. 1(a)–(c) it is shown the result of the numerical integration with the different values of the parameter c.

The internal dissipation parameter c, characterizing the physical-mechanical features of the beam material, not essentially influences on the oscillation frequency, but from the other side essentially influences on the relaxation speed as was shown on the Fig. 1.

The next results (Fig. 2) are obtained for the another model of the nonlinear resistant forces $p = 3$ ($a = -0,001$, $b = -1$, $d = 1000$, $g = 1$, $l = 1$).

The nonlinear index essentially influences on the relaxation speed and the small initial deviation in the nonlinear case doesn't influence on the oscillation amplitude and the oscillation frequency as follows from all studied dependences.

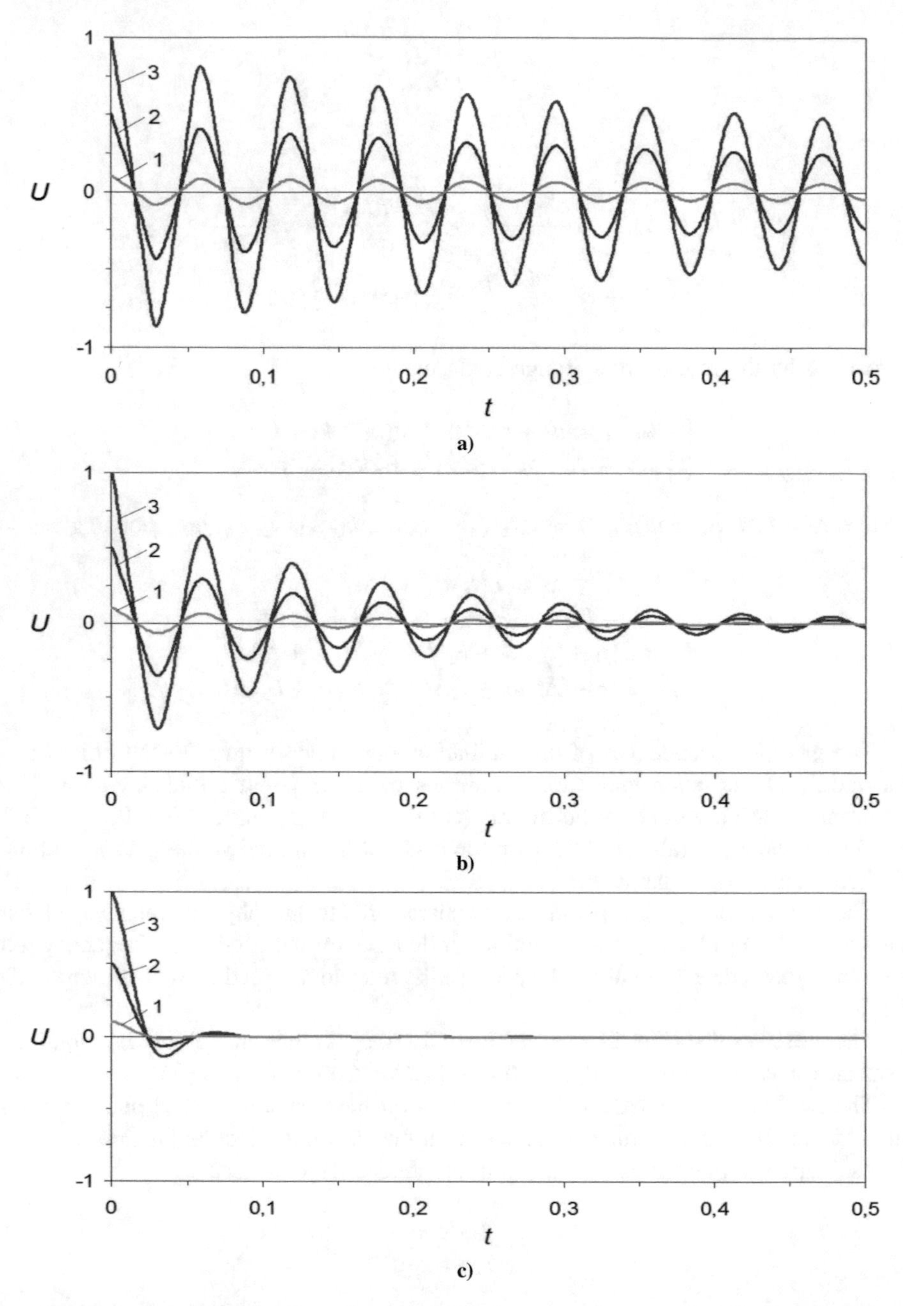

Fig. 1. The graphic dependences of the oscillation amplitude on time with the different values of the internal dissipation parameter: (a) $a = -0,001,\ b = -1,\ c = 0,\ d = 1000,\ p = 2,1,\ g = 1,\ l = 1$; (b) $a = -0,001,\ b = -1,\ c = 1,\ d = 1000,\ p = 2,1,\ g = 1,\ l = 1$; (c) $a = -0,001,\ b = -1,\ c = 10,\ d = 1000,\ p = 2,1,\ g = 1,\ l = 1$.

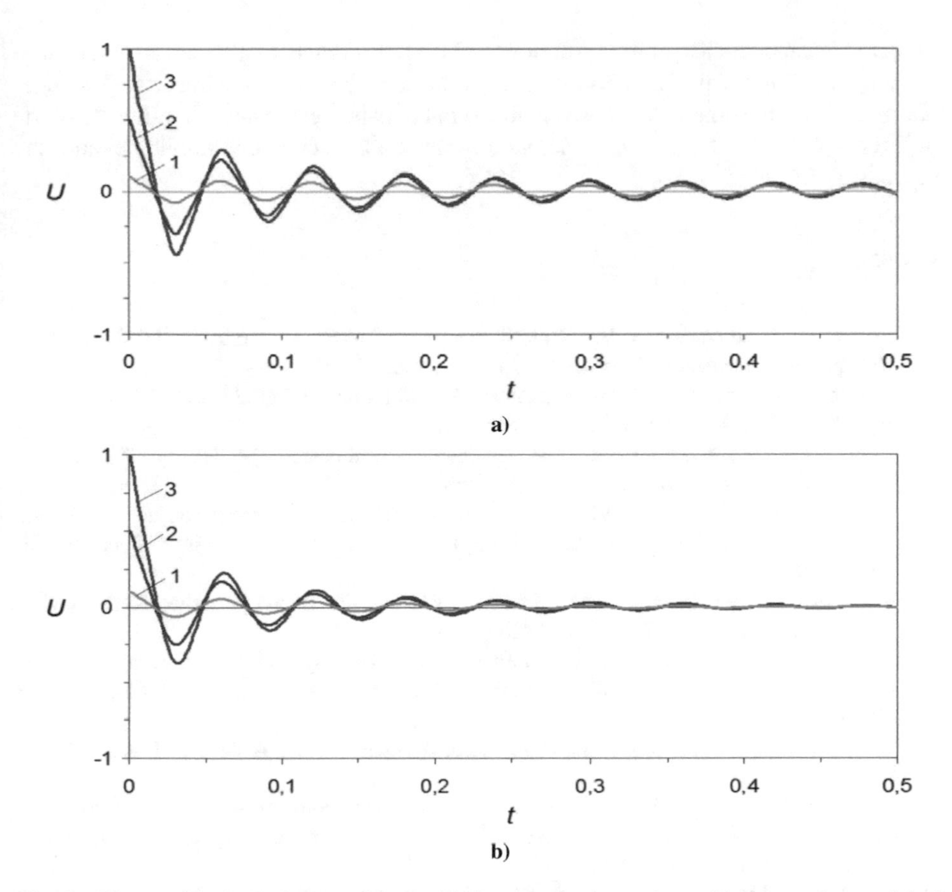

Fig. 2. The graphic dependences of the oscillation amplitude on time with the another model of the nonlinear resistant forces: (a) $c = 0$; (b) $c = 1$.

6 Conclusions

The qualitative results demonstrated in the paper by using Runge-Kutta method and the figures justify: (1) the speed of the oscillation relaxation mainly depends on the non-linearity power of the resistant force; (2) while the essential nonlinearity of the resistant force ($p = 3$) the dynamical process is aperiodic; (3) the influence of the resistant force on the oscillation period while the small values of the parameters g, p and h is miserable. The last fact is proved by the asymptotic integration of the considered differential equations in the case, if it is possible to apply the methods of the nonlinear mechanics.

The conditions of the solution correctness in the mathematical models of the oscillations in the elastic environment under the nonlinear dissipation forces action within Voight-Kelvin theory are obtained in the paper. The obtained qualitative results enable the application to the studied problems Galerkin method and also give the possibility to apply the different numerical methods in the further investigations of the dynamic characteristics of the solutions of the considered mathematical oscillation model. The numerical simulation of the motion fourth-order equations using Runge-Kutta methods in some model cases realized in the paper, estimates the influence of the different physic and mechanic factors on the both oscillation amplitude and the oscillation frequency as well.

References

1. Erofeev, V.I., Kazhaev, V.V., Semerikova, N.P.: Waves in the rods. In: Dispersion, Dissipation, Non-linearity. Fizmatlit, Moscow (2002). [in Russian]
2. Mitropol'skii, Yu., Moiseenkov, B.I.: Asymptotic Solutions of Partial Differential Equations. Vyshcha shkola, Kyiv (1976). [in Russian]
3. Bogolyubov, N., Mitropol'skii, Yu.: Asymptotic Methods in the Theory of Nonlinear Oscillations. Nauka, Moscow (1974). [in Russian]
4. Kagadiy, T.S., Shporta, A.H.: The asymptotic method in problems of the linear and nonlinear elasticity theory. Naukovyi Visnyk Natsionalnoho Hirnychoho Universytetu **3**, 76–81 (2015)
5. Mitropol'skii, Yu.: On construction of asymptotic solution of the perturbed Klein-Gordon equation. Ukr. Math. J. **7**(9), 1378–1386 (1995)
6. Rusek, A., Czaban, A., Lis, M.: A mathematical model of a synchronous drive with protrude poles, an analysis using variational methods. Przeglad Elektrotechniczny **89**(4), 106–108 (2013)
7. Filippov, A.A.: Oscillations of the Deformable Systems. Mashinostroenie, Moscow (1970). [in Russian]
8. Pukach, P.Ya., Kuzio, I.V.: Nonlinear transverse vibrations of semiinfinite cable with consideration paid to resistance. Naukovyi Visnyk Natsionalnoho Hirnychoho Universyte-tu **3**, 82–86 (2013). [in Ukrainian]
9. Pukach, P., Ilkiv, V., Nytrebych, Z., Vovk, M.: On nonexistence of global in time solution for a mixed problem for a nonlinear evolution equation with memory generalizing the Voigt-Kelvin rheological model. Opuscula Math. **37**(5), 735–753 (2017)
10. Il'kiv, V.S., Nytrebych, Z.M., Pukach, P.Y.: Boundary-value problems with integral conditions for a system of Lamé equations in the space of almost periodic functions. Electron. J. Differ. Equ. **2016**(304), 1–12 (2016)
11. Bokalo, T.M., Buhrii, O.M.: Doubly nonlinear parabolic equations with variable exponents of nonlinearity. Ukr. Math. J. **63**(5), 709–728 (2011)
12. Gu, R.J., Kuttler, K.L., Shillor, M.: Frictional wear of a thermoelastic beam. Journ. Math. Anal. And Appl. **242**, 212–236 (2000)
13. Lenci, S., Rega, G.: Axial-transversal coupling in the free nonlinear vibrations of Timoshenko beams with arbitrary slenderness and axial boundary conditions. Proc. R. Soc. A **472**, 1–20 (2016)
14. Denisova, T.S., Erofeev, V.I., Smirnov, P.A.: On the rate of ener-gy transfer by nonlinear waves in strings and beams. Vestnik Nizhe-gorodskogo Universiteta **6**, 200–202 (2011). [in Russian]
15. Lenci, S., Clementi, F., Rega, G.: Comparing nonlinear free vibrations of Timoshenko beams with mechanical or geometric curvature definition. Procedia IUTAM **20**, 34–41 (2017)

16. Bayat, M., Pakara, I., Domairryb, G.: Recent developments of some asymptotic methods and their applications for nonlinear vibration equations in engineering problems: a review. Lat. Am. J. Solids Struct. **1**, 1–93 (2012)
17. Dutta, R.: Asymptotic Methods in Nonlinear dynamics. arXiv:1607.07835.v1[math-ph], pp. 1–20 26 July 2016
18. Chen, L.Q.: Analysis and control of transverse vibrations of axially moving strings. Appl. Mech. Rev. **58**(2), 91–116 (2005)
19. M'Bagne, F., M'Bengue, F., Shillor, M.: Regularity result for the problem of vibrations of a nonlinear beam. Electron. J. Differ. Equ. **2008**(27), 1–12 (2008)
20. Liu, Y., Xu, R.: A class of fourth order wave equations with dissipative and nonlinear strain terms. J. Differ. Equ. **224**, 200–228 (2008)
21. Magrab, E.B.: Vibrations of Elastic Systems with Applications to MEMS and NEMS. Springer, New York (2012)
22. Bondarenko, V.I., Samusya, V.I., Smolanov, S.N.: Mobile lifting units for wrecking works in pit shafts. Gornyi Zhurnal **5**, 99–100 (2005). [in Russian]
23. Andrianov, I., Awrejcewicz, J.: Asymptotic approaches to strongly non-linear dynamical systems. Syst. Anal. Model. Simul. **43**(3), 255–268 (2003)
24. Gendelman, O., Vakakis, A.F.: Transitions from localization to nonlocalization in strongly nonlinear damped oscillators. Chaos Solitons Fractals **11**(10), 1535–1542 (2000)

Methods and Hardware for Diagnosing Thermal Power Equipment Based on Smart Grid Technology

Artur Zaporozhets[1(✉)], Volodymyr Eremenko[2], Roman Serhiienko[1], and Sergiy Ivanov[1]

[1] Institute of Engineering Thermophysics of NAS of Ukraine, Kiev, Ukraine
{a.o.zaporozhets,serhiienko}@nas.gov.ua, teplomer@ukr.net

[2] Igor Sikorsky Kyiv Polytechnic Institute, Kiev, Ukraine
nau_307@ukr.net

Abstract. The article presents methods and devices for diagnosing heat power equipment. A generalized structure of an intelligent distributed multi-level monitoring and diagnostic system for heat engineering equipment is developed, which is consistent with the principles of the Smart Grid concept. Methods for analyzing information signals in frequency-time and amplitude-phase-frequency regions are proposed, which made it possible to conduct a structural analysis of monopulse signals and signals with locally concentrated changes in parameters that are signs of defects in composite materials of heat power equipment. The structure of the measuring module, its hardware and the parameters of the developed prototype of the diagnostic system are given.

Keywords: Heat power equipment · Structure · Diagnostic system
Diagnostic feature · Hardware · Signals · Sensors · Boiler · Smart grid

1 Introduction

Most heat power plants are potentially dangerous for maintenance personnel, environment and public. This is due to the use of water and steam as heat carriers at high temperature and high pressure, fire hazardous substances (oil, solid, liquid or gaseous fuel, etc.), as well as the danger of electrical stress in control, signal and protection systems [1].

More than 90% of the park of operating boilers in Ukraine has worked out the resource and it must be replaced [2, 3]. The accidents at enterprises of the heat power industry become more and more frequent [4, 5]. The main causes of faults in boiler plants are damage of boiler heating surfaces, fuel supply systems, auxiliary equipment, automatics, etc. Heating surfaces are the most common causes of boiler failure (about 80% of cases).

The most dangerous operational factors affecting on the longevity of the elements of heat and power equipment are temperature fluctuations. They lead to a short-term and long-term overheating of the metal and are the reason for changing the properties

N. Shakhovska and M. O. Medykovskyy (Eds.): CSIT 2018, AISC 871, pp. 476–489, 2019.
https://doi.org/10.1007/978-3-030-01069-0_34

and structure, increasing the creep rate, reducing the long-term strength and long plasticity, accelerating the corrosion processes and, as a result, affecting on the intensive development of thermal fatigue. Temperature fluctuations, especially in the region of 450 °C and above, strongly influence the residual creep deformation, which increases the diameter of the pipelines, reduces the wall thickness, and, as a result, leading to cracking and destructing of the pipes.

The technical condition of heat power engineering facilities testifies to the need to ensure operational reliability, durability and safety of heat power equipment. It is necessary to have special monitoring and diagnosing systems that constantly allow to monitor the thermal engineering processes of generation, transportation and consumption of thermal energy; to measure the main parameters of heat power plants, equipment, machinery, mechanisms, etc.; to diagnose and to predict the technical condition of plants and their nodes for solving this problem [6].

Research and development of materials using in industry continue uninterrupted, which lead to the appearance of ever new materials and constant progress in materials science. In our time, there are a large number of various materials that are used to make various designs, equipment, devices. Among them, the most intensively developed materials, called composites.

At present, there is no universal physical method for diagnosing composites, which would allow us to identify all possible types of defects. Taking into account the physical features of objects and the possibilities of obtaining primary measurement information, the most common are specially developed low-frequency acoustic methods and their modifications. Modern instruments and systems for diagnostics of composite materials mainly use deterministic models and the corresponding methods for processing informative signals and making diagnostic decisions that do not provide the necessary noise immunity and reliability of diagnostics results and the classification of defects.

For increasing the reliability of heat power equipment, it is necessary to accumulate and systematize retrospective information on the operation of heat engineering equipment. In electric power systems, this task is solved on the basis of Smart Grid concept, which significantly improves the reliability of power supply and ensures trouble-free operation of the system [7].

Thus, an urgent task in the field of heat power engineering is the development of a system for controlling, diagnosing and monitoring of heat engineering equipment, taking into account the requirements for bilateral information exchange between all elements of the smart network, as well as the decentralization of computing and information resources.

2 Structures of the Diagnostic Object and the Diagnostic System

For solving the problems of diagnosing of large heat power systems it is expedient to use the methodology of the system approach. One of its main provisions is the allocation in the heat energy system of several hierarchical levels. In Fig. 1 showed the hierarchical structure of the thermal power system of a large industrial enterprise.

Elements of the V level are complex installations (for example, a steam turbine) and can be further detailed to lower levels.

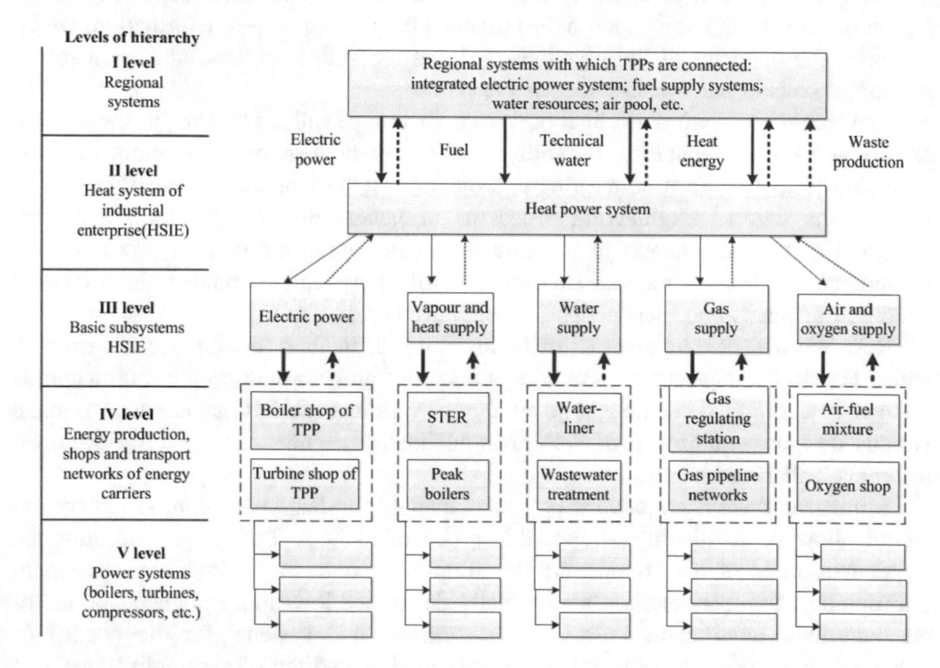

Fig. 1. Hierarchical structure of the thermal power system of a large enterprise

The tasks of hierarchical levels II–IV include such as, for example, the distribution of different types of fuel between individual consumers; choice of composition and profile of the main power equipment; optimization of parameters and type of thermal scheme of thermal power plant (TPP), etc. The tasks of level V and lower hierarchical levels include the selection of optimal thermodynamic and design parameters for specific heat and power equipment with parameters defined at II-IV levels [8].

This approach for the consideration of the heat power system allows the using of Smart Grid technology for diagnosing individual levels.

The essence of the developed system for diagnosing heat and power equipment is to monitor and to make diagnostic decisions at each of the individual hierarchical levels, which allows to identify, to localize and to eliminate defects before the diagnostic objects become faulty.

The emergence and development of the Smart Grid concept is a natural stage in the evolution of the heat and power system, caused on the one hand by the obvious needs and problems of the current heat energy market, and on the other hand by technological progress, primarily in the field of computer and information technologies.

The existing thermal power system without Smart Grid can be characterized as passive and centralized, especially in the part of the last link - from distribution networks to consumers. Exactly in this part of the heat supply chain Smart Grid technology most significantly changes the operating principles, offering new approaches to active and decentralized interaction of system components.

To this date, the structure of the Smart Grid is represented by the following elements:

- Smart Sensors and Devices – intelligent sensors and devices for main and distribution networks;
- IT Hardware and Software – information technologies used in backbone and distribution networks;
- Smart Grid Integrated Communications – integrated control and management systems - complete automation solutions; a certain analog of known ERP (Enterprise Resource Planning) systems in the enterprise;
- Smart Metering Hardware and Software – smart meters in the form of firmware.

Based on the hierarchy of TPP equipment, the system measures diagnostic signals that carry information about the actual state of the equipment nodes, which is diagnosed. Thus, the system can include sensors of those physical quantities that are used to diagnose a specific system. Depending on the diagnostic object, the system may include:

- thermocouples or thermistors – for measuring temperature;
- accelerometers – for measuring vibration parameters;
- measuring microphones – for determining the level of acoustic noise;
- sensors of electrical quantities – for measuring the parameters of the functioning of transformers;
- pressure sensors – for monitoring the depression in the furnace;
- gas sensors – for determining the concentration of harmful substances in the smoke path;
- thermal energy meters – for determining the current operating mode of heat engineering equipment, etc.

Thus, the structure of the diagnostic system that is being developed can be conditionally divided into hierarchical levels, similar to the way it was done above in the heat engineering equipment of the heat power system (Fig. 2).

The distribution of functions between the hierarchical levels of the development system is expediently organized as follows:

- level I (Measuring Transducers (MT)) – primary selection of diagnostic information (measurement of diagnostic signals, amplification, analog filtration, digital conversion);
- level II (LDS) – accumulation, full processing and in-depth data analysis, rapid response to alarms from a lower level, diagnostic decision-making on the diagnostic object as a whole, archiving of statistical data, prediction of reliability and estimation of the remaining equipment life, planning of repair works;
- level III (CDS) – data representation for various users (including geographically remote, for example, through Web technologies) with restricted access rights depending on their official duties.

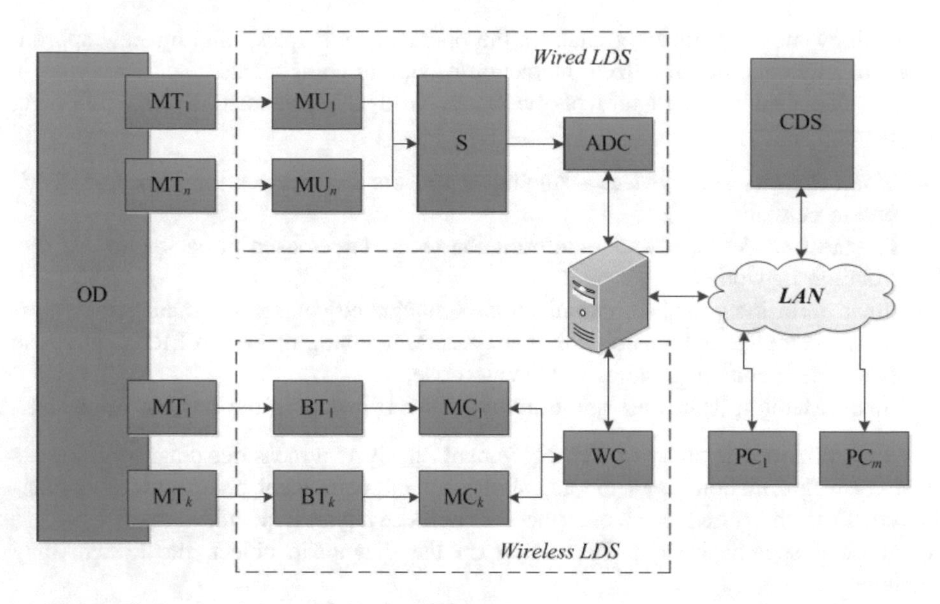

Fig. 2. Structure of the multi-level of diagnostic system of heat engineering equipment

All LDSs are included in an Ethernet-based LAN for displaying information for local users (for example, maintenance personnel), as well as for exchanging information with the central TPS diagnostic system.

The CDS has a connection to the global network (Internet) for enabling the exchange of information with external users (thus can be both people and devices operating outside of this TPS, but integrated into a «smart network»). In connection with this, a number of serious problems arise in ensuring information security and preventing possible terrorist attacks. Special network security hardware is used to solve these problems.

The system for diagnosing heat engineering equipment can work both with wired and wireless LDSs. The wired LDS consists of a matching unit (MU), a switch (S), an analog-to-digital converter (ADC) and PC. Wireless LDS consists of a block of transformation (BT), a microcontroller (MC), a wireless communication (WC) and PC. The use of both wired and wireless LDSs can significantly expand the classes of heat and power equipment that is diagnosed [9].

The consideration of the degree of critical defects at the stage of system development makes it possible to simplify its structure; reduce the amount of information that is processed in the system and transmitted between its hierarchical levels; and ultimately reduce the cost of the system while maintaining its functionality at an adequate level.

3 Models of Formation of Diagnostic Features

For assessing the technical condition of heating surfaces, pipes and other equipment of the boiler, it is necessary to have information about the presence of defects inside them. The construction of a set of diagnostic features can be performed on the basis of a wavelet transform.

Wavelet transform combines two kinds of transforms – forward and reverse, which respectively transform of the function *f(t)* into a set of wavelet coefficient $W\psi(a, b)f$ and vice versa [10]. The direct wavelet transform is performed in accordance with the rule:

$$W_{\psi}(a,b)f = \frac{1}{\sqrt{C_{\psi}}} \int_{-\infty}^{\infty} \frac{1}{\sqrt{|a|}} \psi^{*}\left(\frac{x-b}{a}\right) f(t)dt, \tag{1}$$

where a and b are parameters that determine the scale and displacement of the function ψ, $C\psi$ is a normalizing factor.

The basic, or maternal, wavelet ψ forms with the help of stretch marks and landslides a family of functions $\psi(t - b/a)$. Having a known set of coefficients $W\psi(a, b)f$, we can restore the original form of the function *f(t)*:

$$f(t) = \frac{1}{\sqrt{C_{\psi}}} \int_{-\infty}^{\infty} \int_{-\infty}^{\infty} \frac{1}{\sqrt{|a|}} \psi\left(\frac{t-b}{a}\right) \left[W_{\psi}(a,b)f\right] \frac{da \cdot db}{a^2}. \tag{2}$$

The direct (1) and the inverse (2) transforms depend on some function $\psi(t) \in L^2(R)$ which is called the basic wavelet. In practice, the only restriction on its choice is the condition for the finiteness of the normalizing coefficient:

$$C_{\psi} = \int_{-\infty}^{\infty} \frac{\left|\hat{\psi}(\omega)\right|}{|\omega|} d\omega = 2 \int_{0}^{\infty} \frac{\left|\hat{\psi}(\omega)\right|^2}{|\omega|} d\omega < \infty, \tag{3}$$

where $\hat{\psi}(\omega)$ is Fourier image of the $\psi(\omega)$ wavelet.

This condition satisfies many functions, so it is possible to choose the wavelet type that is most suitable for a particular task. In particular, for analyzing damped harmonic oscillations, it is more expedient to select wavelets, which are also damped oscillations. Diagnostic system examines the MHAT wavelet and the Morlet wavelet [11].

Since the signal of free oscillations of materials is a superposition of the modes of damped oscillations, it was decided to choose the best analyzing wavelet for the example of a model of a multicomponent damped signal with known parameters:

$$S(t) = \sum_{i=1}^{7} A_i \cdot \sin(2\pi f_i t) \cdot e^{-\beta_i t}, \tag{4}$$

where Ai, fi, βi – are the known values of the amplitudes, frequencies, and attenuation coefficients of the *i*-damped component.

The amplitude wavelet function of the signal was calculated using the MHAT wavelet as an analysis wavelet (Fig. 3) and the Morlet wavelet (Fig. 4).

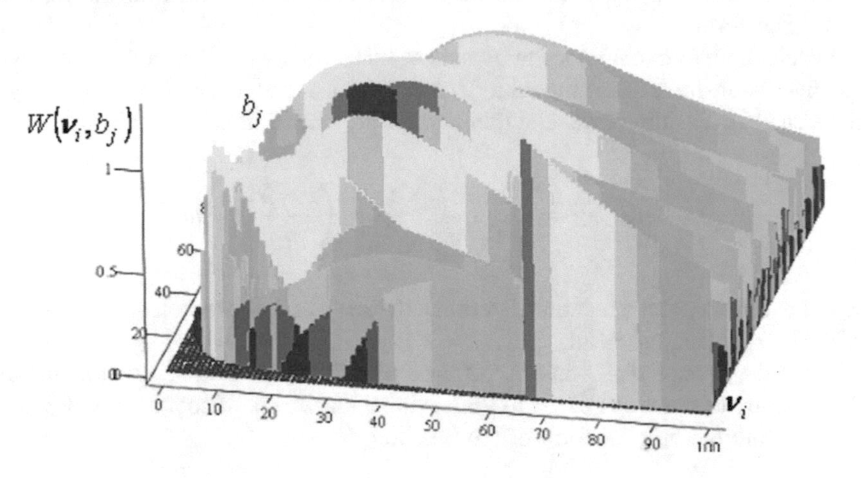

Fig. 3. Amplitude wavelet function of a multicomponent signal obtained using the MHAT wavelet

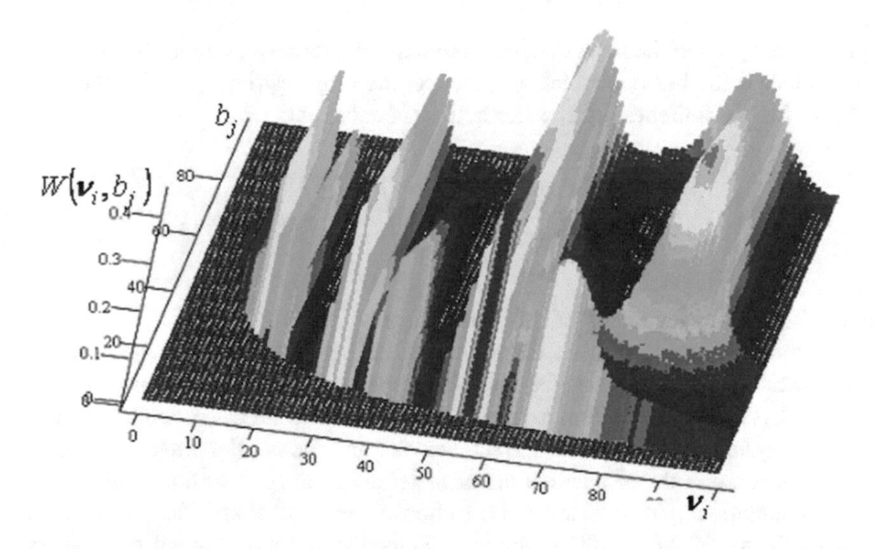

Fig. 4. Amplitude wavelet function of a multicomponent signal obtained using a Morlet wavelet

The wavelet transform of a sinusoidal signal with an MHAT wavelet is a periodic function of the parameter b, and the local energy spectrum behind the Morlet wavelet does not depend on the shift. This explains the shape of the wavelet spectra obtained in

Figs. 3 and 4. These figures show that for visual perception and study of damped components a more suitable wavelet transform is that using the Morlet wavelet. In addition, in Fig. 4 one can clearly see not only the attenuation of each mode, but also the instant of the beginning of the 7th mode, which was set with a delay in time [12].

The wavelet transforms of signals obtained in an intact and damaged panel area of 20 mm thick are considered. The amplitude spectra of these signals are shown in Figs. 5 and 6.

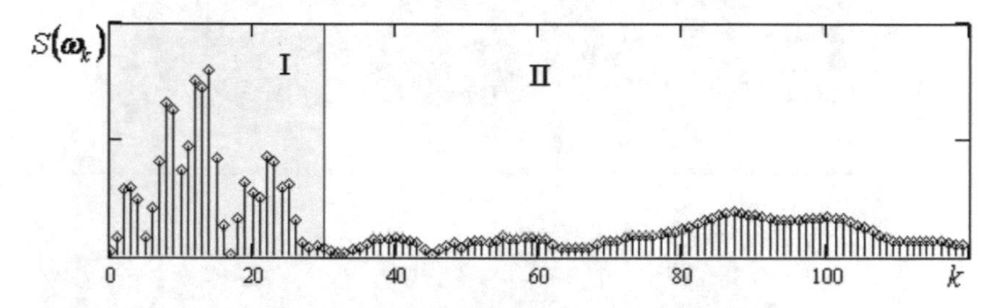

Fig. 5. Amplitude spectrum of the signal of free oscillations of the intact panel zone

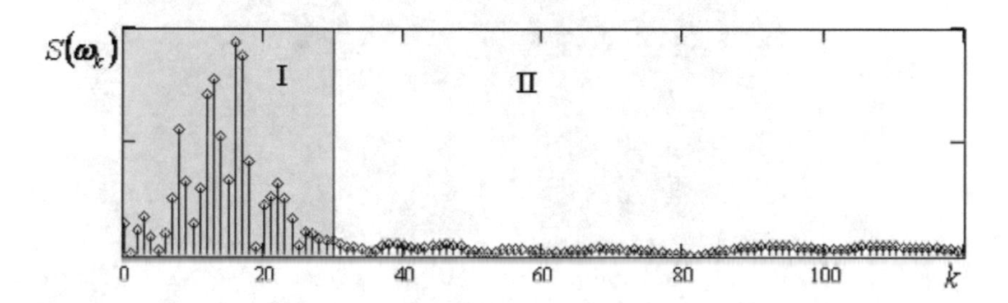

Fig. 6. Amplitude spectrum of the signal of free oscillations of a zone with a defect of a radius of 20 mm of the panel

According to the estimates of the amplitude spectra, the frequency range is preliminarily determined, within which the wavelet transforms will be performed - zone I in Figs. 5 and 6.

Figure 7 displays the graphs of the amplitude wavelet spectra of these signals calculated by the Morlet wavelet in the selected frequency range.

For faster decision making about the presence or absence of a defect in the controlled area of the material, it is proposed to compare the values of the amplitude wavelet spectra of free oscillations of the reference and monitoring zones calculated with the same offset [13]. For example, Fig. 8 (up) shows the amplitude wavelet spectra of the signal of free oscillations of a benign zone with landslides $b_j = b1, j = 30$; $b_j = b2$, $j = 50$; $b_j = b3$, $j = 70$, and Fig. 8 (down) shows similar spectra of free vibrations of a zone with a defect diameter of 20 mm. In other words, these spectra are actually the cross section of the graphs in Fig. 7.

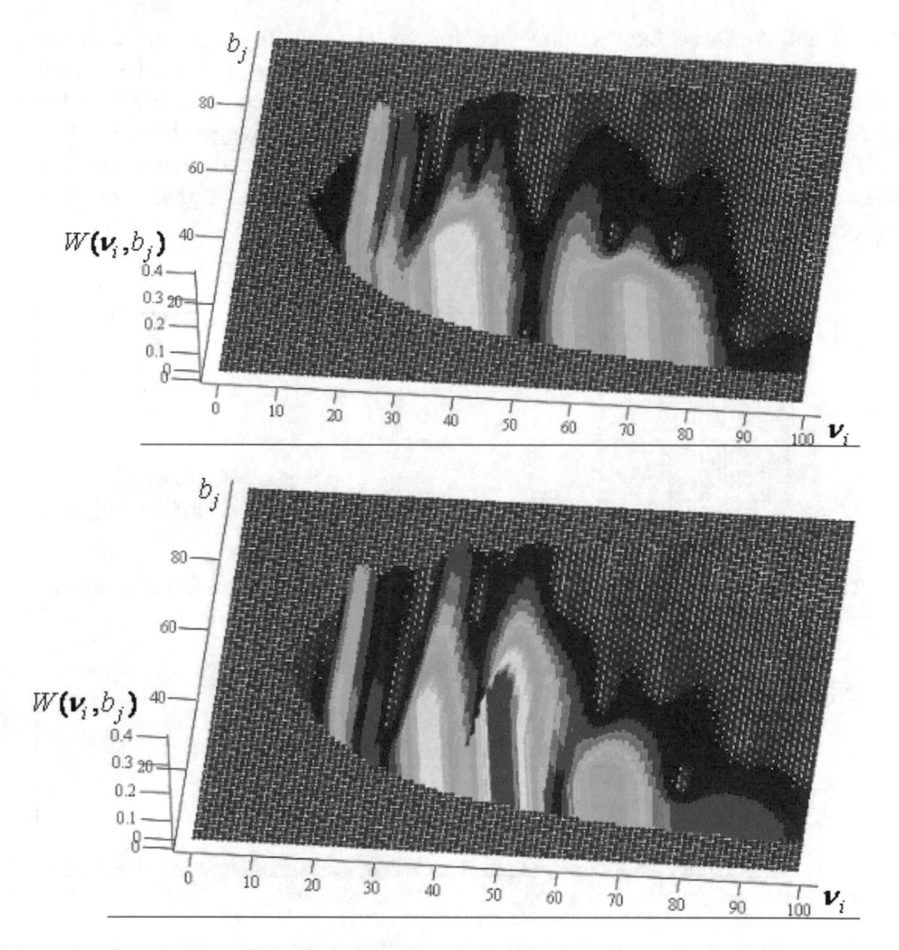

Fig. 7. Graphs of amplitude wavelet functions of free oscillations (*up* – intact area; *down* – zones with damage of radius 20 mm)

The wavelet spectra obtained in Fig. 6 allow us to obtain new diagnostic features that are convenient for visual comparison, and also allow more accurate determination of the frequency range of each individual mode in order to reduce errors in its recovery.

4 Hardware and Software of the Diagnostic System

Table 1 shows the main technical characteristics of the prototype of the system for diagnosing heat power equipment [14, 15].

A special feature of the developed diagnostic system is the measuring modules, which consist of shell (*1*), battery (*2*), printed circuit board (*3*), microcontroller (*4*), sensor (*5*) and transceiver (*6*). You can see all this elements in Fig. 9.

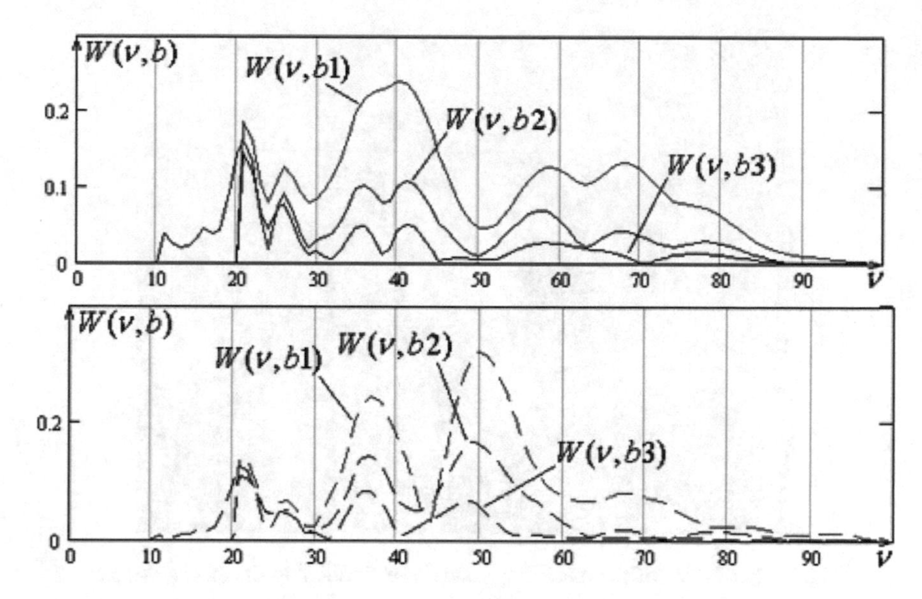

Fig. 8. Graphs of the amplitude wavelet spectra of free vibrations of the intact zone (*up*) and the zone with a 20 mm separation (*down*) at bias $b1 < b2 < b3$

Table 1. Technical characteristics of the prototype of the diagnostic system

Parameters	Values
Circuit board	STM32F103RB (ARM Cortex-M3), 64 pins
Flash memory, kB	128
SRAM, kB	20
Timers (IC, OC, PWM)	4 × 16 bit (16/16/18)
Other timers	2 × WDG, RTC, 24-bit reverse counter
Interfaces	2xSPI/2xI2C 3xUSART7USB/CAN
I/O ports	51
Sensors	Temperature, Acoustic, CO_2, CO
Package	LQFP64
Voltage, V	2…3,6

Various types of sensors are used to measure the performance of heating equipment: thermocouples, accelerometers, measuring microphones, sensors of electrical quantities, pressure sensors, thermal energy meters, gas sensors [16].

The using of accelerometers and the methods of processing information signals that have been observed above can detect defects in heating surfaces, and prevent an emergency situation. We also consider the possibility of using measuring modules in the air quality control system of the environment (based on gas sensors CO, CO2, NOx, SO2). A model of the diagnostic system is shown in Fig. 10. For prototyping, a Nucleo setup card was used based on the STM32f103 microcontroller, for server of developing diagnostic system was used Raspberry Pi 3 (Fig. 11).

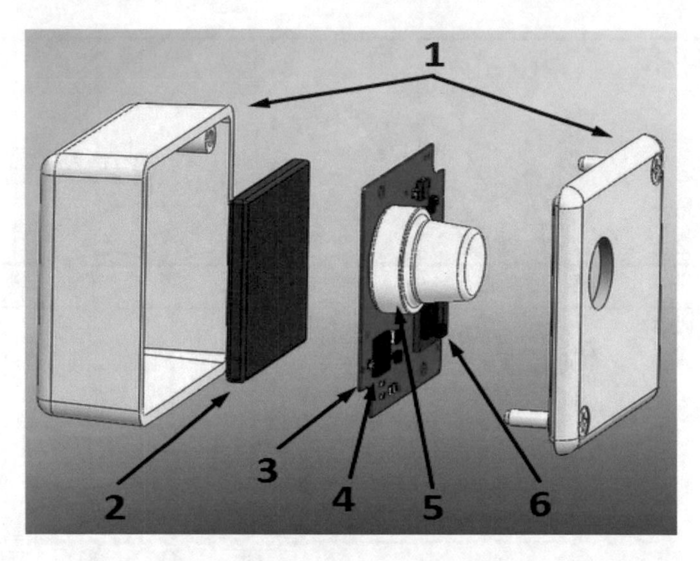

Fig. 9. 3D model of the measuring module of developing diagnostic systems

Fig. 10. The model of developing diagnostic system

In Fig. 12 shows the developed software for the system for diagnosing heat and power equipment. This version of the software product is designed to measure the temperature of heating surfaces, the pressure in the furnace of the boiler, as well as the concentration of CO2 in the flue gases.

In the future, it is planned to improve the developed system by increasing the number of monitoring parameters, as well as carrying out experimental studies.

Fig. 11. Hardware devices used in developing diagnostic system (*left* – STM32f103, *right* – Raspberry Pi 3)

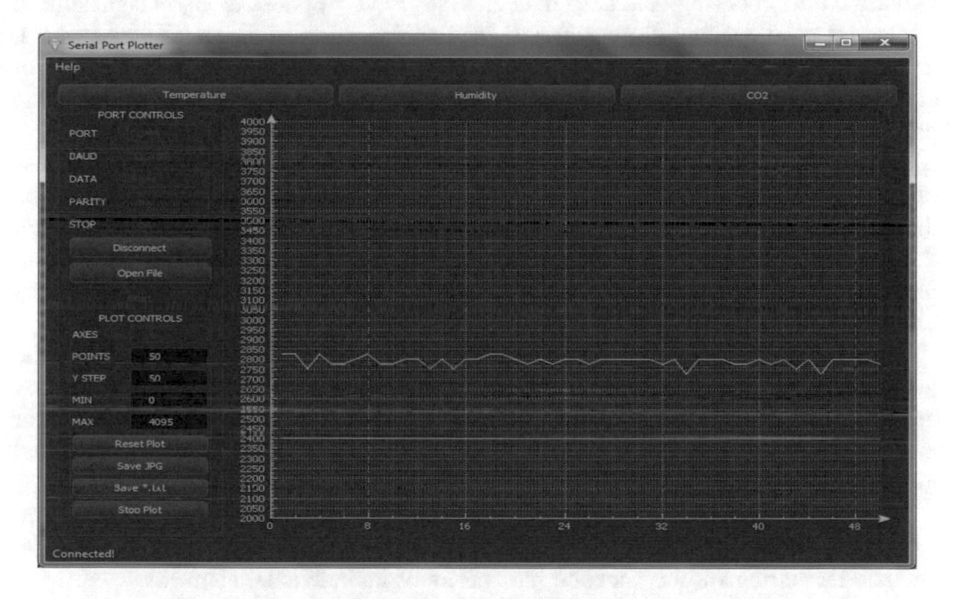

Fig. 12. Screen of software for developing diagnostic system

The main parameters of heat power equipment that can be diagnosed include:

- general parameters – economical factors associated with the factors of the technological process;
- characteristics of the properties of metal structures – hardness, creep, crack resistance, the presence of shells, fissures, the formation of scales of heating surfaces;
- geometric parameters of structures – pipe diameter and thickness, relative displacements of individual components;
- parameters of thermophysical processes – temperature of overheating zones of heating surfaces and steam pipelines;
- parameters of chemical processes – the state of water in cooling media;

- parameters of noise processes – the appearance of acoustic emission signals, acoustic leakage signals, noises of boiling liquid, noise in pipelines, etc.;
- vibration parameters – vibration of the boiler, pipelines, fans, smoke exhausters.

5 Conclusion and Future Developments

The developed methods and technical means allow to actualize the application of the Smart Grid concept in the hierarchical structure of the heat power system.

The structure of a multilevel diagnostic system based on the using of the Smart Grid concept is proposed. The application of the system allows for: primary selection and preparation of diagnostic signals, including digitization; mathematical processing, the adoption of intermediate diagnostic solutions, signaling of possible defects; accumulation, full processing and deep data analysis, rapid response to alarm signals from the lowest level, making diagnostic decisions on the object of diagnostics as a whole, archiving statistical data, predicting reliability and estimating the residual life of equipment, planning repair work; presenting data to different users and ensuring the protection of the system and its information from possible external interventions.

The constructive mathematical model of the information-signal field of the process of diagnosing composite materials was constructed, which made it possible to describe the interaction of the fields of mechanical disturbances in composite materials with defects of various types; use the results of experimental studies for statistical estimation of field characteristics, conduct a wide range of mathematical and computer model experiments.

The methods of primary processing of information signals of acoustic diagnostic methods in frequency-time and amplitude-phase-frequency coordinates have been improved and investigated, which made it possible to conduct a structural analysis of monopulse signals and signals with locally concentrated parameter changes and increase the probability of diagnostics by 20%.

The application of the developed measuring modules provides an opportunity to comprehensively assess the technical state of the heat power equipment by simultaneously measuring various function parameters of its individual elements.

References

1. Babak, V.P.: Information support for monitoring of energy objects (2015). ISBN 978-966-02-7478-5
2. Kostetskyi, V.V.: Prospects of investment and innovation development of housing and communal services of Ukraine. Socio Econ. Res. Bull. **2**, 82–91 (2014)
3. Sigal, O., Boulanger, Q., Vorobiov, L., Pavliuk, N., Serhiienko, R.: Research of the energy characteristics of municipal solid waste in Cherkassy. J. Eng. Sci. **5**(1), 16–22 (2018). https://doi.org/10.21272/jes.2018.5(1).h3
4. Voinov, A.P., Voinov, S.A.: Problems of management of efficiency of use of solid energy fuel in Ukraine. In: New and Non-traditional Technologies in Resource and Energy Saving: Materials of Scientific and Technical Conference, Odessa-Kiev, pp. 31–34 (2014)

5. Lund, H., Moller, B., Mathiesen, B.V., Dyrelund, A.: The role of district heating in future renewable energy. Energy **35**(2), 1381–1390 (2010). https://doi.org/10.1016/j.energy.2009.11.023
6. Babak, V.P.: Hardware-software for monitoring the objects of generation, transportation and consumption of thermal energy (2016). ISBN 978-966-02-7967-4
7. Masy, G., Georges, E., Verhelst, C., Lemort, V., Andre, P.: Smart grid energy flexible buildings through the use of heat pumps and building thermal mass as energy storage in the Belgian context. Sci. Technol. Built Environ. **21**(6), 800–811 (2015). https://doi.org/10.1080/23744731.2015.1035590
8. Babak, V.P., Zaporozhets, A.O., Sverdlova, A.D.: Smart grid technology in monitoring of power system objects. Ind. Heat Eng. **38**(6), 73–83 (2016). https://doi.org/10.31472/ihe.6.2016.10
9. Myslovych, M.V., Sysak, R.M.: On some peculiarities of design of intelligent multi-level systems for technical diagnostics of electric power facilities. Technical Electrodynamics, №1, pp. 78–85 (2015)
10. Daoud, O., Hamarsheh, Q.J., Damati, A.A.: Wavelet transformation method to allocate the OFDM signals peaks. In: 13th International Multi-Conference on IEEE Systems, Signals & Devices, pp. 159–164 (2016). https://doi.org/10.1109/ssd.2016.7473667
11. Eremenko, V.S., Pereidenko, A.V., Rogankov, V.O.: System of standardless diagnostic of cell panels based on Fuzzy-ART neural network. In: Microwaves, Radar and Remote Sensing Symposium (MRRS), pp. 181–183. IEEE (2011). https://doi.org/10.1109/mrrs.2011.6053630
12. He, P., Li, P., Sun, H.: Feature extraction of acoustic signals based on complex Morlet wavelet. Procedia Eng. **15**, 464–468 (2011). https://doi.org/10.1016/j.proeng.2011.08.088
13. Eremenko, V.S., Gileva, O.: Application of linear recognition methods in problems of nondestructive testing of composite materials. In: International Scientific Conference on Electromagnetic and Acoustic Methods of Nondestructive Testing of Materials and Products, LEO TEST-2009 (2009)
14. Ivanov, S.A., Vorobjev, L.Y., Dekusha, L.V.: Information processing in the study of the properties of wet materials by the method of synchronous thermal analysis. Inf. Process. Syst. **131**(6), 75–78 (2015)
15. Zaporozhets, A.O., Bilan, T.R.: Theoretical and applied bases of economic, ecological and technological functioning of energy objects (2017). ISBN 978-966-02-8331-2
16. Isermann, R.: Fault-diagnosis applications: model-based condition monitoring: actuators, drives, machinery, plants, sensors, and fault-tolerant systems. Springer (2011). ISBN 978-3-642-12767-0

Project Management

Managing the Energy-Saving Projects Portfolio at the Metallurgical Enterprises

Sergey Kiyko[1(✉)], Evgeniy Druzhinin[1,2], and Oleksandr Prokhorov[2]

[1] PJSC "Electrometallurgical Works "Dniprospetsstal" named after A.M. Kuzmin", Zaporozhye, Ukraine
kiyko@dss.com.ua

[2] National Aerospace University, Kharkiv Aviation Institute (KHAI), Kharkiv, Ukraine

Abstract. This paper considers managing of energy saving and energy efficiency projects and programs at the metallurgical enterprises. It suggests a multilevel model for energy saving process management at the enterprise. This model allows sequentially analyzing the project to recognize opportunities of tasks implementation, agreeing upon the project plans and enterprise plans realization on different levels of planning, and choosing the most perspective projects that suit the development strategy. This paper suggests the agent simulation analysis model to manage the energy resources at the metallurgical enterprises. This model considers a number of interdependent power flows, requirements, aims and behavior strategies of different divisions, and the manufacturing process dynamics. This model is essential to manage the processes of energy distribution and consumption at intrafactory and intradepartmental mains, and to control the basic energy-intensive equipment operating modes.

Keywords: Energy efficiency · Metallurgical enterprises · Projects portfolio
Energy saving management

1 Introduction

The metallurgical enterprises consume a huge amount of electric and heat energy. Due to this it is vital to develop a comprehensive program and a project portfolio on the principal energy saving and energy efficiency directions with obligatory coordination with the program of the enterprise development.

The need to solve these problems is caused by the obligation to improve the enterprise economic strength, production competitiveness and to reduce the dependency from the energy suppliers. The objects of the energy saving program are the management process, the process of technology, and the community facilities. To reduce the energy consumption it is appropriate to exclude inefficient use of the energy resources, to eliminate the energy resources waste, to increase the energy resources usage effectiveness, and to use and to distribute the energy produced by the enterprise.

It is critical to provide a new methodology of energy saving projects portfolio management. We have to consider the overriding priorities, resources scarcity, and possible risks. The methodology must provide multi-criteria project selection,

N. Shakhovska and M. O. Medykovskyy (Eds.): CSIT 2018, AISC 871, pp. 493–503, 2019.
https://doi.org/10.1007/978-3-030-01069-0_35

considering the variety of tasks on the metallurgical enterprises, planning the process of the project implementation in various timing aspect, interrelations and coordination agreements between the projects, risks and financing mechanisms.

The Dniprospetsstal steel production enterprise has four steel melting shops. The metal powder shop is equipped with induction furnace. The producing steel-melting shop is conducted in the open arc furnace, with the subsequent purge in the argon-oxygen converter and processing on the furnace ladle facility. For the production needs the enterprise uses gaseous and solid fuel. The natural gas is used for production needs as boiler and furnace fuels. The blast-furnace gas is used in mix with natural gas. As far as the enterprise does not own any sources of heat energy (pair and hot water) it receives them from the outside. For steel melting and pouring the enterprise uses argon. The air division products (compressed air, nitrogen, oxygen) are own-produced.

The electro melting production allows reducing expenses by optimization performance of the arc-furnace. Considering the energy balance of the arc furnace we should mention, that a receipt section is defined by electric power received from the network and by heat power from the exothermic reactions.

The electric power share in the balance receipt part makes up to 85% when melting without use of oxygen, and 65–75% with it. In the account part of balance, heat and electric losses make approximately 40%, and the useful expense is about 60%. The greatest impact on the amount of the large arc furnaces heat losses causes the melting cycle time. The main factors influencing electric losses are the oven transformer and the secondary current distributor. The enterprise effective energy management bases on the international standard ISO 50001, which aims to implement the systematical approach to achieve the continual improvement of the power system, including the energy efficiency, energy security and energy consumption. The overall purpose is to increase the energy efficiency at the enterprise. To put it into action the energy saving project portfolio management must be implemented. Thus, the aim is to optimize the energy balance and energy effectiveness, and to minimize the natural gas consumption.

2 The Recent Papers Review

As far as today the process technology in the metallurgical industry requires enhanced specific power consumption, the energy saving and energy efficiency question is serious and important.

The researchers consider energy survey, monitoring and planning to be possible solution of the problem. They emphasize, that the purpose-oriented energy saving programs are implemented considering specific things and features about the particular enterprise. The energy supplies exhaustion arrangements for accounting are considered to be a tactical choice.

At the same time, they emphasize: on the strategic level, the metallurgical enterprises implement the special-purpose energy saving programs considering the characteristics of the particular enterprise; on the tactical level, the enterprises provide energy consumption accounting management at multiple levels.

The data source [1] considers the features of power consumption simulation at the highest levels of management. The authors define the main regularities of the power consumption formation of metallurgical enterprises. The article [2] considers scientific and methodological basis of the energy management at mining and metallurgical enterprises, from the formation of mathematical models of energy consumption on out to the power modes operational management.

The researchers think that the production wastes decrease and effective use of the production wastes, together with use of secondary resources allow cutting down the expenses. The authors of [3] propose some measures to boost energy conservation and energy efficiency. They recommend to introduce the systems for coke dry slaking, and to use steam-gas turbines that employ coke-oven gas or a mixture of gases produced at metallurgical enterprises. The paper [4] considered the ways of using low-potential thermal secondary energy resources of a metallurgical plant.

The paper [5] regards the possibility to use the conservation power plant concept for integrated resource planning and metallurgical processes control. The researchers reviewed the optimization issues occurred at the combined heat and power plants where the secondary power resources of metal manufacturing were recovered to upgrade the fuel usage efficiency.

The complete energy consumption model allows us to estimate the energy saving realization projects efficiency, and to find the share of each energy resource in general stream, that defines the power consumption of the separate production, the shop, and of the whole enterprise. This model also allows correcting the strategy of the energy resources management.

The paper [6] describes how to apply the basic principle of Hybrid Petri net to model and to analyze the Metallurgical Process. This model vividly stimulates the dynamic flow of materials and the real-time change of each technological state in metallurgical process. The paper [7] considers the problem of modeling the production and consumption of electric power in hybrid power systems. The modeling allows defining the optimum quantity and parameters of the renewable electric power-receiving component depending on the predicted consumption needs.

3 The Model of Energy Efficiency and Energy Saving at the Metallurgical Enterprise

Analyzing the structure of energy saving program at the metallurgical enterprises proved that the factors influencing the energy saving policy are:

- inefficient consumption (or considerable losses) of energy resources (natural resources, thermal energy, electricity), and
- control of expenses formation and energy consumption improvement results.

The energy consumption by a metallurgical enterprise has a number of peculiarities, those are:

- a number of electric facilities engaged in production in every division,
- the types and capacity variety of the electric load using equipment, relatively weak interference of the electric load using equipment,

- a number of electric facilities engaged in the production in every division that creates the conditional-constant load, and also depends on the production process intensity,
- the factors influencing the conditions and amounts of energy consumption in a random way,
- predominate influence of the overall production on energy consumption,
- usage hours of maximum electric capacity,
- end-use products high capacity,
- possible operational changes and changes in the equipment configuration of divisions, product mix, and other factors that occur on a regular basis.

The main purpose of the energy saving project is to develop the management systems and to reduce inefficient energy consumption. The criteria of success for the energy saving project are effectiveness, exploitation expenditures, loses, etc. The main difficulty is to evaluate the part of each resource of energy in general stream, the energy consumption for a separate production, shop and the whole enterprise, etc.

The management concept of the energy saving portfolio bases on several interconnected adaptive systems, those are

- planning and formation,
- monitoring, and
- management of changes (Fig. 1).

When setting the priorities for the energy saving projects we consider these factors: organization of account, energy resources regulation, and other system actions; importance of the project for the main production development and its impact on final product cost; payback of the project; improvement of the power supply system reliability; possibility to involve a power service company.

As it was mentioned above, the existence of a complete model of energy consumption allows estimating the efficiency of the selected projects. This allows implementing the energy saving strategies, objectively estimating the part of each resource of energy in the general stream (Fig. 2), defining the power consumption of the separate production, shop and whole enterprise, correcting the strategy of the energy resources management.

We should note, that predicting the energy consumption by a metallurgical enterprise is a difficult multivariable task with a probabilistic part. The actual energy consumption is conditioned not only on management decisions, orders portfolio allocation, main reparation overall production, service maintenance, and the development of in-house sources of power supply, but also on day type (active days or days-off), weather conditions, time of day, etc. The causal relation of energy consumption to any of these parameters is quite difficult and don't have simple formal specification.

These tasks can be solved by applying the simulation model. The use of agent approach in simulation modeling allows implementing the dynamic behavior, autonomy and adaptation of separate components of model. That means that the mechanism of flexible change and coordination of the energy consumption parameters proceeds from the energy resources efficiency of use, management goals, restrictions and requirements, strategy of behavior of certain participants, dynamics of external and

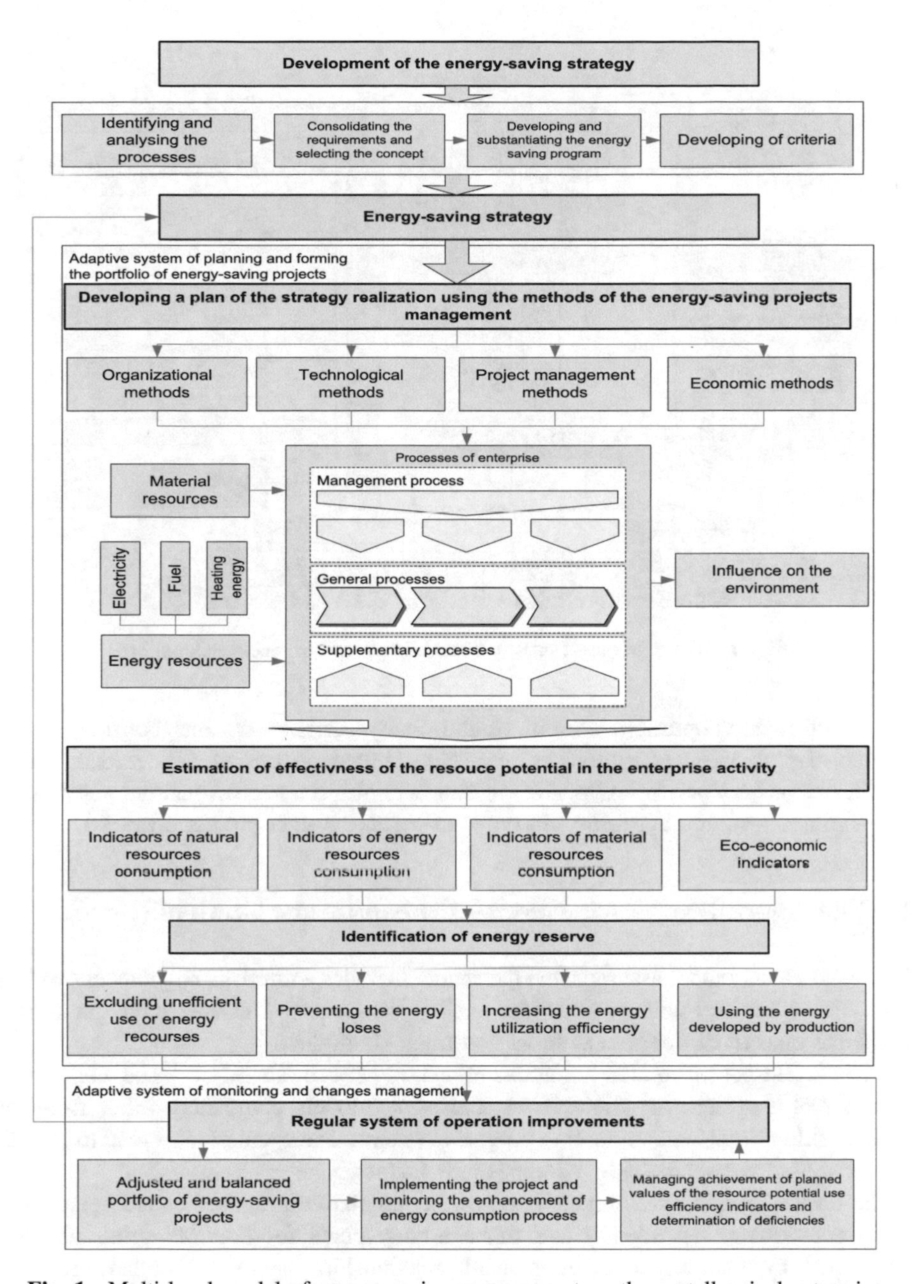

Fig. 1. Multi-level model of energy saving management on the metallurgical enterprise

internal environment. Due to the model, we can solve a number of problems, such as: assessment of rationality and efficiency of the energy consumption structure; predicting the expected levels of energy consumption caused by technology changes, range and quality of production; comparing the technologies and equipment considering the

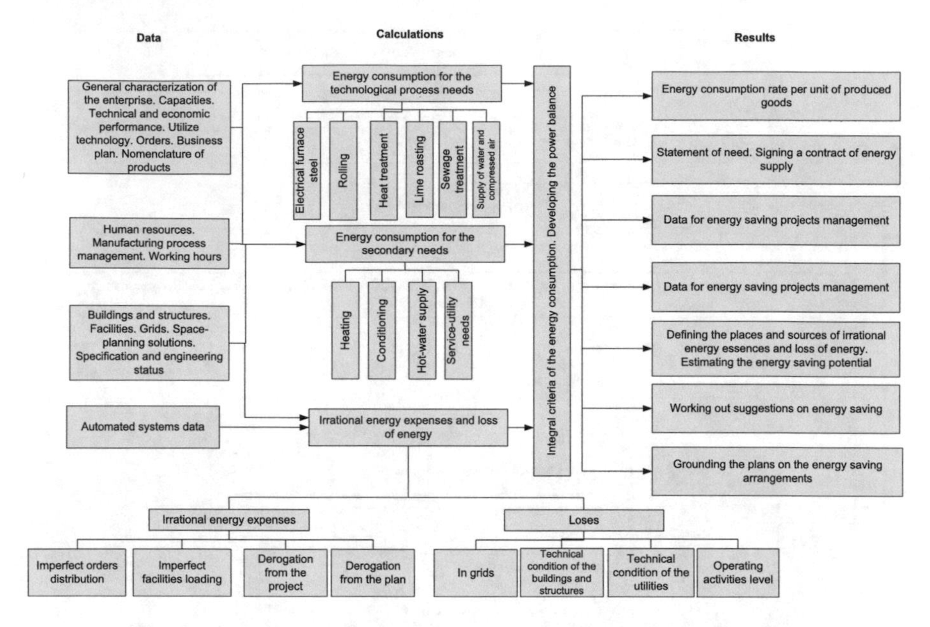

Fig. 2. The scheme of energy consumption effectiveness calculating

energy efficiency, optimum control of the energy carriers streams considering the change of production conditions of, etc. Now, based on this model we can form a number of project portfolios, which we can later subject to the dynamic analysis on financial and resource feasibility, in order to create the final project portfolio.

4 The Agent Model for Energy Consumption Control

The agent-based model assumes that the model includes a number of agents interacting with each other and with environment, called data items. The data items have their goals and objectives, internal statuses and rules of conduct.

The agent-based model is characterized by decentralism, it lacks centralized behavior of a system in whole. Thus, the agent models completely differ from the existent stiffly organized simulation program systems. The agent models tend to be self-managed, that proves to be an essential new feature.

Herewith, certain autonomous parts of the simulative program, called agents, can make decisions independently and make arrangements for possible solutions, they obtain their own activity and can enter into relationships against each other, they start user interaction dialog at the moments of time, those were not predefined, etc. These factors prove that the individual behavior of an agent, and their general behavior is a result of activity and interaction of many agents, where each has its own rules, functions in general environment and interacts with the environment and other agents.

Considering the advantages of the agent-based model at the energy consumption management process modeling, it bears mentioning that:

- the approach of separateness of agents that function together in distributed systems, where a number of interacting and interconnected processes goes simultaneously;
- the presence of individual behavior elements (with simple conditions and restrictions, and the difficult ones, those consider the goals and objectives;
- the agents those are able to learn, to adapt, to change their behavior, and to have dynamic relationships with other agents, that in their turn can appear and disappear in process of functioning.

Applying the multiagent approach in question of simulation modeling of the manufacturing systems requires solving the following tasks:

- analysis and defining the agent roles among the principal components of the simulation modeling system;
- development of the agent distributed knowledge database and the general ontology creation;
- creation of the artificial intelligence (AI) agents with a mechanism providing the solution inference;
- AI agents actions organization and planning;
- development of the agent interaction mechanism, including cooperation, competition, compromise, conformism, interaction avoidance, and developing the agent collective behavior strategies.

An overriding priority at agent modeling of the energy consumption flow processes in the considered environment is to provide a number of alternatives for energy resources production and consumption. This model allows effectively managing the processes of energy distribution and consumption in the intrafactory and intradepartmental mains, and controlling the operating modes of the main energy-consuming equipment.

In this case, the simplest way to manage the multiagent network when solving tasks on the energy distribution management is based on interaction of supplier agents, consumer agents, production agents and transformation agents that searches for matches among the intrafactory energy resources and in the outer energy resources (Fig. 3).

Due to the competition and cooperation when making "deals" to solve the questions together (the agents can use advanced economical mechanisms, such as participation interest, auction sales, etc.), the agents can grant the system new opportunities for self-organization that allows it to adapt permanently to unsteady conditions.

Using the term auction in the agents' communication allows passing the "value" (which is price) from one agent to another. The auction serves as a market mechanism for the processes of self-organization and self-management in the collective behavior.

The auction allows building a sales-chart that will provide the necessary properties of the multiagent system. Some resources, necessary for achieving the goals by several agents come up for sale. These resources are limited so the agents compete at the auction. The access to purchase resources by the agents are also limited, and the purchase usefulness is estimated by the resource utility function, that is the difference between the income from the resource usage and the expenditures for its purchase.

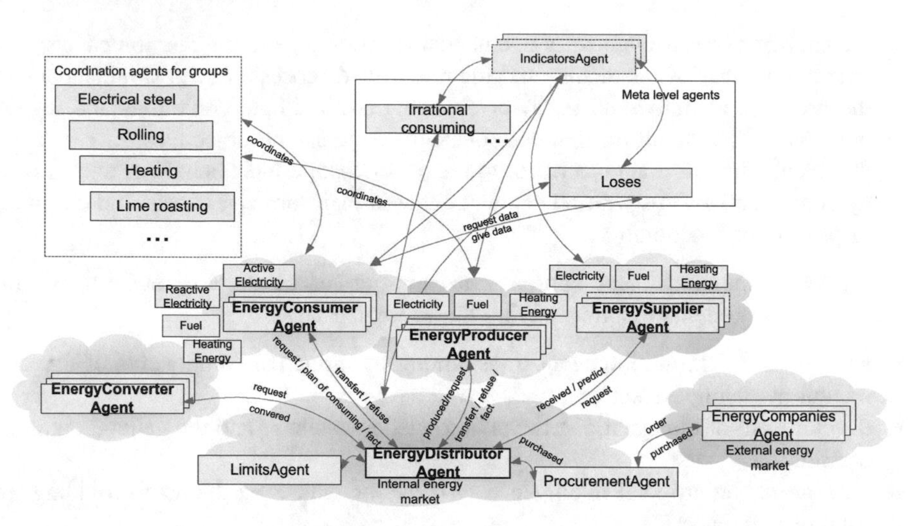

Fig. 3. The agent model for energy consumption control

The agent model hierarchical structure supposes one or several "metalevel" agents that coordinates the tasks and problems solving. Therewith, the agents can form coalitions in order to optimize their expenses (e.g. for resources mobilization).

Thus, the idea of collective behavior leads us to necessity to solve many questions. Among them we can highlight the problem of collective plan formation, the opportunity to take into account the concerns of the agent's partners, coaction synchronization, conflicting goals, competition for the shared resources, conversation organizations, defining the necessity of cooperation, appropriate partner choice, behavioral trainings and ethical rules for groups and communities, task partition and split of responsibilities, shared commitments, etc.

Consider the mentioned characteristics and peculiarities in the developed agent simulation analysis model of energy resources management process at the metallurgical enterprise.

To build an agent representation of the simulation model, we take as a departure point defining the elements with individual behavior.

All of the purchased or produced energy resources are sold to the energy distribution agent at transfer prices. The consumed energy resources are purchased from the energy distribution agent at transfer prices. Herewith, the energy distribution agents provide the consumption agents with a consumption forecast for a stated period. At lack of resources, the agents can purchase them from the distribution agent.

The consumption agents are divided into groups according to the types of consumed resource: active or reactive electric energy, heat energy, or fuel.

For example, every electric energy consumption agent has set parameters: capacity, rated current line, model, cable cross section and cable length, cable electrical resistivity, cable capacity factor. This is the way how the characteristics for electrical equipment for each section is formulated.

The energy distribution agent records the requests in its database, controls the limits and accepts them, in keeping with the condition of resources, the current power balance and the allowable risks. Here is defined the current condition, according to which the energy resources shortage or excess becomes visible. Thereafter, it becomes obvious is the availably energy resources are in use, or if the energy resources are being sold or purchased.

Thus, the day-ahead market works on the principle of exchange house that sets the indicator price for the energy resource. The indicator price is considered by the participants of agreement when signing the agreements and submitting the day-ahead requests.

The resources transfer to the energy consumption agents is carried out in accordance with the priorities, which are meant to manage the goals achievement upon indications of profitability and efficiency.

The energy distribution agent can deny the resource allocation if the sum differs from the sum declared in the request, if there is deficit or if the limits are exceeded. The approved requests are accepted. The accepted deals participate further calculations at modeling, the denied deals can be updated by the agents (there can be changed the amount of energy resources, terms, or other attributes) or they can be deleted. After the energy distribution agent receives information that the operation was executed, the actual energy resources expenditure is registered. If the expenditures go beyond the limit a conflict occurs. To solve it, it's possible to increase the limit, to deny the request or to reconsider the balance of energy, and accordingly to reorganize the energy recourses by taking them from the other agents (or groups).

The energy distribution agent has a regulating function of setting prices for the electrical and heat energy, and the main task is to manage the energy balance structures. The price and rates system must incite cost saving for the producers and energy stretching for consumers. The energy distribution agent also interacts with the purchasing agent in order to pay the energy resources defiance.

Formation of this energy resources home market makes good conditions to form various tariff proposals for consumers. Such as a tariff considering the load demand typical for the consumer, for example in accordance with the time of a day. The main agents' interaction mechanism is conversations that aims to make internal intragroup transactions (deals) on resources attraction and consumption between the agents of consumption, deliveries, production, and distribution.

The suggested model claims that it is vital to unite the agents into the separate activity-specific groups. The criteria to base on when creating these groups are the following: accomplishing close or related operations and services those are related to the process flow, the own activity market, the presence of unit, that manages the activity.

The agent groups are supposed to be created due to the operations carried out by these agents (electric steel, rolling, baking, or heat treatment), to the effective operation according to types of clients (corporate order, government order, investments, etc.) and on the territorial principle. Generally, the agent can belong to a few agent groups. For every group the coordinator agent generation is performed. Multidimensional and detailed distribution of energy consumption indicators within the similar structure gives the chance to estimate efficiency and to influence separate structural divisions, the directions of business and products.

The every agent in the model gets a list of indicators to monitor. Monitoring the mentioned indicators helps to predict risks. When the indicators hit the set limits, it is considered to be a reason to activate various mechanisms and situational scenarios. Thus, when the voltage is lower than the set values the balancing mechanism is on to support it. The same if the losses occur. In this case, the system generates the Indicator agents (their number corresponds with a number of possible solutions) and each of them tries to perform the task in parallel and to work the scenario irrespective of the others and without having ideas of their existence.

That means, that the scores and calculations are kept in the same time for all of the alternative strategies and scenario. Herewith, the strategies and scenario can be changed and modified just when calculating. The every delivery agent, production agent or consumer are responsible for own inbalance, that means off-schedule deviation from the production or consumption plan.

When signing two-sided contract or buying the energy resources at internal or external market one day in advance, both the energy suppliers and energy consumers undertake to provide the consumption and production in certain hours at appropriate level. Herewith, they are parties personally responsible for balance (or enter on a contractual basis into a certain balancing group of agents).

In real life, it is not so easy to provide full accordance with contracts. For example, the weather conditions can get worse, and as a result the customers will demand more electric power in real time or there can occur unforeseen deactivation of the transformer equipment, etc.

Within the developed agent model, the function of the energy efficiency management metalevel is expressed in adjustment of the operating parameters influencing the agents' behavior in process of information exchange course (increase/decrease in production, cut in expenditure, etc.).

The operating influences vector includes adjustment: energy consumption limits; elimination of losses of energy resources; use or sale of the energy developed in the main production, etc.

5 The Results of the Energy Saving Program Implementation at the Enterprise

The developed models allow us to identify, to analyze and to choose the promising energy saving projects in order to choose among them the most viable one. We managed to optimize the portfolio and this allowed us to focus on the most desirable goals.

Approbation of the developed models and computer means in PJSC Dniprospetsstal proved that the effective management of energy efficiency based on the program and portfolio projects management is possible.

The consumption of electricity by the enterprise in 2017 has made 421 168,0 thousand kWh (Fig. 4). Due to implementation of the energy saving program of in 2017 the enterprise saved 1,7 million kWh and 350 thousand m^3 of gaseous fuel.

In comparison with 2011 electricity consumption on steel-melting production has decreased by 30%. Also the share of costs of the electricity for steel-melting production has decreased from 71% in 2011 up to 67,8% in 2017.

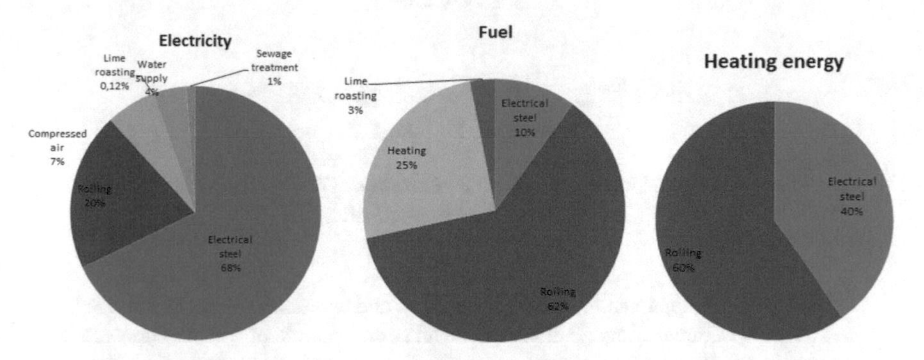

Fig. 4. The balance of energy consumption in PJSC Dniprospetsstal in 2017

References

1. Shemetov, A.: Identification of the electricity consumption of metallurgical enterprises at the highest levels of management, pp. 135–140 (2005)
2. Shemetov, A.N., Fedorova, S.V., Kuznetsov, S.V., Lyapin, R.N.: Modern problems and prospects of model. Forming of energy management at enterprises of mining and metallurgical complex. Elektrotekhnicheskie sistemy i kompleksy **4**(33), 41–48 (2016)
3. Fal'kov, M.I.: Energy conservation and efficiency in Giprokoks designs at Ukrainian ferrous-metallurgical enterprises. Coke Chem. **52**(7), 335 (2009)
4. Shatalov, I., Shatalova, I., Antipov, Yu., Sobennikov, E.: Uilization of secondary energy resources of metallurgical enterprises using heat pump. J. Fundam. Appl. Sci. **9**(7S), 342–352 (2017)
5. Kazarinov, L., Barbasova, T.: Case study of a conservation power plant concept in a metallurgical works. Procedia Eng. **129**, 578–586 (2015)
6. Yujuan, R., Bao, H.: Modeling and simulation of metallurgical process based on hybrid Petri net. In: IOP Conference Series: Materials Science and Engineering, p. 157 (2016)
7. Shcherbakov, M.V., Nabiullin, A.S., Kamaev, V.A.: Multiagent system for modeling the production and consumption of electricity in hybrid power systems in Engineering. Bull. Don **20**(2), 217–221 (2012)

A Method for Assessing the Impact of Technical Risks on the Aerospace Product Development Projects

D. N. Kritsky(✉), E. A. Druzhinin, O. K. Pogudina, and O. S. Kritskaya

National Aerospace University « KhAI», Chkalova 17, 61070 Kharkiv, Ukraine
d.krickiy@khai.edu

Abstract. An approach for the representation and assessment of technical risks in projects of developing aviation equipment as an example of complex products is proposed. The review of risks representation methods is performed. A definition of technical risk is given and for the first time, a method is proposed for assessing technical risk. The method of modeling networks with returns is improved by taking into account the probability of technical risk, which allows obtaining a qualitative characteristic of the project duration. The aim of the work is improving the quality of the planning in projects of developing complex products by taking into account the impact of technical risk.

Keywords: Project risk management · Aviation technology · Technical risk

1 Introduction

The project is considered from the viewpoint of uniqueness in requirements and constraints, which are the quality of the result, the effectiveness and the realization time and cost. Thus, the implementation of the project can be represented by the movement of the system in the phase space from state ω_0 to state ω_k (Fig. 1). The set of works Y determines the shape of the trajectory in the phase space, along which the system moves during the project realization. The same goal or result of the project can be achieved in different ways, each of them correspond to a different trajectory with different time and resource cost values (curves Y_1 and Y_2 in Fig. 1). In addition, under the influence of a number of factors, the system can perform uncontrolled movements in (ω, R, T) space, usually leading to a deterioration in its state. These factors can act constantly or under certain conditions. We are talking about the impact of possible risks on the system.

Therefore, project management should be aimed not only at purposeful improvement of the system state, but also at overcoming these uncontrolled movements, which is risk management. When analyzing project risks, the manager must choose the way for balancing the damage of the occurred risk, and also to allocate dangerous ones. Therefore, in projects of creating complex products, it is necessary to find this kind of mismatch on time and, to perform the required amount of actions in order to eliminate

N. Shakhovska and M. O. Medykovskyy (Eds.): CSIT 2018, AISC 871, pp. 504–521, 2019.
https://doi.org/10.1007/978-3-030-01069-0_36

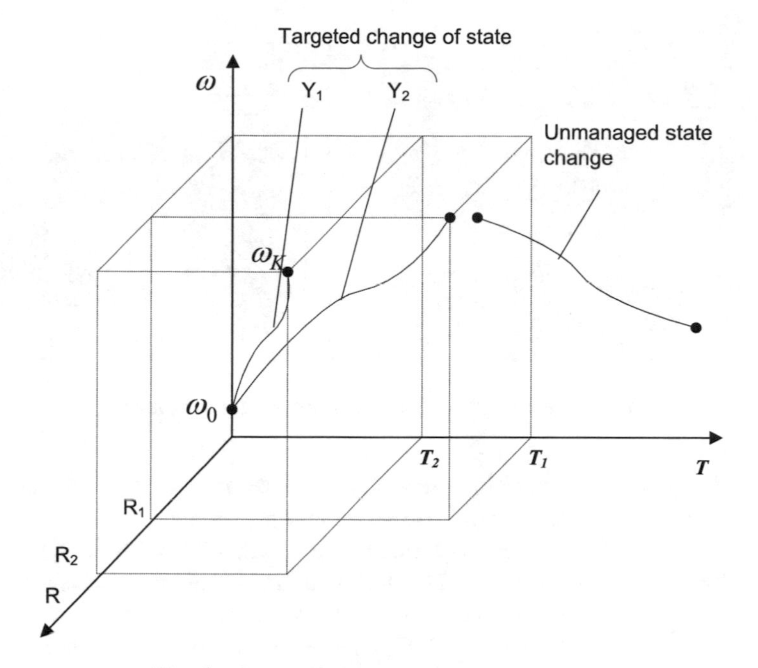

Fig. 1. Geometric interpretation of the project

them by analyzing the situation. In case of occurring risk and mismatch, it is necessary to re-plan the scope of the project, which will lead to additional costs.

In aircraft development projects, one of the ways for finding mismatches is the certification of the aircraft. During the certification process, mismatches can be found between the planned and the actually performed activities. If there is a mismatch, it becomes necessary to determine the scope of the project so that the project goal can be achieved, but the data obtained in the course of the previous works should also be used. Such an approach may be presented in the form of the tree branch. Before starting the project, the project scope is determined from the set of suggested alternatives. A risk is an event or a factor that can cause damage or loss to someone. The risk can be represented as a structure of the form [1]:

$$R\{s, p, x\}, \tag{1}$$

where: s is the risk description, p is the probability or other indicator of the severity (prevalence) of the risk event and x is the weight index, consequence or damage value.

Traditionally, the risk is represented as a probability distribution curve (Fig. 2). For example, in [2], the probability distribution of NPV (net present value) for the confirmed changes of risk factors was given.

In this work, data of distribution functions of NPV terms were collected, and then with the help of calculations by the Mont-Carlo method the final NPV distribution curve was obtained. By analyzing this curve, an estimate of the damage was predicted in the form of insufficient level of project's NPV. In this method, there is no specific

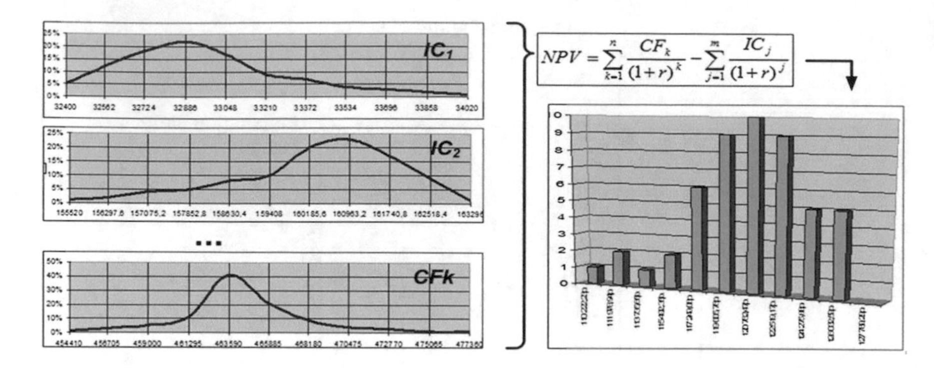

Fig. 2. Probability distribution for project indicators

suggestion for actions, and most importantly, it is necessary to analyze the likelihood of the risk occurrences and their impact on each other separately.

In the GOST R 57272.1-2016 standard, the "Risk Matrix" technique has been suggested for risk analysis (Fig. 3). This technique represents the analysis of a 5 × 5 matrix, which contains the possible combinations of five categories of risk consequences and five degrees of its probability. The five probability degrees include very likely, probable, possible, unlikely and very unlikely.

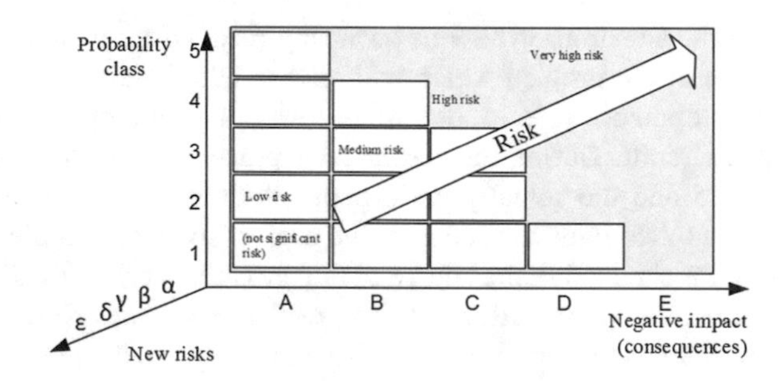

Fig. 3. Risk Matrix method

Categories of consequences include minor adverse effects on human health, living conditions, public services, the economy; limited impact on human health and well-being; moderate impact on human health and well-being; emergency situations with harmful consequences for human health, loss of livelihood and extremely negative events, affecting simultaneously on human health, the environment, the economy, and so on. After filling such tables (matrices), it is suggested to use color coding and analyzing the obtained data to make decisions about the project future. In this technique, the subjective opinion of experts, which assign the appropriate assessments for each risk, has an important influence. This reduces the error in determining the necessary and sufficient design solutions to the error obtained when performing expert evaluation. In [3], each risk event is represented by a circle whose center is located at the intersection point of the probability and the consequence of the risk (Fig. 4).

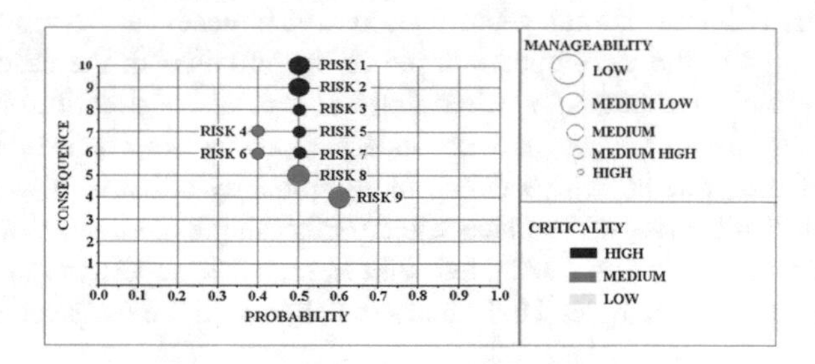

Fig. 4. Safety oriented bubble diagrams in project risk management

Manageability, as a measure of the impact on the consequences of risk, represents the size of the circle. These three components of risk affect the risk's criticality for the project, which is specified by color (black - high criticality for the project, green - medium and yellow - low). From the visibility point of view, it is a very useful tool for the manager, but in innovative projects where there are a lot of risks, using this bubble chart will have difficulties related to overlapping risks to each other and the analysis of the current situation within the project. In [4], a three-dimensional model of risk representation is considered where, in addition to the probability of occurrence and consequences of risk, the axis of the "strength of knowledge" is suggested (Fig. 5). These data are necessary when the probability of occurrence of risk is assessed not statistically, but by expertise.

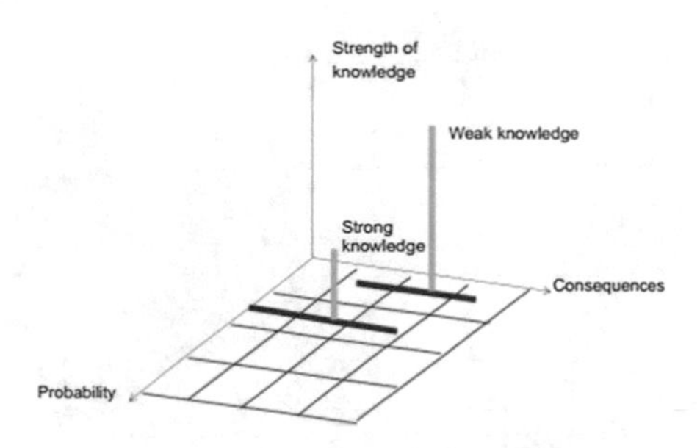

Fig. 5. Three-dimensional model of risk representation

Then, it is important to take into account the coefficient of consensus of the expert group and other factors, which is demonstrated by this axis. The three-dimensional representation of risks is more effective, but in the proposed method, complexity arises with the objective quantitative risk assessments.

When modeling projects of developing complex technical systems, such as aerospace products, it is necessary to take into account specific risks of this industry. To increase the efficiency of implementation of such projects, the concept of requirements management is applied. Therefore, the main risks that are taken into account in project planning are the risks of not achieving the expected quality. In the works of O.K. Pogudina and E.A. Druzhinin the notion of technical risk was represented. The characteristic of this risk type is that when it occurs, it is necessary to make principal or secondary changes in the developed sample of the product. In the case of aerospace products, the principal changes are changing the typical construction of the product, which affects its airworthiness, and secondary changes are changes in the typical construction of the product, which does not significantly affect its airworthiness. The occurrence of technical risk takes place when performing experiments or flight tests of the product and most often is associated with the need to repeat the design activities and recreate the product sample. This causes the change in the duration and the cost of the project.

2 The Proposed Method

Technical risk has an important parameter, which does not exist in the reviewed models of risk representation. It is characterized by the time of occurrence, since it is associated with the results of the project's activities. In planning the development of complex technical systems, experts can evaluate a number of specific risks that arise while performing individual activities or project phases. For example, at the research phase of the formation of the aircraft's configuration, the risk of mismatch between the obtained result (in the form of calculated characteristics) and the requirements of the technical specification is a significant prevalent event. Planning for developing a complex product should be performed with mandatory control of individual risks, as well as their subsequent integration to assess the risk of project failure as a whole (Fig. 6).

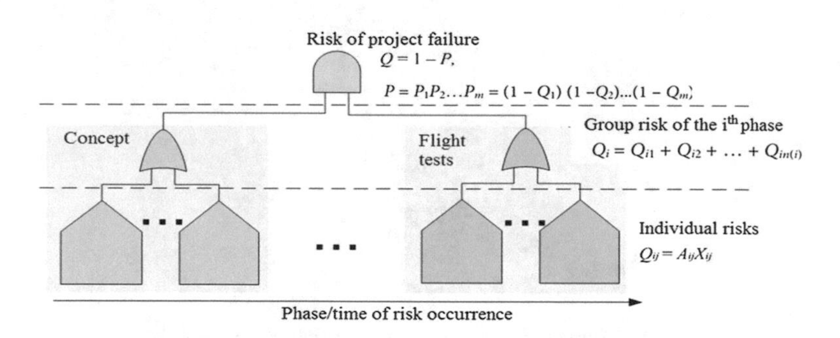

Fig. 6. Calculation of the risk of the project failure

The individual risk R_{ij} assessment Q_{ij} has the form:

$$Q_{ij} = A_{ij}X_{ij}, \tag{2}$$

where A_{ij} is the weight index, for example, an estimate of the economic losses caused by this type of risk and X_{ij} is the severity index (prevalence).

This equation generalizes the known method of risk assessment as the product of the average damage (the mathematical expectation of damage) to the probability of the undesirable event. For the subsequent integration of risks at the level of the individual project phases, the weighting (importance) factors must satisfy the following criteria:

$$A_{i1} \max X_{i1} + A_{i2} \max X_{i2} + \ldots + A_{ik(i)} \max X_{in(i)} = 1 \quad (3)$$

For the project completion, all phases must be completed, therefore, the top-level integration for estimating the likelihood of successful project implementation is carried out by the multiplication $P = P_1 P_2 \ldots P_t$. Accordingly, the risk of project failure will be $Q = 1 - P$.

In order to visualize the individual project risks, it is proposed to use a three-dimensional coordinate system (probability of risk, occurrence time, weight index), and demonstrate the risk in the form of points (Fig. 7). A visual representation will allow analyzing the situation within the project depending on whether the total damage from happening all risks is higher than incomes or not. Depending on the analysis of the data, hedging, diversification, creating special reserves or other methods should be used. The parameters of the analysis will be: the size of the damage, taking into account the probability of occurrence, the expected income, the result of the sum of the integrals at a given time interval as the functions which describe risks.

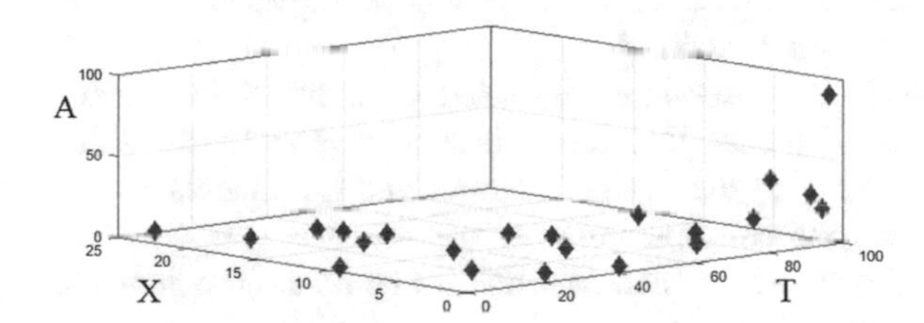

Fig. 7. Three-dimensional coordinate system of risk visualization

In aerospace engineering projects the following states ($V_{i,j}$) can be identified: the requirement state ($V_{1,1}$); formulation of the general problem for the product requirement ($V_{1,2}$); the functional model of system that describes the useful function, connections and relationships with the outside environment ($V_{1,3}$); technical plan ($V_{2,1}$); technical specification ($V_{3,1}$); technical proposal ($V_{4.1}$); application for the issuance of the type certificate ($V_{5.1}$); preliminary design ($V_{6,1}$); layout ($V_{7,1}$); technical project ($V_{8.1}$); detail design documentation for the prototype ($V_{9,1}$); prototype ($V_{10.1}$); the prototype sample submitted for experiments ($V_{10.2}$); typical construction ($V_{11.1}$); certified typical construction ($V_{12,1}$). The works that need to be done for achieving the project goal are divided into two types: preparatory works ($W_{0,P}$, where the first index indicates that this is the preparatory work - 0, and the second number is the work order at this stage) and works aimed to create the product of the project ($W_{i,\ j}$) [5].

Thus, in order to go to the next project state, it is necessary to fulfill the next condition: $V_{0,f} \underset{W_{0,g}}{\longrightarrow} V_{all}$, where: f is the number of the last state and V_{all} is the state of the project, when the documentation needed for creating the specified quantity of the project product is ready.

3 The Model of Forecasting the Results of Projects of Developing Complex Products Based on the Values of Technical Risk

One of the problems, which is solved by the certification center as a project quality control center in the early stages of design is the comparison of selected variants of high-tech product sample for their suitability and sensitivity to refinements in the course of detailed design. Solving this problem involves the application of technical risk criteria in the selection of design solutions. Nominal values of the criteria, which are selected by the results of the "external" design, become a reference point for the subsequent design stages. After choosing reasonable reserves in the process of selecting final variants, the criteria values (mass, cost, the final tactical and technical characteristics) are considered in the project task, as well as the technical specifications for the system, which should be adhered in the course of the aircraft development.

Technical risk is defined as the probability that the obtained values of an aircraft's design parameters in the process of its development goes beyond the limits, which are determined by the design task.

Any technical risk assessment is based on a priori information about possible deviations of the initial data, therefore it is subjective and reflects the designer's views about the unreliability of the design methods and the inadequacy of their real conditions of manufacturing and operation.

At present, the criteria of technical risk is one of the few indicators that reflect the reliability of design.

The method of calculating risk criteria is based on the following assumptions.

1. The values of technical characteristics are determined for each variant of the design solution independently.
2. Distribution functions of the criteria values can be obtained with a high degree of reliability.

For simplified preparatory assessments, it is logical to characterize the effect of limiting deviations of the values of certain data on the magnitude of the criteria by finding the sensitivity of the criteria to the different conditions of the design task.

In the general case, based on the statistical experiments, the correlation of the initial data, the nonlinearity of the objective functions are taken into account and the results of the study are probabilistic indicators of the relationship of the parameters which are realized in the course of the development to the proposed ones in the project.

We consider, as an example, the application of risk criteria by comparing two variants of design solutions. Suppose that at the initial stages of the design, there is the task of choosing the first variant (an innovative design solution, which is optimal by mass, based on using the promising materials and technology, new design solutions, effective but unusual aerodynamic forms, new principles of motion control) or the second variant, by giving up the mass of the design, but retaining the known design principles, unified assemblies, tested and simple aerodynamic configuration, confirmed control schemes and available technological processes.

As the result of modeling, the distribution of possible values for the mass of the structure and the supporting systems, we obtain the graphs shown in Fig. 8. There is a limit value of the mass (M_0) and the probability of exceeding this limit (technical risk) for two variants may be approximately equal, although a comparison by the nominal values shows the advantages of the first variant. When deciding about the further development of the project, it is necessary to assess how technical risk can change if the scope of the project task is expanded.

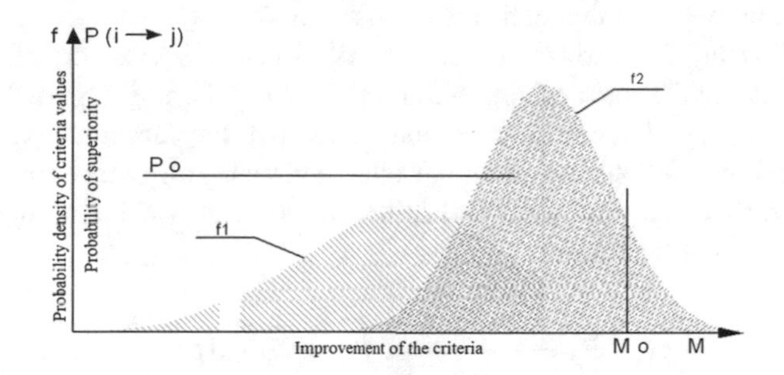

Fig. 8. The probability distribution for various values of mass

The design study should consider how the change in input data and project constraints affects the choice of optimal solutions. By shifting the limit value of M_0 periodically, it is possible to calculate the distribution functions of $P^{(1)}(M^{(1)} < M_o)$ and $P^{(2)}(M^{(2)} < M_o)$ which reflect the probability of successful completion of the design task.

The indicator of the superiority of one variant over another is the probability of a complex event, by which the dominant variant satisfies the project constraints and for the competing variant, it does not.

The ability of stably holding the superiority by probability is visually evaluated as the value of the design constraint is shifted. In this case, the probability of the superiority of the new solution over the traditional one can be calculated as:

$$P(1 \rightarrow 2) = P^{(1)} - P^{(1)}P^{(2)} \tag{4}$$

Similarly, the probability of the situation by which the project task for the variant with traditional solutions is executed and the same task c not be executed for the variant with innovative solutions is obtained from the distribution function of the successful implementation of the constraints:

$$P(2 \rightarrow 1) = P^{(2)} - P^{(1)}P^{(2)} \tag{5}$$

Moving the boundary of M_0 within the value of mass reserve to the payload, we observe that the decrease in the reserve amount is more advantageous for the first variant, while by its increasing the preference will be by the second variant. However, these preferences are doubtful, since the reliability of estimates at the edges of the distribution is low.

Any characteristic of the aircraft (mass, cost, speed, range, etc.) is physically limited. The distribution functions of the values of these parameters by a fair accuracy are determined with the normal distribution [6], and often have unacceptable deviations at the distribution edges.

The certification center, in the process of monitoring, should analyze the ultimate values of the characteristics. Detection of such values is performed by using the mathematical apparatus of the statistics of extreme values.

When analyzing the variants of design solutions, it seems logical to minimize technical risk at fixed values of limitation on the M_0 criterion. Optimal and close to optimal solutions by the criterion of total mass of the structural and supporting equipment, design solutions are obtained after reviewing all competing variants, and for each *i*-th variant, the probability of failure of the task for the comparison is calculated by the following formula:

$$P_p = 1 - \frac{1}{\sigma\sqrt{2\pi}} \int_0^{M_o} e^{-\frac{(\tau - M)^2}{2\sigma^2}} d\tau \tag{6}$$

where M is the mathematical expectation of the characteristic and σ is the standard deviation of the characteristic, for projects of creating technical systems this parameter can be determined on the basis of statistical data of the errors of design or control process.

The dimensionless criterion of "exposure" to risk makes it possible to compare different variants with each other, based on the risk, which corresponds to the variant with traditional and well-examined design solutions $P_T(K_o)$. Hence, the criterion is defined as follows:

$$R_i = \frac{P_i(K_o)}{P_T(K_o)} \tag{7}$$

When $R_i < 1$, there is a "risk redundancy" zone based on the criterion value scale, which shows the profitability of the solution which has the elements of novelty.

The comparison of the variants is also made according to the conditional probability of superiority, which is found, for example, by comparison with the risk inherent in the traditional design decision. The conditional probability of superiority $P_{y,i}$ reflects the probability of the event in which the criterion value obtained during the implementation of the i-th new variant falls within the range of the design task, provided that the unsuccessful coincidence of the circumstances for the realized design solution is known.

In practice, it is necessary to compare variants according to many criteria under the conditions that statistical laws can not be used because of the small accuracy of probability estimates at the edges of the distributions. In addition, there is an interdependency between different criteria, because they are determined by the same set of input data and a single design calculation with the correlation of the values of the aircraft characteristics. Under such conditions, it seems logical to compare the variants by the sum of the relative deviation modules of the limiting values of the initial data. The vector of the limiting state of the data is characterized by the convolution value, which is obtained based on the calculation of the radius of the Heming region in the parameter space when solving optimization problems with discrete variables.

The swing range of the initial data vector is calculated from the equation:

$$r_H^{(q)} = \sum_{i=1}^{n} \left| \frac{q_{i,\lim} - q_{i,cp}}{q_{i,\lim}} \right| \tag{8}$$

where $q_{i,cp}$ is the nominal value of the i-th component of the initial data, $q_{i,\lim}$ is the limiting value of the i-th component of the initial data and n is the dimension of the vector of the initial data.

The represented convolution, summing up the deviations of the data, gives only a qualitative picture of the richness of the design solution by the uncertainty factors. To replenish the same qualitative picture, but not for making quantitative estimates, a convolution can be used that reflects the sum of the limiting deviations of the values of the criteria, calculated by using the methods of statistics of extreme values. Convolution can have the next form:

$$r_H^{(q)} = \sum_{j=1}^{m} \left| \frac{K_{j,\lim} - K_{j,cp}}{K_{j,cp}} \right| \tag{9}$$

where $K_{j,\lim}$ is the maximum deviation of each j-th criterion and $K_{j,cp}$ is the nominal value of each criterion.

The possibility of applying the latter formulas only for the qualitative assessment is explained by the dependence of the criteria values, which are obtained by statistical experiments. That is why the creation of separate resulting histograms for mapping the distribution functions of the values of each criterion does not provide objective information. Only the approximate graphical representation of the spatial multidimensional curve of the distribution density of the values of the vector criterion provides an objective picture that allows to conclude the probability of realizing the project task, which limits the values of two or more design characteristics.

From tens and hundreds of studies of the initial design stage, only a few reach the construction stage, and only some of them cross the start of development and production. Based on these individual implementations, the objective estimates of the reliability of design calculations are constructed.

We suppose that one of the project variants, which was investigated earlier with the help of technical risk criteria, has undergone a number of changes that is provided by detailed modeling calculations, accurate information about the design elements, on-board systems, conditions and flight programs.

At the initial design stage, a priori values of the technical risk is P_p and the probabilities of realizing task is $P_v = 1 - P_p$.

From the histograms obtained at the stage of detailed verification calculations and constructional investigations, a posteriori values of the same probabilities $P_p^{(p)}$ and $P_v^{(p)}$ are determined. Usually there is the next inequality: $P_v^{(p)} < P_v$.

The level of confidence in approximate design models of the initial design stage is reflected by the next dimensionless coefficient:

$$K_d = \frac{P_v^{(p)}}{P_v} \tag{10}$$

4 Examples of the Technical Risk Calculation

In the process of controlling the characteristics of complex equipment, the results of individual works are measured, and in the early stages of the project, as a rule, intermediate results are obtained which are not specified in the product's technical specification. In addition, every method of calculating the characteristics, which is used in developing the complex equipment, is characterized by the magnitude of the error.

For finding the magnitude of the characteristics dispersion, which are specified in the technical specification of the product and associated with the obtained results at intermediate stages of the developing complex products, the momentum method or the Monte-Carlo method are used. The error of the Monte-Carlo method is $N^{-0,5}$, where N is the number of experiments. Let us consider the example of calculating the distribution of the "flight range" characteristic at the stage of the model testing. The flight range depends on the following parameters, each of them is characterized by its distribution law, obtained during the field test of the model:

$$l = L(c_{x_0}, A, c_p, C_{Tp}, L_p, L_{pl}, G_k, G_{ob}, G_{cy}) \tag{11}$$

where c_{x_0} is the drag coefficient in zero lift at the cruise mode, A is the coefficient of the cruise polar, c_p is the coefficient of the engine thrust, C_{Tp} is the fuel consumption during acceleration and climb, L_p is the distance of the acceleration and climbing regime, L_{pl} is the range of planning regime, G_k is the body weight, G_{ob} is the equipment weight, G_{cy} is the weight of the control system.

The result of the calculation by the Monte Carlo method with $N = 5000$ results in the values of $M = 3917$ and $\sigma = 356$ (Fig. 9), while the calculation by the momentum method results in $M = 3917$ and $\sigma = 363$.

For calculating the technical risk parameter, it is necessary to determine the possible limitations specified in the normative documentation or in the technical specification for the project of developing the complex equipment depending on the activities that are being performed. Design activities consider the compliance of the

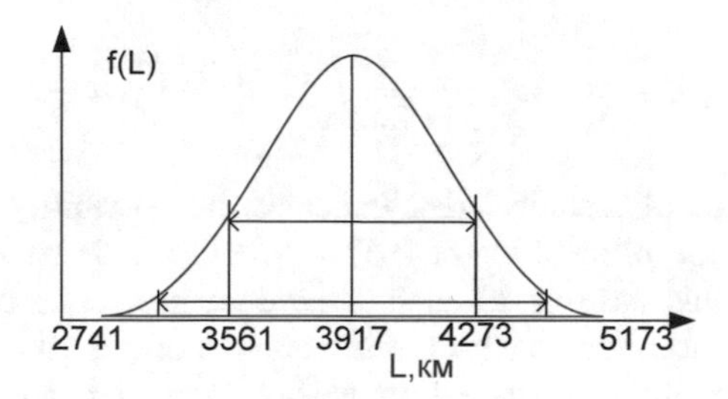

Fig. 9. Density distribution for "flight range" parameter

characteristics with the parameters specified in the technical specification, and certification activities with the certification basis.

For example, if in the technical specification of the aircraft it is stated that the flight range can reach 4700 km, then the risk of non-fulfillment of this condition after the stage of the "Layout" is equal to:

$$P_{\text{flight range}} = 1 - \frac{1}{356\sqrt{2\pi}} \int_{0}^{4700} e^{-\frac{(\tau-3917)^2}{2 \bullet 356^2}} d\tau = 0,014 \tag{12}$$

Let's consider the calculation of the take-off runway length of the aircraft (Fig. 10), this characteristic is contained in the norms of airworthiness. For JAR-VLA [7] it should be up to 500 m and by AP-25 standard [8] it is calculated as 115% of the horizontal distance along the take-off path from the start point to the point, in which the aircraft height above the take-off surface is 10.7 m (specified by the customer).

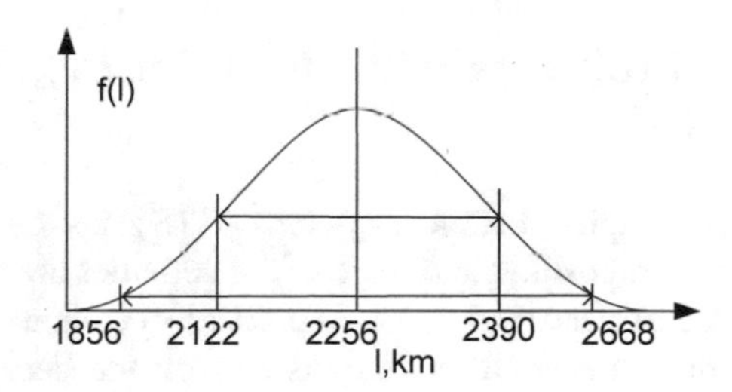

Fig. 10. The distribution density of the "runway length" characteristic

Thus, if this aircraft is to be certified according JAR-VLA standard, then the technical risk is equal to one, it is necessary to repeat the aircraft development steps: the external, general design.

If the aircraft is certified according to aviation regulations AP-25 in the technical specification it is stated that the distance from the start point to the take-off point is 2000 m, which according to the rules of AP-25 corresponds to the limit of the take-off (115%) to 2300 m, so the technical risk will be equal to:

$$P_{\text{runway length}} = 1 - \frac{1}{134\sqrt{2\pi}} \int_0^{2300} e^{-\frac{(\tau - 2256)^2}{2 \bullet 134^2}} d\tau = 0,371 \quad (13)$$

The obtained value of technical risk indicates the possibility of obtaining a sub-standard result with the probability of 0.371. Whether it is necessary to repeat the project activities in this case or in each individual project is decided by practical considerations and in accordance with the importance of the desired project result. To this end, at the project planning stage a number of parameters are defined: NR normal result (the result satisfies the requirements); EM elementary mismatch (there is a slight disagreement with the requirements); SM secondary mismatch (there is a significant disagreement with the requirements); GM global mismatch (the result does not meet the requirements). After this, a piecewise linear interpolation is constructed between the values of technical risk and the values of NR, EM, SM and GM (Fig. 11). Knowing the results, which will be obtained at the end of the next stage allows us to anticipate possible inconsistencies in advance and avoid changes, which means shortening the time of the aircraft development.

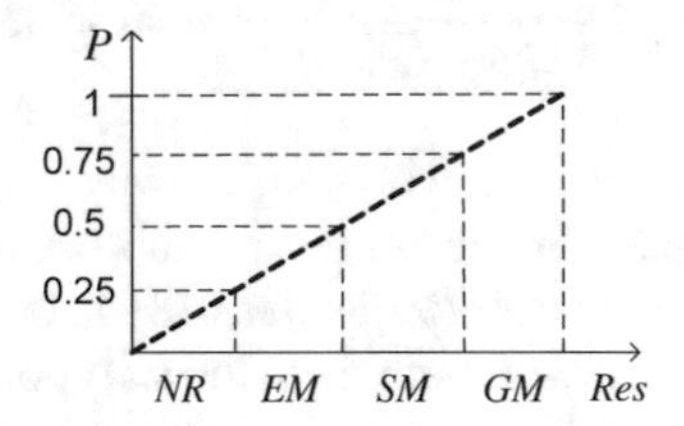

Fig. 11. Interrelation of the technical risk index with the results of the project

5 Scheduling the Project Activities by Taking the Technical Risk into Account

When developing new complex technical systems (CTS), not only the parameters of the project, but also the composition and logic of the activities are subject to significant changes under influence of occurring risks. The set of operations, their sequence and relationships depend on the specific conditions and on the decisions that are made during the development.

Among the features of the CTS development projects it is necessary to highlight these aspects: complexity of the interrelationships between the various stages of activities; the control system; the testing of the various units, blocks and the product as a whole; the repetition of the individual components of the CTS in connection with the technology development and the need for a large number of supplementary and finishing activities. Sources of such activities are mainly certification tests. Insufficient quality of creating individual components of the product can also be the reason for the reworks. Another important source of reworks is the control of technical and working documentation [9]. Working drawings are subject to technological control, as well as standardization control.

If there are any deviations from established norms, non-compliance with standards, incomplete or unreasonable refusal to apply standardized units and parts the drawings are returned back for correction and revision.

The emergence of works that are subject to redesign or refinement, significantly changes the project implementation time and required resources for its implementation. In addition, this kind of additional works can have a rather complex structure and be represented by a subgraph. The model for such case can be the network with return loops (Fig. 12).

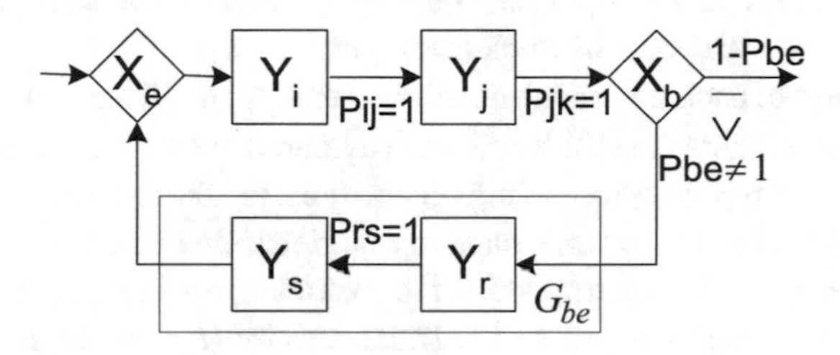

Fig. 12. Fragment of the project network with returns

It is assumed that the graph $G(E, U)$, on the basis of which the model is created is finite, i.e. $|E| < \infty$, and has the following properties:

(1) Has one initial vertex i_o, which is the input of the network for which the $\Gamma_{i_o}^{-} = \varnothing$, where $\Gamma_{i_o}^{-}$ is the set of events, from which the activities emanate to the vertex e and i_n is the exit. The occurrence of event i_n means the end of development process.
(2) The set of graph edges G is inhomogeneous and consists of edges (i, j), which must be fulfilled with a probability equal to 1, and return edges of (b, e), realization of which occurs with the probability of $P_{be}, 0 < P_{be} < 1$.

The set of edges $\{(i, j)\} = U_g$ is a list of edges of the deterministic graph $G_g(E, U_g)$, shows the development process without taking into account the situations that generate the returns. We denote the set of return edges by $U_b = \{(b, e)\}$. Hence $U_b = U \backslash U_g$. Return edge $(b, e) \in U_b$ closes the graph $G(E, U)$ by connecting the vertex b with higher rank to the vertex e with lower rank and generates random contours in it.

(3) The set of vertices of the graph E also is inhomogeneous and consists of the vertices $x \in Y, X$; where $E = Y \cup X$. Here Y is the set of vertices, at the input of which the logical operation "AND" is realized.

In the set X there are all vertices b, from which there is the possibility of return: $\{b\} = X^1$, and, all vertices to which the there is a possible return $\{e\} = X^2$ a $X = X^1 \cup X^2$.

In this case, the vertices $b \in X^1$ are branching vertices that form one of the next two possible generalized outcomes: with the certain probability the development process continues in the directions determined by outgoing edges or as the result of random outcomes of the events, returns back to repeat the already finished development stages.

The vertices $e \in X^2$ realize the logical operation $\wedge$ on the input, which means the occurrence of an event when all deterministic activities which enter to it are realized, or a return occurs into it.

(4) In the simplest case, it is assumed that the return edges $(b, e) \in U_b$ have the following simple structure: $\{(b, e), P_{b,e}\}$, where the value of $P_{b,e}$ is the probability of realization of the return edges (b, e). In more complex modifications of this model, the return edge (b, e) can be deployed with the subgraph G_{be}, having nonzero duration and cost of implementation.

(5) Corresponding to each deterministic edge of the graph (i, j) a vector $A(t_{ij}, S_{ij}, R_{ij})$ is defined which is characterized by the activity parameters i.e. the time t_{ij}, cost S_{ij} and various resource types R_{ij} for its implementation. In addition to these characteristics which are common for network planning on deterministic edges, the parameter α_{ij} (as a rule, $0 < \alpha_{ij} \leq 1$) is determined as the coefficient of change in the duration of the work for the repeated execution, i.e. if the activity (i, j) as the result of returns is realized for k-th time, then its duration will be calculated by the next formula

$$t_{ij}^{k} = \alpha_{ij} t_{ij}^{k-1} \tag{14}$$

For the given contour (b, e) and the probability of passing through it $P_{b,e}$, the probability of the exit from the contour for m-th pass is ξ. Then, the m possible passes within the contour after which we exit from it with the probability of ξ where $\xi = 1 - P_{b,e}^{m}$ will be [10]:

$$m = \frac{\ln(1 - \xi)}{\ln P_{b,e}} \tag{15}$$

For example, for $\xi = 0.95$ and $P_{b,e} = 0.4$ we get $m \approx 3$, i.e. after three times passing within the contour, we exit from it with the probability of 0.95. By determining the allowable time of accomplishing the event j (with the probability of ξ), identified with the exit from the contour, it is necessary to add to the length of the path passing from the event i to the event j, the value of m L(K(i)), where L(K(i)) is the length of contour K(i).

The network model of control processes can be represented as an alternative network with returns. As the probability of $P_{b,e}$, we use the value of technical risk (Fig. 13).

The basic fragments of the alternative network are the following set of activities:

Z2 - the activities of the overall design phase of the aircraft development;
Z3 - the activities of the component design phase of the aircraft development;
Z4 - the activities of detail design documentation;
Z6 - the activities of the overall design for aircraft certification;
Z7 - the activities of the aircraft certification in the “detail design” stage.

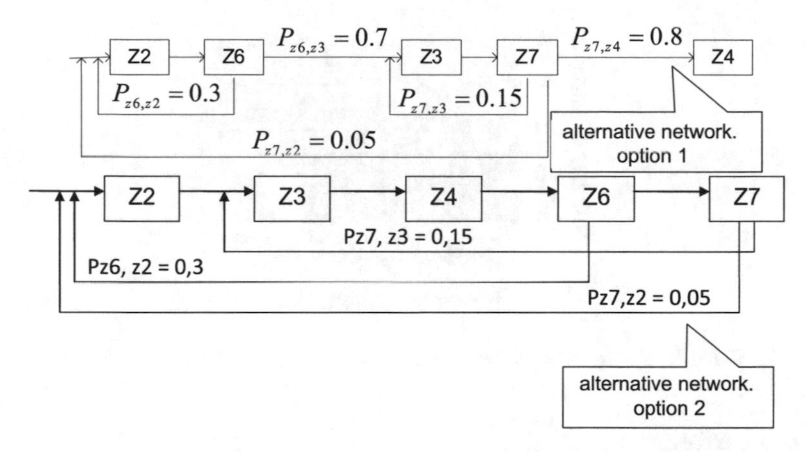

Fig. 13. Fragments of the alternative network

After conducting a series of tests, the following technical risk values were obtained: $P_{z6,z2} = 0.3, P_{z7,z3} = 0.15, P_{z7,z2} = 0.05$.

Alternative network variants for these works will look like as follows:

We define the probability of the exit from the contour as $\xi = 0.95$. Then the number of possible passes m within the contours are:

$$m_{z6,z2} = \frac{\ln(1 - 0.95)}{\ln 0.3} \approx 3; m_{z7,z3} \approx 2; m_{z7,z2} \approx 1.$$

The number of possible project realization variants with considering the occurrence of returns will be $N = 2^{\sum_{i=1}^{l} m_i}$, where l is the number of contours. Thus, for the obtained fragment of the network, it is possible to calculate 64 variants, with different values of project realization time. The minimum time of project implementation is:

$$T = T_{z2} + T_{z6} + T_{z3} + T_{z7} + T_{z4} = 620 \text{ days}$$

The maximum value, for controlling the results immediately after receiving them is

$$T = T_{z2} + T_{z6} + T_{z2} + T_{z6} + T_{z2} + T_{z6} + T_{z2} + T_{z6} + T_{z3} + T_{z7} + \\ + T_{z3} + T_{z7} + T_{z3} + T_{z7} + T_{z2} + T_{z6} + T_{z7} + T_{z3} + T_{z4} = 1712 \text{ days}$$

provided that all activities were executed consequently.

After calculating the project implementation variants, the histogram of the statistical data can be obtained, which shows the project duration. For the considered network and the number of N = 64 cases, the obtained histograms shown in Fig. 14.

The practical value of the obtained model is that its use will allow predicting the progress of developing complex products and for each fixed strategy of creating a new product, helps to answer questions about the relationship between the duration and the probability of achieving the goal.

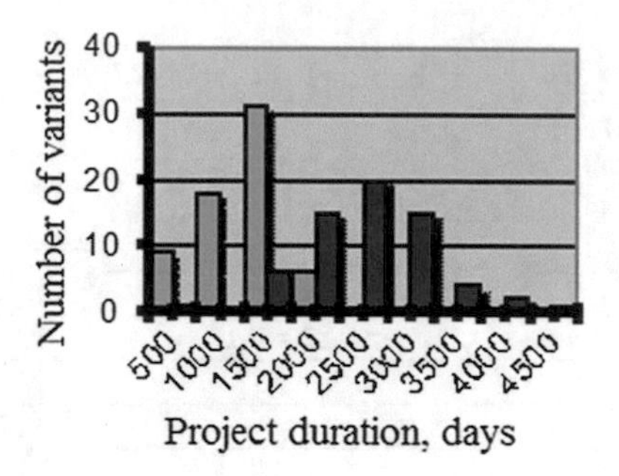

Fig. 14. Histograms of the statistical series of project duration values

6 Conclusions

The analysis of risk management methods showed the requirement not only to visualize the arising risks in the course of the project, but also the requirement of creating a method that would allow for quantitative assessment, on the basis of which further project management is carried out.

In this work, a method was developed for analyzing the technical risk, illustrated by aircraft as the example. For the purpose of quantitative risk assessment, probabilities were introduced, where the variance of the design characteristics is determined by using the Monte Carlo method, statistical models for calculating the confidence intervals for test results.

Obtained values of the technical risk serve as the basis for planning the project activities taking into account the impact of the technical risk. The probability values of obtaining a low-quality result are the initial values for specifying the probability of occurring returns when constructing the alternative network with returns. Multiple calculations of the alternative network with returns, allows obtaining the probabilistic characteristics of the project duration values, taking into account the occurrence of activities that are subject to redesign or revision.

References

1. Mukha, A.R.: Managing the process of developing complex technical systems and processes, features of the FMEA-analysis application. Math. Mach. Syst. **2**, 168–176 (2012)
2. Iskhakov, M.I., Shekalin, A.N., Gorbunov, V.N.: Choice of risk analysis methods for increasing the information content and quality of the economic and mathematical model of the investment project. Modern scientific researches and innovations, vol. 1, no. (2) (2015)
3. Abrahamsen, E.B., Aven, T.: Safety oriented bubble diagrams in project risk management. Int. J. Perform. Eng. **7**, 91–96 (2011)
4. Aven, T.: Practical implications of the new risk perspectives. Reliab. Eng. Syst. Saf. **115**, 136–145 (2013)
5. Chenarani, A., Druzhinin, E.A., Kritskiy, D.N.: Simulating the impact of activity uncertainties and risk combinations in R&D projects. J. Eng. Sci. Technol. Rev. **10**(4), 1–9 (2017)

6. Mendenhall, W., Sincich, T.: Sincich Statistics for Engineering and the Sciences. Chapman and Hall/CRC, London (2016)
7. Joint airworthiness regulation, JAR-VLA. Airworthiness standards: Very light aircraft. Less than 750 kg gross weight, 26.4.90
8. Aircraft specifications. Standards of flight suitability of planes of transport category. Mezhgosudarstvennyi aviatsionnyi komitet LII im. M. Gromova, Moscow (1994). Chap. 25
9. Komiogorov, V.L.: Forecasting the quality of engineering products. The Ural branch 10. Sverdlovsk: UrA of the USSR Academy of Sciences (1991)
10. Voropaev, V.I., Gelrud, Ya.D.: Cyclic alternative network models and their use in project management. http://www.sovnet.ru/pages/casm1.rar. Accessed 17 June 2018

Sustainability and Agility in Project Management: Contradictory or Complementary?

Vladimir Obradović[1(✉)], Marija Todorović[1], and Sergey Bushuyev[2]

[1] Faculty of Organizational Sciences, University of Belgrade, Belgrade, Serbia
obradovicv@fon.bg.ac.rs, todorovicm@fon.bg.ac.rs

[2] Kiev National University of Construction and Architecture, Kiev, Ukraine
sbushuyev@ukr.net

Abstract. This paper aims to analyze the new perspective within the sustainability in project management considering contemporary project management methodologies, practices, knowledge and skills and future trends in this management discipline. One of the main issues at the begging of the 21st century is how to achieve sustainable development. Sustainability as a concept is present at the society level and at the business level as well, therefore there is an increasing effort among researchers and practitioners to integrate project management and sustainable development. In recent years project management discipline is also facing a growing challenge of how to create value and respond to changing the environment, in order to profit. Facing this challenge requires agility. Agile management is now present not only in software development but in other industries too. Based on the analysis of sustainable project management concept, the main challenges of its application, and the key elements of agile project management, the main conclusions of this paper is that sustainability and agility are complementary concepts that help project managers to deal with environment burden.

Keywords: Sustainability · Agile · Project management

1 Introduction

By the end of the 20th century, the concept of sustainable development became one of the most important thoughts for society and in the business world as well. An arising question is how to evolve without limiting future generations to meet their needs. At the society level in recent years there is an increasing development of high-tech housing projects (projects that require a combination of site selection, sustainable materials, proven methods, etc.), sustainable food systems, projects for conversation of cultural heritage projects [1], based on digital services in order to build sustainable infrastructure [2]. Further, some countries, for example, The Republic of Korea, developed a national project to construct a new administrative capital, to relocate regulatory organizations of the central government for improvement of overpopulated capital and regionally balanced development [3]. Those projects are established for the implementation of innovation platform for sustainable regional development.

N. Shakhovska and M. O. Medykovskyy (Eds.): CSIT 2018, AISC 871, pp. 522–532, 2019.
https://doi.org/10.1007/978-3-030-01069-0_37

That fact that sustainability is spread in almost every day life confirms the fact that EU is funding a research project focused on the development and evaluation of strategies aimed to bring household activities that are more sustainable, referring to period 2050 [4].

At the business level, with the overall growing awareness of environmental concerned, being green is becoming a necessity in today business [5]. The fact is that companies have mainly begun to introduce the concept of sustainability into their strategic documents [6]. A large number of studies show that focusing on sustainable development can make the organization more competitive, more flexible, with greater ability to conquer new markets. Corporate Social Responsibility represents the most common way sustainable development manifestation in the organization and represents the basis of company behavior and responsibility for the various influences from society.

Still, the implementation of this concept requires an investment of time, work, and resources, while the results are uncertain and a company needs time to recognize benefits and to measure the effectiveness of sustainability implementation, which is the main reason for companies to drop out this process.

The most common benefits of introducing the sustainability concept in an organization are long-term value creation through greater cohesion in the organization and increasing efficiency and flexibility; improving the reputation and image of the organization [7]. The same author quotes studies that emphasize the extent to which the organization's reputation enhances the organization's profit: the 60% of reputation improvements lead to a 7% increase in stock value.

Project management is the result of decentralizing management and introducing stochastic flexibility into the planning and programming of new ventures. Therefore if we consider projects as a mechanism to implement company's strategy, it would be crucial to integrate and evaluate sustainable development at the operational level [8–11].

This paper aims to analyze the new perspective within the sustainability in project management considering contemporary project management methodologies, practices, knowledge and skills and future trends in this management disciple.

If projects are considered as temporary organizations that result in changes of any kind (in the organization, products, services, assets, etc.) it can be concluded that sustainable development requires projects - projects create the future. In addition to the projectification of societies, which is a measure of the diffusion of project management in all areas of society, as a global trend, future development trends of project management by to 2025 are: dealing with the complexity of projects; trans-nationalization and virtualization of project management, women in Project Management and sustainability [12].

Considering the elementary characteristics of projects such as time and resource constraints, risk, and uncertainty, temporary engagement of team members as well as the presence of significant number of stakeholders, project management influencers, high demands, new technologies, many researchers confirmed that project management requires a transition from traditional to agile management [13, 14]. Findings presented in [15] are focused on verifying the existence of interconnections between agile project management and sustainability.

2 Incorporating Sustainability Practices in Project Management Process

The link between sustainable development and projects was first stated a 30 years ago at the World Commission on Environment and Development, but in the last decade, it became a subject of many research papers and practical implementation. From 2006. we can find a number of papers presented by the leading authors in the field of project management, focused on integrating sustainability into the project management process [8, 16–18]. The highest number of papers in the last ten years are published in (besides the leading journals for project management) in journals for cleaner production, construction management. This is expected having in mind the rapid growth of green building construction phenomenon and sustainable production in the past period. The International Journal of Project Management initiated the special theme aimed to explore how sustainability considerations are integrated into projects and project management.

The practical significance of sustainability in the project management confirms a standard for sustainability in project management has been in use - *The GPM P5 Standard for Sustainability in Project Management*. P5 refers to *People, Planet, Prosperity, Processes, and Product* [19].

Some projects are "green" by its definition – aimed to implement sustainable development, still [5] stated that projects that are not green by definition and primarily are not about sustainability could be run in a more sustainable manner. Silvius and Schipper [20] define sustainable project management from the perspective of all stakeholders - how to plan, monitor and control project outputs and outcomes, considering environmental, economic and social aspects of the life-cycle of all project results, processes, and resources.

The authors emphasize the importance of professional, fair and ethical approach. In project management, sustainability refers to the integration of economic, social and environmental aspects. According to this definition, it is not enough to take into account only the project life cycle, but also the life cycle of the project's results that bring a change in the organizational system, assets, and the behavior. In addition to the life cycle of the project's (product) results, the lifecycle of the resources used during the project should also be taken into account. The authors emphasize that in order to integrate sustainability into the project management process it is necessary to understand all three life cycles, as well as the interactions between these cycles [20].

Increasing attention given to the sustainable development resulted in different conceptual models trying to integrate project management and sustainable development emerging concept "Green Project Management" that emphasizes project management sustainability [5, 21].

The authors in [5] explain that being green means to change the way we think about a project. Even though sustainability is becoming mainstream, not everyone accepts sustainability as a natural part of the project management discipline. If project management is already concerned with reducing costs, increasing value and protecting scarce resources, one can conclude that this fit with being green. Acronym S.M.A.R.T. goals (Specific, Measurable, Attainable, Relevant, and Time-bound) is now supported with S.M.A.R.T.ER (Environmentally Responsible), and the well-known paradigm of project success extends for another one - an environmentally responsible aspect.

Having in mind previously stated, green practices and considerations need to be incorporated into any project and should be applied in every industry, not just in a sustainable one. Traditional project management concepts and methodologies evaluate projects from the perspective of time, money and quality. Because sustainability refers to a more extended period, while projects are time-limited [22] one must emphasize the importance of assessing the impact that project results can have a long time after project completion. Projects can create a change in the system, behavior, assets, while products as accurate results of the project throughout their lifetime can also have different impacts on society and nature.

The main idea of green project management concept is to apply "green thinking" into the existing project management methodologies. Maltzman and Shirley have started from project quality area, developing "greenality" concept and explaining project management processes from the perspective of "greenality" [5]. Also, some different conceptual approaches present the incorporation of sustainability values/criteria in project management process and indicator for project's performances [8, 23].

Green project management concept is more strategically oriented since the original idea was based on ISO (*International Organization for Standardization*) standard 14001:2004. This standard provides a framework established to improve organizational environmental performances, and if the projects are used as a mechanism to implement organization mission and strategy, then they need to comply also with the elements of this standard.

Further, The GPM P5 Standard supports the alignment of projects (programs and portfolios) with organizational strategy oriented toward sustainability and social responsibility. This Standard relies on ISO 21500: 2012 - Guidance on Project Management, according to which a project consists of a group of processes and GPM P5 Standard is evaluating the efficiency in which those processes are contributing sustainability.

The primary recommendation is to create Sustainability management plan as a mechanism to implement sustainability in project initiatives. The purpose of this plan is to translate the sustainability objectives into project objectives, with the explanation of reasons for its implementation, constraints, and conflicts that may evolve and proposed actions. Consequentially this change leads to the changes in project scope and set of activities [19].

Further, to integrate sustainability in project management, a model PRiSM (*Project Integrating Sustainable Methods*) has been developed. It is based on and ISO:26000 - Guidance on Corporate Social Responsibility and P5 Standard and its implementation aims to reduce project risk level, from a social, environmental and economic point of view, while expanding benefits to the organization and society. It includes fundamental principles and workflow based on project processes to envolve sustainability principles [19]. It is essential to state that P5 Standard a PRiSM rely on business agility as the organizational ability to implement changes in the project (program and portfolio) management that are initiated to obtain sustainable project results and create value.

In recent years project management discipline is facing with an increasing challenge of how to create value and respond to changing the environment, in order to profit. This requires 'agility.' The agile concept can increase the flexibility, velocity, learning, and response to change [24]. Agile Project Management (APM) is used where a project goal is clear, but the way to reach that goal is not. It was developed for a software development environment in frameworks: Scrum, Extreme Programming, KANBAN, etc. [25].

In such projects, the traditional approach has proven to be ineffective as it led to a higher level of resource usage and higher costs due to the changes on projects, the dissatisfaction of client and project's team members. Organizations needed a new approach that would allow the delivery of parts or the increment of the project where each iteration would have all the stages from design to the client's feedback. Response to this need has resulted in an agile approach, which provides greater satisfaction for clients and team members, reducing the risk of excessive costs and creating products that can not respond to market needs. APM is suited: for complex and innovative delivery environments [25] to embrace change and learning during the change process; and to contribute to the customer value [26].

However, nowadays APM is not only restricted to software development because the projects in other industries are also challenged by high demands and turbulent environment [13]. The authors investigated management practices related to the APM approach and concluded that there are APM practices that are strongly linked to traditional PM concept. Furthermore, APM could be implemented in different industries with the presence of some APM enablers related to team work, project manager's experience and the level of a new product development formalization process.

If the project is considered as an entity realized in the existing organization, the implementation of agile approach or sustainability in project management process requires adequate changes in a company – its system, structure, staff, and values. Project organization, communication, decision making, reporting system, control and other processes. The analysis of project's context involves identification of project's stakeholders and their impact on the final goals and results of the project that rises the question - how mature a company is to implement agility and sustainability in project management.

To successfully adopt agile methodologies a company needs to reconsider its goals, technologies, managerial and people components [27, 28]. Nerur and Mahapatra define issues for each of those components: for management and organization component the most issues are related to organizational culture, management style, knowledge management, and rewarding system; for people component issues are related to customer relations, competencies and teamwork; the main issue for processes is a shift from process–driven to a feature-driven and people-centric approach; and for technologies, the primary challenge is a new set of skills and reconsideration of tools that were used before in the organization [27]. Through the research provided in the software industry, Boehm and Turnes defined three main barriers to implementing agile methodologies: development process conflicts, business process conflicts, and people conflicts [28].

Patel and Muthu have provided an Agile maturity model (AMM) for software development that has five stages, starting with no process improvement goals to performance management and defect prevention practices at the highest level. The purpose of the Agile maturity model is to demonstrate how could an organization implement agile approach [29].

When we consider sustainability incorporation in project management process there is a lack of research that are focused only on organizational issues that should be considered to adopt this concept successfully. Some papers are emphasizing what should a project manager do (which is more described in the chapter below), but there is no any consistent maturity model for sustainability in projects. Anyhow, if we use as a premise, the logic of agility implementation we can conclude that the implementation of any new concept requires maturity on a company to accept the change.

2.1 Project Manager Competencies and Knowledge Areas

With the mainly present transition from traditional to agile project management and the emerging concept of "green" project management project managers are facing new challenges that consequently change their roles and responsibilities [30]. Project manager competencies are the subject of many research. However only a few have explicitly examined what competences are considered as an essential element of sustainable management, and those researches are mostly conducted in a construction industry.

Researches in this industry argued that project managers need to develop their managerial (contextual) and behavioral skills and knowledge to meet new professional demands [31]. Further, that green project management practices can contribute to a sustainable construction project.

To understand the project manager's competence profile, it is essential to identify in what areas of work (technical competencies) managers need to be competent, together with the behavioral competences and contextual competencies. Individual Competence Baseline [32] defines three competence areas: (1) People competence (a competences to participate in project team and/or to lead a project); (2) Practice competence (specific technical knowledge (tools and techniques) necessary for a project; (3) Perspective competencies (methods and tools used in interaction with the project environment).

Based on literature review we provided a meta-analysis on available research based on what we can conclude that the field of project manager's challenges, role and needed competencies for sustainable project management has not yet been significantly studied. Only studies with this topic have been provided in the construction industry and publicly funded project (health projects). Further, the published research was focused on linking the project manager's knowledge and skills for sustainable project management, based on generally accepted and confirmed knowledge areas by leading certification bodies.

Analyzed research aimed to define what knowledge areas and skills need to strengthen to manage green projects effectively. According to [33] project manager's competencies should be viewed within the context of their operating environment and the constraints under which they work. The essential skills required to manage green construction projects effectively are analytical skills, decision-making, team working, delegation, and problem-solving skills [31].

The integration of project activities to organizational routines formed the strong association with quality, human resource, communications and risk management, as

PMBoK knowledge areas. Results in [34] emphasize the leadership skills that are is related to increased functionality and flexibility of the construction teams in sustainable or green building projects. It is significant conclusion that the authors emphasize the importance of behavioral competencies and contextual competencies.

If projects are providing the change in the organization, as stated through this paper, we need to have a broader picture – beyond the project if we want to incorporate sustainability and understand project manager's role and competencies.

If a project manager improves the key competencies to implement sustainability in every aspect of a project, then it may be a request for other team members and may affect the interactions on the project, communication process, organization, etc. The main role here is on human resource management, to promote and provide professional development with recognition system [35].

2.2 Project Manager's Challenges

Literature review leads us to the conclusion that the main challenges in sustainable project management refer to project initiation, preparation, and planning phases. In the study provided by [36], there are listed challenges related to project initiation: analysis of project benefits, alternatives, and lack of reference point to measure sustainable options. Aarseth et al. [11] argued the importance of a project sustainability strategy and the role of the project and host organization, as a crucial factor for successful integration of sustainability in project initiatives which emphasizes a project manager's contextual competencies.

According to IPMA classification, those competencies are defined as perspective competencies, that refers to "methods, tools, and techniques through which individuals interact with the environment, as well as the rationale that leads people, organization and societies to start and support projects" [32].

Some challenges emphasized the need to straighten the behavioral and contextual competencies, concerning the changing environment, new technologies, high demands, limited resources and a rising number of stakeholders. Additionally, if we adopt the fact that project manager should be the change agent [5] and is responsible for goal setting (which directly affects project scope, project activities, amount of work need and resources), behavioral competencies are recognized as the most important one [33, 34, 37].

3 Agility and Sustainability in Project Management

Concerning the abovementioned challenges for sustainable project management we can conclude that the agility is a necessity of this process. Among other agile principles that are applicable to different frameworks we can extract those who may present a link between sustainability and agility: Agility support dealing with complexity and uncertainty [38–40]; Further, agile welcome change requirements even late in development, emphasize joint work with all partners on a project, build project around motivated team members, promote face-to-face communication and sustainable pace, put continuous attention to technical excellence [41–43].

Different academic research papers are oriented on social aspects in APM: do agile methods support social identity and collective effort [44, 45]; the link between social sustainability and APM [15]. Research presented in [15] was based on socially sustainable elements: diversity, self-organization, capacity to learn, trust, common meaning. These results support the fact that several APM methodological elements have a causal relation to the improvement or degradation of social sustainability factors.

Calvo et al. (2008) investigated if sustainability can be implemented in agile manufacturing process [46]. The authors used a ratio of utility and entropy as measures of sustainability and flexibility and complexity as agility measures. They have defined a framework that presents the utility and its contribution to agility, introduced through system flexibility. The main conclusion of this paper is the systemic approach to sustainability is applicable in agile manufactory process.

Further, a leading conference such as Big Apple Scrum Day, Agile 2017 emphasizes social aspects too and how to engage Human resource team in agile transformation: how to create leadership and engagement at every level.

Those research confirmed that sustainability and agility in project management are overlapping and that agile project management requires the implementation of sustainability aspects [47]. In practice, we can find new courses – Sustainable Agile Project Management - SAPM to ensure sustainable outcomes.

4 Conclusion

The paper showed that in the last decade the development of the project management area had been significantly influenced by changes at the global level in the business world. In all segments of the society, we are striving for sustainable development. Therefore the link between this concept and the implementation of projects is inevitable. A new emerging theory "Green Project Management" emphasizes the fact that even projects that are not green by definition and primarily are not about sustainability can be run in a more sustainable manner. Being green means to change the way we think about a project. In parallel, this discipline is undergoing transition, and there is a growing presence of agile approach in comparison to the traditional project management.

Many research were provided to create a framework that will integrate project management and sustainability. Only a few studies are emphasizing how sustainability principles incorporation can improve agile management, and what is the role of a project manager in sustainable project management and what are the requirements for the competencies of the project manager. Further, there is a lack of research that identify crucial organizational issues that should be considered to adopt sustainability on projects successfully. The main challenges in adoption of agile methodologies are recognized, and they are focused on people, processes and a system. If we use this as a premise, we can conclude that the implementation of any new concept requires maturity of a company to accept the change. Besides the benefits and opportunities brought to a company, either agile approach or sustainability concept implementation (or both) is a long-term process supported by an investment of work and money.

The main conclusions of this paper are (a) sustainability and agility are complementary concepts that help project managers to deal with environmental constraints. The agile concept can increase the flexibility and response to change, enable learning and provide profit, and the integration of sustainability principles should ensure sustainable outcomes; (b) the sustainable project management emphasized behavioral and contextual competencies; (c) the field of a project manager in sustainable project management can be investigated in different industries, not just on a project that is green by its definition in order to see what competences are necessary to obtain project performances on a sustainable manner.

Acknowledgment. This paper is a result of the Project No. 179081 funded by Ministry of Education and Science of the Republic of Serbia: Researching Contemporary Tendencies of Strategic Management Using Specialized Management Disciplines in Function of Competitiveness of Serbian Economy.

References

1. Marotta, A.: Visione sostenibile (Sustainable vision in the conservation of cultural heritage project). Le vie dei Mercanti - XI Forum Internazionale di Studi tenutosi a Aversa/Capri Politecnico di Torino, pp. 1210–1219 (2013)
2. Henshaw, C.: The Wellcome Digital Library: building a sustainable infrastructure. Library Hi Tech. News **28**(1), 18–21 (2011)
3. Seo-Jeong, L., Eung-Hyun, L., Deog-Seong, O.: Establishing the innovation platform for the sustainable regional development: tech-valley project in Sejong city, Korea. World Technopolis Assoc. **6**(1), 75–86 (2017)
4. William Young, C., Quist, J., Green, K.: Strategy for sustainable shopping, cooking and eating for 2050 – suitable for Europe? In: International Sustainable Development Research Conference, pp. 430–437 (2000)
5. Maltzman, R., Shirley, D.: Green Project Management. CRC Press, US: Taylor & Francis Group (2010)
6. Savić, D., Bogetić, Z., Dobrota, M., Petrović, N.: A multivariate approach in measurement of the sustainable development of european countries. Manag.: J. Sustain. Bus. Manag. Solut. Emerg. Econ. **21**(78), 73–86 (2016)
7. Oehlmann, I.: The Sustainable footprint methodology. Delft University of Technology (2010)
8. Gareis, R., Heumann, M., Martinuzzi, A.: Relating Sustainable Development and Project Management. IRNOP IX, Berlin (2009)
9. Obradovic, V., Jovanović, P., Djordjević, N., Beric, I., Jovanovic, F.: Using project management as a way to excellence in healthcare. HealthMED **6**(6), 2100–2107 (2012)
10. Petrović, N., Bošnjak, I., Nedeljković, S.: Disaster risk reduction for sustainable development goals. Eur. Proj. Manag. J. **7**(2), 27–35 (2017)
11. Aarseth, W., Ahola, T., Aaltonen, K., Økland, A., Andersen, B.: Project sustainability strategies: a systematic literature review. Int. J. Project Manage. **35**, 1071–1083 (2017)
12. Schoper, Y.G., Gemünden, H.G., Nguyen, N.M.: Fifteen future trends for Project Management in 2025. In: International Project Management Association, IPMA Expert Seminar, Zurich (2015)

13. Conforto, E.C., Salum, F., Amaral, D.C., da Silva, S.L., de Almeida, L.F.M.: Can agile project management be adopted by industries other than software development? Proj. Manag. J. **45**, 21–34 (2014)
14. Beck, K., Beedle, M., Van Bennekum, A., Cockburn, A., Cunningham, W., Fowler, M., Thomas, D.: The Agile Manifesto. Manifesto for Agile Software Development. http://agilemanifesto.org/. Accessed Nov 2016
15. Albarosa, F., Valenzuela Musura, L.: Social sustainability aspect of agile project management. Umeå School of Business and Economics (2016)
16. Turner, R.: Responsibilities for sustainable development in project and program management. In: International Project Management Association. IPMA Expert Seminar, Zurich (2010)
17. Scally, J.: Sustainability in project management. Journal of Environmental Planning and Management, 57, (2013)
18. Silvius, G.: Sustainability as a new school of thought in project management. J. Clean. Prod. **166**, 1479–1493 (2017)
19. Green Project Management: The GPM P5 Standard for Sustainability in Project Management, USA. https://greenprojectmanagement.org/. Accessed Nov 2017
20. Silvius, A.G., Schipper, P.R.: Sustainability in project management competencies: analyzing the competence gap of project managers. J. Hum. Resour. Sustain. Stud. **2**(02), 40 (2014)
21. Sholarin, E.A., Awange, J.L.: Environmental Project Management: Principles, Methodology, and Processes. Springer (2015)
22. Brent, A., Labuschagne, C.: Social indicators for sustainable project and technology life cycle management in the process industry. Int. J. Life Cycle Assess. **11**(1), 3–15 (2006)
23. Martens, M.L., Carvalho, M.M.: A conceptual framework of sustainability in project management. In: Project Management Institute Research and Education Conference, Phoenix, AZ. Project Management Institute, Newtown Square, PA (2014)
24. Campanelli, A.S., Parreiras, F.S.: Agile methods tailoring – a systematic literature review. J. Syst. Softw. **110**, 85–100 (2015)
25. Cubric, M.: An agile method for teaching agile in business schools. Int. J. Manag. Educ. **11**, 119–131 (2013)
26. Van Waardenburg, G., van Vliet, H.: When agile meets the enterprise. Inf. Softw. Technol. **55**, 2154–2171 (2013)
27. Nerur, S., Mahapatra, R.K., Mangalaraj, G.: Challenges of migrating to agile methodologies. Commun. ACM **48**(5), 72–78 (2005)
28. Boehm, B., Turner, R.: Management challenges to implementing agile processes in traditional development organizations. IEEE Softw. **22**(5), 30–39 (2005)
29. Patel, C., Muthu, R.: Agile maturity model (AMM): A Software Process Improvement framework for agile software development practices. Int. J. Softw. Eng. IJSE **2**(1), 3–28 (2009)
30. Edum-Fotwe, F., McCaffer, R.: Developing project management competency: perspectives from the construction industry. Int. J. Proj. Manag. **18**(1), 111–112 (2000)
31. Hwang, B.-G., Ng, W.-J.: Project management knowledge and skills for green construction: overcoming challenges. Int. J. Project Manage. **31**, 272–284 (2013)
32. IPMA: Individual competence baseline for project, programme and portfolio management. In: International Project Management Association (2015)
33. Mei, C., Dainty, A., Moore, D.R.: What makes a good project manager? Hum. Resour. Manag. J. **15**(1), 25–37 (2005)
34. Tabassi, A., Argyropoulou, M., Roufechaei, M., Argyropoulou, R.: Leadership behavior of project managers in sustainable construction projects. Procedia Comput. Sci. **100**, 724–730 (2016)

35. Tharp, J.: Project management and global sustainability. In: PMI® Global Congress 2012—EMEA, Marsailles, France. Project Management Institute, Newtown Square, PA (2012)
36. Agarval, R.S., Kalmar, T.: Sustainability in project management – Eight principles in practice. Umeå School of Business and Economics, Sweden (2015)
37. Todorovic, M., Obradović, V.: Sustainability in project management: a project manager's perspective. In: Ljumović, I., Ėltető, A. (eds.) Sustainable Growth and Development in Small Economies. Institute of World Economics, pp. 88–107 (2018)
38. Little, T.: Context-adaptive agility: managing complexity and uncertainty. IEEE Softw. **22**(3), 28–35 (2005)
39. Wong, C.Y., Arlbjorn, J.S.: Managing uncertainty in a supply chain reengineering project towards agility. Int. J. Agile Syst. Manag. **3**(3/4), 282–305 (2008)
40. Fernandez, D.J., Fernandez, J.D.: Agile project management - agilism versus traditional approaches. J. Comput. Inf. Syst. **49**(2), 10–17 (2016)
41. Cho, J.: Issues and challenges of agile software development with SCRUM. J. Comput. Inf. Syst. **9**, 188–195 (2008)
42. Cobb, C.G.: Making Sense of Agile Project Management: Balancing Control and Agility. Wiley, Hoboken (2011)
43. Measey, P.: Agile foundations: principles, practices and frameworks. In: Radstad (ed.) Swindon, SN2 1FA, BCS Learning & Development Ltd, UK (2015)
44. Wenying, S., Cecmil, S.: Practitioners' agile-methodology use and job perceptions. IEEE Softw. **35**(2), 52–61 (2018)
45. Whiwort, E., Biddle, R.: The social nature of agile teams. In: Agile Conference (AGILE), IEEE; Washington, DC, USA (2007)
46. Calvo, R., Domingo, R., Sabastián, M.A.: Systemic criterion of sustainability in agile manufacturing. Int. J. Prod. Res. **46**(12), 3345–3358 (2008)
47. Mohrman, S.A., Lawler, E.E.: Generating knowledge that drives change. Acad. Manag. Perspect. **26**(1), 41–51 (2012)

Software Engineering

Smart Integrated Robotics System for SMEs Controlled by Internet of Things Based on Dynamic Manufacturing Processes

Yurij Kryvenchuk[1](✉), Nataliya Shakhovska[1,2], Nataliia Melnykova[1], and Roman Holoshchuk[1]

[1] Lviv Polytechnic National University, Lviv 79013, Ukraine
yurkokryvenchuk@gmail.com,
melnykovanatalia@gmail.com,
nataliya.b.shakhovska@lpnu.ua,
[2] University of Economy, Bydgoszcz, Poland
roman@ridne.net

Abstract. The new technology for Industry 4.0 implementation is proposed. This technology is implemented in one of Ukrainian manufacture. The equipment identification module on the workstation is provided accounting and quality analysis of equipment. The module of the interface part of human-machine interaction during the maintenance of all parts of the production process of the press-shop is developed. The main stages of development and results evaluation is described. The key performance indicators for project management is given.

Keywords: Internet of Things (IoT) · Industry 4.0 · Robotics
Equivalent frequency spectrum components anti-stokes · Raman spectrum

1 Introduction

JV «Spheros-Electron» is Ukrainian manufacture of climate systems for motor transport: variety of liquid heaters for buses, trucks and commercial vehicles, construction and special equipment. The company also offers automakers air conditioners, electromechanical and mechanical ventilation hatches for all types of city and tourist buses, receivers for pneumatic and braking systems of vehicles, fuel tanks of various sizes and capacity, steering columns. Consequently, optimization and improvement of the production line is one of the main tasks. Its realization needs to go beyond the outdated, albeit logical process. The production is based on it and helps the company to implement new ideas. The introduction of innovative processes should be large-scale and fully cover all sectors of the production process for long-term effective development of the enterprise. At the present time, the issues of modernization of the process of assembly and packaging of liquid heaters, the automation of their accounting at all stages of the production process are acute. Now this process is semi-automated, which leads to the risk of errors, unproductive usage of working time and economic inefficiency of the production. Listed above is key issues that need to be addressed to increase profits, and improve the production process.

N. Shakhovska and M. O. Medykovskyy (Eds.): CSIT 2018, AISC 871, pp. 535–549, 2019.
https://doi.org/10.1007/978-3-030-01069-0_38

2 State of Arts

The term "Industry 4.0" was established ex-ante for an expected "fourth industrial revolution" and as a reminiscence of software versioning [1]. According to [1], Industry 4.0 stands for an advanced digitalization within industrial factories, in form of a combination of internet technologies with future-oriented technologies in the field of "smart" objects (machines and products). This enables and transforms industrial manufacturing systems in a way that products control their own manufacturing process. The high importance of digitalization and the internet is also reflected in the discussions about related concepts such as the "Internet of Things" or the "Industrial Internet". Besides the focus on digitalization, Industry 4.0 is expected to be initiated not by a single technology, but by the interaction of numbers of technological advances whose quantitative effects together create new ways of production [2, 13, 14].

The main advantage in comparison with the technological perspective, is the possibility to facilitate tasks that previously required heavy manual work [3], e.g. the connection of suppliers with the ordering company by means of electronic data interchange (EDI) systems.

The HORSE project uses of innovative technologies in order to develop a robust technological framework [4]. More specifically:

- Integrated, Process-oriented management model for control of the production line and automatic resource allocation/dynamic reallocation (BPM),
- OSGI based (IoT) for remote control of production resources (humans, robots) (all resources are accessible in the same manner).

3 The Main Material

3.1 IoT Solution for Data Gathering and Processing

The company has the counters of mechanical stamping stations with information displays, which are used partially. This information is gathering manually, that's why the company hasn't exact information about the count of produced details and stamp's frequency using. Each of the stamp has own life cycle (count of times used). When this count is bigger than planned, stamp creates detail with defect. The management of this company can't predict the period of stamp using and can't organize planning resource. We propose to collect information from sensors which are located near stamps to database through Cloud-solution based on Big data approach [5, 12]. This allows data processing to:

- Stamps' using monitoring,
- Quality management,
- Create complex solution to resource planning.

Each department of this company has own database. Users collect information and create reports, but top management analyzes only information from papers reports. Very important feature of this company is control of production quality, but this control is not organized automatically. The company uses the Ukrainian sensors system MICROTECH. The measurement tolerances of details consist of

- Data measuring and checking of detail's sizes by patterns,
- Comparison of detail's parameters with standards,
- Conclusion forming regarding compliance to quality certificate.

The company uses analogue devices for parameters measuring and saves as printing report. We propose to create user interface for data processing. The system records progress in real time and can monitor planning resource of stamps for future maintenance. Currently, these data are processed manually in form of internal reports. Therefore, errors and delays at all stages are occurred. The aim of proposal is to build the new flexible model of smart JV «Spheros-Electron» factory using IoT solution. Robotics assistance improves worker's safety, quality and production effectiveness. Also we should integrate horizontal and vertical process in factory management because one of the issues is control of working time. The project calls for contributions that validate the framework in Sphere Electron industrial settings involving almost human-robot collaboration. We are planning to provide new functionalities (software components) and create new hardware with sensors (hardware components). The main goals are to increase product quality and reduce work time. The implementation of the above mentioned management model allows us to adopt the IoT paradigm and OSGI middleware, enable and facilitate remote control, organize the production line and resources monitoring. Manufacturing Process Management System is also be used for quality control too.

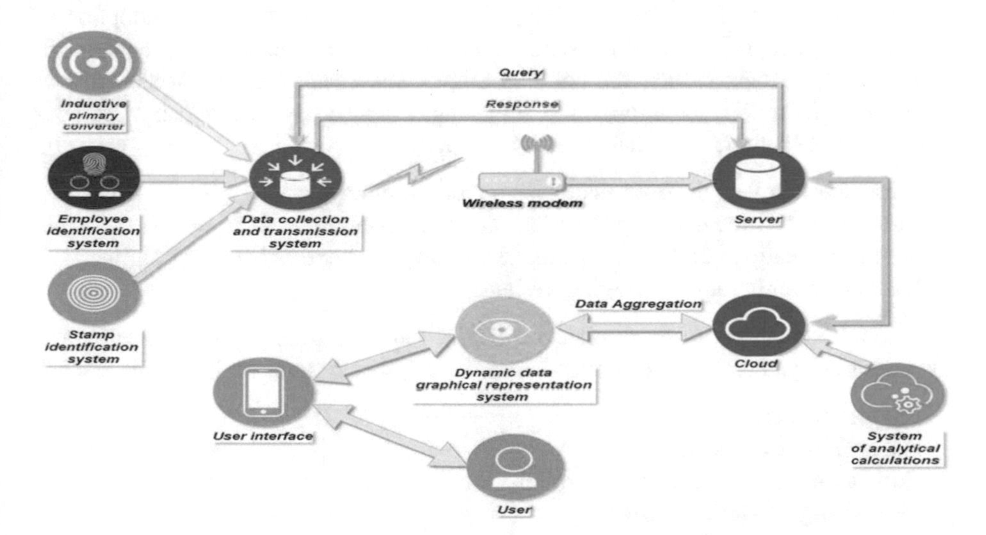

Fig. 1. Typical scheme of Horse framework usage for process control

The typical scheme of the production system is presented in Fig. 1. The system can accept incoming data from the user or the production scheduler. Production is initiated by activating the production process (at level #3 of the hierarchy). As soon as the process reaches the first task from a person or job, the task of the instruction is sent to the robot management system or the user interface. This control system and user interface are part of the HORSE system [4] and are located on the 3rd level of the functional hierarchy.

The current production process consists of the following steps:

- The inductive primary converter (CHE12-10PA-H710) generates an impulse when the press is triggered.
- The generated impulse is recorded using the developed system, the connection between the server and the synthesized system is carried out using wi-fi.
- Identification of stamps is carried out by the district leader after adjustment using the keyboard.
- Identification of employees is done using a personalized card and RFID module.
- The server requests developed system every minute, accumulates the number of triggers, stamp identification, employee identification and provides an answer.
- The accumulated information on the server is processed and a report on the best and worst employees and stamps, the quantity of the manufactured product is provided.

3.2 Technology for Manual Operation Removing

Also we propose to increase human safety. The hardware and software complex is deployed for tracking matrix processing. This allows to control automatically the count of usage of the matrix. As a result, the site operator does not remove the matrix to check its performance. This helps to reduce injuries when setting up the press. For this purposes the new components is developed. Time conditions require the adoption of specific, effective solutions, the usage of the best modern experience, equipment, technology to achieve a breakthrough. To achieve the goals, it is proposed to implement a robotic system for information collecting. Proposed system with new component for HORSE framework has such subsystems:

- Information collecting from machines, presses, employees, etc.
- The server part for handling the transmitted information,
- System of analysis and optimization of working processes of production line,
- Robotic assembly and packaging system.

The robotic system of data collection (RSDC) aims to improve the collection and restoration of data and correct visualization. Issues, that is addressed through RSDC:

- Product accounting;
- Registration of products;
- Conduct the correct data visualization;
- Accounting of working time;
- Employee accounting;
- Accounting of the wear of the machine;
- Accounting of the wear of stamps;
- Identification of the employee;
- Robotic assembly and packaging of finished products.

In fact, the commissioning of a robotic system for collecting information will contribute to the improvement of the production process. Moreover, the introduction of these innovations will lead to improvements in the management system, since obtaining, accounting and analyzing data will help to improve enterprise productivity. The installation of a robotic system for collecting information gives such results:

- Printing of bar-coding and accounting of output of products;
- Automatic generation and accounting of factory numbers based on the results of acceptance tests,
- Control of completeness during assembly and packaging (using bar-codes and comparisons with the specification);
- Automatic accounting and printing of labels on products and boxes, warranty certificates, stickers, etc.,
- Timely writing-off of equipment from the warehouse,
- Tracking the minimum stock products.

The new hardware (Fig. 1) is integrated into the experiment by attaching it to the presses at the working stations. Also a computer network is designed and new software components are developed that allow the server to determine the number of cycles of each matrix and determine the worker that works with the press. All of these data is stored in a local data warehouse. Next, data is integrated and transmitted by a predefined schedule to the enterprise data warehouse. Information is analyzed, relevant text and image reports are formed to support decision-making. The new hardware equipment is planned to design taking into account the integration of human-machine interaction during the production and with using of IOT, robotics, cloud technologies and Data Analysis for

- Registration of production at the workstation of assembly,
- Change the direction of movement of the technical process of assembly,
- Changes in placement of the final assembly of heaters,
- Changing the way of packaging heaters at the site,
- Reading bar-code of the product configuration, writing off from stock.

As a result, this will extend the enterprise's functional capabilities by providing new software components along with human-machine interaction for collecting and analyzing information on the use of equipment or by integrating new robotic applications or other IT technologies for using and detailing the impact in specific parts of the process of production in the region.

The application of the HORSE structure of the company JV «Spheros-Electron» allows to get the following results:

- Streams at the workstation of assembly become sequence,
- The distance and time to move the technological trucks is decreased (saving 10% of the time),
- Introduction of bar-coding and accounting of product release,
- Automatic generation and accounting of factory numbers based on the results of acceptance tests,

- Control of completeness during assembly and packaging (using codes and comparison with the specification);
- Automatic accounting and printing of labels on products and boxes, warranty certificates, stickers, etc.,
- Timely cancellation of equipment from the warehouse,
- Tracking the minimum stock,
- Increase productivity up to 15% and increase quality.

3.3 The Component of Information Technology for Temperature Measuring

It is also important to control the temperature during the manufacture of parts to reduce the defect. There are many methods for measuring temperature, both contact and non-contact. In the process of manufacturing parts that are variables it is expedient to use contactless methods of temperature measurement. The analysis showed that the Raman scattering method is best suited for solving this problem [6, 7].

Based on Raman known at present are two ways to measure temperature. The first and more common method of measuring temperature by Raman intensity is dependent stokes and anti-stokes Raman component. This method is relatively simple to implement, since change with temperature-integrated area anti-stokes and stokes component. This method of temperature measurement by Raman has good sensitivity and accuracy, but has several significant drawbacks. The main drawback is a methodological error that occurs as the result of determining the area of integrated anti-stokes and Stokes components. Spectrophotometer to measure consistently first Stokes then anti-stokes component of Raman spectroscopy, the measurement time of stokes components of the object and is heated by laser heating anti-stokes components that it leads to error. Another way is to measure the frequency shift Raman [8, 9].

For statistical analysis of temperature measuring the information technology was created. All determined values were saved in the cloud. If new measure was created, this data was automatically sent to the database. To measure the temperature, shift frequency Raman enough to determine just anti-stokes component Raman spectroscopy. To measure the temperature, shift frequency Raman frequency is not appropriate to use a spectrophotometer and spectrum analyzer. The peculiarity of the spectrum analyzer is that it measures only anti-stokes component, and the full range of a whole, not just a stepping stone that can reduce the methodological error. In addition, unconditional significant advantage of this method within the temperature measurement by Raman is speed. By comparison, when measuring the temperature integrated area ratio of the maximum speed is 13 s, and the Raman shift frequency of 1 s. By reducing the measurement time is reduced further methodological error caused by heating of the object-studied laser. Therefore, based on this method-conducted research described in the article. The results of experimental studies Raman spectroscopy for H_2O in the temperature range of 18 to 70 °C. Each point temperature for 10 implementations derived components range anti-stokes Raman method of center of mass calculated value equivalent frequency components anti-stokes Raman spectroscopy, and the average value of the equivalent frequency components anti-stokes range and uncertainty determine an equivalent frequency components anti-stokes.

Analytical dependences equivalent frequency components anti-stokes Raman spectrum of temperature. The dependence of error of approximation of the number of coefficients approximating curve for each of the objects, and certainly the best number of factors. Equipment which were used to hold the experiments: laser ν = 532 nm spectrum analyzer MS 3501i, optical circuit using a narrow band filter and prism, studies were conducted under normal conditions.

In the temperature range from 18 °C to 70 °C with step 1 °C for each temperature, 10 realizations of the anti-Stokes component of the spectrum Raman scattering of light (RSL) for H_2O [10, 11]. At each temperature point, for 10 realized spectrum realizations, the centre-of-mass method was used to calculate the equivalent frequency of the anti-stokes component of the spectrum (EFASC), and the average value of EFASC and the uncertainty of the determination EFASC RSL. All calculations were made in the software package Matlab.

Figure 2a shows the spectra of the anti-Stokes component RSL for H_2O at the temperature 18 °C, the corresponding values EFASC RSL calculated by the center-of-mass method and the averaged value EFASC RSL (Fig. 2b).

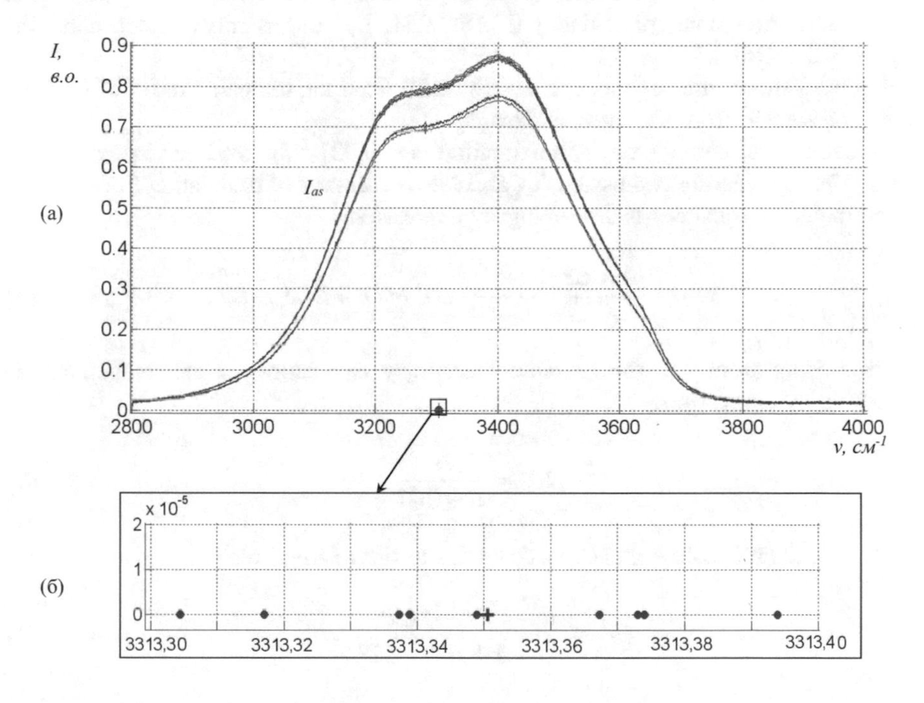

Fig. 2. Research results Raman spectra for H_2O: (a) Raman spectra anti-stokes components of the temperature 18 °C, (b) respective values EFASC Raman and Raman the average value EFASC

Also in Table 1 are presented the results of the dependence study EFASC RSL from the temperature and uncertainty in the definition of mean value EFASC RSL by experimentally obtained spectra of an anti-stokes component of RSL for H_2O with step 4 °C.

Based on the results of the studies (Table 1), the uncertainty in determining the values EFASC RSL in the temperature range from 18 °C to 70 °C for H_2O is less than 0,091 cm^{-1}. Considering the temperature T and the mean values EFASC RSL m_v (Table 1), an interpolation equation describing the dependence EFASC RSL from the temperature:

$$\nu = A + BT + CT^2 \tag{1}$$

where, A = 3309,70, cm −1, B = 0,47, cm −1/ °C, C = −0,01, cm −1/(°C)2, ν – EFASC RSL, cm −1, T – temperature, °C.

By analogy with introduced in the previous formula, in Table 1 denote the following applied: T – temperature at which the spectra of the anti-Stokes component were obtained RSL; m_v – mathematical expectation EFASC RSL; D_v – dispersion EFASC RSL; σ_v – standard deviation EFASC RSL; u_v – uncertainty of determining the value EFASC RSL.

Figure 3 shows the dependence EFASC RSL from the temperature for H_2O and a curve constructed from the interpolation Eq. (1).

Absolute inaccuracy of approximation is 0,021 °C, relative inaccuracy – 0.00052%. Considering expression (1), absolute inaccuracy of calculating EFASC RSL by interpolation equation, is describing by expression:

$$\Delta\nu = \frac{\partial(A + BT + CT^2)}{\partial T} \cdot \Delta T = (B + 2CT) \cdot \Delta T. \tag{2}$$

Proceeding from (2), the absolute inaccuracy calculation of the temperature is describing by expression:

$$\Delta T = \frac{\Delta\nu}{B + 2CT}. \tag{3}$$

Passing from absolute values inaccuracies to the relative, we get:

$$\delta\nu = \frac{\Delta\nu}{\nu} = \frac{BT + 2CT^2}{A + BT + CT^2} \cdot \delta T, \tag{4}$$

then

$$\delta T = \frac{A + BT + CT^2}{BT + 2CT^2} \cdot \delta\nu. \tag{5}$$

Table 1. The dependence of the mean value EFASC RSL from temperature and the uncertainty of determining the mean value EFASC RSL for H_2O

Value EFASC RSL ν, cm^{-1}											mν, cm^{-1}	Dν, cm^{-2}	σν, cm^{-1}	uν, cm^{-1}
T,^{0}C	№ p/p													
	1	2	3	4	5	6	7	8	9	10				
18	3313,31	3313,33	3313,34	3313,34	3313,34	3313,39	3313,38	3313,37	3313,37	3313,37	3313,35	0,000489268	0,022	0,070
19	3313,26	3313,27	3313,28	3313,28	3313,28	3313,25	3313,27	3313,29	3313,28	3313,27	3313,27	0,000109171	0,010	0,033
20	3313,13	3313,15	3313,13	3313,17	3313,19	3313,14	3313,16	3313,20	3313,13	3313,18	3313,16	0,000613342	0,025	0,078
24	3312,40	3312,39	3312,40	3312,41	3312,45	3312,44	3312,44	3312,43	3312,43	3312,45	3312,42	0,0004766	0,022	0,069
28	3311,19	3311,19	3311,22	3311,24	3311,22	3311,18	3311,24	3311,17	3311,18	3311,18	3311,2	0,000574898	0,024	0,076
32	3309,46	3309,50	3309,48	3309,51	3309,50	3309,52	3309,52	3309,48	3309,53	3309,53	3309,5	0,000494628	0,022	0,070
36	3307,34	3307,29	3307,36	3307,31	3307,31	3307,36	3307,35	3307,30	3307,29	3307,29	3307,32	0,000776815	0,028	0,088
40	3304,66	3304,65	3304,67	3304,67	3304,65	3304,68	3304,65	3304,70	3304,64	3304,66	3304,66	0,000275102	0,017	0,053
44	3301,56	3301,51	3301,54	3301,55	3301,53	3301,57	3301,53	3301,57	3301,54	3301,54	3301,55	0,000293285	0,017	0,054
48	3297,93	3297,96	3297,91	3297,94	3297,89	3297,95	3297,95	3297,91	3297,90	3297,96	3297,93	0,000556748	0,024	0,075
52	3293,83	3293,80	3293,81	3293,87	3293,81	3293,87	3293,83	3293,80	3293,82	3293,82	3293,83	0,00065131	0,026	0,081
56	3289,27	3289,28	3289,28	3289,30	3289,26	3289,28	3289,27	3289,24	3289,27	3289,27	3289,27	0,000247454	0,016	0,050
60	3284,21	3284,24	3284,24	3284,20	3284,20	3284,20	3284,19	3284,27	3284,23	3284,23	3284,22	0,000537504	0,023	0,073
64	3278,68	3278,71	3278,71	3278,72	3278,67	3278,71	3278,73	3278,75	3278,68	3278,73	3278,71	0,000560447	0,024	0,075
68	3272,67	3272,69	3272,75	3272,75	3272,72	3272,74	3272,72	3272,75	3272,74	3272,68	3272,72	0,000803419	0,028	0,090
69	3271,16	3271,11	3271,13	3271,12	3271,10	3271,15	3271,17	3271,14	3271,18	3271,12	3271,14	0,000561961	0,024	0,075
70	3269,52	3269,51	3269,51	3269,51	3269,50	3269,57	3269,56	3269,57	3269,57	3269,50	3269,53	0,000823692	0,029	0,091

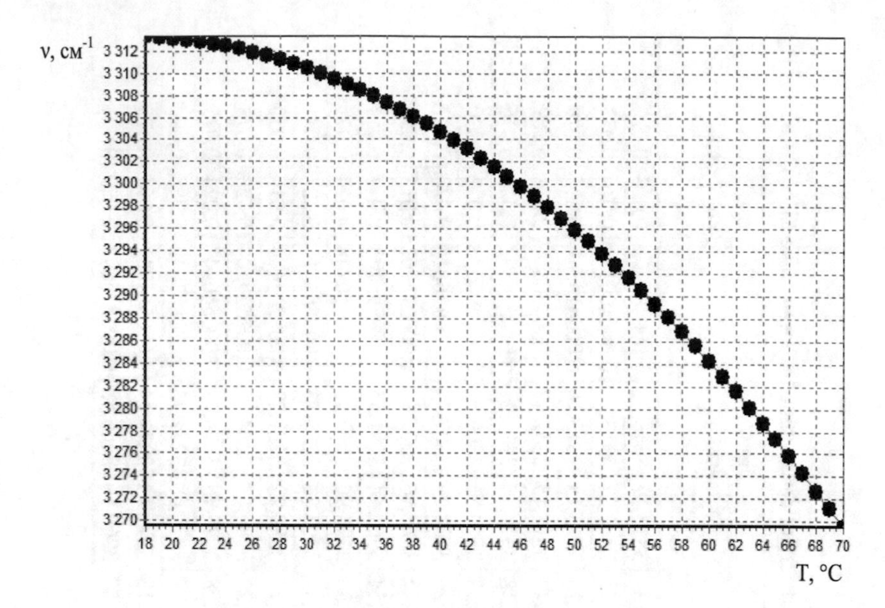

Fig. 3. Dependence EFASC RSL of temperature for H_2O

Passing from uncertainties to uncertainty, expression (5) will have the form:

$$uT = \left| \frac{A + BT + CT^2}{BT + 2CT^2} \right| \cdot uv. \tag{6}$$

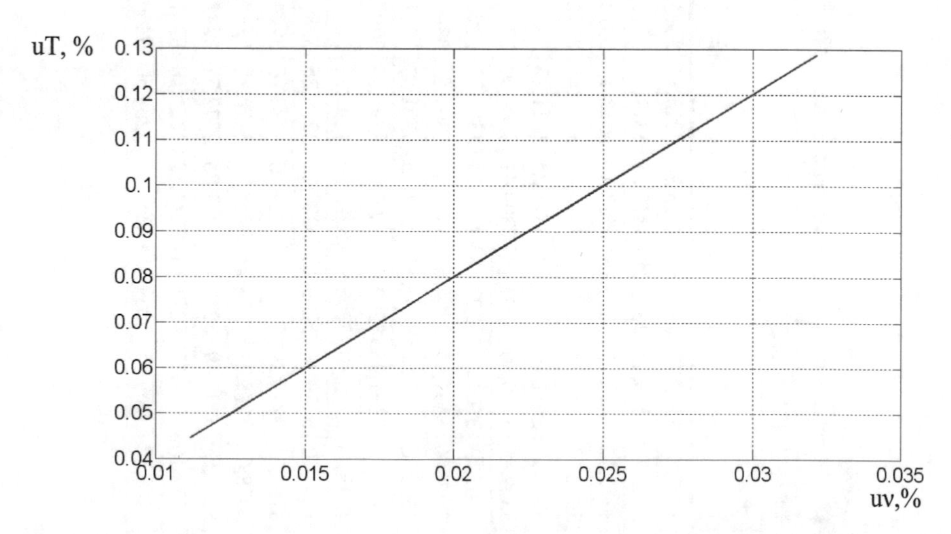

Fig. 4. Dependency uncertainty of determining the temperature of uncertainty Raman shift for H_2O

Figure 4 shows the dependence of the uncertainty of the temperature determination on the uncertainty of finding the equivalent frequency of an anti-stokes component of the spectrum RSL.

For statistical analysis of temperature measuring the information technology was created. The architecture of such system is based on cloud technology. The cloud server consists of database for value saving and business logic. All determined values were saved in the cloud. If new measure was created, this data was automatically sent to the database using sensors in laboratory.

4 The Key Performance Indicators for Project Management

We divide all metrics by classes:

Process Metrics

Percentage of Product Defects: Take the number of defective units and divide it by the total number of units produced in the time frame. This will give you the percentage of defective products (Fig. 5).

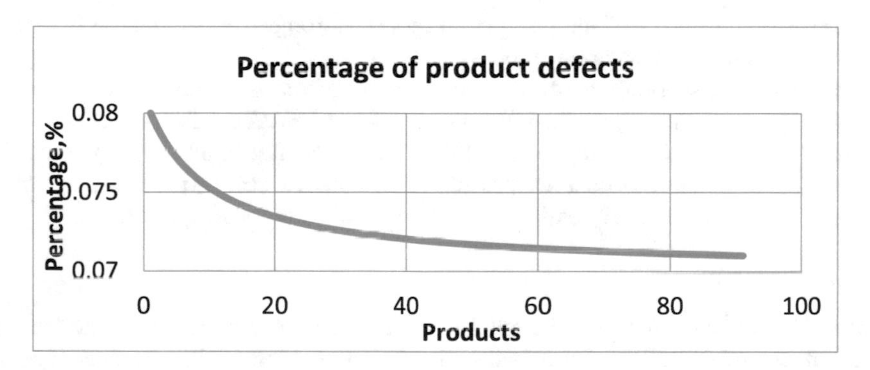

Fig. 5. The dependence between count of products (X) and count of defects (Y)

LOB Efficiency Measure: Measure Spheros-Electron's efficiency by analyzing how many units company has produced every hour, and what percentage of time your plant was up and running.

- Identification of the equipment codes on the workstation will provide accounting and quality analysis of equipment (Fig. 6).
- Minimizing the cost of maintenance and repairs.
- Forecasting and optimization of purchasing needs of production for quality equipment, due to the search of quality equipment by manufacturers, on the basis of comparison of technical indicators.
- Savings of inventory of the enterprise, due to reduction of consumables.

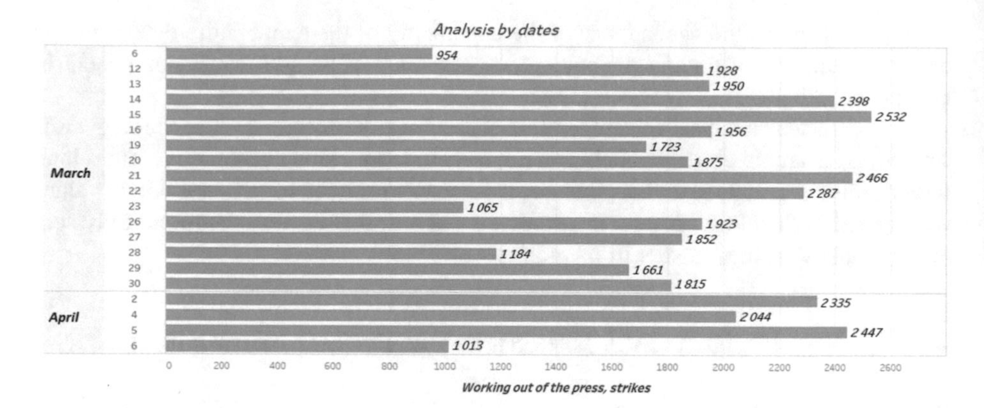

Fig. 6. Analysis of productivity by data

People Metrics

Employee Satisfaction: Measuring employee satisfaction through surveys and other metrics is vital to company's departmental and organizational health.

- Increasing the safety of the production process through the automation of product quality control and stamping process.
- Facilitation of working conditions on the basis of robotic processes of collecting, processing and storing useful information from the process of products stamping.
- Improvement of the quality of employee's work, taking into account the intellectualization of the identification process of equipment stamps, the definition of the types of scheduled work, and the quality control of the work performed.

Financial Metrics

Profit: The usage of intellectual capital will minimize the risks of losing the financial profit of the organization, through the circulation of high-quality products and the reduction of competition in the business market.

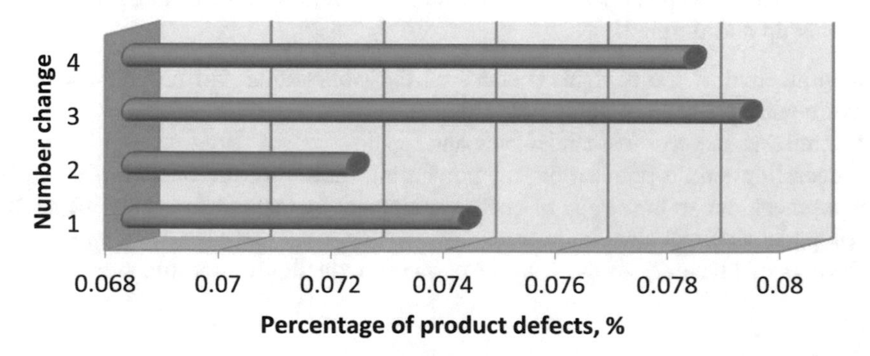

Fig. 7. Defect analysis

Cost: Measure cost effectiveness and find the best ways to reduce and manage costs (Fig. 7). Increase of equipment usage time by determining quality products.

5 Conclusions

The new enterprise information system allows to control all aspects of production, find track defective products and cause its appearance. As a result, user will have a reduction in production losses that were caused by downtime after technical failure of the equipment, minimization of maintenance and repair costs and optimization of material supplies and spare parts. The proposed module for determining the bit number of worked stamps on the press, prevents production from poor-quality products at the press-plant, allows to improve the process of forming products, as well as the choice of type of press equipment, which is determined by the range of products that are produced, as well as the initial physical and mechanical characteristics of raw materials (mineralogical composition, plasticity, etc.).

Proposed approach to the solution of this problem can be applied in various branches of production, namely in the industry at the factories in the manufacture of steel shafts of cars, railway wheels and many other products and machine building, military production. It is also possible to expand in the jewelry case for stamping of precious metal products. Functional properties of the system can be expanded for robotizing press machines to replace equipment matrices, including stopping the exhaust press, replacing it with a new one and starting the pressing process. The expansion of the functional system ensures the continuity of the production process, high performance of the product, increase the level of employee's safety. The equipment identification module on the workstation is provided accounting and quality analysis of equipment. Such rational usage of equipment minimizes maintenance and repair costs, forecast and optimize purchasing needs for quality equipment production as well as material resources of the enterprise.

A promising way to expand the capabilities of the system is to robotize the process of checking the correctness of the usage of a particular equipment for a specific type of work. This will prevent the usage of inappropriate equipment for a particular type of work, which will reduce the risk of poor quality product models. The module of the interface part of human-machine interaction during the maintenance of all parts of the production process of the press-shop is developed. Functional expansion of the system by improving the interface part of the operator's dialogue and press-installation optimizes the process of monitoring by the production process is proposed. It will ensure the product registration and verification of its quality according to the nomenclature requirements, the definition of the stage of matrix development and replacement of equipment, analysis of the quality of equipment types by manufacturers, control for performing operations at the workstation, accounting for employees responsible for the execution of products, determining the timing characteristics of production.

Also we have processed the experimentally obtained spectrums of RSL, certain meaning EFASC RSL at different temperatures. The dependences of the equivalent frequency of the anti-Stokes component of the spectrum RSL from the temperature and uncertainty of the determination of average values EFASC RSL for H_2O is given. Interpolation equations of the temperature dependence of EFASC RSL are obtaining. Also the dependence of temperature detection uncertainty by determining the uncertainty EFASC RSL is mined.

This allows at a particular temperature of measurement uncertainty requirements set to the measurement uncertainty frequency offset EFASC RSL or determining the frequency uncertainty resulting displacement EFASC RSL to calculate the temperature measurement uncertainties.

Acknowledgment. The paper describes the case for Spheros Electron, which was created as result of HULIT project mentored by HORSE (http://horse-project.eu/content/home).

The project calls for contributions that validate the framework in Spheros Electron industrial settings involving nearly human-robot collaboration. We plan to provide new functionalities (software components) and create new machinery with sensors (hardware components). The main goals are to increase the quality of product and reduce working time. These allow to maximize the impact of HORSE on the European manufacturing sector.

References

1. Lasi, H., Fettke, P., Kemper, H.-G., Feld, T., Hoffmann, M.: Industry 4.0. Bus. Inf. Syst. Eng. **6**(4), 239–242 (2014)
2. Schmidt, R., Möhring, M., Härting, R.-C., Reichstein, C., Neumaier, P., Jozinovic, P.: Industry 4.0: potentials for creating smart products: empirical research results. In: Business Information Systems, pp. 16–27. Springer, Cham (2015)
3. Wisner, J.D., Tan, K.–C, Leong, G.K.: Principles of supply chain management: a balanced approach. Cengage, Boston (2015)
4. HORSE project homepage. http://www.horse-project.eu/About-HORSE. Accessed 10 May 2018
5. Shakhovska, N.: The method of big data processing. In: 12th International Scientific and Technical Conference on Computer Sciences and Information Technologies (CSIT), vol. 1, pp. 122–126. Lviv (2017)
6. Stadnyk, B., Yatsyshyn, S., Seheda, O., Kryvenchuk, Yu.: Metrological array of cyber-physical systems. Part 8. Elaboration of raman method. Sens. Transducers **189**(6), 116–120 (2015)
7. Rong, H., Jones, R., Liu, A., Cohen, O., Hak, D., Fang, A., Paniccia, M.: A continuous-wave Raman silicon laser. Nature 725–728 (2005)
8. Grubb, S.G., Erdogan, T., Mizrahi, V., Strasser, T., Cheung, W.Y., Reed, W.A., Lemaire, P. J., Miller, A.E., Kosinski, S.G., Nykolak, G., Becker, P.C., Peckham, D.W.: 1.3 μm cascaded Raman amplifier in germanosilicate fibers. Optical Amplifiers and their Applications Topical Meeting (1994)
9. Michalski, L.: Temperature Measurement, 2nd edn. Wiley, Canada (2012)
10. Zhang, J.X.J., Kazunori, H.: Molecular Sensors and Nanodevices. Springer Science & Business Media, USA (2013)

11. Mulyak, A., Yakovyna, V., Volochiy, B.: Influence of software reliability models on reliability measures of software and hardware systems. East. Eur. J. Enterp. Technol. **4**(9), 53–57 (2015)
12. Shakhovska, N., Vovk, O., Kryvenchuk Yu.: Uncertainty reduction in big data catalogue for information product quality evaluation. East. Eur. J. Enterp. Technol. **1**(2) (2018)
13. Roblek, V., Meško, M., Krapež, A.: A complex view of industry 4.0. SAGE Open **6**(2) (2016)
14. Wieclaw, L., Pasichnyk, V., Kunanets, N., Duda, O., Matsiuk, O., Falat, P.: Cloud computing technologies in 'smart city' projects. In: IEEE 9th International Conference on Intelligent Data Acquisition and Advanced Computing Systems: Technology and Applications, IDAACS 2017, vol. 1, pp. 339–342. Lviv, Ukraine (2017)

Queueing Modeling in the Course in Software Architecture Design

Vira Liubchenko(✉)

Odessa National Polytechnic University, 1 Shevchenko Ave.,
65044 Odessa, Ukraine
lvv@opu.ua

Abstract. The studying of dependencies between architecture design decision and the quality attribute is difficult for students because of the high level of abstractness. The article introduces an idea to use the queueing modeling for examining the software efficiency of the different decisions. Because of study purposes, the only requirement for the model is a similarity to architecture topology; the model parameters are chosen to provide comparability of analyzed decisions. In the article, there are discussed the result of the queueing modeling involvement in the course in Software Architecture Design and defined the directions of future work.

Keywords: Software design · Architecture style · Queueing modeling

1 Introduction

Ukraine owns the fastest-growing number of IT professionals in Europe; it increases by 20% per year. That has been possible due to a new generation of young developers and engineers – during the last four years the number of IT specialists has increased from 42.4 K to 91.7 K IT specialists. By 2020, the number of software developers, engineers, and other IT specialists in Ukraine could attain to 250 K people [1].

Since the industry is rapidly growing, the issue of education has occupied a significant place. The business requires not only modern knowledge but also appropriate comprehension, skills and mindset from graduates. Sometimes for the universities, it is complicated to meet such requirements because of external causes. For example, software engineering is an empirical field of study, based on the experience of engineers. However, some key areas of software engineering provide limited support for teaching purposes. Therefore, teachers need to use the models in the study process to demonstrate the high-level abstractions and dependencies.

The purpose of this paper is an exploration of queueing modeling for examining the software quality characteristics in the course on Software Architecture Design.

The rest of the paper is organized as follows. Section 2 briefly describes the challenges for the course from the modern tendency in software architecture design. Section 3 presents the examples of how to use queueing modeling for analyzing the efficiency of architectural solutions. Section 4 discusses some results of the in-class implementation of the proposed idea.

N. Shakhovska and M. O. Medykovskyy (Eds.): CSIT 2018, AISC 871, pp. 550–560, 2019.
https://doi.org/10.1007/978-3-030-01069-0_39

2 Recent Trends and Teaching Challenge in the Course in Software Architecture Design

Software engineering is an engineering discipline that is concerned with all aspects of software production [2]. Software Engineering Body of Knowledge defines 15 key areas; one of them is Software Design. The result of software design describes the software architecture – that is, how software is decomposed and organized into components – and the interfaces between those components [3]. It represents fundamental decision profoundly affected the software and the development process. The theory of high-level (architecture) design has been developed from the mid-1990s, this gave rise to some exciting concepts about architectural design – architectural styles and architectural tactics. An architectural style provides the software's high-level organization; an architectural tactic supports particular quality attributes of the software. Both – architectural styles and architectural tactics – are kinds of design patterns documented a recurring problem-solution pairing within a given context. A design pattern is more than either just the problem or the solution structure: it includes both the problem and the solution, along with the rationale that binds them together.

The historically first architectural pattern is stand-alone systems, in which the user interface, application 'business' processing, and persistent data resided in one computer. Such architectural styles as layers, pipe-and-filters and model-view-controller were born from this kind of systems. The explosive growth of the Internet cause that most computer software today runs in distributed systems, in which the interactive presentation, application business processing, and data resources reside on loosely coupled computing nodes and service tiers connected by networks. The most known architectural styles for distributed systems are client-server, three-tiers and broker [4].

A recent trend in the area of architectural design is service-oriented architecture and microservices. Instead of building a single monolithic or distributed application, the idea is to split the application into a set of smaller, interconnected services, which typically implement a set of distinct features or functionality. Each microservice is a mini-application that has its hexagonal architecture consisting of business logic along with various adapters. The microservices architecture style features some significant benefits: it tackles the problem of complexity, enables each service to be developed and deployed independently, and enables each service to be scaled independently. Therefore, like every other technology, the microservices architecture style suffers drawbacks: small service size, the complexity of a distributed system, the partitioned database architecture, the complexity of testing and deploying.

Another trendy topic is big data architecture, designed for the systems for processing and analysis of the large and complex collection of datasets [5]. The primary purpose is to develop the data pipelines take raw data and convert it into insight or value. Usually big data architecture bases on pipes-and-filers architectural style [4] and involves such types of workload as batch processing of big data sources at rest, real-time processing of big data in motion, interactive exploration of big data, predictive analytics and machine learning. It is evident every type of workload needs making the own architectural solution. For example, because the size of data sets often a big data solution includes the subsystem for long-running batch jobs to filter, aggregate, and otherwise prepare the data for analysis. Usually, batch processing involves reading source files, processing them, and writing the output to new records. Solution architect should decide how to realize the batch processing.

Instead of recent trend software architect ought to analyze different aspects to take a decision, and often the "old-fashion" architecture styles are more appropriate for the developed application. Usually, the impact on quality attributes and tradeoffs among competing for quality attributes are the basis for design decisions. Professional software architects, who have risen from senior software developer, feel architecture drivers at the intuition level. However, for the students, the lack of professional experience causes misunderstanding of some cause-effect chains.

Curriculum Guidelines [6] insist on teaching course in the architecture design of complete software systems, building on components and patterns for an undergraduate student. The core model for architecture description is 4+1 view model designed for describing the architecture of software-intensive systems, based on the use of multiple, concurrent views [7]. The 4+1 model illustrates the high-level design of the system from different points of view, but it does not provide the possibility of modeling for system dynamic. A primary lesson of the course for students is to learn how to provide quality (or non-functional) attributes in their designs, so they should realize how architectural decisions impact on the quality attributes. It may be the most difficult subjects to teach to students lacking in considerable system experience. Therefore, the teacher needs visual and straightforward modeling tools to analyze quality scenarios for different architectural decision.

One of the appropriate tools is queueing modeling. This kind of models is clear and visual. The only problem concerned the queueing model in the framework of Software Engineering curriculum was the lack of students' knowledge. Therefore, we should have remarkably simplified the model definition.

3 The Examples of Queueing Modeling for Architecture Analysis

Queueing networks are a powerful abstraction for modeling systems involving contention for resources; they are especially useful in modeling computer communication systems. In this model, the computer system is represented as a network of queues, which is evaluated analytically. A network of queues is a collection of service centers, which represent system resources, and customers, which represent users or transactions. The queueing networks are easy to use for students due to many software solutions for queueing modeling. The queueing modeling is usually used when deciding the needed resources, and the core problem is the definition of adequate parameters.

Naturally examining how the architecture style impact on software efficiency does not need a very accurate model. It is enough to provide the possibility of result comparability. Therefore, the task of queueing modeling becomes simpler than in usual operation research case. Let us demonstrate the ideas with two examples.

3.1 Splitting the Monolith

Suppose the software efficiency is examined quality attribute, and the students have to consider how the different levels of software "granularity" influence the efficiency. For the purposes, students compare three architectural designs: monolith – a typical enterprise application consists of three layers (presentation, business logic and data-access), split frontend and backend, extracted services.

Remember, queueing modeling is the demonstration tool only. Having regard to the purpose, we can choose the parameters as simple as possible to focus students' attention on software efficiency. Let us suppose all queues are M/M/1 queue, which represents the queue length in a system having a single server, where a Poisson process determines arrivals and job service times have an exponential distribution. As the controlled attribute of software quality is efficiency, students are recommended to analyze the average time in the system as productivity metric and average utilization as resources utilization metric. All simulation experiments in the course and the article were realized with Simio Personal Edition software.

The first case is queueing model for monolith architecture (see Fig. 1). Suppose, the application works with browser client that runs on a user's local computer, smartphone, or another device.

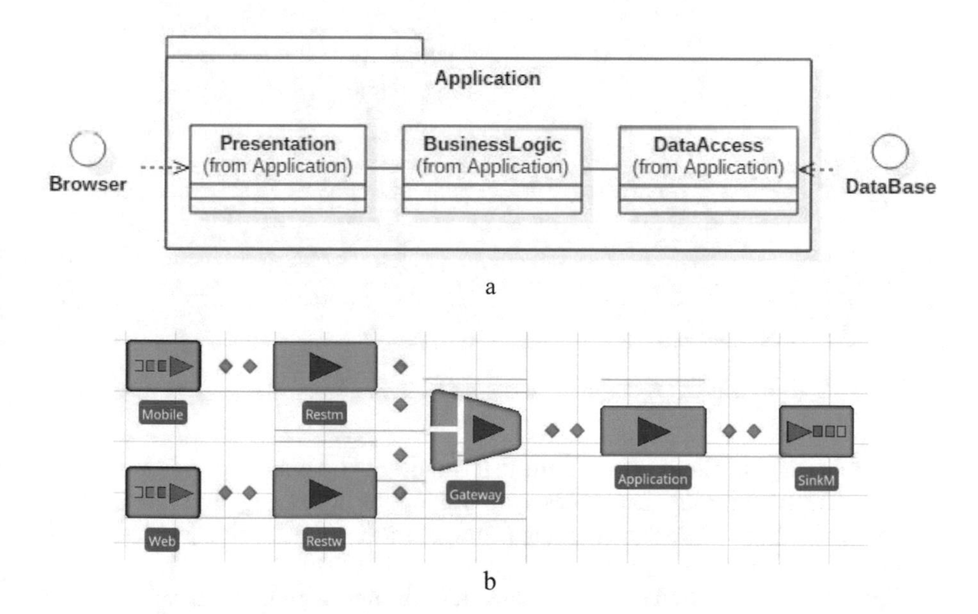

Fig. 1. Monolith architecture: a – UML-model, b – queueing model.

In the queueing model, we separate two types of clients, suppose the average time interval between two arrivals from mobile clients is 4 time units (for example, seconds); the average time interval between two arrivals from web clients is 6 time units. Both clients – mobile and web – retrieve the data by making a single REST call to the application, suppose each remote transaction takes 1 time unit. The Gateway serves as the sole entry point into the Application (like Façade pattern). Suppose each application action takes 2 time units. The application was designed with three-tier architecture pattern, which means the parameter of full processing time distribution is 6 time units.

In this case, the average time in the system is 0.32. To get this result students do not use the complicated calculation, they take the figures from Simio software. For example, at the Fig. 2 the report about Application characteristics is shown.

Server	Application	[Resource]	Capacity	ScheduledUtilization	Percent	99,1924
				UnitsAllocated	Total	4 474,0000
				UnitsScheduled	Average	1,0000
					Maximum	1,0000
				UnitsUtilized	Average	0,9919
					Maximum	1,0000
			ResourceState	TimeProcessing	Average (Ho...	0,2263
					Occurrences	33,0000
					Percent	99,1924
					Total (Hours)	7,4694
				TimeStarved	Average (Ho...	0,0018
					Occurrences	33,0000
					Percent	0,8076
					Total (Hours)	0,0608
		InputBuffer	Content	NumberInStation	Average	34,0191
					Maximum	70,0000
			HoldingTime	TimeInStation	Average (Ho...	0,0567
					Maximum (Ho...	0,1289
					Minimum (Ho...	0,0000
			Throughput	NumberEntered	Total	4 535,0000
				NumberExited	Total	4 474,0000
		OutputBuffer	Throughput	NumberEntered	Total	4 473,0000
				NumberExited	Total	4 473,0000
		Processing	Content	NumberInStation	Average	0,9919
					Maximum	1,0000
			HoldingTime	TimeInStation	Average (Ho...	0,0017
					Maximum (Ho...	0,0146
					Minimum (Ho...	0,0000

Fig. 2. A fragment of the report in Simio software.

From this report, the students learn that the average Application utilized is 0.99. Because of high utilization, the requests are waiting for processing; the average size of the input buffer is 34.02 so that additional component for a buffer of requests should be provided in design.

Highlight that the model, in this case, reflects non-stationary queueing system, which causes the increasing of input buffer size and the average time in the system with increasing the number of arrived requests (Fig. 3).

It could be the additional point for learning by students, but now such assignment is not used in the course because of two reasons. First, the students did not study the queueing systems and do not understand non-stationarity causes. Second, the non-stationarity depends on parameters of the model; the students are not familiar with this

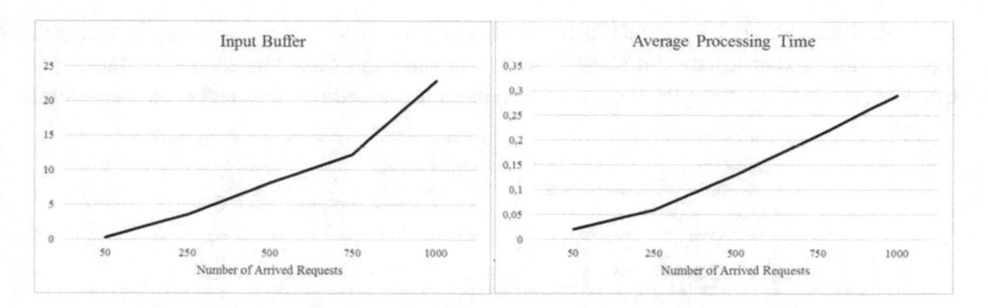

Fig. 3. The demonstration of non-stationarity impact on characteristics of the system.

dependency. Therefore, such assignment needs an additional theoretical base from students.

The second case is the queueing model for split frontend and backend (see Fig. 4). There is usually a clean separation between the presentation logic on one side and the business and data-access logic on the other. It causes splitting a monolith in this way is usually the first step of re-engineering the monolith systems. After the split, the presentation logic application makes remote calls to the business logic application, which should be reflected in queueing model.

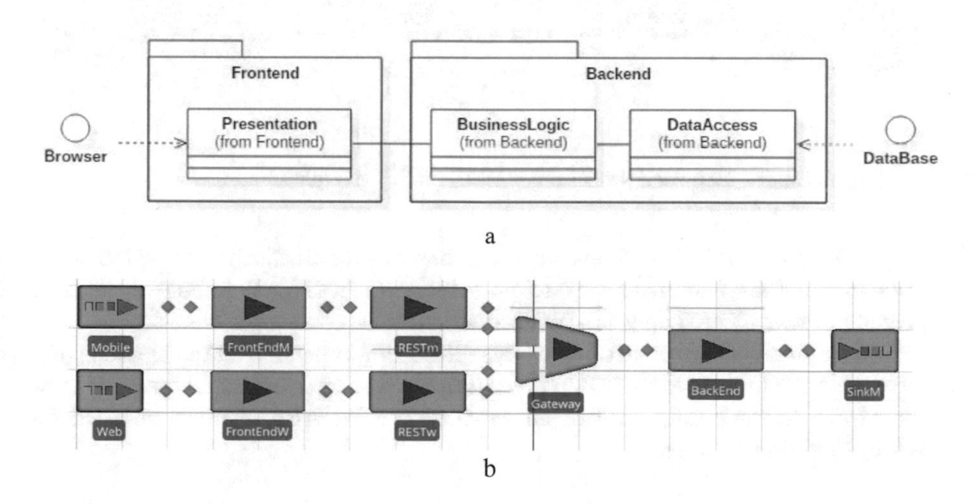

Fig. 4. Split frontend and backend: a – UML-model, b – queueing model.

For comparability of results, both average time intervals for clients remained the same as well as the time for remote transaction and actions of software. The only difference is the topology of components.

Such design decision increases the average time in the system to 0.91. Instead of this utilization of the software components decreases to 0.48 and 0.63 for frontend and backend. The only backend needs the buffer; the average size of the input buffer is 1.2.

The third case is the queueing model for extracted services (see Fig. 5). Suppose one of the existing modules within the monolith was turned into standalone microservice. After the split, the modules make remote calls to connect with each other.

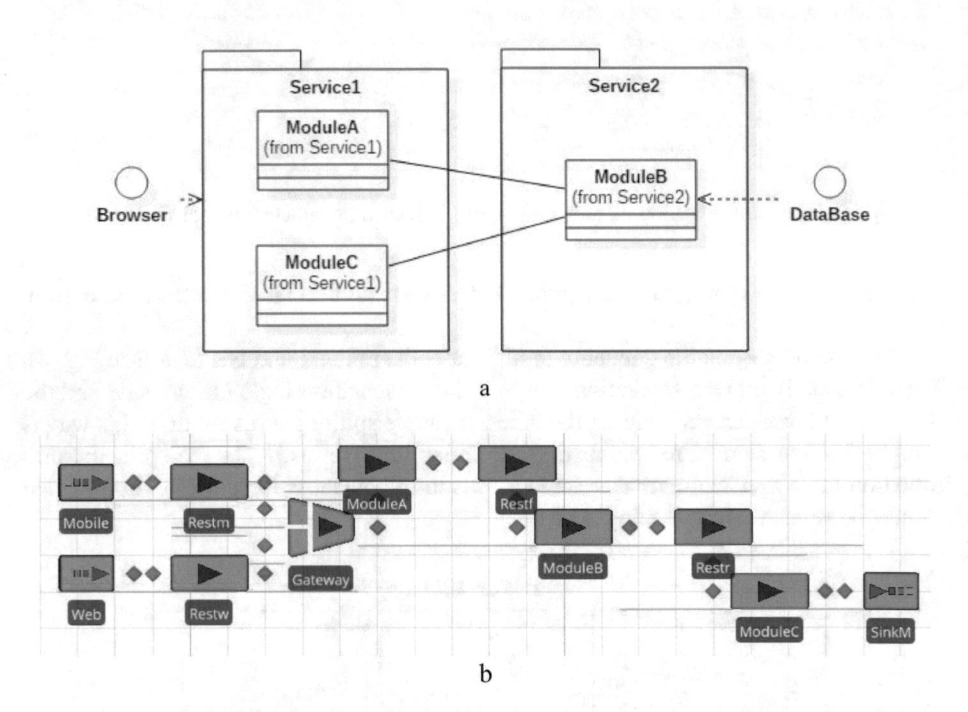

Fig. 5. Extracted services: a – UML-model, b – queueing model.

As in the previous case both average time intervals for clients, the average time for remote transactions and actions were not changed. Additional remote transaction appeared because of modeling the connection between separated services.

From the productivity point of view, this refactoring is better than the previous one, the average time in the system is 0.61, but from the resources utilization point of view is worse, the average utilization of all services is 0.33. Also, there is no need for providing additional buffers.

3.2 Schema of Batch Processing

The queueing modeling support students not only in the comparison of structural solutions but also in the study of the implementation of particular components of the software system. For example, when students are studying Big Data architecture style, they should examine the impact of batch processing schema. Batch processing is a component processed data files using long-running batch jobs to filter, aggregate, and otherwise, prepare the data for analysis.

Suppose the software efficiency is examined quality attribute, and the students have to consider how the different schema of batch processing influence the productivity.

For the purposes, students compare two approaches: MapReduce programming model and user-defined functions of the data warehouse.

MapReduce model is a specialization of the split-apply-combine strategy for data analysis and consist of three operations. Operation Map computes a given function for each data item and emits value of the function as a key and item itself as a value. Operation Shuffle groups the data items based on the output keys. Operation Reducer obtains all items grouped by function value and processes or saves them. The queueing model for MapReduce model is shown in Fig. 6.

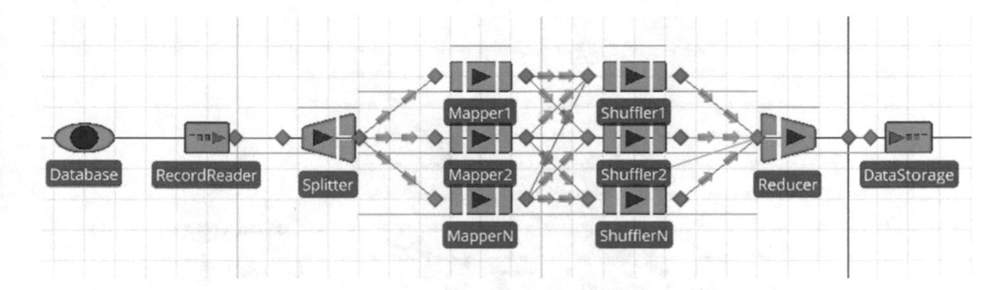

Fig. 6. Queueing model for MapReduce model.

A user-defined function (UDF) implements in program language (e.g., Java) the program logic, which is difficult to simulate in request language (e.g., HiveQL). UDF realizes the logic of data sample processing with all required calculations. Such solution transfers the implementation of particular costly operations on data samples in the data warehouse, which reduces the program execution time, eliminates the necessity of repeatedly data extraction from the repository and recording back, separates the functional duties of the data warehouse and the custom program. The queueing model for UDF solution is shown in Fig. 7.

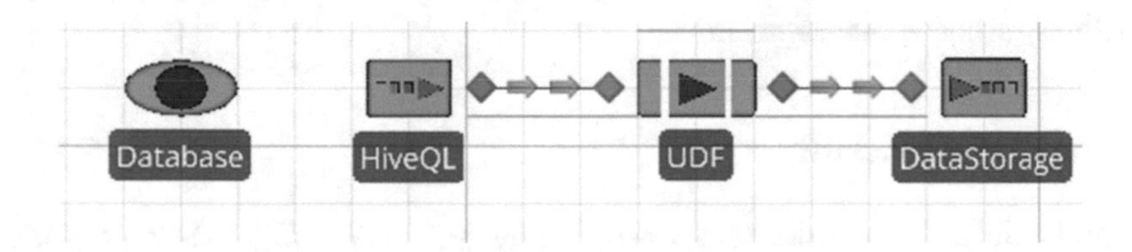

Fig. 7. Queueing model for UDF solution.

In the experiment framework, the students compare the average time in the components built under each of both schemas with different data volumes. Accordingly with the results of simulation UDF-solution is faster than MapReduce-based one, but the gap between two solution decreases with increasing of data volume. The students face the complex trade-off situation for software architect: the implementation of UDF is complicated; the benefit from UDF depends on the data volume, to take decision, abstract "big data analysis" is not enough. Such situation is quite difficult for text illustration; the queueing modeling gives the possibility to realize "what-if" experiment.

Other implementation of the comparison of the same models highlights that the processing time increases for the same sample sizes with the different speed. Such case is more straightforward than the previous one from parametrization point of view. Because students are studying the time increasing rate, they can use the same parameters in the framework of the particular model without providing the comparability of parameters between two models. In Fig. 8 there are shown the results of experiments with models in Figs. 6 and 7.

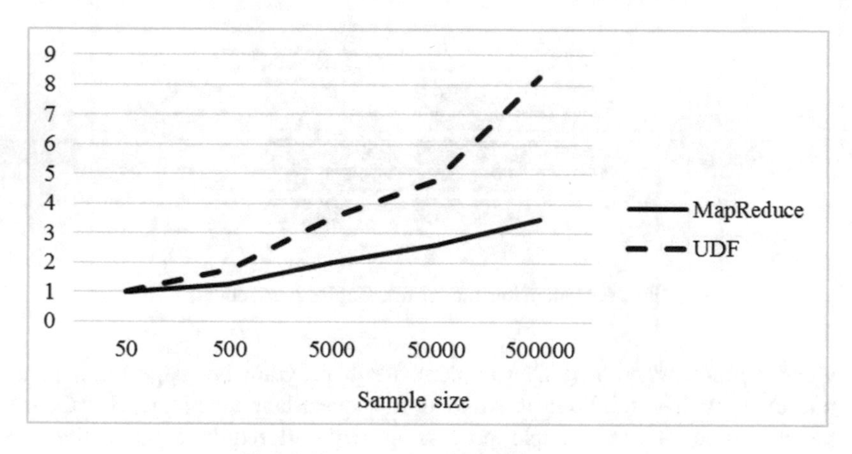

Fig. 8. The results of the experiment.

The students see the critical growth of processing time for UDP solution, while MapReduce solution demonstrates relative stability due to split-apply-combine strategy. Each server works with small data portions, which provides almost regular services utilization. It causes productivity increasing without introduction complex algorithms of queue management.

4 Analysis of Result

Course in Software Architecture Design is taught in the framework of Software Engineering curriculum for 3rd-year students. Traditionally the course is challenging because of the high level of abstractness. The timeframe does not provide the possibility for experiments with software prototypes so that students practice thought experiments to examine the dependencies between architectural decisions and quality attributes.

Introduction the queueing modeling into the course was quite risky because the students were very unfamiliar with such type of modeling. It causes the necessity for the maximum simplicity of the model. Last year we introduced the queueing modeling for efficiency study (two examples of the analyzed scenario is described in Sect. 3) into the course in Software Architecture Design. Let us discuss the results.

To avoid Hawthorne effect [8], the blind experiment was held; the students were not informed about participating in the research. Two different groups of students were involved in the test. The sampling of the students who participated in the experiment was homogeneous accordingly the distribution of their average score.

The following Table 1 gives a summary of students' result demonstrated in homework assessment and a final exam. We consider only those students, who made the task about efficiency characteristic. Groups of 57 and 74 participants were involved, correspondingly, in cases without and with queueing modeling; the same theoretical and practical tasks were used for both groups. Also, we transform the grades for each task into the universal scale.

Table 1. The result of an efficiency study.

Grade	Without queueing modeling		With queueing modeling	
	Assessment	Exam	Assessment	Exam
Perfect	2.1%	0.0%	6.8%	4.5%
Good	60.4%	47.9%	70.5%	63.6%
Satisfactory	20.8%	25.0%	15.9%	22.7%
Unsatisfactory	16.7%	27.1%	6.8%	9.1%
Task success	83.3%	72.9%	93.2%	90.8%
Quality	62.5%	47.9%	77.3%	68.1%

Table 1 shows the queueing modeling led to improving the learning outcome quality by 14.8% for assignment and 20.2% for the exam and growing the task success by 9.9% and 17.9% respectively. Also, the gap in the learning outcomes quality and task success achieved at assignment and exam was reduced, respectively, by 8.0% and 5.4%, so the forgetting effect was reduced.

5 Conclusion and Discussion

The main challenges of Software Engineering curricula are the abstractness and empiricism. The natural tool for teaching are different kinds of the model; the most common of them are the models in UML. Sometimes models in UML remain too abstract for students, and teacher should introduce another means.

The natural model for examining the software architecture design from an efficiency point of view seems to be queueing model. The problem is the students have not studied this class of model. Because of this issue, the parametrization of the queueing model was realized in a primitive way. Instead of inaccuracy of model, it demonstrated the properties of different design decision successfully, and students improved the topic understanding.

The future works ought to be realized in two directions. The first one is introducing the model for other software quality attributes. The second one is developing different analysis scenario based on adequate parametrization schemas.

References

1. Ukrainian it market in numbers and facts. https://blog.softheme.com/ukrainian-it-market-in-numbers-and-facts/. Accessed 01 July 2018
2. Sommerville, I.: Software Engineering, 9th edn. Pearson, London (2010)
3. Bourque, P., Fairley, R.E. (eds.): Guide to the software engineering body of knowledge, Version 3.0. IEEE Computer Society (2014)
4. Buschmann, F., Henney, K., Schmidt, D.C.: Pattern-oriented software architecture 4: a pattern language for distributed computing. Wiley, Hoboken (2007)
5. Sawant, N., Shah, H.: Big Data Application Architecture Q&A. A problem-solution approach. Apress, New York (2013)
6. Joint task force on computing curricula: Curriculum guidelines for undergraduate degree programs in software engineering. Technical Report. ACM, New York, NY (2015)
7. Kruchten, P.: Architectural blueprints – The "4+1" View model of software architecture. IEEE Softw. **12**(6), 42–50 (1995)
8. Gottfredson, G.D.: Hawthorne effect. In: Everitt, B.S., Howell, D. (eds.) Encyclopedia of statistics in behavioral science. Wiley, Hoboken (2005)

Architecture of the Subsystem of the Tourist Profile Formation

Valeriia Savchuk, Olga Lozynska(✉), and Volodymyr Pasichnyk

Information Systems and Networks Department, Lviv Polytechnic National University, Lviv, Ukraine
{Valeriia.V.Savchuk, Olha.V.Lozynska, Volodymyr.V.Pasichnyk}@lpnu.ua

Abstract. This paper is devoted to the architecture of the subsystem of the tourist profile formation. Three main steps of the process of providing the user with personalized content are presented. The model of informational and technological support of the tourist profile formation process is developed. Methods of forming the person's psychological profile such as Q method, Leary's method, Smirnov's method, Eysenck's method are suggested. The analysis of these methods and their comparison is given. The profile of the user is formed on the basis of the following data: the results of the surveys concerning the definition of its psychological type and tourist preferences, information about the previous trips. The tourist profile formation subsystem consists of four interdependence components: user survey, the analysis of past user travel, definition of a psychological type of user and definition of tourist preferences. For informational and technological support of the tourist profile formation the application was developed. The main results of the survey of the people with previous experience of the tourist trips are presented. Tasks that require further research are defined.

Keywords: Petri net · Tourist profile · Leary's method · Q method Smirnov's method · Eysenck's method · Trip support

1 Introduction

A personalized approach to tourists is a component of a tourist trip, which is currently provided only by tour operators and agencies. As a result of the analysis of a wide array of information sources, there was no available software that would take into account the wide range of personality characteristics of the traveler and give him specific recommendations.

That's why it is necessary to develop a system for individual travel support which, unlike the existing ones, provides a personalized approach to the user in accordance with his psychological characteristics and gives recommendations for overcoming possible dangers during the trip.

To improve the quality of the informational content for the particular tourist, it is necessary to determine his/her psychological peculiarities. The methods of forming the

N. Shakhovska and M. O. Medykovskyy (Eds.): CSIT 2018, AISC 871, pp. 561–570, 2019.
https://doi.org/10.1007/978-3-030-01069-0_40

tourist's psychological profile will allow define "comfort zones" of the user, which are an important characteristic when selecting "places of interest".

2 Related Work

The process of providing the user with personalized content consists of three main steps [1]:

Step 1. Collecting the information about the user.
Step 2. Data processing.
Step 3. Forming the content of the information system.

There are two main types of methods for collecting user information: explicit or implicit. The explicit type includes methods when the users provide information about their own traits of character and preferences, the implicit includes the automated collection of information about the interests and the activities of the user.

One of the basic implicit methods for gathering the user information is cookies that consist of background collection and data storage of repetitive actions in the Internet browser [1]. According to the stage of the modern information technology development, this method does not meet the well-known requirements, because in most cases the computer device is used by several users.

The main function of Google Assistant mobile application is to generate relevant informational content to support the user in the daily affairs [2]. In accordance to the goal, the system collects information about the user using implicit methods.

The Google Trips provides the user with a personalized list of tourist attractions according to the current time, location and weather information, and offers a plan for a day that is consistent with the time limits [3].

Among the implicit methods of the personal data processing, separate the collaborative filtering and instant personalization. The method of the collaborative filtration consists in forming groups of the consumers of the information systems according to the similarity of performed actions [1, 4].

The basis of the method of the collaborative filtration is the principle of classification: the assignment of each individual user to a separate group [5–7]. The informational content is selected for non-individual, but for the relevant group.

In most, the modern tourist applications use location information, time and travel archival data when forming recommendations.

3 Main Part

3.1 The Model of Informational and Technological Support of The Tourist Profile Formation Process

During the research the model of informational and technological support of the tourist profile formation process were developed. The model for the identification of the tourist preferences is based on the Petri network, represented by the system

$H = (P, T, I, O)$, where P – the set of the positions, T – the set of the transitions, I – the function of the inputs, O – the function of the outputs (Fig. 1).

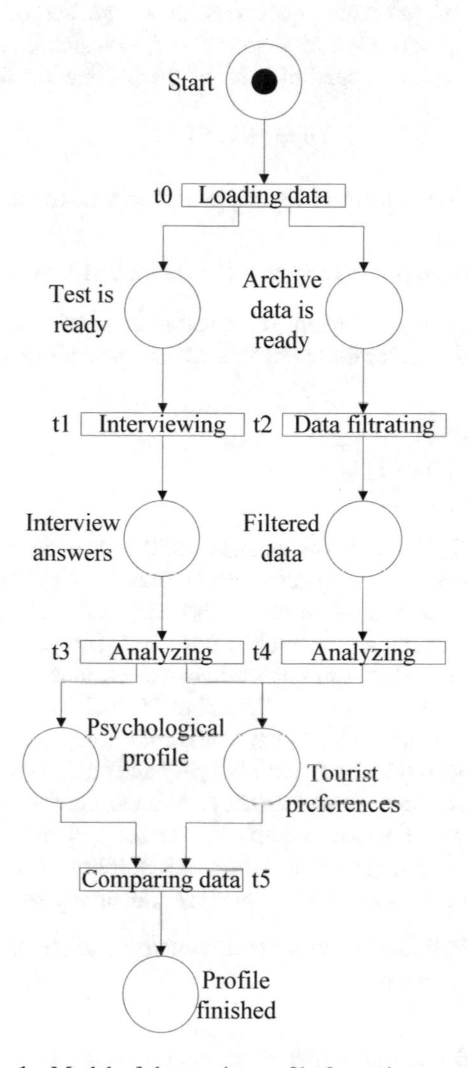

Fig. 1. Model of the tourist profile formation process

It should be noted that the activation of the next transition is impossible without activating at least one previous. The initial marking is $\mu_0 = (1, 0, 0, 0, 0, 0, 0, 0)$.

According to the set theory in mathematics, the model of tourist profile can be formalized with the next formula 1:

$$Tp = \langle Q, A, I, R, D, F, P \rangle, \quad (1)$$

where Q – the set of interview questions, A – the set of answers, I – interview:$I \subset Q \times A$, R – the results of interview: $R \in I$, D – archive data, F – filtered data: $F \in D$. According to the mentioned definitions the affirmation follows:

$$Tp = (R \cup F) \cap P, \quad (2)$$

So, the tourist profile is a set of tourist preferences of the user.

3.2 Methods of Forming the Person's Psychological Profile

To improve the quality of the content for a particular tourist, it is necessary to determine its psychological peculiarities. For this reason, a few methods were analyzed:

- Q method [8, 9];
- Leary's method [4];
- Smirnov's method [10, 11];
- Eysenck's method [10, 12].

The Q method [8, 9] has been known since 1958 and consists in the classification of a person by a number of characteristics in accordance with the results of the survey of the entire group. When passing the survey, it is necessary to determine whether certain statements are true, for example: "I am critical to my friends", "I am avoiding meetings and gathering in the group", "I am not sufficiently restrained in expressing feelings", etc. Among the answers there are "yes", "no" or "I doubt".

In general, 60 statements, which characterize the user versatilely, are formulated. The result is formed by counting the number "yes" and "no" for certain keys and allows to identify the dominant behavioral tendency. A small number of "doubt" answers in various ways is a sign of indecision of the person and his desire to avoid direct responses, while it can be regarded as a sign of flexibility or tact.

The following indicators are used to analyze the survey results:

- dependence or independence from public opinion and generally accepted norms;
- openness to communication;
- rivalry.

These characteristics allow to determine the prevailing psychological peculiarities of the person.

Smirnov's method [4] can detect a number of polar properties of the person's character, for example: balance, flexibility, sincerity, and others.

The method consists in analyzing the answers of 48 questions that may be either false or true, for example: "Do you like noisy companies?" or "Do you fulfill all promises?". Each answer "yes" or "no" is exactly one ball in a specific pair of polar character traits that are eventually added. Depending on the number of points can determine which of the features of the character is inherent in the person. The advantage of the method is the possibility of verifying the results to the truth by counting the points of the corresponding group.

Leary's method makes it possible to investigate the person's perception of himself, and to study relationships in particular groups [4]. In the relationship between people, two main areas can be distinguished, which determine the role of the person in the group: the level of dominance, or vice versa subordination, the level of friendliness to surrounding people and, accordingly, aggressiveness. According to Leary's method, it is believed that these levels are key characteristics of interpersonal relationships, and they are basic knowledge of the type of temperament.

To form the result, person need to determine which of the 128 characters' rice most closely match his/her personality. The disadvantage is that it is impossible to determine the level of truthfulness of the answers provided, but it is simple and understandable.

Leary suggested using a graphical schema for reproduce basic social orientations (Fig. 2).

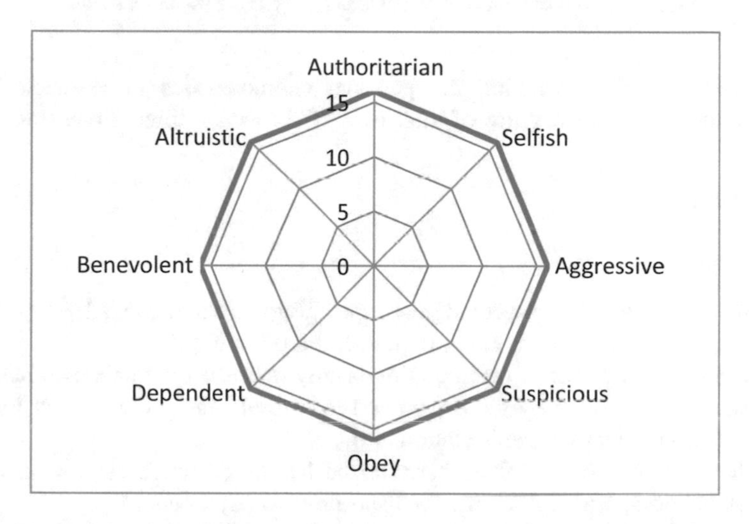

Fig. 2. Conditional diagram of the basic social orientations of the person

The main axis are Authoritarian-Obey and Benevolent-Aggressive, additional – Dependent-Selfish, Altruistic-Suspicious. Thus, the scheme is divided into 8 sectors that determine the basic psychological characteristics. Such division is due to the assumption that the proximity of the result to the center, indicates a closer relationship between these characteristics. The largest ball is 16, the smallest – 0.

In general, it can be argued that the greater the value – the more characteristic is the character trait. If the characteristic corresponds to a value from 0 to 5 points, this indicates an adaptive variant of the trait, from 6 to 10 points – the average level of manifestation, and if the result exceeds 10 points – an indicator of pathology and extreme behavior.

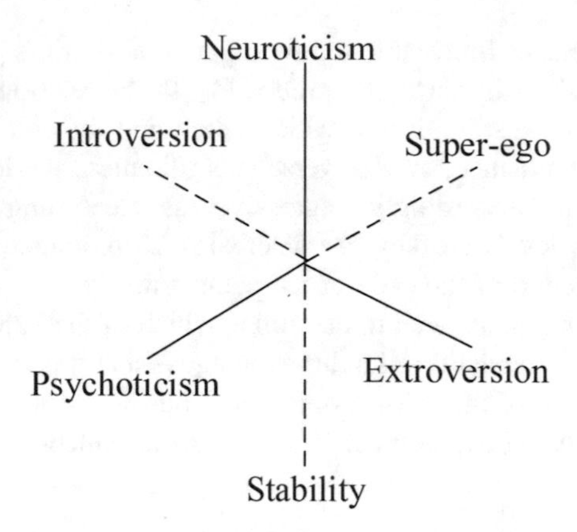

Fig. 3. Three dimensions of personality by the Eysenck's method

The method of determining the personal characteristics of Eysenck helps to determine the following features of their own "I" based on three dimensions [9] (see Fig. 3):

- neuroticism;
- introversion;
- psychoticism.

Neuroticism is a psychological pathology characterized by anxiety, worry, fear, anger, frustration, envy, depressed mood, etc. [8].

Introversion is the state of being predominantly interested in one's own mental self. Introversion characterized by the orientation to himself, the preference is loneliness, desire to observe, creative manifestations [9].

Psychoticism is the state that characterized by the presence of rich imagination, fantasy, selfishness, aggressiveness, predisposition to psychosis [9].

The method consists of 57 questions, for example: "Do you enjoy being among the people?", "How much do you dream?" or "Do you have any jitter?". The response options are "yes", "no" or "possible". The result is formed depending on how many statements correspond to the person, they are assigned 2, 1 or 0 points.

Table 1. Comparative table of techniques

Name	Q method	Eysenck's method	Smirnov's method	Leary's method
Number of questions	60	57	48	128
Average time of passing the test	5 min.	4-5 min.	4-5 min.	6-7 min.
The presence of a mechanism for verifying the sincerity of the user	Missing	Present	Present	Present

The advantage of the method is a powerful mechanism for checking the truthfulness of answers, consisting of 17 questions, for example: “Do you always keep your promises, even if you are not beneficial?” and “Have you ever been late?”. The norm is 4 “false” answers. If there are more these answers (“false”), the test is not objective.

The analysis of the methods and their comparison is given in Table 1.

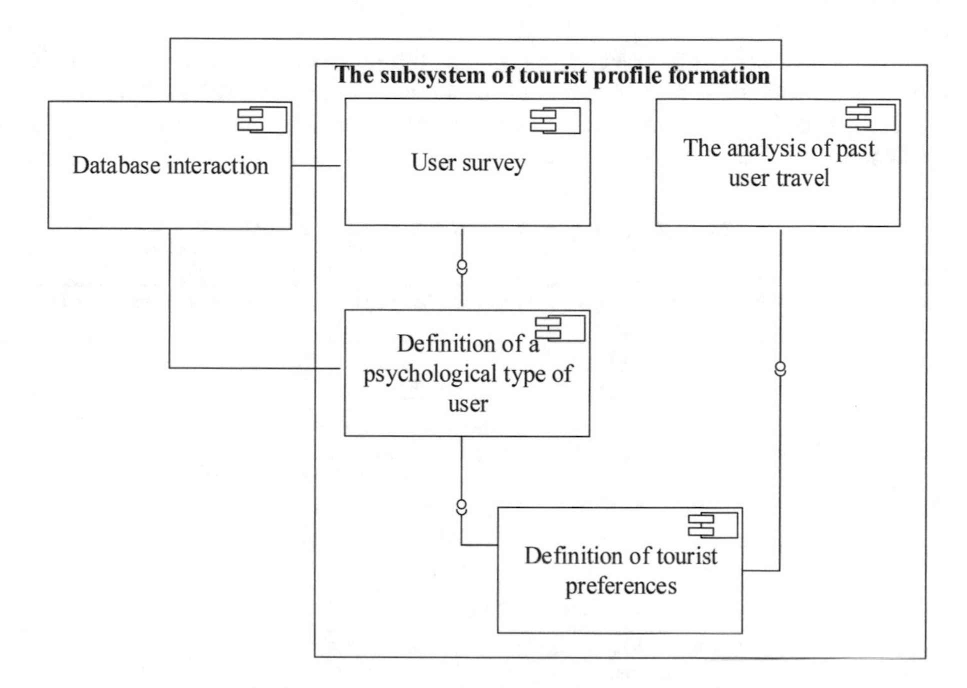

Fig. 4. Architecture of the subsystem of tourist profile formation

3.3 The Architecture of the Subsystem of the Tourist Profile Formation

The architecture of the subsystem of the tourist profile formation is shown in Fig. 4.

The profile of the user is formed on the basis of the following data: the results of the surveys concerning the definition of its psychological type and tourist preferences, information about the previous trips. There are four interdependence components which forms and receives data:

- “User survey” is responsible for querying and writing to the database of the user’s system.
- “The analysis of past user travel” analyzes user ratings of the “places of interest”, which he visited, and defines their features, common and distinctive features.
- “Definition of a psychological type of user” analyzes the results of surveys concerning the psychological type of the user. As a result, we get the probabilities of user belonging to each type (introvert, extravert or phlegmatic).

- "Definition of tourist preferences" analyzes user responses regarding his preferences and the results of the work of the two previous components. As a result, a list of characteristics of tourist objects that should attract the user is formed.

4 Results

For the study sample, the survey of the people with previous experience of the tourist trips was conducted. The main results of the survey are given in Tables 2 and 3.

Each separate tourist preference is considered separately, because most are independent and the user can have several of them or all of them.

Table 2. Survey results using the Leary's method

0	1	2	3	4	5	6	7	8
Number of participant	Authoritarian	Selfish	Aggressive	Suspect	Subordinate	Dependent	Benevolent	Altruistic
1	6	5	11	12	15	14	2	12
2	2	8	3	11	5	7	11	8
3	3	7	2	12	8	7	7	8
4	1	3	4	12	4	4	3	12
5	11	13	0	2	9	10	6	5
6	12	1	10	11	15	13	8	12
7	4	2	5	12	14	11	0	15

Table 3. The survey of personal preferences

Question \user's number	Gender	Camping	Residence in hotels	Hobby	Costs ($)	Architectural sights	Food	Traveling in the company
1	M	No	Yes	Photography	1000	Yes	Restaurants, cafes	Yes
2	W	No	Yes	Hiking in the mountains	500	Yes	Cafe, fast food	Yes
3	M	No	Yes	Traveling	800	Yes	Cafe	Not always
4	M	Yes	Yes	Auto sport	1500	Yes	Restaurants	Yes
5	W	No	Yes	Photography	700	Yes	Cafe	No
6	M	Yes	No	Null	300	No	Fast food	No
7	M	No	No	Cycling	750	Yes	Cafe	Not always

For informational and technological support of the tourist profile formation the subsystem was developed as an application. At the first starting the application, the user must to register and to provide certain personal information by passing a questionnaire (see Fig. 5).

Questions are related to such basic characteristics of a person as the age, the gender, the education, other preferences and also features pointing to the user's psychological profile [13, 14].

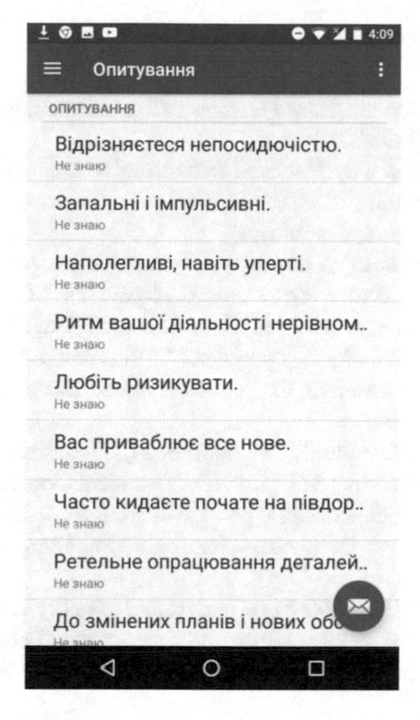

Fig. 5. Mobile app interface for user survey

As a result of the user answers analysis, the subsystem generates its tourist profile. The tourist profile is the basis of personalized user support.

5 Conclusion

The architecture of the subsystem of the tourist profile formation are described in the paper. The model of informational and technological support of the tourist profile formation process is developed.

Methods of forming the person's psychological profile such as Q method, Leary's method, Smirnov's method, Eysenck's method are suggested. The analysis of these methods and their comparison are given.

The profile of the user is formed on the basis of the following data: the results of the surveys concerning the definition of its psychological type and tourist preferences, information about the previous trips. The tourist profile formation subsystem consists of four interdependence components: user survey, the analysis of past user travel, definition of a psychological type of user and definition of tourist preferences.

The main results of the survey are compared to the previous experience of the tourist trips. As a result of the analysis of the user answers, the subsystem generates its tourist profile, which is the basis of personalized user support.

Further research can be focused on improving the model of the tourist profile formation process.

References

1. Sukhorolsky, P.M., Khliboiko, G.P.: Personalization in the Internet and its impact on human rights. Leg. Inform. **4**, 3–9 (2013)
2. Leontiev, S.: What can and why you need an assistant to Google Now. Hi-tech. News. http://hitech.vesti.ru/article/622037
3. Google trips. https://get.google.com/trips
4. Klochko, V.E.: Age psychology. http://medbib.in.ua/vozrastnaya-psihologiya782.html
5. Lytvyn, V., Vysotska, V., Burov, Ye., Veres, O., Rishnyak, I.: The contextual search method based on domain thesaurus. Adv. Intell. Syst. Comput. II **89**, 310–319 (2018)
6. Lytvyn, V., Pukach, P., Bobyk, I., Vysotska, V.: The method of formation of the status of personality understanding based on the content analysis. Eastern Eur. J. Enterp. Technol. **5/2** (83), 4–12 (2016)
7. Naum, O., Vysotska, V., Chyrun, L., Kanishcheva O.: Intellectual system design for content formation. In: 12th International Scientific and Technical Conference on Computer Sciences and Information Technologies (CSIT), pp. 131–138. Lviv, Ukraine (2017)
8. Karvasarskii, B.D.: Clinical Psychology Textbook for High Schools, 4th edn. Peter, St. Petersburg (2004)
9. Pervin, L.A., John, O.P.: Psychology of personality. Theory and research. M. (2001)
10. Ilyin, E.P.: Differential psychophysiology. St. Petersburg: "Peter" (2001)
11. Kulikov, L. Psychological study: methodical recommendations for conducting. SPb: Language (2001)
12. Suslov, V.I., Chumakova, N.P.: Psychodiagnostics. SPb.: SPbSU (1992)
13. Savchuk, V.V., Kunanec, N.E., Pasichnyk, V.V., Popiel, P., Weryńska-Bieniasz, R., Kashaganova, G., Kalizhanova, A.: Safety recommendation component of mobile information assistant of the tourist. In: Proceedings of SPIE – The International Society for Optical Engineering, vol. 10445, pp. 110–118. Wilga, Poland (2017)
14. Shakhovska, N., Vysotska, V., Chyrun, L.: Features of e-learning realization using virtual research laboratory. In: XIth International Scientific and Technical Conference Computer Sciences and Information Technologies (CSIT), pp. 143–148. Lviv, Ukraine (2016)

Formation of Efficient Pipeline Operation Procedures Based on Ontological Approach

O. Halyna Lypak[1,2], Vasyl Lytvyn[1(✉)], Olga Lozynska[3], Roman Vovnyanka[3], Yurii Bolyubash[3], Antonii Rzheuskyi[1], and Dmytro Dosyn[1]

[1] Information Systems and Networks Department, Lviv Polytechnic National University, Stepan Bandera Street, 32a, 79013 Lviv, Ukraine
vasyl17.lytvyn@gmail.com,
antonii.v.rzheuskyi@lpnu.ua, dmytro.dosyn@gmail.com
[2] Zolochiv College of Lviv Polytechnic National University, Zborivsky College of the Ternopil National Technical University named after Puluj, 57, Zboriv, Ternopil region, 47270 Zolochiv, Zboriv, Ukraine
[3] Zolochiv College of Lviv Polytechnic National University, Zolochiv, Ukraine
Olha.V.Lozynska@lpnu.ua, {vovnianka,bol_jura}@ukr.net

Abstract. The approach to the development of computer system for automated construction of the basic ontology is presented. The mathematical support of functioning of intellectual agents of activity planning on the basis of ontologies was developed, which allowed to formalize their behavior in the space of states. Using ontologies allows to narrow the search way from the initial state to the state of the goal, rejecting irrelevant alternatives. Such approach made it possible to reduce the task of planning the activities of the intellectual agent to the problem of dynamic programming, where the goal function is the composition of two functions that specify competitive criteria. Using the developed method, the calculation of the necessary costs for pipeline modernization and the expected economic effect from their application were made.

Keywords: Ontology · Ontological approach · Intellectual agent
Mathematical model · Information systems · Pipelines

1 Introduction

Intensive development of the field of engineering and knowledge generates the need for scientific developments and their testing in the processes of building information systems used to solve these tasks, which require efficient planning and monitoring of activities; forecasting and classification of objects and phenomena, etc.

In the information society, the accumulation of knowledge in various subject areas generates the urgent need to develop new, highly effective methods, means and techniques for presenting knowledge, one of the most popular among which is the ontological approach. Under the "ontology" we will understand the detailed formalization of some subject area, represented using a conceptual scheme. Such scheme usually

N. Shakhovska and M. O. Medykovskyy (Eds.): CSIT 2018, AISC 871, pp. 571–581, 2019.
https://doi.org/10.1007/978-3-030-01069-0_41

consists of a hierarchical structure of concepts, relations between them, theorems and constraints that are adopted in a given subject area [1].

2 Analysis of Recent Researches and Publications

The research on the ontological approach to the construction and functioning of intellectual agents began at the end of the last century. The basic theoretical foundations of formal methods and adequate mathematical models of ontologies are presented by Nehmer and Bennett [2], Potoniec and Ławrynowicz [3], Li, Martínez and Eckert [4], who consider the ontology in the representation of a three-dimensional cortege. The practical aspects of the use of ontologies in applied intelligence information systems are analyzed in the works of Norenkov [5]. The problems of intelligent systems design based on the ontological approach are considered in the works of Munir and Sheraz Anjum [6], Donnelly and Guizzardi [7], Andreasen, Fischer Nilsson and Erdman Thomsen [8], Golenkov [9], Anisimov, Glybovets, Kulyabko, Marchenko, Lyman [10], Gladun and Rogushyna [11].

The analysis of developments in the field of intelligent information systems and Internet services [12] gives grounds to consider that the following software and technical solutions are justified:

- implementation of the ontology synthesis system as a subsystem of the portal service of Internet search;
- applications of OWL as a language of knowledge representation in ontology;
- the application of HTN and OWL-S as the structure and language of the automatic knowledge base planning;
- Java API for Protege-OWL - as a software environment and library for classes, including machine learning (learning support) OWL ontology and knowledge base;
- Link Grammar Parcer - as a mean of grammatical-semantic analysis of English-language text documents in electronic format;
- Apache-PHP-MySQL - as a software for building a user-interface with the web portal architecture;
- Wget - as a web service for automated access to search engines by query generated from keywords;
- SWRL - as the language of rules for the logical output of new knowledge by deductive and inductive methods;
- WordNet is the basic English glossary.

Ontology in the OWL language contains a conceptual apparatus of the upper level and subject domain [13]. The ontology of the upper level provides:

- the logical output of new knowledge,
- the addition of the received messages by the context;
- verification of the truth of the received statements;
- evaluation of the credibility of the sources of messages;
- ensuring the logical integrity of the knowledge base.

Machine learning is implemented with the Java API Protege-OWL. These means contain class libraries, which implement methods for working with OWL structures: their reading, additions. Thus, machine learning means function in interaction with OWL-ontology, taking from it the patterns of grammatically-semantic structures for recognition of statements (predicates of the 1st order logic) in the studied and/or educational texts and adding new elements to it as a result of such recognition.

To do this, Link Grammar Parcer [14] is used, which splits the affirmative sentence, written in grammatically correct English, in a semantically interconnected pair of words. LGP contains in its composition a table of correspondence between grammatical constructions of the English language and types of syntactic-semantic connections between words (concepts). API LGP allows you to link this table to an OWL ontology so that the table can dynamically adapt in the learning process to a given subject area.

The aim of the article is to analyze the capabilities of using an ontology synthesis system with the purpose of achieving the maximum economic effect of exploitation of systems of pipelines of resource networks.

3 Ontological Modeling of Subject Areas

Special attention of researchers deserves activity planning agents. The authors' analysis of the main approaches, methods and means of constructing intellectual activity planning agents convincingly testifies that not all available ontology capabilities are used in these systems, especially as it relates to the stages of modeling the functionality of such systems. The behavior of these systems is reduced mainly to the search for an optimal path in the state space, but in every case it is not obvious what the exact ways this search should be carried out. The search for the optimal path should be based on the rules (laws) that are inherent and set within a specific subject area. In order to effectively implement formalization procedures for such rules, an ontological approach is proposed.

At the same time, it is natural that the researchers intend to develop systems of concepts that are system-specific for groups and complexes of models of many subject areas. The practice of forming large ontologies confirms the necessity of establishing close inter-branch collaboration of researchers. It is expedient to formulate certain hypotheses, assumptions and system predictions in the formation of ontologies of certain subject areas, which would allow them to be easily changed and adapted in the future in accordance with changes in our representations and knowledge of the subject area.

The vast majority of ontology-based practical applications provides the possibility of presenting such description of tasks on formation of target configuration of certain products based on their components in accordance with a given specification and a program that allows the specified configuration to be transformed in spite of the product and specified components. This, in turn, creates conditions for the development of the ontology of components and characteristics of computer systems, which allows us to generate automatically on the basis of this algorithm non-standard configurations of computer systems. The analysis of knowledge on a given subject area implies a declarative specification of a system of terms based on their formal analysis, which in turn provides the preconditions for re-using previously created ontologies and their extension. Generalized ontology is presented as a cortege of four elements of the following categories: concepts, relations, axioms, instances. Formally, the ontology model of O is presented in the following form:

$$O = \langle C, R, F \rangle,$$

where C – finite set of concepts (concepts, terms) of the subject area, which is given by ontology O; $R : C \rightarrow C$ – a finite set of relations between concepts (terms, concepts) of a given subject area; F – finite set of interpretation functions (axioms, constraints) that are defined on the concepts and relations of ontology O.

In an ontological representation, the concept is used to conceptualize certain entities or phenomena. Classes serve as general categories that can be hierarchically organized. Each class describes a group of individual entities that are consolidated on the basis of common properties. Different types of relationships (such as length, location, etc.) can be maintained between entities, which bind them to classes. One of the most common types of relations used in ontologies is the categorization relation, that is, the admission of some entity to a certain category. This type of relationship has several names [1], which are actively used by modern researchers: taxonomic; IS-A; class – subclass; generic; relation a-kind-of. At the same time, it should be noted that these structures do not always provide all the components of ontology.

4 CROCUS System Modules for Building of Ontologies

An ontological presentation of knowledge makes sense only in the part of some intellectual system. The best solution in our opinion is the solution in which such an intellectual system is an information retrieval system, for which an adaptive ontology, on the one hand, is a tool for information retrieval, analysis and classification, and on the other hand it uses search tools to supply new data for its content, the synthesis of new predicates and rules, the teaching of new concepts and semantic relations between them. Such a solution was the intellectual information search system CROCUS, developed on the basis of adaptive ontology, a knowledge base in the field of materials science and a database of scientific publications in this area. The general conceptual scheme of the CROCUS system is presented in Fig. 1.

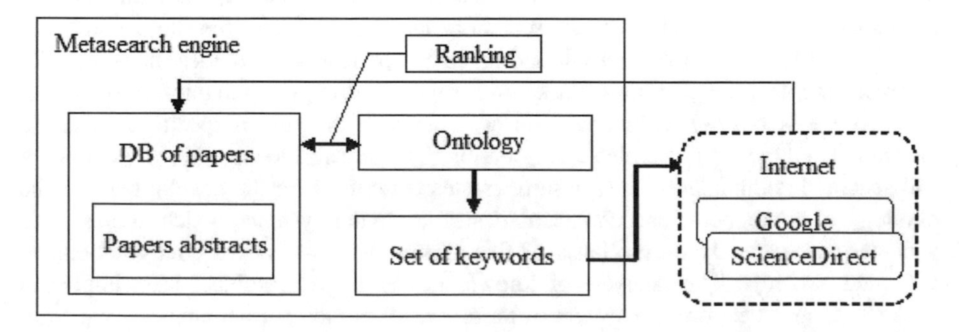

Fig. 1. The conceptual scheme of the CROCUS system

Subsystem of ontology training uses the textbooks of annotations of scientific publications of the database of articles. To fill the database, the system generates a set of keywords that selects from the external source of publications in the Internet (in particular the ScienceDirect, CiteSeer, Wiley Online Library, Springer databases) the main metadata for publications in a given subject domain, their annotations, which form the basis for analysis and learning ontology.

The essence of the method of extracting knowledge [15] from a natural text document is to formulate a strategy of the intellectual agent's activity aimed at recognizing the data allocated in the text document. Strategy and information model - the knowledge base of the intellectual agent - are built in one formal language of presentation of knowledge. The value of information obtained as a result of the recognition of the content of a text document is determined by the increase in expected utility from the implementation of the strategy of the operation of the intellectual agent [16–18].

According to the analysis of annotations, scientific publications are ranked according to their relevance to the user's information needs, that is, according to the ontology, which reflects these needs. For this purpose, an analysis of each annotation as a natural text is performed, its image is constructed in terms of ontology in the form of a set of predicates and rules that are added to the knowledge base of the system. Each time the expected utility of an intellectual agent's activity strategy is calculated. In this ranking, the system gives priority to publications, the submission of metadata which leads to increased utility.

The system is able to adapt to the needs of the user, preserving its preferences in the database. Each user can contribute to the training of their own ontology, the system stores data about this process, maintains statistics of sessions, provides the opportunity to correct mistakes made during training, as well as return to previous versions of ontology. Modules of the CROCUS system are shown in Fig. 2.

Through the user interface, the client has the opportunity to manage the priorities of document rankings, that is, to adjust the order of their placement in the list (relevant to the information needs of the client) and/or to classify them according to certain criteria. In this case, the most important documents are used to study ontology and build effective keyword sets. The metadata for new articles from the Internet is input in the database of publications in conjunction with user preferences and other prerequisites for obtaining a document, first of all the source in which it is stored.

Annotation processing takes place in several stages, which ensure their transformation into a plurality of predicates. With the Link Grammar Parser module, a grammatically-syntactic compilation of annotations and the formation of a model of its context from semantically similar predicates from ontology is performed. Supplemented models are compared with each other to calculate the semantic distance between their centers of semantic weight, and thus the most closely related documents are selected for their further ranking and classification.

The CROCUS system has two main functions:

(1) interactive automated construction of the ontology of a problem area;
(2) search, preservation and classification (ranking) of scientific publications both in interactive semi-automatic and in automatic mode.

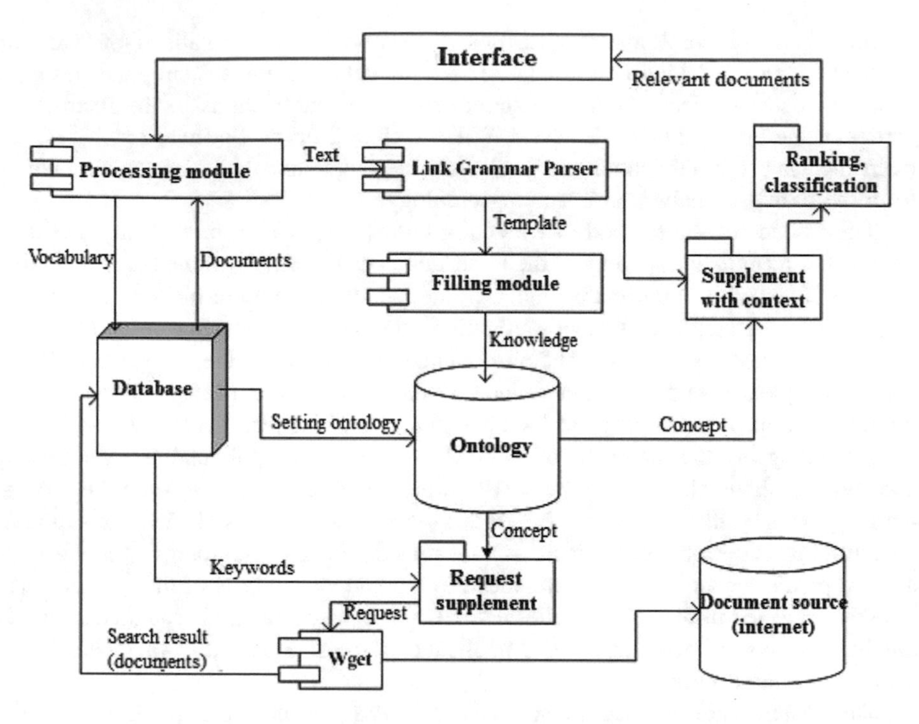

Fig. 2. CROCUS system modules

Each of these functions is implemented by a separate basic set of functional modules, some of which have a dual purpose. The CROCUS system is implemented in the Java programming language for an object-oriented paradigm as a hierarchy of program code classes, instances of which call each other with the parameters specified at the time of call, and/or interact through events and their handlers. Most system modules have a Swing interface and AWT libraries. All connected libraries have an open source status, distributed free of charge. Due to their application, the information system is fully functional and has all the necessary means for successive development (Fig. 3).

The developed architecture of the ontology synthesis system is implemented using tools and software solutions as a module of CROCUS software (Cognition Relations or Concepts Using Semantics).

5 Technologies of Automated Ontology Development

The Protege-OWL Java-based machine learning tools contain a generalized description of the semantic link, which serves as a template for generating new types of semantic links in the learning process and forming them for identification in the text of the appropriate vectors of the signs of these relationships. In this case, the corresponding classes of bonds and their properties are added. Copies of these classes serve to

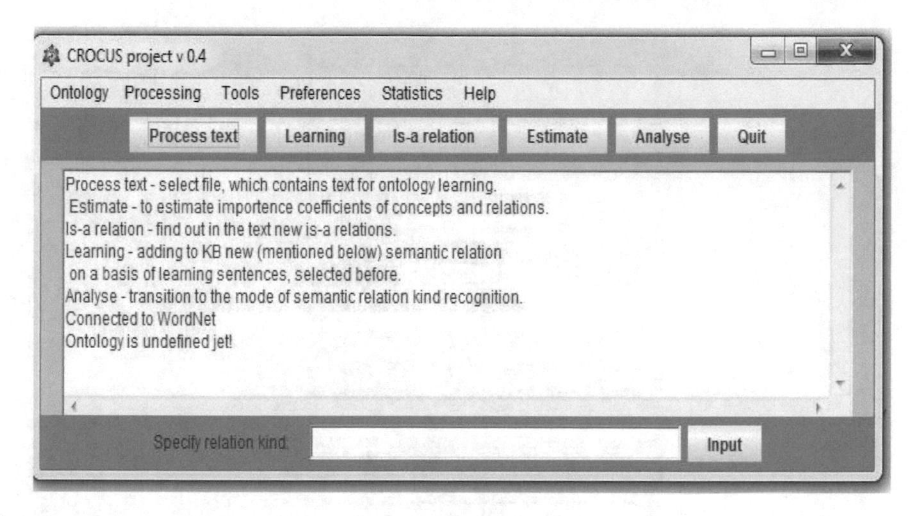

Fig. 3. The main window of the CROCUS user interface

describe the existing and new classes of ontology by using them as predicates of logic 1-st order.

Developing ontologies it should be taken into account that certain classes of concepts impose restrictions on the properties of their instances by means of descriptive logic. Such restrictions are divided into three categories: quantifier restrictions (existence, universality); limit the amount of permissible values (minimum $\leq$, just = , maximum $\geq$); type limit can take value from a plurality.

The system of automated ontology development follows the following sequence of steps:

Step 1. To form a set of information sources U.
Step 2. To calculate credibility σ_i to information source U_i.
Step 3. To calculate the weight W of concept C ontology.
Step 4. To edit the ontology, depending on the weight growth of the concept.

The algorithm for the functioning of the system of automated ontology development is presented in Fig. 4.

6 Construction of the Ontology of the Subject Area of Material Science

In the process of the research on the basis of the ontology built using the developed modules of the CROCUS system the following problem was solved: how to maximize the use of the pipeline for the supply of gas or water at minimum cost, taking into account that: (1) the main restrictive resource factor is electro-chemical corrosion of the pipe; (2) the estimated economic effect that we derive from the exploitation of the pipeline and possible losses from the cessation of operation; (3) the cost of anti-

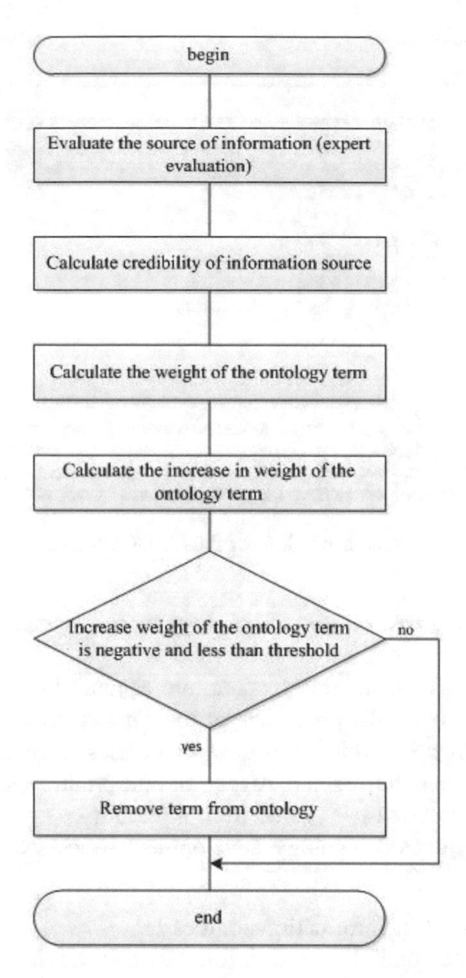

Fig. 4. The algorithm for the system of automated ontology development

corrosion protection is known and determined by the technology of such protection; (4) from the expert assessments, norms, data of non-destructive testing and technical diagnostics, known indicative terms of trouble-free operation of the pipeline are known.

The general rule is that the replacement of the coating restoration is formulated as follows: IF ((It is the term of the coating renewal) OR (There has been an event of damage to the coating) OR (measured parameters exceed the previously established tolerance threshold)) AND (Available resources for updating the coating) THEN (Perform replacement).

The Knowledge Base details this rule through the system of defining production rules. An information retrieval agent is considered to be valuable information that allows you to succeed in solving this problem: about new types of anticorrosion protection that contribute to the extension of the period of trouble-free operation; refined evaluation of the pipeline's resource; more effective coating technology.

The initial state is the state of "unprocessed". Purpose status: "processed". The task is divided into six stages (opening the surface of the pipe, removing the protective coating, degreasing, priming, coating, protection).

Alternative solutions are used to perform each stage. In particular, for the stage "removal of protective coating" one of three alternatives can be used: mechanical, chemical, thermal. The selection of alternatives takes place on the basis of information stored in the relevant ontology.

For the effective functioning of the ontology, axioms of the vocabulary terms, atomic statements about instances of concepts are written, the knowledge base is established. At the same time, all axioms were analyzed in order to remove those that contain incorrect statements.

The ontology constructed contains over 3000 concepts, 40% of the concepts are defined. To obtain metrics for the period of operation and cost of works, the language of requests to the SPARQL ontology is used. For example, we can search for alternatives to the methods for cleaning pipeline surfaces: Hand-operated cleaning, Cleaning by electric tool, Commercial cleaning, Cleaning to almost pure metal, Cleaning to pure metal.

Using the method of functional equations, designed to solve dynamic programming tasks, we obtain an optimal way to find alternatives, which is shown in Fig. 5.

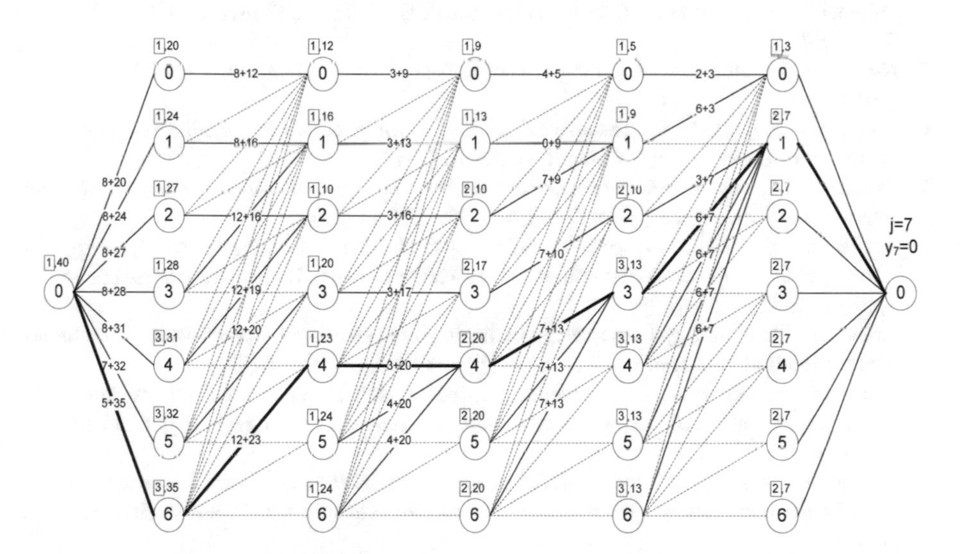

Fig. 5. The way of search for a solution in the space of states

Thus, optimal distribution of funds for pipeline processing is following: priority is given to the first and third processes, the following alternatives are 4 and 6 processes.

7 Conclusions

Thus, the approach to the development of a computer system of automated construction of basic ontology is considered in the paper. The architecture of the ontology synthesis system as a module of CROCUS software (Cognition Relations or Concepts Using Semantics - "recognition of connections and/or concepts according to their semantics") is developed. Using ontologies allows to narrow the search way from the initial state to the state of the goal, rejecting irrelevant alternatives. With using the developed method, the calculation of the necessary costs for pipeline modernization and the expected economic effect (gain) from their application were made.

References

1. Glybovets, M.: Artificial Intelligence. KM Academy, Kiev (2002)
2. Nehmer, R., Bennett, M.: Using mathematical model theory to align conceptual and operational ontologies in FIBO. https://vmbo2018.e3value.com/wp-content/uploads/sites/10/2018/02/VMBO_2018_paper_8.pdf
3. Potoniec, J., Ławrynowicz, A.: Combining ontology class expression generation with mathematical modeling for ontology learning. In: Twenty-Ninth AAAI Conference on Artificial Intelligence, pp. 4198–4199 (2015)
4. Li, X., Martínez, J.-F., Eckert, M.: Uncertainty quantification in mathematics-embedded ontologies using stochastic reduced order model. IEEE Trans. Knowl. Data Eng. **29**, 912–920 (2017)
5. Norenkov, I.: Intelligent technologies on the basis of ontologies. Inf. Technol. (1), 17–23 (2010)
6. Munir, K., Anjum, M.S.: The use of ontologies for effective knowledge modelling and information retrieval. Appl. Comput. Inf. 1–11 (2017)
7. Donnelly, M., Guizzardi, G.: Formal ontology in information systems. In: Seventh International Conference (FOIS 2012), vol. 239, 368 p. (2012)
8. Andreasen, T., Fischer Nilsson, J., Erdman Thomsen, H.: Ontology-based querying. In: Larsem, H.L., et al. (eds.) Flexible query answering systems. Recent Advances, pp. 15–26. Springer (2000)
9. Golenkov, V.: Ontology-based design of intelligent systems. In: Open Semantic Technologies for Intelligent Systems (OSTIS-2017), pp. 37–56. Minsk (2017)
10. Anisimov, A., Glibovets, M., Kulyabko, P., Marchenko, O., Liman, K.: The method of the automated expansion of the pre-toxic ontological bases of knowledge. Comput. Sci. **99**, 50–53 (2009)
11. Gladun, A.: Methodology for the development of terminology of information resources as a basis for the formation of ontologies and thesauri for semantic search. Softw. Eng. **1**, 41–52 (2014)
12. Hatzi, O., Vrakas, D., Bassiliades, N., Anagnostopoulos, D., Vlahavas, I.: The PORSCE II framework: using ai planning for automated semantic web service composition. Knowl. Eng. Rev. **02**(3), 1–24 (2010)
13. Dosyn, D., Kovalevych, V., Lytvyn, V., Oborska, O., Holoshchuk, R.: Knowledge discovery as planning development in knowledgebase framework. In: Modern Problems of Radio Engineering, Telecom-munications and Computer Science (TCSET'2016), pp. 449–451. Lviv-Slavske, Ukraine (2016)

14. Link Grammar – Carnegie Mellon University. http://bobo.link.cs.cmu.edu/link
15. Kut, V., Kunanets, N., Pasichnik, V., Tomashevskyi, V.: The procedures for the selection of knowledge representation methods in the "virtual university" distance learning system. In: Advances in Intelligent Systems and Computing, pp. 713–723 (2018)
16. Lytvyn, V., Vysotska, V., Pukach, P., Vovk, M., Ugryn, D.: Method of functioning of intelligent agents, designed to solve action planning problems based on ontological approach. East.-Eur. J. Enterp. Technol. **3/2**(87), 11–17 (2017)
17. Burov, Ye., Mykich, Kh.: Uncertainty in situational awareness systems. In: Modern Problems of Radio Engineering, Telecommunications and Computer Science (TCSET'2016), pp. 729–732. Lviv-Slavske, Ukraine (2016)
18. Kravets, P.: Game model of dragonfly animat self-learning. In: Perspective Technologies and Methods in MEMS Design (MEMSTECH 2016), pp. 195–201. Lviv Politechnic Publishing House, Lviv-Polyana, Ukraine (2016)

Distributed Malware Detection System Based on Decentralized Architecture in Local Area Networks

George Markowsky[1(✉)], Oleg Savenko[2], and Anatoliy Sachenko[3,4]

[1] Missouri University of Science and Technology, Rolla, USA
markov@maine.edu

[2] Khmelnitsky National University, Khmelnitsky, Ukraine
savenko_oleg_st@ukr.net

[3] Kazimierz Pulaski University of Technology and Humanities, Radom, Poland
sachenkoa@yahoo.com

[4] Ternopil National Economic University, Ternopil, Ukraine
as@tneu.edu.ua

Abstract. The paper proposes the architecture of a distributed malware detection system based on decentralized architecture in local area computer networks. Its feature is the synthesis of its requirements of distribution, decentralization, multilevel. This allows you to use it autonomously. In addition, the feature of the components of the system is the same organization, which allows the exchange of knowledge in the middle of the system, which, unlike the known systems, allows you to use the knowledge gained by separate parts of the system in other parts. The developed system allows to fill it with subsystems of detection of various types of malicious software in local area networks. The paper presents the results of experiments on the use of the developed system for the detection of metamorphic viruses.

Keywords: Distributed system · Malware · Local area networks
Decentralized architecture · Structure Kripke

1 Introduction

According to data from [1], malicious software is one of the main tools of cybercrime. Distribution of computer systems and information technologies in various fields and fields, their integration into the global Internet network, as well as increasing opportunities for obtaining financial returns that appear at the same time, motivate developers of malware to increase and spread them [2, 3]. Distribution of malware in information systems of local area networks creates problems for users. Existing means of its detection for today do not meet the needs of users. This is especially true of the task of detecting malware prior to it, at the stage of its initial distribution. As a rule, detection of malware occurs after it has been spread over a period of time and has undergone destructive actions. A variety of antivirus tools that detect malware at different stages of its lifecycle are not known to provide high authenticity of its detection [2]. A special place is taken by antivirus [4–8] tools that execute the removal of malware in local area networks. They allow you to take advantage of an organization with more computing

N. Shakhovska and M. O. Medykovskyy (Eds.): CSIT 2018, AISC 871, pp. 582–598, 2019.
https://doi.org/10.1007/978-3-030-01069-0_42

power than individual computer systems. Creating such malicious software removal systems in local area networks based on their effective organization will increase the level of authenticity of the detection.

2 Related Works

Known classical methods (signature analysis method, method of checksums, method of heuristic analysis, etc.) detection of malware are mainly focused on applications in end-user computer systems. For network-based antivirus tools, methods are developed, which are only possible on the server or on corporate or local networks. Most of these methods are developed using technologies and components of artificial intelligence. As a rule, modern malware detection systems contain sets of many methods and their combinations, which is influenced by the growth of malicious software varieties. Let's consider more detailed systems and methods for detecting malware.

A system of identification and classification for network cyberattacks is proposed in [9]. To implement the system, different methods of machine learning are used, namely, neural networks, the immune system, neurophysical classifiers. Moreover a method of reference vectors is proposed. A distinctive feature of the proposed system is the multi-level analysis of network traffic, which enables to detect attacks by signature as well as combine a set of adaptive detectors.

In [10] the static system for detecting the malicious software is developed. It's based on the principal components method for extracting data and classifiers SVM, J48 and Naive Bayes. To eliminate the disadvantages of known antivirus tools, methods of static analysis are used to generate features derived from information about the Windows PE header, DLL libraries, and API calls inside each DLL library. To reduce the resulting set of characteristics, the method of the main components is used.

The system for detecting cyberattacks on the basis of the involvement of neural network immune detectors is presented in [11]. The developed system consists of two parts. The first is implemented hardware and it works constantly in real time. The second part is represented by software on a dedicated computer, which is used to analyze current attacks and create appropriate security features. The decision on the possible influence of SPS is carried out with the involvement of a system of neural network detectors based on the algorithm of Mamdani.

Another approach to detecting malicious software is to isolate the characteristics based on information about the flow of the program. A proposed system [12] based on constructing a graph for malware flow control, which is converting into the vector space afterward.

In addition to forming a control flow graph for detecting metamorphic viruses, signs based on the API tracking program executing the program are used [13]. In the work, the authors proposed a static approach to the formation of a signature of metamorphic virus, based on the calculation of the number of corresponding API calls. The conclusion about the presence of a metamorphic virus is formed on the basis of finding the similarity of the formed signature with the base of signatures using similar metrics.

The method of detecting metamorphic viruses on the basis of analysis of information flows is described in [14]. The proposed method of detection is based on the analysis of program values during its implementation (value set analysis). It implies that each program at the time of execution can be represented by a set of values of memory cells and registers. An unknown program is placed in a protected environment, after which for each API call the analysis of the status of registers before and after the API call is performed. Based on the change of values in the registers, a vector of signs for each register is formed. The proposed approach showed high efficiency, with a percentage of false positives at 2.9%. However, the presented method does not take into account the technique of evasion from the emulation.

The analysis showed that for detecting malicious software known systems carry out analysis of network traffic, audit files, packets transmitted over the network, checking the configuration of open network services. To define the fact of LAN work violation various methods are used, namely, neural networks, artificial immune systems, reference vector method, Bayesian networks, fuzzy clustering [9–19]. At the same time, the main disadvantage of known systems is their host-oriented approach for detecting the malicious software.

The authors are proposing therefore the advanced system for detecting malware described below. The use of the developed distributed system is provided in the local area network. Its task is to detect such malicious software: file viruses, program bookmarks (local and remote exploits), botnet.

3 Architecture of the Distributed System for Detecting the Malware in Local Area Networks

The effectiveness and reliability of detecting malicious software through a variety of tools essentially depend on, among other things, the architecture of such tools, as well as their positioning and placement in computer systems of local networks. Taking into account that the process of detection of malware will be conducted on local networks, the choice of the model of system operation should involve the inclusion of information from all computer systems of the local network, that is, placement in all computer systems of the system. This is necessary to increase the efficiency and reliability of detection by taking into account information on the state of other computer systems for decision-making in a particular computer system. These basic requirements that a system should be placed on the network in each computer system, affect the choice of model of its architecture. Also important for such systems is that the decision-making center of the system is not presented and identified unambiguously, since its detection will lead to an attack on it to remove the entire system from the working state. The system should be constructed so that its components located in the computer systems of the local area communicate effectively with each other to exchange information about the state of the computer systems in order to provide additional information for decision-making. In addition, the malware detection system should be structured accordingly in order to be able to grow and increase it should not slow down the detection process.

Considering that the system will function on a local computer network and solve many different problems in detecting malicious software, its determinants will be distributed and multilevel. The main function of the system is to check existing software and running processes in the computer systems of the local network to be able to attribute to malware. The achievement of the system's compliance with the specified characteristics and its functions in detecting malicious software creates requirements for it, the main ones being as follows: distribution; decentralization; autonomy in decision making; multi-levelness; self-organization; adaptability According to the analysis of the task, the requirements for malware detection system, on the basis of system functions and its characteristics, the architecture of the developed system can be synthesized as a set of the following components: models of collective intelligence, multi-agent systems, distributed systems, decentralized systems, self-organized and adaptive systems. Taking into account such components in the system model is the basis of its architecture and will increase the reliability of the functioning and survivability of the system in the local network. Integrate the basic requirements into a model of a developed system, whose architecture is depicted by a generalized scheme of the main components in Fig. 1.

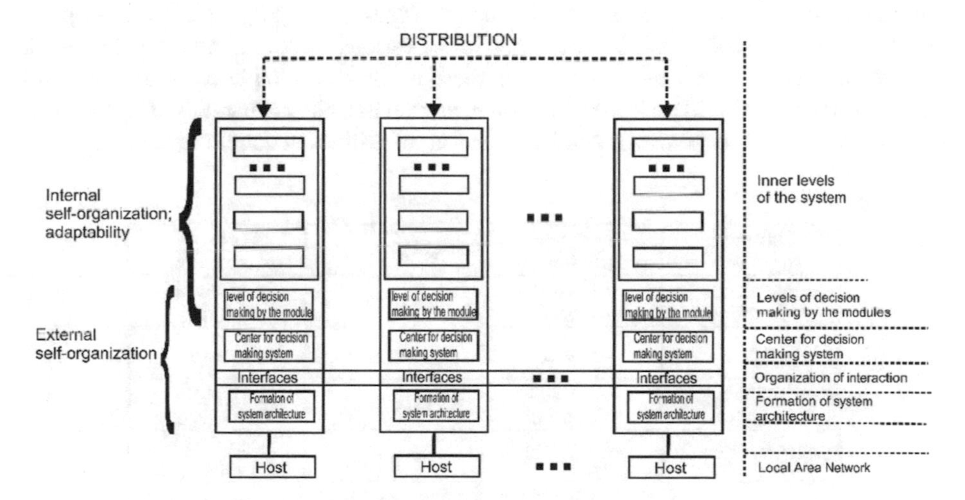

Fig. 1. A generalized scheme of system architecture with main components

Gradually, this distributed architecture will be filled with subsystems, which will implement other models included in the system. The system is distributed in space and according to the characteristic requirements should be decentralized, that is, the system should not have a single control center of all its parts and the possibility of making decisions depending on the change of the state of any computer systems of the network. In this context, the decentrality and independence of decision making by the system module located in a particular computer system are not identified. In the concept of decentralization of the system we will put the function of the higher level of abstraction, whose task should be the ability of the system of specific computer systems to decide on the beginning of its implementation, the transition between the identified levels on the basis of information received from other levels of the system, the decision to communicate with other parts of the whole systems, receiving information from other parts of the system from different computer systems and transferring this information to the corresponding levels, the completion of work. For components of the system located in computer systems, we introduce the concept of an autonomous software module of the system. Autonomy in decision making by an autonomous program module of the system includes the following possibilities: decision-making about the state of the threat of malware for computer systems on the basis of integrated information from other levels of the program module and transfer of this value to the level of decentralization; transfer to the level of decentralization the level of safety; determining the number and involvement of the appropriate levels of the program module for the investigation of malware. Thus, the level of decentralization will differ from the level of decision-making by the fact that at its level decisions are made on the general organization of the functioning of the software module in the structure of the whole system, and at the decision-making level - on the direct execution of the very basic tasks of the investigation of malware.

The multilevelness (Fig. 2) of the system will allow separating processes that relate to the functioning of the system as a whole, the processes of the program module's operation, and divide the task levels into the detection of various types of malware and network attacks and the necessary utilities to solve these problems. In addition, multilevel will allow the system to build up a new functional, as well as a clear separation of the various tasks of parts of the system will allow the sharing of utilities at certain levels.

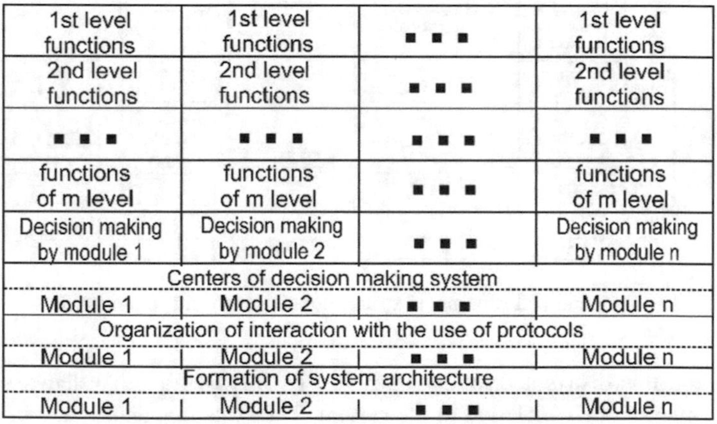

1st level functions	1st level functions	■ ■ ■	1st level functions
2nd level functions	2nd level functions	■ ■ ■	2nd level functions
■ ■ ■	■ ■ ■	■ ■ ■	■ ■ ■
functions of m level	functions of m level	■ ■ ■	functions of m level
Decision making by module 1	Decision making by module 2	■ ■ ■	Decision making by module n
Centers of decision making system			
Module 1	Module 2	■ ■ ■	Module n
Organization of interaction with the use of protocols			
Module 1	Module 2	■ ■ ■	Module n
Formation of system architecture			
Module 1	Module 2	■ ■ ■	Module n

Fig. 2. Multi-level scheme of the system

The functional, which will be responsible for the self-organization of the autonomous software module, is placed in a separate level of the program module, which will interact with the decision maker module. In addition, certain levels of program modules will be able to carry out self-study and will be based on a change in the security status. Self-organization at the level of the whole system will be implemented in the sub-block of the decentralization block and will include the operation of the entire system and its transition to different states, depending on external changes in computer systems and networks. The property of self-organization of the system is closely related to adaptability and will be manifested in changing the structure of its system and level of organization in the course of its life cycle as a result of accumulation of data, stored information in memory. The accumulated experience will be expressed in changing the parameters important for the purpose of the system, which will change the way the system works, and will express the property of self-study.

The formation of the system architecture in the network will be carried out throughout its life cycle and will consist of storing the previous formats for analysis. Part of the levels of autonomous software modules of the system will have the properties of adaptability to increase the effectiveness of the implementation of the tasks of detecting malicious software. Adaptability of the system will be manifested in the automatic change of the algorithms of its functioning and, if necessary, its structure in order to maintain or achieve an optimal state when changing external conditions. System interfaces are divided into the following: administrative, daily report, critical situation report, stand-alone intermodule.

Distributed system in the local area network according to its model is represented by the same software modules, which are located in each computer system. That is, no matter how many computer systems on a local area network, each of them contains an autonomous software module. If in the process of using computer systems it turns out that not all of them are on the network, then the system consists of those that are active. Independent software modules of active computer systems in the local network form a directly distributed system. This will allow even the presence of the work of two computer systems to support the execution of system tasks. Each stand-alone software module contains functionality at appropriate levels to detect a particular type of malware or attack.

Distributed system contains in each software module a level that is responsible for communication between autonomous software modules of the whole system, that is, provides the operation of communication channels. With this level, a connection is established between the program modules of the system as a whole. An important function of the system's operation is the identification of one specific standalone software module of the remaining stand-alone software modules for activating and operating the integral system. To do this, when installing a program into a specific computer system activates a functional that reads the system information about the hardware and software of computer systems, stores this information in its internal repository, forms on its basis the identification of this particular module, stores it as identifies an autonomous software module and uses it in packages that will be sent to other distributed module software modules. When installing all autonomous software modules of the system on the network they activate them to collect all identifiers. Each separate stand-alone software module in the system after the complete installation of the system will contain the characteristics and identifiers of all autonomous program modules of the system.

The number of levels of the standalone software module and their content will depend on the tasks that will be solved. The main tasks of the system are to conduct an analysis of executable files in order to detect the malicious functionality and behavior of computer software executed in the presence of malicious actions. Both of these tasks combine the need for analysis of software behavior, that is, the discovery through behavior. This will allow for the consideration and investigation of a sufficiently wide variety of file malware, by separating the malicious behavior for analysis into its component. That is, the system will include tools for researching software hosted on computer systems, which will be in active state or in external memory, the common element of which will be tracking and analyzing their behavior.

Given that as a result of functional subsystems accumulate large volumes of heterogeneous information, for the efficient operation of the distributed system requires subsystems of autonomous software modules to account for the work of specific computer systems and to optimize the arrays of information in the system as a whole.

The distribution of tasks in the system software modules will include the states in which the program module will be located on specific computer systems in the course of their lifecycle. A distributed system model that takes into account the architectural constituents and their states and relationships between them will be represented as follows:

$$M_A^s = \langle S, G_A \rangle, \quad (1)$$

where S is the set of states of the system, G_A is an oriented graph of the generalized states for the distributed system (Fig. 3).

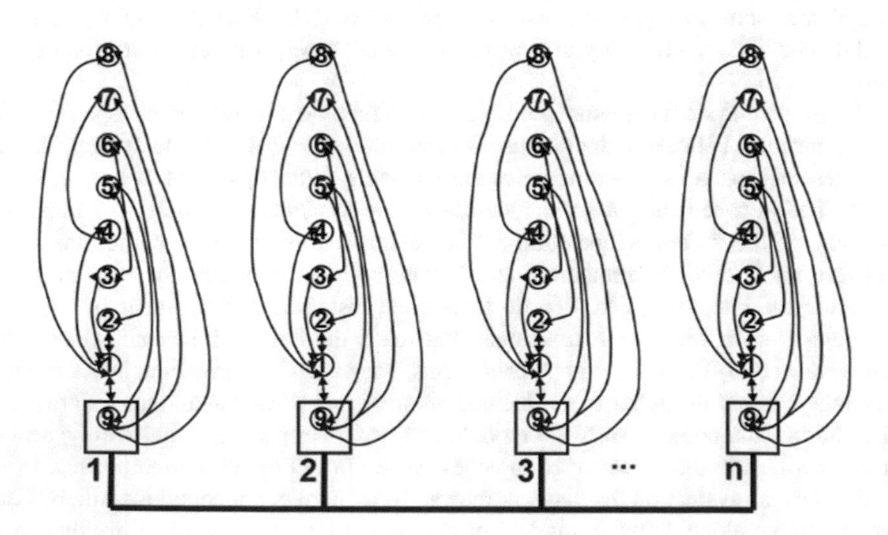

Fig. 3. Oriented graph of connecting the generalized states autonomous software modules for the distributed system

In accordance with Fig. 3 for each state the following functions are defined with the following notation:

1 - basic state, monitoring of computer systems, definition of transition to the state 2, 3, 4, 8;
2 - check executable files, definition of transition to the state of 1, 5;
3 - verification of running processes and network activities of computer systems, definition of transition to the state of 1 and 6;
4 - check executable files and running processes, definition of transition to the states 1 and 7;
5 - verification of executable files using other software modules, return to level 2;
6 - verification of running processes and network activities and their comparison with other computer systems of the local area network, return to level 4;
7 - processing executable files and running processes using other autonomous software modules of the system;
8 - processing and optimization of information from the base of packages of the software module with the use of information from other software modules of the system, return to level 1;
9 - providing communication with other software modules of the system.

Each autonomous software module of the system has the same structure and it is divided into four levels depending on the function assignment and grouped tasks:

(1) Level 1 includes monitoring events and defining transitions to the following levels, as well as processing information from other standalone software modules;
(2) Level 2 includes checking executable files, checking running processes and network activity without the involvement of information from other system software modules;
(3) Level 3 involves performing level 2 tasks with the use of information from other autonomous software modules of the system;
(4) Level 4 carries out processing, optimization and extraction of information from the base of the autonomous software module of the computer system.

It is necessary to provide a communication of Levels 1, 3 and 4 with other program modules of the system. The program module, as a rule, is in level 1, where it monitors events. If changes occur in computer systems, then depending on the results of monitoring at its level, decisions are made to move to vertices 2, 3 or 4, ie to Level 2. At Level 2, the tasks of Level 1 are investigated by methods and means without involving other parts of the system. If the result of the research turns out to be negative, that is, malware is not found, then it is determined that a deeper check is required and a transition to Level 3 is carried out, which is used to identify other parts of the system on the network. Using this level of resources involves distributed online stand-alone software modules and, as a result, will increase the effectiveness of detection of malware. At the fourth level, optimization of information from the base of the software module with the involvement of other components of the distributed system, as well as after receiving information from all other autonomous software modules of the decision-making system on the state of the system as a whole.

Thus, the first level contains means that ensure the autonomy of the program module of the system. At the third level, a deeper analysis of the objects under study is achieved in comparison with the second level, which increases the efficiency of the system. The fourth level solves part of the tasks of self-organization of the system, associated with the optimization of the accumulated during the work of information. Each level and its corresponding generalized subsystems in turn, too, are represented by sets of sublevels that impose the implementation of certain functionals.

The image of the relationship of sublayers is presented in Fig. 4 with a graph with vertices corresponding to the purpose of sublevels.

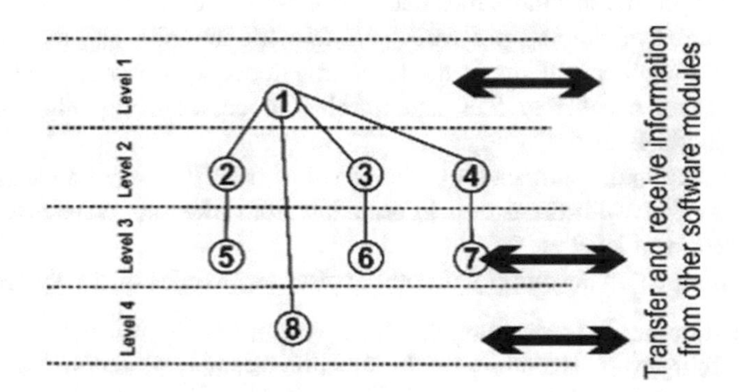

Fig. 4. Graphic diagram of sublevels interaction for system software modules

The time model of the behavior of a distributed system through its behavioral models of standalone software modules is considered not as a linear sequence of the set of computations, but as a tree of possible computations, that is, an extended time pattern with feedback bonds. Therefore, we introduce a distributed system model based on the structure of the Kripke:

$$M_{RS} = \langle S, S_0, R, F_{RS} \rangle, \tag{2}$$

where S is a finite set of states of a distributed system, S0 is the set of initial states, R is the set of transitions between states, TA is the set of atomic states related to states and is true only in these states, FRS is a function representing each state from a set S into a subset of TA atoms that are true in the displayed states. Since the system is distributed, then the set of states of the system can be represented through subsets of states belonging to the program modules Ai, i = 1, 2, …, n, namely: $S = \bigcup_{i=1}^{n} S_i$, that is, the plurality of states of autonomous program modules will form a plurality of states in which the system will be. Similarly, the set of initial states of a distributed system $S_0 = \bigcup_{i=1}^{n} S_{0i}$, and among their elements can not be the same. For any state s ∈ S, there is at least one transition to another state, that is, the relation $R \subseteq S \times S$ is valid, which means that for any state in the set of states there is a corresponding state from the same set. If the set of atomic statements TA is complete, then the set of its subsets 2^{T_A} will be the FRS map for the set S into those subsets of atoms that will be true in $s \in S$.

Similarly, we represent the structural component of a distributed system, an autonomous program module into submodel:

$$M_{A_i} = \langle S_i, S_{0i}, R_i, F_{A_i} \rangle, \tag{3}$$

where Ai is the i-th programm module of the distributed system i = $1, 2, \ldots$, n, R_i is the set of transitions between states $s_{ij} \in S_i$ is the number i of those states, F_{A_i} is the display function of the set states S_i into the set of subsets of the set T_{A_i}, i.e. $2^{T_{A_i}}$. Each state of module necessarily has a connection with some other states. Switching from one state to another, if there is a connection between them, will display the sequence $s_{ij}s_{ip}$, where i is the number of the program module, j and p are the states of the same module. Then, the sequences $s_{ij}s_{ip}s_{ij}s_{ih}s_{iy}s_{iu}s_{ie}s_{ik}$..... will mean transitions of an autonomous software module from one state to another during its operation. Distributed system in the process of operation will be characterized by a plurality of sequences of transitions from state to state of program modules. The Kripke structure for an autonomous software module according to its transition diagram is shown in Fig. 5.

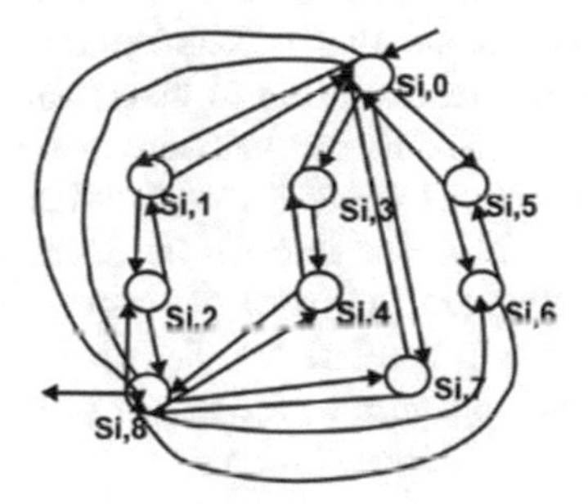

Fig. 5. The Kripke structure of the A_i software module

R_i is the set of transitions between the states of the i-th module on sets of variables and is defined as follows:

R_i={								
	0000i	0001i,	0001i	0000i,	0001i	0100i,	0100i	0001i,
	0100i	1000i,	1000i	0100i,	0000i	1000i,	1000i	0000i,
	0000i	0010i,	0010i	0000i,	0010i	0101i,	0101i	0010i,
	0101i	1000i,	1000i	0101i,	0000i	0011i,	0011i	0000i,
	0011i	0110i,	0110i	0011i,	1000i	0110i,	0110i	1000i,
	0000i	0111i,	0111i	0000i,	1000i	0111i,	0111i	1000i }

The function F_{A_i} to represent the set of states S_i in the set of subsets of the set T_{A_i} is given tabularly. A fragment of the coded transition from state to state is presented in the Table 1.

Self-replication of the distributed system is carried out at the commencement of the work of each of the software modules, the addition of each new software module, the removal of a specific software module. These events save the system and allow you to

Table 1. Coded transition from state of module to state.

Property	Coded transition	State of module
$p_{i,0}$	0000i	$s_{i,0}$
$p_{i,1}$	0000i, 0001i	$s_{i,1}$
…	…	…
$p_{i,47}$	0111i, 1000i	$s_{i,47}$

solve the tasks set on it. This level of consideration of the distributed system represents this property as belonging to the existence of the system. The described events in the organization of communication between autonomous software modules show the principle of the formation of a distributed system and its self-reproduction in the life cycle of its parts. Minimum number of working computer systems, where standalone software modules are installed, at least two. But the use of a very small number of components of the system is not effective, since then to increase the reliability of malware detection, the amount of information from different computer systems of the local area network is insufficient. The presence of only one stand-alone software module within a distributed system, that is, one system-enabled computer, will not allow the use of opportunities for transition to other states for the purpose of more detailed study of the behavior of malicious software, and will only use capabilities at the level of single-user antivirus software. Self-replication of the distributed system at the level of its components will occur through the transfer of autonomous software modules, which for a long time are in one of the states, to the base state of module. Also, the distributed system will analyze the statistics of the autonomous software modules of computer systems to optimize the transfer of tasks in order to attract the most active computer systems to their solution.

Independent software modules of the system while staying at the third and fourth levels for solving the tasks will involve self-learning technologies that will affect the change of their work in solving the same tasks at the next stages of the life cycle; thus they, passing tasks for processing to other modules, and the results of their implementation will affect other software modules; in general, the impact of modules, as structural parts of the system, on other modules will allow the evolution of a distributed system.

The distributed system at the level of its structural parts of the components will carry out self-monitoring, which will be manifested in the periodic verification of the completeness of the system, the analysis of the availability of stand-alone software modules, which for a long time are in the same state and require the automatic removal of current tasks for execution and transfer to another state, processing module bases for optimization and distributing modules to several groups according to the analysis of their states over a long period of time.

An important element of self-organization of the system is the development of mechanisms in it for the formation of its own goals. For such purposes we shall include the following: dynamic formation of the system; distribution and correlation of all structural units by groups of load, processing of critical events in the system, collective execution of tasks solved by one module, processing and optimization of accumulated statistical data.

The basis of the constructed model of the distributed system is its structural parts, which are represented by autonomous software modules, which can be in different states. The transition between the modules states is based on a defined set of transitions. Interaction and communication between autonomous software modules is based on their presence in certain states during operation. Distributed system is a responsive system that will monitor certain events. Each program module contains a resident mechanism, the motive mechanisms for the transition between states, the transitions between which are given subsets of transitions, the data for which will be formed using artificial intelligence technologies.

Picture of the window forms of the software implementation of the distributed system is shown in Fig. 6.

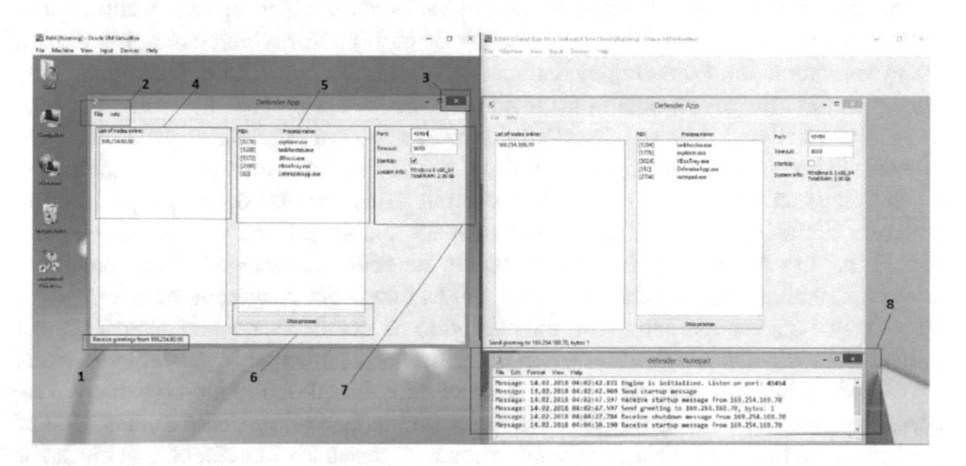

Fig. 6. Interface windows of the developed distributed system consist of the following 8 components:

1. Issuance of informative messages.
2. Main menu. File/Quit - complete, Info/About - show additional information: title, short description, author, version.
3. When closing the main window, the program rolls out in the system tray and continues its work. To complete the work, use File/Quit, or PCM on the program icon in the system window and Quit.
4. List of machine IP addresses, which also runs an instance of the program. Updates automatically when starting/finished programs on other machines.
5. List of processes in the system running in computer systems. Updated according to a given timer interval.
6. The command "stop the process selected from the list".
7. Settings panel. Port is the port number for communication between instances of programs on a network, Timeout - a timer interval for monitoring system processes,

StartUp - on/off. Startup program at OS startup, System info - System characteristics.
8. Log file. Created in a directory with executable module. Collects message/warning/program errors.

4 An Example of Using a Distributed System and the Results of Experiments

As an example of the application of a distributed system, consider the subsystem of detecting metamorphic viruses in a local area network. The use of the network is dictated by the presence of, in addition to obfuscation techniques, anti-emulation that hinders the implementation of the implementation emulation. Using an emulator is one of the main methods for detecting metamorphic viruses, which in turn results in low detection efficiency. Therefore, the detection of highly effective metamorphic viruses using anti-emulation technologies is impossible with the means of one computer system, therefore, the involvement of a local area network is proposed.

In order to detect the suspicious activity on each host, a program suspicion analyzer is used. Its main task is to track the flow of ARI calls that are carried out during the execution of an unknown program. In the event of application of the program to the software confusing the code, its API calls remain unchanged, only the parameters and values that are returned by the corresponding function are changed. Each individual suspicious activity represented by an API call feature is not dangerous when executing an unknown application. However, the execution of a certain sequence of such actions may indicate a possible risk of infection by malware, in particular a metamorphic virus.

The detection is based on ARI tracking of calls that describe the potentially dangerous behavior of a metamorphic virus and a comparison of the disassembled code of the functional blocks of the metamorphic virus with the code of the functional blocks of its modified version. To create an altered version of the metamorphic virus on the LAN hosts, modifiable emulators are installed that provide a variable execution environment. In order to increase the overall effectiveness of detecting metamorphic viruses, the system involves finding matching between the functional blocks of the metamorphic virus and its modified version. To form the conclusion about the similarity of a suspicious program on a metamorphic virus, a system of fuzzy logical conclusion is used. In case of maladministration and increasing the level of reliability for detecting a metamorphic virus, other network hosts are involved. More about the implemented method in [19, 20].

Let's consider the example of implementing the developed system which enables to detect metamorphic viruses through the investigation of suspicious code by modified emulators located in various LAN components. To perform the classification, signs were obtained based on the search and comparison of equivalent functional blocks between applications in the Damaerau-Levenstein metric [19]. As a result of the comparison, we received such features as the number of insertion, deletion, permutation, and matching of operating codes, as well as the Damaerau-Levenstein distance.

Several experiments were carried out to determine the effectiveness of the developed system. For this purpose, a university network consisting of 40 computer systems was involved. Each computer system was equipped with a virtual environment based on Qemu.

The metamorphic generators NGVCK, PS-MPC, VCL32 and G2 [21] were used to obtain test data to verify the effectiveness of detecting metamorphic viruses. All metamorphic versions created using the specified generators were compiled with anti-debugging and anti-emulation options. Using these generators, 100 samples of each type of harmful code (total of 400 samples) were generated and installed. During the experiment, the threshold values were set at levels 0.5, 0.6 and 0.7.

Experiments included the definition following performance indicators (for different threshold values of similarity of equivalent functional blocks):

(i) True Detection Rate (TP Rate) –the percentage of correctly detected metamorphic viruses:

$$TP\,Rare = \frac{TP}{TP + FN}; \tag{4}$$

(ii) False Alarm Rate (TP Rate) –percentage of false-identifiable useful applications:

$$FP\,Rare = \frac{FP}{FP + TN}; \tag{5}$$

(iii) Precision – the share of malicious software belonging to this class relative to all the test samples that the system attributed to this class:

$$P = \frac{TP}{TP + FP}; \tag{6}$$

(iv) Recall – the proportion of specimens belonging to the class in relation to all samples of metamorphic verses of this class in the test sample:

$$R = \frac{TP}{TP + FN}, \tag{7}$$

where TP is the number of correctly detected metamorphic viruses, FN is the number of false metamorphic viruses, TN is the number of correctly identified utility programs, FP is the number of useful programs that are incorrectly classified as metamorphic viruses.

(v) F-Measure - average harmonic values P and R:

$$F = 2 * \frac{P * R}{P + R}. \tag{8}$$

The results of evaluating the effectiveness of detecting metamorphic viruses from the threshold value of the similarity of equivalent functional blocks are given in Table 2.

Table 2. Estimation of perfomance indicators for metamorphic viruses' detection

	TP Rate	FP Rate	Precision	Recall	F-Measure	Class
Threshold value 0,5	0,854	0,074	0,920	0,913	0,917	NGVCK
	0,917	0,038	0,960	0,967	0,964	PS-MPC
	0,923	0,053	0,946	0,921	0,933	VCL32
	0,946	0,032	0,967	0,941	0,954	G2
Weighted avg.	0,910	0,049	0,948	0,936	0,942	
Threshold value 0,6	0,912	0,059	0,939	0,944	0,942	NGVCK
	0,934	0,044	0,955	0,947	0,951	PS-MPC
	0,928	0,048	0,951	0,958	0,954	VCL32
	0,978	0,027	0,973	0,974	0,974	G2
Weighted avg.	0,938	0,045	0,955	0,956	0,955	
Threshold value 0,7	0,894	0,088	0,910	0,919	0,915	NGVCK
	0,927	0,042	0,957	0,912	0,934	PS-MPC
	0,907	0,051	0,947	0,927	0,937	VCL32
	0,934	0,031	0,968	0,962	0,965	G2
Weighted avg.	0,916	0,053	0,945	0,930	0,938	

The results of experiments confirmed that the highest level of detecting the metamorphic viruses is 97.8% which corresponds to G2 (with the level of false positives was 2.7%). The highest metamorphic virus detection rates were recorded with a similarity threshold of 0.6 for all types of test samples (see Table 2).

Experimentally it was proved that the level of manifestation of metamorphic properties increases with the increase in the number of hosts in the network. In addition, the choice of equivalent functional units for comparison can reduce the level of false positives compared with previous studies. Thus, the use of a distributed system of detection of such a class of malicious software as metamorphic viruses has allowed to increase the authenticity of the detection. The benefits of this approach are achieved by provoking a possible manifestation of a metamorphic virus in different parts of the distributed system on the network. In this case, stand-alone software modules were filled with modified emulators to create a modifiable environment research.

5 Conclusion

The developed distributed system malware detection in local area networks. It is built on the basis of decentralized architecture and allows it to be filled with various functionality to detect malicious software. Distributed system refers to reactive systems, which will continuously monitor the running processes and executable programs in the computer systems of the local network. Objects for research on the system side are the testing of existing software and running processes in computer systems on the LAN to the ability to attribute to malicious software.

The basis of the architecture of the distributed system is autonomous software modules with the same architecture, but each of them can independently take decisions based on various data collected from different computer systems of the network.

For the effective operation of the system, it is necessary to develop methods and models of interaction and coordinate the work of different software modules among themselves and their respective levels, the detailed structure of its states and filling the subsystems of detection of various types of malicious software.

The use of the developed distributed system to detect metamorphic viruses has shown the ability to reduce the level of false positives by attracting components of the system located in various computer systems of the local area network.

Promising areas of research are the development of new effective methods for detecting malicious software, taking into account the peculiarities of its functioning in local area networks, as well as further filling up the system with formalized behavioral signatures of malicious software in order to increase the authenticity of its identification.

References

1. INTERPOL. https://www.interpol.int/Crime-areas/Cybercrime/The-threats/Malware,-bots,-botnets
2. Security response publications. Monthly threat report. https://www.symantec.com/security_response/publications/monthlythreatreport.jsp
3. McAfee labs threat report. https://www.mcafee.com/us/resources/reports/rp-quarterly-threats-dec-2017.pdf
4. Overview of symantec endpoint protection 12, Part 2. https://www.anti-malware.ru/reviews/Symantec_Endpoint_Protection_12_2
5. Palo Alto Networks. https://www.paloaltonetworks.com/
6. Malwarebytes endpoint security. https://ru.malwarebytes.com/business/endpointsecurity/
7. Cisco NAC Appliance (Clean Access). https://www.cisco.com/c/en/us/products/security/nac-appliance-clean-access/index.html
8. Comodo cybersecurity. https://www.comodo.com/
9. Branitskiy, A., Kotenko, I.: Hybridization of computational intelligence methods for attack detection in computer networks. J. Comput. Sci. **23**, 145–156 (2017)
10. Baldangombo, U., Jambaljav, N., Horng, S.: A static malware detection system using data mining methods. Int. J. Artif. Intell. Appl. (IJAIA) **4**(4), 113–126 (2013)
11. Bezobrazov, S., Sachenko, A., Komar, M., Rubanau, V.: The methods of artificial intelligence for malicious applications detection in Android OS. Int. J. Comput. **15**(3), 184–190 (2016)
12. Eskandari, M., Hashemi, S.: A graph mining approach for detecting unknown malwares. J. Vis. Lang. Comput. **23**(3), 154–162 (2012)
13. Kaushal, K., Swadas, P., Prajapati, N.: Metamorphic malware detection using statistical analysis. Int. J. Soft Comput. Eng. **2**(3), 49–53 (2012)
14. Ghiasi, M., Sami, A., Salehi, Z.: Dynamic malware detection using registers values set analysis. In: Information Security and Cryptology, pp. 54–59 (2012)

15. Komar, M., Golovko, V., Sachenko, A., Bezobrazov, S.: Intelligent system for detection of networking intrusion. In: Proceedings of the 6th IEEE International Conference on Intelligent Data Acquisition and Advanced Computing Systems: Technology and Applications, IDAACS 2011, Prague, Czech Republic, pp. 374–377 (2011)
16. Lysenko, S., Savenko, O., Kryshchuk, A., Kljots, Y.: Botnet detection technique for corporate area network. In: Proceedings of the 2013 IEEE 7th International Conference on Intelligent Data Acquisition and Advanced Computing Systems, pp. 363–368 (2013)
17. Shafiq, M.Z., Tabish, S.M., Mirza, F., Farooq, M.: PE-Miner: Mining structural information to detect malicious executables in realtime. In: International Workshop on Recent Advances in Intrusion Detection, pp. 121–141 (2009)
18. David, B., Filiol, E., Gallienne, K.: Structural analysis of binary executable headers for malware detection optimization. J. Comput. Virol. Hacking Tech. **13**(2), 87–93 (2017)
19. Savenko, O., Lysenko, S., Nicheporuk, A., Savenko, B.: Metamorphic viruses' detection technique based on the equivalent functional block search. ICT in Education, Research and Industrial Applications, Integration, Harmonization and Knowledge Transfer, vol. 1844, pp. 555–569 (2017)
20. Savenko, O., Lysenko, S., Nicheporuk, A., Savenko, B.: Approach for the unknown metamorphic virus detection. In: 9th IEEE International Conference on Intelligent Data Acquisition and Advanced Computing Systems. Technology and Applications, Bucharest, Romania, pp. 453–458 (2017)
21. VX Heavens. http://vxheaven.org/

Method of Reliability Block Diagram Visualization and Automated Construction of Technical System Operability Condition

Yuriy Bobalo[1], Maksym Seniv[1], Vitaliy Yakovyna[1,2(✉)], and Ivan Symets[1]

[1] Lviv Polytechnic National University, Lviv, Ukraine
vitaliy.s.yakovyna@lpnu.ua
[2] University of Warmia and Mazury in Olsztyn, Olsztyn, Poland

Abstract. This paper describes the method of reliability block diagram visualization and automated construction of technical system operability condition, which is based on the reliability block diagram analysis with splitting into segments. The developed method includes RBD traversal to examine the topology of the diagram with subsequent identification of scheme's elements connection type for further construction of a technical system working condition. To verify the correctness of the developed method of automated construction of the operability condition of technical systems, the comparison of its results with the operability condition obtained using the recursive algorithm was carried out. The performance of the implemented software was studied as well using parallel and sequential combinations with different number of modules in the diagram. When the number of modules is 200, the performance of the developed method is higher about 25% comparing to the recursive algorithm.

Keywords: Reliability · Reliability analysis · Operability condition

1 Introduction

The problem of estimation of reliability criteria in modern technical and particularly computer systems of responsible designation is becoming especially acute, as far as their failure is unacceptable due to possible human casualties or enormous material losses. Therefore the matters of submission and estimation of reliability indicators are essential yet at the very beginning of a designing process of these systems, when their architecture is being created, which corresponds to the functioning structure and algorithm. The structure of the study subject is revealed as a set of single-function nodes (modules) and the required interrelations between them. The functioning algorithm displays the character and sequence of interoperability between the nodes. The description of the functioning algorithm of a technical system with regard to reliability starts with the development of its Reliability Block Diagram (RBD).

An RBD carries out system reliability and availability analyses on large and complex systems by means of block diagrams to display network relations.

N. Shakhovska and M. O. Medykovskyy (Eds.): CSIT 2018, AISC 871, pp. 599–610, 2019.
https://doi.org/10.1007/978-3-030-01069-0_43

The structure of the RBD determines the logical interconnection of failures within a system which are needed to support system operation [1, 2].

Reliability block diagrams can be arranged on different levels of disaggregation (indenture level): maintainable component level, maintainable unit level and system level. RBD is basically applied at design concept stage to register reliability on various levels of disaggregation.

During the first stage of a complex process design, an engineering plant or any facility a block diagram is built up in this diagram and each block represents one of the facilities component systems, sub-systems or assets/equipment [2]. An illustrative block diagram displays the way in which the assets are physically connected, whereas a functional block diagram reveals the flow of power, material, etc., through the system, with the interaction between input and output defined for each block. This diagram provides the conceptual design of the system and requires approval prior to the accomplishment of any detailed engineering design. In a similar manner, the estimation of full system reliability can be built up by the construction and analysis of an RBD [1, 2]. In an RBD the connections between the assets, symbolize the ways in which the system will operate as needed and will not necessarily point out the actual physical connections.

The components within a system can be related to each other in two basic ways: sequential configuration or parallel configuration. Sequential and active parallel RBD represent the simplest building blocks for reliability analysis. In sequential configuration all the components must function so that the system is able to operate. However, in a parallel or redundant configuration at least one component must function for the operating of the system [1, 2].

Figure 1 displays the samples of RBD of some simple configurations. The operability conditions can be easily revealed from the displayed RBD: for instance, the system (a) is operable, in case if all the elements are operable; the system (b) is operable, in case if any of the elements is operable; the system (c) is operable, if the first element is operable and any of the two elements of the second group, and any of the three elements of the third group of elements [2–5] are operable.

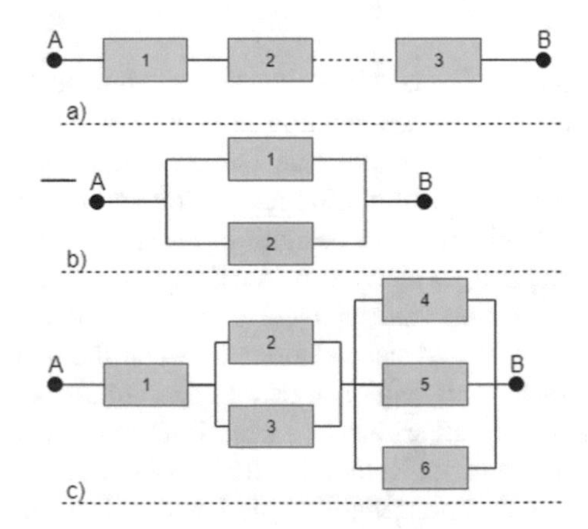

Fig. 1. An example of RBD for basic cases (a – sequential connection; b – parallel connection; c – mixed scheme).

Talking about the complex RBD structure, the task of operability conditions formulation can appear much more complicated because the real complex computer systems can be made of hundreds of thousands elements. That is why; the processing of the automated formulation method for the operability condition is a challenge number one at the present moment. It will provide an opportunity to perform the RBD automation development reliability analysis of complex technical systems.

2 Related Works

By application of an RBD you can create the corresponding Markov model for reliability estimation of the technical system represented by the specified diagram. Markov analysis is a vigorous and flexible method for estimation of reliability measurements of safety instrumented systems, but manual creation of Markov models is a troublesome and long-lasting process. A number of papers bring up the issues of automatic Markov model creation from the RBD [5–7].

Paper [5] introduces a new method of automatic creation of Markov models for reliability evaluation of safety related systems. Such safety related criteria as failure modes, self-diagnostics, and restorations, common cause and voting, are added into Markov models. In paper [5] a framework is primarily simulated on the basis of voting, failure modes and self-diagnostic. Afterwards restorations and common-cause failures were included into the framework for construction of a complete Markov model. Ultimately, state merging can make the created Markov models simpler [5]. In paper [6] the authors have developed the method of automated generation of Markov models on the basis of a verbal description of the system under study. The method, described in [6] provides an opportunity to design a graph of reliability behavior for the given RBD and then to create the systems of Kolmogorov – Chapman equations for the respective Markov model.

In paper [3] major attention is paid to RBD generating for providing of reliability prediction on the first steps of industrial system processing. The specific case study was centered on gas turbine auxiliary systems [3]. System design created on the basis of reliability evaluation provides an opportunity for project engineers to reduce time-delivery as well as the time for introduction of advancements, achieve reliability goals and provide availability performance to the customers. More over a new approach to standby redundancy architectures was introduced in [3] for the purpose of achievement of reliability evaluation for complex systems, commonly used in Oil and Gas applications.

In combination with optimization, the model-based design allows engineers to identify design options sufficiently and in automatic manner. Offering higher reliability and safety requirements, reliability-based design is applied in multidisciplinary design optimization more and more by the day. "Multidisciplinary" refers to various aspects which are to be included in a system design. In [4] a technique of sequential optimization and reliability evaluation for multidisciplinary design optimization was suggested in order to improve the productivity of reliability-based multidisciplinary design optimization. The main idea of [4] is to separate the reliability analysis from multidisciplinary design optimization.

Popular software products for reliable design [8–11], such as Advanced Specialty Engineering Networked Toolkit, ReliaSoft Synthesis Master Suite, Reliability Workbench, RAM Commander and others afford an opportunity to automate the process of reliability assessment, but it is also necessary to accomplish "manual" construction of a reliability behavior graph.

In spite of substantial functionalities and flexibility of software systems reliability analysis during the research process and in-depth study of these software products, some disadvantages have been displayed:

- Requirement for certain "manual" calculations of input data (criteria of set of states performance, etc.)
- Deficiency in definition for system operability
- Complexity of training and maintenance service of software tools (specific qualification is required in the field of reliability theory, which allows one to work with the software)

With a purpose of elimination of these disadvantages the authors have developed particular algorithms for graphic RBD design and automated formation of technical system space with a limited number of restorations (papers [12, 13]), and their program implementation was realized in the form of Windows Form Application based on these algorithms.

When using this software, the user must first enter the details of the system structure and formulate its operability condition, which will be tested by the program in the simulation process in each of the possible states of the system. In the case of a complex system, this can increase time expenditures and lead to errors. That is why, the issue of RBD visualization method development of the technical system, which enables automated formulation of its operability condition, is highly-demanded.

3 Algorithmic Background for the Method of RBD Processing

The condition of operability of any technical system can be determined visually by the appearance of its RBD. For automation of this process one can use a mathematical tool of logical algebra and graph theory. For this purpose it is necessary to accomplish a graph traversal, which basically corresponds to RBD and to apply logical operations and/or in case of sequential and parallel connection of elements on the scheme, respectively [6]. The solution of a problem of automated construction of complex technical system operability condition with its RBD requires the development of algorithms for scheme traversal and identification of a connection type between the elements of a scheme. The method of automated construction of technical system operability condition should be designed on the basis of such algorithms.

In the previous paper [12] the authors designed a software subsystem of RBD visualization, which, however, did not provide an opportunity to receive the condition of operability and technical system failure.

One of the authors has developed a recursive algorithm for RBD traversal and determination of technical system operability condition, which is simulated by the following scheme [13]. This algorithm makes it possible to determine parallel and sequential RBD connections; however, as a result of its recursive character it contains a number of defects and limited capacities. In particular, this algorithm cannot be applied for the determination of operability conditions of a particular type diagram as far as it does not contain any data of full diagram topology and it only designs the operability condition in the course of diagram bypass. Along with that such algorithm has the ability to operate correctly only upon the condition, that the starting and the ending nodes of parallel subsystems are clearly arranged on the scheme. For instance, in a three parallel subsystem RBD all the branches coming from one node should be attributed to the same node, in other cases such algorithm will reveal incorrect condition of system working capacity.

For the elimination of the abovementioned defects and development of an effective method of automated determination of technical system operability condition based upon their RBD, there have been processed [14] the enhanced algorithms of graph traversal, which corresponds to RBD, the identification of parallel and sequential connections on the scheme, as well as the automated creation of operability condition. In particular, the paper [14] introduces the development of three algorithms: the algorithm of scheme traversal (see Fig. 2) and two algorithms of scheme connection type definition (sequential and parallel respectively).

The subsystem of RBD visualization and the algorithms of RBD traversal with a simultaneous identification of its connection types [14], developed in the paper [12] serve as a basis for RBD visualization method of complex technical systems and automated definition of their operability condition, revealed in this paper. The explanation and the scheme of the developed method are provided in Sect. 4 alongside with the description of its software implementation. Section 5 is devoted to the case study of application of the described method exemplified by the system of 200 elements in its RBD and its comparison with the recursive algorithm of scheme traversal [13].

4 The Method of Visualization of the Structural Schemes of Reliability and System's Working Condition Construction

The method of RBD visualization and the automated construction of the operability conditions of technical systems consists of the following main components:

- RBD traversal to examine the topology of the diagram and identify the set of sections, each of them being either a sequential combination of elements or a single element;
- Identification of scheme's elements connection type to reveal the set of elements connected in parallel or sequential way for further construction of a technical system working condition.

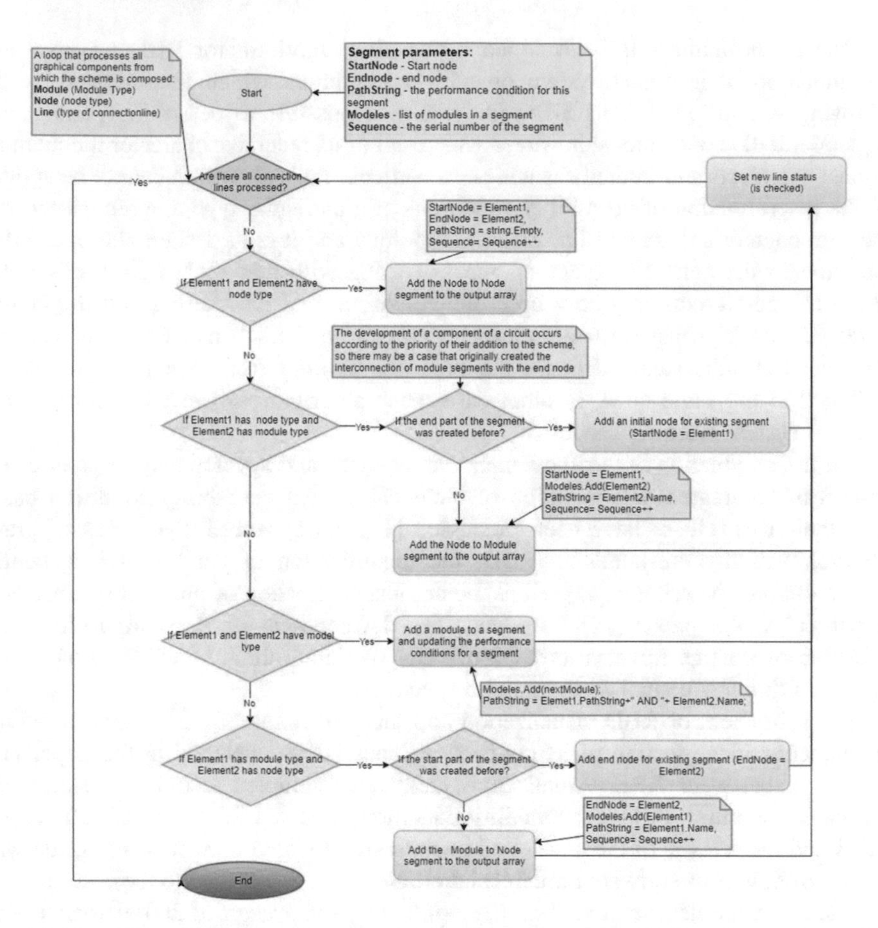

Fig. 2. Algorithm of RBD traversal for further connection type identification [14] (© 2018 IEEE. Reprinted, with permission, from XIV-th International Conference on Perspective Technologies and Methods in MEMS Design (MEMSTECH), Lviv, 2018.).

Figure 3 depicts the representation of the developed method of RBD visualization and construction of operability condition of a technical system.

To apply the method, it is needed to complete the following steps:

Step 1. Analysis of the topology and definition of the array of segments of the diagram

Analyze the topology of a graph which represents a studied RBD to identify all its nodes and branches. This task is performed by using the RBD traversing algorithm (Fig. 2). This algorithm splits the topology of a RBD into segments consisting of either a sequential combination of elements or a single element.

Step 2. Sequential and parallel connection of segments with partial operability conditions

The next step after splitting the RBD topology into segments, is identification of segments connections types (both sequential and parallel) and following construction of partial operability conditions of the studied system, until the RBD convolutes into the single segment and hence the single operability condition remains.

Step 3. Search and connect the parallel RBD segments
If some segments have identical starting and ending nodes, then such segments are considered to be parallel, and hence they will be included into the operability condition using the OR operator.

Step 4. Search and connect the sequential RBD segments
After processing all parallel connections, it is needed to use the sequential algorithm to build the operability condition. If the endpoint of any segment of the array is the only endpoint in the array, the starting point of this segment is the only initial point in the array, so they are connected sequentially. In this case the operability condition is constructed using the logical AND operation.

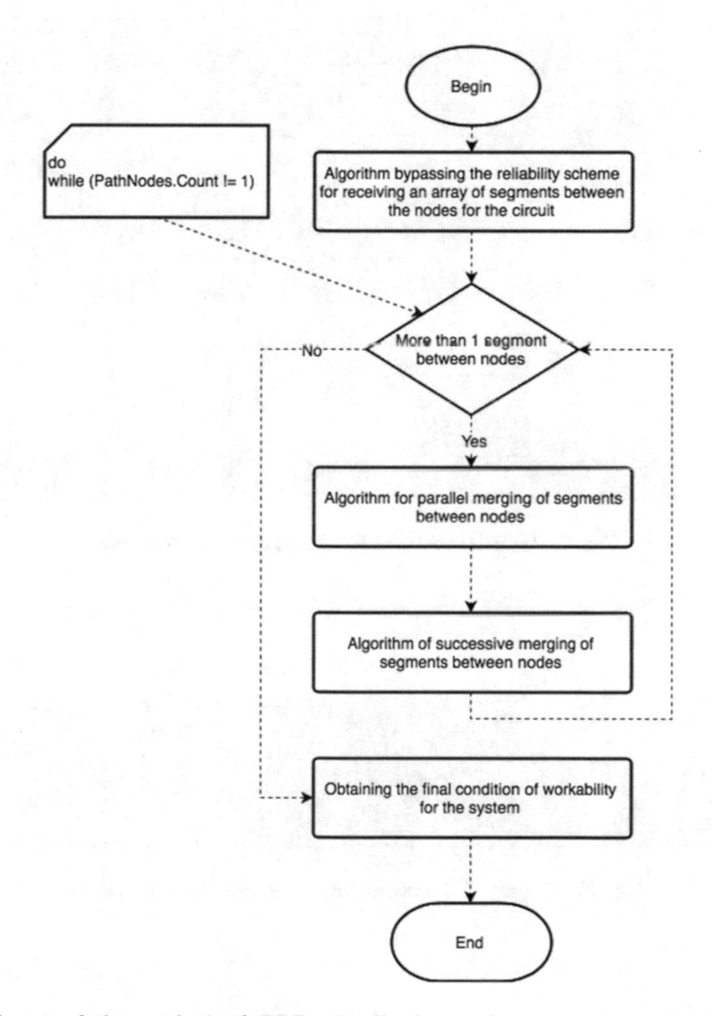

Fig. 3. Scheme of the method of RBD visualization and automated construction of the operability condition of a technical system.

Step 5. Checking for the final operability condition
If the single element remains in the segments array, then the convoluted operability condition is obtained and the method is finished, if not, then go back to step 2.

For practical usage of the proposed method as well as for it verification and comparison with previously developed recursive RBD transversal, its software implementation has been developed [15]. The input of the developed software is the RBD of a technical system; the output is the operability condition of the technical system studied.

Figure 4 depicts user interface of the mentioned RBD visualization software, while Fig. 5 represents the system operability condition output determined using the developed software.

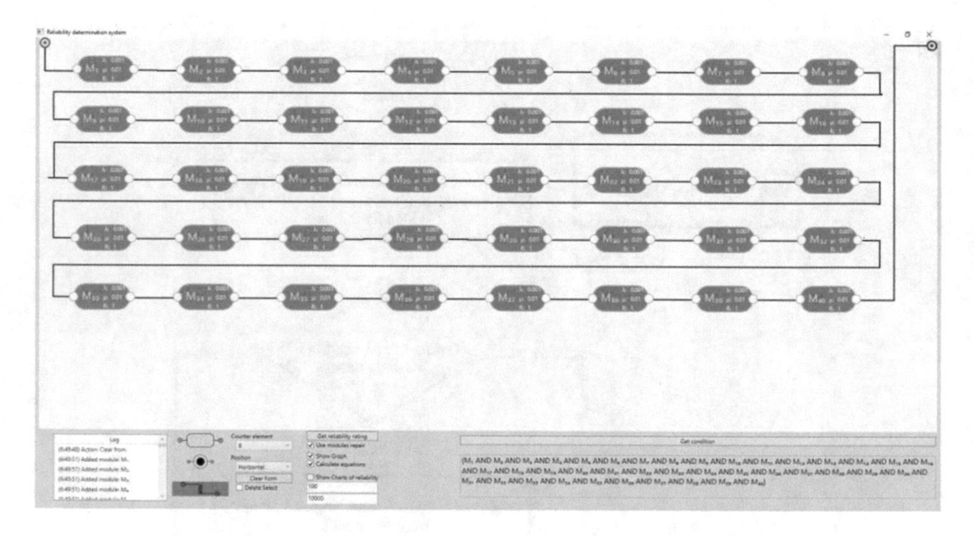

Fig. 4. User interface of RBD visualization software.

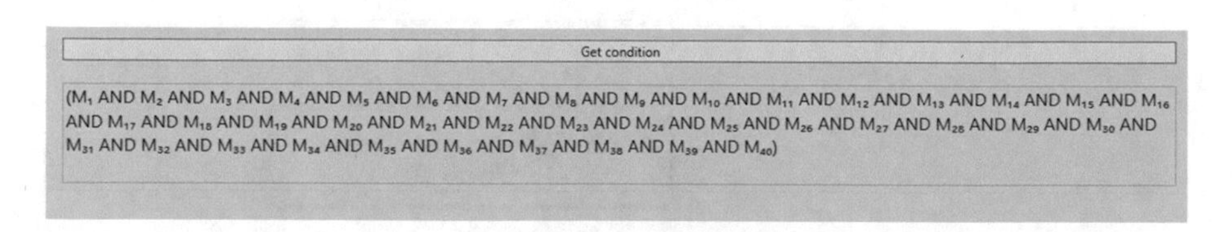

Fig. 5. Output of the operability conditions of the system

5 Case Study and the Verification of the Method of RBD Visualization and Automated Construction of Technical System Operability Condition

To verify the developed method the implemented software was used to construct the operability condition for different technical test systems consisting of up to 200 components and having different types of RBD structure. The types and number of interconnections between the modules is an important factor that results in the operation velocity of the developed software. RBD of real safety related technical systems usually contains mixed types of interconnections (Fig. 6).

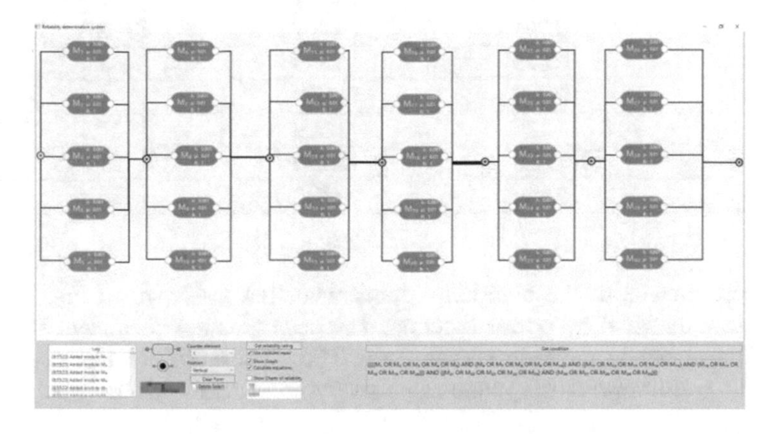

Fig. 6. RBD of the technical system used for testing the developed method.

The case study was run on a PC running Windows 10 OS with the following hardware configuration: Intel Core i3-3220 CPU @ 3.3 GHz, and 8 GB RAM. To verify the correctness of the developed method of automated construction of technical system operability condition the comparison of its results with the operability condition obtained using the recursive algorithm [13] was used.

The performance of the implemented software was studied as well using both sequential and parallel combinations with various number of RBD components. Table 1 shows the comparison of the average speed of both algorithms for different types of connection. As can be seen from the Table 1 the recursive algorithm with small number of elements in the system is faster than the developed, however, with the number of elements greater than 50, the speed of construction of the system becomes larger for the algorithm [14], which is the basis of the developed method, and continues to grow to reach values of about 25% for the investigated configuration. This is due to the fact that the developed method is based on the whole diagram analysis and then the to construction of a condition begins [14, 15], whereas the recursive algorithm begins to construct an operability condition during the RBD bypassing. Hence, at an increasing number of modules, the developed in this paper method based on the splitting RBD into segments shows a significant calculation speed increasing compared with the recursive algorithm [13].

Table 1. The comparison of the performance of algorithms for the construction of the operability condition.

Number of elements	Processing time, ms		Difference, %
	Developed algorithm	Recursive algorithm	
5	0.309	0.259	−19.3
10	0.378	0.308	−22.7
15	0.416	0.396	−5.0
20	0.460	0.441	−4.3
30	0.491	0.482	−1.9
40	0.663	0.660	−0.4
50	0.728	0.721	−1.0
60	0.791	0.807	2.0
70	0.823	0.843	2.4
80	0.949	0.999	5.0
110	1.120	1.520	26.3
150	1.620	2.220	27.0
200	2.120	2.820	24.8

The performance of the algorithms implementation is shown in Fig. 7. Experimental data were fitted by power function. The determination coefficient R^2 was not less than 0.993, while the reduced Chi^2 was 0.005 and 0.002 for the recursive and developed algorithm processing time respectively. It is clear from the graph that both

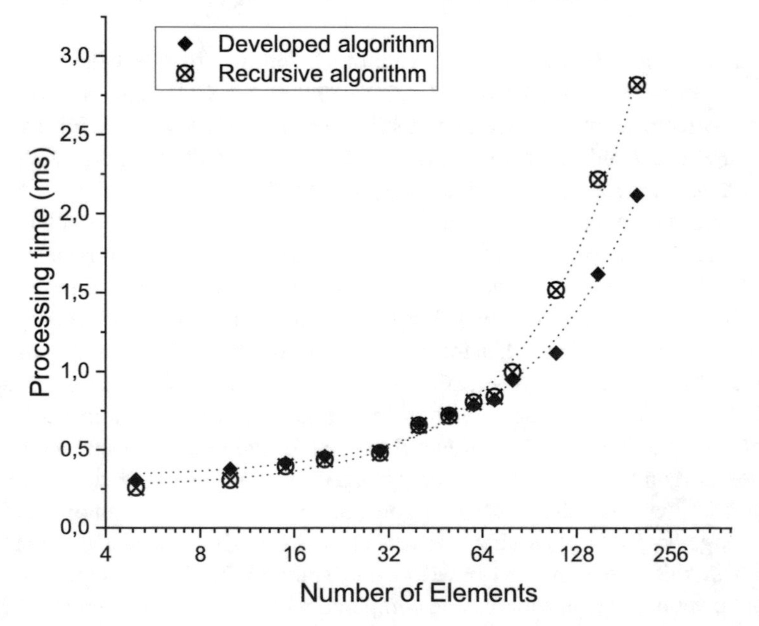

Fig. 7. The processing time of the developed method vs. recursive algorithm.

dependencies are well described by the power expression with the value of the power index 1.28 ± 0.08 for the recursive algorithm and 1.16 ± 0.07 for the developed algorithm of constructing the system operability condition. When the number of components is 200, the time gain is about 25%.

6 Conclusion

This paper presents the method of RBD visualization and automated construction of technical system operability condition, which consists of:

- Algorithm of RBD traversal to examine the topology of the diagram and identify the set of sections, each of them being either a sequential combination of elements or a single element;
- Algorithms for identification of scheme's elements connection type to reveal the set of elements connected in parallel or sequential way for further construction of a technical system working condition.

The developed method was implemented into software tool for automated construction of the operability condition of complex technical systems and is able to process and visualize various types of RBD.

To verify the correctness of the developed method of automated construction of the operability condition, the comparison of its results with the operability condition obtained using the recursive algorithm was carried out. The performance of the implemented software was studied as well using various combinations with variable quantity of components in the diagram. To verify the developed method the implemented software was used to construct the operability condition for different technical test systems consisting of up to 200 components and having different types of RBD structure. It is shown that while the number of components increases, the developed method based on the RBD splitting into segments yields an increase up to 25% in computational speed comparing to the previously developed recursive algorithm. Experimental data were fitted by power function. Both dependencies are well described by the power expression with the value of the power index 1.28 ± 0.08 for the recursive algorithm and 1.16 ± 0.07 for the developed algorithm of constructing the system operability condition.

References

1. Ebeling, C.E.: An Introduction to Reliability and Maintainability Engineering. McGraw-Hill, New York (2004)
2. Modarres, M., Kaminskiy, M., Krivtsov, V.: Reliability Engineering and Risk Analysis. Marcel Decker, New York (1999)
3. Catelani, M., Ciani, L., Venzi, M.: Improved RBD analysis for reliability assessment in industrial application. In: Proceedings of 2014 IEEE International Instrumentation and Measurement Technology Conference (I2MTC), Montevideo, Uruguay, pp. 670–674. IEEE (2014). https://doi.org/10.1109/i2mtc.2014.6860827

4. Du, X., Guo, J., Beeram, H.: Sequential optimization and reliability assessment for multidisciplinary systems design. Struct. Multidiscip. Optim. **35**(2), 117–130 (2008). https://doi.org/10.1007/s00158-007-0121-7
5. Guo, H., Yang, X.: Automatic creation of Markov models for reliability assessment of safety instrumented systems. Reliab. Eng. Syst. Saf. **93**(6), 829–837 (2008). https://doi.org/10.1016/j.ress.2007.03.029
6. Volochiy, B.Y.: Technology of Modeling Algorithms for Behavior of Information Systems. Lviv Polytechnic Publishing House, Lviv (2004). (in Ukrainian)
7. Gorbatyy, I.V.: Investigation of the technical efficiency of state-of-the-art telecommunication systems and networks with limited bandwidth and signal power. Autom. Control. Comput. Sci. **48**(1), 47–55 (2014). https://doi.org/10.3103/S0146411614010039
8. Relex Software. http://www.relexsoftware.it. Accessed 15 June 2018
9. RAM Commander. http://www.aldsoftware.com. Accessed 15 June 2018
10. Reliability Workbench. http://www.armsreliabilitysoftware.com/software-solutions. Accessed 15 June 2018
11. Raytheon Company. https://www.raytheoneagle.com/asent/index.htm. Accessed 15 June 2018
12. Mandziy, B., Seniv, M., Yakovyna, V., Mosondz, N.: Software realization of the improved model of reliability of the technical reserve system with limited number of recoveries. Bull. Lviv. Polytech. Natl. Univ. **826**, 43–51 (2015). (in Ukrainian)
13. Seniv, M., Mykuliak, A., Senechko, A.: Recursive algorithm of traversing reliability block diagram for creation reliability and refuse logical expressions. In: Proceedings of Perspective Technologies and Methods in MEMS Design Conference (MEMSTECH 2016), Lviv – Polyana, Ukraine, pp. 199–201. IEEE (2016). https://doi.org/10.1109/MEMSTECH.2016.7507542
14. Bobalo, Yu., Seniv, M., Symets, I.: Algorithms of automated formulation of the operability condition of complex technical systems. In: Proceedings of Perspective Technologies and Methods in MEMS design Conference (MEMSTECH 2018), Lviv – Polyana, Ukraine, pp. 14–17. IEEE (2018). https://doi.org/10.1109/MEMSTECH.2018.8365692
15. Seniv, M., Yakovyna, V., Symets, I.: Software for visualization of reliability block diagram and automated formulation of operability conditions of technical systems. In: Proceedings of Perspective Technologies and Methods in MEMS design Conference (MEMSTECH 2018), Lviv – Polyana, Ukraine, pp. 191–195. IEEE (2018). https://doi.org/10.1109/MEMSTECH.2018.8365731

Author Index

N. Shakhovska and M. O. Medykovskyy (Eds.): CSIT 2018, AISC 871, pp. 611–612, 2019.
https://doi.org/10.1007/978-3-030-01069-0

Printed in the United States
By Bookmasters